BRAKING OF ROAD VEHICLES

AF750899

10 083 063 01

UNIVERSITY OF NORTHUMBRIA
AT NEWCASTLE LIBRARY
WITHDRAWN

1996

1997

BRAKING OF ROAD VEHICLES

I Mech E CONFERENCE PUBLICATIONS 1976 - 5

Conference sponsored by the
Automobile Division of the Institution of Mechanical Engineers
in association with the Institute of Road Transport Engineers

Loughborough University of Technology
23 - 25 March 1976

Published by
Mechanical Engineering Publications Limited for
The Institution of Mechanical Engineers
LONDON AND NEW YORK

First published 1977

This publication is copyright under the Berne convention and the International Copyright Convention. Apart from any fair dealing for the purpose of private study, research, criticism or review as permitted under the Copyright Act 1956, no part may be reproduced, stored in a retrieval system, or transmitted in any form or by any means, electronic, electrical, chemical, mechanical, photocopying, recording or otherwise, without the prior permission of the copyright owners. Inquiries should be addressed to: The Managing Editor, Books and Academic Journals, Mechanical Engineering Publications Limited, PO Box 24, Northgate Avenue, Bury St Edmunds, Suffolk IP32 6BW

© The Institution of Mechanical Engineers 1977

This volume is complete in itself.

There is no supplementary discussion volume.

ISBN 0 85298 356 5

NEWCASTLE UPON TYNE
POLYTECHNIC LIBRARY

629-11-59

Printed in Great Britain at The Lavenham Press Limited, Lavenham, Suffolk
Bound by The Burlington Press, Foxton, Royston, Hertfordshire

CONTENTS

CONTENTS

C22/76

BASIC PRINCIPLES

T. P. NEWCOMB, DSc, CEng, FIMechE
Loughborough University of Technology, Leics.

The MS of this paper was received at the Institution on 1 October 1975 and accepted for publication on 16 October 1975

SYNOPSIS A broad view is presented of the main factors that influence the braking behaviour of road vehicles. First the general principles involved in decelerating a vehicle are examined; these include road-tyre adhesion, steady state and transient load transfers and the complications resulting from the many types of vehicles and their loadings. Methods of improving stability are also briefly mentioned. Braking controls and the means of achieving the braking force are then outlined. A final section discusses the brake mechanism and the parts played by the main components and a brief comparison is made of the overall performance from the various brakes.

INTRODUCTION

1. The general requirements that a vehicle braking system should meet are self-evident. The brakes should be capable of stopping the vehicle safely within the shortest distance possible on all types of road surfaces and in all motoring conditions. During braking the vehicle should not deviate to one side or the other, and should be completely under the control of the driver, and the brakes must not interact with other vehicle components.

2. The vehicle braking system should be completely reliable and unaffected by temperature, water and road debris, and its performance must not deteriorate if parts wear. Brakes and braking systems should also require as little maintenance and adjustment as possible.

3. The contribution of the driver to braking control is also of vital importance. For driver comfort the deceleration should be uniform for a given pedal effort throughout the application and in different applications should be the same for the same pedal effort exerted by the driver. The deceleration should also increase with pedal effort in a regular way and high decelerations should not require excessive effort. The overall response time of the system should also be uniform and as small as possible.

4. Bearing in mind the above requirements the basic principles involved in vehicle braking will be described from the standpoint of the designer, who has to cope initially with the various technical problems.

Road-Tyre Adhesion

5. If the braking force at the road tyre interface is greater than the frictional force, that is, the adhesion between tyre and road, the wheel will lock and any further increase in pedal effort will cause no further increase in deceleration. Also the locked wheel will not develop a sufficient cornering force and so directional control over the vehicle will be lost.

6. In many accidents with all classes of vehicles, loss of control occurred, caused by the braking force exceeding the adhesion force.

7. The retarding force exerted by a tyre is developed as the wheel slips as well as rolls and the force builds up to a peak value before decreasing to a lower value as the wheel locks in the fully-sliding condition. The ratio of the tangential force transmitted by the wheel to the normal load carried by the wheel is termed the coefficient of adhesion μ_A. In normal motoring the driver can utilise values of μ_A up to the maximum value whereas the lower locked wheel value corresponds to the skidding condition.

8. When the tyre is subjected to braking and cornering forces the peak braking force is reduced and occurs at a larger percentage slip. In the fully-locked wheel condition, there is virtually no sideways force available for cornering.

9. The main factors that affect μ_A are the material and tread pattern of the tyres, the nature of the road surface and the water depth. Secondary effects arise from other factors such as size, inflation pressure and loads.

©IMechE 1976

10. Some of the important factors that influence the μ_A between tyre and road are illustrated in Figs. 1 and 2. Fig. 1 (Ref.1) shows the locked wheel μ_A at different speeds for various types of truck tyres including a smooth tyre compared with a car tyre. These curves show that wide differences in μ_A behaviour of patterned truck tyres can exist and these differences are larger than those between patterned car tyres. Fig. 2 (Ref.2) shows the effect of water depth on the grip with a fully-patterned tyre and these results suggest that the tyre road-surface combination gives least μ_A around 4mm water depth.

11. The tyre tread material influences μ_A and a considerable increase in μ_A can be obtained by making the tread of high hysteresis rubber. Benefits are obtained if the surface deforms the rubber even in the presence of a water film.

12. The tread pattern on a tyre has little effect on μ_A on a dry road surface but has a profound influence on the grip on a wet road surface. In the worst possible case, where the μ_A is zero at high speed because of aquaplaning, a suitably patterned tread can increase μ_A to 0.1 approximately. At lower speeds μ_A can approach the values obtained on a dry surface.

13. In the dry condition, provided there is no loose grit, there are only small differences in the μ_A of different road surfaces but in wet conditions large variations in μ_A exist especially at higher vehicle speeds. The grip on wet surfaces can be much increased by giving it a harsh texture to help penetrate the film and to provide drainage from actual contact regions. Despite recent advances in our knowledge of wet friction we still do not know how best to match road surface and tread patterns. Further developments in tyre technology will however improve the frictional characteristics of tyres.

14. In general the brake designer must make provision to cope with widely-varying road adhesions. For maximum deceleration of a vehicle on a given surface all wheels must be on the point of locking simultaneously and the maximum deceleration cannot exceed $\mu_A g$. To ensure this condition the braking force at each wheel must be in proportion to the load being carried.

LOAD TRANSFER

15. One complication influencing braking performance is that the static loads carried on each axle vary depending on whether the vehicle is loaded or not. Another complication is load transfer that causes the wheel loads to vary with the degree of braking. Load transfers in turn can be divided into steady state load transfer in which the longitudinal braking forces are independent of time and transient load transfer in which the braking forces are dependent on time.

Steady State - Two-axle Vehicles

16. Considering steady-state load transfer on a two-axle vehicle, the deceleration when the front wheels are about to lock, is given by

$$\frac{a}{g} = \frac{\mu_A d\, W_F}{W(dF_F - \mu_A h)} \quad .. \quad .. \quad .. \quad .. \qquad (1)$$

and when the rears are about to lock during straight-line braking

$$\frac{a}{g} = \frac{\mu_A d\, W_R}{W(dF_R + \mu_A h)} \quad .. \quad .. \quad .. \quad .. \qquad (2)$$

where h is the height of the centre of gravity above the road surface, d is the wheel base, F_F is the proportion of braking at the front and F_R is the proportion at the rear and $F_F + F_R = 1.0$.

17. Equations (1) and (2) are illustrated graphically in Fig.3 where lines ab and cd represent the maximum braking forces that can be applied at the front and rears respectively, expressed as a proportion of the vehicle weight when the coefficient of adhesion $\mu_A = 1.0$ and the deceleration is a/g. These results are based on a car of weight 1360kg which has 0.54 of its static weight carried by the front axle (driver only condition) and for which h/d is 0.2. The lines ef and gh depend on the braking ratio of the vehicle which is assumed to be 70:30 and it is also assumed there is no variation of μ of the friction material or other factors to change the brake torques. The points i and j where the two sets of lines intersect give the solutions to equations (1) and (2) and the lower value determines the maximum braking; in this example at j the rears would lock first and the maximum deceleration would be 0.92g. The other two lines shown in Fig.3 give the front and rear braking performance when μ_A = 0.8. Similarly, simple proportioning of the μ_A = 1.0 values will enable the braking performance to be determined for other values of μ_A.

18. The ratio of the useful axle weight (weight plus transferred weight) is

$$\frac{W_{FT}}{W_{RT}} = \frac{W_F + W\mu_A h/d}{W_R - W\mu_A h/d} \quad .. \quad .. \quad .. \quad .. \qquad (3)$$

and as the deceleration will be a maximum when all wheels are about to lock simultaneously for a given μ_A, this ratio should be the same as that of the braking force F_F/F_R.

19. Generally the braking ratio is fixed and on passenger vehicles it is usual to fix the ratio such that the rear wheels will not lock before the fronts, until a high

© IMechE 1976

value of μ_A is attained. In Fig.3 both wheels will lock simultaneously at a μ_A of 0.8.

20. A further complication occurs with split or dual braking systems in ensuring that the greatest possible deceleration can be obtained and that vehicle stability is retained when only half the brakes are working.

21. The behaviour of systems split in various ways is illustrated in Fig.4 for the car mentioned previously when all wheels lock simultaneously at an adhesion of 0.8 in the driver only condition. The simple split systems will produce rear wheel locking at decelerations of approximately 0.3g if the rear brakes only are operative. Consequently the simple front-rear split or diagonal split can cause problems and if decelerations of greater than 0.3g are required some braking must be applied to at least three wheels (Ref.3).

22. Premature locking of either front or rear wheels will result in a deceleration less than the maximum for that surface where both fronts and rears lock simultaneously, as well as affecting the directional control of the vehicle.

Adhesion Utilisation

23. A measure of the interaction of adhesion and braking ratio is found by determining the ratio $\eta = a/\mu_A g$ where η is the coeff.of adhesion utilisation and a is the actual deceleration at which one axle locks and $\mu_A g$ is max. possible deceleration.

24. If the front wheels are about to lock

$$\eta = \frac{W_F}{W(F_F - \mu_A h/d)} \quad \dots\dots\dots\dots \quad (4)$$

and if the rears lock first

$$\eta = \frac{W_R}{W(F_R + \mu_A h/d)} \quad \dots\dots\dots\dots \quad (5)$$

and either equation (4) or (5) is used for η up to unity.

25. A study of adhesion utilisation (Ref. 4) shows to what extent the braking system is able to make use of the available μ_A. Fig.5 shows a typical plot of η against μ_A for a passenger car where maximum efficiency occurs when $\mu_A = 0.8$. Below this value the front wheels lock first and the vehicle continues to move in a straight line. Above $\mu_A = 0.8$ the rear wheels lock first and the vehicle tends to turn about its c. of g. By choosing $\eta = 1.0$ at high decelerations rear wheel locking can be avoided but the loss of maximum deceleration is greatest at low values of μ_A and becomes much less at high values of μ_A.

26. An extreme case of load transfer occurs on two-wheeled vehicles and the maximum deceleration is only about 0.36g when the rear wheel only is braked and $\mu_A = 1.0$.

Steady-state - Three-axle Vehicles

27. Vehicles with three or more axles can be treated by an analysis similar to that outlined for two-axle vehicles. Articulated vehicles are specially important because of the large differences in axle-loading that can exist.

28. The simplest form of solution of the equations for articulated vehicles is when all the axles lock simultaneously. The basic equations (Ref.5) show that the deceleration of the vehicle is $\mu_A g$, the trailer load transfer is as for a car, modified by the difference in king-pin and vertical distance from c. of g. The tractor load transfers are in two parts, that owing to the mass of the tractor which is as for a car, together with that owing to the trailer which acts like an additional point mass at the king-pin location. On a typical articulated vehicle the ratio of tractor plus trailer load to tractor front axle load can vary from 2.5 to 1 in the fully-laden vehicle, to 5 to 1 in the unladen state. Thus a fixed braking system is unable to ensure that all axles lock together for a range of μ_A and vehicle load and is the fundamental problem with vehicles of this type.

Transient Load Transfers

29. During braking a pitching moment is applied to the vehicle and this imparts pitch and vertical motions to the vehicle body. The resulting effects may be critical as the vertical axle loads can vary and in the critical situation cause axle locking and affect the vehicle stability.

30. The complex case of an articulated vehicle has been studied (Ref.6) and Figs. 6 and 7 show typical transient load transfers on the front axle, tractor rear axle and trailer axle for the vehicle unladen and laden respectively for various brake inputs.

31. These results show that step brake inputs can result in bigger variations of load than exponential inputs. This can be critical if there is a large reduction in load on the tractor rear axle. Exponential inputs reduce the excitation as the slower response system produces less oscillatory motion.

32. In the laden vehicle time delays in the trailer brake operation cause large excitation of the oscillatory motions of the trailer axle load. The effect of brake delay also gives a transient drop in rear axle load during the period when the trailer brakes are not operating.

©IMechE 1976

33. The existence of transient load changes is thus a potential source of danger on a vehicle as these can increase the probability of locking under critical adhesion conditions.

Improving Stability

34. Some means of adjusting the braking ratio to the road tyre adhesion and loading on each wheel would help give the shortest braking distance without loss of control.

35. Variable braking ratios can be achieved by sensing by the pipe line pressure or deceleration and providing accordingly restricted braking to the rear wheels in such a way that as deceleration is increased more braking is applied to the fronts and so reduce the risk of the rear wheels locking. The braking ratio can then be made to approximate to the ideal for a fixed static loading as can be seen in Fig.8.

36. Load-sensing proportioning valves are practically essential for trucks, or buses which can experience large changes in the load distribution. These units have been made to respond to static loading and to deceleration. On articulated vehicles two valves, one placed on the tractor rear axle and the other on the trailer axle, can cover most of the requirements of load and deceleration changes.

37. In practice variations in brake torque for a given input limit the effectiveness of brake proportioning arrangements. For example, a designer might arrange his system so that the braking ratio is optimised at 65/35 for a particular set of conditions on a drum-braked car. If the brakes, however, should be heavily worked the front brakes will become hotter than the rears and if they should fade the braking ratio may fall to say 50/50 with the result that in an emergency stop the rear wheels instead of the fronts will lock first and braking distance will be increased. Of course, one way to maintain the ratio constant is to fit more stable brakes and brake stability will be discussed later. Another and more general way to cope with changing conditions is to fit anti-lock devices.

38. Anti-lock devices all work on the principle of determining the deceleration of the sensed wheel and if this is above a certain value, that is, if the wheel is on the point of locking, the brake actuating force is reduced, the wheel speeds up and the actuation force is re-applied, and the cycle is then repeated as long as the wheel is on the point of locking. These devices give improved stability and controllability in emergencies. Studies (Ref.7) suggest that anti-lock systems would have a definite benefit in 4 to 7.5% of accidents.

BRAKING CONTROL

Ergonomics

39. Pedal effort during a stop is invariably modulated and the driver calls up a particular programme of changing pedal effort, probably with a very strong proprioceptive element, for a particular situation and monitors his situation visually throughout the stop. The more skilled the driver the less visual feedback he requires and he works to a firmly imprinted programme that is difficult to perturb. Braking skills are mostly acquired at low pedal efforts involving decelerations of approximately 0.2g and high decelerations on average are attained once every 16000km in the U.S.A. but more frequently in Europe More knowledge of the factors that influence the driver decision to brake is required if the vehicle and environment is to be best adapted to realise the capabilities and overcome the limitations of the driver.

40. The actuating system provides the force to apply the brakes and the torque developed depends on the type of brake mechanism. The gain that has to be designed into the braking system is influenced by the pedal effort a driver can exert and this depends on anthropometric factors. Most drivers, if reasonably fit, can exert a force 2.3 times this body weight for a short time and even higher forces in an emergency situation. However there is poor correlation between maximum force and body weight as there is between age and maximum force. Females are capable of exerting a maximum force of average about 55 per cent that of males (Ref.8). Pedal efforts of 356-400N are considered reasonable for the weakest car drivers to produce a deceleration of 1.0g. Greater pedal efforts are tolerated on commercial vehicles and 445N pedal effort typically develops a 0.6g deceleration.

41. Similarly for handbrakes a 222N effort is considered satisfactory to hold a car on a 1 in 4 gradient and allowance has to be made at the design stage for the poor efficiency of a mechanical system. Travel of the force control should not be excessive - maximum pedal and lever movements should preferably not exceed about 60mm and 90mm respectively for cars.

Actuating Systems

42. The actuating system connects the pedal with the brake and provides the gain in force to operate the brake which is approximately three times the load on the wheel if reasonably high decelerations of the wheel are to be attained. This is achieved by hydraulic, air or mixed systems.

43. Hydraulic systems have small light com-

ponents which can be readily adapted to incorporate a servo unit. They are used with and without a servo unit on vehicles weighing up to 5000kg. On vehicles up to 10000kg weight a servo unit is essential. On heavier vehicles up to 13000kg weight air/hydraulics are used fitted with a hydrovac servo.

44. Direct air systems can provide very large brake actuating forces and are used on heavier vehicles with spring brake actuators for the secondary and parking brake. Air brake systems on bigger vehicles have to be quite complex in order to meet safety requirements. Systems and their components have been highly developed and designed to minimise time delays.

45. The system and size of components are selected together with the size and type of brake to give not only the required torque but to cater for the performance and operating duties of the vehicle.

BRAKE MECHANISMS

Brakes

46. Brakes are almost exclusively of the friction type and classified as drum and disc. The torque developed for a given input depends on its design and particularly the coefficient of friction between lining and drum or disc. In comparing brakes it is convenient to derive the non-dimensional quantity S, the shoe factor, defined as the total frictional force on a shoe or pad divided by the applied load actuating the shoe or pad.

47. In general S can be expressed in the form

$$\frac{1}{S} = \left(\frac{a}{\mu} \mp \frac{b}{c}\right) \quad \ldots \quad (6)$$

where a, b and c are constant and the minus sign refers to the leading shoe and the plus sign refers to the trailing shoe. Expressions have been derived (Ref.9) to determine S for all the well known brakes. Each shoe factor multiplied by the actuating force and the drum radius or the mean radius for a pad gives the brake shoe torque. In determining the actuating force allowance must be made for the efficiency of the system which is approximately 0.9 for a hydraulic system and can be half this value for a mechanical system.

48. The torque of most drum brakes increases in a non-proportional manner with μ as a result of self-energisation. The S cam and disc brake output torques are proportional to μ. A comparison of the brake torque developed by the most commonly used brakes is shown in Fig. 9.

49. In practice it is essential to ensure that not too great a shoe factor is used as this may give rise to excessive sensitivity s. As a general guide sensitivity can be defined as the percentage change in torque either for a given change in μ (Ref.10) or for a percentage change in μ (Ref.11). Either definition is acceptable and if the latter parameter is adopted experience suggests that an s value below 2 gives a reasonably stable brake for most purposes.

50. The conventional analysis for drum brakes assumes the shoe and drum to be rigid and the lining perfectly elastic, and the calculated torques are directly proportional to the brake shoe actuating force. In practice shoe flexibility and distortions and differential geometry between drum and lining result in the torque varying non-linearly with the shoe actuating loads (Ref.12). At higher loads the torque increases linearly with load, the curves being parallel but displaced to the curves obtained when complete contact occurs. Greater displacements are observed when initial contact occurs towards the ends of the lining. In a well bedded-in lining torques calculated from the simplified analysis are in reasonably good agreement with measured values.

Brake Components

51. The components of the brake, the drum and friction material, also influence the stresses that a brake can support. Thermal expansion of the drum can alter the shoe factor and therefore the torque; the torque tends to fall when the drum is expanding and to increase when the drum is contracting when cooling. The brake will therefore grab and give an erratic performance and may also be noisy.

52. Expansion, deflections, lining compression and clearances also affect the shoe-centre lift which is the radial displacement of the shoes at the centre of the shoe and limits the effectiveness of the brake. Any losses must be allowed for in the available shoe-centre lift if useful lining wear is to be obtained. The available shoe-centre lift in turn must be catered for in the design of the actuating mechanism together with the fluid displacement requirement in the brake system.

53. With discs thermal expansion effects are smaller than with drums as the expansion is towards the pads but pad compressibility can cause considerable lost motion in the brake system.

54. Automatic adjusters to cater for wear of the friction material are an essential requirement of any brake to maintain a high mechanical advantage and this is especially important for integral handbrakes with disc brakes. There is therefore a need for a cheap and reliable automatic wear adjuster.

©IMechE 1976

55. The metal member may craze from cracks formed by overheating and repeated stress cycling from compression to tension as the surface is heated and cooled.

56. If large amounts of heat are dissipated over small areas of cold metal member the resulting thermal gradients are high and metallurgical transformations can take place at these areas giving rise to small discoloured hard "hot spots". Hot spots may cause local cracking and as they are slightly above the metal surface they machine away the friction material. Cracked metal surfaces can also give rise to brake judder.

57. Asbestos-resin based friction materials incorporating inorganic and other fillers continue to be developed to withstand higher temperatures and to exhibit lower wear rates. These materials have the advantages of good performance and low wear over the wide range of temperatures experienced in service while still remaining cheaper than other types of friction materials. When considering the friction pair, linings are assessed on their performance against cast iron which is exclusively used as the mating surface. Recent work (Ref.13) however has shown that the presence of very small amounts of other elements in the cast iron can cause remarkably large changes in μ and wear. This is an area open to greater exploitation. Good matching of the friction pairs is essential to retain the balance of braking on all wheels of a vehicle. Brake defects that cause different torques on each side of the vehicle produce lateral instability that is particularly dangerous at high speed and during braking and cornering (Ref.14).

Choice of Brakes

58. Weight and performance determine the type of drum brake to be fitted. On lighter cars of medium performance drum brakes of two leading shoes or leading/trailing shoe brakes are fitted to the front wheels and leading/trailing shoe brakes on the rear wheels. On heavier cars, duo servo brakes are used and these are very suitable for the rear brakes to give a good handbrake performance without power assistance and disc brakes are fitted to the front wheels.

59. Overall, disc brakes are more stable than drum brakes and can run at higher temperatures. As performance depends on temperature and the cooling rates of the front and rear brakes are different, it is easier to maintain a constant braking ratio by using disc brakes, which are less sensitive to temperature, on all four wheels, rather than fit disc brakes on front and drum brakes on rear wheels. There is therefore, a trend towards disc brakes being used on all wheels of cars.

60. On heavy vehicles drum brakes have the advantage, as conventional discs are more exposed to road debris than drums and score badly and pad life is less than lining life. As ample power is available with most systems used on heavy vehicles leading/trailing shoe brakes are normally fitted although two leading shoe brakes are also used for the front wheels with leading/trailing shoe brakes at the rear wheels. The shoes are generally forced apart by a cam or by a wedge mechanism.

SUMMARY

61. Some of the more important factors involved in the braking of road vehicles have been presented here to enable the problems to be appreciated at the design stage. Any design should always be finalised in consultation with specialist component manufacturers, as developments are continually taking place in order to make components safer and more reliable and easier to maintain.

62. Safety has been further emphasised by the many international braking regulations now in force; these have hastened progress towards improved systems and better braking performance and the development of methods of giving improved stability to the vehicle during braking under emergency conditions.

REFERENCES

1. WILLIAMS, T., MEADES, J.K., and RILEY, B.S. The wet road grip of lorry tyres. Road Res.Rept. LR544 (Road Research Laboratory, Crowthorne, Berks.) 1973.
2. STAUGHTON, G.C. and WILLIAMS, T. Tyre performance in wet surface conditions. Road Res.Rept. LR355 (Road Research Laboratory, Crowthorne, Berks.) 1970.
3. INGRAM, B. and OPPENHEIMER, P. Safer braking systems. Proc.Instn.Mech.Engrs. 1973, 187, 10.
4. ROUSE, A.J. The distribution of braking in road vehicles. Symposium on Control of Vehicles During Braking and Cornering (Inst.Mech.Engrs.) 1963.
5. HALES, F.D. Commercial vehicles (Inst.Mech.Engrs.) 1967, Chapter 6.
6. NORTH, M.R. and OLIVER, S.J. Transient load transfers in an articulated vehicle under braking conditions. Rept. No. 1967/7 (MIRA) 1967.
7. MACKAY, G.M. The effectiveness of vehicle safety designs in accidents and injury reduction. Conference on Vehicle Safety Legislation - Its Engineering and Social Implications (Inst.Mech.Engrs.) 1973.
8. MORTIMER, R.G., SEGEL, L., DUGOFF, H., CAMPBELL, J.D., JORGESON, C.M. and MURPHY, R.W. Brake force requirement study. Driver-vehicle Braking Performance as a Function of Brake System Design

Variables. Report No. PB-193-258. National Highway Traffic Safety Administration Contract FH-11-6952. Highway Safety Research Institute, University of Michigan, 1970.

9. NEWCOMB, T.P. and SPURR, R.T. Braking of road vehicles. (Chapman and Hall Ltd.) 1967, Chapter 3.

10. ODIER, J. Contribution to the study of brakes and the rational use of friction materials. Bulletin Société Française des Mécaniciens, 1955, No. 16.

11. FAZEKAS, G.A.G. Brake design. A detailed investigation of fade sensitivity. Auto. Engr., 1951, 41, 53.

12. MILLNER, N. and PARSONS, B. Effect of contact geometry and elastic deformations on the torque characteristics of a drum brake. Proc.I.Mech.E. Auto.Div., 1973, 178, 317.

13. HATCH, D. Cast iron brake discs. J.Auto.Eng. 1972, 3, 38.

14. ODIER, J. Contribution à l'etude de l'instabilité en freinage des véhicules automobiles. Journal de la Sociéte des Ingénieurs de l'Automobile 1960, 33, No.10.

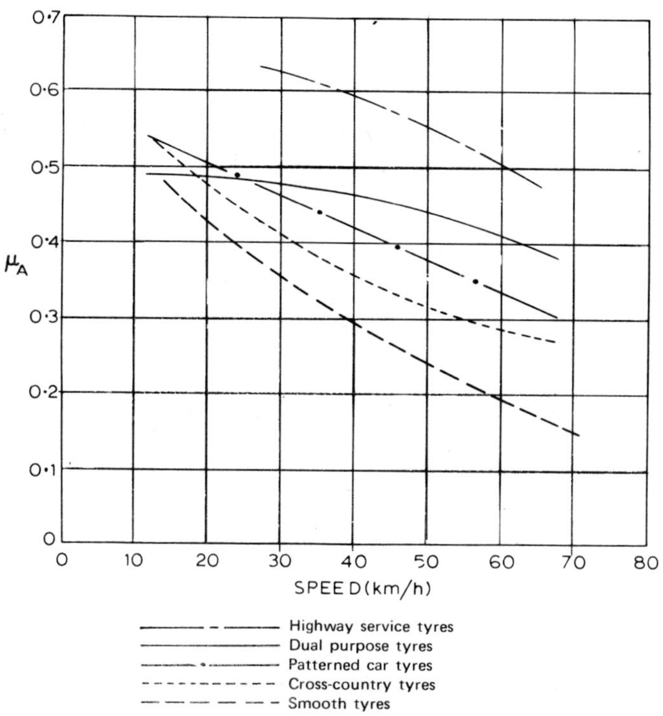

Fig. 1: Locked wheel μ_A on a wet, smooth concrete surface.

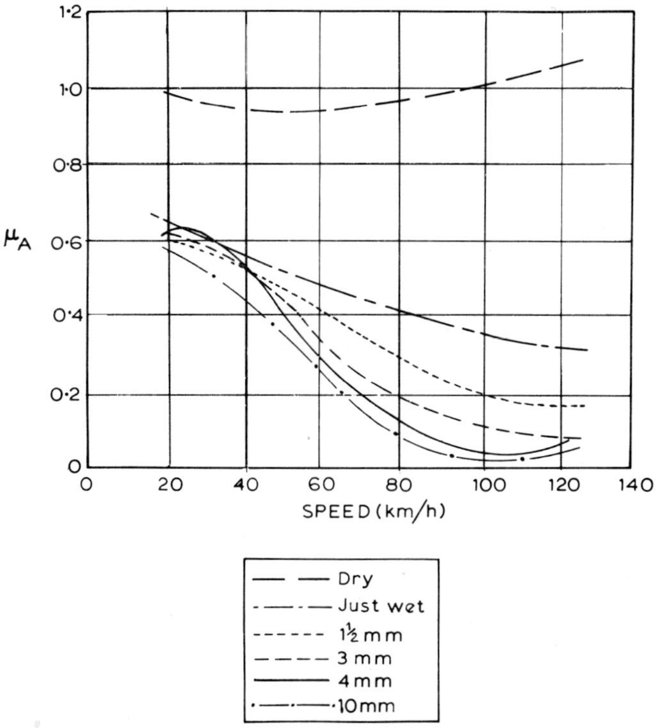

Fig. 2: Effect of water depth on μ_A with speed, smooth concrete surface

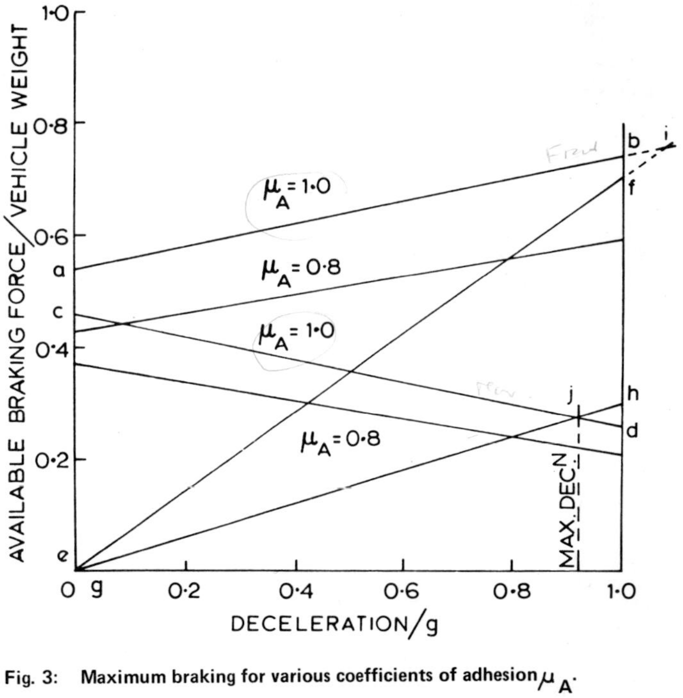

Fig. 3: Maximum braking for various coefficients of adhesion μ_A.

©IMechE 1976

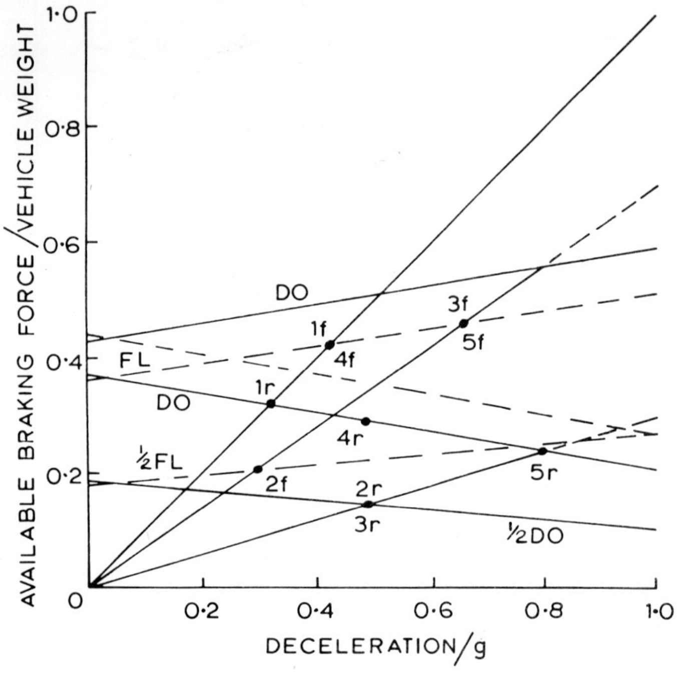

Fig. 4: Braking performance of different types of dual-circuit systems

1 – front-rear split, 2 – diagonal split, 3 – triangular split, 4 – duplicated front single rear, 5 – fully-duplicated (dual) system

FL = fully-laden vehicle
DO = driver only condition

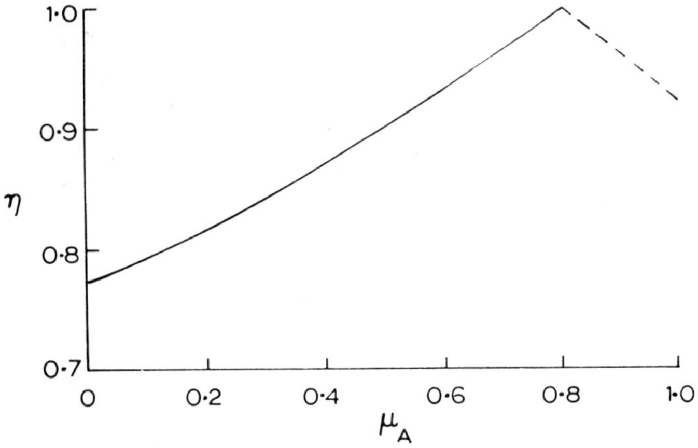

Fig. 5: Typical variation of η and μ_A for a passenger car.

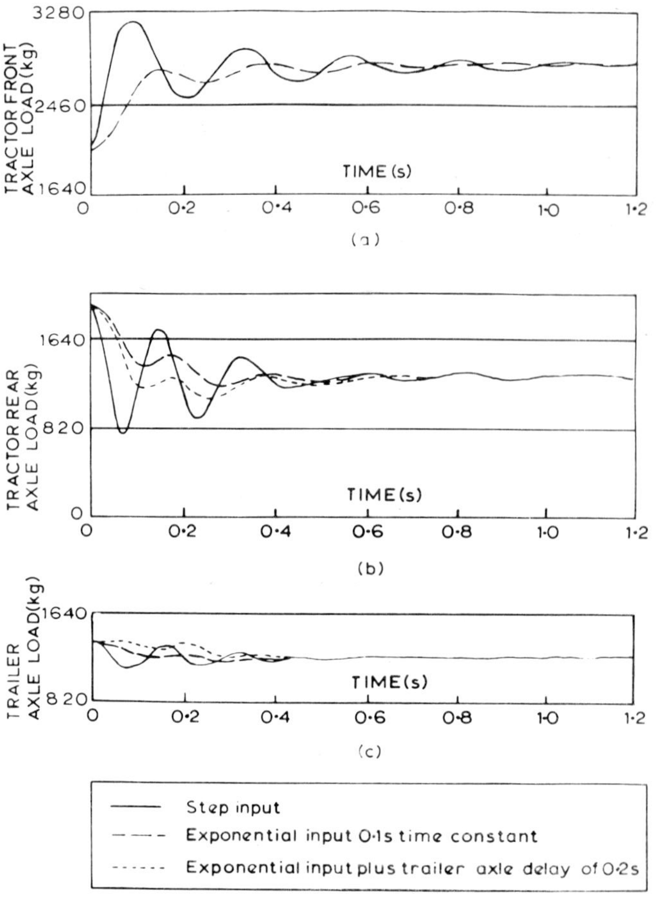

Fig. 6: Axle loadings on articulated vehicle (unladen) for step and exponential brake inputs

a) Front axle load
b) Rear axle load
c) Trailer axle load

©IMechE 1976

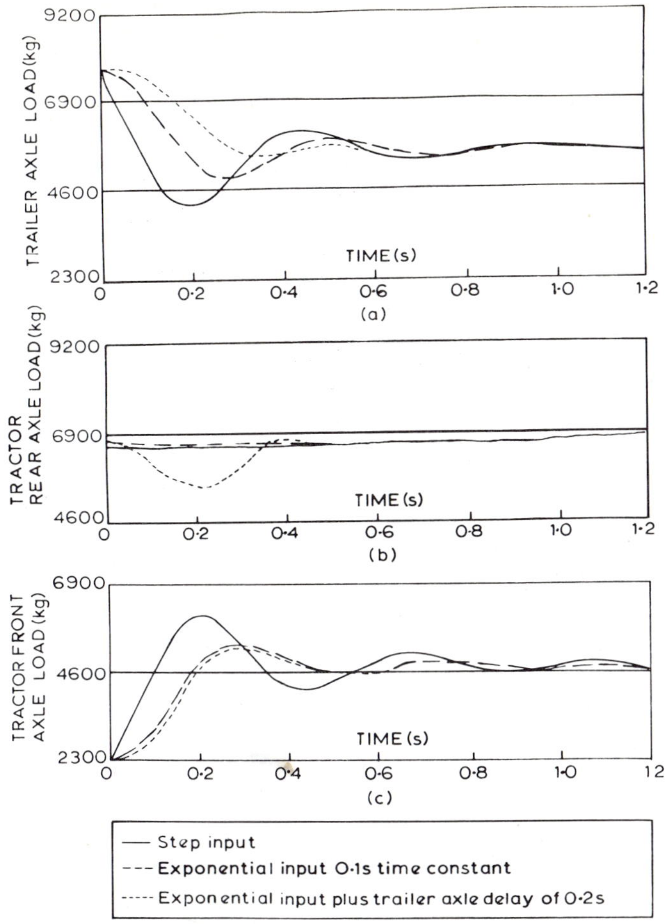

Fig. 7: Axle loadings on articulated vehicle (unladen) for step and exponential brake inputs

a) Trailer axle load
b) Tractor rear axle load
c) Tractor front axle load

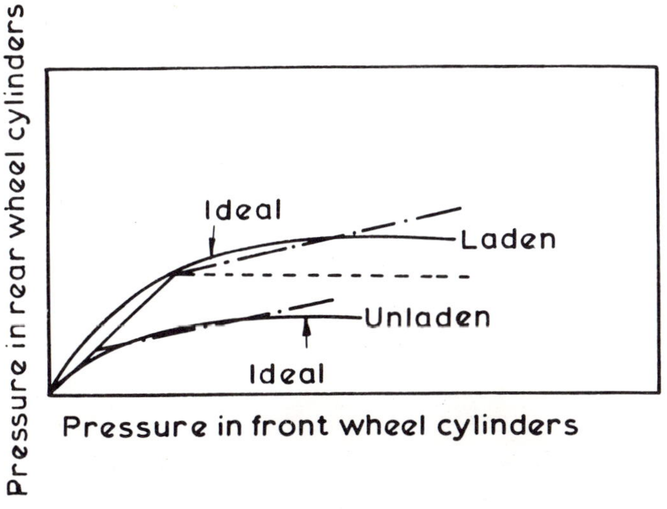

Fig. 8: The way in which the pressure in the rear-wheel cylinders varies with respect to that in the front-wheel cylinders. Dashed line is pressure when pressure-limiting valve is fitted. Chained lines is pressure when pressure-reducing valve is fitted.

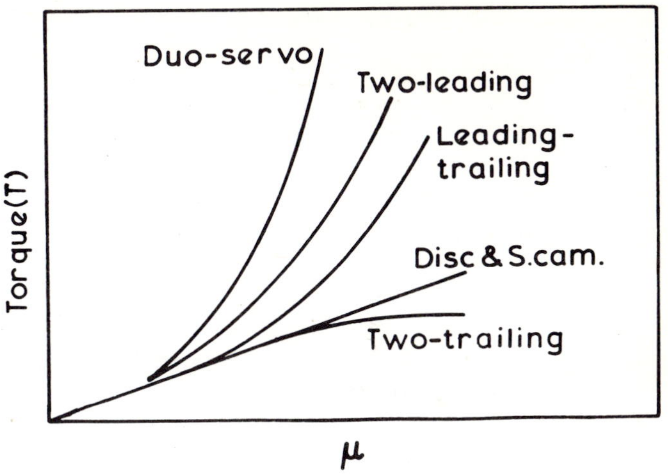

Fig. 9: Variation of torque with μ for different brakes.

© IMechE 1976

C49/76

BRAKING SYSTEMS FOR RIGID VEHICLES

B. INGRAM, CEng, MIMechE and D. PEASLEY
Girling Limited, Kings Road, Tyseley, Birmingham

The MS of this paper was received at the Institution on 18 November 1975 and accepted for publication on 2 December 1975

INTRODUCTION

This presentation outlines the considerations involved in selecting a braking system for a rigid vehicle, these are basically that any braking system should be capable of being operated in safety for a reasonable period of time without maintenance, ensuring an acceptable pedal effort and travel, bearing in mind the purpose for which the vehicle is to be used. It is relevant to note that the requirements of the countries in which the vehicle has to operate must also be met although it is not part of the scope of this paper to deal with legislation in detail. Many of the methods well known to brake engineers involved in prescribing brake system are discussed in ways which it is hoped will prompt engineers always to question motives for the approach used to solve any braking problem.

So far as the driver of a rigid vehicle is concerned the braking system starts with the tyre on the road and ends with his foot on the brake pedal. This theme is used through this paper with sections dealing with the way in which the level of braking is developed at the wheel, the selection of the wheel brakes to cope with the energy and torque created as a result, and finally, the means to enable the requisite force from the foot, suitably augmented and apportioned, to be transmitted to the brakes.

Basic Limitations of a fixed ratio system

To achieve maximum deceleration or 'ideal braking' front and rear wheels must approach locking point simultaneously. At this point the brake force requirements at any wheel are equivalent to the wheel load related to the tyre to road adhesion, frequently referred to as the "effective weight".

As the wheel load changes as a result of weight transfer, different braking force requirements will be required to achieve the ideal over the tyre to road adhesion range.

Because it is common practice to use 'fixed' ratio systems, the braking distribution cannot be arranged to give ideal braking conditions for varying brake force requirements.

Some compromise is necessary and in practice, a condition for optimum performance is selected and the system pressures are modulated, where possible, with valves to produce satisfactory braking over the range See fig. 1.

Braking Distribution for Optimum Performance

As with a basic fixed ratio system, optimum braking can only be achieved for one tyre to road adhesion level (see fig. 1) this condition is selected to be in the safest usage range, say 0.5 mu for trucks and 0.7 mu for cars, and the braking distribution established accordingly.

Taking moments about the rear tyre to ground contact point (see fig. 2)

$$M_0 = 0$$

$$R_F L - Wy - \frac{Wah}{g} = 0$$

$$R_F = \frac{Wy}{L} + \frac{Wah}{gL} \quad \text{..................................} 1$$

At optimum braking (Front and Rear Lock)

$$ayg = \mu \text{ and } \mu R_F = B_F$$

$$\therefore B_F = \frac{Wy\mu}{L} + \frac{W\mu^2 h}{L}$$

and as $B_F + B_R = \mu W$ (effective weight) then the braking proportions or braking ratio is identified.

Significance of Wheel Lock

Apart from stability effects which have been discussed elsewhere (1) wheel locking points are basic to any performance analysis as they ultimately limit braking potential.

© IMechE 1976

The locking characteristics of a vehicle can be presented graphically by plotting the effective weight against deceleration and superimposing the braking proportions. At the intersection, wheel lock occurs such that for the truck represented in fig. 3, on a 0.5 surface, simultaneous front and rear lock occurs (selected optimum) while the rears lock at 0.84g on a 1 μ tyre to road adhesion.

To express the single axle lock condition mathematically, from fig. 3 by similar triangles:-

$$\frac{a}{g} = \frac{\text{front axle weight}}{\text{front braking proportion}} = \frac{R_F}{P_F}$$

$$R_F = \frac{P_F a}{g} \quad \ldots\ldots\ldots\ldots\ldots\ldots\ldots\ldots 2$$

Equating (1) and (2)

$$\frac{P_F a}{g} = \frac{Wy}{L} + \frac{Wah}{gL}$$

and resolving for any tyre to road adhesion to express the locking deceleration of a front wheel.

$$a = \frac{gWy\,\mu}{P_F L - Wh\,\mu}$$

There is legislation in force in certain countries, such as the EEC Directive 75/524/EEC, which controls locking relationships and this is a typical technique used in examinations of that type.

Selection of Wheel Brakes

The selection of brakes for a vehicle is considered to be influenced by the following:-

a) Vehicle Weight

The ultimate strength requirement of a brake is primarily determined by the torque created by the load on the tyre and the adhesion between it and the road. In this respect safety margins (or factors of safety) are especially relevant as it is this developed torque which is transmitted to the brake parts and can be well in excess of that determined from nominal values of lining friction/brake factor and line pressure. For instance, failures of brake components have been experienced even when the ultimate strength has been in excess of twice the deceleration capability of the actuation/brake package, based on nominal performance values. Spike testing is now typically a part of most brake proving programmes.

Weight is also the key factor in terms of energy input levels during hill descents or for that matter on any single brake application.

b) Engine Power

Ultimately engine power determines two things, the time between successive brake applications and the maximum speed at the commencement of the brake application. Brake usage on main roads bears a relationship to maximum speed and, for all practical purposes, a direct relationship to engine power on cross country routes.

In reality until recently, the power to weight ratio was only significant in car usage as trucks, with around 8 BHP/TON have their brake sizes influenced more by weight than engine power, but some modern vehicles now boast 20 BHP/TON and it is evident that this increase, in future, will have to be reconciled in any brake specification.

c) Disc or Drum Brakes

Passenger Cars

The drum brake obtained a bad reputation on medium/heavy cars of the late 50's because of the continual uprating in power to weight ratio's and improved road holding which reflected in excessive levels of duty at the brakes. Adopting the disc brake reconciled many of the problems on these vehicles but, it is relevant to note, this change resulted in an increase in rubbed area of 30 to 40%.

Trucks

This choice is somewhat limited because of the availability of suitable disc brakes. This situation is a legacy left by the disappointments of the early 1960's.

One operator, at least, has persisted over the past 10 years or so with a fleet of single decker buses so equipped, which totalled over 400 at the peak, demonstrating that careful selection of vehicle and usage can give an acceptable compromise between smoothness of operation, fade resistance and lining life. The whole problem facing disc brakes, on heavy commercial vehicles, centres on the difficulty of getting a sufficiently large disc into the wheel. Because of this restriction it is necessary to utilize discs at high specific energy levels which results in excessive lining/disc wear and even disc crazing problems. Multiple discs are an obvious development but, because of the attendent mechanical sophistication necessary to provide adequate solutions, current market forces do not make them commercially attractive when compared with drum brakes. The drum brake maintains its precedence by the simple fact that wheels which diametrically limit disc brakes, will accommodate drums of 8" in width. It is also relevant to note that drum brakes have the ability of self-limit, a feature which cannot be over-emphasised and is one reason for their continued use on heavy vehicles in spite of the disc brake's superior fade and smoothness characteristics.

d) Brakes in addition to mechanical suitability have to be judged on the following operational criteria.

i) "one off" stop from maximum speed - no sensible heat loss - heat sink capability/temperature rise, main limiting characteristic.

ii) Driving down a long hill descent - energy input dependent on grade and gross vehicle weight - ability to reject heat at a rate sufficient to avoid excessive temperatures, main requirement.

iii) Driving on the level - energy input governed by engine power - ability to reject heat at a rate sufficient to avoid excessive temperatures, main criteria.

i) and ii) can fairly readily be assessed but iii) is more complex because the energy input level is most influenced by the driver.

© IMechE 1976

We can, however, make some fairly simple judgements based on various researchers and backed by our own experience, for example:-

Brakes used on both Passenger Cars and Commercial Vehicles are significantly effected by long hill descents such as the type 1 ECE/EEC 6 Km - 7% grade considered typical of the alpine pass. So far as commercial vehicles are concerned, the effects of such a decline are extremely serious unless a low gear is selected - there is no substitute for a retarder if such conditions are a regular occurrance. It has also been shown, by MacKenzie, Newcombe and Watton (2) that even passenger cars, unless they are capable of exceeding 100 mph, are unlikely to input more energy on the level than on such declines. The passenger car has an advantage in that the relationship between rubbed area and GVW tend to be superior, particularly since the incorporation of disc brakes, such that the ratio is in the order of 2 to 1 in favour of the passenger car.

The Significance of Brake Types

In any one torque range there is normally a choice of brakes available including disc brakes and various types of drum brakes.

Types of Brakes	Typical Design Characteristics
1. Disc Brakes	
a) Opposed Piston b) Single Sided	Equal work per pad - No geometry effects
2. Drum Brakes	Geometry Effects
a) Leading Trailing	Equal work per shoe - e.g. fixed cam brakes or Equal input per shoe e.g. the Girling GNS brake
b) Two Leading Shoe	Equal work and equal input per shoe
c) Two Trailing Shoe	Equal work and equal input per shoe
d) Duo Servo	Equal input per shoe

There are various ways of comparing brake designs of a given torque capacity for example, by 'PV' analysis, one very relevant method well understood by the brake engineer in which the input force (P)/travel (V) relationship is compared with the resultant torque output. In this respect, an obvious comparison is between the disc brake and the duo servo drum where the 'PV' requirement is in the order of $1\frac{1}{2}$ to 1 respectively to the advantage of the duo servo. On the other hand comparing a leading trailing equal input per shoe drum with a duo servo drum the 'PV' requirement is $2\frac{1}{2}$ to 1 again in favour of the duo servo. On this basis why are there any other sorts of brakes?

We need to look at stability characteristics to find the answer - fade sensitivity of a brake is influenced by the rate of change of brake factor with lining friction - which is a means of indicating the fade resistance capability of various brake designs - good fade resistance being synonomous with small rate of change in brake factor. Hence some brakes such as disc brakes should be good in this respect and duo servo on the other hand poor. If we consider the relationship between lining friction and brake factor, which can be deduced using methods ascribed to Acres, et alia (3) (see fig. 4), then by knowing the variation of friction with temperature, (see fig. 5) it is possible to predict the fade/temperature characteristics for each brake type being considered. Fig. 6 shows the increase in % fade with fall off in friction coefficient for various brake designs.

On this basis of evaluation it can be shown that the disc brake and 'S' cam are evidently the best as they show the least fade. The following table shows the brake types in descending order of fade resistance deduced on this basis.

Brake Type	Order of Fade Resistance	
Disc Brake	1	good
Cam Brake	2	
L/T Brake	3	
Two Leading Shoe	4	
Duo Servo	5	poor

In reality it is difficult to demonstrate this comparison quite so clearly between some designs as brakes are complex mechanical devices. Although the input/output characteristics can be considered to be defined by the geometry effect, this "definition" can be seriously affected by a whole range of these factors.

The 'S' cam for example is simple - rugged - and is therefore popular throughout the world because it can be easily understood, even though it has various characteristics which don't endear it to the purist e.g. its inability to provide good shoe/drum conformity and high friction losses which give rise to its consequent poor definition. The more sophisticated two leading shoe "Twinstop" brake on the other hand, has good definition which, together with its sliding shoe, enable the use of very high friction linings with a consequent reduction in actuation 'PV' requirements.

Factors affecting "definition of a brake:-

a) Friction losses in actuating mechanism.

b) Variation in reaction forces, in comparison with brake geometry, because of deformation of shoe tips and abutment friction.

c) Variation in lining/drum conformity because of drum/shoe deflection through physical and thermal stresses.

d) The actuating means may be unable to provide adequate travel to meet the whole range of mechanical/thermal deflections.

Practical tests carried out on the dynamometer and vehicle show the degree of fade which exists is often different to that which can be accounted for by using the theoretical brake factor v lining friction coefficient relationship, even having taken into assessment the friction/temperature characteristics of the particular lining material, see fig. 7. For instance, investigation into pertinent dimensions of the drum

© IMechE 1976

designs used revealed that variations in the rim thickness from 0.44" to 0.768" had very little effect on the maximum mean rubbing surface temperature see fig. 8. This observation is in agreement with work undertaken by Fazekas (4). The thermal aspects of rim thickness were, therefore, considered to be of a low order of significance and the deflection and strength characteristics of the rim section were considered to be more relevant see fig. 9. Hence the position of the rib, its shape, the relative strength of the flange and the general thickness become important.

It was also evident from the work carried out that there was a relationship between fade resistance and specific rubbed area, in that fig. 10 shows the drum temperature reached during a particular fade test procedure and the way in which this varies with the specific rubbed area of the drum. From this work it was deduced that the fade resistance of a particular duo servo brake could, for example, be equated on a particular test routine with a two leading shoe brake if the specific rubbed area of the duo servo was 30% greater. This increased area would keep the duo servo temperature down, thereby balancing out the effect of its greater fade sensitivity as shown in fig. 11.

It was interesting to note from the test work that the average temperature recorded on the vehicle under equivalent fade test energy levels. The vehicle tests showed that the rear brakes, during fade test conditions, typically ran 20% hotter than those on the front axle at equivalent energy input levels per ft^2 of rubbed drum area. Refering back to fig. 10 it can be appreciated that a small increase in the size of brakes fitted to a particular vehicle weight (i.e. an increase in specific rubbed area) can result in considerable improvements in fade resistance and lining life. In addition small differences in the spectrum of vehicle usage can produce equally dramatic changes in lining life as shown by fig. 12 which relates the significance of rubbing temperature to lining life and duty level.

Another feature which is significant, though its relative importance cannot readily be determined without a very detailed knowledge of the work spectrum, is "conditioning" i.e. improving or modifying a lining's fade behaviour by say, working it hard; fig. 13 shows the result of repeating a medium duty fade test on a particular lining. Care must therefore be exercised in pre-judging results quoted at random from a series of tests without understanding the ramification of the work spectrum defined by the test procedure.

Actuation

It can be argued that the process of selecting the actuating media from vacuum, air, hydraulic or a combination is associated with identifying the losses within a braking system. There are of course significant characteristics associated with each actuation media, but essentially two are compressible and one not. In this respect, losses are on the one hand immediately interpreted into having adequate volume reservoirs with acceptable depletion rates, and on the other, limiting ergonomic factors that dictate pedal effort/pedal travel involvements. Legislation is, therefore, in evidence to control depletion rates and pedal effort levels. As will become apparent, designing a braking system is all to do with the conservation of PV, (P) referring to pressure or effort and (V) referring to volume or travel.

Losses in a brake system are basically of two sorts:-

a) losses incurred by manufacturing tolerances and running clearances.

b) losses incurred by deflection as a result of physical stresses.

By way of example, consider a simple hydrostatically operated hydraulic braking system on a passenger car as follows:-

Basic Data

Braking Ratio	65% Front 35% Rear	Effective lining friction	0.3
Vehicle GVW	3000 lbs	Front Brakes	Disc
Pedal Ratio	4.5 to 1	Rear Brakes	Drum

The following table shows how the losses build up to equate to the 3 inches travel found to be necessary to achieve 0.87g.

Item	Losses: Manufacturing and Running Clearances	Losses: Mechanical Deflection (Δ Mech)
1. Master Cylinder valve cut off	2½%	
2. Lost travel in pedal gear	6%	
3. Caliper seal roll back	14%	
4. Caliper seal knock back	20½%	
5. Drum shoe to drum clearance	15%	
6. Hoses and bundy tube expansion		12½%
7. Drum Elastic Deflection		10½%
8. Caliper Elastic Deflection		16½%
9. Fluid Compression		2½%
Totals	58% +	42%=100%

Now consider the significance of increasing the effective friction level of the brake linings:-

As for practical purposes it can be simply considered that:-

Mechanical Deflection

$$(\Delta_{mech}) \propto \frac{1}{\text{effective lining friction} \quad (u)}$$

If we increase 'μ' from 0.3 to 0.4

$$\Delta\, mech = k \frac{1}{\mu} \qquad \text{hence } k = \frac{42}{100} \times 0.3$$

Therefore, if

$$\mu = 0.4 \text{ then } \underline{mech} = \frac{42 \times 0.3}{100} \times \frac{1}{0.4} = 31.5\%$$

© IMechE 1976

In addition, Pedal Effort (or System Pressure)

$$\propto \frac{1}{\text{effective lining friction} \quad (\mu)}$$

hence there will also be a lowering of pedal effort to $\frac{3}{4}$ of the original value. So what is the cumulative effect of the consequent reduction in mechanical deflection?

It will either reduce the pedal travel by (42% - 31.5%) = 10.5% or if the travel is retained at 3 inches by modifying the pedal ratio, the pedal effort will be further reduced to 2/3 of the original value. We can see from this why the brake engineer's dream is a 0.5 mu lining with zero fade.

One way of artificially increasing effective lining friction is to increase the geometric self-servo effect of the brake design, however the engineer must, in addition to remembering the effects on fade resistance, be careful not to introduce travel losses in the process.

A similar effect can be achieved in system terms, by making say, the caliper stiffer, also automatic adjusters can be introduced to offset and control the manufacturing losses in both disc and drum brakes.

The hydrostatic hydraulic system, when applied to vehicles in excess of 3000 lb (GVW), tends to have pedal efforts which are unacceptable by present day fashion, and typically one then finds the addition of servo devices to augment personal effort. The servo is essentially a device to make up losses and in this respect the vacuum hydraulic servo was typically found in the U.K. on rigid vehicles up to 24,000 lb GVW. However, this type of servo has, over recent years, lost favour because of the inception of divided circuit systems which introduced problems of pressure equalization.

The air system, in its various forms, now dominates the commercial vehicle actuation scene where it augments hydrostatic hydraulic systems and also operates mechanical brakes directly by way of air actuators. Fundamentally air pressure is superior to vacuum being less affected by altitude and capable of higher pressure thresholds. Even the packaging of air systems presents problems, hence the use of air over hydraulic systems in conjunction with high definition/high factor hydraulic brakes gives a realistic compromise.

Stored Energy (or power hydraulic systems) have, more recently, been used on cars and commercial vehicles. But there are two fundamental characteristics which must always be remembered:-

a) "Cliff-Edge" Characteristic

The driver of a stored energy system can be lured into a false sense of security because the choice of gas pre-charge pressure in the accumulator is relatively high and gives a realistic compromise deceleration, the safety of the system being completely reliant on the driver heeding warning lights and buzzers. The air system depletes slowly and ultimately to zero there being no "sudden death" situation as shown in fig. 14.

b) Header-Tank Volume

The air system draws its basic fluid from the atmosphere and can, therefore, make up for small leaks on an indefinite basis. By comparison, however generous the header tank is on the Stored Energy system, it is nevertheless of finite volume. This situation might be likened to the hydrostatic hydraulic braking system, but only providing there is a divided reservoir and a two pump system.

Whether or not the hydraulic power system comes into more general usage because of its packaging convenience, these two simple facts will continue to harass the brake engineer on the subject of its inherent safety.

So far in this presentation philosophical considerations have been avoided, but this short section on actuation has shown that at some stage the engineer, with the ultimate responsibility for a vehicle braking system, must develop his philosophy sufficiently to reason out a sensible balance between conflicting requirements. It must be borne in mind that human beings have to operate these machines when others are standing in close proximity, none of whom necessarily have the benifit of our insight into brake system design concepts.

Adhesion Utilization

As a result of the anomalies of fixed ratio systems, even valved by the most sophisticated techniques commercially available, the overall braking performance is still a compromise and ultimately the level of compromisse must be quantified. This can be achieved by using the adhesion utilization curve where the adhesion available at any road condition is compared with limiting deceleration produced by wheel locking.

Until recently, the extent of the compromise was the responsibility of the engineer and identified with his philosophy, there being very limited statutory requirements.

The adoption of Directive 75/524/EEC now requires specific locking characteristics, mainly associated with avoiding rear wheel locking by increasing the front braking proportion, and although it is desirable to regularise a system that is open to abuse by the unscrupulous, does it improve the specification?

By the natural course of events and motivated by the need to progress, valves were becoming a more universal fitment. However, the presence of a valve is not essential to 75/524 and compliance can be achieved with a fixed ratio. The attached compliance plots for a truck will demonstrate this:-

Fig. 15A and 15B a code of practice type system with valve (system X)

Fig. 16A and 16B 75/524/EEC without valve (system Y).

System X is a fairly conventional system with generally accepted levels of braking front and rear and displaying good adhesion utilization across the range. System Y has a 66/34 braking ratio and levels of adhesion utilization which are unlikely to be considered consistent with good braking philosophy. Yet all are satisfactory within the terms of 75/524/EEC. Legislation has the effect of removing the responsibility from the individual, but this example shows that it does not necessarily provide a satisfactory substitute!

© IMechE 1976

The Data Sheet

Various aspects of braking systems have so far been discussed, including many of the problems, no attempt having been made to draw all the apparently conflicting requirements together. One way in which the brake manufacturer does this is by the use of the vehicle data sheet. Although this document is merely pages of numbers, it tells a comprehensive story including a 'PV' analysis, an energy absorption analysis, choice of apportioning valve and its affect on distribution, including, in the case of commercial vehicles, compatability with the towing of trailers, but most significant of all, it spells out the basic design of the brake system. A typical passenger car and commercial vehicle data sheet are included by way of examples. It is not within the scope of this paper however, to examine, justify, or necessarily quote the parameters chosen in order to uphold them as representing good practice. Nevertheless, the inclusion of these data sheets provides a focal point, demonstrating that no one single aspect makes a braking system and that it is necessary to carry out a careful balancing act to achieve the best practice in brake system design with the cold hard facts of legislative and commercial pressures.

CONCLUSIONS

The subject of braking has been variously called a science and an art. This presentation continues to show that both of these descriptions are apt, especially as legislation has now complicated the judgement process. Vehicle improvements, resulting in better handling, in turn enable higher power/weight ratios to be utilized, thereby placing an even greater burden on the braking system. Emphasis is placed in this report on the significance of the area of braking surface which is rubbed by the brake lining and, in this respect, it is recommended that vehicle designers ensure adequate provision is made for a satisfactory braking package.

ACKNOWLEDGEMENTS

Much of the supporting data used in this paper is the result of painstaking research work by our colleagues and we should like to gratefully acknowledge their support and assistance which we received in the course of its preparation. We should also like to thank the Directors of Girling Limited and Joseph Lucas Limited for permission to publish this paper.

REFERENCES

1. Ingram B. Oppenheimer P.
1973 Proc. Auto. Div. Instn. Mech. Engs.
"Safer Braking Systems"

2. Mackenzie D.K. Newcombe T.K. Watton B.
1962-63 Proc. Auto. Div. Instn. Mech. Engs.
"Car Brake Usage and Design of Test Schedules"

3. Acres F.A.S.
1946 Proc. Inst. Auto. Eng. Vol. 41
"Some Problems in the Design of Braking Systems"

4. Fazekas G.A.G.
Trans. S.A.E. Polytechnic Institute of Brooklyn 1953
"Temperature Gradients and Heat Stresses in brake drums".

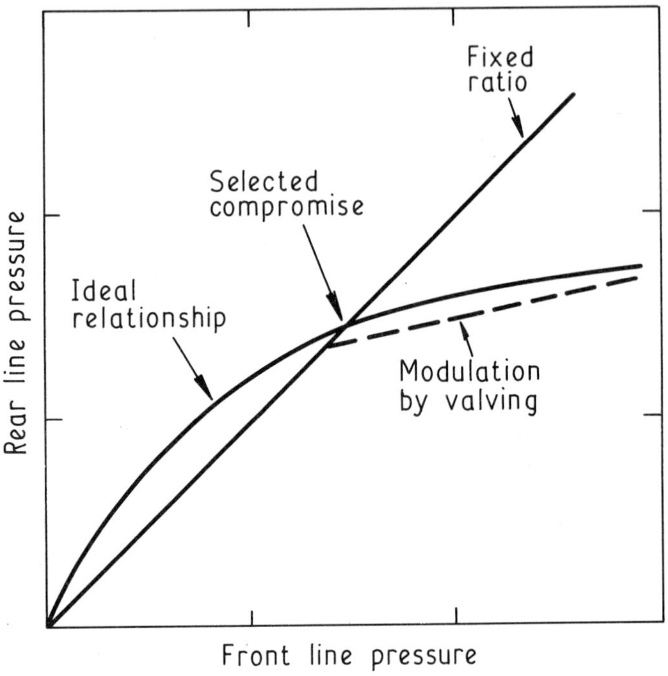

Fig. 1

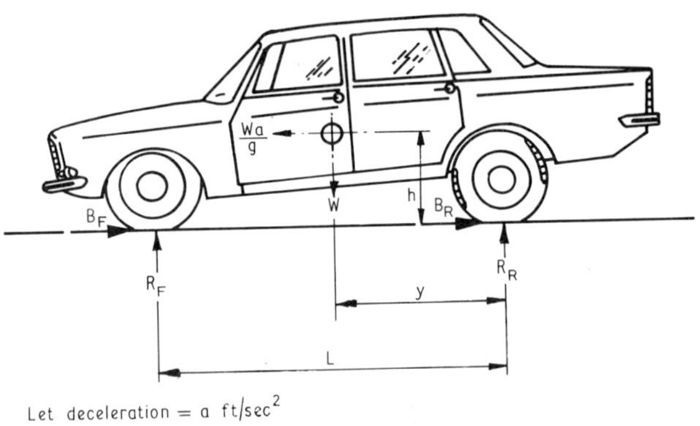

Fig. 2

© IMechE 1976

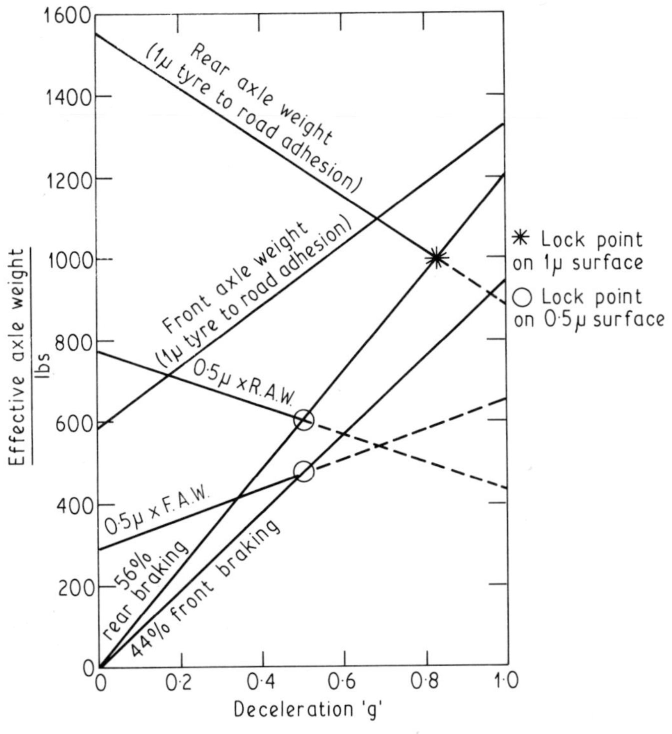

Fig. 3

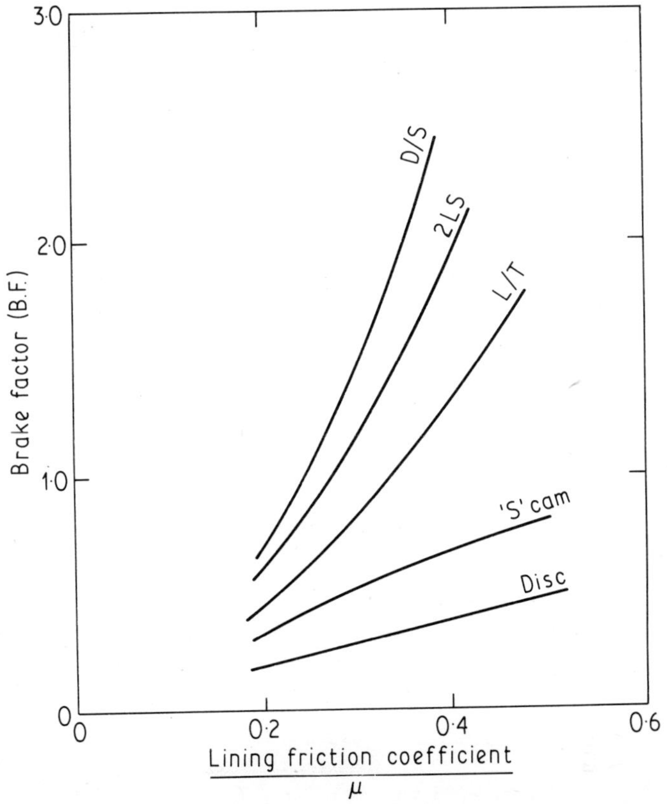

Fig. 4

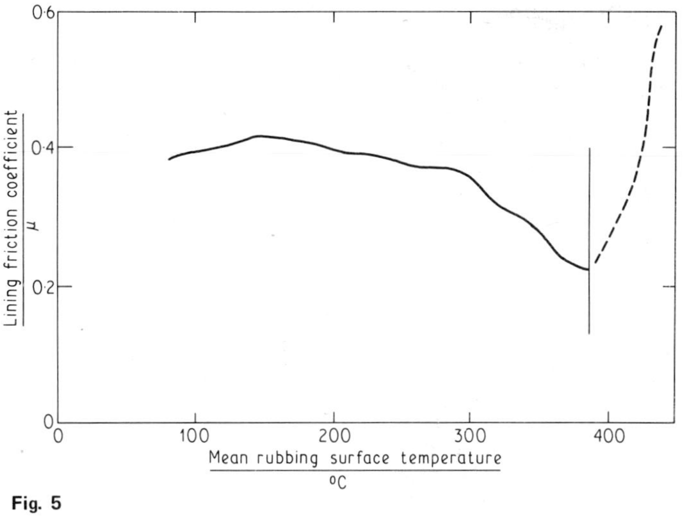

Fig. 5

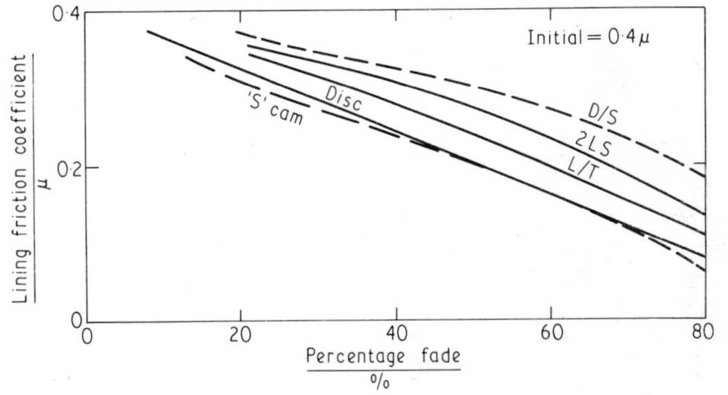

Fig. 6

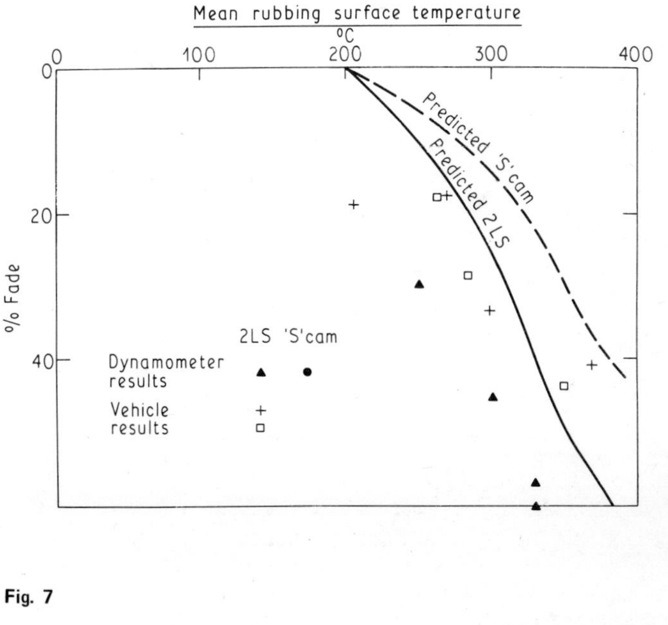

Fig. 7

© IMechE 1976

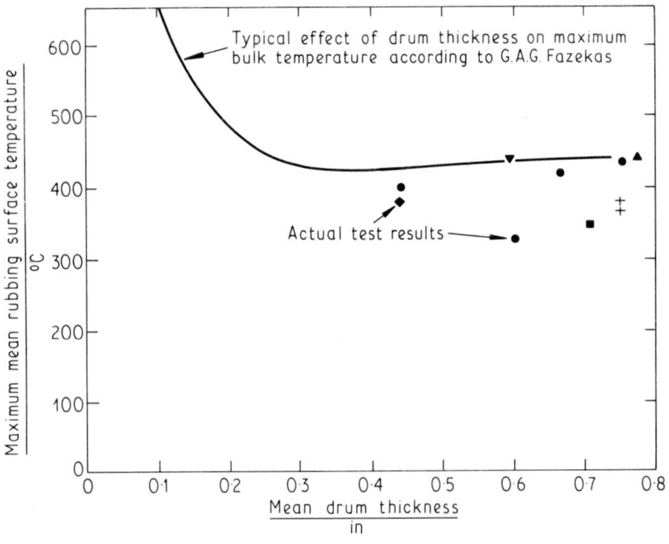

Fig. 8

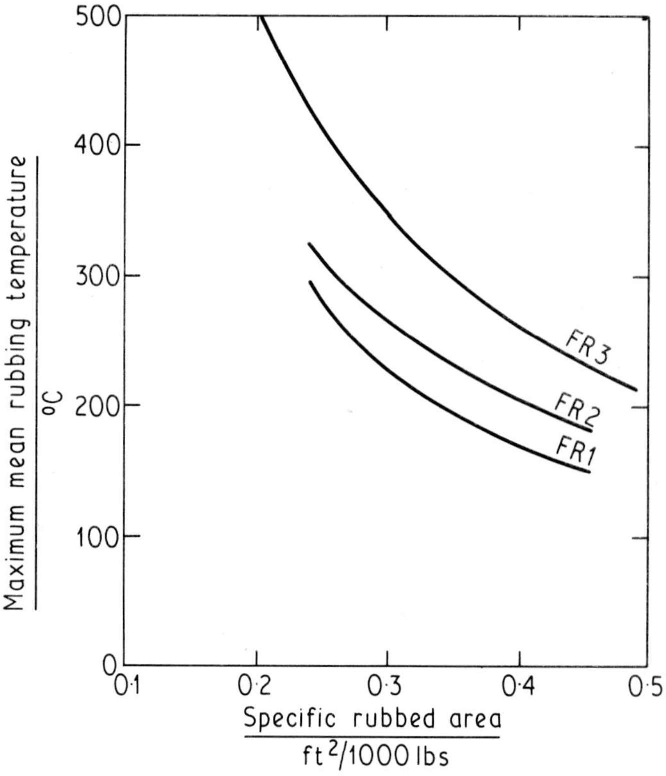

Fig. 10

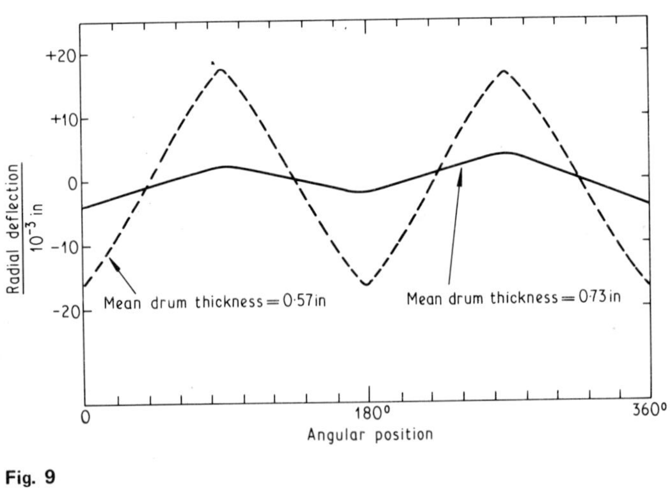

Fig. 9

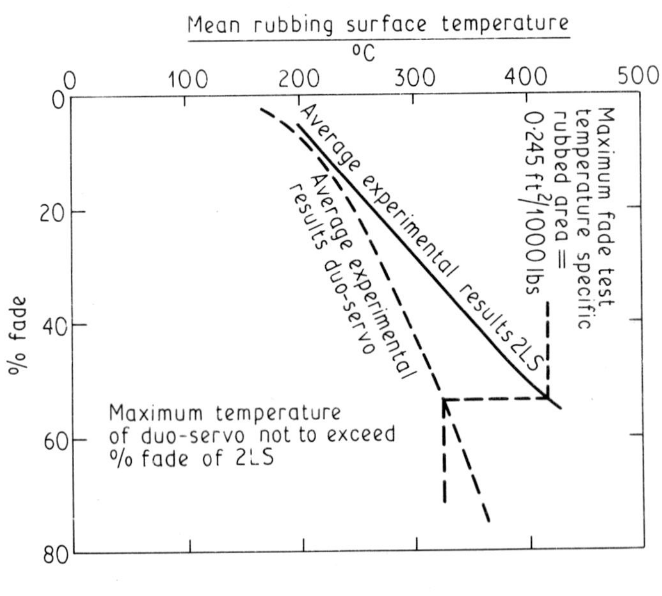

Fig. 11

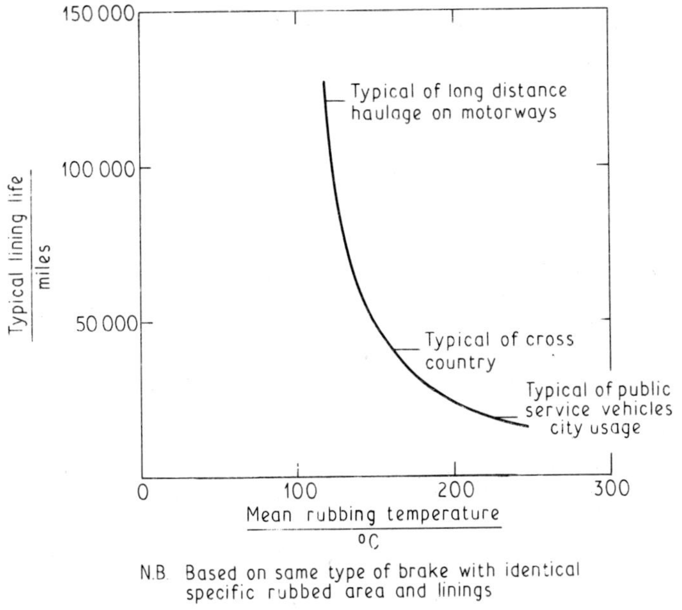

Fig. 12

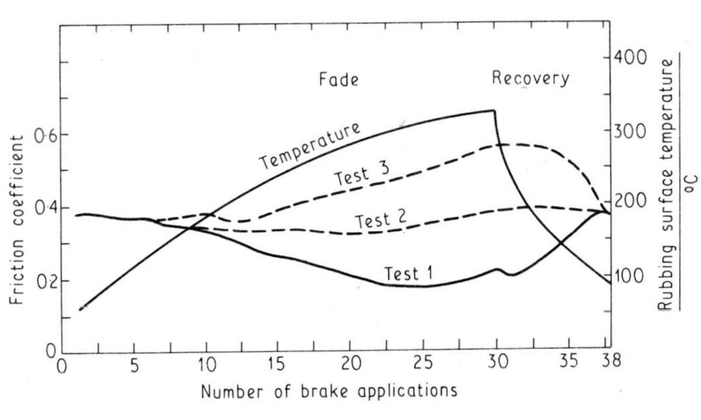

Fig. 13

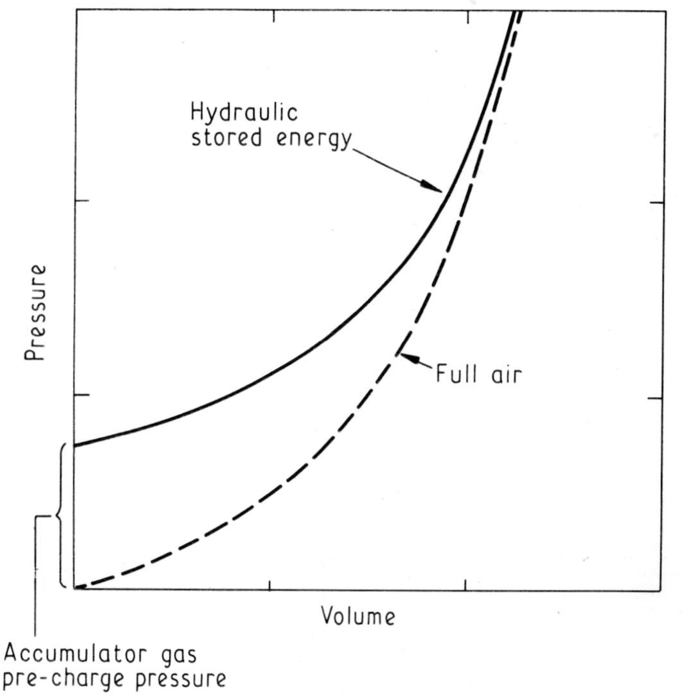

Fig. 14

© IMechE 1976

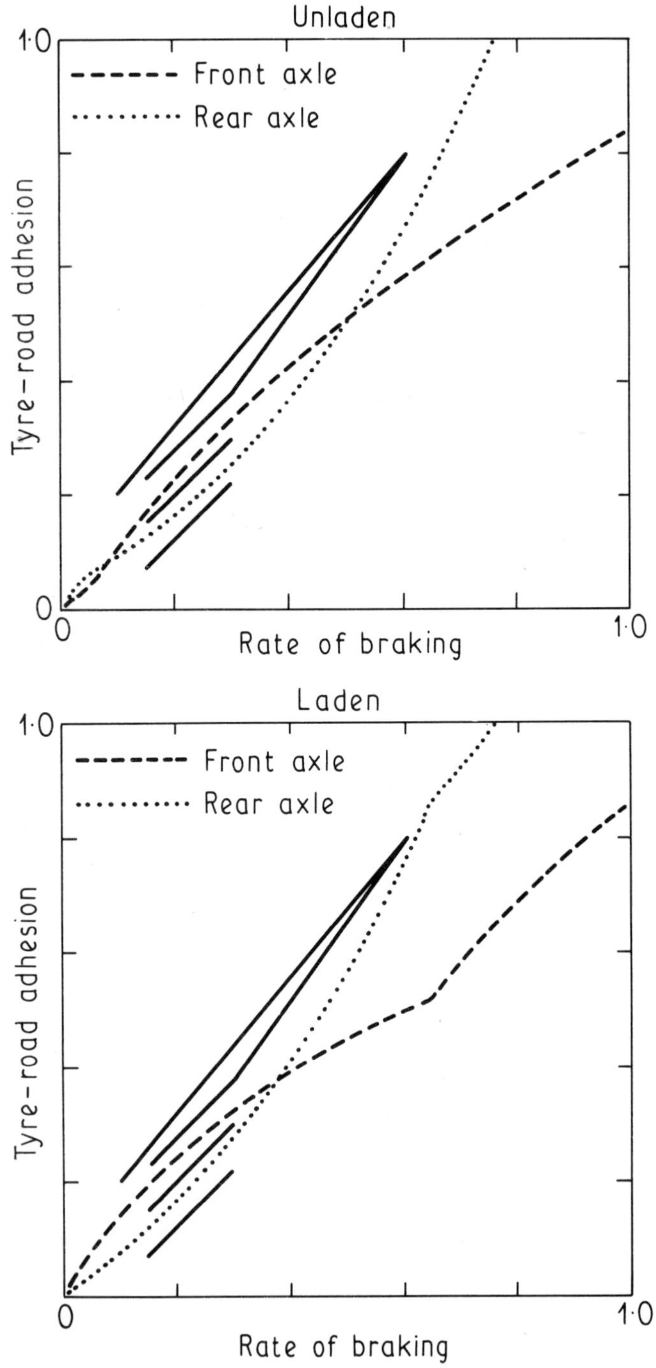

Fig. 15A

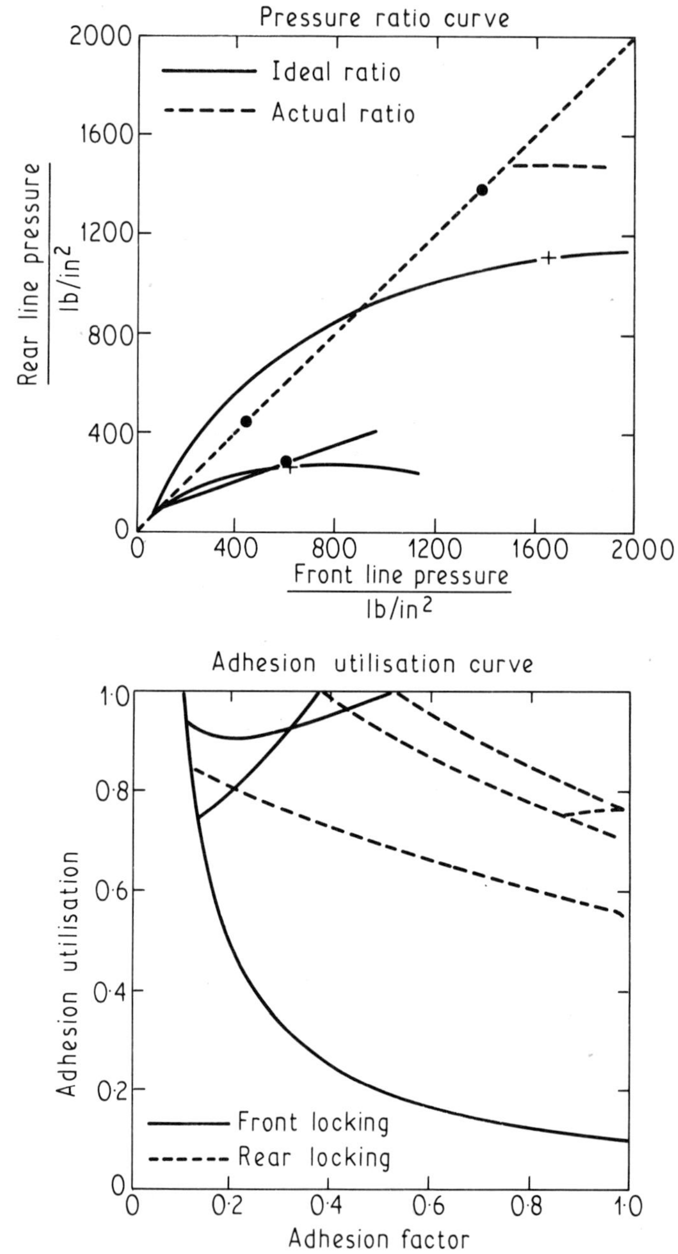

Fig. 15B

© IMechE 1976

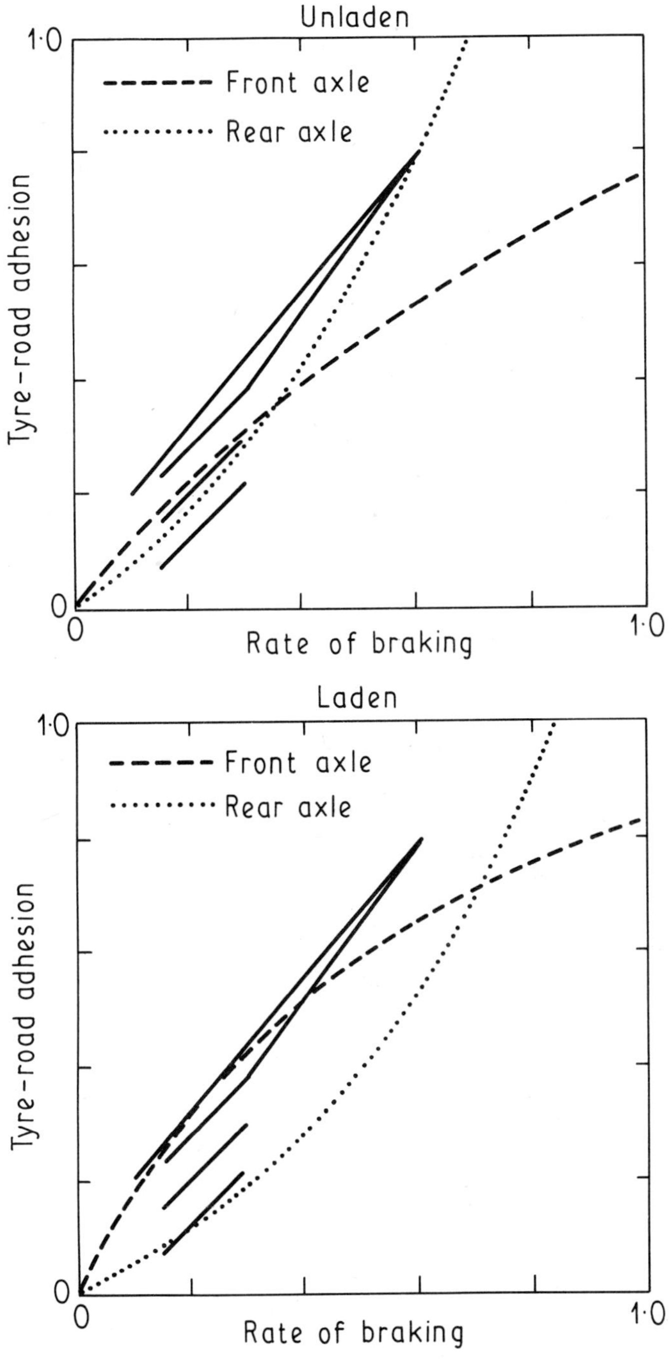

Fig. 16A

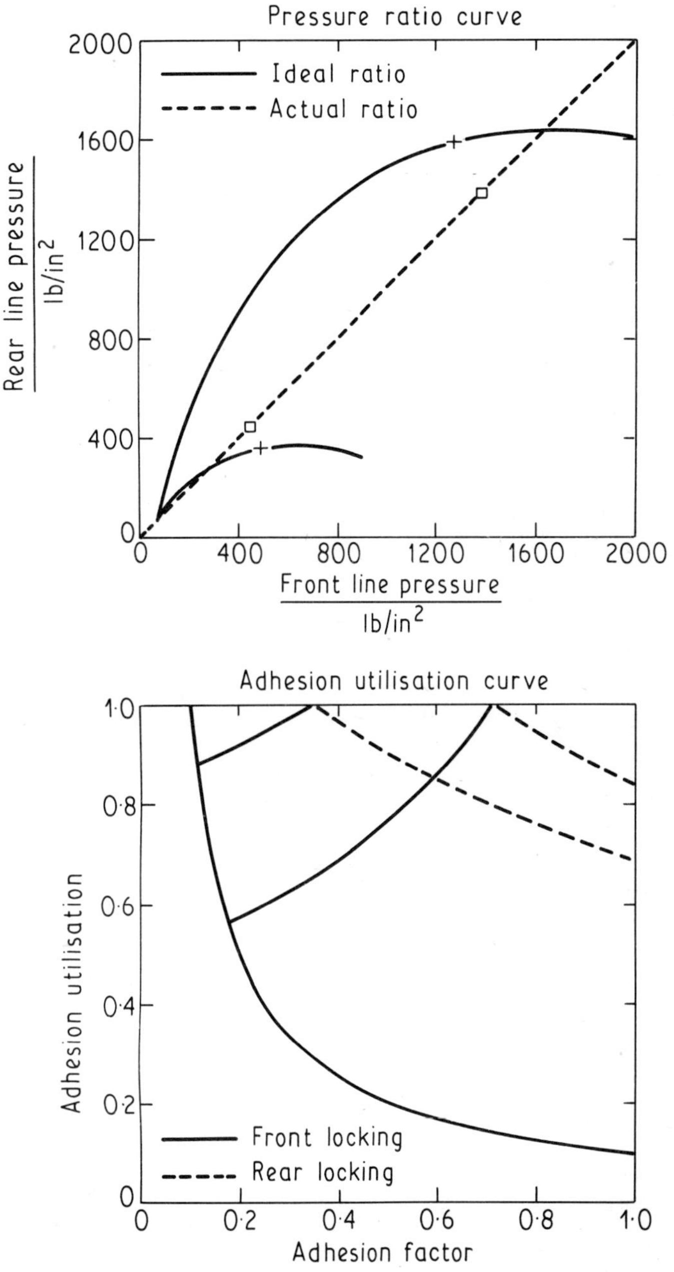

Fig. 16B

© IMechE 1976

```
GIRLING LIMITED, KINGS ROAD, BIRMINGHAM B11 2AH

COMMERCIAL VEHICLE BRAKE PERFORMANCE PREDICTION

CUSTOMER AND MODEL: TYPICAL TRUCK 4X2  WITH TRAILER
                                   PURPOSE:
REQUEST NO. ARN        0

VEHICLE SPECIFICATION FROM QUESTIONNAIRE NO    DATED

GROSS TRAIN WT. 88174  LB
GROSS TRLR. WT. 52344  LB
HEIGHT OF PICK-UP 33.0 IN


HT C.G. UNLADEN  41.3 IN
HT C.G. LADEN    63.8 IN



                                        208   IN

LADEN WEIGHT        35830 LB    14555                   21275
UNLADEN WEIGHT      18150 LB    11221                    6929
TYRE SECTION                    12.00                   12.00
RIM SIZE                        22.50                   22.50
ROLLING RADIUS            IN    20.75                   20.75
ENGINE POWER             BHP      335
POWER / WEIGHT       BHP/TON    20.94
MAXIMUM SPEED            MPH       76

   MK.III TWIN STOP BRAKE SPECIFICATION
DIAMETER                  IN    15.50                   15.50
WIDTH                     IN    7.874                   6.889
WEDGE ANGLE              DEG     6.25                    7.50
LINING MU                        0.42                    0.42
LINING AREA / AXLE    SQ.IN     426.0                   372.7
DESIGN RATING TORQUE LB.IN     159320                  137192

   ACTUATION SPECIFICATION
MANUFACTURER                    M.G.M.
SERVO TYPE                     T 9 /T 9S               T 9 /T 9S
SPRING GEOMETRY                XHD L/T                 XHD L/T
VALVE                                                     NONE


   ACTUATION SYSTEM OPERATING DIAGRAM
```

Fig. 17a

```
GIRLING LIMITED, KINGS ROAD, BIRMINGHAM B11 2AH

COMMERCIAL VEHICLE BRAKE PERFORMANCE PREDICTION

                                    FRONT                  REAR
                             1ST AXLE  2ND AXLE    1ST AXLE  2ND AXLE

   SERVICE BRAKE AT .60G DECELERATION LADEN
EFC BRAKING AT .61G       %    46.66                53.34
ACTUAL BRAKING RATIO      %    54.58                45.42
BRAKE OUTPUT TORQUE  LB.IN    121738               101303
LINING DRAG             PSI     73.7                 70.1
BRAKE CHAMBER EFFORT     LB      793                  802
AIR PRESSURE            PSI     86.2                 86.2
SERVICE S.C.L.           IN     .080                 .073
CHAMBER TRAVEL (100 C)   IN    0.748                0.575
POWER ABSORPTION:
    - LINING       HP/SQIN      2.78                 2.64
- RUBBED AREA      HP/SQIN      1.54                 1.47
FACTOR                         21.13                17.49
THRESHOLD EFFORT         LB       49                   55
MAX AVAILABLE S.C.L.     IN     .189                 .228

   SECONDARY BRAKE AT .30G OFF THE TREADLE (TRUCK ONLY)
BRAKE GEOMETRY                                        2LS
BRAKING RATIO             %                        100.00
S.C.L.                   IN                          .076
BRAKE CHAMBER EFFORT     LB                           878
CHAMBER TRAVEL (100 C)   IN                         0.601
AIR PRESSURE            PSI                          94.6
MAX AVAILABLE S.C.L.     IN                          .228
FACTOR                                              17.49

   SECONDARY PERFORMANCE OFF SPRINGS (TRUCK ONLY)
FORWARD     57.1 %G    ADH.     0.52                 0.64
PLUNGER EFFORT           LB     1197                 1277
PLUNGER TRAVEL (100 C)   IN    1.423                1.121
FACTOR                         12.66                10.50

   GRADIENT PARKING PERFORMANCE OFF SPRINGS (TRUCK ONLY)
FORWARD     60.7 %G    ADH.     0.74                 0.80
PLUNGER EFFORT           LB     1170                 1060
PLUNGER TRAVEL (100 C)   IN    1.523                1.106
FACTOR                         14.90                12.36
REVERSE     43.7 %G    ADH.     0.80                 0.37
PLUNGER EFFORT           LB      878                 1289
PLUNGER TRAVEL (100 C)   IN    1.184                1.074
FACTOR                         11.32                 9.39

   LOCKING POINT DECELERATIONS ON A 0.8 ADHESION  SURFACE (%G)
SERVICE     LADEN                                    67.9
            UNLADEN                                  49.8
EMERGENCY   LADEN                                    38.1
            UNLADEN                                  26.4
```

Fig. 17b

© IMechE 1976

```
GIRLING LIMITED, KINGS ROAD, BIRMINGHAM B11 2AH

COMMERCIAL VEHICLE BRAKE PERFORMANCE PREDICTION

                                     FRONT                    REAR
                              1ST AXLE   2ND AXLE      1ST AXLE   2ND AXLE

    SECONDARY TRAIN BRAKING AT .30G OFF THE TREADLE
      (NO ASSISTANCE FROM TRAILER)
BRAKE GEOMETRY                 2LS                      2LS
BRAKING RATIO          %      54.60                    45.40
S.C.L.                 IN      .089                     .081
BRAKE CHAMBER EFFORT   LB      964                      974
CHAMBER TRAVEL (100 C) IN     0.832                    0.633
AIR PRESSURE           PSI    105.3                    105.3
FACTOR                         21.13                    17.49

    SECONDARY TRAIN PERFORMANCE OFF SPRINGS
FORWARD     23.2 %G     ADH.   0.57                     0.57
PLUNGER EFFORT         LB      1197                     1277
PLUNGER TRAVEL (100 C) IN     1.423                    1.121
FACTOR                         12.66                    10.50

    GRADIENT PARKING TRAIN PERFORMANCE OFF SPRINGS
FORWARD     26.7 %G     ADH.   0.65                     0.72
PLUNGER EFFORT         LB      1170                     1256
PLUNGER TRAVEL (100 C) IN     1.523                    1.199
FACTOR                         14.90                    12.36
REVERSE     19.4 %G     ADH.   0.80                     0.35
PLUNGER EFFORT         LB      1044                     1289
PLUNGER TRAVEL (100 C) IN     1.271                    1.074
FACTOR                         11.32                    9.39


    TRAIN SERVICE LOCKING DECELERATION ON A 0.8 ADHESION SURFACE
      (TRAILER BRAKING RATIO = 59.36 %.)
REAR AXLE LOCKS FIRST AT 67.9 %G

    TRAIN SECONDARY LOCKING DECELERATION ON A 0.8 ADHESION SURFACE
      (TRAILER BRAKING RATIO =   0.00 %.)
REAR AXLE LOCKS FIRST AT 30.7 %G


    REMARKS

THE ABOVE INSTALLATION MEETS ANNEXE 10 LADEN
SPRINGS OPERATE ON TRUCK ONLY
```

Fig. 17c

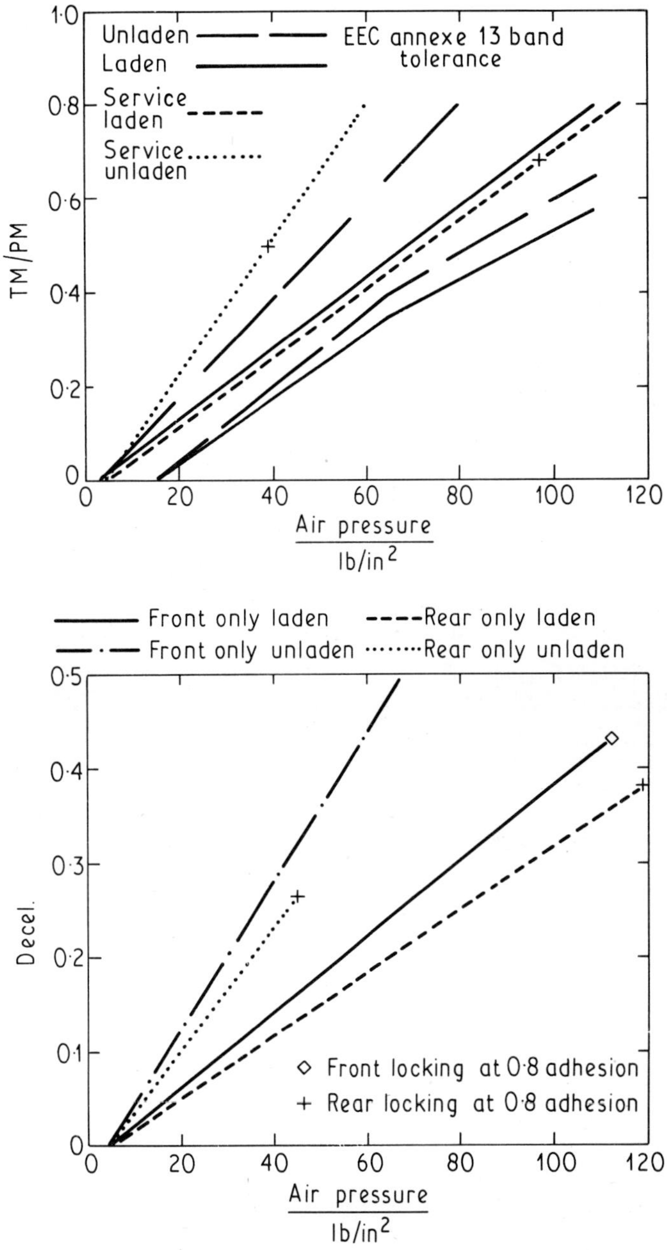

Fig. 17d

© IMechE 1976

```
GIRLING LTD, KINGS RD, BIRMINGHAM B11 2AH.                    DATE 9-10-75

CAR BRAKING PERFORMANCE PREDICTION                            NO. 18 A
CUSTOMER AND MODEL:TYPICAL SALOON
REQUEST:0                     PURPOSE:ENGINEERING INFORMATION

VEHICLE SPECIFICATION FROM QUEST. NO B64/36
  LADEN WT.   TOTAL    LB       2888          WHEELBASE          IN        98.0
  LADEN WT.   FRONT    %        47.3          HT.OF CG.LADEN     IN        20.5
  D.O.WT.     TOTAL    LB       2163          HT.OF CG.D.O.      IN        21.0
  D.O.WT.     FRONT    %        55.1          TYRE SIZE          -      155SR13
  PEDAL  RATIO         -        3.85:1        ROLLING RADIUS     IN        11.04
  PEDAL  TRAVEL  AVAIL.IN       5.50   *      ENGINE POWER       BHP       80
  PEDAL  SPRING LOAD   LB        1.0          MAX.SPEED          MPH       95.0
  PEDAL  SPRING RATE   LB/IN     0.5          VACUUM DRAW        IN.HG.    20.0
  PEDAL  M.A.          -        3.62          P'BRAKE M.A.       -         1.00
  PEDAL  V.R.          -        4.04          P'BRAKE V.R.       -         1.00

BRAKE SPECIFICATION                        FRONT        REAR           PARKING (R)
  TYPE                                     14LFI        HASC           HASC
  EFF.RAD:DIA & WIDTH          IN          3.890        8.000X 1.50    8.000X 1.50
  DISC:DRUM THICKNESS          IN          0.375        0.295          0.295
  LINING MU                    -           0.32         0.40           0.40
  LINING ARC                   DEG         -            110-110        110-110
  CYL DIA:LEVER RATIO          IN:-        -            0.700          2.77:1
  LINING AREA/BRAKE            SQ.IN       8.700        23.00          23.00
  SWEPT AREA/BRAKE             SQ.IN       91.42        37.70          37.70

ACTUATION SPECIFICATION
  MASTER CYLINDER TYPE         -     AS/AS +PDWA        SYSTEM SPLIT F.R.
  DIAMETER                     IN    0.750
  STROKE                       IN    1.358              SPLIT 62-38
  SERVO TYPE                   -     38 SUPVAC          RATIO 2.20:1
  VALVE TYPE                   -     PCRV   + BYPASS    RATIO 2.35:1

ESTIMATED LADEN PERFORMANCE
                      DECEL G    P.E LB      P.T IN       %FRONT      FLP PSI.
  1.01 ADH            0.87       107         3.47         75.8        1512
  SERVO K.P           0.74        77         3.12         74.9        1279
  VALVE CUT IN        0.32        34         2.35         66.9         500
  MEDIUM BRAKING      0.50        52         2.69         71.9         829
  CHECK BRAKING       0.30        33         2.31         67.0         468
  SERVO FAIL          0.30        76         2.44         67.0         468
  PARKING ON) EFFORT LB:BRAKE FWD  144,REV  173:CAB FWD  144,REV  173
  25% GRAD  ) TRAVEL IN:BRAKE FWD 0.00,REV 0.00:CAB FWD 0.00,REV 0.00

IDEAL FRONT BRAKING AT 0.8 G :-  LADEN 64.0%  D.O. 72.2%
WHEEL LOCK ON 0.8 MU SURFACE: LADEN 0.66G FRONT: D.O. 0.79G FRONT

SPLIT SYSTEM PERFORMANCE ON .80 MU SURFACE
  SYSTEM   LADEN DECEL : P.E : P.T        D.O. DECEL : P.E : P.T
    F       .30G :  46LB :  4.10IN         .30G :  36LB :  3.94IN
    R       .30G :  81LB :  5.09IN         .30G :  62LB :  4.83IN

TECHNICAL INFORMATION, LADEN, .87G:-                 I FLUID RES, LADEN:-
    BRAKE   LINING   ENERGY     TEMP   BRAKE HOSE    I         SYSTEM
    TORQUE  LOAD     ABSORP'N   RISE   FLUID FLUID   I FULL.87G    HALF.30G
    LB.IN   PSI      HP/SQ.IN   DEG.C  CU.IN CU.IN        IN          IN
  F 10512   311      13.9       407    0.144 0.042      0.011       0.014
  R  3357   36.5     2.51       193    0.065 0.023      0.033       0.019

REMARKS.    (  ) ESTIMATED VALUE.
            * 5.80 IN PEDAL TRAVEL REQUIRED


CHIEF ENGINEER U.K PRODUCT GROUP              PERFORMANCE ENGINEER
```

Fig. 18a

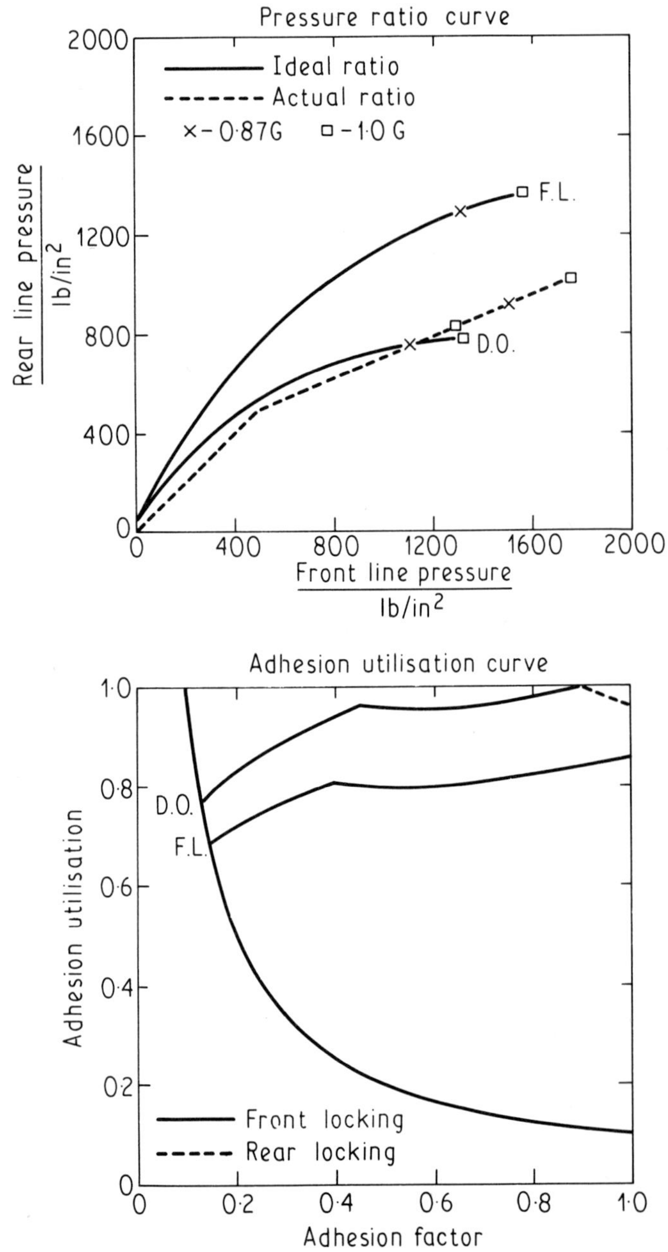

Fig. 18b

C26/76

NON-RIGID VEHICLE BRAKING SYSTEMS

B. R. SHILTON, BSc, CEng, MIMechE
Clayton Dewandre Co. Ltd, Lincoln

The MS of this paper was received at the Institution on 24 November 1975 and accepted for publication on 2 December 1975

SYNOPSIS The paper discusses the problems associated with providing a braking system for a non-rigid vehicle, which has to cater for the articulation of the vehicle, for the separation of the prime mover from the towed vehicle and for interchangeability between towing and towed vehicles. The flexibility of operation which this last factor allows is one of the principal economic reasons for the existence of commercial vehicle trailers and semi-trailers, and safety considerations dictate that compatibility of many aspects of braking performance is considered, as well as the more obvious one of mechanical interchangeability.

A number of brake systems are described, including over-run or inertia brakes, mechanical systems, vacuum systems and air pressure systems, and the means by which these meet the fundamental requirements are discussed.

Recent developments in non-rigid commercial vehicle braking systems are described, and future trends and developments are discussed.

1. NON-RIGID VEHICLE TYPES.

Most non-rigid road vehicles belong to one of three types, and the braking system fitted to such a vehicle is largely dependent upon the category to which it belongs. On all non-rigid vehicles, except very lightweight trailers, the towed portion of the outfit is required by legislation to be fitted with brakes, and these must be automatically actuated when the brakes on the towing vehicle are applied by the driver.

The three main categories of vehicles are:

(1) Passenger cars towing luggage trailers or trailer caravans.

The towing vehicle is usually driven without the trailer attached for the majority of its life, and the design of its braking system takes no account of the trailer requirements.

(2) Load carrying trucks towing drawbar trailers.

The truck may frequently run without the trailer, and its braking system must be fundamentally suited to solo usage, but it must take account of and be adapted to cater for the requirements of the trailer. The trailer braking should be self sufficient, except insofar as the energy required to operate the brakes must be initially derived from the towing vehicle.

(3) Tractor and semi-trailer combinations, in which a significant proportion of the semi-trailer weight is carried by the tractor.

The braking system of the tractor must take account of this, and be capable of braking more than the mass of the tractor itself, and similarly the trailer brakes will be required to brake less than the total mass of the trailer. The tractor will most probably be required to run solo on occasions, and the braking must obviously be safe in this condition.

A small number of special purpose non separable tractor/semi-trailers are in use, the articulation allowed by this configuration aiding manoevrability. Providing suitable braking for such outfits is generally somewhat less problematical than for combinations which can be separated.

2. BRAKING SYSTEM REQUIREMENTS.

The braking system of a non-rigid vehicle, or of a non-rigid outfit consisting of two or more vehicles coupled together, must obviously allow for relative articulation of the vehicles comprising the outfit. The greatest relative movement will take place in yaw when the vehicles are cornering or manoeuvring, but some movement will take place in pitch, when travelling over undulating ground.

A second requirement is that disconnection of the towed vehicle from the prime mover should leave the prime mover driveable, and capable of being safely braked by the driver. In addition, accidental separation of the trailer from the towing vehicle, which may occur on the road, should result in the trailer brakes being applied automatically.

Thirdly, for combinations of vehicles which are required to be interchangeable,it must be possible to disconnect and reconnect any braking devices in a simple manner, and with the minimum chance of incorrect and potentially unsafe connections being made.

The design of the braking systems must be such that when connected and running together, the combination of vehicles must have a satisfactory overall braking performance, without one part of the combination having to provide an unnecessarily high proportion

© IMechE 1976

of the total braking. This aspect of compatibility of braking performance is currently receiving a great deal of attention in the industry.

3. INERTIA BRAKING SYSTEMS.

Probably the simplest form of braking applied to a non-rigid vehicle is the inertia or overrun brake, which is used on most light trailers and caravans which can be towed by passenger cars. As the name implies, this functions by using the compressive force which develops in the coupling or tow bar when the trailer overruns the prime mover, due to the prime mover being braked, to apply the brakes on the trailer. In its most widely used form, this system utilises a telescopic member in the trailer coupling, which, when compressed, applies the trailer brakes via a lever and a conventional mechanical rod mechanism.

A ratchet handbrake acting on the linkage is provided as a trailer parking brake.

A small proportion of the total braking force required to decelerate the trailer - generally between 10% and 20% - must be provided by the compressive force in the tow bar, and therefore by the brakes on the towing vehicle. In addition, the method inherently requires that the towing vehicle's brakes are effective before the trailer's, and therefore the proportion of the total braking force required to decelerate the outfit which is provided by the towing vehicle must change as the brakes are being applied, until a steady state situation is achieved. Hydraulic damping of the telescopic member is generally provided, to prevent trailer brake snatch and to prevent the trailer brakes from being applied due to short term compressive forces in the drawbar, during ride motions of the outfit, and this inevitably delays the onset of trailer braking.

For these reasons, overrun systems are generally limited to relatively light trailers which can be coupled on a casual basis to towing vehicles, the braking systems of which require no modification.

They are, however, virtually self sufficient, and a trailer fitted with such a system is inherently compatible with the towing vehicle as far as braking performance is concerned.

To cater for accidental separation of the trailer from the towing vehicle, a simple safety chain or cable is used to connect the two vehicles, this applies the trailer handbrake mechanism should the towing coupling fail or become disconnected.

One problem which is peculiar to overrun braking systems is that the trailer brakes are automatically applied when attempts are made to reverse the outfit. A number of satisfactory solutions exist, including a solenoid operated hydraulic unit, incorporated in the brake linkage, which is normally rigid but which will freely extend when the solenoid valve is energized, and a recently introduced drum brake in which the direction of the drum rotation is sensed at the brake shoe abutments, and the brake itself rendered virtually ineffective in the reverse direction.

4. MECHANICAL SYSTEMS.

A number of articulated tractor/semi-trailer combinations have been fitted with mechanical devices for transmitting brake linkage forces from the tractor to the semi-trailer, either as the sole means of braking the trailer or in addition to vacuum or air pressure systems. An air pressure actuator or hydraulic cylinder on the tractor, actuated by the same control that applies the tractor brakes, pushes axially on a king pin positioned concentrically with the tractor/semi-trailer coupling. The operation is thus not affected by articulation of the outfit. The king pin thrust is transmitted to a lever mounted on the semi-trailer and thence through a mechanical linkage to the trailer brakes.

5. VACUUM SYSTEMS.

The simplest vacuum systems used on commercial vehicles consisted of brake actuators on the trailer, connected by a vacuum pipe to a control valve on the towing vehicle. This valve was actuated in conjunction with the towing vehicle's brakes and produced a vacuum output level proportional to the driver's footbrake effort. A version of this system is available for fitting to light trailers and caravans, to replace the overrun brake. Later commercial vehicles employed inverted vacuum control. In this system a vacuum reservoir was fitted to the trailer and this was evacuated through a pipe from the towing vehicle. This pipe was subjected to vacuum when the brakes were in the "off" position, and the vacuum level was reduced in order to apply the trailer brakes. Automatic braking in the event of trailer breakaway was thus achieved, and the provision of a reservoir on the trailer speeded up trailer brake application. The system required the use of special valves on the trailer to invert the signal in the pipeline, and a similar valve was often fitted to the towing vehicle to provide the correct signal to the trailer, based upon the level of vacuum in the truck's own brake vacuum servo unit.

6. AIR PRESSURE SYSTEMS.

The majority of trailers and semi-trailers drawn by commercial vehicles are now fitted with air pressure braking equipment, which derives its energy from a source on the towing vehicle. Over the years, differences in national customs and regulations have led to the evolution of different systems, but the trend in Europe is now towards a unified approach, based upon the legislative requirements of the E.E.C. Directive on The Braking of Road Vehicles.

Figures 1 and 2 illustrate truck and drawbar trailer air pressure braking systems designed to conform to the E.E.C. Directive, and typical of modern practice. When the truck is operating without a trailer, it functions as a normal single vehicle, the two air line couplings to which the trailer pipes can be connected being automatically sealed off.

The truck braking system consists of an engine driven air compressor supplying air via a small sensing reservoir to four main storage reservoirs, through system protection valves. Two reservoirs are connected to a dual brake valve operated by the brake pedal, the brake valve supplying independent circuits to the front and rear axle brakes. In the event of any single system or component failure which prevents effective application of the brakes

© IMechE 1976

on one axle, the system protection valves will ensure that the reservoir supplying the brakes on the other axle will continue to be charged by the compressor up to a predetermined pressure level. By ensuring that this pressure is sufficient to decelerate the vehicle at secondary brake performance level, the legal requirements for both service and secondary braking can be met from the footbrake valve.

A third reservoir supplies air via a hand control valve and a relay valve to spring brakes on all wheels of the truck, which are used for parking purposes. To prevent the foundation brakes from being overloaded when both the footbrake and the hand control valve are operated simultaneously, anti-compounding valves in the form of change-over valves are connected across the service brake and spring brake lines. A change-over valve allows the higher of two input pressures through to the output, whilst blanking off the lower input pressure. Thus, when air pressure is applied to the spring brakes to hold them off, any air pressure delivered from the footbrake valve cannot pass through the change-over valve. When the spring brakes are exhausted of air and therefore applied, however, air pressure delivered from the footbrake valve will apply the service brakes and simultaneously pass through the change-over valve to the spring brakes, so reducing the force applied to the brakes by the springs.

The fourth reservoir supplies air to the emergency coupling for the trailer via a relay valve which is triggered by air pressure from the spring brake reservoir. By this means, any system failure which depletes the spring brake reservoir and applies the spring brakes will also exhaust the emergency line to the trailer, and so cause an automatic application of the trailer brakes through the trailer relay emergency valve. This same reservoir also supplies the air to a triple relay valve, the output of which is fed to the trailer service line coupling. The three signal or trigger lines to the relay valve comprise one from each of the front and rear brake lines from the footbrake valve and one from the hand control valve. Thus, should any fault on the tractor prevent the service brakes on one axle from being applied, operation of the footbrake will ensure that the trailer brakes are applied in addition to the service brakes on the remaining truck axle.

In addition, operation of the hand control valve which applies the truck spring brakes will signal the triple relay valve and thus apply the trailer brakes. In order to combine this function with the requirement that air pressure should not be permanently applied to the brakes when the outfit is parked, the hand control valve has a multiple output function. When moved from the "OFF" to the "ON" position, the valve delivers a graduated reducing pressure to apply the spring brakes, and simultaneously a graduated increasing pressure to the triple relay valve to apply the trailer brakes. Then, when moved from the "ON" to the "PARK" position, the pressure delivered to the relay valve is reduced to zero, thus releasing the trailer brakes but leaving the truck spring brakes fully applied.

The dump valve which is incorporated in the truck system ensures that in the event of a leak in the service line to the trailer preventing the trailer brakes from being applied effectively, the emergency line will be quickly exhausted and the trailer brakes applied through the trailer relay emergency valve. It does this by sensing triple relay valve input and output signals, and reacting to an abnormal pressure differential.

The system on the trailer appears to be extremely simple when compared with that on the truck, but it does incorporate the relay emergency valve feature, which can be regarded as being the heart of a trailer system. That illustrated in Figure 2 includes a load sensing valve incorporating a relay emergency valve on each axle of the trailer. This configuration lends itself to a neat installation and reduces delays in brake application.

7. RELAY EMERGENCY VALVES.

Relay emergency valves have been in widespread use on air pressure braked trailers and semi-trailers for many years. They combine the normal relay valve function of speeding up brake application times with the emergency breakaway function of automatically applying the brakes of the trailer should it become detached from the prime mover.

A typical valve is illustrated diagramatically in Figure 3. When the trailer is coupled to a prime mover, compressed air passes to the trailer along the emergency line past the non-return valve to the trailer reservoir, and simultaneously through the inlet valve which is held open by the spring in the base of the unit, to the trailer brake chambers. When the pressure in the trailer reservoir and brakes reaches approximately 4 bar, the emergency piston which incorporates the inlet valve seat moves down against the action of the emergency spring, the inlet valve closes and the exhaust valve opens, so releasing the trailer brakes. In this way the vehicle is prevented from being driven away until sufficient pressure exists in the trailer reservoir to make an effective brake application.

Once in this normal running condition, the valve will allow the trailer reservoir to be charged up to full pressure.

When the brakes on the towing vehicle are applied, driver controlled air pressure passed to the trailer through the service line depresses the relay piston, which closes the exhaust and opens the inlet valve. This allows air under pressure to flow from the trailer reservoir to the brake chambers. The amount of air which is supplied through the service line from the towing vehicle is therefore only that which is required to operate the relay piston.

Should the trailer break away from the towing vehicle and the emergency line become detached, the loss of pressure above the emergency piston will allow the spring to force it upwards, carrying the inlet valve with it until the exhaust valve is closed and the inlet valve opens.

The trailer brakes will thus be fully applied, and the non-return valve will prevent loss of air from the trailer reservoir through the disconnected emergency line.

8. COMPATIBILITY OF TRUCK AND TRAILER BRAKE PERFORMANCE.

In addition to the matching of braking systems and

© IMechE 1976

components to allow physical interchangeability of trucks and trailers, it is essential that the braking performances of the truck and the trailer are matched to each other.

If, at any driver applied force on the footbrake, the retarding forces produced by the truck brakes and by the trailer brakes are not in proportion to the respective masses of the truck and the trailer, then a horizontal force will be produced in the drawbar. This force causes one part of the outfit to be underbraked, and the other to be overbraked. The limit of tyre to road adhesion will then be reached on the overbraked vehicle before the brakes on the underbraked part have produced their optimum retarding force, and premature wheel locking may result. In addition, the horizontal force in the drawbar, acting at some distance above ground level, will alter the weight distribution between the axles on both the truck and the trailer. For example, if a truck is fitted with more effective brakes than a trailer which it is towing, under braking conditions the drawbar will be under compression, and the truck wheels will tend to lock before the trailer wheels. As far as the ground to tyre normal reaction forces are concerned, the drawbar compressive force will increase the truck front axle and trailer rear axle forces and reduce the truck rear axle and trailer front axle forces. It can be seen that in this case the truck rear axle is likely to be the most severely overbraked axle, and therefore the most likely to lock first; such a situation is likely to lead to the outfit jack-knifing under emergency braking.

The dangers of unmatched truck and trailer braking have been recognized for many years, and the United Kingdom Code of Practice for Air Pressure Braking on vehicles requires that trucks and trailers must be designed so that when fully laden a deceleration of 0.6g must be achieved at a brake line air pressure of 85 ± 7 lbf/in^2. For tractors and semi-trailers the requirements are modified slightly to take account of the fact that when decelerating, the weight transfer experienced by the semi-trailer reduces the weight carried by the semi-trailer wheels and adds to the weight carried by the tractor wheels. For the semi-trailer therefore the 85 ± 7 lbf/in^2 air pressure must produce a retarding force at the semi-trailer wheels of 0.5 X P, where P is the static ground to tyre normal reaction force.

This aspect of the Code of Practice ensures that trucks and trailers achieve an acceptable standard of performance compatibility when laden, but, of course, it does not cater for cases when one or other of the parts of the outfit is in the part laden or unladen condition. For example, if a laden truck is towing an unladen trailer and the trailer braking force to brake line air pressure relationship is such as to match the truck brake performance, the trailer will be severely overbraked and is likely to suffer from premature wheel locking when the outfit is braked at all heavily.

The magnitude of this problem is best illustrated by referring to Figures 4, 5 and 6, which show "adhesion utilization curves" for a 16 ton GVW truck coupled to a 16 ton GVW trailer, in the laden/unladen and unladen/laden states respectively. The adhesion utilization curves are plots of μ against Z for each axle, where μ is the coefficient of tyre to road friction actually used by each axle, and is therefore equal to the braking force divided by the dynamic normal reaction force, and Z is the deceleration of the vehicle, measured as a proportion of the acceleration due to gravity. The physical data of the truck and trailer used for this analysis are listed in Appendix I.

Figure 4 shows that in the laden/laden condition, when the truck and trailer brakes are well matched, at decelerations below 0.5g the adhesion utilization curves for all axles are close to each other and to the theoretically ideal $\mu = Z$ line. This means that on road surfaces having limiting μ values of up to 0.5, decelerations approaching the theoretical maxima of μg should be attainable without wheel locking.

Figure 5 however shows that in the laden/unladen condition, on a road surface have a limiting μ of 0.5, the trailer front wheels will lock at a deceleration of less than 0.2g, and Figure 6 shows that in the unladen/laden condition, on the same road surface the truck rear wheels will lock at only 0.21g.

In order to achieve ideal compatibility between a truck and a trailer with conventional brake actuation systems under any condition of loading, it is necessary for the relationships between the deceleration and brake line air pressure to be independent of the state of loading of the vehicles and to be the same for each.

An attempt to legislate such a requirement was made in the ECE (Economic Commission for Europe) Regulation No. 13 on Vehicle Braking, with some tolerance in the deceleration pressure relationship being permitted.

The theoretical advantages and practical difficulties of this regulation were argued at great length between Government and Industry representatives from many European countries, and eventually what was considered to be compromise solution was issued, which became part of the E.E.C. (European Economic Community) Directive on Braking.

The compromise involved defining a band within which the air pressure to deceleration characteristics of laden trucks and trailers must lie, and a separate somewhat lower pressure band within which the characteristics of the unladen vehicles must lie. The widths of the bands constitute tolerances, and the implication of having different laden and unladen requirements is that outfits must be compatible in the laden/laden and unladen/unladen conditions, but that some measure of incompatibility is permissible in the laden/unladen and unladen/laden conditions.

The required relationship between the applied air pressure and the resultant deceleration is achieved on a laden vehicle by specifying appropriately sized basic actuation equipment. This equipment must also proportion the braking forces between the axles of the vehicle so as to prevent premature front or rear wheel locking under emergency braking. When unladen, a conventional truck requires a higher proportion of the braking to be applied to the front axle to prevent premature rear wheel locking. This is generally achieved by fitting a load sensing valve to the rear axle. This aspect of braking performance also forms one of the requirements in the EEC Braking Directive.

A rear axle load sensing valve, by attenuating the pressure supplied to the rear brakes, increases the

© IMechE 1976

brake valve delivery pressure required to achieve a given unladen vehicle deceleration. If the truck has a large payload capacity relative to its unladen weight, however, this increase in pressure requirement will not be sufficient to bring the characteristics within the band required by the Directive. In such a case it is necessary to further attenuate the braking when the vehicle is unladen. To achieve this, whilst maintaining the required distribution between the axles, may require a front brake pressure attenuating valve, which is sensitive to the state of vehicle loading, together with a higher degree of attenuation at the rear axle.

The effects of fitting appropriate equipment to the truck and trailer described in Appendix 1 are illustrated by the adhesion utilization curves in Figures 7 and 8. The truck has been assumed to be fitted with a rear axle load sensing valve set at a 4/1 ratio in the unladen condition, together with a 2/1 front brake pressure attenuating valve, whilst the trailer is fitted with 4/1 load sensing valves on both axles.

By comparison with Figure 5, Figure 7 shows that on the laden/unladen outfit, the trailer overbraking has been corrected and the risk of premature trailer wheel locking considerably reduced. Similarly, Figure 8 illustrates that in the unladen/laden condition, the gross overbraking of the truck rear axle shown in Figure 6 has been eliminated, and that on a road surface having a limiting μ of 0.5 a deceleration of over 0.45g can be achieved before the truck front wheels lock.

Whereas for trucks and drawbar trailers ideal compatibility under braking implies zero tensile or compressive force in the drawbar, for tractors and semi-trailers the requirements are different. Ideally, for these combinations, braking should induce a horizontal compressive force at the king pin equal to $R \times Z$, where R is the vertical force at the king pin and Z is the deceleration. Due to the weight transfer experienced by the semi-trailer, the value of R increases with increasing deceleration, and therefore the proportion of total braking which is supplied by the tractor should increase as the deceleration increases.

The E.E.C. Directive recognizes this, and requires that a tractor be considered with mathematical simulations of laden and unladen semi-trailers imposed upon it, and that prescribed conditions of adhesion utilization and brake pressure/deceleration relationships must be met. The requirement for semi-trailers is that the braking force, measured as a proportion of the semi-trailer static axle load, should bear a prescribed relationship with the applied air pressure. The prescription takes account not only of the unladen condition, but also of such factors as the axle position and the height of the centre of gravity of the semi-trailer. It requires, for example, that a short, high semi-trailer, which suffers considerably more weight transfer than a long, low one, should have less effective brakes, so as to prevent the premature wheel locking and resultant trailer swing which might otherwise occur.

It has generally been found that satisfactory laden and unladen combination performance and braking distribution can be achieved by fitting load sensing valves to the tractor and semi-trailer axles only. If these valves can respond to the weight transfer experienced dynamically, there will be an automatic increase in the proportion of the total braking contributed by the tractor as the deceleration increases.

9. FUTURE TRENDS AND DEVELOPMENTS.

On heavy commercial vehicles, air pressure braking systems have almost completely displaced vacuum systems, and full power hydraulic systems, using accumulator energy storage and closed-centre valve control units, are coming into use on some rigid vehicles. Among the advantages of hydraulic systems are reduced component size, due to the considerably higher pressures employed, reduced brake lag, due to the use of a virtually incompressible fluid, and reduced susceptibility to atmospheric contamination, due to the closed circuit nature of the fluid system. These advantages also apply to non-rigid vehicle systems, but there are obvious operational disadvantages in changing from one system to another if truck and trailer interchangeability is a particular fleet requirement.

A similar problem existed several years ago during the changeover from vacuum to air pressure braking systems. For an interim period, some air pressure braked tractors were equipped with compressor/exhauster units, to enable them to tow existing semi-trailers fitted with vacuum brakes.

An analogous solution exists to enable air pressure braked trailers to be towed by trucks fitted with power hydraulic brakes, requiring the use of a combined hydraulic pump/air compressor unit. A layout of such a system is shown in Figure 9.

In this instance, the truck braking system comprises two entirely separate hydraulic circuits, supplied from a duplex pump, and a hydraulic spring brake for parking. The dual hydraulic/air pressure reaction valve ensures that the trailer brakes can be applied after a failure of either hydraulic circuit on the truck, and the single hydraulic/air pressure reaction valve provides automatic emergency braking of the trailer should a depleted parking accumulator result in the spring brake being applied.

When considering a powered hydraulic system on a trailer, the most obvious approach is to simulate air pressure techniques, and to connect the vehicles through hydraulic service and emergency pipes, equipping the trailer with an accumulator and a hydraulic relay emergency valve. This requires the use of connecting hoses which neither leak fluid nor ingest dirt when uncoupled and recoupled, and which can be supplied at an economic price.

Other approaches entail mounting a hydraulic pump on the trailer. This can be driven either by the motion of the trailer or by an electric motor. A road wheel driven pump requires that in some circumstances the trailer must be started from rest before pressurized fluid is available for a brake application. This may not be a significant disadvantage in practice, because the likely precharge pressure characteristics of the hydraulic accumulator will result in adequate pressure being available quickly. Nevertheless, a system incorporating an electrically driven pump could incorporate similar safety features to those provided by a relay emergency valve.

With either method of driving the hydraulic pump, an electrical control line, providing a graduated signal for applying the trailer brakes, can be used.

©IMechE 1976

In order to provide the trailer breakaway feature, the control would have to be such that an open circuit would cause the trailer brakes to be applied; this could be achieved by using an inverted signal. The magnitude of the variable control signal could be derived from the pressure delivered to the brakes of the drawing vehicle, or from other sources indicative of the amount of braking required, such as the deceleration of the prime mover's road wheels and the force in the drawbar. By such means, ideal compatibility between the vehicles could be achieved.

The concept of using an electrical signal to control the hydraulic pressure applied to the brakes leads to the consideration of integrating the normal actuation with an antilock system. Solenoid operated hydraulic inlet and outlet valves could apply braking in accordance with the signal from the driver, overriding this if he demanded a higher deceleration that the road surface could provide.

Such a system may incorporate fewer passive antilock components than is usual, with a corresponding reduction in the amount of systematic exercising and monitoring of antilock functions being required.

These techniques need not of course be restricted to hydraulic brake systems, electro-pneumatic control being equally feasible.

APPENDIX 1

Truck and Trailer Data.

Truck

	Laden	Unladen
Mass	16000 kg	6000 kg.
Front axle load	6000 kg	3500 kg.
Rear axle load	10000 kg	2500 kg.
Height of c of g	1.70 m	0.95 m.

Wheelbase	5.00 m.
Braking Distribution	54/46.
Laden Brake Performance	0.6g deceleration at 6 bar air pressure.

Trailer

	Laden	Unladen
Mass	16000 kg	4000 kg.
Front axle load	8000 kg	2000 kg.
Rear axle load	8000 kg	2000 kg.
Height of c of g	1.50 m	0.90 m.

Wheelbase	5.50 m.
Braking Distribution	60/40.
Laden Brake Performance	0.6g deceleration at 6 bar air pressure.

© IMechE 1976

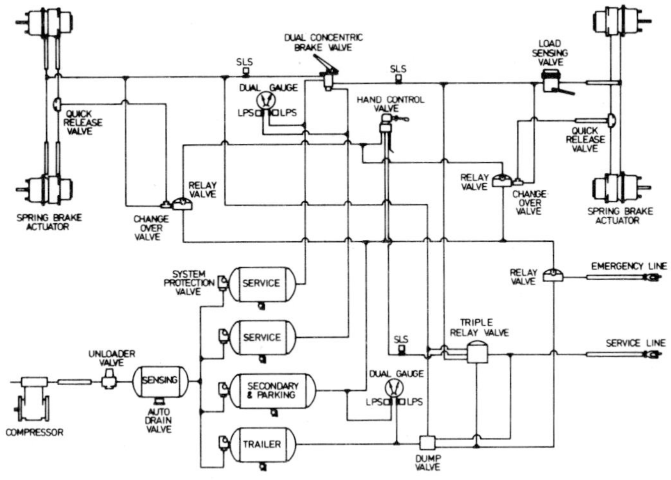

Fig. 1: Typical 2 Line Truck System to E.E.C. Regulations

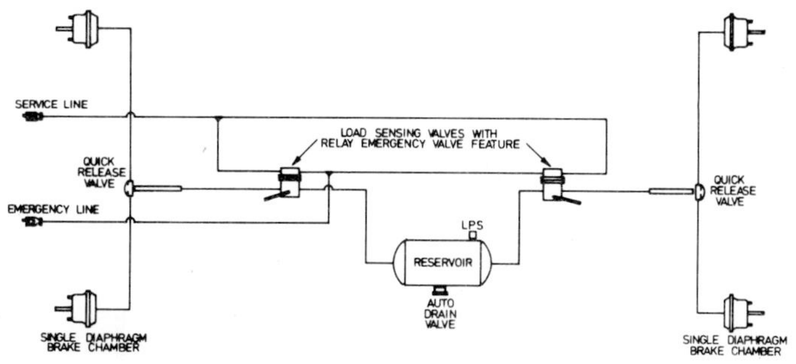

Fig. 2: Typical Trailer System to E.E.C. Regulations

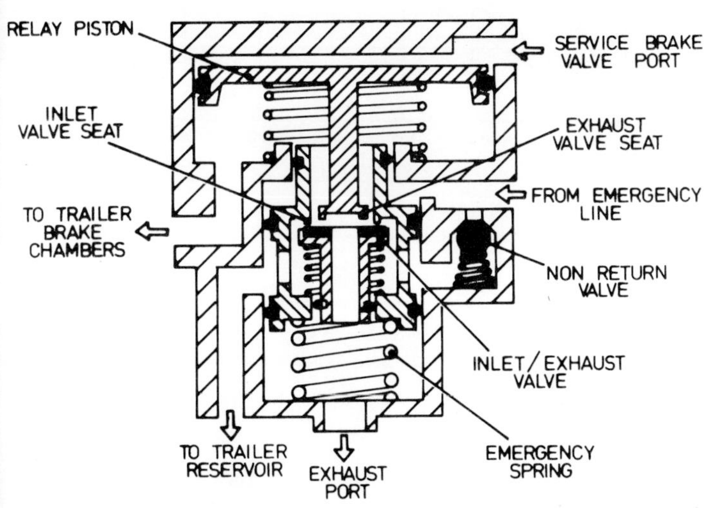

Fig. 3: Relay Emergency Valve

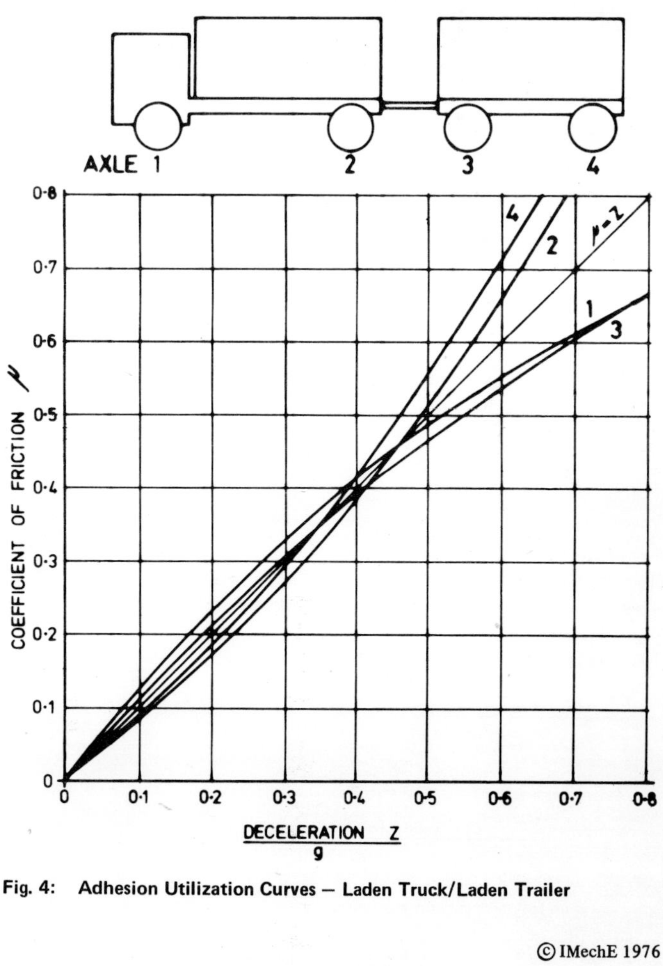

Fig. 4: Adhesion Utilization Curves – Laden Truck/Laden Trailer

© IMechE 1976

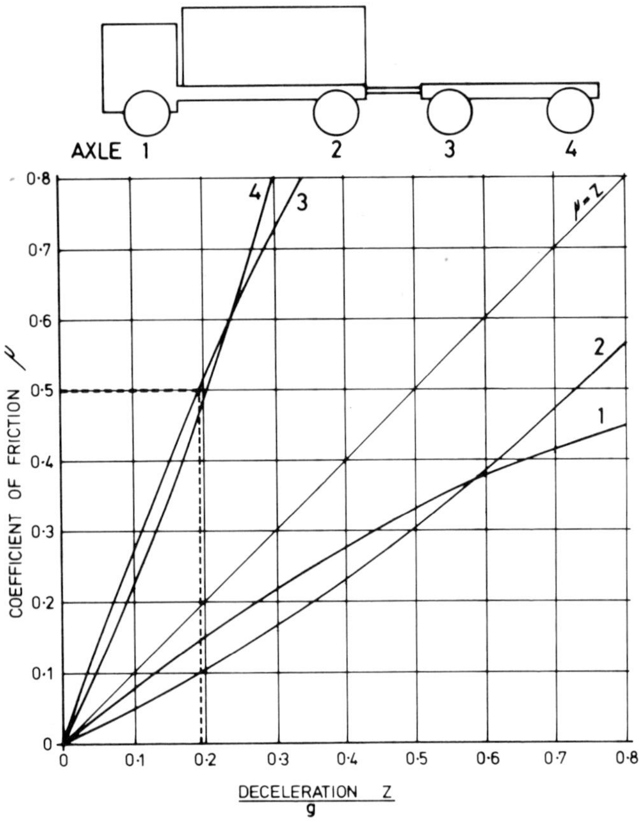

Fig. 5: Adhesion Utilization Curves — Laden Truck/Unladen Trailer

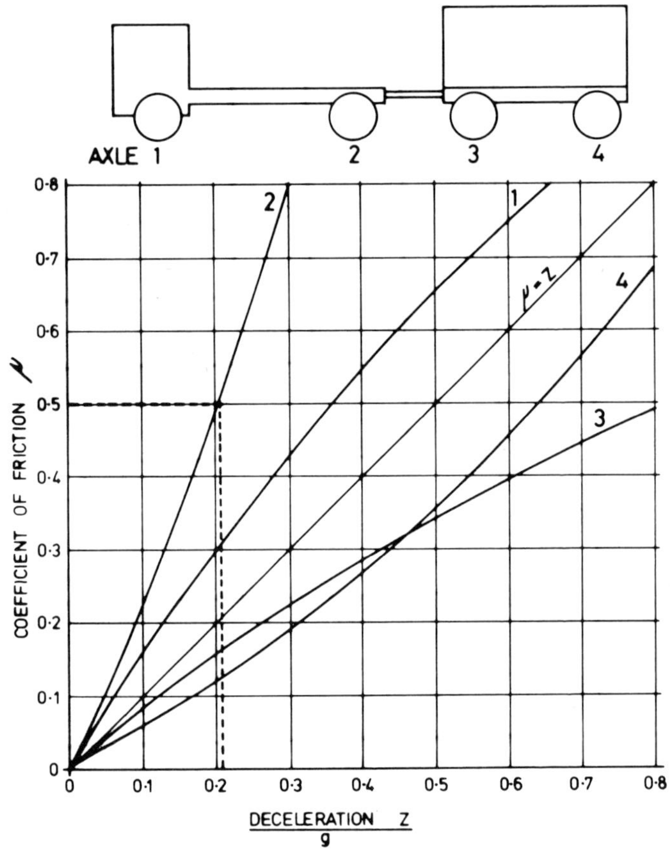

Fig. 6: Adhesion Utilization Curves — Unladen Truck/Laden Trailer

© IMechE 1976

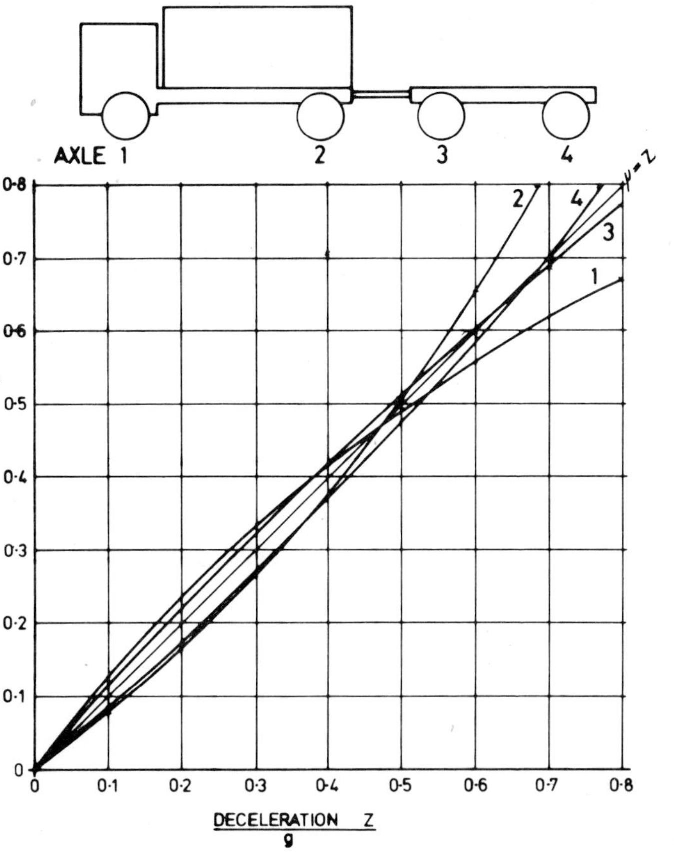

Fig. 7: Adhesion Utilization Curves – Laden Truck/Unladen Trailer – Fitted with Load Sensing Equipment

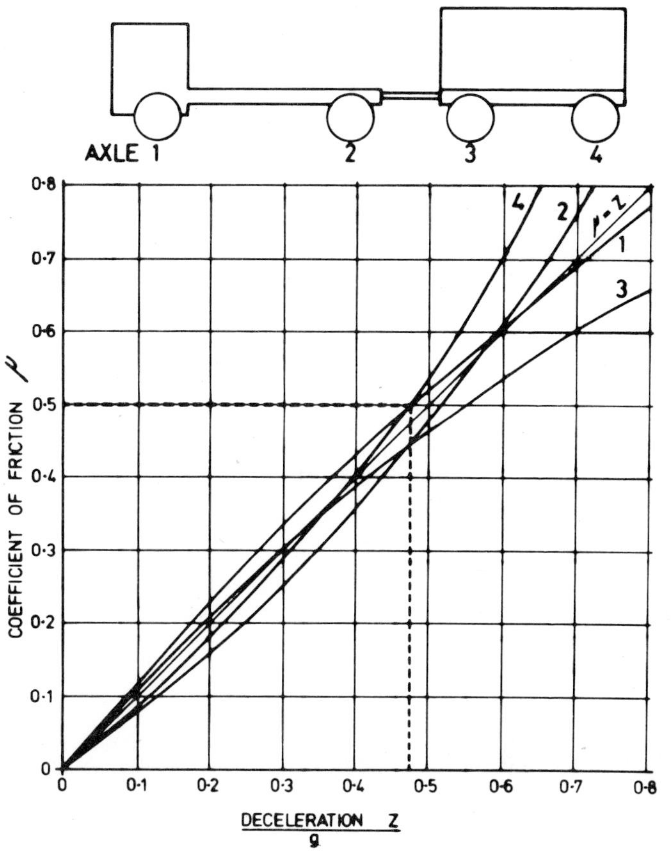

Fig. 8: Adhesion Utilization Curves – Unladen Truck/Laden Trailer – Fitted with Load Sensing Equipment

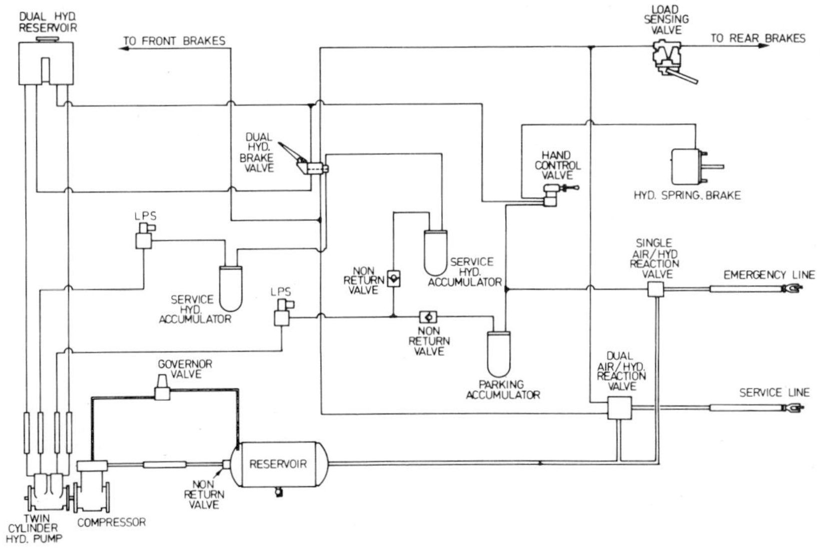

Fig. 9: Full Power Hydraulic Truck with Air Pressure Trailer Braking

© IMechE 1976

C23/76

LATERAL STABILITY OF COMMERCIAL ROAD VEHICLE TRAINS UNDER BRAKING CONDITIONS

F. D. HALES, PhD, CEng, MIMechE
Loughborough University of Technology, Leicestershire.

The MS of this paper was received at the Institution on 14 September 1975 and accepted for publication on 17 October 1975

SYNOPSIS Road trains are defined as a tractor plus an arbitrary number of trailers. Commercial road vehicle trains are defined as two types; truck + full trailer(s) and tractor + semi-trailer + full trailer(s). A range of six road trains, with one, two and three load units for each type are specified.

The equations of lateral motion while braking for a generalised road train are outlined and a method of linearising them proposed so that eigen values may be obtained to provide data on the dynamical characteristics.

The equations are solved for the six laden road trains, unbraked, braked without axle locking, with single axle locking, and with two axles locking, and the results analysed and discussed.

It is concluded that there are strong similarities between the two types of road train and that the major deteriorations in lateral stability are associated with locking of the second and third axles of any road train.

INTRODUCTION

1. Road vehicles have had braking problems since the early days when the pioneer automotive engineers first fitted brakes to the back wheels of their vehicles. In the light of current technology these early difficulties are perhaps now viewed by some with a sense of superiority. Braking problems are of many types and one of the most intriguing and long-lived has been in the area where under braking conditions the lateral characteristics of the vehicle are altered. Sometimes the changes are small but on occasions they are both large and dramatic and uncontrollable lateral motions occur. It is with this aspect of braking that this paper is concerned.

2. The family of vehicle types that are considered are commercial road vehicle trains. A commercial road vehicle train is defined as a tractor unit plus an arbitrary number of trailers designed to carry goods. A general approach is possible because such vehicle systems may always be reduced to a tractor plus a number of semi-trailers.

3. The approach used is to determine the lateral stability of the various vehicles without braking and then to introduce braking, both as longitudinal forces and skidding axles and to assess the changes in behaviour that occur.

COMMERCIAL ROAD VEHICLE TRAINS

4. Commercial road vehicle trains have a number of potential types. In practice only two are normally used; they are distinguished easily (see Fig. 1).

Type (a) Load-carrying tractor (truck) + n. full trailers
$n = 0,1,2$

Type (b) Non-load carrying tractor + semi-trailer + n full trailers
$n = 0,1,2$

This provides a range of six vehicles, with a type (a) and type (b) example each carrying 1, 2 or 3 load units.

5. The single load unit vehicles (truck and articulated vehicle) are familiar vehicles. Trucks have essentially similar lateral behaviour, due to braking, as automobiles, and their lateral motion during braking seems to have attracted little attention from engineers. Articulated vehicles, on the other hand, have always had a lateral motion problem under braking - jack-knifing. This behaviour, having been identified and picturesquely named, has attracted a great deal of attention from both engineers and the press. References 1, 2 and 3 were an early series of papers and in recent years several excellent review papers (e.g. reference 4) have appeared.

© IMechE 1976

6. Considering the amount of effort that has been put into solving the artic equations of motion, the paucity of comparative studies between articulated vehicles and trucks is surprising but reference 5 attempts to put the problem into perspective.

7. The two-load unit vehicles are only beginning to appear on United Kingdom roads, but have of course been common for many years in Continental Europe and the United States of America. Little or no published information exists on the effects of braking on lateral stability of such vehicles but some work is appearing on their lateral stability (e.g. references 6,7,8).

8. The three-load unit vehicles are, perhaps, no more than a dream in the eyes of transport operators and no literature, other than that of reference 8 seems to exist at present. Vehicles of the general type however do exist in the United Kingdom as fairground vehicles.

9. In order to determine the behaviour of the family of road trains, representative parameters for the four components of which they are composed, need to be fixed. The four components are:

Truck
Tractor
Semi-trailer
Full trailer

where the full trailer is composed of a 'dolly' + semi-trailer, see Figure 1. To ensure that variations are due only to the differences in the two types of road train and the number of load units the following assumptions are made:

1. all load units are identical
2. all semi-trailers are identical
3. all dollies are identical
4. tyre coefficients are based on the same assumptions for all vehicles.

With these assumptions a set of parameter values was generated and is given in full in Table 1, the symbols used are defined at the end of the paper. Naturally the vehicles thus obtained are unlikely to be identical with an existing vehicle, but comparisons between differing types are made practicable.

EQUATIONS OF MOTION AND THEIR SOLUTION

10. The derivation of the equations of motion is based on that given in reference 9, with the addition of terms due to longitudinal braking forces. Because all trains are composed of a tractor + varying numbers of semi-trailers only a tractor unit, and the ith semi-trailer, need to be considered. The equations are derived relative to a moving axes system located at the C. of G. of the tractor unit and moving with the tractor. The axes and notation follow normal vehicle stability practice with x-forward, y-laterally (to starboard) and z-down to form a right-handed triad.

11. The process of deriving the equations is only outlined (see reference 9 for full details). They are obtained by equating rates of change of linear and angular momenta to the external forces and moments and then eliminating the longitudinal and lateral hinge forces. The equations of motion may then be written as:

$$\sum_{j=1}^{n} m_j\dot{U} = X_{1F} + \sum_{j=1}^{n} X_{iR} \qquad (1)$$

$$\sum_{j=1}^{n} m_j(\dot{v}_j+(r_1+r_j)U) = Y_{1F} + \sum_{j=1}^{n} Y_{jR} \qquad (2)$$

$$C_1\dot{r}_1+x_{12}\sum_{j=2}^{n} m_j(\dot{v}_j+(r_1+r_j)U)$$

$$= x_{1F}Y_{1F}+x_{1R}Y_{1R} + x_{1R}\sum_{j=2}^{n} Y_{jR} \qquad (3)$$

and for $1 < i \leq n$

$$C_i(\dot{r}_1+\dot{r}_i)=x_{i,i-1}m_i(\dot{v}_i+(r_1+r_i)U-U\dot{\theta}_i)$$

$$-(x_{i,i-1}-x_{iR})(Y_{iR}-X_{iR}\theta_i)$$

$$+(x_{i,i-1}-x_{i,i+1})\sum_{j=i+1}^{n}(m_j(\dot{v}_j+(r_1+r_i)U-U\dot{\theta}_j)$$

$$-(Y_{jR}-X_j\theta_j) \qquad (4)$$

where the lateral tyre forces may be written:

$$Y_{1F} = -k_{1F}\frac{v_1 + x_{1F}r_1}{U}$$

$$Y_{1R} = -k_{1R}\frac{v_1 + x_{1R}r_1}{U}$$

$$Y_{iR} = -k_{iR}\frac{v_1 + x_{iR}(r_1 + r_i)}{U} \quad \text{for} \quad i > 1$$

.... (5)

and from the geometrical constraints on the system the lateral velocity v_i of the ith component for $1 < i \leq n$ is:

$$v_i = v_1+x_{12}r_1 - \sum_{j=2}^{i-1}(x_{j,j-1}-x_{j,j+1})(r_1+r_j)$$

$$- x_{i,i-1}(r_1+r_i)-\theta_i U \qquad (6)$$

and

$$\frac{d}{dt}\theta_i = r_i \qquad (7)$$

By substituting for the external forces from (5) and eliminating the dependent variables from (6) the set of equations (1) - (7) are reduced to 2n+1 1st order non-linear differential equations. The non-linearities occur due to the U-term in the denominator of the external tyre force terms and the product terms with θ_i in equation group (4).

12. The method of solution frequently adopted for such a system is essentially that of calculating the motion throughout a specified braking sequence. However a more sophisticated approach is required in this case, where comparisons of basic dynamic behaviour are required for a number of vehicle types for a range of conditions. A possible approach is to simplify the equations by assuming:

1. X_{1F}, X_{iR} $(i=1,\ldots,n)$ are known, and constant
2. The stability of the vehicles varies only slowly with speed.

The first assumptions imply that $\dot{U}$=const. and eliminate the non-linearities due to $\dot{U}$, and X_{iF}, and X_{iR}. The second assumptions mean that if U is taken as a constant, and results calculated over a range of speeds, (thus eliminating the non-linearities associated with 1/U), the stability will not be seriously affected. With these assumptions the equations are reduced to 2n 1st order linear equations in the variables $(v_1 r_1,\ldots r_n,\theta_2,\ldots\theta_n)$. In particular their stability can be investigated by standard eigen value techniques.

LATERAL STABILITY OF ROAD TRAINS UNDER BRAKING

13. To appreciate the effects of braking on lateral stability a datum must be established for the unbraked condition. This is obtained by calculating the eigen values for the equations of motion with the longitudinal force terms omitted. Such a situation has been investigated in some detail in reference 8 for the same set of vehicles as those defined in Table 1. Figures 3 and 4 present results for a range of speeds for the six vehicles. The real parts of the roots only are presented, but the natural frequencies at 30m/s are included. Points of interest are that after the low speed real roots become oscillatory the characteristics of the motion change comparatively slowly, with reducing damping as speed increases. The low damping of the full trailer modes is also of interest.

14. Braking forces can be accommodated in the equations and a strategy developed for making a comparison practicable, but a number of questions concerning the level of braking, its distribution and the effect of braking and incidence of axle locking on the lateral tyre force coefficients, remain to be decided. Any effects due to braking can reasonably be expected to increase with increasing braking; therefore a high level of 0.6g was taken for these studies. The distribution of braking forces takes two forms, in time, and in space (i.e. brake forces may be applied at different times and have different magnitudes on the various axles). The significance of variations in time will be considered in the discussion below but the distribution of braking forces along the train must be decided before calculations are made. The essentially simple strategy of allocating brake forces proportional to axle loads, ignoring load transfers, was adopted.

15. The effect of braking, and incidence of wheel locking, on the lateral tyre force coefficients is in detail a most complex phenomenon. However simple upper and lower limits for the effects may be easily defined. The lower limit is given by braking forces having no effect on lateral tyre forces. The upper limit is given when braking forces reduce the lateral tyre coefficients to zero. The lateral tyre forces being essentially unchanged by braking is associated with conditions where the longitudinal forces are small compared with the adhesion limit between the tyre and road. The lateral tyre forces being essentially zero while braking is associated with conditions where the longitudinal forces are close to the adhesion limit. The process of axle locking which results when the demanded brake torque is in excess of the torque produced by the road-tyre forces is, due to the negative slope (near axle locking) of the road-tyre force-v-wheel-slip curve that often occurs, frequently an unstable process. Therefore, although intermediate conditions can and do occur, for practical purposes the approximate upper and lower limits outlined above are the predominant operating conditions. In the calculations below, these upper and lower limits are associated with axle locked, or axle unlocked, states.

16. Figure 5 gives the eigen values at 30m/s for the unbraked and braked conditions without axle locking for all six vehicle types. Of particular interest is the fact that the introduction of braking forces has a negligible effect on the lateral dynamics when axle locking does not occur. Of more general interest, it may be seen that the degree of dynamic coupling induced by adding full trailers is fairly low. The full trailer modes are very similar for both sorts of road train (i.e. compare 5(c) and (d), and 5(e) and (f)).

17. Figure 6 gives the eigen values at 30m/s for single axle locking cases, for the six vehicle types. A comparison of Figure 6 and Figure 5 indicates that the introduction of axle locking produces major changes to the lateral stability.

18. Considering firstly the single load unit vehicles (Figs. 6(a) and (b)), locking the first (front) axle maintains the stable motion of Figures 5(a) and (b). Locking the second axle produces in both cases a violent exponential divergence of very similar magnitudes ($\lambda \approx 2.7$), the articulated vehicle motion is known as jack-knifing, but the nearly equally violent truck motion has escaped such a descriptor. Locking the third axle, for the articulated vehicle, produces a divergence (trailer swing) with a magnitude an order of magnitude less than

©IMechE 1976

that of jack-knifing.

19. For the two load unit vehicles (Fig. 6(c) and (d)) again locking of the first axle maintains stability. Locking the second axle again produces the exponential instability. The truck case remaining constant, but the tractor case worsening by some 20%. Both vehicle types now have a third axle and the effect of locking it is similar; an exponential divergence, but the truck case is twice as severe ($\lambda \approx 0.5$) as for the artic. Fourth axle locking produces in both cases a near neutral oscillatory motion, just stable for the truck and just unstable for the tractor (indicated in Fig. 6(d) by the 4 being in the unstable region) case. Locking the fifth axle, for the articulated vehicle, produces another just unstable oscillatory motion.

20. The three-load units cases (Figs. 6(e) and (f)) are again stable with the first axle locked. Locking of the second and third axle produces virtually identical results to those of the two load case, stable oscillatory motion for the truck case and a just unstable oscillatory motion for the sixth (and last) axle. For the tractor case locking of the fourth, fifth, sixth and seventh axles all produces just stable oscillatory motions with reducing frequencies.

21. Before leaving the single axle locking cases the effect of speed (as a constant) will be considered. Since without axle locking the effects of braking are negligible, there is little reason to question the validity of the second assumption concerning the effect of speed variation. However the single axle locking cases show considerable differences from the unbraked condition and need further study. Figure 7 gives the eigen values at 15m/s for single axle locking cases, for the six vehicle types. A comparison of Figure 7 and Figure 6 indicates a very similar dynamic situation at the two speeds. That for the truck trains is identical, while that for the tractor trains has a minor modification in that the minor exponential divergence present at 30m/s has not yet appeared at 15m/s and is replaced by a near-neutral stable low frequency oscillation. From the overall viewpoint, however, that difference cannot be regarded as a major one. Clearly this evidence is not able to validate the second assumption but it does not contradict it, in that under the assumption the changes in stability with speed that occur are small.

22. In a situation in which an axle may lock, it is pertinent to investigate the effect of locking more than one axle. Table 2 catalogues the unstable or near unstable eigen values that occur for the range of vehicles with pairs of axles locking. Broadly speaking no new instabilities occur with pairs of axles locking. Indeed the major effect of locking pairs of axles is the profound stabilising effect of locking the front axle, which virtually eliminates the serious exponential instability arising from locking axle number 2. A smaller effect is the suppression of the third axle divergence by the second axle divergence, in the studies presented a near-neutral stable oscillation occurs.

DISCUSSION

23. Perhaps the two most interesting features of the lateral behaviour of road trains under braking are:

1. The strong similarity in the lateral stability of type (a) and type (b) road trains under braking.
2. The non-appearance of further serious instabilities, comparable with jack-knifing.

This first feature means that it is possible to consider the behaviour of road trains collectively. Without axle locking serious instabilities due to braking would not be expected. The effects of locking a single axle are that locking the front axle is stabilising, locking any other axle is de-stabilising. Locking the second axle of any road train produces a serious divergent instability.

24. Locking the third axle normally produces a moderate divergent instability, alternatively a near-neutral oscillation. Locking the fourth, or higher number axle normally produces a near-neutral oscillation but a mild divergence could occur. The effects of locking pairs of axles are generally a quasi-linear superposition of the effects due to single axle locking. Exceptions occur when the front axle is locked since its stabilising effect suppresses almost entirely the instabilities due to locking the other axles.

25. A feature of the brake systems on road trains is that there are delays in operation of the brakes as the axle concerned recedes from the brake pedal. With no axles locking stability is likely to be unaffected by braking, thus brake delays are not expected to be important either.

26. If axle locking occurs, then the road train becomes a dynamic system that changes when any axle locks. For example let the ith axle lock t_i seconds after initiation of a braking sequence and the jth axle lock t_j seconds after initiation. Then the road train has three dynamic states:

1.	$0 \leq t < t_i$	all axle unlocked
2.	$t_i \leq t < t_j$	$t_i{}^{th}$ axle locked
3.	$t_j \leq t$	$t_i{}^{th}$, $t_j{}^{th}$ axles locked

and the behaviour of the system at any time will be that appropriate to its dynamic state, subject to the initial conditions when it entered that state.

27. From the stability point of view the occurrence of near-neutral oscillatory modes might be overlooked as of little

©IMechE 1976

consequence. Unfortunately the vehicle driver could well be embarrassed by the motions, happening towards the rear of the train, which are difficult for him to perceive and over which he has little direct control.

28. The principal conclusion from these braking studies is that the dynamics of truck and tractor-based road vehicle trains under braking are very similar. The major deteriorations to the lateral stability of road trains are associated with locking of the second and third axles in such systems and that either eliminating such locking of ensuring that the first axle is also locked would substantially eliminate such effects.

NOTATION

Suffices

F, R Front and rear
i, j Vehicle component number
. Differentiation w.r.t. time

In the case of a double suffix the first identifies which vehicle is the train, the second identifies the other location.

C Yaw inertia
k Lateral tyre coefficient
m Mass
r Yaw velocity
U Forward speed
v Lateral velocity
x Longitudinal distance
X Longitudinal force
Y Lateral force
θ Yaw angle of trailer relative to tractor

REFERENCES

1. HALES, F.D. The lateral stability of a simplified articulated vehicle. I.Mech.E. Symposium on Control of Vehicles, 1963.

2. HALES, F.D. Lateral stability problems of simply articulated vehicles, I.U.T.A.M., Paris, 1964.

3. HALES, F.D. et al. The handling and stability of motor vehicles: Part 3 - The lateral stability of articulated vehicles under braking conditions. MIRA Report 1966/3.

4. DUGOFF, H. et al. The dynamic performance of articulated vehicles, A review of the state-of-the-art. SAE 710223.

5. FARR, B.N. A survey into the accident rate of articulated and rigid commercial vehicles. TRRL Report LR 197, 1968.

6. HAZEMOTO, T. Analysis of lateral stability for doubles. SAE 730688, June 1973.

7. NORDSTRÖM, O. et al. Tunga Fordmskombinationers Dynamiska Stabilitet. Statens Vag-och Trafikinstitut, 1972.

8. HALES, F.D. The rigid body dynamics of road vehicle trains. I.U.T.A.M., Delft, 1975.

9. HALES, F.D. The vertical motion and lateral stability of road vehicle trains. TT 7211, Loughborough University of Technology, October, 1972.

©IMechE 1976

Table 1

Parameter values for various road train components (after ref.8)

Vehicle	Truck	Tractor	Dolly	Semi-Trailer
Quantity (units)				
m_i (kg)	21000	6000	1750	16500
C_i (kgm^2)	57750	15600	1750	41340
x_{i_F} (m)	2.8	1.25	--	--
x_{i_R} (m)	-0.95	-2.00	-0.1	-2.25
$x_{i,i-1}$(m)	--	--	2.0	2.75
$x_{i,i+1}$(m)	-3.45	-1.675	-0.05	-3.35
k_{i_F} (N/rad)	240900	211900	--	--
k_{i_R} (N/rad)	956300	505200	501000	528300

©IMechE 1976

Table 2

Unstable and near-unstable eigen values for axles locking in pairs (30m/s)

TRUCK

Axle No.1		2
1	-	0.0 +0 R
2	0.0 +0	2.76

TRUCK, FULL

Axle No.1		2	3	4
1	-	-0	0.11	-0
2		2.67	2.64 +0	2.68 +0
3			0.51	0.48 -0
4				-0

TRUCK, FULL, FULL

Axle No.1		2	3	4	5	6
1	-	-0	0.14	-0	-0	+0
2		2.62	2.56 +0	2.62 -0	2.62 -0	2.62 -0
3			0.69	0.69 -0	0.68 -0	0.69 -0
4				-0	-0 -0	-0 -0
5					-0	+0 +0
6						+0

TRACTOR, SEMI

Axle No.1		2	3
1	-	-0	+0
2	-0	3.45	3.42 +0
3	+0	3.42 +0	0.11

TRACTOR, SEMI, FULL

Axle No.1		2	3	4	5
1	-	-0	-0	-0	-0
2		3.42	3.39 +0	3.42 -0	3.42 -0
3			0.22	0.27 -0	0.22 -0
4				+0	+0 +0
5					+0

TRACTOR, SEMI, FULL, FULL

Axle No.1		2	3	4	5	6	7
1	-	-0	-0	-0	-0	-0	-0
2		3.39	3.36 -0	3.39 -0	3.39 -0	3.39 -0	3.39 -0
3			0.27	0.34 -0	0.27 -0	0.27 -0	0.27 +0
4				+0	+0 +0	+0 -0	+0 -0
5					-0	-0 -0	-0 -0
6						-0	-0 -0
7							-0

©IMechE 1976

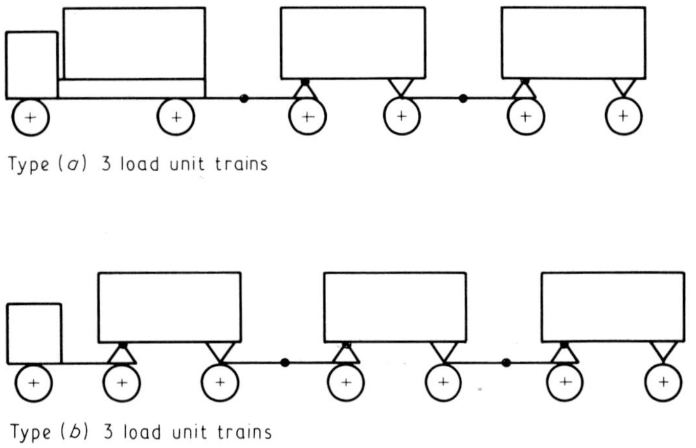

Fig. 1: Road train types

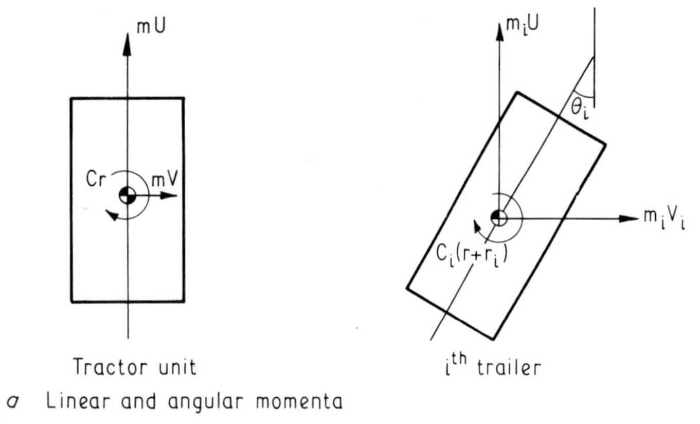

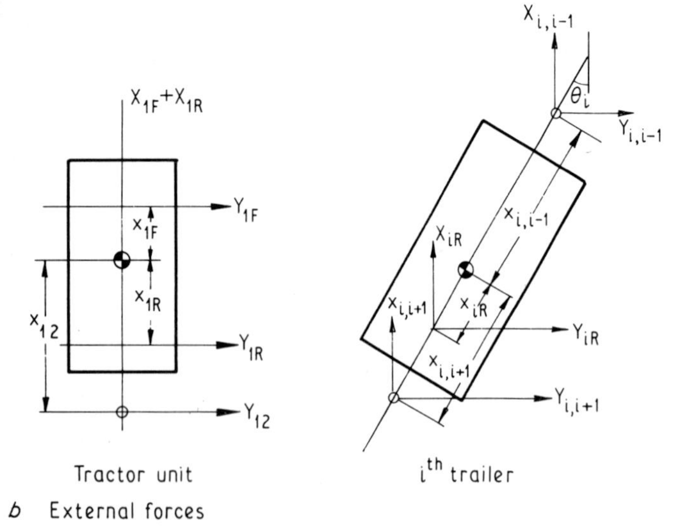

Fig. 2: Linear and angular momenta and external forces

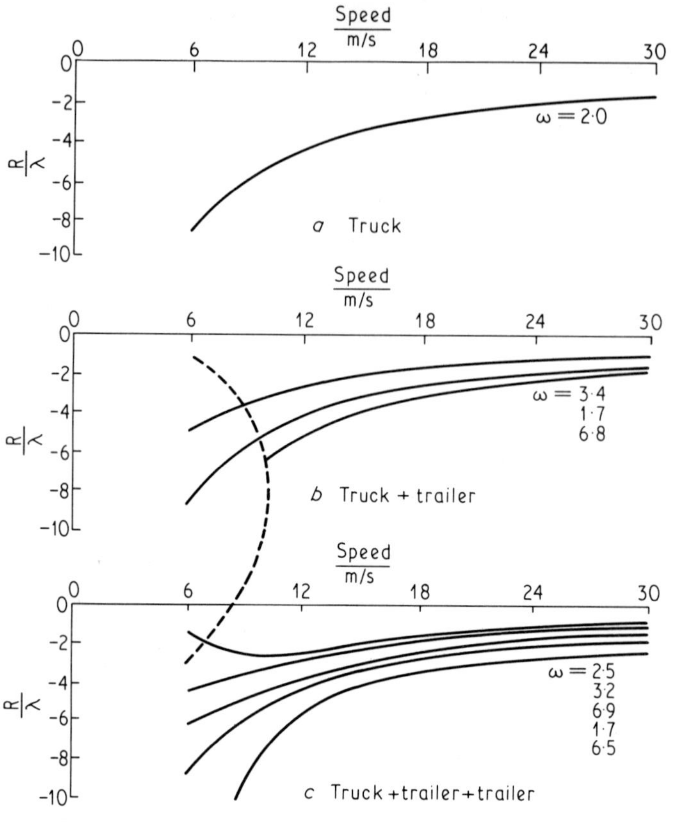

Fig. 3: Truck trains, real parts of lateral eigen values for datum cases (λ Hz given at 30 m/s, dotted lines indicate real roots, full lines indicate complex roots)

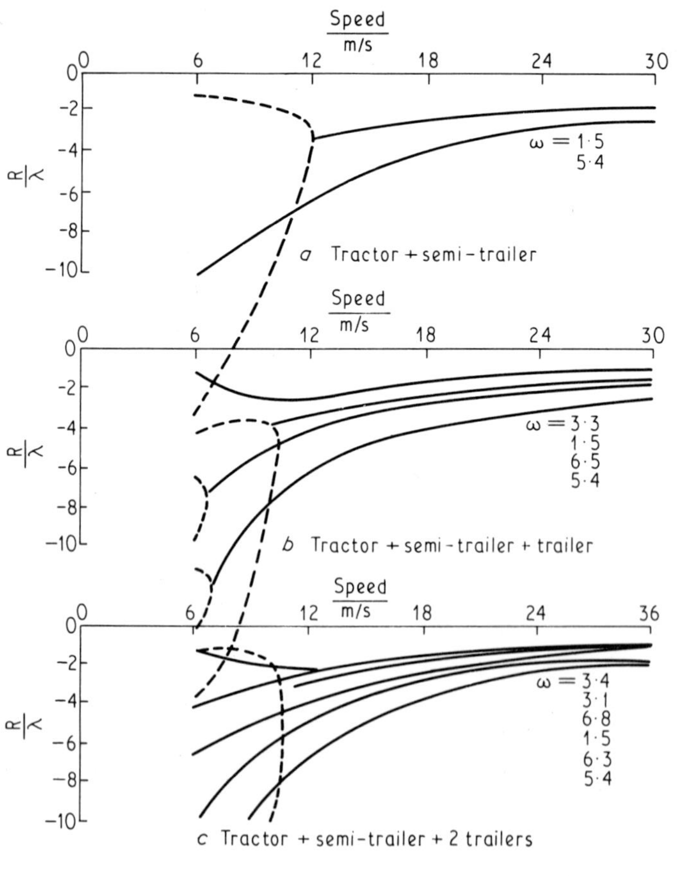

Fig. 4: Tractor trains, real parts of lateral eigen values for datum cases (λ Hz given at 30 m/s, dotted lines indicate real roots, full lines indicate complex roots)

© IMechE 1976

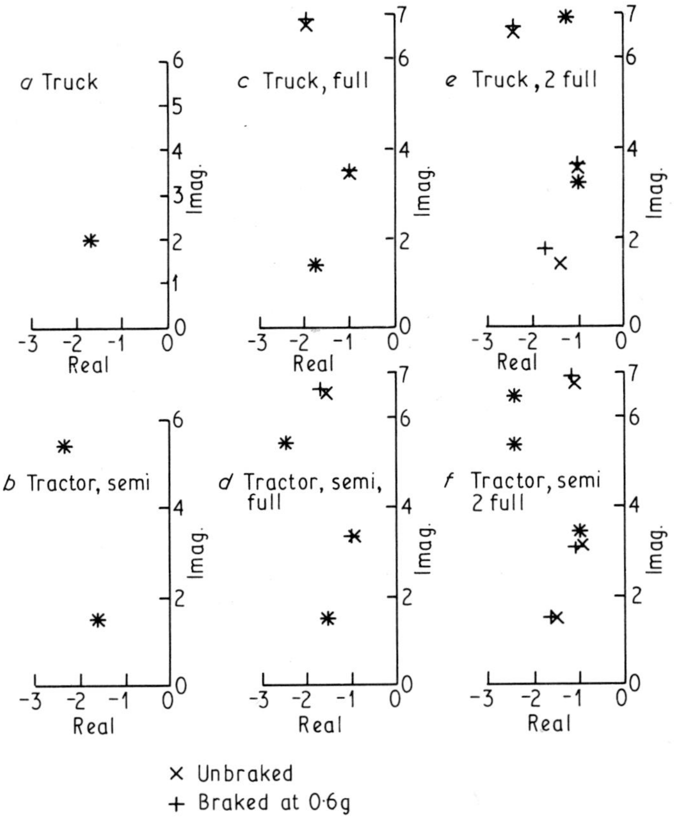

Fig. 5: Effect of braking on lateral stability, no axle locking, (30 m/s)

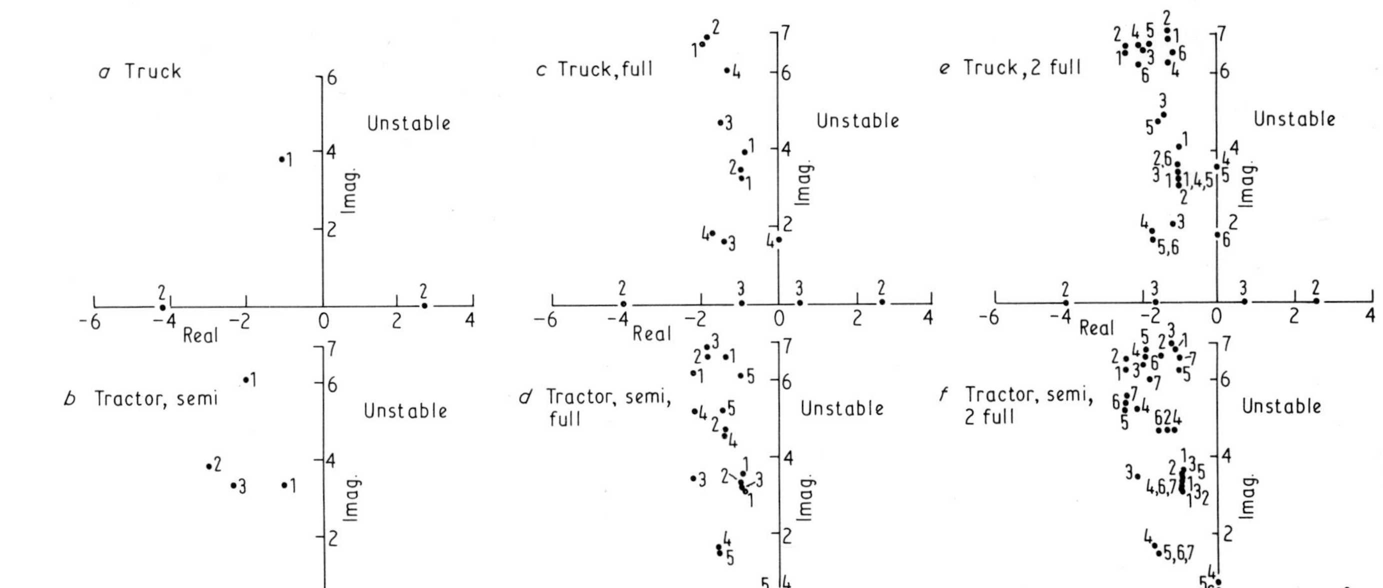

Fig. 6: Effect of braking on lateral stability, single axle locking one, two and three load units (30 m/s, the location of the locked axle is indicated by integers, numbered from the front)

© IMechE 1976

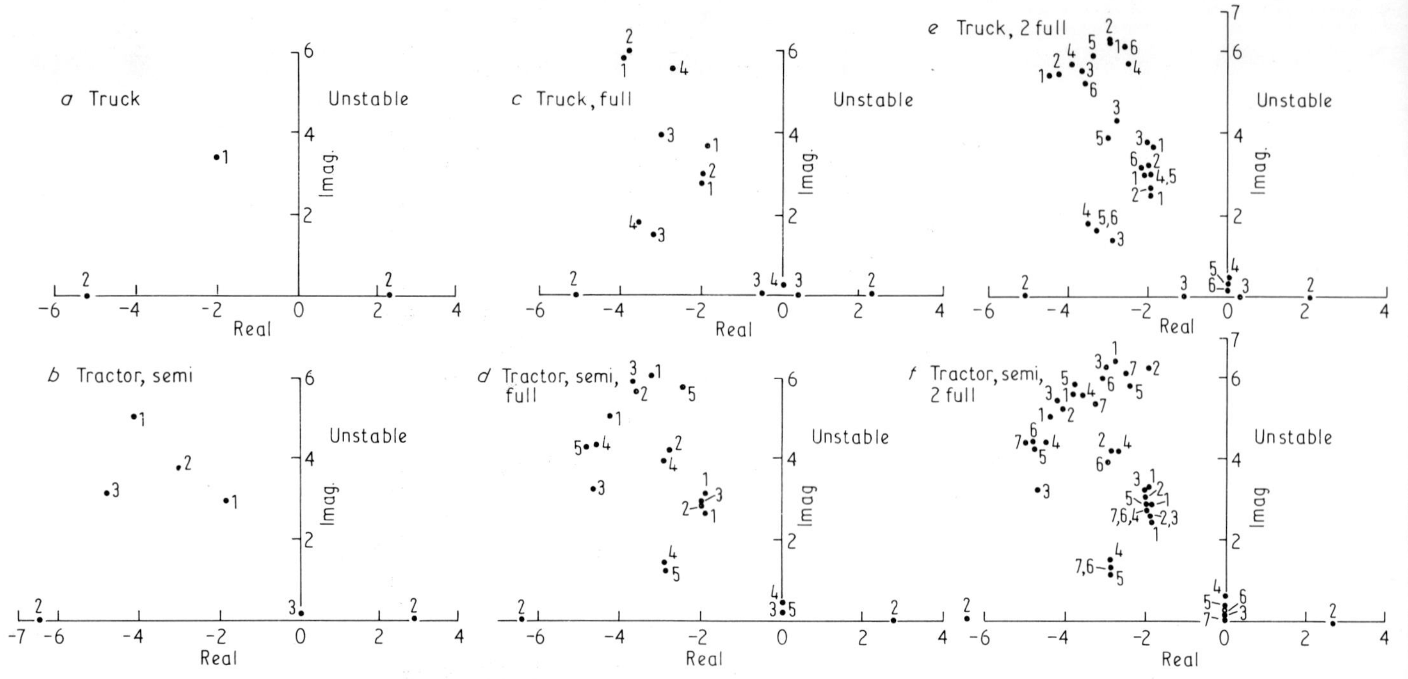

Fig. 7: Effect of braking on lateral stability, single axle locking one, two and three load units (15 m/s, the location of the locked axle is indicated by integers, numbered from the front)

C24/76

THE STABILITY OF MOTORCYCLES IN ACCELERATION AND DECELERATION

R. S. SHARP, BSc, MSc
Department of Mechanical Engineering, University of Leeds

The MS of this paper was received at the Institution on 18 August 1975 and accepted for publication on 17 October 1975

SYNOPSIS The effects of accelerating and braking on the stability characteristics of a motorcycle and rider are calculated through a modification of a previously published analysis. The individual influences present are enumerated and their contributions considered. Some practical consequences are discussed.

INTRODUCTION

1. Theoretical studies of motorcycle steering behaviour have concentrated heavily on the constant forward speed case. When linear equations of motion describing small perturbations from straight running have been employed to find stability characteristics, constant forward speed has been assumed (Refs 1,2,3,4), and when digital solution of non-linear equations, including that for the forward motion of the machine, has been employed, only constant speed results have been presented (Refs 5,6).

2. In studying the jack-knifing of articulated vehicles, essentially a phenomenon of the decelerating vehicle, Hales (Ref.7) has successfully included only the inertia effects of deceleration, and has avoided the necessity for solving the longitudinal equation of motion simultaneously with the lateral equations.

3. This paper is concerned with the effects on the lateral stability characteristics of motorcycles of a changing forward speed, results being obtained by regarding the longitudinal equation of motion as uncoupled from the lateral equations, and treating the longitudinal acceleration as a parameter of the lateral motion. The base model is that employed by Sharp (Ref.1) with some refinement of the tyre description as utilised by Eaton and Segel (Ref.4). This model has had some success in describing the motions of an experimental motorcycle (Ref.4).

NOTATION

4. b,e,h,j,k,l,R_f,R_r,t linear dimensions (Figure 1)

C_{f1}, C_{r1}	front and rear tyre cornering stiffnesses
C_{f2}, C_{r2}	front and rear tyre camber stiffnesses
C_{f3}, C_{r3}	front and rear tyre aligning moment to sideslip angle constants
C_{f4}, C_{r4}	front and rear tyre aligning moment to camber angle constants
C_{f5}, C_{r5}	front and rear tyre rolling moment to camber angle constants
C_{rxy}	rear frame product of inertia with respect to reference axes
g	acceleration due to gravity
I_{fx}, I_{fy}, I_{fz}	principal inertias of front frame
I_{rx}, I_{ry}, I_{rz}	rear frame inertias about axes through the mass centre forwards, to starboard, and downwards respectively
i	polar moment of inertia of engine flywheel
i_{fy}	polar moment of inertia of front wheel
$i_{rx} = i_{rz}$	camber inertia of rear wheel
i_{ry}	polar moment of inertia of rear wheel
K	steering damper coefficient
M_f	mass of front frame
M_r	mass of rear frame
U	forward velocity of reference point A (Figure 1)
$\dot{y}_1$	lateral velocity of reference point A
Y_f, Y_r	lateral tyre forces
M_{zf}, M_{zr}	tyre yawing moments
ε	steering head angle
λ	gear ratio between rear wheel and engine flywheel
σ_f, σ_r	tyre relaxation lengths
ψ	yaw angle
ϕ	roll angle
δ	steer angle
Z_f	front tyre load (negative)

THE ANALYSIS

5. The model employed is a development of that of reference 1. The variable forward speed gives rise to some "inertia force" terms in the equations of motion which were previously zero. These terms are obtained by differentiation of the kinetic

© IMechE 1976

energy expression given in reference 1. Also the dependence of tyre sideforce on path curvature, the dependence of tyre aligning moment on sideslip angle, camber angle, and path curvature, and the dependence of rolling moment on camber angle (Ref.4) are now included. The aligning moments contribute to the external yawing moment and that from the front tyre contributes to the steering moment. The rolling moments arise from the circular cross-section of the tyres, and do not depend on tyre distortion. They are therefore taken to be in phase with the camber. The sideforce and self aligning torques, being distortion dependent, are considered to lag the motion variables as in reference 1.

6. The linearised equations of motion, describing small perturbations from straight running, with signs chosen to make all the tyre constants positive, are:-

$$(M_f + M_r)\ddot{y}_1 + M_f k\ddot{\psi} + (M_f + M_r)U\dot{\psi} + (M_f j + M_r h)\ddot{\phi} + M_f e\ddot{\delta} - Y_f - Y_r = 0$$

$$M_f k\ddot{y}_1 + (M_f k^2 + I_{rz} + I_{fx}\sin^2\varepsilon + I_{fz}\cos^2\varepsilon)\ddot{\psi} + M_f k\,U\dot{\psi} + \{M_f jk - C_{rxz} + (I_{fz} - I_{fx})\sin\varepsilon\cos\varepsilon\}\ddot{\phi} - \{i_{fy}/R_f + (i_{ry} + \lambda i)/R_r\}U\dot{\phi} - \{M_f j + M_r h + i_{fy}/R_f + (i_{ry} + \lambda i)/R_r\}\dot{U}\phi + (M_f ek + I_{fz}\cos\varepsilon)\ddot{\delta} - (i_{fy}/R_f)\sin\varepsilon\,U\dot{\delta} - lY_f + bY_r - M_{zf} - M_{zr} = 0$$

$$(M_f j + M_r h)\ddot{y}_1 + \{M_f jk - C_{rxz} + (I_{fz} - I_{fx})\sin\varepsilon\cos\varepsilon\}\ddot{\psi} + \{M_f j + M_r h + i_{fy}/R_f + (i_{ry} + \lambda i)/R_r\}U\dot{\psi} + (M_f j^2 + M_r h^2 + I_{rx} + I_{fx}\cos^2\varepsilon + I_{fz}\sin^2\varepsilon)\ddot{\phi} + \{C_{f5} + C_{r5} - (M_f j + M_r h)g\}\phi + (M_f ej + I_{fz}\sin\varepsilon)\ddot{\delta} + (i_{fy}/R_f)\cos\varepsilon\,U\dot{\delta} + \{tZ_f - M_f eg + C_{f5}\sin\varepsilon + (i_{fy}/R_f)\cos\varepsilon\,\dot{U}\}\delta = 0$$

$$M_f e\ddot{y}_1 + (M_f ek + I_{fz}\cos\varepsilon)\ddot{\psi} + \{M_f e + (i_{fy}/R_f)\sin\varepsilon\}U\dot{\psi} + (M_f ej + I_{fz}\sin\varepsilon)\ddot{\phi} - (i_{fy}/R_f)\cos\varepsilon\,U\dot{\phi} + (tZ_f - M_f eg + C_{f5}\sin\varepsilon)\phi + (I_{fz} + M_f e^2)\ddot{\delta} + K\dot{\delta} + \{(tZ_f - M_f eg)\sin\varepsilon - M_f e\cos\varepsilon\,\dot{U} + C_{f5}\sin^2\varepsilon\}\delta + tY_f - \cos\varepsilon\,M_{zf} = 0$$

$$(C_{f2}R_f/U^2)\ddot{y}_1 + (C_{f1}/U)\dot{y}_1 + (C_{f2}R_f\ell)/U^2)\ddot{\psi} + \{(C_{f1}\ell + C_{f2}R_f)/U\}\dot{\psi} - C_{f2}\phi - (C_{f2}R_f t/U^2)\ddot{\delta} + \{(C_{f2}R_f\cos\varepsilon - C_{f1}t)/U\}\dot{\delta} - (C_{f1}\cos\varepsilon + C_{f2}\sin\varepsilon)\delta + (\sigma_f/U)\dot{Y}_f + Y_f = 0$$

$$(C_{r2}R_r/U^2)\ddot{y}_1 + (C_{r1}/U)\dot{y}_1 - (C_{r2}R_r b/U^2)\ddot{\psi} + \{(C_{r2}R_r - C_{r1}b)/U\}\dot{\psi} - C_{r2}\phi + (\sigma_r/U)\dot{Y}_r + Y_r = 0$$

$$(C_{f4}R_f/U^2)\ddot{y}_1 - (C_{f3}/U)\dot{y}_1 + (C_{f4}R_f\ell/U^2)\ddot{\psi} + (C_{f4}R_f - C_{f3}\ell)/U\}\dot{\psi} - C_{f4}\phi - (C_{f4}R_f t/U^2)\ddot{\delta} + \{(C_{f4}R_f\cos\varepsilon + C_{f3}t)/U\}\dot{\delta} + (C_{f3}\cos\varepsilon - C_{f4}\sin\varepsilon)\delta + (\sigma_f/U)\dot{M}_{zf} + M_{zf} = 0$$

$$(C_{r4}R_r/U^2)\ddot{y}_1 - (C_{r3}/U)\dot{y}_1 - (C_{r4}R_r b/U^2)\ddot{\psi} + \{(C_{r4}R_r + C_{r3}b)/U\}\dot{\psi} - C_{r4}\phi + (\sigma_r/U)\dot{M}_{zr} + M_{zr} = 0$$

7. The datum parameter values are those of reference 1, except that subsequent measurements have shown the values of I_{fx} and I_{fz} used to be too small. These have been increased and values for C_{f1}, C_{f2}, C_{f3}, C_{f4}, C_{f5}, C_{r1}, C_{r2}, C_{r3}, C_{r4}, C_{r5}, σ_f and σ_r have been estimated from the values given by Eaton and Segel (Ref.4). A full parameter list is given in appendix 1.

8. The "inertia force" terms containing the longitudinal acceleration are three in number. They are $-\{M_f j + M_r h + i_{fy}/R_f + (i_{ry} + \lambda i)/R_r\}\dot{U}\phi$ in the yaw equation, $(i_{fy}/R_f)\cos\varepsilon\,\dot{U}\delta$ in the roll equation, and $-M_f e\cos\varepsilon\,\dot{U}\delta$ in the steer equation. Employing the appropriate numerical values, it is easily shown that the last two of these terms make very small contributions to the coefficients of δ, and the primary inertia effect arises from the first term, representing a yawing moment from roll angle.

9. Also longitudinal acceleration changes the wheel loads so that Z_f, and C_{f5} and C_{r5}, which depend directly on the wheel load (Ref.8) are affected. The other tyre constants will also, in general, be changed, and results have been obtained for two limiting cases. In one, the tyre constants $C_{f1} - C_{f4}$, $C_{r1} - C_{r4}$ are unchanged from the datum (steady speed, static weight distribution) values. In the other, the changes to $C_{f1} - C_{f4}$ and $C_{r1} - C_{r4}$ are in proportion to the changes in wheel load. Longitudinal force production at the tyres would be expected to influence the lateral tyre characteristics, but for non-limiting conditions, the effects will be small and have been ignored.

RESULTS

10. Stability characteristics were obtained by reducing the equations of motion to first order ones by the introduction of new variables, and assuming solutions of the form

$$\dot{y}_1 = X_1 e^{\lambda t},\quad \dot{\psi} = X_2 e^{\lambda t}\quad \text{etc.}$$

Substitution of these assumed solutions back into the first order equations yielded a set of algebraic equations which were expressed in the matrix form $\{B\}\lambda\{X\} + \{A\}\{X\} = 0$ in which $\{A\}$ and $\{B\}$ were square matrices, while $\{X\}$ was a column matrix of unknown amplitudes X_1, X_2, etc. Pre-multiplying by $\{B\}^{-1}$ and rearranging $-\{B\}^{-1}\{A\}\{X\} = \lambda\{X\}$ and for non-trivial solutions for $\{X\}$, values of λ must be eigenvalues of

©IMechE 1976

$-\{B\}^{-1}\{A\}$ (Ref.9). The real parts of the eigenvalues indicate the damping and the imaginary parts the natural frequencies of the normal mode motions. A numerical procedure was employed for calculating the elements of the {A} and {B} matrices, forming $-\{B\}^{-1}\{A\}$ and finding its eigenvalues (Ref.10).

11. The results of references 1, 2 and 4 show that the physically significant normal mode motions for perturbations from straight running of the motorcycle are three in number, and they can be referred to as the capsize, weave and wobble modes. Over most of the speed range of a motorcycle, the capsize mode is a slow divergence, while the weave and wobble modes involve oscillations which may damp out or build up depending on the machine design and the forward speed.

12. Eigenvalues which relate to the capsize mode are presented in Figure 2, those relating to the weave mode in Figures 3 and 4, and those relating to the wobble mode in Figure 5. With no changes to $C_{f1} - C_{f4}$, $C_{r1} - C_{r4}$ (assumption (a)), the changes to the capsize mode stability due to acceleration are less marked than when all the tyre constants are adjusted (assumption (b)). In both cases acceleration improves the stability while deceleration makes it worse, the effect being particularly pronounced near to 10 m/s forward speed where the capsize mode stability is normally at its worst, and where most machines are capable of rapid acceleration. Subjectively motorcycles seem to handle better when accelerating than when braking, and this difference in capsize mode stability may be the main reason for this.

13. According to assumption (a), the effect of acceleration on the weave mode is negligible, but according to (b), under some circumstances the weave mode damping can be seriously impaired by accelerating, while the wobble mode damping is increased. The large change in weave damping found under assumption (b) for high acceleration levels is associated with a lowering of the wobble mode natural frequency. This occurs mainly through lowering of the front tyre cornering stiffness and aligning moment to slip angle constant as a result of weight transfer. Where the large changes in damping occur, the weave and wobble mode frequencies cross over and loss of damping of one mode and gain of damping of the other when the frequencies come very close together is not entirely unexpected. This result is consistent with the normally observed behaviour, that slowing down is the best course if a weave oscillation starts, but that it is often possible to alleviate a wobble oscillation by accelerating, even though the steady speed characteristics will invariably show the stability to be worsening with increasing speed.

14. The question of how the motorcyclist, having accelerated out of a wobble oscillation, can safely return to low speed travel, is immediately posed, and there appear to be two basic possibilities. The first is that extra steering damping may be provided by adjustment of the machine, or by muscular effort on the part of the rider. The other is that by avoiding a disturbance of any magnitude while decelerating, the friction level normally present will be sufficient to prevent a wobble building up.

15. Changes to the wobble mode due to acceleration are negligible except in the case already discussed in which the weave mode damping was much reduced.

CONCLUSION

16. Acceleration of a typical large motorcycle has a significant stabilising influence on the capsize mode and deceleration acts conversely. This may account for the better "feel" of motorcycles when accelerating.

17. Under some circumstances, acceleration can bring a large reduction in weave mode damping, together with a corresponding increase in wobble mode damping. This situation appears to be associated with reduction of the wobble mode frequency to that of the weave mode, so that the two modes, to some extent, lose their identity.

18. Any design features which tend to separate the weave and wobble mode frequencies, mainly by increasing the latter, will tend to prevent this confusion of the modes, and its undesirable consequences. Among these features are low steering inertia, high trail, and high values of the front tyre cornering stiffness and self aligning moment to sideslip angle constant.

APPENDIX 1

Parameter values for datum machine

M_f	30.7 kg	C_{f1}	9800 N/rad
M_r	218 kg	C_{r1}	14120 N/rad
Z_f	- 1010 N	C_{f2}	835 N/rad
I_{rx}	31.2 kgm^2	C_{r2}	1200 N/rad
I_{rz}	21.0 kgm^2	C_{f3}	238 Nm/rad
C_{rxz}	1.76 kgm^2	C_{r3}	342 Nm/rad
I_{fx}	2.88 kgm^2	C_{f4}	6.49 Nm/rad
I_{fz}	1.34 kgm^2	C_{r4}	9.35 Nm/rad
i_{fy}	0.719 kgm^2	C_{f5}	36.6 Nm/rad
$i_{ry} + \lambda i$	1.06 kgm^2	C_{r5}	53.0 Nm/rad
b	0.478 m		
e	0.0244 m		
h	0.616 m		
j	0.466 m		
ℓ	0.933 m		
R_f	0.305 m		
R_r	0.305 m		
t	0.116 m		
σ_f	0.092 m		
σ_r	0.152 m		
ε	0.472 rad		
K	6.78 Nm/rad/sec		

© IMechE 1976

APPENDIX 2

REFERENCES

1. Sharp, R.S. "The stability and control of motorcycles" J.Mech.Eng.Sci., 1971, 13(No.5), 316-329

2. Sharp, R.S. "The influence of frame flexibility on the lateral stability of motorcycles" J.Mech.Eng.Sci., 1974, 16(No.2), 117-120

3. Weir, D.H. "A manual control view of motorcycle handling" Proceedings of the Second International Conference on Automotive Safety, San Francisco, July 1973

4. Eaton, D.J. and Segel, L. "Lateral dynamics of the uncontrolled motorcycle" Proceedings of the Second International Conference on Automotive Safety, San Francisco, July 1973

5. Roland, R.D. "Computer simulation of bicycle dynamics" ASME Meeting, Mechanics and Sport, Nov. 1973, Detroit, 35-83

6. Roland, R.D. "Simulation study of motorcycle stability at high speed" Proceedings of the Second International Conference on Automotive Safety, San Francisco, July 1973

7. Hales, F.D. "Lateral stability problems of simply articulated vehicles" Proceedings of IUTAM Symposium on Recent Progress in Linear Mechanical Vibrations, Paris 1965

8. Eaton, D.J. "Man-Machine Dynamics in the Stabilisation of Single Track Vehicles" Doctoral Dissertation, Highway Safety Research Institute, The University of Michigan, Ann Arbor, 1973

9. Crandall, S.H. "Engineering analysis: a survey of numerical procedures" 1956 (McGraw-Hill Inc., New York and London)

10. Sharp, R.S. "Eigenvalues and eigenvectors in linear vibration theory" Proceedings of the First British Conference on the Teaching of Vibration and Noise in Higher Education, Sheffield Polytechnic, July 1975

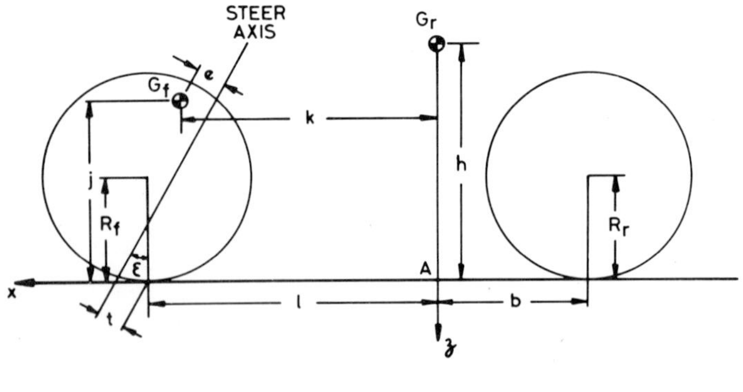

Fig. 1: Motorcycle dimensions

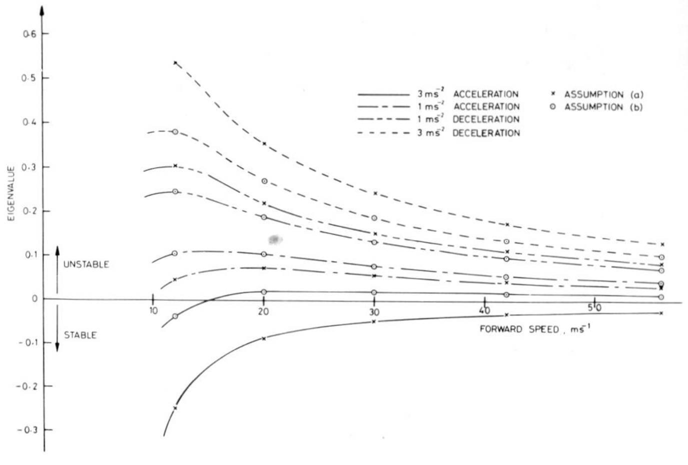

Fig. 2: Capsize mode eigenvalues as a function of forward speed for different accelerations

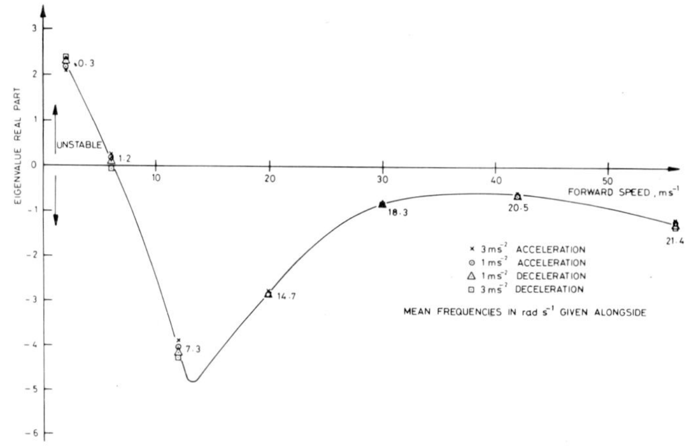

Fig. 3: Weave mode eigenvalues as a function of forward speed for different accelerations (using assumption (a))

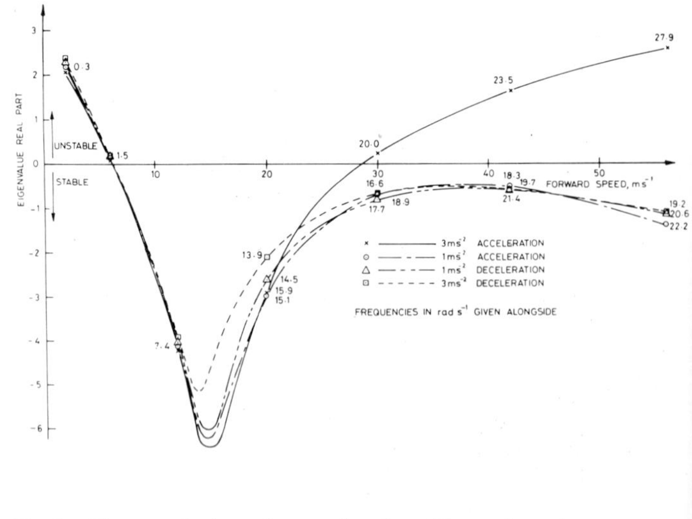

Fig. 4: Weave mode eigenvalues as a function of forward speed for different accelerations (using assumption (b))

©IMechE 1976

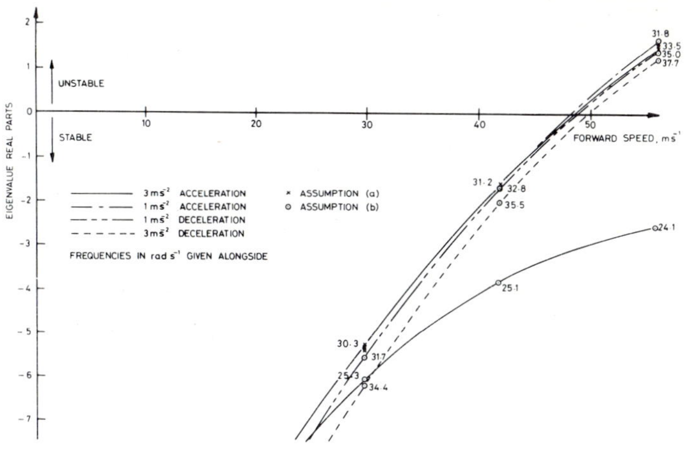

Fig. 5: Wobble mode eigenvalues as a function of forward speed for different accelerations

©IMechE 1976

© IMechE 1976

C27/76

THE DYNAMICS OF WHEEL BRAKING

M. B. PACKER, CEng, MIMechE
Automotive Products Limited, Leamington Spa

The MS of this paper was received at the Institution on 12 November 1975 and accepted for publication on 1 January 1975

SYNOPSIS An analysis of the dynamic behaviour of a wheel during transient brake application has been undertaken. Emphasis is placed on the use of simple computer models to present fundamental data of importance to brake system design.

Dynamic weight upon the wheel, rate of brake application, the brake force slip diagram shape and the gear engaged are shown to have a significant influence on wheel motion.

The use of brake pressure as an indication of effective brake torque is shown to be misleading, it being possible to develop very high structural brake torques on slippery surfaces. Wheel deceleration in isolation is demonstrated to be an inadequate parameter to define skid corrective action.

The effect of combined steering and braking, suspension compliance and weight transfer are discussed.

INTRODUCTION

Until relatively recently it has been sufficient for most brake calculations to assume that the effective torque reacted by the road is proportional to the applied brake effort, making suitable allowances for changes in lining coefficient. Little attention appears to have been given, or indeed to have been necessary, to considering the fundamental dynamic behaviour of a wheel subjected to transient brake forces. The advent of antilock systems initiated a quest for more fundamental factors influencing the dynamic behaviour of wheels under the influence of braking. In particular, a great deal of effort has been expended upon the relationship between tyre slip and the resultant force acting on the road (Ref.1,2). The work has yielded a variety of braking and steering slip curves for many situations. The relationships established are generally highly non-linear and interactive between several variables.

Engineers charged with the responsibility of developing brake systems clearly have a problem in utilizing the available data to good effect. Evidence in the form of papers, publicity material and patents suggests that most brake control systems evolve from ideas associated with a typical slip diagram. Good theoretical systems can undoubtedly be proposed to suit the idealised conditions but their practical execution will often fall short of theory and involve considerable ad hoc development.

During the development of advanced braking systems over a considerable period of time, it has been found to be advantageous to study the dynamic behaviour of a wheel under transient brake applications using very simple computer models. The approach leads to a basic understanding of many fundamental characteristics that can otherwise be lost, or incorrectly diagnosed, in the complexity of a model with many degrees of freedom incorporating a control system. This paper collates some of the more interesting results that give an insight into problems confronting the brake control system designer, without making reference to any specific devices. No attempt is made to cover a wide spectrum of variables, but the graphical presentation used highlights significant parameters influencing dynamic behaviour, enabling design and development engineers to identify potential problem areas.

DYNAMIC BEHAVIOUR OF A SINGLE BRAKED WHEEL

Throughout this section the first principles of wheel braking dynamics are established by considering the torsional equilibrium of a simple single wheel model governed by the relationship:-

> The torque accelerating a wheel equals the difference between the torque reacted by the road surface and the brake torque applied to that wheel.

The applied brake torque is basically under driver control but devices interposed between the brake pedal and the brake assembly frequently produce a complex relationship between pedal effort and applied brake torque at the wheel. The reacted brake torque is governed by conditions at the tyre to road interface. It is normal to express the relationship as a brake slip diagram, in which the tangential brake force at the road is plotted as a function of the percentage difference between the actual and free rolling wheel velocity.

The equation for the system investigated is, therefore:-

$$I\ddot{\theta} = f(ws) - f(t)$$

where I = wheel inertia
$\ddot{\theta}$ = angular acceleration of wheel
$f(ws)$ = reacted brake torque as function of weight and slip.
$f(t)$ = applied brake torque as function of time.

Before investigating the consequences of varying parameters influencing the individual terms in the equation, it is possible to establish many important principles by solving the equation for the following simple arbitary case:-

V	Vehicle speed (constant)	50 Km/h
W	Weight upon wheel (constant)	4453 N.
I	Wheel Inertia	1,2 $Kg.m^2$
R	Wheel Radius	0,3 m.
$\dot{\theta}_R$	Free rolling wheel velocity	46 rad/s.
$\dot{\theta}_C$	Wheel velocity at critical slip	38 rad/s.

© IMechE 1976

Brake applied linearly at 2174 Nm/s until wheel locks and then released at the same rate.

Reacted brake torque defined by slip diagram in which peak brake force coefficient of 0,4 occurs at 17,5% slip and the locked wheel coefficient is 0,32.

Note: The wheel slip coinciding with the maximum reacted brake torque is said to be critical slip.

Computation of the equation for the above case, using digital simulation techniques, yields the results presented graphically in Fig.1. Wheel acceleration, velocity, applied brake torque and road reacted brake torque are plotted against time. The form of the slip diagram has been superimposed to demonstrate how reacted brake torque is completely defined by wheel velocity.

A wheel operating on the positive slope of the slip diagram, that is before critical slip, is generally considered to be a stable situation. Any increase in slip is accompanied by an increased reacted brake torque. However, Fig.1 shows a dynamic lag between applied and reacted brake torque and illustrates that, in the absence of remedial action, wheel lock was inevitable before the reacted brake torque reached a maximum. Since the vehicle is travelling at constant speed and a rigid system has been assumed, the wheel deceleration is entirely attributable to the torque difference between applied and reacted torque until the wheel establishes a stable slip velocity. At critical slip the excess of torque decelerating the wheel is considerable. As wheel slip increases further the fall in the reacted torque associated with the slip diagram, coupled with increasing applied brake, produces a highly unstable condition and the wheel rapidly locks. When the wheel locks, the applied brake torque equals the locked wheel torque, regardless of the applied pressure.

Recovery from the locked wheel condition commences when the applied brake torque falls to the locked wheel reacted torque. The wheel recovery is subsequently governed by the difference between the falling applied brake torque and the reacted brake torque as defined by the slip diagram. In this particular example, the peak reacted torque is achieved when the applied brake torque is virtually zero. The recovery from peak brake torque slip to zero slip, i.e., free rolling, is associated with an extremely rapid reduction of reacted brake torque.

Finally, an important characteristic of the diagram is that the area between the applied torque and reacted torque in the period up to lock point is equal to the area between the curves during recovery. This can be readily explained. The torque accelerating a rotational mass equals Inertia times Acceleration. The area between the two curves is, therefore, proportional to the integral of wheel acceleration, i.e. velocity. Since the vehicle velocity was constant, the velocity into the skid equals the velocity out of a skid and the areas must be identical.

It follows from the above reasoning that the total area under the applied brake torque curve equals the area under the reacted brake curve.

PRACTICAL IMPLICATION

When stressing brake assemblies, the input load is normally proportional to pedal effort with corrections made for non-linearities; for instance, limit valves and finite power sources. It is sometimes reasoned that, although excessive pedal effort can cause structural problems in the actuation system, the maximum torque developed by the brake is limited by the tyre to road adhesion. This is clearly falacious since the brake torque retains its normal relationship to brake input up to the instant of wheel lock. Directly actuated high factor brakes are normally characterised with a fast response and light pedal effort. It is, therefore, possible to subject the brake to torsional stresses associated with the maximum pedal effort before the applied brake torque is limited by wheel lock.

Brake energy dissipation requirements during hill descents are characterised by long periods of brake application at relatively low levels of retarding force. This condition is frequently simulated on test tracks by a series of high deceleration brake applications to give the same mean rate of brake energy dissipation. Due to a relatively high level of wheel slip at high brake torques, a significant amount of energy is dissipated by the tyre compared with the modest brake torques associated with hill descents. For strict accuracy, an allowance should be made for wheel slip when compiling a test schedule from road test results.

Most brake system engineers will have encountered, through inventions submitted or discussions with their colleagues, antilock schemes that basically depend upon arbitrarily pumping the brakes on and off. If the equivalance of area between applied and reacted torque is considered for a complete cyclic brake application, it is clear that, in the periods between wheel lock, the effective mean torque will equal the applied brake torque. If this mean applied brake torque is low, there will be a loss of stopping distance and, if high, the wheel will tend to hover close to 100% slip. Consequently, the pumping device must take account of wheel load and road surface coefficient for effective operation.

The magnitude of wheel deceleration plays an important role in many wheel skid control systems. It is frequently stated 'When the wheel deceleration is greater than that associated with the maximum vehicle deceleration, typically 1 g., a skid must be imminent'. The existence of wheel deceleration from free rolling to a slip velocity associated with a perfectly stable wheel condition is clearly of importance in anti-skid logic design. Consequently, it is desirable to establish all the factors influencing the magnitude of this extraneous acceleration. Unfortunately, the permutations of the variables are numerous and, because of the nonlinear relationships involved, it is virtually impossible to obtain generalized solutions. In practice, this is not a serious restriction since, by expanding the case already considered, it is possible to put the various aspects in perspective. Throughout the study the system parameters remain as in the case already considered unless otherwise stated.

THE SLIP DIAGRAM

A knowledge of the relationship between reacted brake torque and tyre slip, i.e. the slip diagram, has been shown to be fundamental to the study of wheel braking dynamics. The exact form of the brake slip diagram normally provokes lively discussion among researchers but, in practice, brake control devices have to work within an operational envelope embracing many characteristics that are likely to make even quite large discrepancies in the form of the slip diagram of little consequence. To test this hypothesis the dynamic behaviour of a wheel is computed for a particular case. The computation is then repeated using fundamental

©IMechE 1976

changes in the form of the slip diagram to test their relative significance.

It is normal to express the slip diagram as a plot of Brake Force Coefficient (BFC) v. % wheel slip (s); where BFC = (Tangential brake force (F) per unit wheel load (W), i.e. BFC = F/W.

If the BFC is redefined as:-

BFC' = $F/W\mu$
μ = Coefficient of friction at critical slip,

the basic form of most slip diagrams can be classified by specifying the slip at which peak brake force occurs and the percentage reduction in the BFC' at 100% slip.

Using this technique, the influence of slip diagram characteristics has been studied for a particular situation using three alternative basic slip diagrams (Fig.2), defined as follows:-

Curve A: BFC' max. at 17,5% slip & 0,8 at 100% slip
Curve B: BFC' max. at 17,5% slip & 0,5 at 100% slip
Curve C: BFC' max. at 12% slip & 0,5 at 100% slip

The wheel and loading parameters of the previous example have been retained, but the applied brake torque has been modified to represent an emergency application to a level just in excess of that necessary to lock the wheel, followed by brake release at a rate typical of an antilock device.

Probably the most striking feature of the results of this exercise (Fig.3) is the occurrence of a wheel deceleration maxima prior to peak brake force, followed by a minimum deceleration at peak brake force. Although the shape of the slip diagram influences the amplitude of the peak initial deceleration, it is caused by, and is coincident with the knee point in the brake application curve. Factors affecting this phenomenum become clear when considering rates and amplitude of brake application.

For the particular example shown the presence of peak brake force at a low slip has little practical effect on the dynamic behaviour of the wheel within the accuracy of typical control circuits. The difference between peak and locked wheel BFC' has the readily predictable effect of proportionally increasing the maximum deceleration into a skid. Due to the fast brake release and the common magnitude of peak BFC', the maximum acceleration when recovering from a skid is identical for all three cases. The high reduction in BFC' between peak and sliding, however, does delay recovery and modify the acceleration v. time history during recovery.

FORM OF BRAKE APPLICATION

The influence of brake application on dynamic wheel behaviour has been studied for various amplitudes of brake application and rates of application and release. The study is confined to slip diagram A of the previous study, i.e. 17,5% critical slip with 20% fall off in BFC' at locked wheel.

Fig.4 shows the results of applying various amplitudes of brake torque at a common rate. The amplitude of applied brake torque has been expressed as a percentage of that necessary to just achieve peak reacted brake torque.

The reacted torque and wheel velocity curves resulting from overbrake show the readily anticipated reduction in time to wheel lock as the overbrake increases.

In the case of the application to 70% peak, the wheels assume a stable steady state slip and effective brake torque as defined by the slip diagram. The transient deceleration in achieving this state is in excess of 3g. even though the vehicle speed is constant.

If the brake torque is just in excess of 100%, the torque causing the wheel to lock is principly created by the BFC' reduction between peak and locked wheel condition. As the amplitude of application increases there is a marked reduction in the time to lock and the influence in the form of the slip diagram diminishes. The locus of wheel acceleration at peak brake force superimposed on Fig.4 shows that it is coincident with a minimum value of deceleration except for the 130% brake application where peak brake force is characterised by an inflexion in the curve.

Extension of the study to vary the rate of brake application while limiting the brake application to 130% peak reacted torque assists with the understanding of the phenomena. Fig.5 gives the results of varying the nominal application time between 0,1 and 0,5 seconds. The locus of peak brake force and knee point of application is constructed on the diagram.

From these results it is evident that the coincidence of a minimum deceleration and maximum BFC is conditional upon the limit of brake application being achieved before peak reacted torque. The variation in wheel acceleration at peak brake torque emphasises that wheel deceleration, or rate of change of deceleration, cannot be used in isolation to define precisely the imminence of a skid.

Variation in the rate of brake release to permit locked wheel recovery, Fig.5, shows a consistent sudden reduction in reacted brake torque if the applied brake torque is low at critical slip. In a modern compliant suspension this rapid reduction in reacted brake force can excite a torsional suspension oscillation as a wheel recovers from a skid. The frequency is typically 10 c.p.s. and the magnitude is sufficient to trigger a skid sensing system unless special precautions are taken.

BRAKING IN GEAR ON A SLIPPERY SURFACE

If the brakes are applied to driven wheels without disengaging the clutch, the difference between applied and reacted brake torque not only acts on the wheel inertia but also upon the engine. Since the effective engine inertia increases as the square of the total transmission gearing, its effect can be very significant. From evidence in the previous examples, it is clear that the degree of excess brake application plays a significant role in controlling the dynamic characteristics of a wheel tending to lock. The force accelerating the wheel during skid recovery, however, is limited by the road adhesion even if the brakes are completely released. Consequently, an adverse case occurs when braking in gear on a slippery surface.

This case has been computed for the single wheel model assuming slip curve 'A' with a peak brake force coefficient of 0,15, using a panic brake application in which the torque reaches twice the peak reacted torque in 0,2 seconds. This is followed by a rapid brake release. The engine inertia and gear ratios are typical of a six-cylinder saloon car and it is assumed that the inertia load of the engine is shared equally between the driven wheels. The effect of engine power on a closed throttle has been ignored. The results of this computation plotted in Fig.6 demonstrate the high attenuations in the dynamic wheel motion that occur as a consequence of leaving the clutch engaged when braking. Of particular significance is the fact that the deceleration into a skid in third gear approximates to a peripheral wheel deceleration of 10 m/s^2 which is below the

©IMechE 1976

theoretical sensing level of many antiskid systems. In practice, vehicle deceleration can normally be added to this deceleration and it is fortuitous that one wheel tends to lock before the other so alleviating the engine inertia load on the first wheel to lock.

The time for a wheel to recover from a skid in gear on a slippery surface is potentially very important in an emergency situation. If a vehicle with rear wheel drive locks all its wheels, with the clutch engaged, and then releases the brakes, the front wheels rapidly recover but the rear wheels remain in a high state of slip. This is a highly unstable condition and the ease with which it can lead to loss of vehicle control can be readily demonstrated on a skid track. Tests on a 16 tonne truck showed that, once the rear wheels were locked in fourth gear on a slippery surface, the engine would remain stalled even through the brakes were released. In some cases, as one wheel recovered the other wheel skidded in reverse.

VARIATION IN VERTICAL WHEEL LOAD

Although a road input profile to a wheel may be random, a frequency analysis of body acceleration shows a predominance of motion at the wheel bounce and body natural frequencies. The consequences of weight variations at wheel hop frequencies have been studied by superimposing an arbitrary ± 20% sinusoidal weight variation at 10 Hz on a single wheel model during a brake application to 80% of the available adhesion. The results shown in Fig.7 reveal a cyclic variation in reacted brake torque that peaks above the applied brake torque. If one ignores the initial brake application, the area rule established in Fig.1 is seen to apply. Had the computation included brake release, the rule would have applied without reservation. The cyclic reacted torque is accompanied by a low amplitude variation in velocity but, because of its relatively high frequency, the associated peripheral acceleration of the wheel exceeds $\pm$ 19,62 m/s^2 ($\pm$ 2g.)

This is an extremely significant result in antiskid system design. Any attempt to follow the wheel acceleration would result in a much sought after high frequency operation in circumstances where brake release was totally unnecessary. It is also suspected that the sensitivity to weight variations and the associated fast transients is the principal reason for experimental error in establishing the exact form of slip diagrams.

If the single wheel model used a cyclic variation in the surface coefficient of friction, it would produce an identical result due to the use of the product (weight) (surface coefficient) in the modified form of BFC.

The effects of varying the load on the wheel at body frequencies is not so dramatic. The rates of change in wheel load are relatively slow and their effect is almost analogous to slow rates of brake application. Consequently, wheel deceleration levels due to dynamic weight transfer changes are relatively low, but problems can arise with antiskid systems exciting a pitch resonance initiated by brake torque reduction during a period of reducing vertical force on the tyre.

COMBINED BRAKING AND CORNERING

In publicity type of material it has been common to superimpose upon the brake force slip diagram a curve representing the sideways force coefficient available. A casual analysis of this construction implies that a significant cornering force is available while maintaining maximum brake force. Holmes and Stone (Ref.1) made a study of tyre forces during combined braking and cornering. Plotting their results of brake force coefficient and side force coefficient against percentage wheel slip for a typical case, yields a family of curves showing characteristics of the form reproduced in Fig.8. The curves are based upon the conventional definition of BFC for a maximum road surface coefficient of 0,4. Analysis of the diagram shows that above 10% slip the vectorial sum of the side and braking force coefficients approximately equals the BFC at zero steer angle. This result implies that there is a slip dependent limit to the absolute adhesion available for either cornering or braking.

The form of the brake force coefficient at high cornering slip angles is particularly interesting since it maintains a positive slope up to 50% brake slip. Therefore, up to this point the wheel is able to react more brake force, to the detriment of cornering force, without going unstable. This fact is of great interest in the design of antilock sensing logic so the simple one-wheel computer model was extended to include tyre to road interface characteristics shown in Fig.8. A simple limited ramp function has been used for the steering slip angle input. No account has been taken of dynamic lags between steering wheel input and the true tyre slip angle since they are only meaningful when applied to a particular vehicle configuration.

The two basic sequences of braking preceding steering input, and vice versa, have been selected to demonstrate the important dynamic principles involved within the limits of the simple model. The brake application is sufficient to achieve 80% adhesion utilisation and the steering slip angle input of 10° is well in excess of the 4° necessary to produce maximum cornering force. The interaction of cornering force and brake force for each input sequence is shown in Fig. 9. The associated wheel velocity and acceleration histories are included

The most obvious feature of the results is the priority of braking over cornering for both cases. Although the vectorial sum of the adhesion demanded exceeds the capability of the surface, the demanded steady state brake torque is achieved after the initial transients and any residual adhesion is available for cornering.

The wheel accelerations associated with transients in steering slip angle originate from the need to establish a new wheel slip equilibrium commensurate with the combined brake and steering input (Ref. Fig.8). In the example shown the wheel deceleration is considerably in excess of that commonly associated with skid sensing systems, even though no skid, (100% slip) was imminent. However, limited brake release in this case would clearly be beneficial to give a compromise between cornering and braking.

The magnitude of the wheel acceleration during steer angle transients will depend upon the form of the brake/steer tyre to road slip characteristics and the rate of steer angle build-up. Since the road characteristics and the driver steer input are substantially outside the control of a brake system designer, the use of wheel deceleration as a skid sensing parameter usually results in variability in the compromise between steering and braking.

DISCUSSION AND CONCLUSIONS

Formal experimental evidence to obtain correlation with the theoretical work has not been undertaken. The theoretical results presented, however, do offer a rational explanation to practical situations that have been experienced in advanced brake

©IMechE 1976

system development. Many of these have already been discussed in the text but it is pertinent to briefly review the results in relation to skid sensing techniques.

The problems of using deceleration as the sole criteria for anticipating a skid are self-evident, as are the difficulties likely to be encountered if any attempt is made to use the derivative of deceleration for phase advance. In practice, system lags and controlled delays play an important role in skid sensing logic to avoid unnecessary brake release. Complete immunity from spurious brake release, particularly due to dynamic wheel loading, would either involve a very complex system, in which many system parameters were monitored and processed, or excessive filtering and consequently sluggish response and poor steering capability. The design of vehicle suspensions to attenuate the torsional response of the wheel hub can have a significant effect in reducing the skid sensing filtering requirements, consequently improving the performance of an antilock system.

Little has been said about the problems of sensing the ideal point for brake re-application. The results of this paper constantly show a peak acceleration as the wheels recover to critical slip velocity. Unfortunately the use of this point for initiating brake re-application is not usually acceptable since it will, in most circumstances, be followed by a rapid reduction in reacted brake torque. Consequently, it is normal to devise logic to initiate brake re-application before the wheel has recovered to the critical slip velocity. The quality of the re-application logic plays an important role in the characteristics of an antilock system. Early initiation of brake re-application can result in premature wheel lock and late re-application is characterised by jerky operation and the generation of suspension knock.

A beneficial aspect of late brake re-application is the high cornering force potential available at the point of skid recovery. However, it is difficult to generalise on steer characteristics since they are dependent upon a total vehicle system.

In conclusion, it is recognised that the use of particular examples to demonstrate the dynamic characteristics of a wheel during transient brake application inevitably results in limited coverage of the variables involved. Nevertheless the analogue presentation of simple cases has proved extremely useful in conveying fundamental principles to design and development engineers as an aid to creative design and problem diagnosis. Once these principles are accepted it is frequently possible to anticipate problems and the consequences of design changes without resorting to the use of a computer except for quantitative information.

ACKNOWLEDGEMENT

The author wishes to thank the Management of Automotive Products Ltd. for permission to publish this paper.

REFERENCES

1. Holmes K.E. and Stone R.D. - Tyre Forces as Functions of Cornering and Braking Slip on Wet Surfaces. - Symposium, 'Handling of Vehicles Under Emergency Conditions', 1969 (I.Mech.E.)

2. Kullberg G. and Kinlgren B. - Investigation of Friction Properties of Winter Tyres and Anti-skid Devices (Report 36, Statens Vaginstitut, Stockholm).

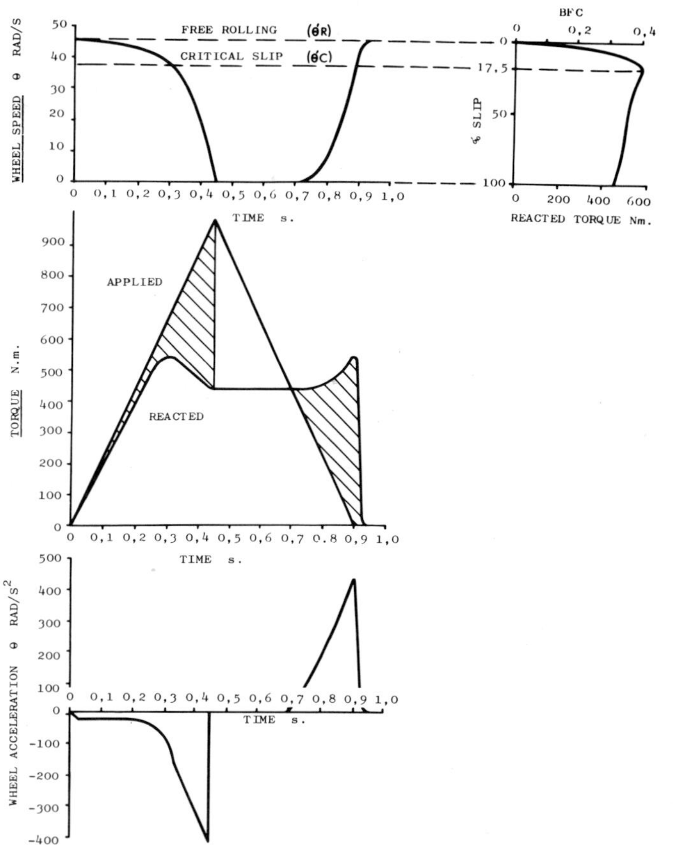

Fig. 1: **Dynamic wheel response to a simple brake application and release sequence.**

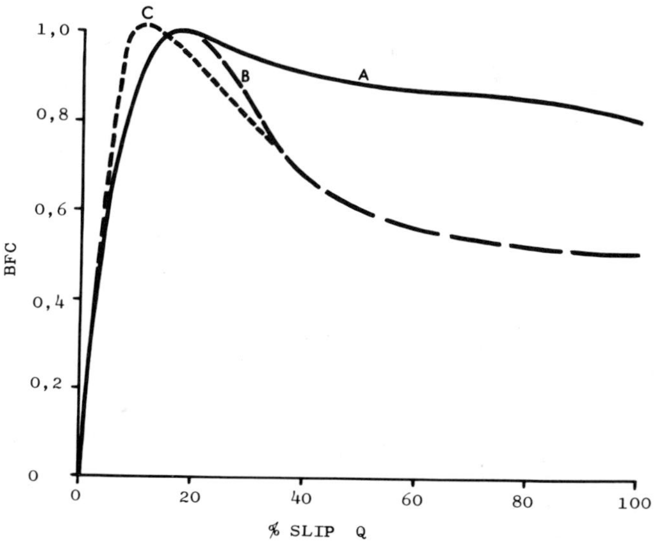

Fig. 2: **Alternative basic forms of brake slip diagrams A, B and C, using modified brake force coefficient BFC' = F/μW.**

©IMechE 1976

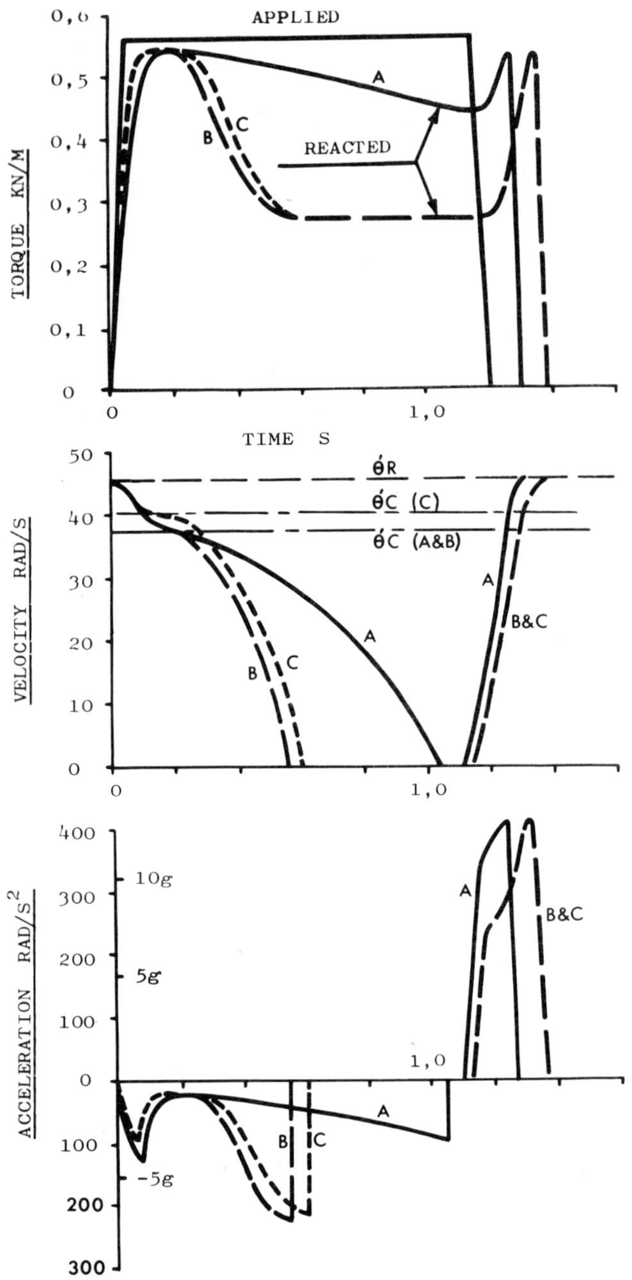

Fig. 3: Wheel response to brake application with alternative slip diagram characteristics A, B or C.

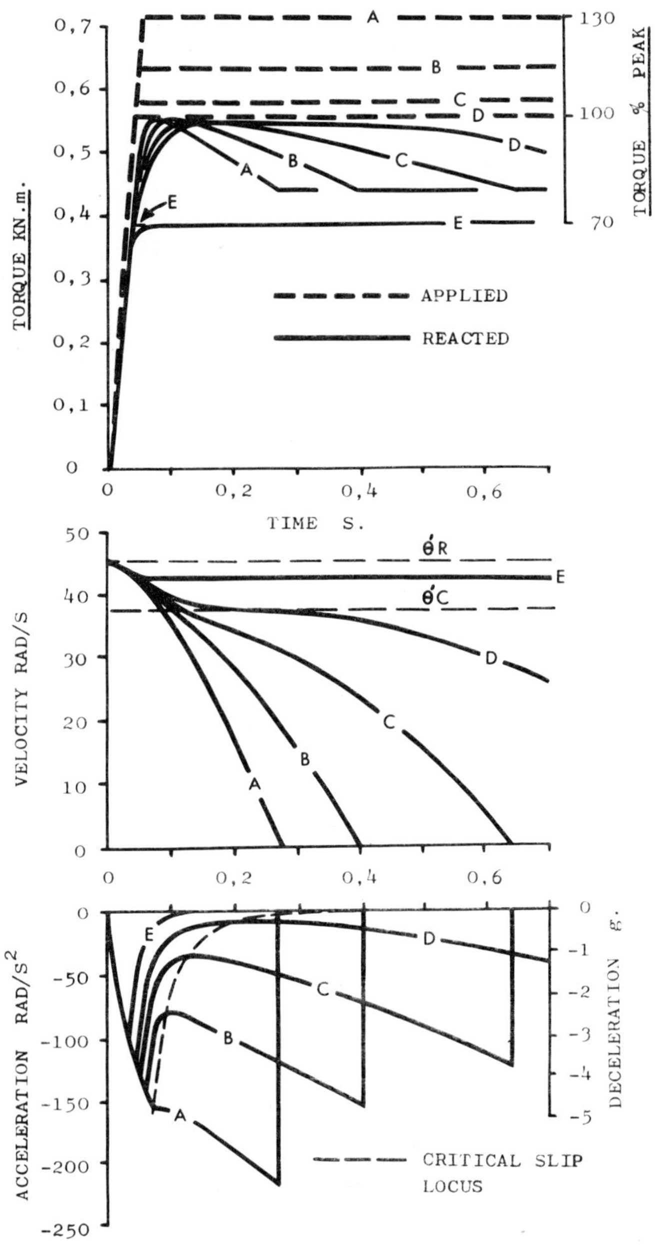

Fig. 4: Wheel response to various amplitudes of brake application.

© IMechE 1976

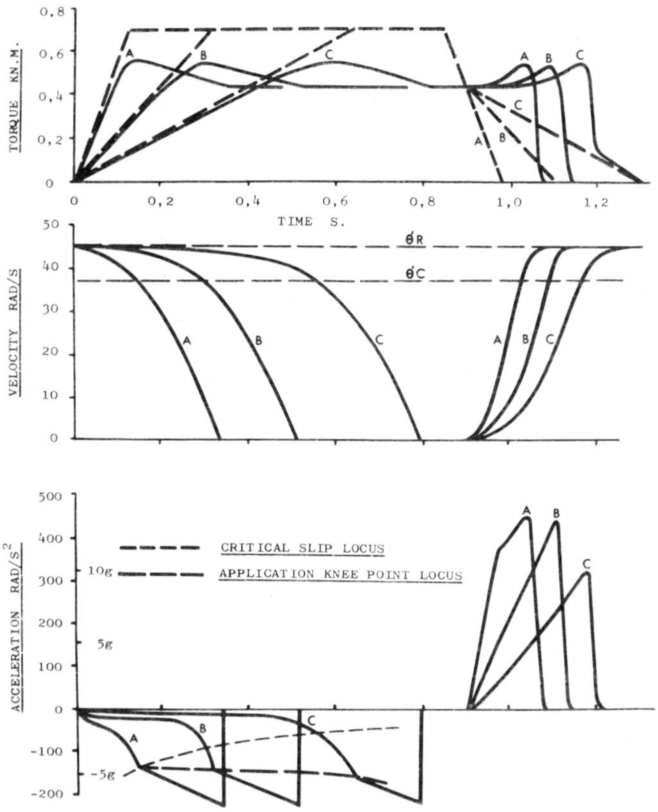

Fig. 5: Wheel response to various rates of brake application and release.

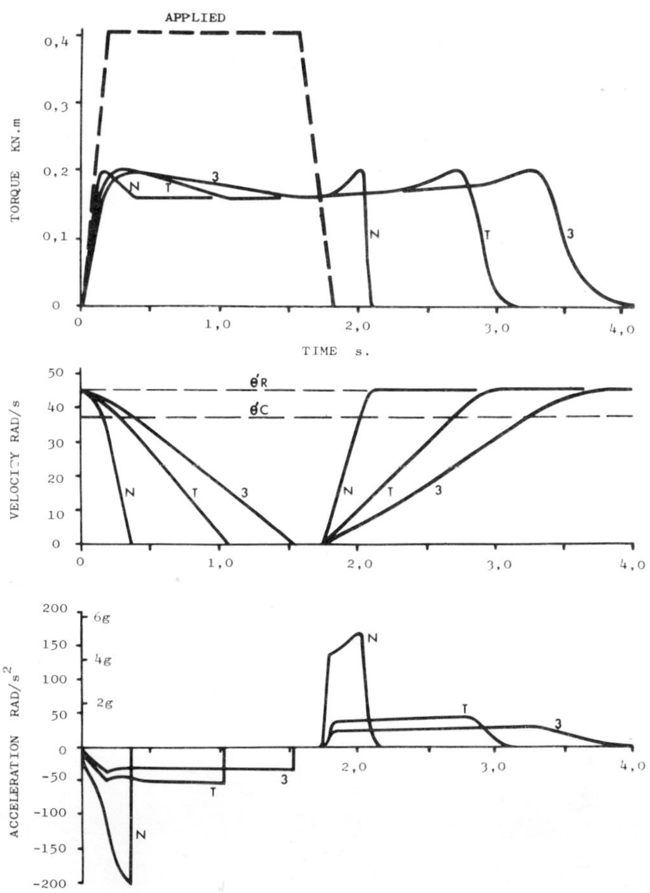

Fig. 6: Wheel response when braking with clutch engaged on slippery surface (μ = 0,15)

N Neutral T Top gear 3 Third gear

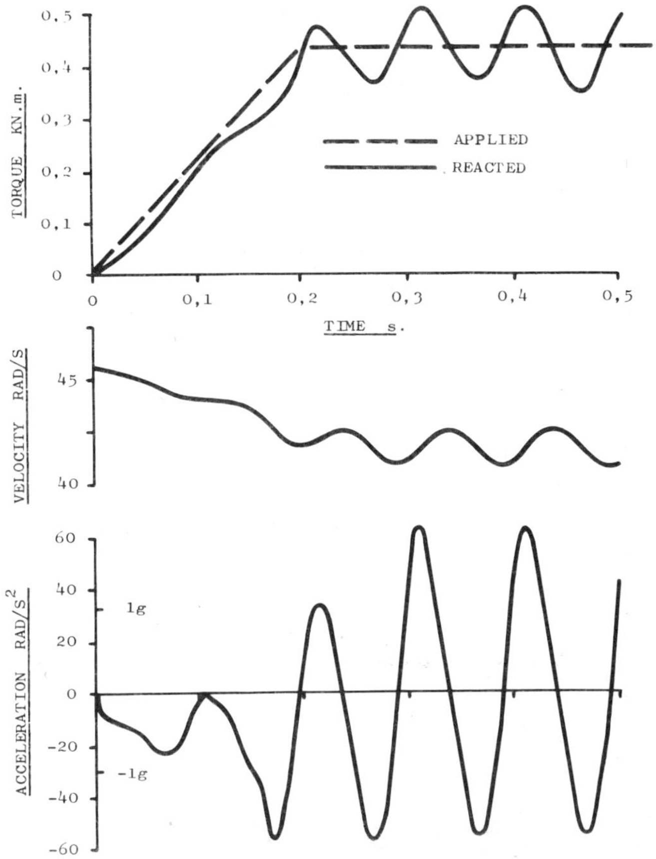

Fig. 7: Wheel response to $\pm$ 20% vertical force variation at 10 Hz while braking.

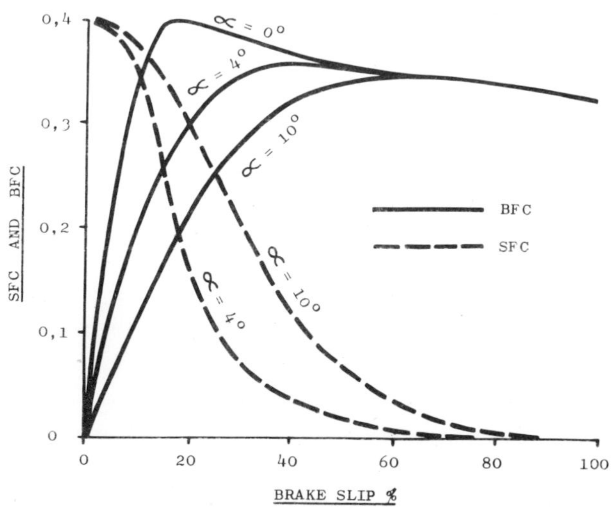

Fig. 8: Typical characteristics of brake slip diagram in presence of steer angle.

© IMechE 1976

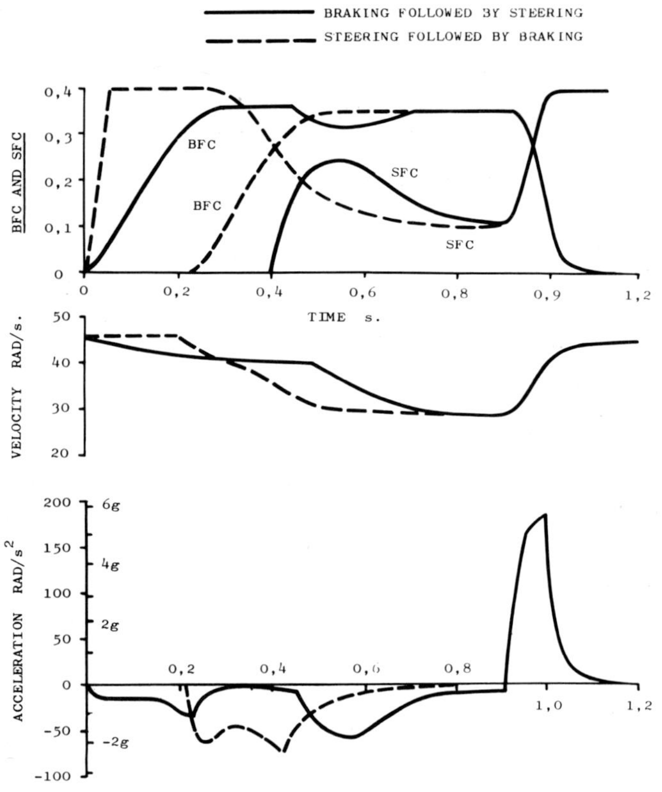

Fig. 9: Wheel response to combined braking and cornering.

© IMechE 1976

C28/76

THE INFLUENCE OF WHEEL SLIP CONTROL DYNAMICS ON VEHICLE STABILITY DURING BRAKING AND STEERING

E. ELGESKOG, Tekn. lic. and S. BRODD, Civ.ing.
AB Volvo, Technological Development Dept, Gothenburg, Sweden

The MS of this paper was received at the Institution on 7 July 1975 and accepted for publication on 17 October 1975

SYNOPSIS The dynamics of a wheel during braking is controlled by applied braking torque, which is generated by braking devices and balanced by wheel inertia and reaction forces between tyre and ground. These forces depend mainly on wheel slip and slip angle, wheel velocity and normal force between wheel and road surface. The vehicle movement is influenced by the interaction of the tyre forces of each individual wheel, and thus the vehicle stability depends on the dynamics of wheel locking.

The paper describes a mathematical wheel model, based on dynamical tyre and vehicle tests on ice and on asphalt. The wheel locking properties are studied in hybrid simulations during panic manoeuvres with special consideration of vehicle stability. Different principles of anti-skid-systems are simulated and general requirements on wheel locking behaviour for vehicle stability are formulated.

Most of the work was performed within the Swedish joint government industry ESV-project "Steerability during emergency braking". The paper describes some extensions and continuations of this work.

1. INTRODUCTION

The Swedish Experimental Safety Vehicle (ESV) project "Steerability during emergency braking" started in March 1972 with the objectives to investigate the safety benefits of steerability during emergency braking, to recommend test procedures, and to suggest performance standards. The project was a joint effort with representatives from the Swedish Government, represented by the Swedish Board for Technical Development, and from the Swedish car manufacturers Volvo and Saab-Scania. The total fund amounted to Sw. Crs. 2,6 millions.

The program was divided into the following tasks:

Skid accident analysis

Field testing, Performance tests and Statistical tests

Simulation

Measurements of tyre-road characteristics

These tasks were completed and reported in June 1974 (Ref. 1).

The simulation activities were concerned with the development of a mathematical model for a complete passenger car (Ref. 2), the validation of this model through comparisons with field test results (Ref. 3), and simulation studies of the braking and steering behaviour of different car types on various road surfaces (Ref. 4). Special care was devoted to the development of a detailed and accurate model for the tyre forces, based on measurements of tyre characteristics for relevant tyres on actual road surfaces (Ref. 5).

In this paper the mathematical model and some simulation results are presented, in particular as they concern the vehicle stability. The wheel slip control strategy for stability is discussed and the influence of wheel load characteristics on the stability is pointed out.

2. VEHICLE MODEL

The simulation model described here was developed for the Swedish ESV-project. The objectives of the simulations were (Ref. 1):

- To complement the field tests
- To perform a systematical investigation of the influence of typical anti-lock parameters.

The model has been reported in detail (Ref. 2). Only a short description will be given here. The model is validated (Ref. 3), see appendix A.

The model is very flexible. Various vehicles with various anti-lock system configurations can easily be simulated during various normal or severe steering and braking manoeuvres with various tyre-road conditions.

The mathematical formulation of the model is such that the following principal parts and phenomena are included:

Sprung mass

Suspension links

Wheel movements

Steering gear

Tyre forces

Anti-lock system

The sprung mass of the vehicle is free to move in longitudinal, lateral and vertical directions as well as to rotate in roll, pitch and yaw directions. However, very extreme pitch and roll rotations, that are of no interest in this project, such as "roll-over" cannot be simulated.

The forces, acting on the sprung mass, described by the model, are suspension link forces, spring forces and damper forces.

©IMechE 1976

The assumption of a known and fixed roll axle has been avoided. A consequence of this is that the directions of the suspension forces must be described by the model. The way the link directions depend on the roll and pitch mode and the loading of the vehicle is formulated in a simplified (linearized) manner. Effects such as anti-lift and anti-dive can thus be simulated.

The mathematical formulation of the suspension forces is such that it is easy to change between different suspension geometries.

In the project, all manoeuvres were carried out on smooth road surfaces. The vertical motions of the wheel hubs were therefore assumed to be small and the translational motions of the wheel hubs were limited to a horizontal plane. The model has however been further developed in this respect since the ESV-project was terminated. The unsprung mass movements in vertical and roll directions are now included in the model, so driving on any uneven surface can now be simulated.

Each wheel can rotate individually around its axle. This is of primary importance when braking manoeuvres are simulated on vehicles with individual anti-lock systems.

Furthermore each wheel can move around a vertical axle. This makes it possible to steer each wheel individually. The steering of each wheel can be due to the driver's action (turning the steering wheel) or the rolling of the vehicle (roll-steering). Both front and rear wheels can be steered, which is of importance as roll steering effects often occur at both front and rear wheels.

The driver's action on the steering wheel is transferred via a stiff gear to the front wheels. This means that, in this model, there is no elasticity in the steering system. There can, however, be different gear ratios between the steering wheel and the two front wheels.

Whenever vehicle dynamics are to be simulated, it is of fundamental importance that the tyre forces are adequately described. The tyre force model is described in more detail in paragraph 2.3.

In order to simulate braking manoeuvres, it is necessary to apply different braking torques on the wheels. The braking torque is considered proportional to the brake pressure at the wheel. The brake pressure can be individually regulated for each wheel via simulated brake valves.

The opening and closing of the brake valves can be controlled by an anti-lock control system or directly by the brake pedal. The models of the anti-lock systems are built up in accordance with the description of the system logic given by the manufacturers. In general, the anti-lock systems use the wheel rotational velocities and accelerations as "inputs". The "outputs" from these systems are signals that control the brake valves.

It is possible to simulate not only different types of anti-lock systems (for example systems from different manufacturers) but also different principles such as individual wheel regulation and axle regulation (for instance "select low"). Theoretical systems of different kinds can further be simulated in order to study e.g. different wheel slip control strategies. Results from such a study are given in section 3 of this paper.

The model is implemented on a hybrid computer (EAI Pacer 600) as shown in Fig. 1.

2.1 Interactions and stability

Excluding aerodynamic and gravity forces, all changes in a vehicle's state of motion are accomplished by the combined effect of the tyre forces at the wheels. These forces cause accelerations and rotations (inertia forces) of the vehicle which in turn influence the geometry and the wheel loads, causing changes in the tyre forces etc., as indicated by the heavy closed loop in Fig. 1.

The vehicle stability is very sensitive to the interactions of these forces. One should expect the stability to be strongly dependent on the forces on each wheel in relation to the forces on all the other wheels. This means that in order to study vehicle stability, you must have a detailed knowledge and representation of the tyre forces and the way you control them.

That is the reason why special care has been taken in this project to develop a detailed and accurate tyre model. It is also the reason why this model (illustrated by the flow chart in Fig. 2) is described in more detail in this paper. The model is believed to be a significant contribution to the state of the art.

2.2 Characteristics of tyres

Curves showing longitudinal and lateral friction numbers (μ_L and μ_A) as functions of the wheel slip (λ) for different tyre-ground combinations are frequently given in the literature. The load, velocity, slip angle, camber angle and inflation pressure of the tyre are parameters in such curves (Ref. 2).

Such curves give good information about the dynamic variations of the tyre forces. Fig. 3 shows tyre characteristics for Michelin ZX-165-SR 15, which was used in the project. The influences of speed, load and side slip angle on μ_L and μ_A on dry asphalt are shown in the figure. Full details of the measured characteristics for this tyre on asphalt and on ice and full details for Michelin XM+S with 120 studs on ice have been reported (Ref. 5).

It was found during the simulation studies that the way the tyre forces depend on the wheel load is of significant importance for the stability of a vehicle as discussed in section 4. Therefore this dependence had to be measured in more detail than was planned at first. These measurements showed that the load dependence of the studded tyres on ice differed significantly from the other tyre-road combinations (summer tyre on ice and on dry and wet asphalt). An example is shown in Fig. 4. The minima in the curves for μ_L and μ_A at a wheel load of about 2000 N is significant for this tyre on ice. The friction numbers for the other measured tyre-road combinations are in general decreasing with increasing load or approximately constant (Ref. 5).

2.3 The tyre model

The longitudinal and lateral contact forces between tyre and ground are (for a given tyre/ground combination and a given inflation pressure) dependent on:

- wheel slip (λ)
- wheel slip angle (β)
- wheel hub velocity (v)
- wheel load (F_z)

The tyre model used in this project describes the tyre contact forces with respect to these variables. Other variables, e.g. the camber angle, can also influence the contact forces but are not considered in this model.

©IMechE 1976

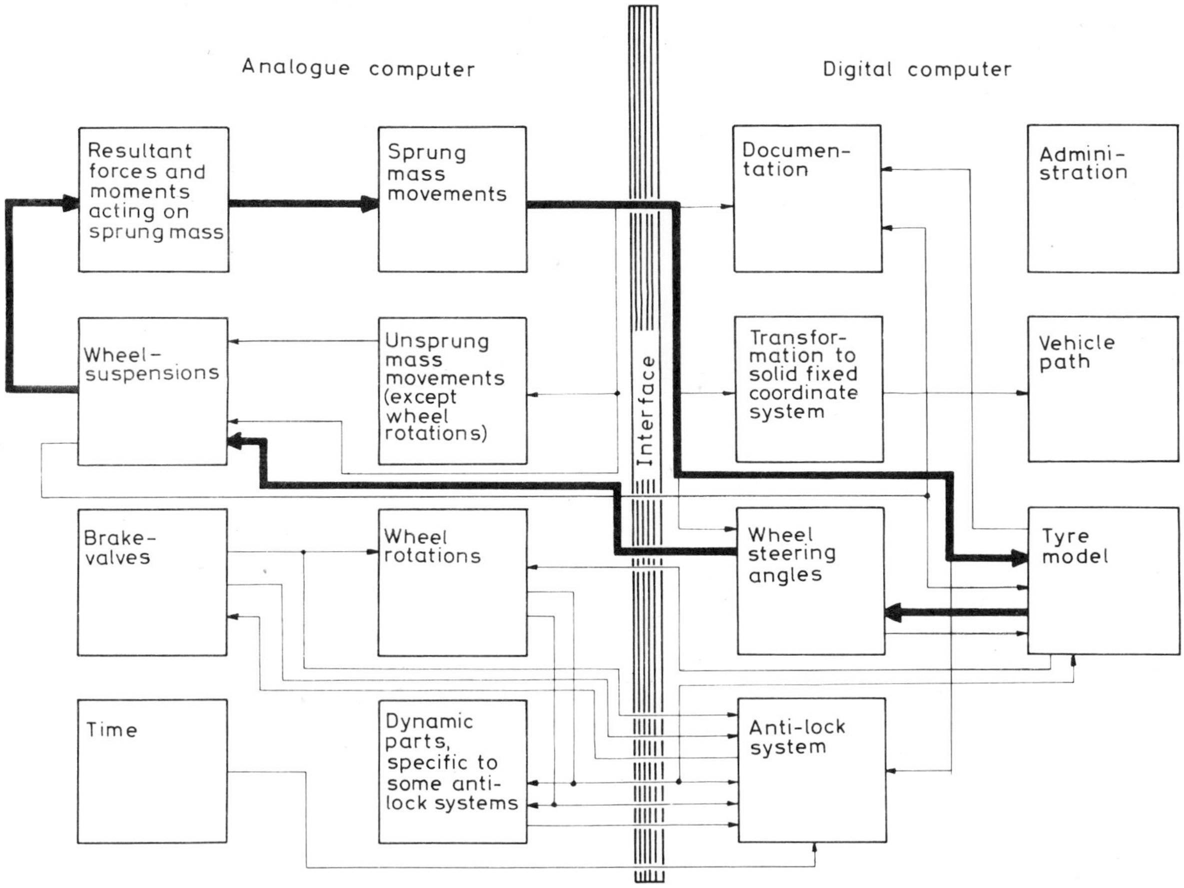

Fig. 1: Implementation of simulation model.

The model is empirical, i.e. it is built from measured data of tyre forces as functions of the above quantities and not from theories including material characteristics. This means that

$$F_L = F_L (F_z, v, \lambda, \beta), \quad (1)$$

$$F_A = F_A (F_z, v, \lambda, \beta), \quad (2)$$

where F_L is the longitudinal and F_A the lateral force for one wheel.

As mentioned in paragraph 2.2, measured μ-λ -curves (μ = friction number) are often found in the literature. Further, The National Swedish Road and Traffic Research Institute has facilities for measuring such curves, which thus could be mearused for those tyre/road combinations of interest in the project (Ref. 5).

The tyre model is built in such a way that curves of this kind could easily be used as input data.

Two quantities, μ_L and μ_A, have been defined:

$$\mu_L = F_L/F_z, \quad (3)$$

$$\mu_A = F_A/F_z. \quad (4)$$

From equations (1) - (4) it follows that

$$\mu_L = \mu_L (F_z, v, \lambda, \beta), \quad (5)$$

$$\mu_A = \mu_A (F_z, v, \lambda, \beta). \quad (6)$$

The measured curves for μ_L and μ_A as functions of slip (when the other variables are constant) are reproduced as exactly as possible in the model. Therefore, a set of nominal values of velocity and tyre load is chosen. Coefficients defining the slip angle influence on μ_L and μ_A at this nominal condition are tabulated for a set of slip values. Furthermore, coefficients defining the influence of variations in velocity and normal load are tabulated for the same set of slip values. The number of slip values is so chosen that sufficiently smooth curves will result from a linear interpolation between the tabulated slip values. (The number of slip values and their spacing can be chosen as desired.)

This means that the simulated curves can be fitted as close as desired to the measured curves for a set of F_z-, v-, and β-values at a wheel.

The tabulated values described here are prepared for every tyre-road combination that is to be simulated. These values are then used during the simulations in order to calculate the tyre forces that correspond to the instantaneous slip, slip angle, velocity and tyre load.

The tyre forces are calculated in the digital part of the simulator. All analogue variables that are used in the tyre model are read from the analogue part 300 to 500 times per second. The corresponding longitudinal and lateral tyre forces for each wheel are calculated with this frequency and written into the analogue part. The calculations are carried out through the following steps, see also the flow chart in Fig. 2. Wheel number k (used as a subscript) is considered.

A. The values F_{zk}, v_k, λ_k and β_k are determined. F_{zk} is read from the analogue computer, v_k and β_k are calculated from the vehicle translational and rotational velocities and the wheel steering angle δ_k

© IMechE 1976

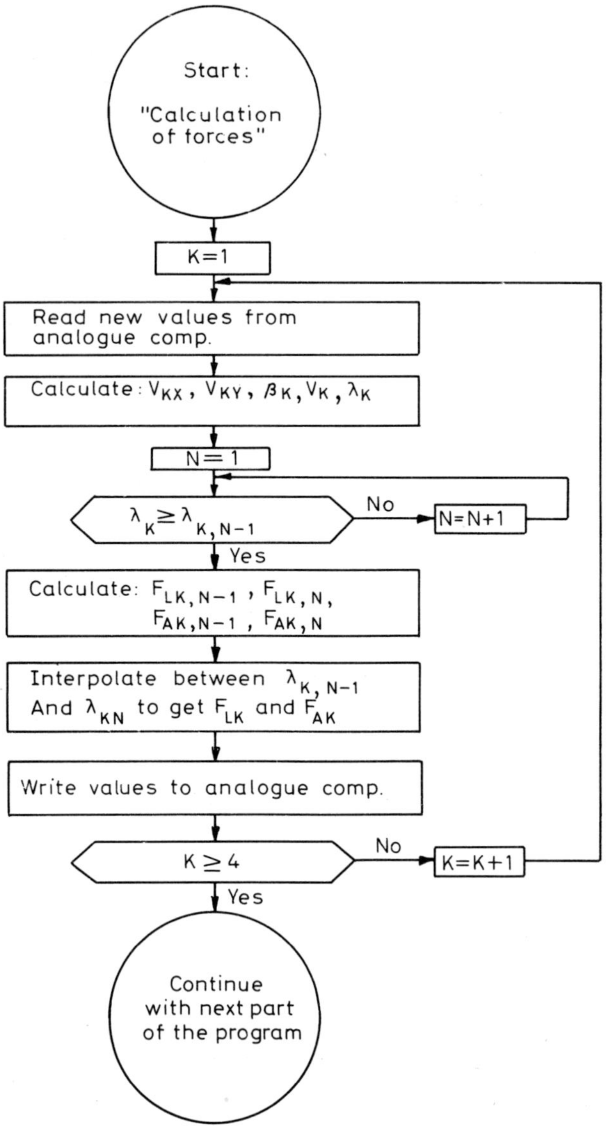

Fig. 2: Tyre model. Flow chart.

(including roll steering)

$$v_k = v_{kx} \cdot \cos \delta_k + v_{ky} \cdot \sin \delta_k, \tag{7}$$

$$\beta_k = \text{arctg}\ (v_{ky}/v_{kx}) - \delta_k. \tag{8}$$

v_{kx} and v_{ky} are the wheel hub velocities in the x- and y-directions of the vehicle attached coordinate system (Fig. 5).

The slip is defined as

$$\lambda_k = 1 - \omega_k \cdot R_r/v_k, \tag{9}$$

where ω_k is the wheel rotation velocity and R_r is the wheel rotation radius.

B. The next step is to determine in which slip interval the calculated λ_k -value lies. This must be done in order to read the corresponding tabulated values. Therefore the tabulated values $\lambda_{k,n}$ and $\lambda_{k,n-1}$, are determined, for which the following relation holds:

$$\lambda_{k,n-1} \leq \lambda_k < \lambda_{k,n}. \tag{10}$$

C. The μ_L and μ_A -values are now determined at the points $\lambda_{k,n-1}$ and $\lambda_{k,n}$. This is done in the following way:

A factor $f_{1L,n}$ (β_k) is calculated from the following empirically chosen interpolation formula

$$f_{1L,n}(\beta_k) = b_{1,n} + b_{2,n} \cdot \beta_k + b_{3,n}/(\beta_k + b_{4,n}). \tag{11}$$

All subscripted quantities $b_{i,j}$ are tabulated constants.

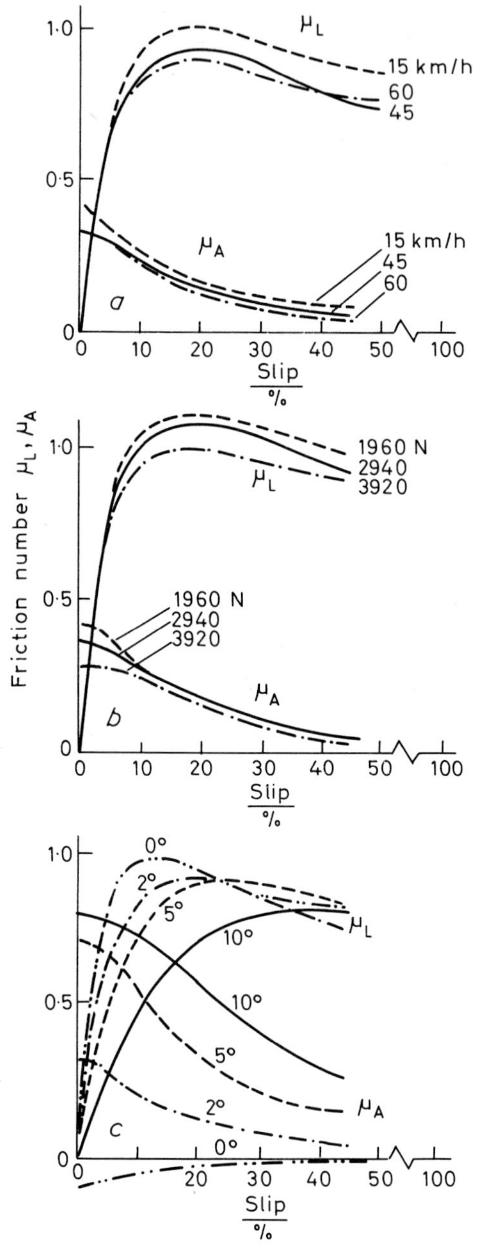

Fig. 3: Tyre characteristics for Michelin zx-165-SR-15 on dry asphalt.

a speed variation for slip angle 2° and wheel load 2940 N.

b wheel load variation for slip angle 2° and speed 30 km/h.

c side slip angle variation for wheel load 2940 N and speed 45 km/h.

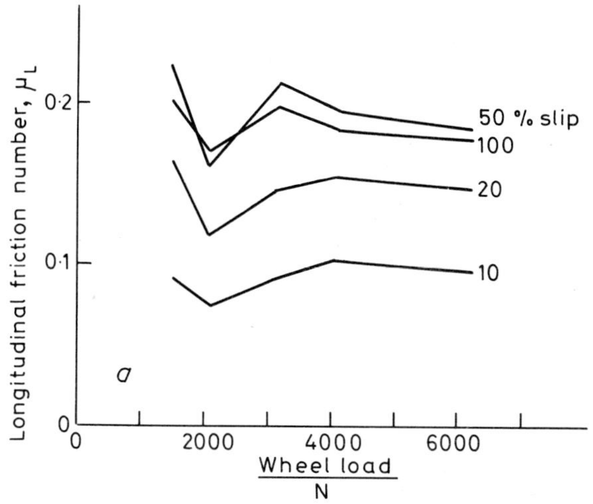

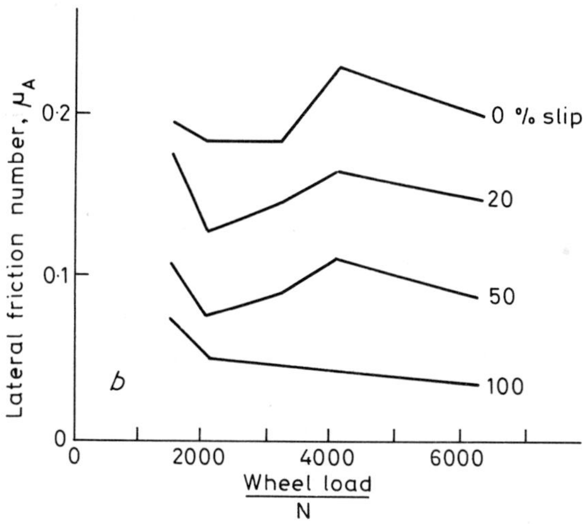

Fig. 4: Friction number versus wheel load for Michelin XM+S with 120 studs on ice.

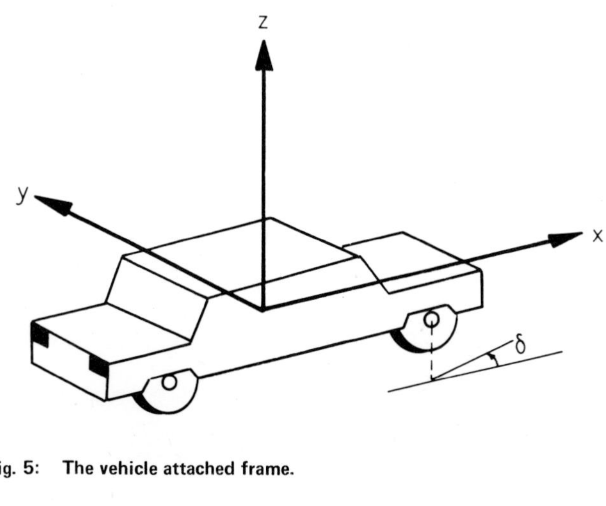

Fig. 5: The vehicle attached frame.

Equation (11) gives the μ_L-dependence of slip angle β_k for wheel number k, when its slip is $\lambda_{k,n}$. The same procedure is then followed for the slip value $\lambda_{k,n-1}$, giving a factor $f_{1L,n-1}$ (β_k).

Similar factors - $f_{1A,n}$ (β_k) and $f_{1A,n-1}$ (β_k) - are also calculated for μ_A.

Corresponding velocity factors $f_{2L,n}$ (v_k), $f_{2L,n-1}$ (v_k), $f_{2A,n}$ (v_k) and $f_{2A,n-1}$ (v_k) are calculated for the same λ_k-values. The following formula is found to give a good description of the velocity dependencies and is therefore used for these calculations:

$$f_{2L,n}\ (v_k) = 1/(1 + b_{5,n} \cdot v_k), \tag{12}$$

and analogously for the other factors.

The way μ_L and μ_A depend on the wheel load is described by f_{3L} and f_{3A}, given by the following relation:

$$f_{3L,n}\ (F_{zk}) = b_{6,n} + b_{7,n} \cdot F_{zk} \tag{13}$$

and corresponding for $f_{3L,n-1}$ (F_{zk}), $f_{3A,n}$ (F_{zk}) and $f_{3A,n-1}$ (F_{zk}). It must be pointed out that for both factors f_{3A} and f_{3L} different sets of constants b_6 and b_7 are used for different F_z-intervals.

This means that the μ_L- and μ_A-dependence of the wheel load is described by straight lines with different slopes in different F_z-intervals and the various wheel load characteristics mentioned in this paper can thus be simulated. The results given below refer to a model with two such F_z-intervals.

The μ_L- and μ_A-values are now calculated from

$$\mu_{L,n} = f_{1L,n}(\beta_k) \cdot f_{2L,n}(v_k) \cdot f_{3L,n}(F_{zk}) \tag{14}$$

and analogously for $\mu_{L,n-1}$, $\mu_{A,n}$ and $\mu_{A,n-1}$. The values μ_{Lk} and μ_{Ak} for the actual slip value λ_k are now determined by linear interpolation between the values for $\lambda_{k,n}$ and $\lambda_{k,n-1}$.

The longitudinal and lateral forces $F_{L,k}$ and $F_{A,k}$ are finally calculated from equations (3) and (4), giving

$$F_{L,k} = \mu_{L,k} \cdot F_{z,k}, \qquad F_{A,k} = \mu_{A,k} \cdot F_{z,k} \tag{15}$$

The tyre model is described in more detail in (Ref. 2).

3. WHEEL SLIP CONTROL STRATEGY FOR STABILITY

In order to study how anti-lock systems can influence the vehicle brakeability, steerability and stability in a panic situation, a drastic manoeuvre was chosen: the steering wheel was rapidly turned to a stop corresponding to minimum free rolling radius for the actual parameter combination (velocity, tyre-road condition, vehicle weight etc.). The brakes were fully applied one second later. The brakeability, steerability and stability are defined as follows (Ref. 1):

©IMechE 1976

Brakeability = $\bar{a}_{curve}/\bar{a}_0$,

where $\bar{a}_{curve}$ is the mean deceleration during braking in a curve and $\bar{a}_0$ is the mean deceleration during braking straight ahead with locked wheels on the same surface.

Steerability = Y/S,

where Y is the lateral deviation from brake application to standstill and S is the braking distance along the vehicle path.

Stability = the ability of the vehicle to correct deviations due to disturbances and to decrease these deviations. It is expressed by the maximum vehicle side slip angle during the combined braking and steering manoeuvre (β_{max}).

From field tests as well as from simulations it has been concluded that stability is the most significant quality in panic situations (Ref. 1).

By simulation techniques, the influence on vehicle stability from variations in brake and anti-lock system parameters has been studied for one front driven and one rear driven car. It was found that the vehicle stability (especially on ice) was extremely sensitive to the choice of certain parameter values, and it was concluded that the anti-lock system had to be well adapted to the car (see Ref. 4 for full details).

It was also found that these sensitive parameters had one quality in common: they influenced the wheel slip in a significant way. This influence could be separated into three parts: mean slip level, (periodic) variations from this mean level and slip distribution between front and rear wheels. It is therefore of interest to study the vehicle stability as a function of the slip control strategy. Results from such a study are given below (see also Ref. 4).

In this study two vehicles (A and B) are simulated when the real anti-lock systems are replaced by a theoretical system. This system controls the slip of the individual wheels and permits variations of

- mean slip levels,
- amplitude and frequency of slip oscillations,
- relations between mean slip levels on front and rear wheels.

Vehicle A is front-driven with mass distribution front/rear = 60/40 and vehicle B is rear-driven with mass distribution front/rear = 53/47 at minimum weight.

Two tyre-road combinations are used, namely

- Michelin ZX on dry asphalt
- Michelin XM+S with 120 studs on ice

Minimum and maximum weights are studied. The velocities are 90 km/h on asphalt and 50 or 30 km/h on ice. The results given in this section are valid for these above defined conditions.

3.1 Slip levels

As expected, the slip levels at the wheels are found to be very important parameters. This is clearly demonstrated by figure 6, which gives typical results from the simulation study. The slip levels are in this case constant.

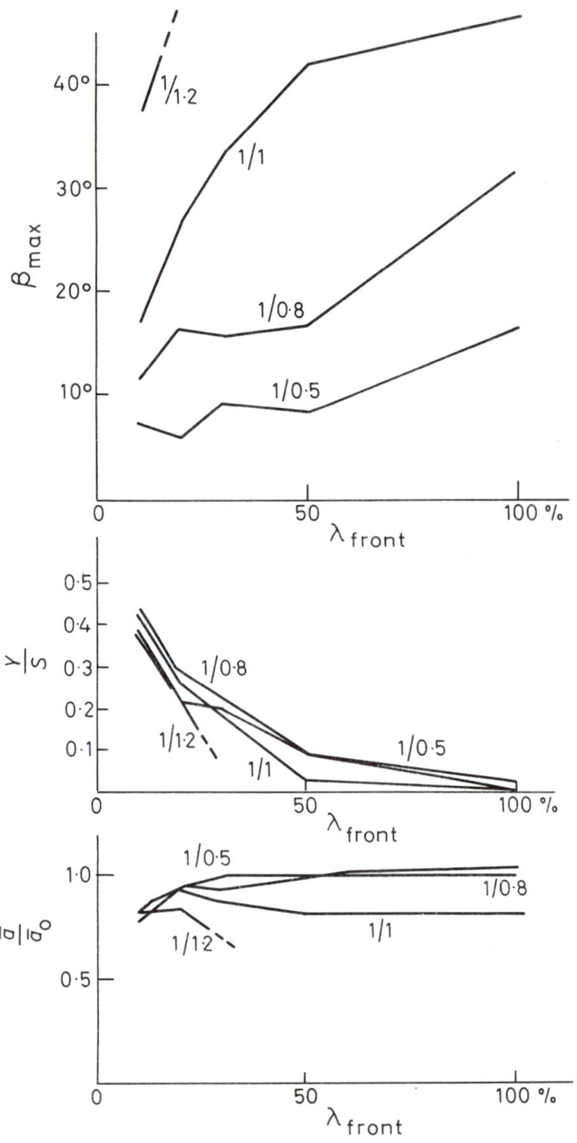

Fig. 6: Influence of wheel mean slip level and ratio between front and rear slip levels on the vehicle stability (β_{max}), steerability (Y/S) and brakeability (a/a_0).

The following conclusions can be drawn:

- The brakeability in general increases with increasing slip levels, for slip levels lower than approximately 20-30%. The brakeability is not much effected for slip levels above 20-30%.
- The steerability is decreasing with increasing slip levels within the whole slip region (0-100%). The change is largest for small slip levels.
- The stability is much affected by variations in mean slip levels. In general, the stability is decreased (β_{max}) with increasing slip levels.

©IMechE 1976

3.2 Ratio between front and rear slip levels

This parameter is also, as expected, found to be of fundamental importance for the vehicle behaviour in this drastic braking and steering situation. The following conclusions can be drawn from the simulation results, which are exemplified by figure 6. The slip level ratios front/rear are 1/1.2, 1/1, 1/0.8 and 1/0.5. These ratios are indicated at the corresponding curves in the figure.

- The stability of the vehicle is very significantly decreased when the slip ratio front to rear is changed towards higher rear slip levels relative to the front level.
- The steerability and brakeability are only slightly affected by variations in the slip ratio front to rear when mean slip level is kept constant.
- Mean front slip levels around 20% give good steerability as well as good brakeability. It is possible to get, in addition, good stability, provided the rear slip level is lower than the front slip level (for instance 10-15%).

3.3 Slip level oscillations

The oscillation frequencies that have been used are 0, 3, 5 and 10 Hz. The results from the simulations as well as the field tests of the real systems indicate that these frequencies are representative with the exception of 0 Hz. The case 0 Hz is however also included, as it is of principal interest.

The oscillation amplitudes have been chosen up to 50%, and 0 and 180 degrees phase difference between front and rear wheels has been used. The slip oscillations have been periodic and sinusoidal.

The results indicate that these variables influence the vehicle behaviour only slightly. The tendencies are the same for the different vehicle-road combinations, so the example in Figure 7 will be representative for all the studied cases. This example corresponds to vehicle A on ice.

The following conclusions can be drawn:

- Oscillations in slip levels influence only slightly the steerability and brakeability of the vehicles.
- Oscillations in slip levels stabilize the vehicles slightly when the mean slip levels are small (less than 30%). This increase in stability seems to occur in the frequency interval 0-3 Hz.
- The phase shift between front and rear slip oscillations does not seem to affect the vehicle behaviour significantly in the frequency interval 3-10 Hz.
- Simulations with no oscillations in the slip levels (i.e. the majority of the simulations) thus give a good approximation to the more realistic case where oscillations occur.

A more detailed study is planned to be carried out for the frequence interval 0-3 Hz.

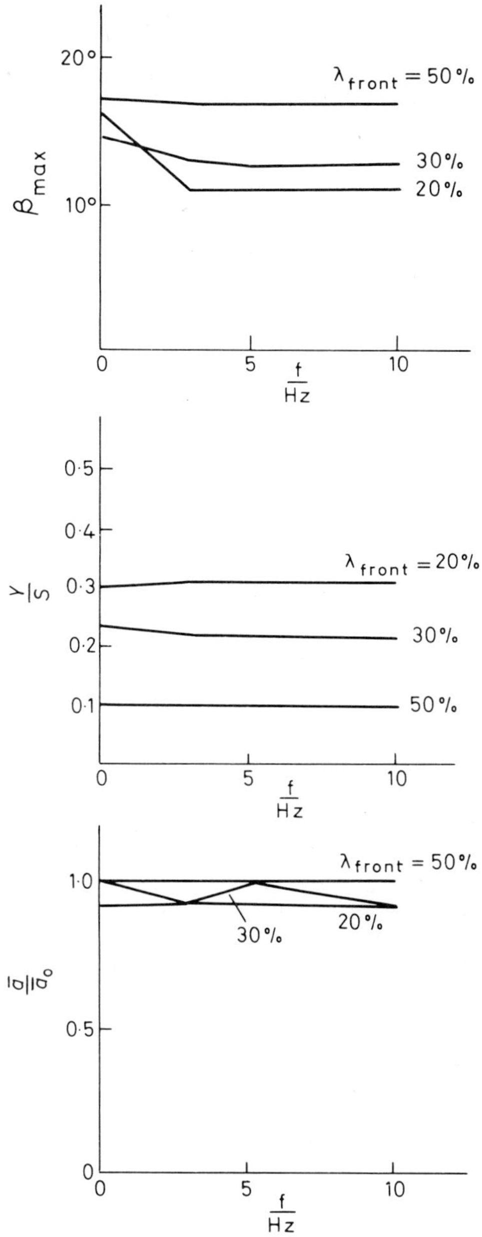

Fig. 7: Influence of periodic slip level oscillations on the vehicle behaviour.

4 INFLUENCE OF WHEEL LOAD CHARACTERISTICS

As mentioned in paragraph 2.2, it was found that the way the friction numbers μ_L and μ_A depend on the wheel load is of great importance for the stability of a vehicle. It has been found during testing of vehicles that the stability of a vehicle in the previously described braking and steering manoeuvre can be significantly different on ice (with studded tyres) compared to wetted sealed asphalt (with summer tyres) even if the friction numbers for these tyre-road combinations are of the same magnitude. A more stable behaviour on wetted sealed asphalt than on ice has been observed.

The different tyre force versus tyre load characteristics, that were measured within the ESV-project, exemplified by figures 3 and 4, can contribute to the explanation of such differences in stability.

It has also been found that the stability of vehicles with different load distributions front/rear

©IMechE 1976

show differences that can not be explained unless these tyre characteristics are considered.

In order to illustrate this, a basic study has been carried out. Two principal cases are considered (Fig. 8 a):

Case I A linear relation between side force and normal load

Case II A nonlinear relation between side force and normal load

Results from simulations with the two real vehicles A and B are given in Fig. 8 b. Vehicle A has a weight distribution, front/rear, of 60/40 and vehicle B 53/47.

The wheel load variations shown in Fig. 8 a refer to variations at the front wheels and the rear wheels respectively during a run from initial forward driving with constant speed (50 km/h) until steady state is reached after a step steering input. The maximum friction number is approximately 0.2. Note that the side forces in these simulations have also been dependent on velocity and slip angle in accordance with the tyre model described in paragraph 2.3.

Fig. 8 b shows the maximum slip angles during the simulated runs for the two vehicles and the two cases. It can easily be seen in this figure that the two different side force characteristics affect the vehicle stability considerably. This is most critical for vehicle A, which is quite stable in case II, but will probably be judged unstable in case I. Vehicle B would probably be judged unstable in both cases.

This example indicates the importance of having detailed and adequate data for the tyre model. It is thus one motivation for the extensive measurements of tyre-road characteristics which were carried out within the project (Ref. 5).

It is well known that different tyre-road combinations have different tyre force characteristics. Rather unexpected characteristics can also occur as is clearly demonstrated by the measured data for studded tyres, which is exemplified in paragraph 2.2. It is therefore recommended that characteristics for such tyre-road combinations that can be present when using a car should be measured in detail in order to make it possible to analyse correctly vehicle stabilities.

The field tests in this project have been carried out on real ice and on asphalt. However, the skid and stability behaviour of vehicles are often tested on wetted sealed asphalt. Tyre force characteristics in combination with such surfaces do not seem to be sufficiently well measured in general. The known measurement results indicate however that they differ significantly from realistic roads. It is therefore recommended that, when tests on wetted sealed asphalt are to be carried out, the tyre/wetted sealed asphalt combinations will be mapped out in more detail in order to obtain more relevant conclusions from tests on such surfaces than is possible at present.

5 CONCLUSIONS

A mathematical model for the simulation of a complete vehicle is described. The tyre model is of paramount importance for the vehicle stability and it is therefore described in detail.

The way the tyre forces depend on the wheel load has a great influence on vehicle stability, as demonstrated by the behaviour of vehicles with studded tyres on ice.

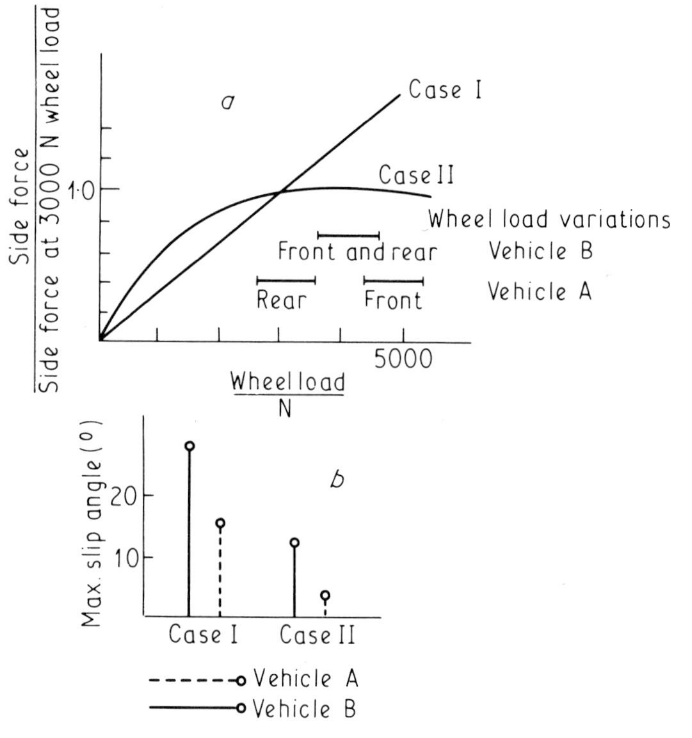

Fig. 8: Example showing how the relations between side force and wheel load influence the stability of the vehicles A and B in a drastic steering manoeuvre.

Care should be taken when testing vehicle stability on skid pads (with wetted sealed asphalt). The behaviour might be different from that on the "real" surface the skid pad is trying to simulate.

There is in general a need for more complete measurements of tyre characteristics on real and simulated surfaces. The use of "friction number" is not enough to characterize a tyre-road combination when studying vehicle stability.

In panic situations the vehicle stability is the most important single quality. It is concluded from the studied cases, which are representative for realistic situations on low-μ as well as on high-μ that the anti-lock systems must be well adapted to the vehicle in order to insure stability. The system should then control the mean slip levels to about 20% at the front wheels and 10-15% at the rear wheels. This will also give good steerability as well as good brakeability.

6 REFERENCES

The Swedish ESV-project (selection)

(1) Rundkvist S., Steerability during emergency braking: Summary report, Swedish Experimental Safety Vehicle program, Report 1-01, June 1974, Göteborg, 37 pp.

© IMechE 1976

(2) Brodd S., Simulation: Mathematical model, Swedish Experimental Safety Vehicle program Steerability during Emergency Braking, Report 5-01, June 1974, Göteborg, 57 pp.

(3) Brodd S., Simulation: Validation of mathematical model Swedish Experimental Safety Vehicle program Steerability During Emergency Braking, Report 5-02, June 1974, Göteborg, 60 pp.

(4) Elgeskog E., Brodd S., Simulation: Low-μ, High-μ, Swedish Experimental Safety Vehicle program Steerability During Emergency Braking, Report 5-03, June 1974, Göteborg, 63 pp.

(5) Nordström O., Nilsson A., Nilsson B., Measurements: tyre-road characteristics: Low-μ, Variable-μ, High-μ, Swedish Experim.Safety Veh. prog. Steerability During Emergency Braking, Report 6-01, June 1974, Göteborg, 46 pp.

Vehicle models

(6) Ellis J.R., Vehicle Dynamics, Business Books Ltd, London, 1969.

(7) Gnadler R., Umfassendes Ersatzsystem zur theoretischen Untersuchung der Fahreigenschaften von vierrädigen Kraftfahrzeugen Automobil - Industri 4/71, pp 75-84.

(8) McHenry R.R, DeLeys N.J., Vehicle Dynamics in Single - Vehicle Accidents - Validation and Extensions of a Computer Simulation, Cal Report No. VJ-2251-V-3, Buffalo, N.Y., 1968.

(9) McHenry R.R., DeLeys N.J., Automobile Dynamics - A Computer Simulation of Three - Dimensional Motions for Use in Braking Systems and of the Driving Task. Cal Report No. VJ--2251-V-7, Buffalo, N.Y., 1970.

(10) Kohno T., Tsuchiya S., Komoda N., On the Vehicle Dynamic Response to the Steering Control - Experimental Evaluation of the Response and Analytical Approach to the Design of the Performance with Seven Degrees Model. SAE Paper 690488, 1969.

(11) Mitschke M., Dynamik der Kraftfahrzeuge, Springer Verlag, Berlin Heidelberg New York, 1972.

(12) Mitschke M., Helm H.-J., Niemann K., Die Auswirkung unterschiedlicher Abbremsung der Räder eines Achse (Schiefziehen). Deutsche Kraftfahrtforschung und Strassenverkehrstechnik, Heft 219, 1971.

(13) Nakatsuka T., Tabanami K., Cornering Ability Analysis Based on Vehicle Dynamics System. SAE Paper 700368, 1970.

(14) Piziali R.A., Dynamics of Automobiles During Brake Applications - Validation of a Computer Simulation. Cal Report No. VJ-2251-V-9, Buffalo, N.Y., 1971.

(15) Rompe K., Verhalten von Kraftfahrzeugen bei Kreisfahrt im Grenzbereich. Deutsche Kraftfahrtforschung und Strassenverkehrstechnik, Heft 227, 1972.

(16) Saibel E., Tsao M.C., Further Investigations in Vehicle Dynamics. SAE Paper 700173, 1970.

(17) Vehicle Handling - A Simulation Study of the Handling Performance of Thirteen Configurations of an Intermediate Size American Passenger Car. (Bendix Corp.). Report PB 200880.

Tyres

(18) Fancher Jr P.S., Dugoff H., Ludema K.C., Segel L., Experimental Studies of Tire Shear Force Mechanics. Report PB 201547, 1970.

(19) Holmes K.E., Braking Force/Braking Slip. Measurements over a range of Conditions between 0 and 100 percent slip. Road Research Laboratory Report LR 292, 1970.

(20) Holmes K.E., Stone R.D., Tyre Forces as Functions of Cornering and Braking Slip on Wet Road Surfaces. Road Research Laboratory Report LR 254, 1969.

(21) Lugner P., Numerische Erfassung von Reifenkennfeldern zur Berechnung von Fahrzeugbewegungen. ATZ vol 74, 1972 Nr 1, pp 17-23.

(22) Nordeen D.L., Analysis of Tire Lateral Forces and Interpretation of Experimental Tire Data. SAE Paper 670173, 1967.

(23) Weber R., Reifen auf Glatteis. ATZ vol 74, 1972, Nr 1, pp 1-9 and ATZ, vol 74, Nr 4, pp 146-150.

(24) Automobildynamische Versuchsstelle, Kantonales Technikum, Versuchsbericht Nr 1, Biel, 1968.

(25) Mechanics of Pneumatic Tires, N B S Monograph 122, edited by S.K. Clark, Washington, D.C., 1971.

APPENDIX

Some comparisons between simulations and field tests

A lot of field tests have been carried out within the project. The test manoeuvre was drastic: The steering wheel was rapidly turned to a stop corresponding to minimum free rolling radius for the actual parameter combination. One second later, the brakes were fully applied.

Two vehicles, A and B (A=front driven, B=rear driven) equipped with anti-lock systems were tested. For validation purpose, the following parameter combinations were chosen:

- summer tyre on dry asphalt for minimum and maximum weight. Velocities: 90 and 110 km/h

- summer tyre on ice for minimum weight. Velocity: 30 km/h

- studded tyre on ice for minimum weight. Velocities: 30 and 70 km/h.

This gives seven parameter combinations for each vehicle. In general five to ten field tests were carried out for each combination. These field test results are compared with the results from simulations of the same manoeuvres and parameter combinations. The qualitative and quantitative correspondence between simulations and field tests is satisfactory. Fig. A1 gives as an example the yaw angle velocity (which was found to be a very sensitive variable) as a function of time for vehicle A, equipped with studded tyres on ice, with a

©IMechE 1976

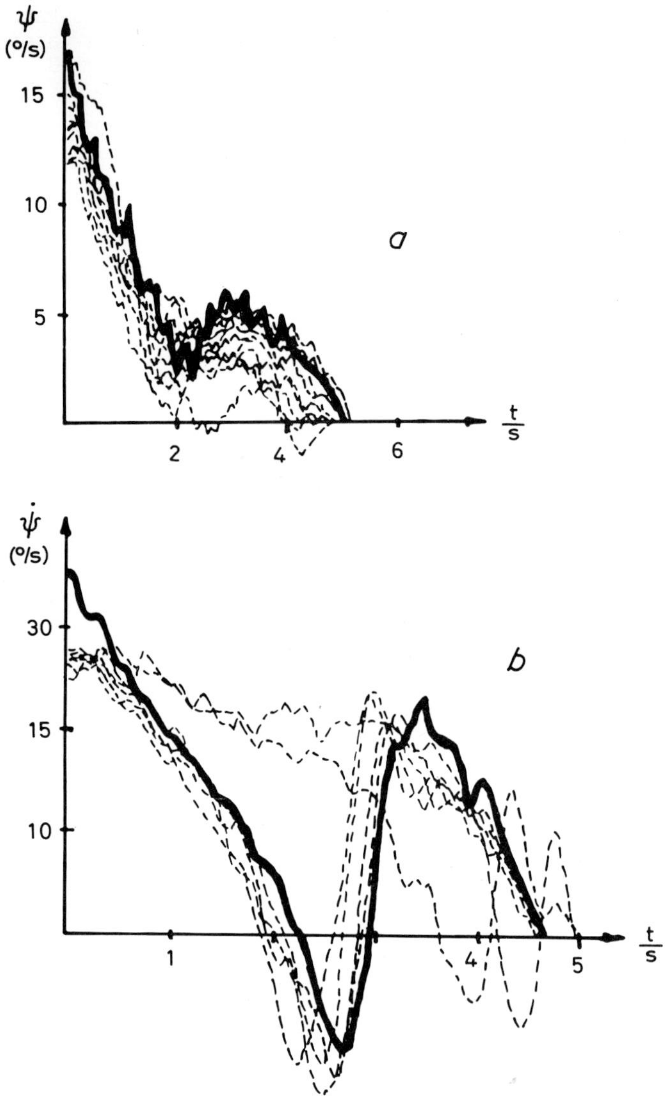

Fig. A1: Yaw angle velocity ( ) as a function of time (t) from simulations (continuous curves and field tests (dotted curves).

Fig. A1a: Vehicle A, 30 km/h, minimum weight, studded tyres on ice.

Fig. A1b: Vehicle B, 110 km/h, maximum weight, summer tyres on asphalt.

---

speed of 30 km/h and minimum weight (Fig. A1 a) and vehicle B, equipped with summer tyres on asphalt, with a speed of 110 km/h and maximum weight (Fig. A1 b).

Full details of the validation procedure and results are given in Ref. 3.

C25/76

THE TYRE/ROAD INTERFACE – ITS EFFECT ON BRAKING

A. R. WILLIAMS, PhD, MSc, MInstP
R. BOND
J. H. PENNELLS
Dunlop Ltd, Birmingham

The MS of this paper was received at the Institution on 4 September 1975 and accepted for publication on 12 November 1975

SYNOPSIS Recent advances in the understanding of the respective parts played by the road surface and tyre in the interface problem are critically reviewed and the current situation from the practical aspect is discussed.

Road accident statistics under wet conditions confirm the need for further improvements in the vehicle frictional characteristics despite advances made in road surface, tyre and brake technology over recent years.

The required overall performance is dependent on the ability of the tyre to create adequate friction at the interface with the road surface; this is determined to a large extent by the characteristics of the road surface and vehicle brakes.

The variables of road surface type and condition, tyre construction and tread compound composition, and some aspects of brake performance have been studied and related to the performance of the vehicle under braking and cornering manoeuvres.

Some results of the practical application of the study of road surfaces with particular emphasis on tyre-to-road interaction are given to support research into the required geometry of road surface texture for optimum wet skidding resistance and minimum tyre to road noise generation.

The paper concludes with a review of legislation involving the wet frictional performance of both tyres and roads.

INTRODUCTION.

The tyre is the intermediary between the vehicle braking system and the road surface; its efficiency is dependent on the performance properties of the vehicle brakes and braking system and on the road surface. The latter becomes most critical under wet conditions where accidents involving skidding account for 27.8 per cent of all accidents (Ref 1).

The process of vehicle braking is a friction process where energy is dissipated mainly within the brakes or mainly at the tyre to road interface. Under maximum braking conditions with the tyre rolling then 95 per cent of the vehicle's kinetic energy is dissipated within the brakes and only 5 per cent at the tyre to road interface. Under non rotating wheel conditions 95 per cent of the energy is dissipated at the tyre-to-road interface.

This observation relates directly to the prolonged debate concerning the origin of the frictional force which was initially considered to be associated with internal losses in the polymer of the tyre tread compound and is now more recently associated with energy losses at the molecular or atomic level occurring at the surface, Gough (Ref 2), Lupton and Williams (Ref 3) and Hegman (Ref 4 and 5). The manifestation of bulk hysteresis in the form of severe tyre abrasion damage known by the misleading term 'reversion' and the minimal influence of this form of abrasion in the generation of a 'safe' level of wet skidding resistance has been discussed by Holmes et al (ref 6).

This controversy is an example of the continuing failure to recognise the tyre's responsibilities and limitations within its operating environment. The aim of this paper is to review the recent advances of tyre technology relating to the tyre's performance under braking and relate these to some operating restraints.

REVIEW.

Summary.

The respective roles of the tyre tread pattern, tread compound formulation, road surface, and vehicle speed have been summarised by Gough and Badger (Ref 7), Fig. 1. The conclusions from this and similar studies are:-

a) An aggregate microtexture harshness is essential for high wet skidding resistance at all speeds.

b) A macrotexture geometric configuration efficient for bulk water drainage is essential at high speeds.

c) The road surface is the dominating influence in the provision of 'safe' levels of wet skidding resistance.

d) That the rating of tyres for specification

©IMechE 1976

purposes requires an accurate specification of the test conditions and that ratings may change in merit order with small changes in these conditions.

e) When considering the difference in performance of tread compounds with different polymers the surface with the greatest ability to differentiate between variants is one possessing a harsh microtexture and an open macrotexture.

f) The maximum benefits of high friction polymers under straight ahead braking are obtained under the peak rolling condition on harsh microtextured, open macrotextured surfaces.

g) The condition of locked wheel sliding where the need for drainage via the road surface and tread pattern is a maximum is the condition where tread pattern variants designed to enhance water drainage are most effective.

h) Tyre performance in the attainment of the maximum braking force coefficient (BFC) is at a maximum under the 'peak rolling' condition rather than the locked wheel sliding condition.

The Tread Compound.

The optimum physical properties of the tread compound for wet skidding resistance depend on the road surface characteristics and the operating condition of the tyre. For example under the peak rolling condition, polymers of inherently high friction give their maximum values with low loadings of carbon black compared with the locked wheel condition where a high compound modulus minimises the closure of the tread pattern elements and aids the breakdown of the pertaining lubrication conditions.

The ratings with regard to wet skidding resistance of polymers suitable for use in tyres has been given by many authors including Rieger (Ref 8). In general terms the second order glass transition temperature (S.O.T.T.) of a polymer increases as the percentage of atoms in the side chains increases (Ref 9). Increases in wet friction may be achieved by increasing the styrene content of a styrene-butadiene co-polymer (Ref 10). As the percentage of styrene is increased and the S.O.T.T. is increased wear resistance will reduce and wet skidding resistance increase.

The S.O.T.T. of styrene-butadiene co-polymers (S.B.R.) is also dependent on the microstructure, i.e. the relative fraction of cis, trans and vinyl groups in the butadiene (Ref 9 and 10). With recent shortages of styrene associated in part with the oil crisis the use of high vinyl content solution S.B.R's has been examined.

Polyisobutylene is an exception to the rule relating S.O.T.T. to the rating of the polymer's wet skidding resistance.

Peterson (Ref 9) has presented an extensive review of tread compounding in relation to wet skidding resistance relating all compounding ingredients to their effect on not only wet skidding resistance but also wear and physical properties of compounds.

The use of carbon black is basically to enhance resistance to tread wear and has only a minor influence on wet skidding resistance, the type and amount of added carbon black for optimum performance being dependent on the polymer. The peak BFC has been found to pass through a maximum with varying levels of carbon black content. The value of the carbon black content at maximum BFC decreases with increasing S.O.T.T. of the polymer and increases with the vehicle speed. The former is a rubber friction phenomenon and the latter a lubrication phenomenon. Thus at 50 km/h the friction of butyl decreases with carbon black content within a practical range of black loading whilst that of polybutadiene increases.

The requirements for type and amount of oil to be added are again governed by the selected polymer. Whilst there are optimum levels of oil and carbon black content for a given polymer and a given operating condition the optimum skid resistance for polymers of high inherent friction is, in general achieved with a relatively low loading of high structure index carbon black and a high oil content (Ref 11).

Theories of Rubber Friction.

Moore and Geyer (Ref 13) have reviewed the adhesion theories of visco-elastic materials. Schallamach (Ref 14) by distinguishing between the mechanisms of bond formation and bond rupture was able to predict the visco-elastic peak in adhesional friction, this was confirmed by the experiments of Grosch (Ref 15). Savkoor (Ref 16) and Bartenev (Ref 17) assumed that molecular adhesion is due to weak van der Waal's forces. In Savkoor's theory each bonding site acquires stored energy due to shear stress and local adhesion is assumed to fail when this energy exceeds the van der Waal potential.

Only Kummer (Ref 18) to date, has attempted to present a unified theory by combining the molecular and macroscopic concepts of adhesion and hysteresis and no one has yet attempted to take this concept further into the more practical but specialized area of tyres and roads in order to produce a generalized theory of tyre to road interaction.

Gough (Ref 2) has questioned the extensive use of the Williams Landel Ferry transform, in which sliding velocities are on a logarithmic scale, in respect of whether the transform should be taken as an exact law or just a very good approximate law. The value of the transform when consideration of compounding variants in relation to friction and abrasion are made, together with the failure of the existing theories to predict improvements, indicates a yet unaccomplished aim, that of completely understanding the theories of visco-elastic materials and their application to tyres running on an unspecified substrate - the road surface.

Tyre Construction and Tread Wear Resistance.

The rate of improvement in the wet skid resistance of tyres has become much slower over recent years, there being few major developments in tread compounding for truck and passenger car tyres. However over the period from the late 1950's to the early 1970's the wet skidding resistance of car tyres has been approximately doubled through tyre developments and compounding technology.

©IMechE 1976

Improvements in the roadhold performance of tread compounds are usually accompanied by a predictable loss in wear resistance. However, the hastening trend for car tyre manufacture to move away from cross-ply construction towards textile breaker and steel breaker radial ply tyres and in part a trend to a lower aspect ratio (ratio of tyre section width to section height), has increased the projected mileage of a tyre (Ref 12). The broken lines in Fig. 2 added to the original graph by Brenner indicate the potentially reduced influence of the tread compound formulation on tyre wear. The greater wear resistance of the radial tyre may allow the tyre designer to further develop tyre tread compounds of higher wet skidding resistance.

The difference in the braking performance of a car and heavy vehicles has been progressively reduced by advances in truck tyre design and the increased efficiency of truck braking systems.

However, on polished close textured road surfaces there remains a considerable disparity between the respective stopping distances of these vehicles. The influence of tyre wear on this disparity is shown in Fig. 3.

THE TYRE UNDER LONGITUDINAL BRAKING.

Tyre Variables.

Under straight ahead driving conditions longitudinal slip occurs at the tyre-to-ground interface during braking as a consequence of the inextensible band theory; there is also the influence of the compressibility or extensibility of the tread band, and the influence of the tread rubber longitudinal elasticity (Ref 19). The latter may be a function of the tread pattern or tread compound dynamic modulus and the inherent frictional properties of the tread compound.

The initial slope of the Braking Force Coefficient (BFC) against slip curve has been shown to increase with the change from cross ply to steel radial car tyres (Ref 20) and with the degree of tyre wear, a worn tyre having the steeper slope (Ref 21). There is also a trend towards a lower slip at peak B.F.C. with lower aspect ratio tyres (Ref 22).

The percentage slip at peak B.F.C. is partly dependent on initial vehicle speed when tests are conducted on high wet skidding resistance surfaces but less speed dependent on low friction surfaces.

The lower the inherent wet skidding resistance of the surface the lower the percentage slip at peak B.F.C.

Tread compound formulations of higher inherent friction produce the maximum B.F.C. value at a lower percentage slip than other compounds of equal dynamic modulus (Fig. 4). There is a direct relationship between the visco-elastic nature of rubber friction and its dependence on the sliding velocity at the tyre road interface at the point of maximum friction.

The Influence of Brake Application Rates.

The influence of brake application rate of peak values of B.F.C. has been the subject of a recent survey by the American Society for Testing Materials (ASTM) (Ref 23). This has indicated that results are strongly dependent on the method of test.

When related to the techniques of tyre testing on instrumented vehicles used by Dunlop Limited (Ref 24) the rate of brake application under longitudinal braking has been found not only to influence the peak B.F.C. but also the percentage slip at which the peak is reached.

Automotive brakes give a nonlinear slip characteristic, with the time to peak B.F.C. being the only linear position, thereafter the slip increases very abruptly.

The slope of the B.F.C. against slip curve may be altered with rate of brake application. For the period up to the peak B.F.C. the optimum rate of pressure application tends towards the longer time when tests are made over a range of times to peak brake pressure of 0.1 to 0.5 seconds. When a high peak B.F.C. is obtained a correspondingly higher locked wheel B.F.C. is obtained because the speed is then lower. In addition the distance to wheel locking is increased and the total stopping distance reduced when the optimum brake application rate is used.

The influence of brake application rate is most noticeable on surfaces of low wet skidding resistance.

The trends of tyre and brake variables as discussed in relation to peak B.F.C. are summarized in Fig. 4.

Within the range of radial tyre cross sectional size between 145 and 185 the effect on tyre cross section on peak and locked wheel values of B.F.C. is small with the trend for the most narrow section to give the inferior results. For new tyres the influence of tyre aspect ratio on peak and locked B.F.C. when varied between 60 and 80 per cent is also small.

THE TYRE UNDER CORNERING AND CORNERING WITH BRAKING.

The influence of the tyre tread compound formulation on the cornering coefficient (C.F.C.) for a range of slip angles is shown in Fig. 5(a). The cornering force coefficient is the force at right angles to the direction of motion. The considerable enhancement in the value of C.F.C. that can be achieved through tread compound technology is illustrated.

The effect on the value of the B.F.C. of braking under a cornering condition is shown in Fig. 5(b). The value of the peak B.F.C. reduces and the percentage slip at which the peak B.F.C. value occurs increases with an increase in the slip angle.

In Fig. 5(c) tyres of identical construction but with tread compounds of equal modulus but differing friction properties are compared for B.F.C. measurements under wet conditions. The trend discussed in section 3 of this paper for the tread compound formulation possessing the higher frictional properties to achieve the peak value of B.F.C. at a lower percentage slip than tread compounds of lower friction properties is again illustrated.

The influence of the tyre's static longitudinal

©IMechE 1976

stiffness in the circumferential direction and static lateral stiffness in the lateral direction on the tyre's braking and cornering performance is shown in Fig. 5(d).

The terms circumferential stiffness (C_s) and cornering stiffness (C_α) have previously been defined (Ref 25) as:-

The circumferential stiffness $C_s = \left.\frac{\partial F_x}{\partial s}\right|_{s=0}$

The cornering stiffness $C_\alpha = \left.\frac{\partial F_y}{\partial \alpha}\right|_{\alpha=0}$

Where α = slip angle

S = circumferential slip parameter

F_x = longitudinal traction force

F_y = lateral force during cornering

To obtain the curves shown in Fig. 5(d) the terms C_s and C_α have been redefined as the slope of the B.F.C. against slip curve at each value of slip between 1 and 10 per cent and the slope of the C.F.C. against slip angle curve at each value of slip angle between 1° and 10°.

For the purpose of plotting the curve the respective values at 1 percent slip and 1 degree slip angle have been equated.

It is seen that two of the four tyres tested have broken away under the test condition and that there are significant differences up to the break-away or near break-away point.

The greater the stiffness of the tyre (C_s or C_α) the greater the potential peak traction force or cornering force, however this potential value needs to be maintained in the case of straight ahead braking over the greatest range of percentage slip.

There appears to be a significant variation in the form of the B.F.C. against longitudinal percentage slip curves for car and truck tyres. The former in general shows a peak at approximately 15 per cent slip (Ref 21), whilst the latter sustains the peak value to in excess of 40 per cent slip (Ref 26) when both are tested on low friction surfaces in wet conditions.

WATER FILM THICKNESS AND TREAD PATTERN DESIGN.

The concept of the three zone contact patch has been widely reported (Ref 24) and there have been various studies of tyre to road lubrication phenomenon. The study of thin water films under smooth rubber contact has been conducted by Roberts (Ref 27) and Browne (Ref 28). By use of Moiré contouring the variation in the depth of water film between a loaded tyre and a smooth surface may be investigated.

If the gap of interest is filled with a fluid of refractive index μ the contour interval Δf is given following the expression due to Browne (Ref 28).

$$\text{Contour interval } \Delta f = \frac{P}{\alpha_i + \alpha_v}$$

$$\text{where } \alpha_\theta = \frac{\tan\theta}{\mu}\left(1 + \left(1 - \frac{1}{\mu^2}\right)\tan^2\theta\right)^{\frac{1}{2}}$$

Thus all fluids having $\mu > 1$ (i.e. water $\mu = 1.333$) have the effect of desensitising the method so increasing the contour interval. Examination under thin water film conditions of a slowly rotating tyre (forward velocity 7 mm/sec) indicates that the promontary areas of the zig zag pattern of a rib are susceptible to water film penetration. The band of thicker water film increases from the leading edge of the promontory and reduces towards the rear of the contact patch under low rolling speeds, Fig. 6.

The extent of the thick water film penetration into the rib increases with load and is explained in terms of the lateral stress and normal pressure distribution at the localised areas across a rib. The possibility of slippage at the rib edge is high as the effective friction (lateral stress/normal pressure) is low.

Under conditions of low normal pressure at the rib edge lateral movement occurs; this is increased on a low friction substrate and the condition exists for the water film to penetrate the rib edges. This is confirmed by photographs of the tyre contact patch under high speed rolling conditions once the grooves are flooded.

Studies of the importance of the microtexture in obtaining a high value of wet skidding resistance (Ref 29) indicate a minimum level of texture is required of $5 - 10 \times 10^{-3}$. Measurements of thin water films (Ref 27 and 30) and a theoretical approach by Moore (Ref 31) indicate that a film of water between rubber and a smooth surface may be as thin as $1.5 - 5 \times 10^{-3}$ mm, depending on its purity.

Without the rupture of this thin viscous water film, rubber to asperity contact is not made and friction values are low. It has been postulated that under the contact pressure exerted by car tyre tread elements the film between the rubber and a smooth surface such as a glass plate remains intact. Moiré contouring has been used in an attempt to establish the minimum film thickness under a car tyre rolling on a glass surface. It was found that no dry areas were detected when the rubber to grating gap was reduced to 25×10^{-3} mm.

THE ROAD SURFACE.

The Microtexture.

For adequate resistance to skidding at any speed a road surface's constituent aggregates should possess a specific level of microtexture. Previously reported studies of microtexture, from both theoretical and practical approaches, indicate minimum levels of microtexture for the attainment

© IMechE 1976

of 'safe' levels of wet skidding resistance (Ref 22 and 31). Aggregates of all sizes within a surface mix design should possess and maintain throughout the service life of the material a micro-texture harshness of between 10 and 100 x 10^{-3} mm.

The upper limit of harshness recommended in practice is determined from the relationship of substrate harshness to tread wear and real contact area.

The Macrotexture.

A road surface should also present to the tyre a geometric configuration of its macrotexture to enhance water drainage from under the contact patch. The demand for water drainage increases with vehicle speed and it follows that the macro-texture of roads should therefore vary depending on the traffic speed of a particular site in order to maintain the maximum potential real contact area between tyre and road surface. Bond et al (Ref 32) have suggested a theory for sub-tyre drainage using a simplified model of the road surface and tyre contact area based on hexagonally packed spheres.

The conclusions from this theory indicate that there exists an optimum value of spacing between road aggregates, and an optimum value of the maximum size of aggregate for the attainment of the highest value of the ratio

$$\frac{\text{actual area of contact within contact patch}}{\text{apparent area of contact}}$$

and the minimum value of the time taken to remove water from under the contact patch. Fig. 7(a) shows the dependence of these variables on particle radius for a given absolute texture depth of 2.0 mm for various values of spacing between spheres of 0.0mm to 5.0 mm.

The effectiveness of the proposed macrotexture geometry in drainage and the close agreement of the practical condition and drainage theory is shown in Fig. 7(c). The figure illustrates the existence of a constant level of contact area with increase in vehicle speed when seen through a transparent material with the macrotexture geometry of hemisphere diameter 10 mm, absolute texture depth 1.5 mm and centre to centre distance between the hemispheres in the hexagonal close packing arrangement of 11 mm.

The fall off in potential contact area with speed on a smooth surface is also shown. A close correlation has been achieved between the theory, practical tests on road surfaces of different macrotexture geometries and the outflow method of road surface texture evaluation (Ref 32).

The Pressure Distribution within the Tyre Contact Patch.

The drainage theory also shows up the importance of the contact patch pressure distribution. If one considered a smooth surface and a smooth tyre having a contact patch with a high shoulder pressure, the theory predicts the curve shown in Fig. 7(b) where A_w represents the apparent water free area in the contact patch and A_o the apparent total contact patch area. The shape of the curve has been verified experimentally as shown. If the friction at the road surface/tyre interface is considered to be proportional to the actual area of contact in the contact patch, the curves shown in Fig. 7(b) can represent the friction/speed curves produced for different surfaces and tyre contact patch pressure distributions. If the smooth tyre had a uniform contact patch pressure distribution, the drainage characteristics would be improved as shown for a smooth surface. For a road surface with good drainage characteristics, e.g. a new BS 594 (Ref 33), Delugrip, pervious macadam, or grooved concrete, the pressure distribution in the contact patch does not play such an important role in the water drainage since the macrotexture of the road surface is capable of removing the bulk water as soon as the tyre rolls over it.

THE OPTIMISATION OF TYRE AND ROAD INTERACTION.

Walker and Major (Ref 34) have reviewed the factors that influence tyre to road noise generation and shown them to be analogous to wet skid resistance factors in so much as the road surface is the dominant factor. Their evidence suggests that the reduction of power train noise by 10dB(A) which is thought to be within the range of existing technology may not be fully realised unless the rolling noise can be reduced by about 5dB(A) (Ref 35). Tyre to road noise generation increases with vehicle speed and is a dominant factor for trucks at speeds above 72 km/h and for cars at speeds above 50 km/h.

In March 1975 Hammersmith fly-over on the A4, London was resurfaced with a wearing course material designed to a rational method of aggregate grading* (Ref 36). The material has been designed to comply with the microtexture and macrotexture requirements as discussed above. The macrotexture is achieved through mix design principles to give adequate sub-tyre drainage for the average traffic speed for the site. The benefits of this approach are illustrated by the value of the sideways force coefficient (S.F.C) against speed for a trial area of the same material laid in 1973 also on the A4 at Hammersmith, London (Fig. 8 (a)). The values of S.F.C. have been taken using the Sideway Force Coefficient Routine Investigation Machine (S.C.R.I.M) (Ref 37).

Walker and Major give a clear indication of the need to control the void spacing of the aggregates on the road surface in order to minimise the frequency of sound generated within the frequency band 500 Hz to 1.5 Hz. This is necessary in order to reduce the possibility of noise generated at the interface adding to the natural tyre resonances that occur within the same frequency band (Fig. 8 (b)).

The spacing of the aggregates dictated by drainage requirements as discussed above, tend, depending on vehicle speed, to generate frequencies at a higher level at an area of the sound pressure level curve where the value of sound pressure is reducing.

By use of the Quantimet image analysing computer it is possible to measure the aggregate void spacing and geometry for a representative area of a rationally designed road surface; Fig 8 (c) illustrates the distribution of spacing of the upstanding aggregates.

---

* The material is known under the Dunlop Limited Trade Mark as 'Delugrip'.

©IMechE 1976

Fig. 8 (d) gives a clear indication of the positive advantages achieved in reducing the overall level of traffic noise by changing the road surface on the Hammersmith fly-over. A reduction of 3dB(A) has been achieved from the values of noise monitored initially on the British Standard 594 hot rolled asphalt with 19 mm precoated chippings.

This study indicates the considerable advantages to be gained from road surface developments brought about by the close liaison between the tyre and road industries. Once the nature of the road surface required for optimum performance of the tyre has been specified, (Table 1), and the performance of the road surface over a period of time has been shown to be durable the future could produce two major developments: firstly a reversal of the trend to reduce the allowable recommended values of wet skidding resistance on roads as currently being proposed, (Ref 38). This paper and others have shown that the road surface is the dominant factor in the attainment of a 'safe' level of wet skidding resistance.

Secondly the value of producing a quiet truck will be fully realised only if 'quiet' road surfaces are produced. Furthermore when such a road surface is specified (Table 1) the important step in the sharing of the joint responsibilities of the tyre and road in the optimisation of their respective performance will have been achieved.

repeated skidding accidents (Ref 38).
Giles (Ref 40) showed that the risk of an accident on a surface having a S.F.C. at 50 km/h of 0.60 was exceedingly small, becoming measurable at a S.F.C. within the S.F.C. range of 0.55 - 0.60 and increasing sharply by more than 20 times as the coefficient falls to 0.40 - 0.55 and by 300 times when the coefficient is between 0.30 and 0.35.

Under the newly proposed system for a wet skidding resistance specification such roads as motorways may be allowed to have a S.F.C. value of 0.30, whilst such a site may never be a scene of repeated skidding accidents such a surface condition could be not only a cause, but the governing factor in the severity of an accident in high traffic density conditions.

As this and other papers have indicated, without the appropriate assistance from the road surface the tyre alone cannot be relied upon to take the responsibility for provision of a 'safe' level of wet skidding resistance. The development and testing on roads of newly designed wearing course and surface dressing materials i.e. open textured macadams (Ref 41), Delugrip high friction dense asphalts, and Shellgrip (Ref 42) does offer hope for up-grading the suggested specifications. The impetus for such improvements has been given by the clear advantages of providing skid-resistant road surfaces in central London through the studies of Hatherley (Ref 43).

Table 1.

PAVEMENT TEXTURAL CHARACTERISTICS REQUIRED

FOR OPTIMUM TYRE PERFORMANCE

TYRE PROPERTY	PAVEMENT PROPERTY	SPECIFICATION
ABRASION	MICROTEXTURE (All Aggregates in Mix)	$10-100 \times 10^{-3}$ mm
WET FRICTION	MICROTEXTURE (All Aggregates in Mix); MACROTEXTURE	AGGREGATE Particle Size 6-12 mm Dia.
RIDE/COMFORT	MACROTEXTURE	INTER PARTICLE SPACING 1-4 mm
NOISE	MACROTEXTURE	ABSOLUTE TEXTURE DEPTH 1-3 mm (Dependent on Aggregate Size)

LEGISLATION.

The Road Surface Wet Skidding Resistance.

The Department of the Environment in the U.K. have proposed limits of road surface wet skidding resistance that are dependent on not only the classification of the road but also on the accident risk rating of the particular site (Ref 38). This is an advance on the Marshall Committee recommendations (Ref 39) where only site classification and speed were taken into account.

However, figures have been obtained from accident statistics which have made it possible to correlate wet skidding resistance of the site and the possibility of the site becoming a scene of

The Tyre Traction Performance.

On the question of legislation relating to road holding most governments have so far avoided any attempts to get involved in the technical intricacies of this subject, contenting themselves, in many instances, with inscribing into their individual national legislation a requirement that tyres shall not be used after the tread pattern depth has worn below a specified value, usually 1 mm; it is of interest to note that one European government - Austria - has recently increased this minimum permissible tread pattern depth to 1.6 mm for tyres whose tread pattern design includes treadwear indicators at 1.6 mm (as required for the U.S. market) and 2.0 mm for tyres not furnished with treadwear indicators.

©IMechE 1976

It is possible that other European governments may use this precedent to increase the minimum tread pattern depth at some future date, as more government-inspired wet braking test work is undertaken (Holland, Germany). If this happens then the question of all-round increases in minimum tread pattern depth will almost certainly be raised again in the E.E.C. Working Party on Road Transport (WP 29); at previous discussions on the merit of an E.E.C. Recommendation on this subject no conclusion could be reached, largely due to the inability of any government to provide a tenable socio-economic justification for the increase.

In the particular case of road holding under winter conditions the E.E.C. is currently studying proposals for recommendations governing the use of studded winter tyres and snow chains or similar tractive devices.

The situation in the U.S.A., at the present time, is broadly similar to that in Europe but is about to change significantly.

Up to now federal laws require tyres for use on public highways to be equipped with treadwear indicators (1.6 mm) to show when tyres are worn to this depth, but do not specify how these are to be used or interpreted which is a responsibility of the State Legislatures or State Police Departments.

The significant change mentioned above arises from the fact that the U.S. government is shortly to introduce legislation which will require tyres for passenger cars to carry symbols indicating among other performance properties, their wet braking or "traction rating". This requirement was written into the U.S. "National Traffic and Motor Vehicle Safety Act of 1966" and was required to be effective within two years. The fact that this requirement has still not yet been implemented nine years later is clearly indicative of the degree of difficulty (and contention) which will always arise when governments attempt to legislate in highly technical areas. Wet road holding, as the tyre and vehicle industries know from long experience, is an extremely complex subject embracing many parameters and is, therefore, a subject which cannot be simplified to the extent proposed in the forthcoming U.S. legislation and still serve a valid and useful purpose. It remains to be seen what the benefit, to the American consumer, will turn out to be but we in Europe are extremely sceptical on this point.

ACKNOWLEDGEMENTS.

The authors wish to thank Dunlop Limited for permission to present this paper and their colleagues within the Company, and Dr. G. Lees of the Department of Transportation and Environmental Planning, University of Birmingham for their help in the experimental work reported, particularly Mr. R. G. Clifton for his contribution on the subject of Legislation.

Grateful acknowledgements are also due to the Greater London Council for permission to report test results.

****

REFERENCES.

1. SABEY B. E. Accidents, Their Cost and Relation to Surface Characteristics. November 29th 1973. Paper presented at a symposium organised by the Cement and Concrete Association, England.

2. GOUGH V. E. A Tyre Engineer Looks Critically at Current Traction Physics. Paper presented at a symposium The Physics of Tyre Traction, held at the General Motors Research Laboratories, October 1973.

3. LUPTON G. N. WILLIAMS T. Wet Road Skidding Resistance. Transport and Road Research Laboratory report 86 UC, 1974.

4. HEGMAN R. R. Contact Temperatures of a Skidding Tyre, Wear. June 1975 p 45 - 64.

5. HEGMAN R. R. HENRY J. J. Thermal Analysis of a Skidding Tyre. Wear, 24, 1973, p 361 - 380.

6. HOLMES T. LEES G. WILLIAMS A. R. A Combined Approach to the Optimisation of Tyre and Pavement Interaction, Wear, 20, 1972, p 241 - 276.

7. GOUGH V. E. BADGER D. W. Tyres and Road Safety, Inter. Road Federation, 5th World Meeting, London September 1966.

8. RIEGER H. Experimentelle und Theoretische Untersuchungen Zur Gummireibung in Einem GroBen Geschwindigkeits - Und TemperaturBereich Unter Berucksichtigung der Reibungswärme. Thesis - Technischen Hochschule, Müchen, 1968.

9. PETERSON R. F. ECKERT C. F. CARR C. I. Tread Compound Effects in Tyre Traction. Paper presented at a symposium. The Physics of Tyre Traction, held at the General Motors Research Laboratories, October, 1973.

10. NORDSIEK K. H. Kautschuk und Gummi, 25, 1972, p 87.

11. SCOTT C. E. BURGESS K. A. CHIRIO V. E. MICEK E. G. Carbon Black Usage for Treadwear and Traction: A Summary, American Chemical Society 99th meeting, April 1971.

12. BRENNER F. C. KONDO A. The Elements in Road Evaluation of Tyre Wear. Presentation to A.S.T.M. Committee E9, November 11th 1971, Akron, Ohio, U.S.A.

13. MOORE D. F. GEYER W. A Review of Adhesion Theories for Elastomers, Wear, 22, 1972, p 113 - 141.

14. SCHALLAMACH A. A Theory of Dynamic Rubber Friction. Wear 6, 1963, p 375 - 382.

15. GROSCH K. A. The Speed and Temperature Dependance of Rubber Friction and its Bearing on the Skid Resistance of Tyres. Paper presented at a symposium. The Physics of Tyre Traction, held at the General Motors Research Laboratories, October 1973.

16. SAVKOOR A. R. Some Aspects of Friction and Wear of Tyres Arising from Deformations and Slip Stresses at the Ground Interface. Wear 9, 1966, p 66.

©IMechE 1976

17. BARTENEV G. M. On the Theory of Dry Friction of Rubber. Dokl. Akad. Nauks USSR. 96, p11 16 - 4; Rubber and Plastics Research Association Translation No. 925.

18. KUMMER H. W. A Unified Theory of Rubber and Tyre Friction, Pennsylvania State University Research Bulletin B94, July 1966.

19. PACEJKA H. B. Tyre in Plane Dynamics, N.B.S. Monograph 122, Mechanics of Pneumatic Tyres, Edited by S. K. CLARK, 1971, p 695 - 757.

20. HOLMES K. E. STONE R. D. Tyre Forces as Functions of Cornering and Braking Slip on Wet Road Surfaces. Transport and Road Research Laboratory, Report LR 254, 1969.

21. HOLMES K. E. Braking Force/Braking Slip: Measurements Over a Range of Conditions between 0 and 100 Per Cent Slip. Transport and Road Research Laboratory, Report LR 292, 1970.

22. BOND R. LEES G. WILLIAMS A. R. An Approach Towards the Understanding and Design of the Pavements Textural Characteristics Required for Optimum Performance of the Tyre. Paper presented at a symposium, The Physics of Tyre Traction, held at General Motors Research Laboratories, October 1972.

23. A.S.T.M. Survey on the Effect of Rate of Braking Application on Peak Braking Coefficient. May 1975. (Private Communication to A.S.T.M. Members).

24. ALLBERT B. J. WALKER J. C. Tyre to Wet Road Friction, Proceeding Inst. Mech. Engs. 1965 - 1966, 180 (2A) (4), p 105 - 21.

25. TIELKING J. T. FANCHER P. S. WILD R. E. Mechanical Properties of Truck Tyres. Society of Automobile Engineering 730183 January 1973.

26. ERVIN R. D. FANCHER P. S. Preliminary Measurements of the Longitudinal Traction Properties of Truck Tyres. Society of Automotive Engineers 741139, November 1974.

27. ROBERTS A. D. The Extrusion of Liquids between Highly Elastic Solids, Proc. R. Soc. A 325, 1971, 323 - 345.

28. BROWNE A. L. Fluid Film Thickness Measurement with Moire Fringes. Applied Optics, October 1972, Vol. 11, No. 10, p 2269 - 2277.

29. WILLIAMS A. R. LEES G. Topographical and Petrographical Variation of Road Aggregates and the Wet Skidding Resistance of Tyres. Quart. J. Eng. Geol. 2 (3), 1970, p 217 - 235.

30. DERJAGUIN B. Z. Phys. 1933, 84, p 657 - 70.

31. MOORE D. F. An Elastrohydrodynamic Theory of Tyre Skidding, VII Congress Int. Des Techniques De L'automobile, Barcelona, May 1968.

32. BOND R. KATEKHDA I.e.D. LEES G. WILLIAMS A. R. Road Surface Characteristics in Relation to the Tyre Performance. Institute of Highway Engineers (to be printed).

33. British Standards Institution BS 594; 1961 Rolled Asphalt (hot process).

34. WALKER J. C. MAJOR D. J. Noise Generated at the Tyre-Road Interface. Stress Analysis Group, Inst. of Physics, Conference at University of Aston in Birmingham September 1974.

35. HARLAND D. G. Rolling Noise and Vehicle Noise, Symposium on Noise in Transportation, University of Southampton July 1974.

36. LEES G. SHARIF R. L. The Design and Field and Laboratory Testing of High Friction Dense Asphalts. The Highway Engineer, June 1973, p 4 - 20.

37. Sideway Force Routine Investigation Machine, Transport and Road Research Laboratory, Crowthorne, England.

38. SALT F. G. SZATKOWSKI W. S. A Guide to Levels of Skidding Resistance, Transport and Road Research Laboratory Report LR 510, 1973.

39. Ministry of Transport, Report of the Committee on Highway Maintenance, London 1970 (H.M. Stationery Office).

40. GILES G. G. The Skidding Resistance of Roads and the Requirements of Modern Traffic. Proc. Inst. Civ. Eng. 6, 1957, p 216 - 249.

41. MARTIN E. R. JUDGE R. F. A. Civil Engineer. December 1966 p 1495.

42. SHELL-MEX and B.P. LIMITED, Shellgrip and Safety, 1968.

43. HATHERLEY L. W. LAMB D. R. VI World Highway Conference. Canada 1970.

****

©IMechE 1976

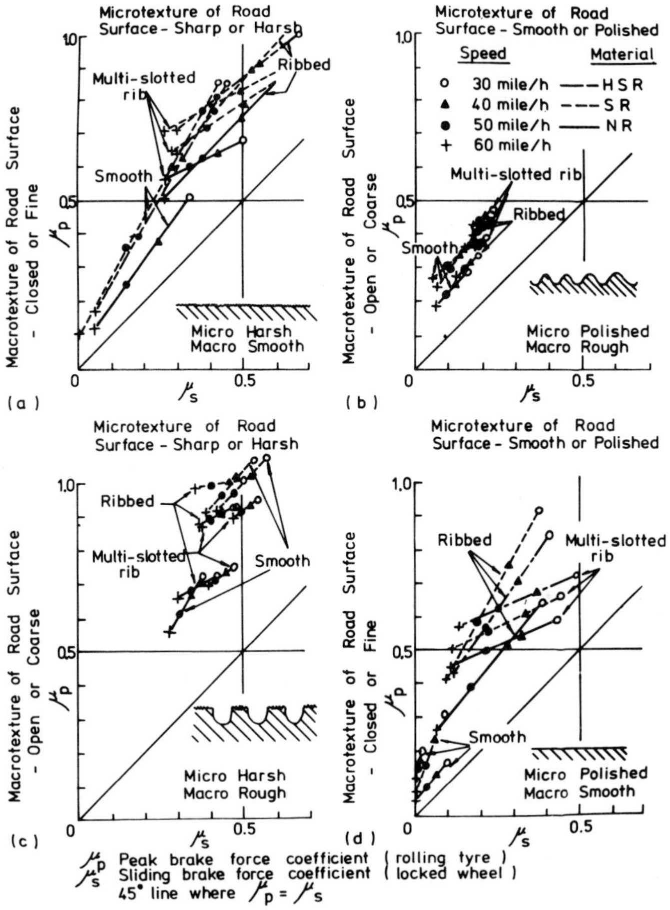

Fig. 1 (a - d): Peak (rolling) and slide (locked) brake force coefficients at four speeds of tyres of three types of tread pattern and three tread materials on four contrasting road surfaces showing the influence of micro and macro-texture of road surface and how they interact with tyre variables. (HSR = High Styrene Butadiene, SBR = Styrene Butadiene, NR = Natural Rubber).

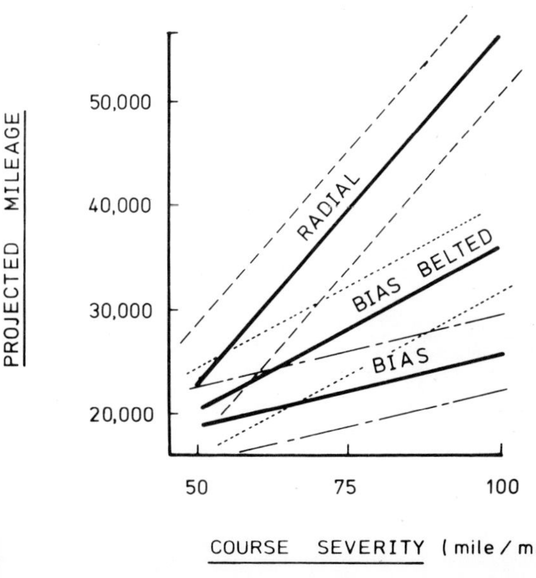

Fig. 2: The influence of tyre construction and tread compound formulation on the wear resistance of tyres under varying conditions of wear severity.

TYRE	MAX. B.F.C.	STOPPING DISTANCE	
New	0.55	23 m	Car
Half worn	0.48	26 m	Car
Worn	0.28	45 m	Car
New	0.20	61 m	Truck
Half worn	0.16	78 m	Truck
Worn	0.14	91 m	Truck

Fig. 3: A comparison of the performance under wet conditions of car and truck tyres under different conditions of wear.

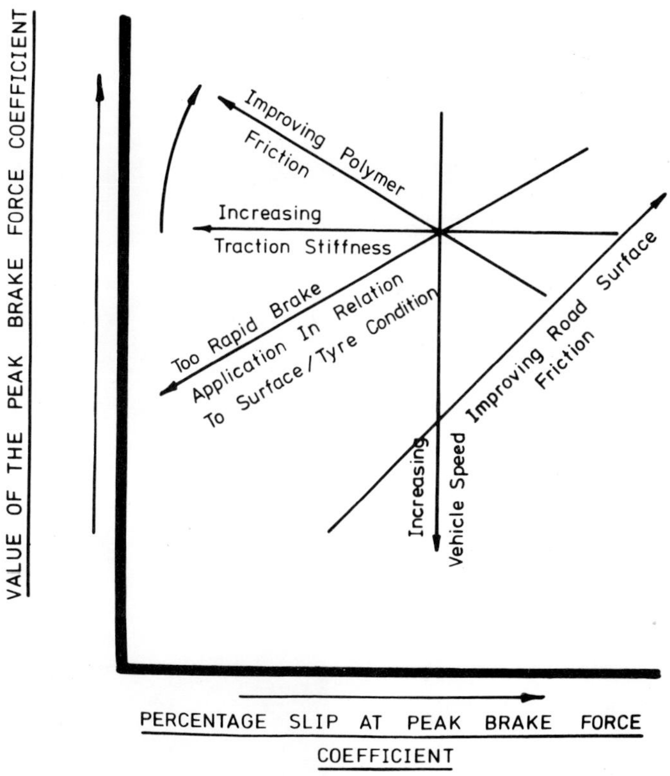

Fig. 4: The influence of tyre and test variables on the value of the peak brake force coefficient and the percentage slip at which this is attained.

© IMechE 1976

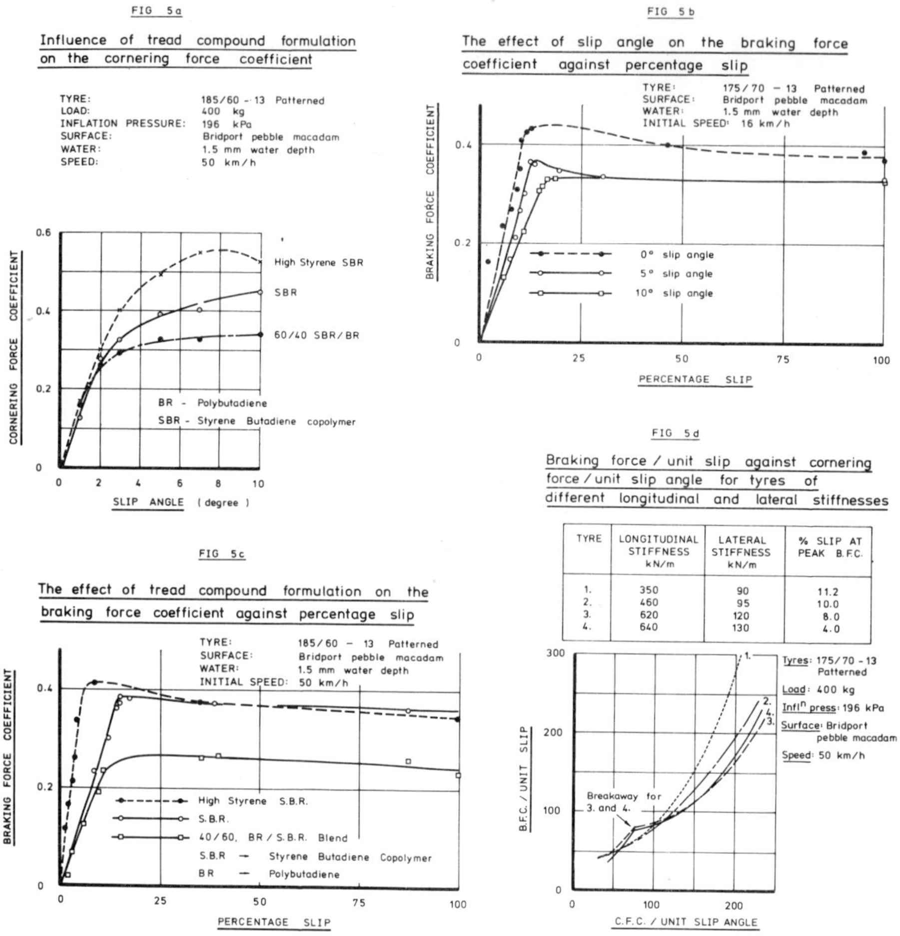

TYRE	LONGITUDINAL STIFFNESS kN/m	LATERAL STIFFNESS kN/m	% SLIP AT PEAK B.F.C.
1.	350	90	11.2
2.	460	95	10.0
3.	620	120	8.0
4.	640	130	4.0

Fig. 5 (a - d): The influence of tyre and test variants on the peak (rolling) brake force coefficient and cornering force coefficient.

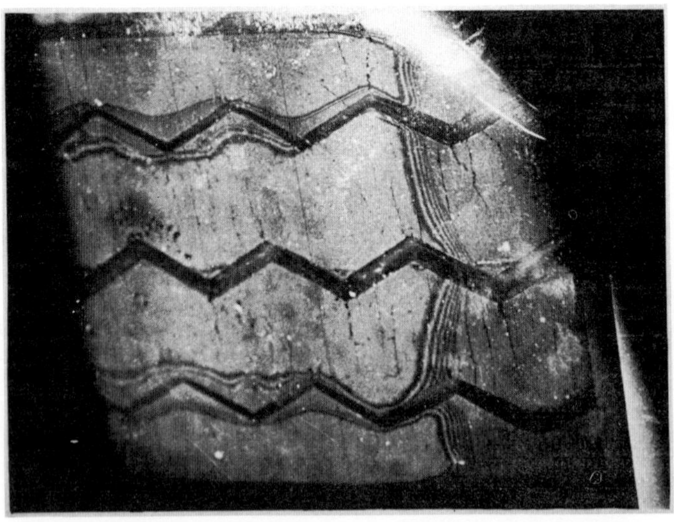

Fig. 6: The penetration of the thin water film under low speed rolling conditions.

© IMechE 1976

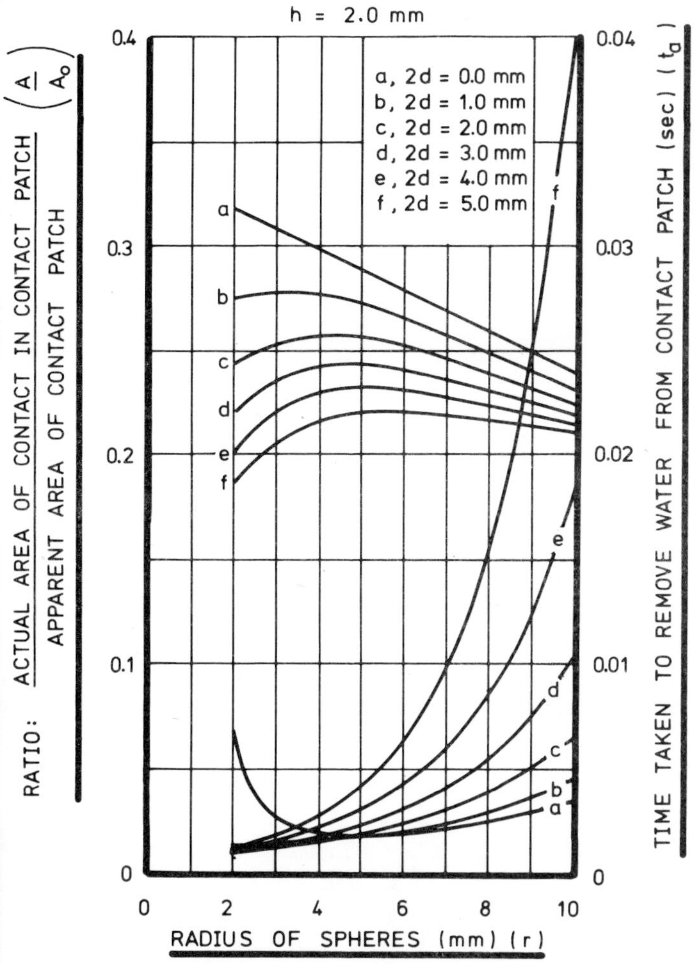

Fig. 7a: Optimisation of road surface drainage properties for the maximum potential contact area (higher sets of curves) and the minimum drainage time of water from under the contact patch (lower set of curves).

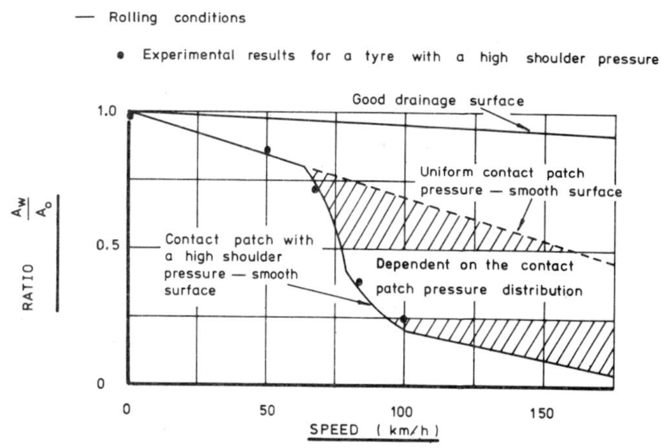

Fig. 7b: A comparison of the theoretical and experimental results for the drainage characteristics of a road surface and their dependance on the tyre contact patch pressure distribution.

PLAIN TREADED TYRE ROLLING OVER A GLASS PLATE AT 70 km/hr IN 1.5mm OF WATER

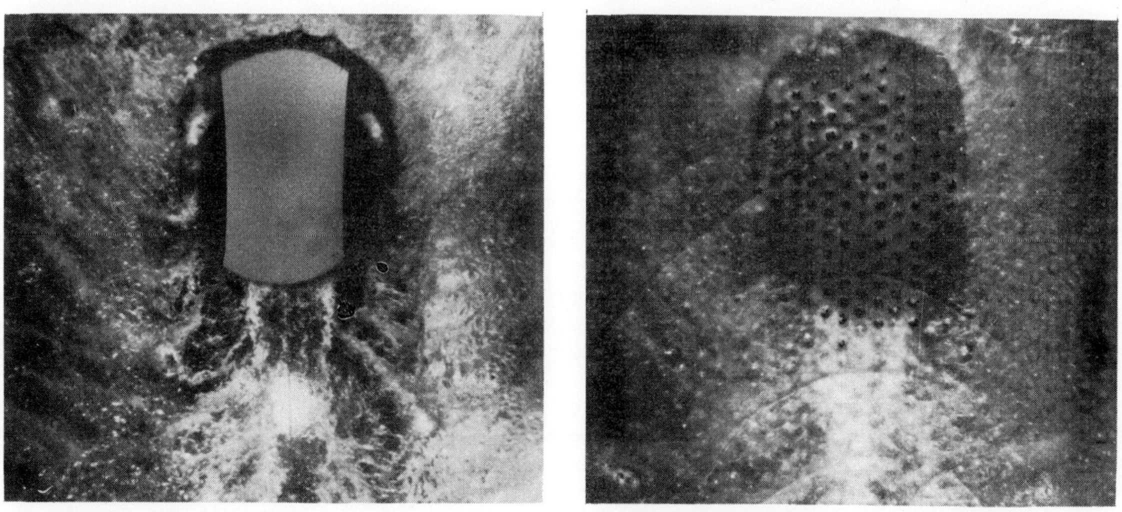

(a) Smooth glass plate

(b) Model surface of hexagonal close packed hemispheres
- Diameter 10 mm
- Separation between centre 11 mm

Fig. 7c: A comparison of the water penetration within the contact patch of a rolling tyre on a smooth surface and a textured surface made to the dimensions of the theoretical model.

© IMechE 1976

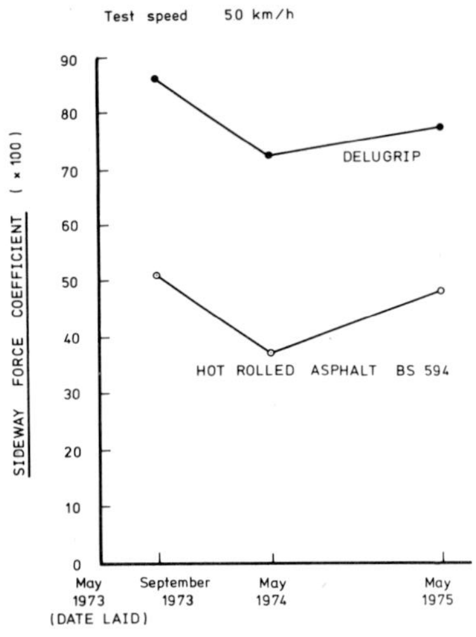

Fig. 8a: **The value of the sideway force coefficient under wet conditions for two surfaces on the A4 road, London for the first two years of trafficking.**

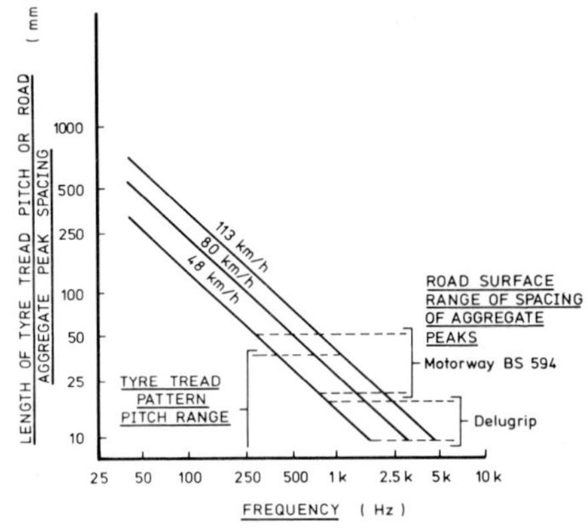

Fig. 8b: **The relationship between the frequency of the noise generation and the spacing of tyre tread pattern elements and road surface.**

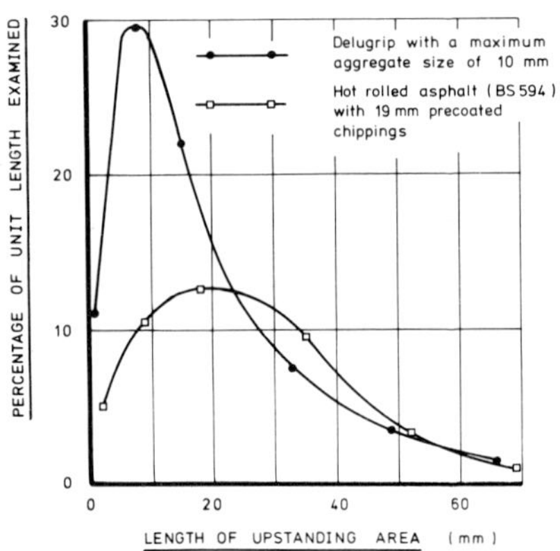

Fig. 8c: **The distribution of aggregate upstanding areas for a typical sample of high friction dense asphalt road surfaces ('Delugrip') and a sample of hot rolled asphalt (BS 594) obtained through analysis using the Quantimet image analysing computer.**

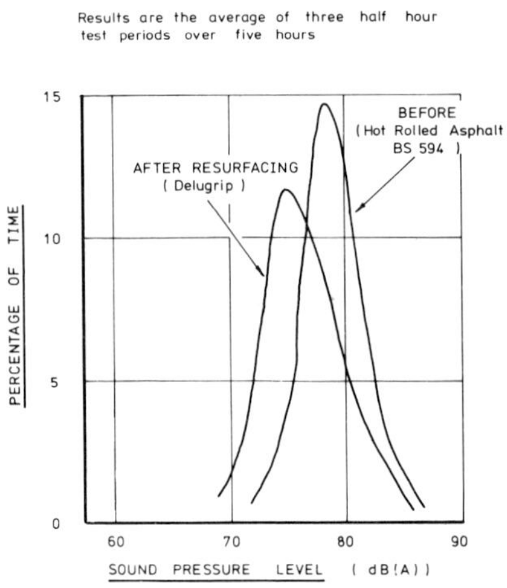

Fig. 8d: **A comparison of the noise spectrum for the Hammersmith flyover, London before and after re-surfacing with a high friction dense asphalt surface.**

© IMechE 1976

C29/76

WET FRICTION – TYRE AND ROAD

T. WILLIAMS, CEng, MRAeS, FIHE
Transport and Road Research Laboratory
Department of the Environment, Crowthorne, Berkshire

The MS of this paper was received at the Institution on 27 August 1975 and accepted for publication on 17 October 1975

SYNOPSIS One of the limiting parameters in braking performance is the friction between the tyre and road surface and considerable reduction in performance can occur when the road is wet. This paper deals with factors affecting wet road friction. The need for drainage in both the tyre and road surface becomes more important with increasing speed and of the two, the road surface can be most effective at providing adequate drainage and hence improved grip. Tyre tread materials also influence the level of wet road skid-resistance and contribute to improved braking, traction and cornering characteristics of vehicles. Commercial vehicle tyres in general use tread materials which are more resilient than those used for car tyres, they also have less effective tread drainage, and both features, polymer and drainage are determined to some extent by the problems of heating and loading. Tests on simple straight grooved lorry tyre patterns suggest that the ratio of the groove to rib width is less important than an effective distribution of grooves across the width of a tyre. Adequate levels of wet road skid-resistance are provided by the correct properties at the interface with the tyre. Road surface wearing courses when new provide these properties, but regular monitoring of both skid-resistance and texture depth is needed to establish when routine maintenance should be carried out.

INTRODUCTION

1. Road surfaces are required to provide a comfortable ride in combination with the tyres and suspension of a vehicle. Impact damage increases with surface irregularities and its severity depends on the type of vehicle and suspension; (Ref. 1) therefore strength and wear characteristics must be carefully considered. The road surface must also provide sufficient grip in both wet and dry conditions to cater for the tractive, cornering and braking requirements of vehicles. These factors must be considered in the design of the road, taking into account durability, sight lines, levels, drainage and road geometry. Other environmental considerations have become more important as vehicle speeds have increased over the years, particularly noise generation and spray. Recent events have also directed more effort into reducing energy consumption and the general efficiency of vehicles is being given more emphasis and in this context the effect of tyre/road combinations on general wearing properties of the road surface and the tyre and also fuel consumption must be taken into account.

2. The requirements of a road surface can vary depending on the type of road, and the likely predicted traffic speeds and flows. In working towards a balance between the various factors and taking into account the cost of any improvement it can be seen that the frictional problems, although extremely important, cannot be considered in isolation as the characteristics of the road surface may well affect other parameters.

3. Annual accidents in Great Britain involving death or injury total about a quarter of a million and additionally an unknown number of "damage only" accidents occur. A breakdown of a group of 1164 local accidents showed that the road factors were contributing in 27% of accidents, vehicle factors 20% and human factors 88%: (Ref. 2). Although the contribution attributed to human factors is high it is possible that the sequence of events could have in some cases been remedied by engineering design. Table 1 shows a comparison of National accidents where skidding has occurred for wet and dry conditions. The figures over the three year period 1972-1974 remained fairly constant for skidding rate.

ROAD SURFACE TEXTURE

4. The major part of the research into the skid-resistance of wearing courses of road surfaces throughout the world has been concentrated on maintaining wet road friction as in almost all cases the wet values are lower than those obtained in the dry and much more speed-dependent. In the wet situation, the requirement of the road surface is to achieve rapid dispersal of the surface water and allow high contact pressures to be established between the individual asperities in the road surface and the tyre. On extremely smooth textured surfaces very thin water films will produce low friction results, and it is on surfaces like these that tyre pattern effects show up to advantage. However, at very low speeds below 2 km/h if sufficient time is available to disperse the water film, this type of surface can provide better grip than surfaces accepted to have good friction properties. Any conclusions drawn from measurements made over a limited speed range or at one speed must be used with some reservation unless some other factor, such as depth of texture is taken into consideration at the same time.

5. In flexible pavements the composition and thickness of the various layers in the structure will govern the traffic carrying capacity and design

© IMechE 1976

life. The subgrade, sub-base, road base and surfacing courses all contribute to the total strength of the pavement. The wearing course also provides the skid-resisting properties. In pavements of this type the aim has been to provide by some method drainage paths in the wearing course of the road surface. The large scale asperities in the surface of the material or macrotexture provide these paths for the bulk drainage of the water. The skid-resistance once the bulk water is removed will then be influenced by the fine scale or microtexture and is dependent on the type of aggregate used and its resistance to polishing. On runway surfaces grooving techniques have been used to improve frictional properties and on both roads and runways pervious surfacings which allow drainage through the wearing course have proved effective in improving friction and reducing spray (Refs. 3 and 4).

6. On rigid concrete pavements the macrotexture is usually provided by some form of grooving or brushing and one of the most important constituents in improving the low speed wet friction is the nature of the fine aggregate (Ref. 5). There are also concrete surfaces designed to introduce larger aggregates in the contact area of the tyre (Ref. 6).

7. On both rigid and flexible surfaces the removal of the bulk water brings into play the available microtexture.

8. The problem of excessive water films particularly at high speeds has been the subject of much investigation (Refs. 7 and 8). The phenomenon of aquaplaning of aircraft tyres emphasised the problems of speed and pressure in the tyre contact area (Refs. 9 and 10). These investigations of the hydrodynamic effects on tyres showed the importance of the surface drainage properties particularly at high speeds. In the aircraft field various methods of improving runway wet skid-resistance have been used. Some of those surfaces having poor friction properties have been grooved or roughened. A further problem peculiar to the aircraft investigations is the ingestion of water into engines at certain speeds. This led to the development of a pervious bituminous surfacing by the Air Ministry of Works Directorate known as the "Friction course" surface (Ref. 4). This type of surface when used on roads in the 1940-50 period tended to close up and become dense under heavy traffic conditions. More recent experiments on roads with pervious surfacings have shown that although some silting and compaction occurs the surfaces have remained in good condition and are still effectively reducing splash and spray (Ref. 3). The skid-resistance of most wet surfaces is to some extent speed-dependent, decreasing with speed. The ability to remove water under the hydrodynamic effects of a vehicle tyre relies on the texture of the road surface and its drainage capability. Methods of measuring the run-off capabilities of surfaces have shown the likely water depths related to the intensity of rainfall (Ref. 11). Various types of outflow meters are used to rate the drainage potential of a surface (Refs. 6 and 12). It is appropriate to mention that the aircraft friction problem differs in some respects to the road vehicle. The majority of accidents to road vehicles occur at relatively low speeds and much of the loss of skid-resistance in the wet is due to heavy trafficking by commercial vehicles polishing the aggregate in the road surfaces. Figure 1 demonstrates the effect of traffic volume over a period of years on the SFC (sideway-force coefficient) as measured by the SCRIM test vehicle (sideway-force coefficient routine investigation machine) which uses a rolling test wheel angled at 20° to the direction of travel on a wetted road, (Figure 2). It can be seen that after about 1 year of traffic an equilibrium level of skidding-resistance is achieved (based on a mean summer value) which does not continue to decrease with life. On roads where there is a change in traffic flow a change in SFC can be expected. Figure 3 shows the result of a decrease in vehicle flow due to a diversion of traffic because of the opening of a motorway, with a consequent increase in the SFC (Ref. 13). The PSV (polished stone value) is a value given to an aggregate after subjecting it to an accelerated polishing test and it is measured with an adapted portable skid-resistance tester (Refs. 14 and 15). Additionally there is seasonal change in resistance to skidding which shows a trend to decrease in the drier summer months due to polishing of the road stones followed by a weathering action in the winter (Refs. 13 and 16). It is therefore necessary to take account of these effects when assessing the performance of road surfaces and recognise that under heavy traffic conditions on adverse sites such as bends or junctions, specialised surfacing, may be required using aggregates with a high PSV such as Calcined Bauxite (Refs. 13 and 17).

9. Work at the Transport and Road Research Laboratory and many other places has shown that the effect of water on road surface skid-resistance is particularly speed dependent. Figure 4 demonstrates this effect for three of the surfaces shown in Table 2, surfaces 2, 4 and 8, covering a wide range of texture depths as measured by the sand patch method (Ref. 15). The three surfaces were a quartzite macadam, rolled asphalt with precoated chippings and a concrete deliberately treated to produce a smooth surface finish (Ref. 18). The water depths were measured relative to the tips of the surface asperities at fixed datum points, and the water retained in rubber troughs. The results were obtained with a full patterned car tyre and the BFC (locked wheel braking force coefficient) measured over a speed range 50-120 km/h. As a comparison the results obtained with a smooth tyre are shown dotted for the concrete surface only. At the lower speed (50 km/h) the changes due to water depth are relatively small but there is a larger change due to tyre pattern. As speed increases (Figs. 4b,c, 80 and 120 km/h) particularly in the 0-2 mm water depth range the surface differences become more apparent. The differences between surfaces, and tyres on one surface, are shown cross-hatched. Large differences appear when comparing the worst condition of the smooth tyre on the smooth concrete at 120 km/h, with the quartzite macadam, producing differences of 0.4 - 0.5 BFC in the 0-2 mm water depth range.

10. Figure 5 demonstrates the effect of a range of surfaces more clearly and shows the results of BFC measurements for all the surfaces described in Table 2. These results are not directly comparable with the results shown in Figure 4 because of differences in tyre polymer and pattern and time of test but both sets of results show the same trends, and are the mean values for five patterned tyres of different tread polymers (Table 3) on each of the eight surfaces. The water depth was maintained around $\frac{1}{2}$ mm. The scale has been divided into logarithmic for low speed and linear for high speed. At the very low speeds the surface that is generally regarded as slippery, the Bitumastic emulsion

© IMechE 1976

produced the highest BFC but rapidly changed order to the lowest BFC at a speed of 10 km/h. In the mid speed range 50 km/h, the effect of the micro-texture predominates and the mean SRV (skidding resistance value with portable tester) which represents a test at this speed, shows that a high SRV is also associated with a high BFC at this speed. Looking at the high speed, in the 140 km/h range, surfaces having a high SRV and large TD (texture depth) give the best results and a surface like the Bridport macadam can compensate to some extent for its lack of SRV by its texture depth. The surfaces having good SRV and low TD tend to be more speed-dependent, falling into the range of surfaces with good macrotexture and poor SRV. With worn tyres the wet skid-resistance values of surfaces with little macrotexture will be lower than those shown in Figure 5 at other than very low speeds. Figure 5 indicates the wide range of skid-resistance that can be obtained by the surface characteristics of the road.

TYRE TREAD MATERIALS

11. Classical laws of friction which state that the coefficient of friction is independent of load, sliding speed and nominal contact area and is dependent on the nature of the materials in contact do not apply to the viscoelastic and hydrodynamic properties of a rubber tyre sliding over a wet road surface. There are a number of components in the total frictional force and it is suggested that the main divisions are (Ref. 19):-

$$F_{total} = F_{adhesion} + F_{deformation} + F_{viscous} + F_{tearing}$$

The primary component is the molecular adhesion between the contacting surfaces. As the tyre is not purely elastic the asperities on the road surface cause energy to be lost by hysteresis in the stress-strain cycle. The behaviour of both these components is viscoelastic, depending on the frequency of the process and the temperature of the materials. The viscous component takes into account the effect of a layer of liquid between or adsorbed by the surfaces. Tearing accounts for a proportion of the friction when particles are torn from the rubber surface.

12. One part of the programme investigating tread polymer effects measured the BFC of a number of polymers for a range of surfaces. The surfaces are described in Table 2 and the resilience of the tread polymers in Table 3. A mean value of BFC was obtained for the eight test surfaces for each polymer, (Figure 6) using smooth test tyres. The tyres were tested over a speed range 1 to 140 km/h. The curves are displayed in a similar way to Figure 5 (logarithmic for low speed, linear for high). The different polymers form a fan shaped system of curves converging with increasing speed. The order of separation suggests some viscoelastic behaviour throughout the speed range with the more resilient polymer, the BR (Polybutadiene) having a lower BFC throughout. Individually some surfaces gave greater separation between polymers but the mean value for all eight surfaces gives a more general appreciation. Laboratory tests show that viscoelastic adhesion and hysteresis components produce frequency dependent peaks. The separation between polymers appears to be related to resilience effects though no real indication of peaks could be determined, possibly because of the random nature of the surface asperities and the hydrodynamic effect as the speed increased. At the highest speed 140 km/h, the range was 0.18 to 0.25 BFC. The mean for patterned tyres would increase the BFC by about 0.1 BFC giving a range of 0.18 to 0.35 BFC. Referring to Figure 5 the range of BFC for the surfaces was about 0.1 to 0.65 showing the importance of the surface in relation to the tyre in providing good skid-resistance. Figure 6 also demonstrates that in the low speed range where a large proportion of accidents occur, tread polymer can make a significant contribution to skid-resistance values.

COMMERCIAL VEHICLE TYRES

13. Much of the loss of skid-resistance due to polishing of road stones can be attributed to commercial vehicles. The commercial vehicle tyre, because it operates under more adverse conditions of loading and temperature than the car tyre, tends to use more resilient tread polymers and less sophisticated tread patterns than cars. There are a range of lorry tyres available to suit different applications and they can be divided into Highway Service, Dual Purpose and Cross Country, the second two categories designed to have 'off road' capabilities. In cornering and braking tests on smooth wet surfaces the Cross Country tyres were found to be little better than a smooth tyre (Ref. 20). In the front wheel braking tests to determine the BFC values difficulties were often experienced in locking the wheels on high coefficient surfaces because of insufficient braking capacity.

14. To examine the effect of tyre drainage sets of smooth tyres were circumferentially grooved in three configurations, three, five and seven ribs. The tyres were cut in such a way that the ratio of the groove width to rib width could be increased by recutting the grooves. The tests were limited to a single surface the smooth concrete, an average tyre tread depth of about 7 mm and a water depth of about 1/2 mm. The peak and locked wheel braking force coefficient was measured over a speed range 20-90 km/h for a range of groove/rib ratios 0.2, 0.4, 0.6 and 0.8.

15. Figure 7 shows the locked braking force coefficient (BFC) and peak braking force coefficient (PBFC) for two speeds, 30 km/h and 80 km/h. For ratios above 0.2 there was little change in PBFC or BFC. At the higher speeds there were distinct differences for the rib configurations, suggesting that the number of ribs was more significant in improving the skid-resistance than the size of the drainage channel above a ratio of 0.2. There are practical difficulties in producing lorry tyres with large numbers of grooves however. Figure 8 suggests that configurations of 7 ribs and over would produce better results from the skid-resistance requirement. Tests varying the groove/rib ratio across a 7 rib tyre with large outer to small inner grooves and the reverse situation (0.6, 0.4, 0.2 and 0.2, 0.4, 0.6) made little difference to the PBFC or BFC measurement when compared to a seven rib tyre all of 0.4 groove/rib ratio. The tyres used in these tests had a flat profile and therefore a fairly even pressure distribution across the tyre. Differences of the order of 2 : 1 between BFC and PBFC at the higher speeds

© IMechE 1976

could be utilised by an efficient anti-locking braking system. Investigation of loss of skid-resistance with tread wear on surfaces 2, 3 and 4 showed that the decrease with wear was gradual for BFC. The PBFC tended to fall more sharply with about 6 mm tread remaining on the fine textured surfaces (Refs. 21 and 22). Added to the fact that generally the lorry tyres produce lower wet skid-resistance values than car tyres the decision on what is a minimum safe tread depth is difficult to make and may be influenced by the position of the tyres on the vehicle.

FURTHER STUDIES

16. In selecting wearing course properties of road surfaces account must be taken of durability, grip, wear and noise level. Tests on a range of surfaces to study the effect of tyre wear showed the wear on a rough (macro) harsh (micro) textured surface was approximately three times as severe as on a rough polished surface, the harshness being of major importance and the roughness a slightly modifying factor (Ref. 23). Increased speed also materially affects tyre life. Efforts to devise standard methods of comparing tyre wear over roads with a range of surfacings have proved difficult because of the variation between vehicles and climatic changes. In many cases it is difficult to determine in the wet condition the degree of wetness or whether a surface that is nominally dry looking is affected by humidity.

17. Much of the laboratory work on the frictional properties of rubber uses the dynamic strain method to determine the viscoelastic properties of different materials. Additionally low speed sliding tests up to 2 m/s (7 km/h) on a wetted rubber track with metal indentors indicated the importance of shape (Ref. 24), with the more angular shapes producing higher values of skid-resistance. An extension of this work up to speeds of 34 m/s (120 km/h) showed that with four shapes, wedges, cones, spheres and cylinders, a great range of wet skid-resistance could be achieved, giving BFC values up to 3.0 using the wedge shapes, down to extremely low values with the cylinders. The results were load dependent in most, though not all cases, tending to increase skid-resistance with increase in load. Increased penetration due to change in angularity or size also tended to increase the skid-resistance. These shapes were produced with a polished finish with little microtexture. Photographs of the shapes after use showed definite evidence of erosion on the trailing edges in the form of pitting of the surface, and a rounding off on the sharper edges. This erosion is similar to that found in high pressure hydraulic systems particularly on the port edges of high precision selectors and is common in systems with poor filtration. This may be part of the process causing the seasonal variation in skid-resistance, where contaminated fluid is brought into contact with the surface at relatively high pressure, perhaps assisting the improvement in microtexture.

18. Some pilot scale tests to produce a range of shapes suitable for wearing courses made of high-PSV materials in about a 12 mm size illustrated some of the practical problems. Many of the better shapes rolled flat when applied to a surface tending to present their flat faces rather than an edge to the tyre. To prevent the cylinders standing end upwards their ends were angled to make them fall over. Possibly a hexagon or similar shape with angled ends would produce a reasonable compromise.

STANDARDS OF SKIDDING RESISTANCE

19. It is clearly important to provide both the Highway Engineer and the Vehicle Engineer methods of categorising surface characteristics within some acceptable tolerances by means that are easily applied. With the increased use of sideway-force measurement (using SCRIM) and the correlation of results with accident data it is possible on highways to apply selectively a standard of skid-resistance suitable to the individual needs of the particular site (Ref. 25). A contactless sensor primarily developed for profilometer measurements gave a resolution of 0.25 mm and indicates a potential for measuring surface texture from a mobile vehicle. This opens up the possibility of making both SFC and TD measurements from the same vehicle (Ref. 26).

20. To standardise test procedures in the vehicle manufacturing industry there is a trend to categorise a surface by a single skid number so that engineers in different areas can carry out acceptance tests on vehicles. For dry testing with cars this is not unreasonable, but as has already been shown the speed dependence of skid-resistance varies greatly with different surfaces. Figure 9 shows the results of three measurements on a range of surfaces SRV, TD and the mean deceleration from a BFC measurement. The results are shown for a smooth OESBR tyre for two speed ranges, 2-50 km/h (full line) and 100-150 km/h (dotted line), the low and high speed end of the range. These results have been deliberately extrapolated beyond the range of the test surfaces used to show the general trends and emphasises the importance of the microtexture as measured by the SRV in the low speed condition; an increase of 0.1g for about 10 increase in SRV over the whole range. At the higher speed changing the SRV has little effect on the skid-resistance at lower TD values and progressively improves as the TD increases.

CONCLUDING REMARKS

21. Wet road friction is often the critical factor in braking. This review of the current situation discusses the part played in wet road grip by road surface materials, drainage which is a function of both surface and tyre and finally the influence of tyre tread materials.

22. Roads tend to attain a long-term equilibrium level of low speed skid-resistance (based on a mean of summer measurements) and this is dependent on the traffic density and the polish resistance of the stone used. This level can be measured using SCRIM (sideway-force coefficient routine investigation machine) and if it falls below a nominated standard, maintenance can be implemented using a higher quality of stone.

23. High speed resistance to skidding requires in addition a good surface macrotexture but this will gradually deteriorate with time due to wear of the aggregate particles and to their embedment. Texture can be measured using a simple sand-patch test but a more sophisticated (and much more rapid) method is being developed using a laser technique;

© IMechE 1976

this will enable a monitoring procedure to be carried out for maintenance planning purposes.

24. Tyre tread polymer materials are already making a significant contribution to improved wet road grip for the car tyre, but these benefits cannot be fully utilised for lorry tyres because their working conditions are more arduous. Tests on grooved lorry tyres suggest that well spaced multi-grooved designs may offer the only improvements still possible.

ACKNOWLEDGEMENT

25. The work described in this Paper forms part of the programme of the Transport and Road Research Laboratory and the Paper is published by permission of the Director.

NOTATION

SFC	=	Sideway-force coefficient (SCRIM)	F_y/F_z
PSV	=	Polished stone value	
SRV	=	Skidding resistance value (Portable skid-resistance tester)	
TD	=	Texture depth	
BFC	=	Locked wheel braking force coefficient	F_x/F_z
PBFC	=	Peak braking force coefficient	F_x/F_z
IIR	=	Butyl rubber	
OENR	=	Oil extended natural rubber	
OESBR	=	Oil extended styrene butadiene rubber	
NR	=	Natural rubber	
BR	=	Polybutadiene rubber	
cv/day	=	Commercial vehicles per lane per day.	

REFERENCES

1. AASHO Road Test, Report 7, Summary report, Highway Research Board Special Report 61G. Washington DC, 1962 (National Academy of Sciences - National Research Council).

2. SABEY, B.E. Accidents: their cost in relation to surface characteristics. Symposium. Cement and Concrete Association, Birmingham, Nov. 1973.

3. BROWN, J.R. Pervious bitumen-macadam surfacings laid to reduce splash and spray at Stonebridge, Warwickshire. Department of the Environment, TRRL Report LR 563. Crowthorne 1973. (Transport and Road Research Laboratory).

4. MARTIN, F.R., and R.F.A. JUDGE. Airfield pavements. Problems of skidding and aquaplaning. Civil Engineering. Dec. 1966.

5. WELLER, D.E., and D.P. MAYNARD. The influence of materials and mix design on the skid-resistance value and texture depth of concrete. Ministry of Transport, RRL Report LR 334. Crowthorne, 1970. (Road Research Laboratory).

6. LEES, G., and K.K. TAM. The design of non-grooved high friction concrete surfacings. Symposium on the influence of the road surface on skidding. University of Salford, April, 1975.

7. GILES, C.G., and F.T.W. LANDER. The skid-resisting properties of wet surfaces at high speeds. Journal of the Royal Aeronautical Society. Feb. 1956.

8. ALLBERT, B.J., J.C. WALKER, and G. MAYCOCK. Tyre to wet road friction, Institution of Mechanical Engineers. Proceedings 1965-66, Vol. 180.

9. HORNE, W.B., and R.C. DREHER. Phenomena of pneumatic tire hydroplaning. NASA Technical Note, NASA TN D-2056 Nov. 1963.

10. GRAY, W.E. Measurements of aquaplaning height on a Meteor aircraft and photos of flow pattern under a model tyre. Royal Aircraft Establishment. Tech. Note No. Aero 2855. Nov. 1962. (Royal Aircraft Establishment).

11. ROSS, N.F., and K. RUSSAM. The depth of rain-water on road surfaces. Ministry of Transport, RRL Report LR 236. Crowthorne, 1968. (Road Research Laboratory).

12. SMITH, T.R. An assessment of the outflow meter for measuring the drainage capability and macrotexture channel size of open graded runway surfaces. Royal Aircraft Establishment Tech. memo EP 579. (Royal Aircraft Establishment).

13. SZATKOWSKI, W.S., and J.R. HOSKING. The effect of traffic and aggregate on the skidding resistance of bituminous surfacings. Department of the Environment, TRRL Report LR 504. Crowthorne, 1972. (Transport and Road Research Laboratory).

14. BRITISH STANDARDS INSTITUTION. Method for sampling and testing mineral aggregates and fillers. British Standard BS 812: London, 1967 (British Standards Institution).

15. ROAD RESEARCH LABORATORY. Instructions for using the portable skid-resistance tester. Ministry of Transport, Road Note No. 27. London, 1969 (HM Stationery Office).

16. HOLMES, T., G. LEES, and A.R. WILLIAMS. A combined approach to the optimisation of tyre and pavement interaction. Symposium on "Tread wear and traction" American Chemical Society. April 1971.

17. HATHERLEY, L.W. et al. Skid resistance of city streets and road safety. J. Inst. Highway Engineers, 1969 (April) 7-14.

18. SABEY, B.E., T. WILLIAMS, and G.N. LUPTON. Factors affecting the friction of tyres on wet roads. International Automobile Safety Conference Compendium P.30. June, 1970. (Society of Automotive Engineers Inc.).

19. LUDEMA, K.C. Friction of rubber. Mechanics of pneumatic tires. National Bureau of Standards 1971.

20. WILLIAMS, T., J.K. MEADES, and B.S. RILEY. The wet road grip of lorry tyres: a comparison of three types. Department of the Environment, TRRL Report LR 544. Crowthorne, 1973. (Transport and Road Research Laboratory)

© IMechE 1976

21. WILLIAMS, T., and J.K. MEADES. The effect of tread pattern depth and grooving on lorry tyre skidding resistance. Department of the Environment, TRRL Report LR 687. Crowthorne, 1975 (Transport and Road Research Laboratory).

22. DIJKS, A. Wet skid resistance of car and truck tyres. Tyre Science and Technology, Vol. 2, No. 2 May 1974.

23. LOWNE, R.W. The effect of road surface texture on tyre wear. Ministry of Transport, RRL Report LR 265. Crowthorne, 1969. (Road Research Laboratory).

24. SABEY, B.E. Pressure distribution beneath spherical and conical shapes pressed into a rubber plane, and their bearing on coefficient of friction under wet conditions. Proc. Physical Society 1958.

25. SALT, G.F., and W.S. SZATKOWSKI. A guide to levels of skidding resistance for roads. Department of the Environment, TRRL Report LR 510. Crowthorne, 1973. (Transport and Road Research Laboratory).

26. COOPER, D.R.C. Measurements of road surface texture by a contactless sensor. Department of the Environment, TRRL Report LR 639. Crowthorne, 1974. (Transport and Road Research Laboratory).

Crown copyright 1975. Any views expressed in this Paper are not necessarily those of the Department of the Environment. Extracts from the text may be reproduced, except for commercial purposes, provided the source is acknowledged.

© IMechE 1976

Table 1

Injury Accidents

Year	Dry		Wet		Total *		Skidding Rate %		
	Skid	Non Skid	Skid	Non Skid	Skid	Non Skid	Wet	Dry	All
1972	19367	151966	22246	65996	45658	219750	25.2	11.3	17.2
1973	19646	160000	19252	56298	43839	218802	25.5	10.9	16.7
1974	16014	144402	18934	61819	37143	207202	23.4	10.0	15.2

* total includes icy roads and conditions not known.

Table 2

Details of test surfacings

	Details of surface	SFC(1) (1971 only)		SRV(2)	Texture(3) depth µm
		50 km/h	80 km/h		
1 (M)	Bitumastic Emulsion Mastic asphalt covered with thin layer of bitumastic emulsion	0.29	0.15	17 to 37	130
2 (C)	Concrete Monolithic granolithic concrete flooring, artificially polished	0.47	0.20	57 to 61	228
3 (FTA)	Fine cold asphalt Fine cold asphalt to BS1690 with blastfurnace slag aggregate	0.73	0.48	61 to 75	533
4 (A+P)	Asphalt with precoated chippings Asphalt wearing course with 20 mm precoated chippings	0.63	0.56	57 to 62	1295
5 (NB)	Bridport Macadam (new) Bitumen macadam carpet with 10 mm Bridport gravel aggregate	0.40	0.35	34 to 38	1346
6 (NQ)	Quartzite (new) Bitumen macadam carpet to BS1621 with 10 mm crushed quartzite aggregate	0.80	0.73	66 to 71	1397
7 (Q/B)	Mixed Aggregate Bitumen macadam carpet with mixed Bridport gravel and crushed quartzite aggregates in roughly equal proportions	0.51	0.49	44 to 51	2133
8 (OQ)	Quartzite (old) Specification as for 6 above	0.83	0.79	69 to 71	2590

1. SFC = sideway-force coefficient
2. SRV = skid-resistance value; portable skid-resistance tester
3. TD = texture depth using sand patch technique

© IMechE 1976

Table 3

Resilience of the test tyres

Resilience *		OESBR	NR	BR	IIR	OENR
Resilience % - Temp °C (9.5 mm thick dample)	0	16	28	48	8	14
	10	21	33	50	11	19
	20	27	39	51	15	24
	30	30	42	52	20	29
	40	33	44	53	26	35
Resilience % - Tyre No. (test tyre on spokes, 20oC)	1	34	48	51	20	34
	2	-	45	53	21	34
	3	35	49	52	21	34
	4	33	48	51	20	34
	5	34	48	52	19	34
	6	-	48	53	20	32

* Measured with a modified B.S. Lupke Resiliometer

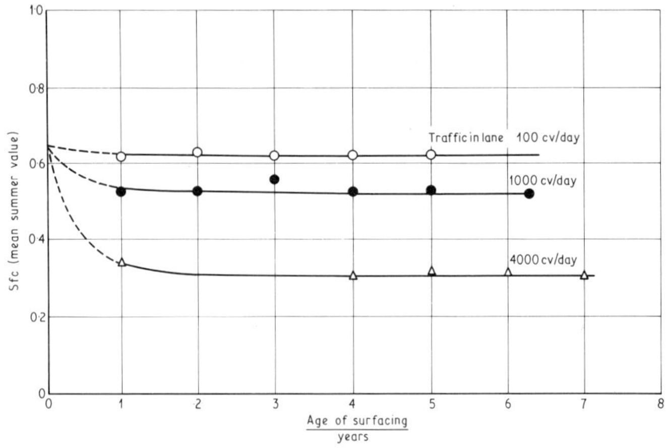

Fig. 1: Effect of Traffic on Skidding Resistance of a Typical Motorway – Standard Surfacing (Rolled Asphalt with Precoated Chippings of PSV 58 - 60)

Fig. 2: Sideway Force Coefficient Routine Investigation Machine

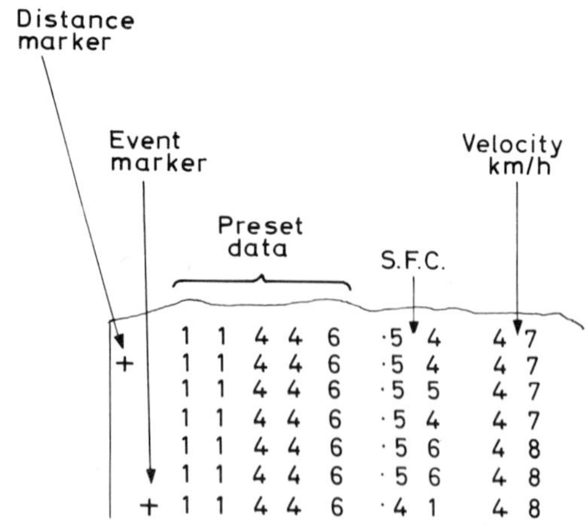

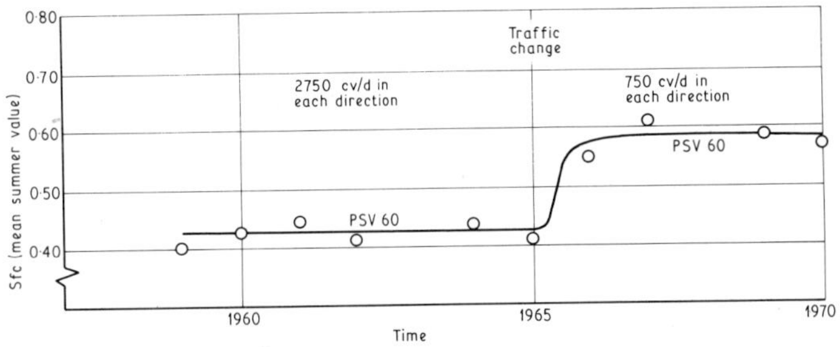

Fig. 3: An Increase in the Level of Skidding Resistance Recorded on Trunk Road A4, Colnbrook By-pass, when Traffic Decreased due to the Opening of a Motorway

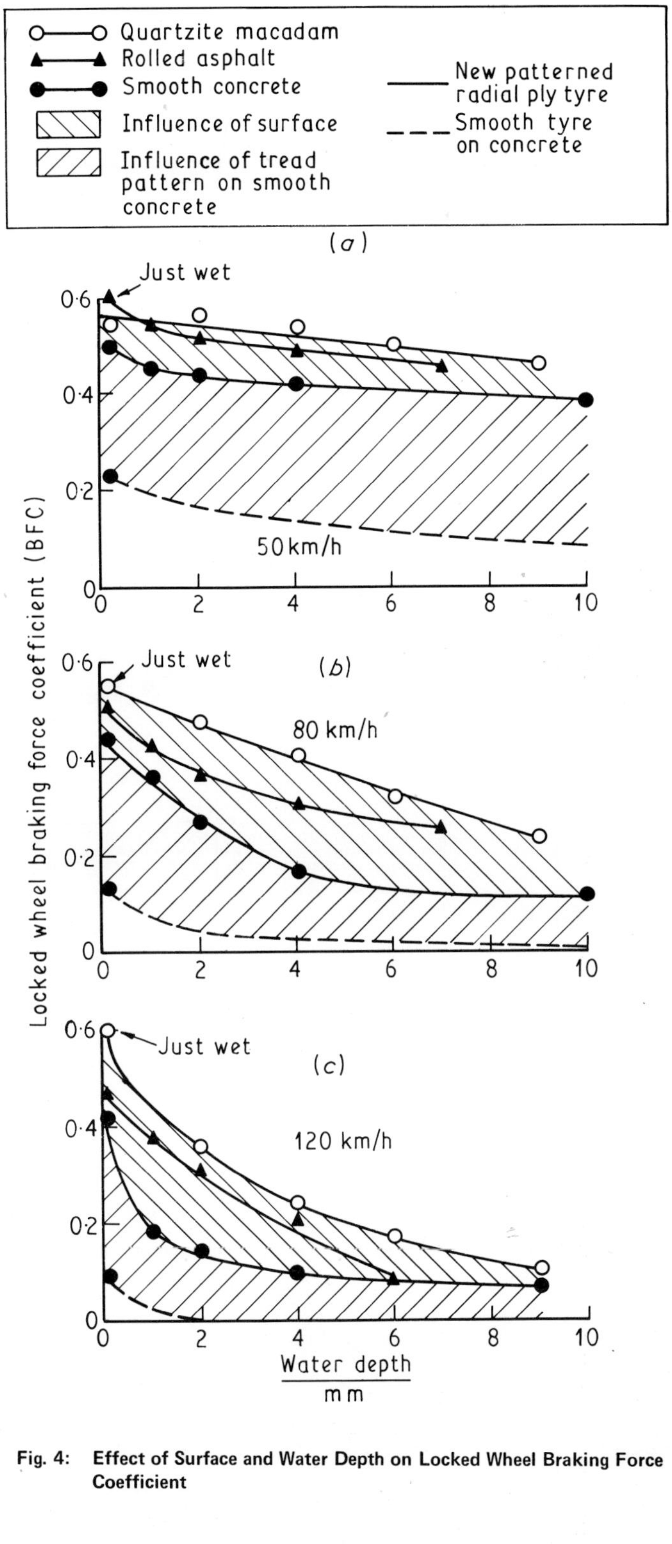

Fig. 4: Effect of Surface and Water Depth on Locked Wheel Braking Force Coefficient

© IMechE 1976

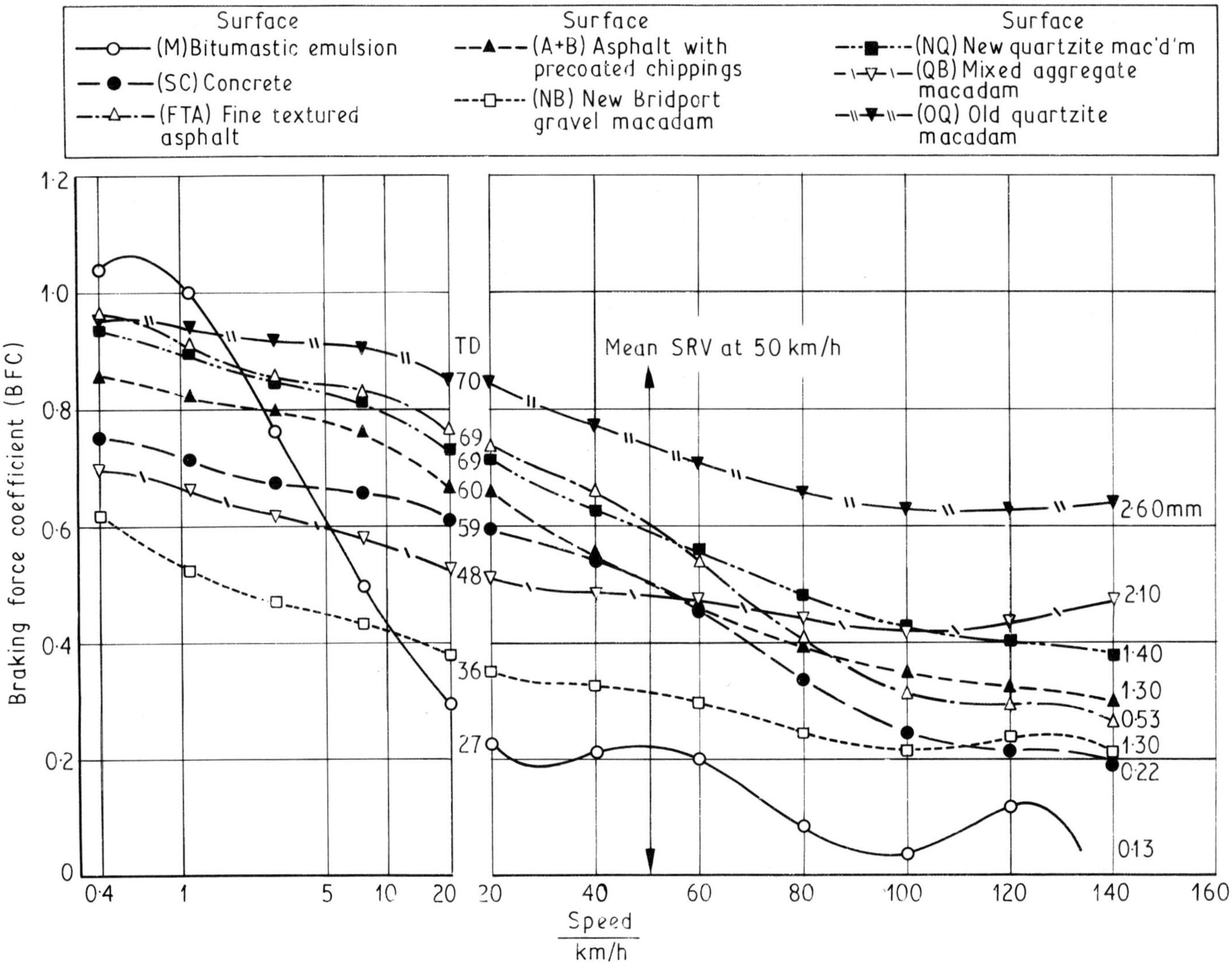

Fig. 5: **Relation between Braking Force Coefficient and Speed-mean of Five Patterned Tyres of Different Tread Polymers on Eight Surfaces (1972)**

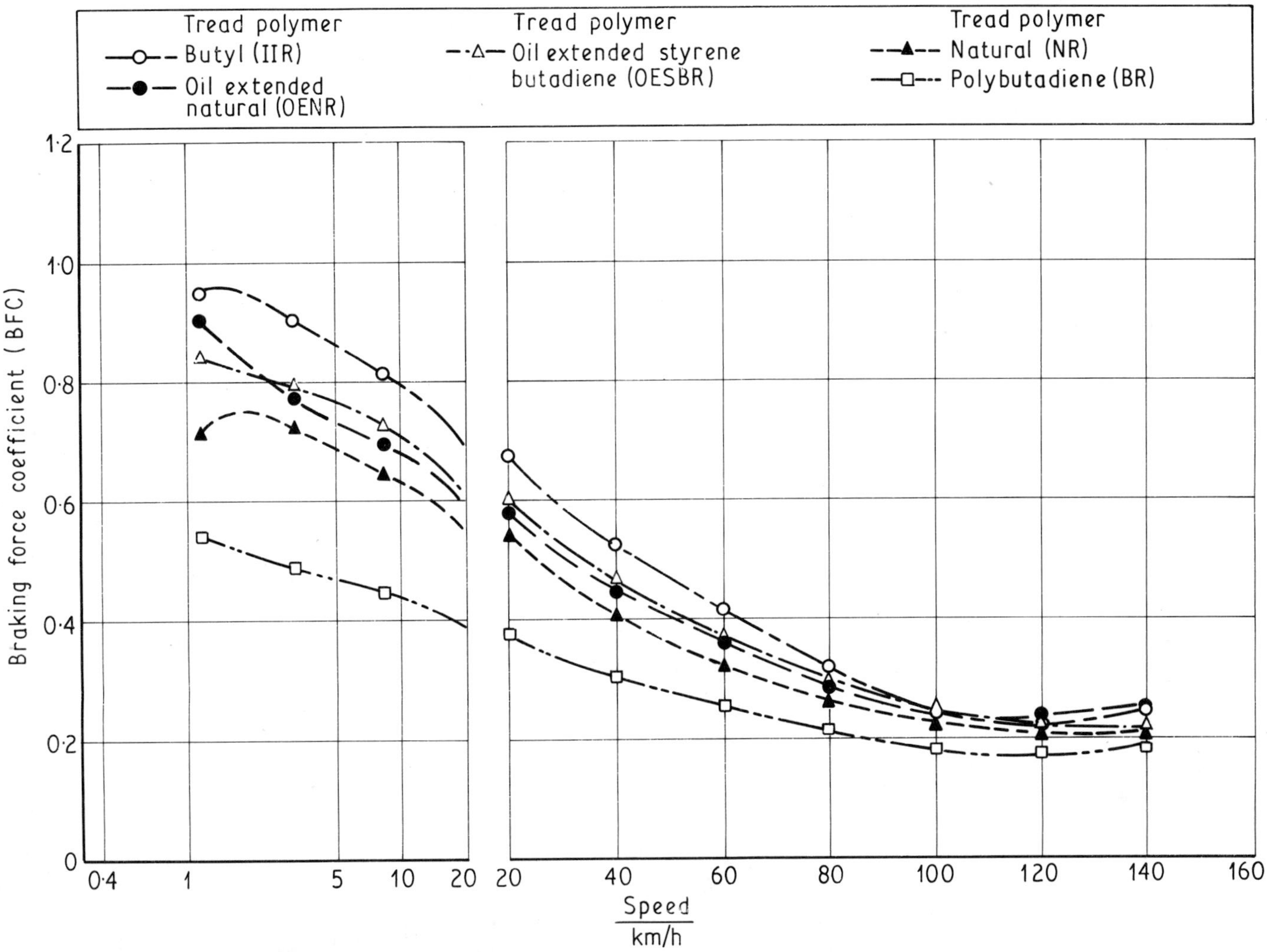

Fig. 6: **Relation between Braking Force Coefficient and Speed-mean for Eight Surfaces Using Five Smooth Tyres of Different Polymers (1971)**

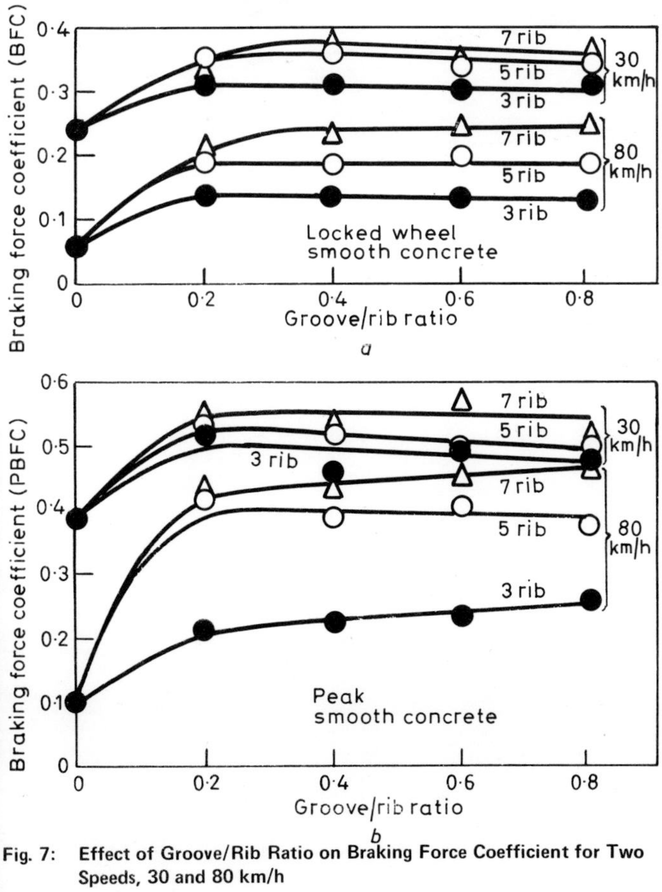

Fig. 7: **Effect of Groove/Rib Ratio on Braking Force Coefficient for Two Speeds, 30 and 80 km/h**

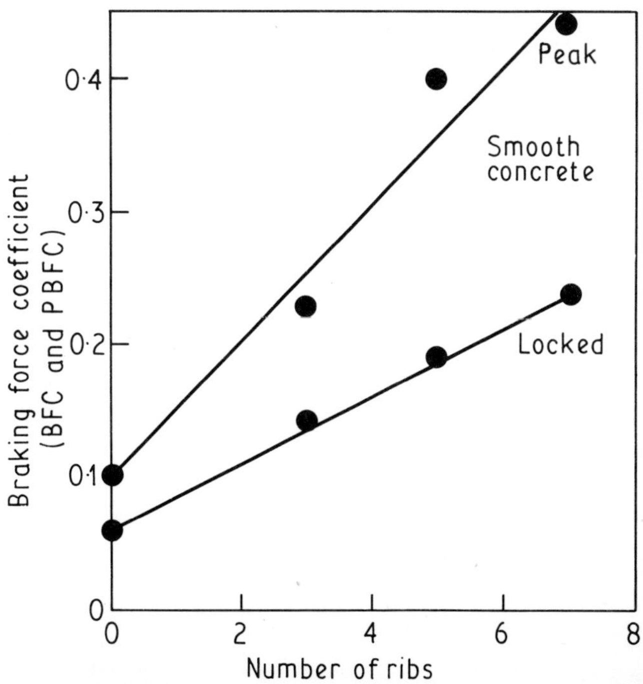

Fig. 8: **Effect of Number of Ribs for Groove/Rib Ratio 0.4 on Wet Road Grip at 80 km/h**

© IMechE 1976

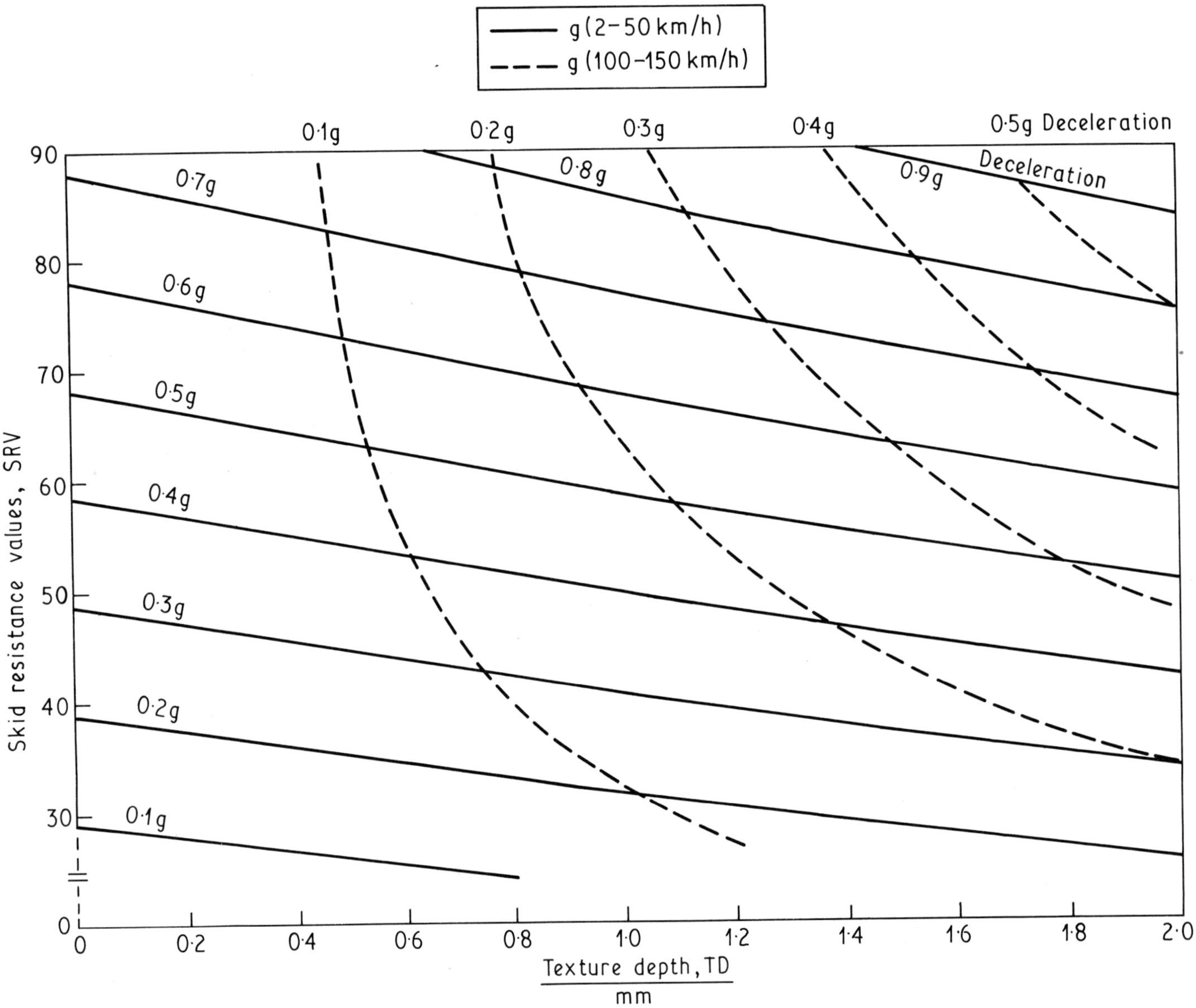

Fig. 9: Mean Deceleration Compared to SRV and TD for Two Speed Ranges 2-50 km/h and 100-150 km/h

© IMechE 1976

C30/76

MEASUREMENTS OF THE LONGITUDINAL AND LATERAL TRACTION PROPERTIES OF TRUCK TIRES

R. D. ERVIN, MS
Highway Safety Research Institute, The University of Michigan, Ann Arbor, Michigan

The MS of this paper was received at the Institution on 22 September 1975 and accepted for publication on 17 December 1975

SYNOPSIS A mobile apparatus has been developed to permit measurement of the longitudinal and lateral shear force behavior of commercial vehicle tires on actual paved surfaces. The apparatus is discussed and data are presented describing various traction properties of truck tires. Data relevant to heavy truck braking cover the longitudinal force response to longitudinal slip, as influenced by velocity, tire load, and pavement selection. Data relevant to cornering cover the side force response to angular slip, as influenced by velocity and vertical load. The experimental repeatability of these measurements is discussed, together with certain confirmation of these findings as obtained through laboratory measurements.

INTRODUCTION

1. This paper presents experimental findings which have been made using a mobile device for measuring the "over-the-road" traction behavior of heavy truck tires. The device itself was designed and developed at the Highway Safety Research Institute (HSRI) of The University of Michigan, U.S.A.

2. While numerous contributions have been made to the literature on the traction mechanics of passenger car tires, there has existed a dearth of information extending this knowledge to heavy truck tires. Although the obvious structural features are shared by both tire types, the range of loading experienced by the truck tire together with its markedly higher level of contact pressure would suggest that specific attention to this class of tires is warranted. Nevertheless, only very recently has the technical capability been developed for conducting precision measurements of the friction-limited shear force properties of heavy truck tires. Reportings of this capability have been made by Bradisse (ref. 1), Dijks (ref. 2), and by Ervin (ref. 3, 4) all of which have described longitudinal force measurement trailer devices. It is also known that three other organizations in the U.S. (Firestone Tire & Rubber Co., Uniroyal, Inc., and the Transportation Research Center of Ohio) operate mobile devices capable of making over-the-road measurements of both the longitudinal and lateral traction characteristics of truck tires at highway velocities.

3. In addition, certain investigators, such as Williams (ref. 5), have measured the inertial reactions of instrumented trucks to deduce tire properties over the road. It should be noted that the above-cited apparatuses are conceptually distinguished from other devices such as have been employed by Tielking (ref. 6) and Ginn (ref. 7), in which low-speed flat-bed machines were applied to truck tire testing, and by Freudenstein (ref. 8) who employed a low-speed mobile device in measurements of cornering traction. The conceptual distinction derives from the inability of a low-speed machine to adequately assess those traction characteristics which are influenced by the sliding velocities prevailing at the interface during the slip process. Such characteristics are determined by frictional mechanisms which sustain shear forces at high values of slip rather than by the structural deflection mechanisms which determine traction behavior at small levels of slip. Thus, low-speed machines have been most useful for characterizing low slip properties. A sophisticated test facility compromising between low-speed laboratory testers and high-speed, over-the-road machines has been reported by Bird and Schuring (ref. 9). This facility places the tire against a steel belt to effect precision measurements at highway-realistic velocities, although it lacks a demonstrably representative test surface.

4. The high-speed over-the-road tester, such as that presented in this paper, constitutes a complement to laboratory apparatuses, providing a maximum realism while endeavoring to cope with climatological variances and non-planar test surfaces.

MOBILE TRACTION MEASUREMENT APPARATUS

5. The HSRI mobile dynamometer consists of a tractor, semi-trailer vehicle which exposes heavy truck tires to controlled conditions of vertical load, velocity, and lateral or longitudinal slip while traveling over selected pavements. The system, shown in Figure 1, permits measurement of longitudinal properties by way of a trailer-configured dynamometer which is towed and serviced by the instrumented tractor. Mounted on the same tractor is a lateral traction measurement apparatus.

6. The trailer device mounts a single test tire on its centerline and supports the sample by way of a parallelogram linkage suspension. The test wheel is braked by a conventional dual wedge air brake whose actuation is controlled through a pneumatic circuit providing an adjustable brake onset time history. The vertical load condition is controlled through pressure regulation of an

©IMechE 1976

overhead air spring assembly. The air spring provides not only an adjustable loading element, but also a low rate coupling between the trailer structure and the test wheel. Since the minimization of tire load fluctuations is a primary concern, the soft coupling provided by the loading spring is augmented with the use of air suspension systems on both the trailer rear axle and tractor rear tandem axle. Not only do these suspensions provide a cushioned ride, but their automatic leveling systems serve to maintain the design attitude and elevation of the test trailer despite major redistributions in axle loading as the tire sample becomes loaded.

7. The measurement of the applied load, F_z, and the tractive force resulting from braking, F_x, is accomplished through use of a multicomponent strain-gauged load cell which is mounted just inboard of the test wheel. Wheel angular velocity, ω, is measured using a DC tachometer. Signals from these transducers are recorded on magnetic tape in the instrumented module mounted on the tractor.

8. The lateral force dynamometer mounts two identical tires on either side of the tractor. The tire samples are supported by a common cross axle which is articulated at two "kingpin" assemblies, permitting "steering" of the two tires. The tires are steered together, in a "toeing" fashion, by an electrohydraulic servo mechanism which responds to programmed sequences of slip angle, α. The cross axle is loaded through a single longitudinal pin at its center, thereby permitting an equal sharing of applied load by the two tires, and permitting a pivoting freedom of the axle as each tire runs over surface irregularities which are uncorrelated, side to side. The axle's center pin is, in turn, fastened within a vertically-sliding cage which is ultimately loaded by a set of pressure regulated air springs.

9. The traction response of only one of the two tires is measured by the instrument system, the other tire serves merely to provide a lateral force balance on the tractor. The instrumented spindle incorporates a cantilever-mounted strain-gauged load cell which permits measurement of the vertical load, F_z, the lateral force, F_y, and the aligning moment, M_z.

MEASUREMENTS OF LONGITUDINAL TRACTION

10. The trailer dynamometer has been employed in a program of baseline experiments intended to establish the first-order sensitivities of the response of truck tires to longitudinal slip. In this, and in subsequent programs, a simple matrix of vertical load and velocity conditions has been examined according to a sequence which includes the periodic repeat of a reference condition to permit later assessment of the statistical quality of the test. At a given condition, the test sample is subjected to a series of six "lockup cycles" while load and velocity are held constant. At the occurrence of the locked-wheel condition, the brake is automatically released (after about .15 seconds) to avoid the localized wear, or flat-spotting, which has been seen to seriously disturb the vertical load condition on the rolling tire.

11. The time histories of F_x, F_z, and ω are recorded on FM analog magnetic tape and later processed through a computerized reduction scheme. The reduction process provides, for each six-lockup test, a single plot of normalized longitudinal force, F_x/F_z, versus longitudinal slip, s, for which s is formally defined by the relationship:

$$s = \left(1 - \frac{\omega R_e}{V}\right) \times 100\% \qquad (1)$$

where V is the translational velocity of the wheel center

and R_e is the effective rolling radius of the non-slipping tire.

12. In HSRI's constant velocity experiments slip is actually calculated using an initially determined value of $\omega = \omega_o = V/R_e$ for the freely rolling tire, thus reducing the calculation of instantaneous slip to:

$$s = \left(1 - \frac{\omega}{\omega_o}\right) \times 100\% \qquad (2)$$

Plots of F_x/F_z versus s, each of which represents the average of the six lockup cycles, have been gathered in Figure 2 to illustrate typical influences of load and velocity on the (F_x/F_z versus slip) curve shapes. As is well known, the shape of this relationship is of primary importance in the design of automatic anti-wheel-locking systems. The most notable feature of these data, which represent dry surface behavior, is the large drop in F_x/F_z which occurs between the peak value of F_x/F_z and the value at the locked-wheel condition, 100% slip. These data represent a typical 10.00 x 20/F truck tire (the "/F" designation refers to the load range as rated by the U.S. Tire and Rim Association, T & RA), which has been inflated to its rated "cold" inflation pressure of 5.9×10^5 Pa but which operated throughout testing at an equilibrium "hot" inflation of about 7×10^5 Pa.

13. Looking at the influence of vertical load on F_x/F_z behavior, as shown on the left side of Figure 2, it is apparent that the heavy truck tire suffers a first-order loss in normalized tractive capability as load is increased from 0.4 to 1.6 times the tire's T & RA rating. Since such an increment in load is easily achieved by a truck during severe braking, the load sensitivity of heavy truck tires must be recognized as a significant mechanism in limiting truck stopping performance.

14. A significant anomaly of the load sensitivity data concerns an influence not directly evident in the normalized force presentation of Figure 2. The property termed "longitudinal stiffness," C_s, defined by the relationship,

$$C_s = \left.\frac{\partial F_x}{\partial s}\right|_{s=0} \qquad (3)$$

represents the linear behavior of the tire under braking and is directly influenced by vertical load. The increasing C_s with load, as listed for a typical 10.00 x 20/F tire in Table 1, shows a characteristic influence of the increasing length of the tire's "contact patch" or "footprint," with

©IMechE 1976

Table 1

Vertical Load, F_z N	Longitudinal Stiffness, C_s N/Unit Slip	Contact Length cm
10,000	132,000	21.6
24,500	242,000	26.7
40,500	308,000	31.2

load. As the contact length increases, each tread element must spend a longer time interval in contact with the "slipping" pavement during each revolution, thus accumulating a larger net rearward deflection during contact and generating an accordingly increased force. Measurements of C_s, made over the 8 to 88 km/h range, confirm "flat-bed" laboratory measurements reported by Tielking, et al. (ref. 6), and further show the expected result that this elastic property is unaffected by velocity.

15. It is interesting to note that the curve shapes deriving at different loads are virtual scale models of one another while the influence of velocity on the F_x/F_z versus s relationship, as shown on the right side of Figure 2, involves marked changes in the shape of the traction curve. It is seen that any given F_x/F_z versus s curve can be rather accurately modeled by use of an approximately linear relationship between slip velocity ($s \times V$) and friction level, such as has been employed in mathematical modeling of passenger car tire traction (ref. 10). As with passenger car tires, however, the truck tire's sensitivity to slip velocity, per se, is markedly influenced by the overall translational velocity of the wheel center. We see, for example, that data shown in the right-hand curves in Figure 2 indicate normalized force sensitivities to slip velocity (in the regime of slip between the peak of the curve and s = 100%) that range from -0.043 (units of F_x/F_z per km/h slip velocity) at V = 5 km/h to -0.005 units/Km/h at V = 88 km/h. Data taken on typical passenger car tires (ref. 11) indicate values of the same parameter ranging from -.031 units/km/h at V = 8 km/h to -.0001 units/km/h at 65 km/h and generally even turning positive, to +.0012 units/km/h at 97 km/h. Another means of comparing the slip velocity sensitivities of truck tires and passenger car tires is by the illustration of Table 2. Here, the so-called "peak-to-slide ratio," $(F_x/F_z)_{peak}/(F_x/F_z)_{s=100\%}$, is given for the bias-ply passenger car tire of reference 11, together with a common bias-ply truck tire as characterized on the HSRI mobile dynamometer. The notable feature of these data is that peak-to-slide ratios representing car tire behavior are decreasing over the velocity range while the corresponding truck tire numeric is increasing with velocity.

16. The F_x/F_z versus s format has been reduced in Figure 3 to a cross-plot of "peak and slide" numerics versus vertical load for a sample of six different truck tires configured with rib-type tread patterns. The sample provides a mixture of comparisons, including three tires of the common 10.00 x 20/F size but differing with respect to the manufacturer, a fourth tire which is the tubeless equivalent of the 10.00 x 20/F, a high load rating tire of conventional cross-section profile, and, finally, a high load rating tire of the so-called

Table 2

Peak/slide ratios as influenced by velocity for typical car and truck tires of bias-ply construction (both tires operated at their rated loads on dry asphalt surfaces).

	Velocities				
Tire	8	16	32	64	88
Passenger Car Tire	1.23	1.24	1.20	1.08	1.00
Heavy Truck Tire	1.20	1.31	1.46	1.64	1.62

"wide base single" variety. Although representing an admittedly small selection, these data serve to indicate a fact which faces the truck engineer who desires improved limit braking capability, namely, that "tire-ing up" in hopes of accruing improved traction benefits from tires of a higher-than-required load rating may not provide the desired improvement. Indeed, certain higher load rating tires may be less suitable for operation at light loads than tires which are rated for lower loads.

17. Taking a broad view, these data clearly indicate a peak force capability for truck tires on dry pavements which compares quite favorably with passenger car tire performance. Thus, if the truck tire is deserving of its reputation for poor traction quality, it can only be for dry pavement cases in which locked-wheel braking occurs, or as indicated by Dijks' measurements (ref. 2), on wetted surfaces for which both peak and slide performance is reported to be inferior.

18. A cross-plot of peak and slide values versus velocity is presented in Figure 4 for the six-tire sample. These data serve to further emphasize the influence of slip velocity, (sV), on longitudinal force as indicated by the marked contrast between the sensitivity of peak values and slide values. Also, it is interesting to note the dramatic low velocity increase in the slide numerics—a phenomenon which should, in part, explain the violent terminal behavior of heavy trucks during limit stopping.

19. As in the Figure 3 data, the H-rated tires are seen to produce a reduced level of normalized traction capability, although it should be recognized that each tire here was subjected to the velocity sensitivity tests while loaded to its rated value.

20. In the U.S., federal regulations requiring trucks to achieve stopping distances (with wheels unlocked) less than or equal to a specified distance have caused considerable concern because of their inability to effect a normalization of results to account for the prevailing test pavement's frictional potential. To assess the nature and severity of this problem, a limited set of tests was performed, yielding the peak and slide data shown in Figure 5. Samples of each of two tires were tested on each of four dry pavements comprising facilities which are employed in the stopping performance testing of heavy trucks. The velocity sensitivities of Figure 5 not only indicate as much as a 15% change in each tire's traction performance over the four selected pavements, but the complexity of the pavement's

©IMechE 1976

interaction with innate tire properties is also evidenced by the changing relationship between the two subject tires. Thus, it would appear that the (dry) longitudinal traction of truck tires is influenced by differences in the composition and wear state of dry pavements to an extent which is significant to the maximum stopping capability of heavy trucks.

21. To characterize the statistical repeatability of data describing longitudinal traction properties, standard deviations of the peak and slide values of F_x/F_z gathered from test runs repeating a reference load and velocity condition were calculated, showing typical σ's of .007 to .012. In tests conducted on four selected surfaces, one pavement showed σ's on peak traction values as high as .033. It was subsequently discovered that a pavement grinding operation had been performed over various, unmapped portions of the facility thereby explaining the relatively large variations in data. Indeed, data gathered to date appear to indicate that the repeatability of measurements of truck tire longitudinal traction depends more upon pavement uniformity than upon the stationarity of innate tire properties.

22. Taking this observation one step further, it also appears that no significant preliminary "working" of truck tire specimens is necessary to assure stable and representative longitudinal traction behavior on dry pavements. Indeed, it has been routinely observed that the new truck tire, with whatever degree of surface contamination as commonly results from final manufacture, delivers a longitudinal traction performance on its first (unprepared) lockup cycle which performance is statistically indistinguishable from its performance in later test cycles up to approximately 25% loss in tread depth. It is felt that an observation such as this is appropriate commentary in a field for which no standard test practices have yet been formulated.

MEASUREMENTS OF LATERAL TRACTION

23. The HSRI tractor-mounted dynamometer has been employed in a limited series of baseline experiments intended to examine the lateral traction properties of heavy truck tires on dry surfaces. These tests were conducted in a fashion similar to that described for longitudinal measurements, with a standard matrix of load and velocity conditions which is periodically interrupted for conduct of the reference condition run. Each test run is characterized by constant velocity and load as before, but instead of repeated lockup cycles, the tire is subjected to a "staircase" time history of slip angle, α, as provided by the servo steering system. Typically, the slip angle sequence includes "pauses" at the α values of -1°, 0°, +1°, +2°, +4°, 8°, 12°, 20°, followed by return to the 0° position. Between pauses the servo slews at a rate of 10 degrees per second. A pause duration of one second has been selected to assure the gathering of a sufficiently long data record to establish steady-state performance. Of course, it is recognized that the principal determinant of the dynamic transient is related to distance traveled rather than elapsed time but, even at the lowest test speeds examined, one second has been found an acceptable value.

24. The continuous time histories of F_z, F_x, M_z, α, and V are recorded on magnetic tape and later processed by computer to yield filtered and averaged response numerics at each value of α for which a pause was employed. All filtering techniques have incorporated no-phase-lag digital routines.

25. A typical set of (F_y/F_z versus α) plots are shown in Figure 6, illustrating the basic range of load sensitivity for three types of 10.00 x 20 tires. The data represent behavior typical of three radial ply, rib tread tires, four bias ply, rib tread tires, and four bias ply, lug tread tires selected from among the production of major U.S. manufacturers of truck tires. The data indicate the typically steeper and more abruptly-saturating side force behavior of the radial tire—a characteristic common to many radial passenger car tires as well. Surprisingly, the lug tread tires showed a higher force capability at lighter loads than did the rib tread selections.

26. Again, as with longitudinal characteristics, the truck tire compares favorably in side force capability to passenger car tires. An important characteristic which is indirectly evident here is the influence of load on cornering stiffness, C_α, defined as the slope of the F_y/α relationship through the origin:

$$C_\alpha = \left.\frac{\partial F_y}{\partial \alpha}\right|_{\alpha=0} \qquad (4)$$

27. As with the corollary longitudinal stiffness, C_s, the cornering stiffness of truck tires increases with vertical load as the longer tread contact length increases the net deflection experienced by tread elements as they pass through the contact patch. With many truck tires the rate of increase of C_α with load has been found to be so large over the entire load range as to assist markedly in maintaining the path curvature gain or understeer characteristics for trucks in both their empty and fully loaded conditions. This apparently fortuitous result derives from the fact that as a truck is loaded, and thus experiences a relocation of its center of mass, the front and rear tires undergo vertical load changes which beget a C_α redistribution tending to compensate for the mass center shift. This characteristic contrasts markedly with the sensitivity to loading which typifies most passenger cars as a result of the rather flat C_α versus F_z relationship exhibited by passenger car tires.

28. On balance, however, the truck tire's C_α sensitivity to load is only part of the truck yaw stability picture, and should not be viewed as an influence overriding all other mechanisms.

29. Although the mobile apparatus employed in these tests is more elaborate than what is required to measure the elastic properties of a tire, such as C_α, these measurements did confirm the insensitivity of tire cornering stiffness to velocity and pavement type. Indeed, the mobile machine has routinely correlated its C_α measurements within 10% of those made on the same tire using the HSRI flat-bed machine.

©IMechE 1976

30. The velocity sensitivity of the F_y/F_z versus α relationship is shown in Figure 7, as typifies the behavior of eleven radial and bias ply tires which were tested on a dry concrete surface. The relatively small influence of velocity on normalized side force, up to a slip angle of 20°, is comparable to the small influence of velocity on peak values of F_x/F_z which were shown earlier. This observation is explained by considering the relatively low absolute slip velocities experienced by tread elements when the tire generates a peak value of F_x/F_z (s ≅ 20%, or absolute velocity of the sliding elements ≅ .2V) and operates at 20° slip angle (a condition at which the velocity of sliding elements is V tan 20 = .36V).

31. With regard to the repeatability of lateral force measurements, examination of data taken during the repeated reference condition runs has revealed that a significant change in behavior can accompany the accrual of test-induced wear. Confirming findings made for passenger car tires (ref. 12), the data in Figure 8 show that the peak side force generated on the fifth repeat run (with a substantial degree of wear at the outermost rib) is significantly higher than that exhibited by the new tire. These data are typical of most tires in the sample. In addition to the peak force increase, an offset at zero is observed as a consequence of the "artificial camber" effect of an asymmetric wear pattern.

32. Having anticipated that degradation of truck tire samples might take place during testing, the load sensitivity experiments were conducted at 32 km/h and were performed first in the sequence, following the first reference condition run. Velocity sensitivity tests, on the other hand, obviously required that large amounts of abrasion be experienced at the needed higher velocities. Thus the velocity sensitivity data shown in Figure 7 represent a more worn sample with the 88 km/h data than with the 32 km/h data. Accordingly, the absolute magnitude of the F_y/F_z reduction with velocity may be a few percent larger than that indicated due to the accumulated influence of wear on the test sample.

33. While the tire wear phenomenon clearly poses severe constraints to any high speed measurements of lateral traction, the phenomenon also influences vehicle performance testing and, consequently, is receiving considerable research attention in the U.S.

CONCLUDING REMARKS

34. It must be emphasized that data presented herein represent limited investigations with small samples of tires on dry surfaces only. While it is felt that illustrations of "typical" behavior will hold up under further scrutiny as research in this area broadens, many facets of truck tire traction mechanics remain to be explored. In particular, of course, the need for an extensive examination of wet traction performance is elemental to safety-oriented research, and reference 2 constitutes a worthy starting point in this area.

35. Although existing dynamometers appear to be adequate for the task of measuring truck tire traction performance, a substantial period of experimental activity will be necessary before a thorough technology can be assembled to address this crucial component of the commercial vehicle.

ACKNOWLEDGEMENT

36. The HSRI mobile dynamometer apparatus was developed under support of a grant from the Motor Vehicle Manufacturers Association.

REFERENCES

1. BRADISSE, J.E., RAMSEY, A.F., and SACIA, S.R. Truck tire traction trailer. Society of Automotive Engineers Paper No. 741138, November 1974.

2. DIJKS, A. Wet skid resistance of car and truck tires. Tire Science and Technology, TSTCA, Vol. 2, No. 2, May 1974, pp. 102-116.

3. ERVIN, R.D. and FANCHER, P.S. Preliminary measurements of the longitudinal traction properties of truck tires. Society of Automotive Engineers Paper No. 741139, November 1974.

4. ERVIN, R.D., MACADAM, C.C., and FANCHER, P.S. The longitudinal traction characteristics of truck tires as measured on dry pavements. HSRI, Univ. of Michigan, Ann Arbor, Mich., Report No. UM-HSRI-PF-75-3, February 1975.

5. WILLIAMS, T., MEADES, J.K., and RILEY, B.S. The wet road grip of lorry tyres: A comparison of three types. Transport and Road Research Laboratory Report LR 544, Crowthorne, Berkshire, England, 1973.

6. TIELKING, J.T., FANCHER, P.S., and WILD, R.E. Mechanical properties of truck tires. Society of Automotive Engineers Paper No. 730183, January 1973.

7. FREUDENSTEIN, G. Pneumatic tires when running at an angle or on curves. Deutsche Kraftfahrforschung und Strass enverkehrstechnik, Heft 152, 1961.

8. GINN, J.L. and MARLOWE, R.L. Road contact forces of truck tires as measured in the laboratory. SAE Transactions, Vol. 76 (1967), Paper No. 670493.

9. BIRD, K.D. and SCHURING, D.J. Truck tire testing on TIRF. Proceedings of Symposium on Commercial Vehicle Braking and Handling, Univ. of Michigan, May 1975.

10. DUGOFF, H., FANCHER, P.S., and SEGEL, L. An analysis of tire traction properties and their influence on vehicle dynamic performance. 1970 International Automobile Safety Conference Compendium, Society of Automotive Engineers Paper No. 700377, 1970.

11. HARNED, J.L., JOHNSTON, L.E., and SCHARPF, G. Measurement of tire brake force characteristics as related to wheel slip (anti-lock) control system design. General Motors Engineering Publication No. 3845, November 20, 1968.

12. ERVIN, R.D., GROTE, P., FANCHER, P.S. Vehicle handling performance. Final Report, U.S. Dept. of Transportation Contract DOT-HS-031-1-159, HSRI, Univ. of Michigan, Ann Arbor, Mich., 1972.

©IMechE 1976

Fig. 1: HSRI mobile traction dynamometer for truck tires

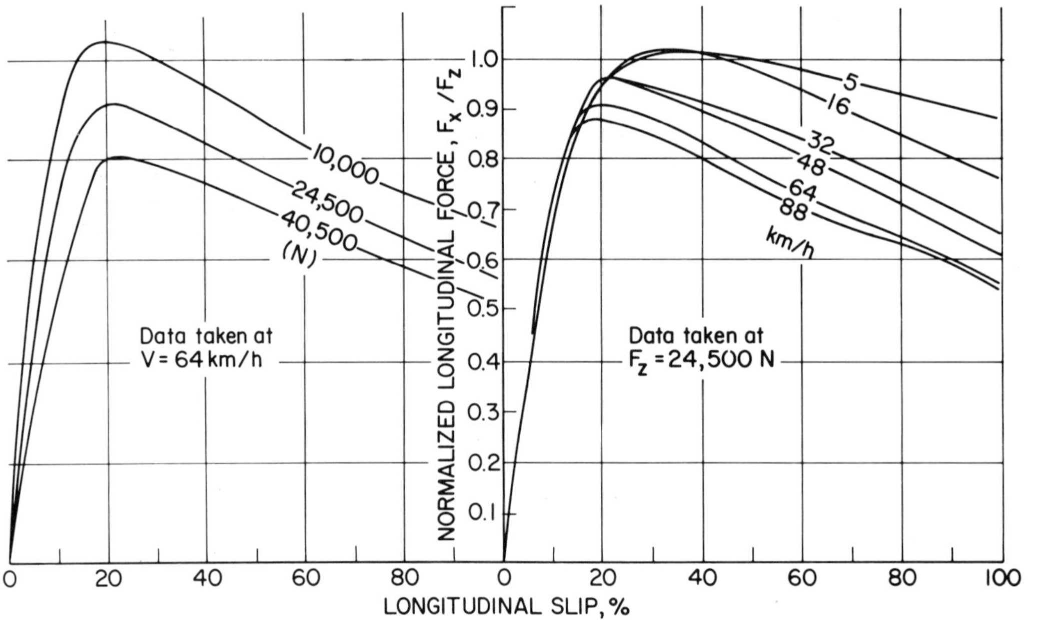

Fig. 2: Typical load and velocity influences on the F_x/F_z versus slip behavior of a 10.00 x 20/F tire, as measured on a dry asphalt surface.

©IMechE 1976

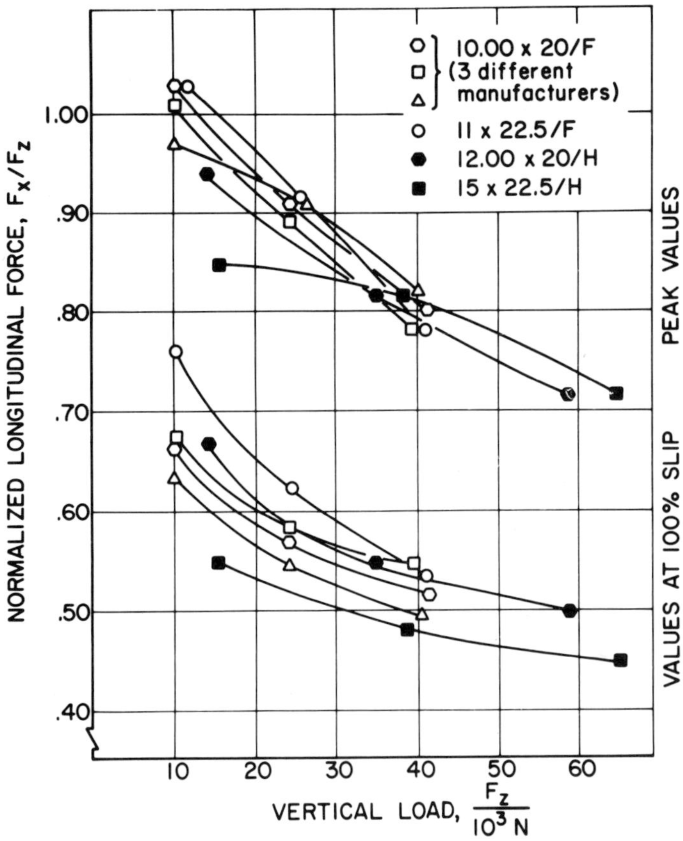

Fig. 3: Load sensitivity in the peak and slide traction of a six-tire sample on dry asphalt. All tests run at 64 km/h.

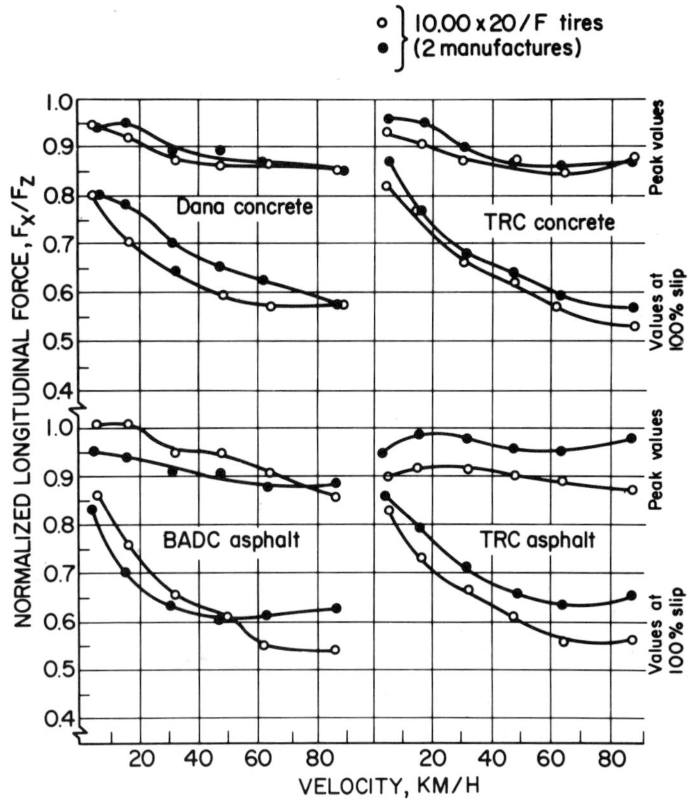

Fig. 5: The differing influence of pavement surface on the velocity sensitivities of two tires.

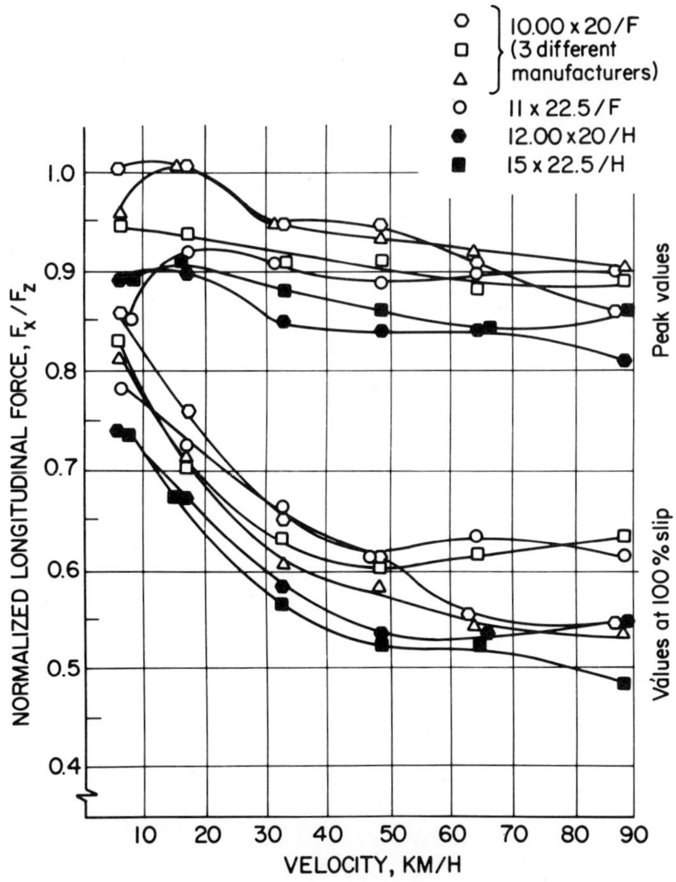

Fig. 4: Velocity sensitivity of the peak and slide traction values for a six-tire sample on dry asphalt. All tires operated at their respective T & RA rated load.

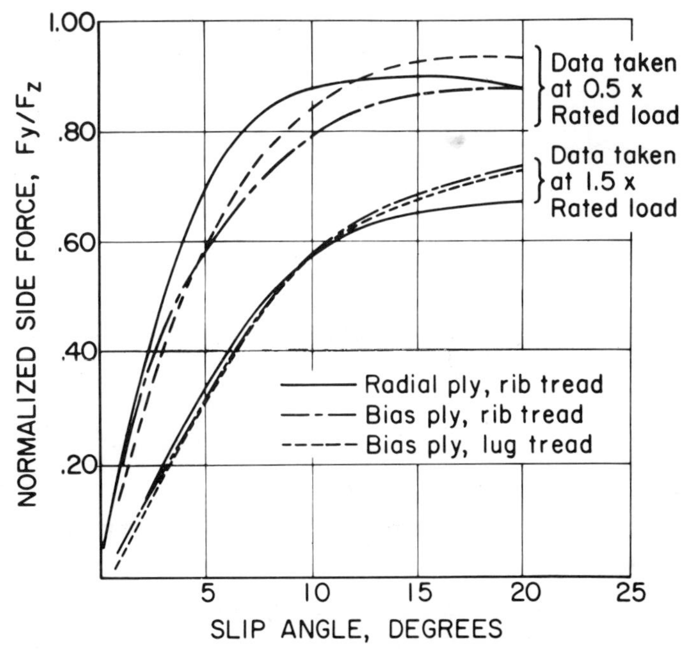

Fig. 6: Typical load sensitivities in the side force response of a sample of 10.00 x 20 tires tested at 32 km/h on a dry concrete surface.

©IMechE 1976

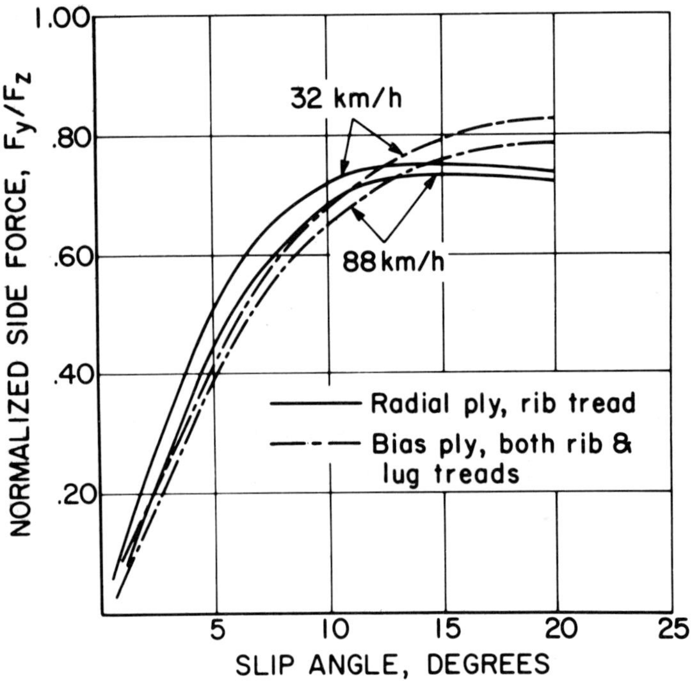

Fig. 7: Typical velocity sensitivities in the side force response of a sample of 10.00 x 20 tires tested at the T & RA rated load on a dry concrete surface.

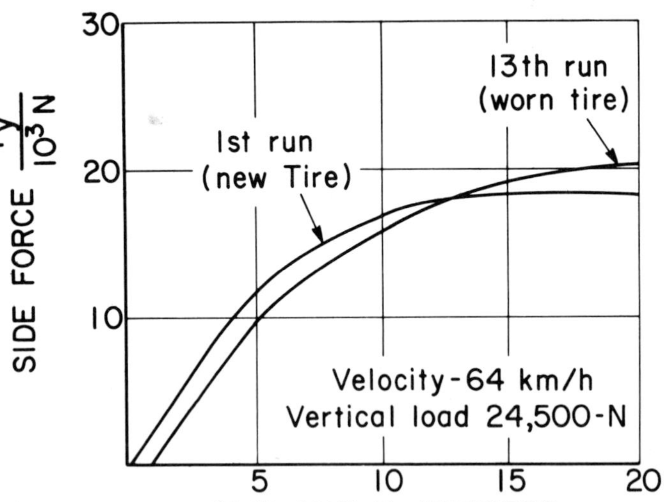

Fig. 8: Changing side force behavior of a 10.00 x 20 tire as a result of test-induced wear.

©IMechE 1976

C31/76

EVALUATION OF THE BRAKING PERFORMANCE OF PASSENGER CARS WITH FIXED AND WITH 'KINKED' BRAKE-FORCE DISTRIBUTIONS

Dr. -Ing. M. BURCKHARDT, Test Department, Passenger Cars, Daimler-Benz AG, Stuttgart
Dr. -Ing. E. C. GLASNER VON OSTENWALL, Test Department, Commercial Vehicles, Daimler-Benz AG, Stuttgart

The MS of this paper was received at the Institution on 14 April 1975 and accepted for publication on 31 July 1975

Preface

The investigation of the braking performance of a motor vehicle calls for the establishment of a basis of evaluation which permits the discussion of all brake systems and brake-force distributions in a single diagram. This requirement is met by the brake-force diagram. The subsequent paper describes all necessary relations and deductions material for establishing the brake-force diagram [1, 2, 4, 27, 32, 36] .

The investigation of a number of papers [1, 2, ... 23] leaves the impression that vehicles which are equipped with brake-force proportioning devices at their rear axles will in any event be capable of avoiding rear-axle lock. This antilock feature is allegedly due to their superior utilization of the available coefficient of friction.

This paper endeavours to clarify the advantages as well as the shortcomings of "kinked" brake-force distribution of a brake-pressure dependent change-over point in contrast to the fixed-distribution type [33] . The above brake-force distribution diagram serves as the basis for evaluation.

Notation

a	Retardation, i.e. the ratio of deceleration to gravitational constant
$\bar{a}$	Specific lateral acceleration, i.e. the ratio of lateral acceleration to gravitational constant
a^*	Ratio of the sum of all brake-forces to vehicle weight
B	Brake-forces causing vehicle deceleration
B^*	Brake-forces produced by the brake system
B_M	Specific engine brake force, rear axle
f	Coefficient of friction, tyre-to-road
G	Weight force (vehicle weight)
G'	Radial force at wheel or axle (dynamic axle load or wheel load)
G_N	Power-to-weight ratio of a vehicle
i_v	Pressure reduction ratio, brake-force proportioning device
p_s	Brake pressure at change-over point of proportioning device
R	Resulting maximum friction force
S	Lateral force
χ	Relative height of centre of gravity, i.e. the ratio of height of centre of gravity to wheelbase
ψ	Rear-axle load share, i.e. the ratio of rear-axle load to vehicle weight
Φ	Share of rear-axle brake force, i.e. the ratio of rear-axle brake force to total brake force

Subscripts

VA	Front axle
HA	Rear axle

©IMechE 1976

1 Introduction

It is the tyre which is the element transferring the brake force onto the road. According to Coulomb's law, the transferable brake force equals the dynamic load of axle or wheel muliplied by the coefficient of friction tyre-to-road. Its magnitude depends on the quality of the tyre as well as on the road surface. When designing a brake system, therefore, the influence of a variation of the coefficient of friction onto the braking performance of a vehicle has to be investigated. The effect of such a variation can be discussed by means of the brake-force distribution diagram [36] . This diagram is derived subsequently. This is not place for a complete derivation of the mathematical relations, the final equations being given instead. A comprehensive deduction of the pertinent equations can be obtained from a study of papers [32] and [36] .

2 Structure of the brake-force distribution diagram

2.1 Parabola of the so-called "ideal brake-force distribution" 1)

Considering the forces acting upon a braked, two-axle vehicle, figure 1, the condition of the dynamic balance in the direction of travel is

$$\frac{B_{VA}}{G} + \frac{B_{HA}}{G} = a \qquad (1)$$

From a consideration of the moments, the dynamic axle-load distribution is for the front axle

$$\frac{G'_{VA}}{G} = \frac{l - l_{VA}}{l} + a\frac{h}{l} \qquad (2)$$

and similarly for the rear axle

$$\frac{G'_{HA}}{G} = \frac{l_{VA}}{l} - a\frac{h}{l} \qquad (3)$$

Using the terms "relative height of the centre of gravity χ " and "rear axle load share ψ ", defined above, and taking into consideration that in the ideal case of an identical utilization of the brake force at front and rear axles, the coefficient of friction tyre-to-road will be equal to the maximum obtainable retardation 2), then the referred brake force at the front axle is

$$\frac{B_{VA}}{G} = a\left(1 - \psi + a\chi\right) \qquad (4)$$

and at the rear axle

$$\frac{B_{HA}}{G} = a\left(\psi - a\chi\right) \qquad (5)$$

Eliminating the parameter"a"from these equations means:

$$\frac{B_{HA}}{G} = \sqrt{\left(\frac{1-\psi}{2\chi}\right)^2 + \frac{1}{\chi}\cdot\frac{B_{VA}}{G}} - \frac{1-\psi}{2\chi} - \frac{B_{VA}}{G} \qquad (6)$$

As can be seen, this is a parabola. It i called the "parabola of the ideal brake-force distribution"1) or more correctly "parabola of the dynamic brake-force distribution". This curve has already been mentioned in many papers [1, 2, 4, 5, 8, 9 - 14, 17, 18, 22, 27, 32, 36, 38

An illustration is shown in figure 2. Th shape of the curve is determined by the position of the centre of gravity, that is by χ and ψ. It is of significance only in the first and third quadrants an passes through the origin of coordinates

For additional statements, the lines of constant retardation and of constant coefficients of friction of front and rear axles as well as the installed brak force distribution still have to be dete mined.

2.2 Curves of constant retardation

The lines of constant retardation in equation (1) are straight lines which are inclined by -45 degrees.

2.3 Curves of constant coefficient of friction tyre-to-road

Finally, an indication of the conditions existing between tyre and road, represented by the lines of constant coefficient of friction tyre-to-road, are stil required. As axle load and brake-force are linearly coupled to each other these have to be straight lines.

The lines of constant coefficient of friction tyre-to-road are derived, by analogy, to [32] and [36] .
The equation of the front axle reads

$$\frac{B_{HA}}{G} = \frac{1 - f\chi}{f\chi}\left(\frac{B_{VA}}{G}\right) - \frac{1-\psi}{\chi} \qquad (7)$$

and the one of the rear axle

$$\frac{B_{HA}}{G} = -\frac{f\chi}{1 - f\chi}\left(\frac{B_{VA}}{G} - \frac{\psi}{\chi}\right) \qquad (8)$$

2.4 Installed brake-force distribution

For characterizing a vehicle, one also needs to know certain "interior parameters". This is not the place to explai their fixation during the design phase but their effect only is considered here

---

1) The term "ideal brake-force distribution" quoted in many papers frequently causes misunderstandings. This curve is the border line between the stable and instable ranges of braking when the coefficient of friction tyre-to-road is fully utilized at the pertinent axle. The brake-force distribution installed into the vehicle, therefore, should never exceed this line in the appropriate range of friction having regard to all disturbing factors occuring during operation.

2) This is valid only for a rectilinear stop on a level ground and when neglecting all exterior forces.

©IMechE 1976

When rating a brake system, use is made either of the so-called fixed distribution - which means that there is a linear relation between the braking moment acting upon the front and rear axles - or brake-force proportioning devices are installed which vary the pressure ratio of front axle to rear axle as a function of the brake pressure, of the vehicle's retardation, its dynamic axle load or of some other auxiliary element. In such a case, the parabola of the dynamic brake-force distribution in the first quadrant is more or less perfectly approximated by sections of straight lines. In both instances, this is called installed brake-force distribution. As regards the fixed distribution, it reads in parametric form:

$$\frac{B_{VA}}{G} = a(1-\Phi) \qquad (9)$$

$$\frac{B_{HA}}{G} = a \cdot \Phi \qquad (10)$$

For the working procedure, shown in section 3, a fixed distribution is assumed for the brake-force distribution diagram in order to simplify the diagram facilitating its understanding.

3. Evaluation of the brake-force distribution diagram [32] , [36]

As a rule, the fixed distribution of brake systems is so rated that the parabola of the dynamic brake-force distribution is intersected by the straight line of the installed brake-force distribution in the range from a = 0.9 to 1.0. In figure 3 that part of first quadrant, which is of practical interest and can be realized, is shown to a larger scale.

In the range above the parabola of the dynamic brake-force distribution, the rear axle locks when fully utilizing the pertinent coefficients of friction tyre-to-road (as, e.g., fixed distribution 2, point B). In the range below, the front axle will lock first (as, e.g., fixed distribution 1, point A). Locking of the wheels occurs whenever the straight line of the installed brake-force distribution intersects the correlated, valid straight line of the coefficient of friction. The straight line of constant retardation, passing this point, represents a measure for the vehicle retardation obtained.

When the driver of a vehicle - equipped with distribution 1, the front axle of which locks first (point A) - depresses the brake pedal further, the brake-force distribution moves along the line of constant coefficient of friction (tyre-to-road) until the parabola of the dynamic brake-force distribution is reached (point C). In this point, the second axle, too, locks. In this instance the maximum retardation, physically obtainable under the given conditions, occurs [3).]

The performance of the vehicle with failure of the brake force at one axle is easily discussed. When, for instance, the brake force at the rear axle fails, the installed brake-force distribution moves along the abscissa only. The front axle thus locks at the intersection of the straight line f_{VA} = const. with the axis of the abscissa (point F). In case of failure of the front-axle brake force, the installed brake-force distribution analogously moves along the ordinate only. The rear axle, therefore, locks at the intersection of the straight line f_{HA}= const. with the ordinate axis (point G).

It has been shown above that the brake performance of a vehicle can be clearly outlined in the so-called brake-force diagram. Above all, this diagram makes possible the clear representation of the stability performance and of the obtainable retardations. It thus permits a comprehensive evaluation of the brake system of a vehicle.

4. Brake-dynamical analysis of vehicles with fixed and with "kinked" brake-force distribution [34, 36]

4.1 General remarks

The following investigations refer to so-called conventional vehicles, that is with engine at the front and propulsion at the rear. Other vehicle constructions, such as front-wheel drive, result in different aspects. For engine outputs above 100 kW, the conventional construction has the best overall properties. With smaller engine output, the vehicle with front-wheel drive might possibly be more at advantage [29] .

The most important criterion for rating of a brake system is the braking stability [26]. In a simplified form, figure 4 shows that a vehicle with locked front axle is in a stable driving condition because a yawing reaction of the vehicle, produced by disturbing forces, is compensated by the counterclockwise rotation of the lateral forces at the rear axle. When the rear axle locks, the yaw response around the vertical axis of the vehicle is amplified by the moment of the lateral forces at the front axle since the rear axle - in accordance with Kamm's circle of friction [24], figure 5 - can no longer build up lateral guidance. The vehicle begins to swerve.

[3)] The above is valid only when assuming that during lock of wheels the coefficient of friction tyre-to-road will not change or only slightly. In order to arrive at easily foreseeable conditions, the assumption of a constant maximum coefficient of friction tyre-to-road is nevertheless maintained. The resultant error is of no significance for the design of brake.

©IMechE 1976

When, on the other side, the maximum physically transferable brake force at the rear axle is utilized to an insufficient degree for the vehicle under consideration, this will result in an excessive stopping distance. The rating of a brake system represents, therefore, at all times a compromise between stopping distance and driving stability during the braking manoeuvre. The manner of actuation of the brake by the driver is of particular importance. A multitude of tests has shown that the average driver is not in a position to actuate the brake so that the optimum retardation is obtained at once. During an average braking, one slowly moves towards the optimum brake pressure which is determined by the maximum utilizable coefficient of friction between tyre and road. The panic stop, in which case the brake is abruptly actuated too strongly, is not the standard case. It equally does not represent the most critical condition since in this instance - and this will be particularly true in the case of vehicles with brake-force boosters and thus relatively low pedal forces - all four wheels lock simultaneously. The vehicle then continues its way without steering control, it is true, but nevertheless straight forward on a level road [33, 34, 35] .

A brake application with a slow increase of the pressure - so-called progressive partial braking - is much more critical. With locked rear axle, the vehicle has sufficient time to rotate in an uncontrollable manner until all four wheels lock by the continuing increase in brake-pressure. This is particularly true for vehicles of a short wheel base and small moment of inertia around the vertical axis [33, 34, 35].

From the above, it becomes evident that the loss of driving stability during braking is considerably more aggravating than a possibly slightly longer stopping distance. Particularly in the range of medium coefficients of friction, that means on a wet road, a sufficiently large reserve of stability is essential [26, 38] .

4.2 Definition of the braking system under considerations

The following considerations are based on two installed brake-force distributions:

(1) a fixed distribution with an intersection of the parabola of the dynamic brake-force distribution at $a_{crit.} = 0.953$ (brake system No. 1 in figure 6);

(2) a "kinked" distribution with a proportioning device of a change-over pressure of 34 bar and a pressure-reduction ratio of $i_v = 0.41$. There is an identical value for $a_{crit.} = 0.953$ (brake system No. 2 in figure 6).

Both distributions are allegedly rated stably as in both instances first the front axle locks for a coefficient of friction of $f = 0.8$, obtaining retardations of $a_1 = 0.76$ for the fixed-distribution system and of $a_2 = 0.785$ with a proportioning device.

Depressing the brake pedal further, the actual distribution moves along line $f_{VA} = 0.8$ up to the intersection with the parabola of the dynamic brake-force distribution. The then resulting maximum retardation to be obtained amounts to $a = 0.8$ in both cases.

4.3 Braking on a level ground under consideration of the engine-braking moment and of the inertia moments

The clear indications of figure 6 apply to stopping on a level ground with the engine disengaged. High-power vehicles which are increasingly equipped with automatic transmissions, however, require amendments to the diagram. While it might be assumed that the driver of a vehicle with manual transmission disengages the clutch during a critical braking manoeuvre, the intervention of the driver cannot be taken for granted in the case of vehicles with automatic transmissions. Thus, it becomes necessary to analyze the influence of the rotating masses on the brake-force distribution. The same will hold true for a vehicle with manual transmission the driver of which, however, does not disengage the clutch. The result is shown in figure 7. The fixed distribution is considered first. The brake force of the engine at the rear wheels, transformed into the brake-force distribution diagram, yields a curve the shape and position of which depends on the driving speed and on the selected gear. The influence of the engine is greater, the higher the speed, the lower the gear and the lower the coefficient of friction. The deviation from the installed brake-force distribution is low for high retardations, increasing, however, considerably toward the smaller retardations.

The displacement of the installed brake-force distribution leads to a reduction of $a_{crit.}$ to 0.848 for a fixed distribution (figure 7, bottom). The front axle locks at $f = 0.8$ for a retardation of $a_2 = 0.79$. The retardation is in this case slightly higher than without the engine-braking moment. Below $a_{crit.} = 0.254$ -corresponding to an equally large coefficient of friction between tyre and road - the vehicle brakes in an unstable manner. Between $a_{21} = 0.848$ and $a_{22} = 0.254$, however, the dynamic brake-force distribution is well approximated.

© IMechE 1976

For a vehicle with a proportioning device (figure 7, top), it reveals that it has become unstable in the entire operating range. With a coefficient of friction of f = 0.8 and for a retardation of a_2 = 0.76, the rear axle locks first and the vehicle begins to swerve. Summing up, one could say that at first sight the fixed distribution appears to be markedly inferior when compared to the "kinked" brake-force distribution. This holds true for a straightforward stop in the medium range of coefficients of friction with a disengaged engine. Closer examination, however, reveals it to be definitely superior under practical conditions, that is with an engaged engine [33, 38] .

4.4 Braking on positive and negative grades

Any motor vehicle must be capable of being stopped not only on level ground but on positive and negative gradients as well. The latter case deserves no special attention when proper rating for the level ground can be assumed since stability of the braking manoeuvre is improved by the downhill driving force when compared to the level plane. Figure 8 shows the effect of upgrades and downgrades on the dynamic brake-force distribution. Above all, one should differentiate between the terms "a*" and "a".

"a*" indicates the fraction of the weight acting as a brake force at one axle or at the vehicle whereas "a" expresses the actual retardation, that is the deceleration with reference to the terrestrial gravity. The values of "a*" and "a" are equal as long as only the brake forces are effective which act upon the vehicle and are transferred by its wheels. Aerodynamic forces or a downhill drive change "a" in comparison to "a*".

The influence of a negative grade is suitably represented by an upward displacement of the origin of coordinates for the retardation along the dynamic brake-force distribution.

For showing the influence of a positive grade, the origin of the coordinate system is downward displaced on the parabola. The stability of the vehicle is improved for a given "a" by increasing the rear-axle load share ψ .

Figure 9 shows a braking manoeuvre on a negative slope of 15 degrees together with the influence exerted by the engine braking moment (for a given speed). It reveals that a vehicle with proportioning device (figure 9, top)and a coefficient of friction f = 0.8, a retardation of a_2 = 0.58 will result in which case the rear axle locks first. The vehicle with a fixed distribution (figure 9, bottom) is stable in this instance and permits a maximum retardation of a_2 = 0.627.

4.5 Braking in a curve

Besides a straightforward downhill braking taking into consideration the influence of the engine - which, as has been shown, substantially deviates from the straightforward braking on a level ground - as the only one considered customarily, there is no doubt that todays traffic attributes extraordinary significance to braking in a curve [38]. It often happens on motorways and similar roads that the driver is compelled to apply the brakes although his vehicle will be subject to material lateral accelerations due to the simultaneously ongoing ride in a curve. The subsequent numerical data may serve as an illustration:

The radius of the curve being 100 mtrs, the lateral acceleration for a speed of 80 km per hr amounts already to $\bar{a}$ = 0.5. For a speed of 180 km per hr, a lateral acceleration of $\bar{a}$ = 0.25 results already from a radius of 1.000 mtrs. For a given, installed brake-force distribution and a definite coefficient of friction tyre-to-road, the locking point of the critical wheel can be found by equivalent means as for the downhill braking, that is via the brake-force distribution diagram.

Figure 10 shows the brake performance of both vehicles for a lateral acceleration of $\bar{a}$ = 0.25. While the installed fixed distribution (figure 10, bottom) remains stable in the range under consideration, the brake system with a proportioning device (figure 10, top) is unstable by reason of its locking potential in regard to the locking of the rear axle in the applicable range of friction. For a maximum coefficient of friction of f = 0.9, the vehicle with fixed distribution comes to lock of the front wheel at the inner side of the curve for a retardation of a = 0,756. The brake performance is stable. The vehicle with the "kinked" brake-force distribution however, initiates swerving already for a retardation of a = 0.74.

In figure 11, the obtainable retardation of the vehicle up to locking of the most critical wheel or axle has been plotted as a function of the lateral acceleration. Diagram No. 1 represents the vehicle with fixed distribution and diagram No. 2 the vehicle with a proportioning device of the characteristic described in figure 6. It can be taken from the diagram that the vehicle with a fixed distribution is capable of achieving rather large retardations even for relatively high lateral accelerations without losing control whereas the vehicle with a proportioning device already becomes unstable for a materially lower retardation.

The diagram also contains the so-called critical points at which the transition

©IMechE 1976

from locking of the front wheel to locking of the rear wheel takes place. It is evident that the vehicle with a proportioning device becomes unstable already above a lateral acceleration of $\bar{a} = 0.175$ at the rear axle whereas this condition does not occur in the vehicle with fixed distribution prior to above $\bar{a} = 0.325$. Much more significant for the average driver is the fact that for a lateral acceleration of $\bar{a} = 0.5$ which occurs already for rather large curve radii when driving fast, the vehicle with a proportioning device swerves already for a retardation of $a_2 = 0.37$ at the rear axle. The vehicle with the described fixed distribution, however, still permits a retardation of $a_1 = 0.46$. The vehicle with the fixed distribution thus displays a markedly low rotational effect during braking and has a stability gain of 24 percent in contrast to the vehicle with "kinked" brake-force distribution.

4.6 Influence of fluctuations of the coefficient of friction of brake linings

Unfortunately, the described theoretical regularities of the brake-force distribution are hard to achieve in large-series production. In the field, the actual brake-force distribution deviates from the installed brake-force distribution. This is due to the brake linings. The above considerations are based altogether on the assumption that the coefficient of friction between disc and brake pad corresponds under all circumstances to the mean value used for calculation or that a variation of the coefficient of friction at the front axle is followed by a corresponding change of the coefficient of friction at the rear axle. This assumption is not valid at all times [33, 34] .

The expert knows that the coefficient of friction of a given lining is subject to a considerable scatter ($\pm$ 10 percent minimum, figure 12) even when the exterior conditions are kept constant. In figure 13, the installed brake-force distribution of a vehicle is influenced negatively by the fluctuations of the coefficient of friction of the linings. While instability can still be avoided with the vehicle of the fixed distribution (figure 13, bottom) by reason of the larger reserve in comparison to locking of the rear axle (front axle locks for $a_2 = 0.79$), the vehicle with a proportioning device (figure 13, top) starts swerving for a retardation of $a_2 = 0.785$ because the fluctuations of the coefficient of friction cannot be compensated by a brake-force distribution device.
The above considerations emphasize the need for fitting of proper brake linings which are adapted to the vehicle. The problems of brake linings have been dealt with in a separate paper [34] .

5. Utilization of the maximum coefficient of friction tyre-to-road

On a theoretical basis, a term has been established called "Braking Efficiency" [23] . Attempts are even being made to introduce it into the Directives of the European Community on brake systems [39]. "Braking Efficiency" is the ratio of the retardation obtainable up to locking of a wheel to the maximum coefficient of friction tyre-to-road [3, 5, ... 37] .

Comparing the above results to it, the conclusion is reached that the vehicle with a fixed distribution has a worse "Braking Efficiency" than the vehicle with a "kinked" distribution. Vice versa, the vehicle with the higher "Braking Efficiency" displays a substantially worse brake performance.

For this reason, it might be considered whether there would be any meaning in defining "Braking Efficiency" legally, as is demanded today by some organizations. This is doubtful since otherwise the vehicle manufacturer might be compelled to design vehicles against his better judgement the brake performance of which - for instance, due to fluctuations in the coefficient of friction, on downgrades with engaged engine and in a curve - would be substantially worse than necessary for meeting legal specifications only, based on the theoretical considerations which take into account only a straightforward braking on a level ground with disengaged engine. In addition, one should keep in mind that the retardation obtained by a locking axle is not necessarily be identical with the maximum retardation physically obtainable. It is for these reasons that the "Braking Efficiency" cannot represent a measure for evaluating the performance of a brake system [23, 36, 37] .

6. Summary

Basing on the arguments presented, it can be concluded that brake-force proportioning devices are not safe methods of preventing the locking of the rear axle. The approximation of the dynamic brake-force distribution by an installed, "kinked" brake-force distribution under simultaneously higher utilization of the coefficient of friction - called here an advantage - is seen to be occasionally a serious disadvantage when closely examined, in particular when discussing situations which deviate from the simple consideration of a straightforward braking on level ground [33] . Rating a brake system solely in accordance with the factors valid for straightforward stops can therefore no longer be defended [36] .

References

[1] Tönnies, C.: Die Blockier- und Idealverteilung der größten Bremskräfte eines Fahrzeuges (The Locking and Ideal Distributions of the Maximum Brake Force of a Vehicle); ATZ 57 (1958) 8, p.235 - 237

[2] Tönnies, C.: Das Varia-Prinzip für die Bremsung von Kraftfahrzeugen (The "Varia" Principle for the Braking of Motor Vehicles); ATZ 60 (1958) 4, p. 99 - 105

[3] Rouse, J.A.: The Distribution of Braking on Road Vehicles; IME, Symposium on Control of Vehicles, 1963, p. 80 - 90

[4] Oberthür, H. and Baum, F.: Bremskraftverteilung für hydraulisch betätigte Fahrzeugbremsen (Brake-Force Distribution for Hydraulically Actuated Vehicle Brakes); ATZ 66 (1964) 8, p. 222 - 225

[5] Strien, H.: Bremskraftverteilung bei Personenwagen (Brake-Force Distribution of Passenger Cars); ATZ 67 (1965) 8, p. 240 - 245

[6] Beuchle, F.: Moderne Bremsen in Personenwagen (Modern Brakes in Passenger Cars); ATE Special from Automobil Revue, Bern, Dec. 1966, No. 51

[7] Mitschke, M.: Kurshaltung und Kraftschlußbeanspruchung während des Bremsens (Directional Stability and Utilization of Tyre-to-Road Coefficient of Friction During Braking); ATZ 69 (1967) 3, p. 73 - 77

[8] Strien, H.: Zweikreis-Bremsanlagen (Dual-Circuit Brake Systems); ATZ 70 (1968) 10, p. 309 - 343

[9] Vallin, E.G.: Legislative Effect on Braking Design; SAE-Paper No. 680 017, 1968

[10] Vallin, E.G.: Bremsen und Fahrsicherheit (Brakes and Driving Safety); Lecture at the 8th Technical Conference on Motor Vehicles in September 1968 at Karl-Marx-Stadt

[11] Strien, H.: Bremsenentwicklung (Development of Brake Systems); Lecture on Occasion of the Traffic Courts' Meeting at Goslar, 1970

[12] Puleo, G.: Automatic Brake Proportioning Devices; SAE-Paper No. 700 375, 1970

[13] Limpert, R. and Warner, C.W.: Proportional Braking of Solid Frame Vehicles; SAE-Paper No. 710 047, 1971

[14] Mitschke, M.: Dynamik der Kraftfahrzeuge (Motor-Vehicle Dynamics); Springer Verlag, Berlin, Heidelberg, New York, 1972

[15] Strien, H.: Bremswegbeeinflussende Faktoren beim Kraftfahrzeug (Motor-Vehicle Factors Influencing the Stopping Distance); ATZ 64 (1962) 7, p. 208 - 210

[16] Rompe, K.: Ausfall eines Bremskreises während einer Vollbremsung (Failure of One Brake Circuit During a Panic Stop); ATZ 73 (1971) 1, p. 12 - 16

[17] Mitschke, M. and Bisimis, E.: Ausfall von Zweikreisbremsanlagen (Failure of Dual-Circuit Brake Systems); Deutsche Kraftfahrtforschung und Straßenverkehrstechnik, VDI-Verlag, 1971, Part 211

[18] Bisimis, E.: Bremskreisausfall bei Zweikreisbremsanlagen (Brake-Circuit Failure of Dual-Circuit Brake Systems); ATZ 73 (1971) 6, p. 197 - 204

[19] Ingram, B. and Oppenheimer, P.: Safer Braking Systems; IME, Automobile Division, 1973, p.87 - 97

[20] Troost, A.: Zur Dynamik der Fahrzeugbremsung (Dynamics of Vehicle Braking); ATZ 56 (1954) 1, p.1-5

[21] Murphy, R.W.: A Procedure for Evaluating Vehicle Braking Performance; HSRI-Report 71/127 Oct. 1971

©IMechE 1976

[22] Buschmann, H. and Koeßler, P.: Taschenbuch für den Kfz-Ingenieur (Pocket Book for the Motor-Vehicle Engineer); Deutsche Verlagsanstalt, Stuttgart, 8th Edition, 1973

[23] Klaue, H.: Bremswerkuntersuchungen am Kraftfahrzeug (Studies on Brake Systems of Motor Vehicles); Deutsche Kraftfahrzeugforschung, 1937, part 13

[24] Kamm, W.: Kraftwagen und Straße in ihrer Wechselwirkung (Motor Vehicle and Road in their Interaction); VDI-Verlag, Berlin, Motor Vehicle Conference, 1934

[25] Wagenführer, H.: Reibwert und Abrieb bei Festkörperreibung (Coefficient of Friction and Wear of Solid State Friction); ATZ 70 (1968) 10, p.344-350

[26] Essers, U. and Hörz, E.: Statistische Auswertung von Unfällen, die mit Bremsen bzw. Gasgeben, Versagen in der Bremsbetätigung und Schleudern infolge fahrwerksbedingter Instabilität in Zusammenhang stehen (Statistical Evaluation of Accidents Connected with Brake Application, Acceleration, Failure of Brake Actuating Systems and Skidding due to Instability of Chassis); Forschungsinstitut für Kraftfahrwesen und Fahrzeugmotoren, Universität Stuttgart, Report No. 29/1973

[27] Burckhardt, M. and Kruse, W.: Beiträge zur Frage der Zweikreisbremse (Contributions to the Problems of the Dual-Circuit Brake System); ATZ 71 (1969) 7, p.231-238

[28] Glasner von Ostenwall, E.C.: How Good is the Braking Efficiency as a Criterion for Evaluation of a Brake System? NHTSA, Report on the Fourth International Technical Conference on Experimental Safety Vehicles, Kyoto, Japon, March 1973

[29] Burckhardt, M.: Der Einfluß der Straßenführung (Steigung, Gefälle, Kurven), des Motorbremsmomentes und der Drehmasse auf die Bremskraftverteilung und die Bremskreisaufteilung (Influence of Road Characteristics (Slope, Inclination, Bends), of Engine Braking Moment and Rotating Mass onto Brake-Force Distribution and on Brake-Circuit Splitting); Lecture on the Occasion of VDI/ATG Meeting, Stuttgart, March 1973

[30] Glasner von Ostenwall, E.C.: Consequences for Two-Wheel Dual Brake Systems Due to the Intensification on the Safety Regulations in the USA; Publication of Daimler-Benz AG, Stuttgart 1972

[31] Glasner von Ostenwall, E.C.: How Good is the Braking Efficiency as a Criterion for Evaluation of a Brake System? Publication of Daimler-Benz AG, Stuttgart 1972

[32] Burckhardt, M. and Glasner von Ostenwall, E.C.: Beitrag zur Beurteilung des Beschleunigungs- und Bremsverhaltens eines Kraftfahrzeugs (Contribution to the Evaluation of Acceleration and Braking Performances of a Motor Vehicle); ATZ 76 (1974) 4, p. 103 - 107

[33] Burckhardt, M. and Glasner von Ostenwall, E.C.: Bremsdynamische Untersuchung von Fahrzeugen mit fester und abgeknickter Bremskraftverteilung (Study on Braking Dynamics of Vehicles with Fixed and Kinked Brake-Force Distribution); ATZ 76 (1974) 7, p. 230 - 234

[34] Burckhardt, M. and Glasner von Ostenwall, E.C. and Näumann, E.: Der Bremsbelag - ein wichtiges Konstruktionselement für das Kfz. (The Brake Lining - An Important Design Element of the Motor Vehicle); ATZ 76 (1974) 11, p. 357 - 365

©IMechE 1976

[35] Burckhardt, M. and Glasner von Ostenwall, E.C. and Krohn, H.: Möglichkeiten und Grenzen von Antiblockiersystemen (Capabilities and Limits of Anti-Lock Systems); ATZ 77 (1975) 1, p. 13 - 18

[36] Glasner von Ostenwall, E.C.: Beitrag zur Auslegung von Kraftfahrzeugbremsanlagen (Contribution to the Design of Motor-Vehicle Brake Systems); Dissertation, University of Stuttgart, 1973

[37] Sorsche, J.H., Enke, K and Bauer, K.: Some Aspects of Suspension and Steering Design for Modern Compact Cars; SAE-Paper No. 741 039, 1974

[38] Limpert, R., Gamero, F.E. and Boyer, R.: An Investigation of Brake Balance for Straight and Curved Braking; SAE-Paper No. 741 086, 1974

[39] Official Gazette of the European Communities Council Directive No. 71/320/EEC of 26th July 1971, published on 6th September 1971 amended by Directive No. 74/132/EWG.

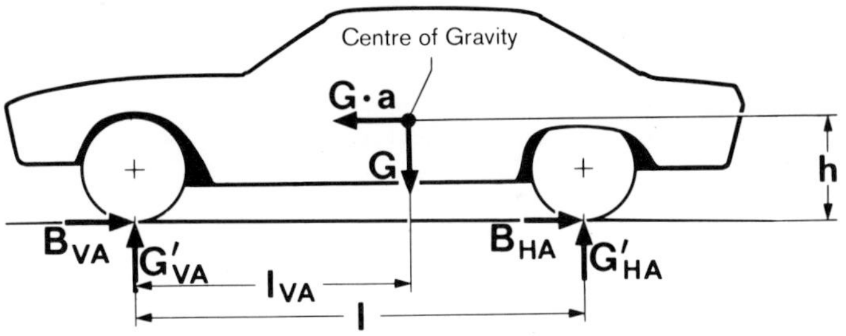

Fig. 1: Forces Acting upon a Braking Vehicle

B_{VA} = Brake force, front
B_{HA} = Brake force, rear
G = Weight force (vehicle weight)
G'_{VA} = Radial force at axle (dynamic axle load), front
G'_{HA} = Radial force at axle (dynamic axle load), rear
a = Retardation, the ratio of deceleration to gravitational constant
h = Height of centre of gravity
l = Wheelbase
l_{VA} = Distance between centre of gravity and front axle

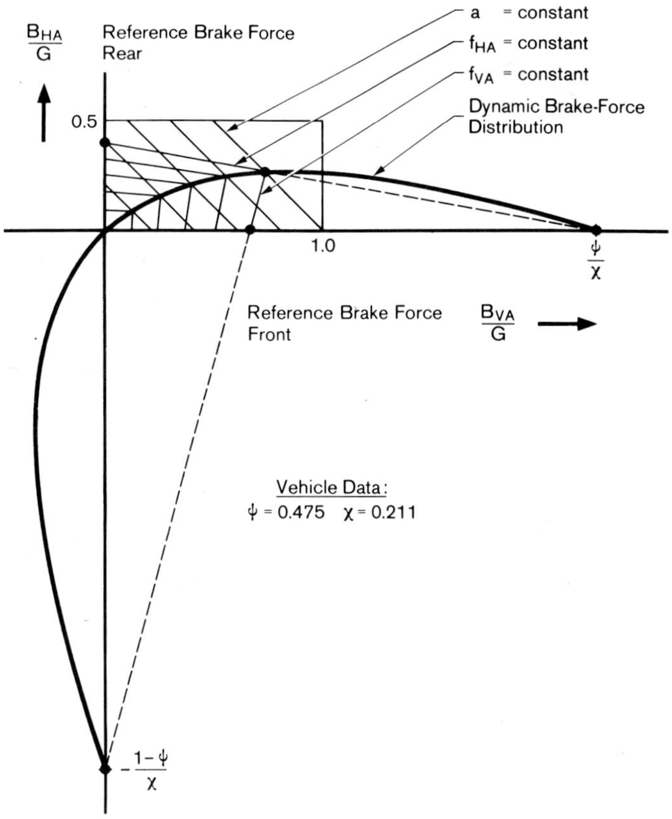

Fig. 2: Dynamic Brake-Force Distribution

a = Retardation, the ratio of deceleration to gravitational constant
f_{VA} = Coefficient of friction, front
f_{HA} = Coefficient of friction, rear
χ = Ratio of the height of centre of gravity to wheelbase
ψ = Ratio of the rear-axle load to vehicle weight

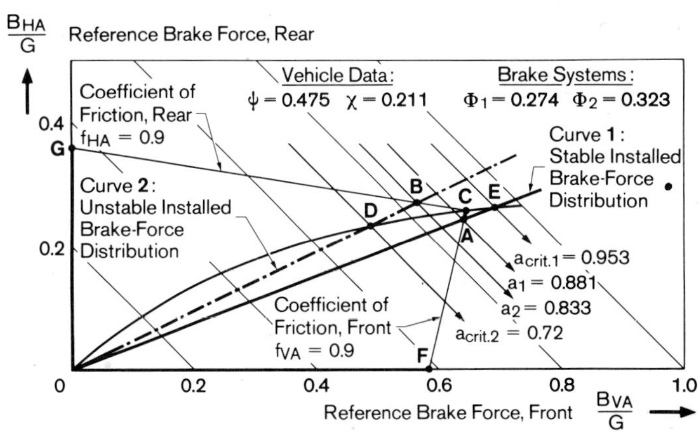

Fig. 3: Brake-Force Distribution Diagram

a = Retardation, the ratio of deceleration to gravitational constant
Φ = Ratio of the rear-axle brake force to total brake force
χ = Ratio of the height of centre of gravity to wheelbase
ψ = Ratio of the rear-axle load to vehicle weight

©IMechE 1976

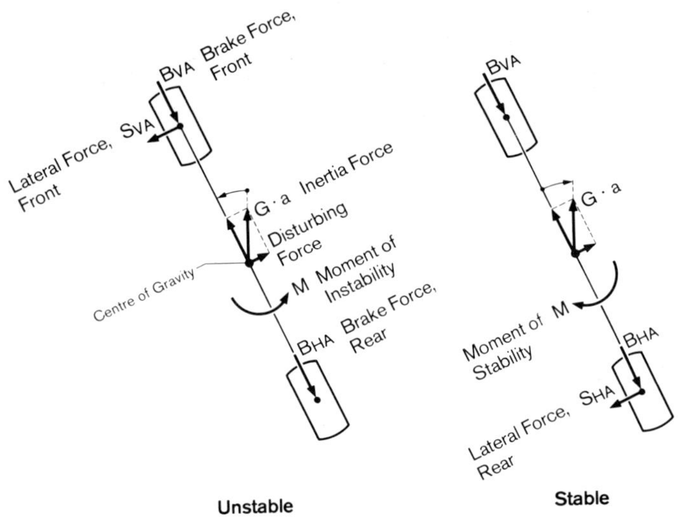

Fig. 4: Definition of Stable and Unstable Brake Performances of Vehicles

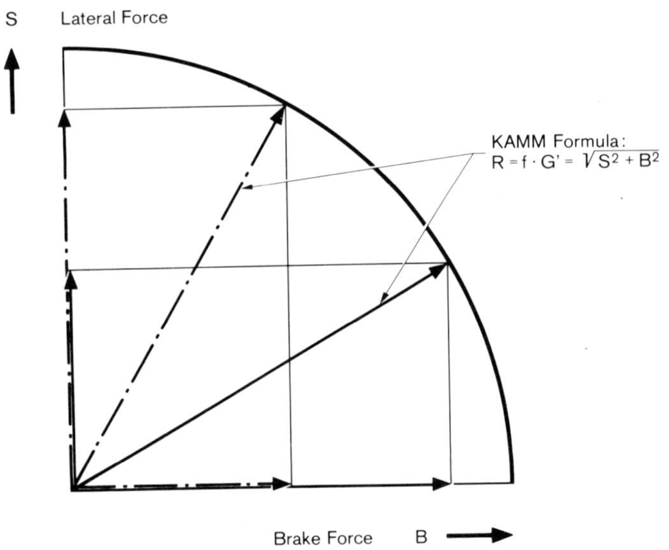

Fig. 5: Derivation of the KAMM Formula

G' = Radial force at axle (dynamic axle load)
R = Resulting maximum friction force
f = Coefficient of friction

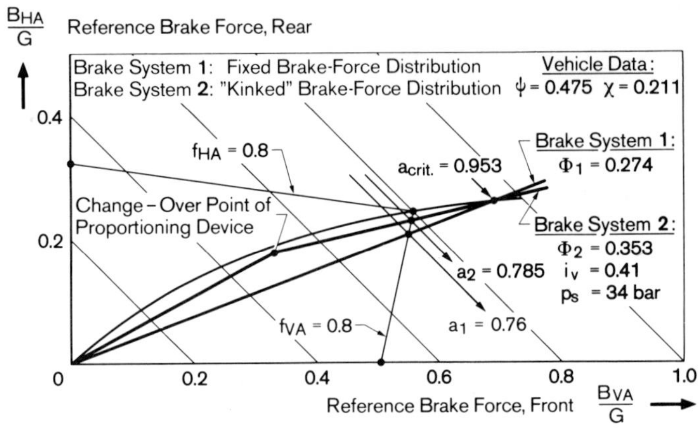

Fig. 6: Parabola of the Dynamic Brake-Force Distribution on a Level Ground with One Fixed and One "Kinked" Installed Brake Force Distribution

a = Retardation, the ratio of deceleration to gravitational constant
f_{VA} = Coefficient of friction, front
f_{HA} = Coefficient of friction, rear
i_v = Pressure reduction ratio
p_s = Brake pressure at change-over point of proportioning device
Φ = Ratio of the rear-axle brake force to total brake force
χ = Ratio of the height of centre of gravity to wheelbase
ψ = Ratio of the rear-axle load to vehicle weight

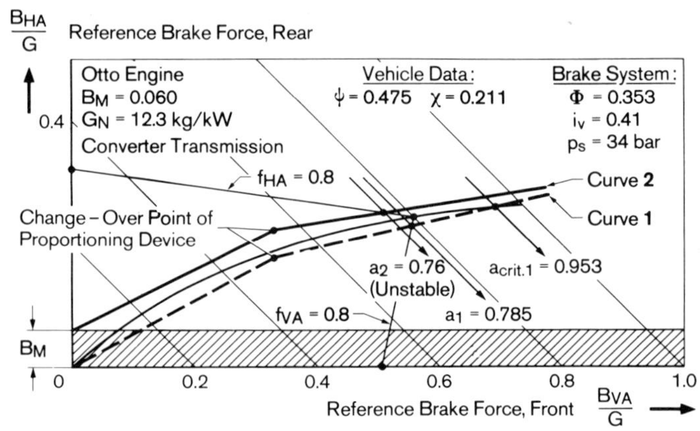

Curve 1: Installed Brake-Force Distribution
Curve 2: Actual Total Distribution Considering the Negative Shares of the Rotating Axle Masses and of the Engine-Braking Moment

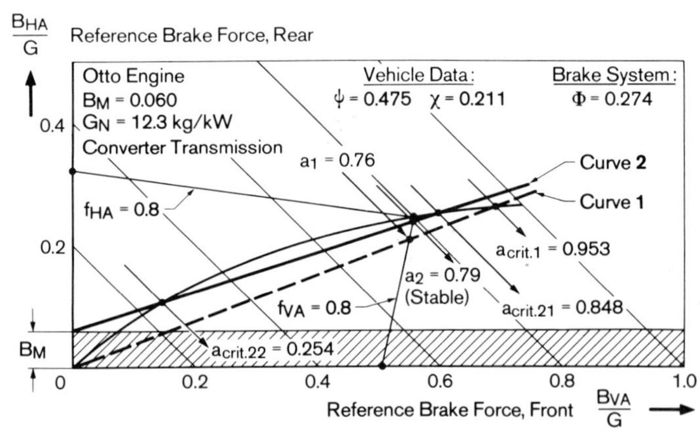

Fig. 7: Displacement of the Fixed and of the "Kinked" Installed Brake-Force Distribution by the Effect of the Inertia Moments and by an Engine-Braking Moment Acting upon the Rear Wheels

B_M = Specific engine brake force, rear axle
G_N = Power-to-weight ratio of a vehicle
a = Retardation, the ratio of deceleration to gravitational constant
f_{VA} = Coefficient of friction, front
f_{HA} = Coefficient of friction, rear
i_v = Pressure reduction ratio
p_s = Brake pressure at change-over point of proportioning device
Φ = Ratio of the rear-axle brake force to total brake force
χ = Ratio of the height of centre of gravity to wheelbase
ψ = Ratio of the rear-axle load to vehicle weight

© IMechE 1976

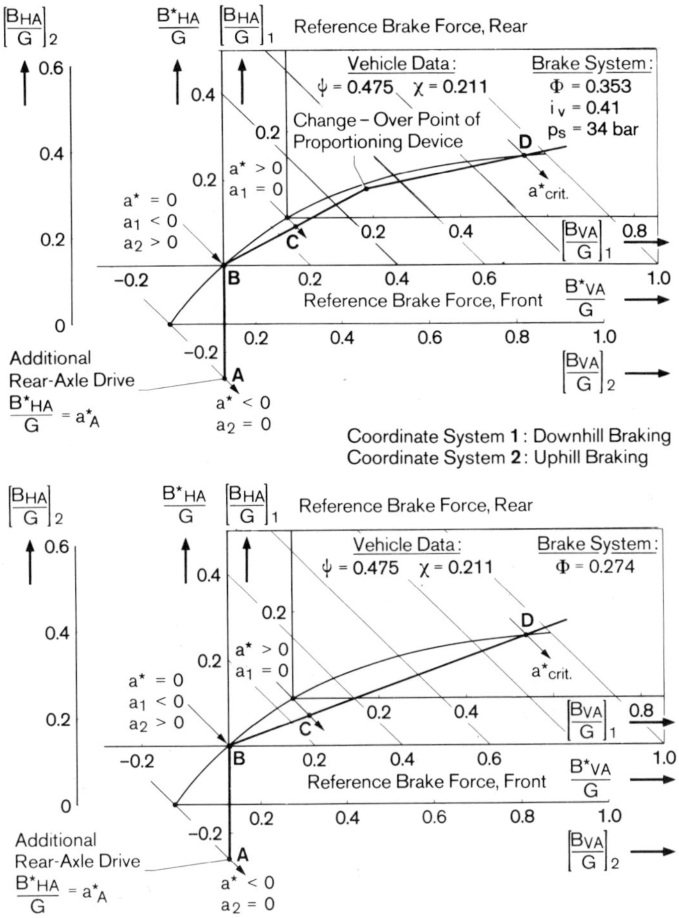

Fig. 8: Brake-Force Distribution Diagram of a Vehicle with Fixed and of a Vehicle with "Kinked" Distribution on Positive and Negative Grades

B = Brake forces causing vehicle deceleration
B* = Brake forces produced by the brake system
a = Retardation, the ratio of deceleration to gravitational constant
a* = Ratio of the sum of all brake forces to vehicle weight
i_v = Pressure reduction ratio
p_s = Brake pressure at change-over point of proportioning device
Φ = Ratio of the rear-axle brake force to total brake force
χ = Ratio of the height of centre of gravity to wheelbase
ψ = Ratio of the rear-axle load to vehicle weight

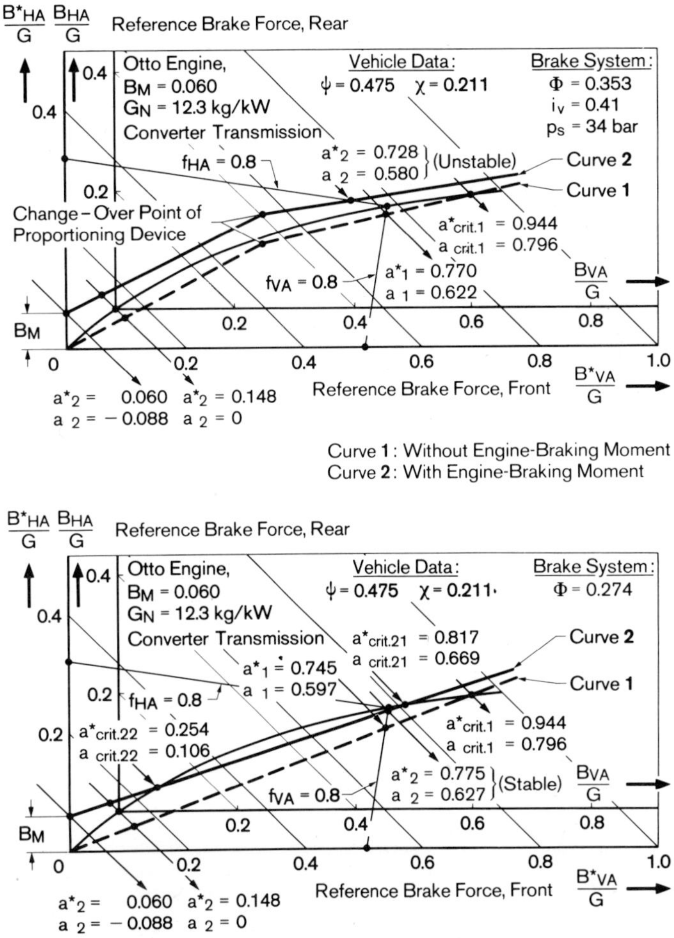

Fig. 9: Braking on a Negative Grade of 15 Percent with the Influence of the Engine-Braking Moment and of the Inertia Moments on a Vehicle with Fixed Distribution and on a Vehicle with "Kinked" Distribution

B = Brake forces causing vehicle deceleration
B* = Brake forces produced by the brake system
B_M = Specific engine brake force, rear axle
G_N = Power-to-weight ratio of a vehicle
a = Retardation, the ratio of deceleration to gravitational constant
a* = Ratio of the sum of all brake forces to vehicle weight
f_{VA} = Coefficient of friction, front
f_{HA} = Coefficient of friction, rear
i_v = Pressure reduction ratio
p_s = Brake pressure at change-over point of proportioning device
Φ = Ratio of the rear-axle brake force to total brake force
χ = Ratio of the height of centre of gravity to wheelbase
ψ = Ratio of the rear-axle load to vehicle weight

© IMechE 1976

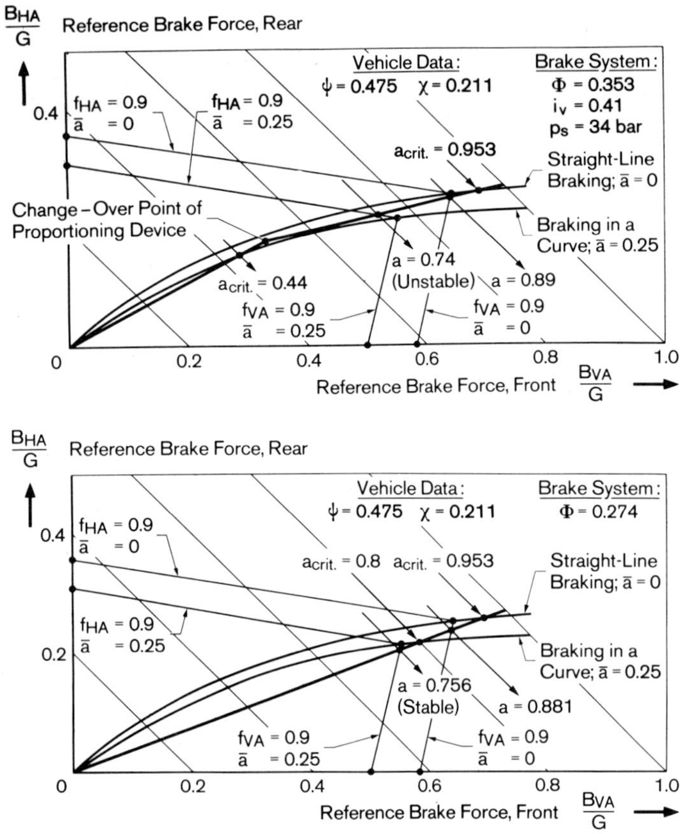

Fig. 10: Displacement of the Dynamic Brake-Force Distribution by a Lateral Acceleration a = 0.25 and its Influence on a Vehicle with Fixed Distribution and on a Vehicle with "Kinked" Distribution

- a = Retardation, the ratio of deceleration tp gravitational constant
- ā = Specific lateral acceleration, the ratio of lateral acceleration to gravitational constant
- fVA = Coefficient of friction, front
- fHA = Coefficient of friction, rear
- iv = Pressure reduction ratio
- ps = Brake pressure at change-over point of proportioning device
- Φ = Ratio of the rear-axle brake force to total brake force
- χ = Ratio of the height of centre of gravity to wheelbase
- ψ = Ratio of the rear-axle load to vehicle weight

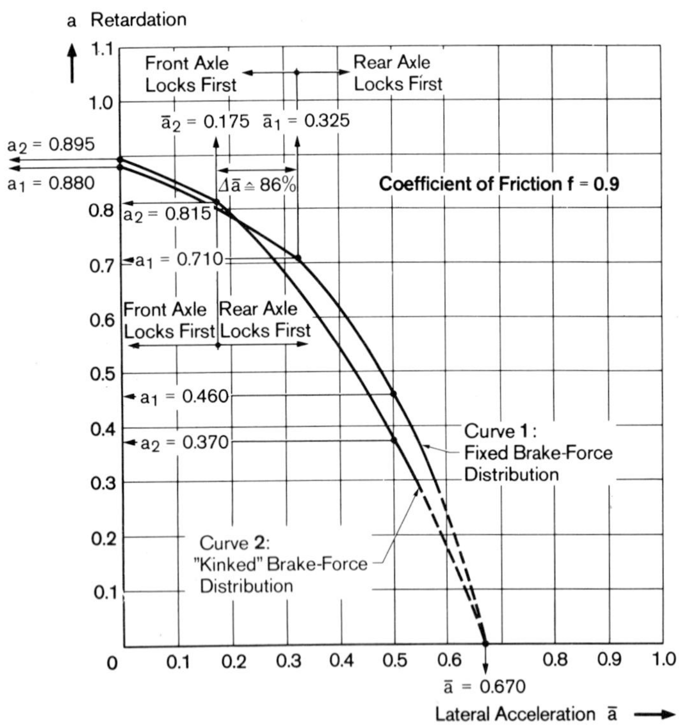

Fig. 11: Obtainable Retardation at the Slip Limit of the Most Critical Axle as a Function of the Specific Lateral Acceleration for a Vehicle with Fixed Distribution and for a Vehicle with "Kinked" Distribution

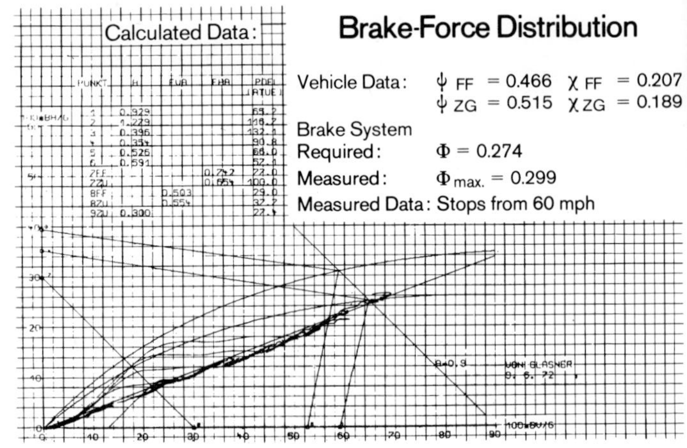

Fig. 12: Recorded Installed Brake-Force Distribution Related to the Design Data (Original Strip Chart Retardation)

- Φ = Ratio of the rear-axle brake force to total brake force
- χ = Ratio of the height of centre of gravity to wheelbase
- ψ = Ratio of the rear-axle load to vehicle weight

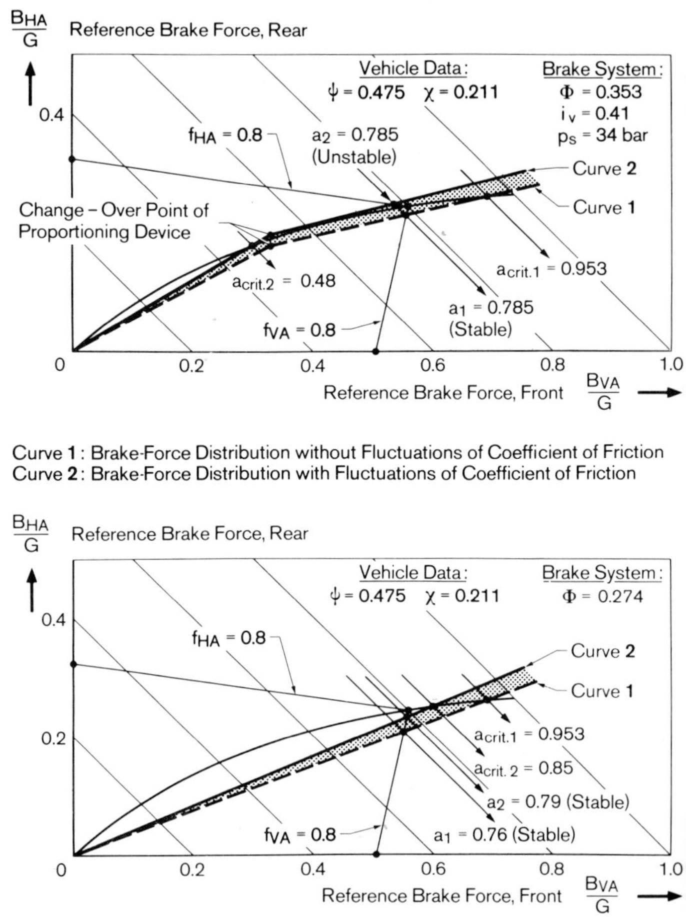

Fig. 13: Displacement of Fixed and "Kinked" Installed Brake-Force Distributions by Ten-Percent Fluctuations of Coefficient of Friction

- a = Retardation, the ratio of deceleration to gravitational constant
- fVA = Coefficient of friction, front
- fHA = Coefficient of friction, rear
- iv = Pressure reduction ratio
- ps = Brake pressure at change-over point of proportioning device
- Φ = Ratio of the rear-axle brake force to total brake force
- χ = Ratio of the height of centre of gravity to wheelbase
- ψ = Ratio of the rear-axle load to vehicle weight

© IMechE 1976

C32/76

COMPUTER ANALYSIS OF ANTILOCK SYSTEM PERFORMANCE IN THE BRAKING OF COMMERCIAL VEHICLES

P. S. FANCHER MSE and C. C. MACADAM, MSE
Highway Safety Research Institute, The University of Michigan, Ann Arbor, Michigan

The MS of this paper was received at the Institution on 2 October 1975 and accepted for publication on 30 October 1975

SYNPOSIS The use of computer simulation to analyze the braking performance of commercial vehicles equipped with antilock systems is examined herein. A technique for measuring the pressure modulation characteristics of antilock systems is delineated. A subprogram for synthesizing antilock system performance within the framework of a large-scale digital computer simulation of the braking of trucks and tractor-semitrailer vehicles is described. Comparisons between measured and calculated vehicle performance are made to indicate the validity of the computerized analysis. Findings concerning the sensitivity of the braking performance of a truck to tire, brake, suspension, and antilock properties are discussed.

INTRODUCTION

1. In an effort to improve highway safety, the United States Department of Transportation has established rules which specify maximum stopping distances with only momentary, controlled wheel lockup for various initial velocities, vehicle loading conditions, and surface-friction characteristics for vehicles equipped with air brake systems (ref. 1). An effect of this government regulation has been the incorporation of antilock braking systems into the design of most heavy vehicles produced in the United States (ref. 2). Consequently, design-analysis methods which account for antilock system performance are now needed to predict and understand the braking performance of currently manufactured heavy vehicles. Accordingly, this paper illustrates a computer-based method consisting of antilock system testing, vehicle testing, and vehicle simulation for analyzing (1) the stopping characteristics of specific vehicles and (2) the influence of changes in tire, brake, suspension, and antilock system parameters on braking performance.

2. In this method, antilock system testing is done in the laboratory on a stationary vehicle with the aid of an analog computer circuit in which left and right wheel dynamics are simulated to supply a pair of wheel speed signals to the truck's antilock module. Pressure transducers, installed in the truck's corresponding brake chambers, return to the analog computer electrical signals which are proportional to brake pressure, thereby providing inputs for simulating brake torque as a function of pressure. (See Figure 1.) During a typical test, an operator stationed in the truck's cab fully depresses the air brake to simulate a panic stop. During each test, time histories of wheel speed, wheel acceleration, and brake pressure are recorded for use in devising a model for synthesizing the operational characteristics of the antilock system under test.

3. This technique of interconnecting antilock system hardware with an analog or hybrid computer is similar to the approach which has been used by brake manufacturers in developing antilock systems for automobiles and trucks (ref. 3, 4).

4. To analyze braking performance, the measured and derived operational characteristics of the antilock system are entered into a large digital simulation of a vehicle. The simulation used in this study was developed expressly for this purpose (ref. 5). A block diagram of the simulation is presented in Appendix 1. The vehicle model employed in the simulation contains the equations of motion for straight-line braking plus detailed means for representing tire shear force characteristics, brake system characteristics, tandem suspension properties, and antilock system characteristics (ref. 6).

5. This paper emphasizes that portion of the simulation associated with modeling the antilock system. Once the method used for representing an antilock system is described, comparisons between test results for a specific vehicle and computed performance are presented to indicate the accuracy of the computer model and the associated parameter measurement activity. The simulation is then used to develop findings on the influence of tire, brake, and suspension factors on antilock braking.

NOTATION

a, b, c, d, e, f, g, h	Constants in the antilock control laws
p	Gauge air pressure
t	Time
T	Brake torque
T_i	"Unfaded torque"
θ	Temperature (in degrees Farenheit)

©IMechE 1976

θ_F	Brake fade factor used in the simulation
V	Velocity
V_R	Reference velocity signal generated in the antilock model
ω	Wheel angular velocity
α	Wheel angular acceleration
t_o	Time of the last brakes-off command from the antilock system

DIGITAL ANTILOCK SIMULATION

6. The digital antilock program used by the main truck simulation provides a versatile, general-purpose framework within which to model the widely varying characteristics exhibited by many of today's antilock systems (ref. 7). The antilock simulation focuses on three areas common to most systems:

1) wheel speed sensor characteristics,
2) control logic, and
3) pressure modulator characteristics.

One antilock system is allowed for each axle, with the same pressure being returned to the brakes on either side of the axle. The following options are available for controlling antilock cycling: (1) worst wheel (the wheel with the lowest angular velocity); (2) best wheel (the wheel with the highest angular velocity); and (3) average wheel speed.

7. The antilock simulation is programmed by inputting various numerical parameters which define the detailed operating characteristics for a given antilock system. A catalog of sixty parameters and variables most commonly associated with antilock operation is provided for programming use. The combining of flexible numerical input with a catalog of numerous parameters and variables provides a highly versatile framework in which to effectively implement on the computer the brake control conditions and logical operations used in the antilock system to be modeled. An example illustrating the type of numerical expressions and logical decisions which can be used appears in the next section. (Details of this program are given in ref. 7.)

ANALYSIS OF ANTILOCK LABORATORY RESULTS

8. A suitable set of numerical input used by the digital antilock program must be obtained in order to adequately describe the operating characteristics for a given system. If the necessary information and parametric data cannot be obtained directly from an antilock manufacturer, an alternative is to extract such information from a series of convenient laboratory tests. Such has been the approach taken in this paper and for other recent studies (ref. 7).

9. Observation of different laboratory time history traces, such as the example shown in Figure 2, allows one to obtain directly from such traces several different parameters and characteristics required for the digital antilock program (e.g., solenoid pressure time delays, pressure rise and fall characteristics). Parameters and characteristics such as those related to the antilock logic scheme are less apparent and must be extracted by a trial-and-error procedure wherein a particular set of rules for prediction of impending wheel lock and reselection of brake pressure is initially assumed, exercised, and then compared with laboratory results.

10. The particular antilock logic model derived by this method for the vehicle under study is defined by the following basic set of inequalities and logical relationships:

11. A prediction, or solenoid-pressure OFF command, is generated if

$$\alpha \leq -a \quad (1)$$

and

$$b\,V_R - \omega + c\,\alpha + d \geqq 0 \quad (2)$$

or if

$$\omega \leq 0 \text{ (wheel lock)}, \quad (3)$$

a, b, c, and d being positive constants.

12. A reselection, or solenoid-pressure ON command, is generated if

$$\alpha \geq e \quad (4)$$

and

$$\omega - V_R + f \geqq 0 \quad (5)$$

or if

$$\alpha \geq g, \quad (6)$$

or if

$$t - t_o \geq h \quad (7)$$

e, f, g, h being positive constants.

13. The form of the reference velocity signal, V_R, is illustrated in Figure 3. The signal V_R is generated by a time ramp, which decreases at a prescribed rate until the wheel angular velocity exceeds the reference velocity signal. Then, V_R reinitializes near the local maximum value of angular velocity for starting another decreasing ramp.

14. The remaining parameters for this system, which pertain to the wheel sensor and pressure modulator characteristics, could be easily estimated or obtained directly from the laboratory test results.

15. Examples of the agreement between laboratory test results and the full-scale simulation are shown in Figures 2 and 4. In these cases, the parametric values used in the full-scale vehicle

© IMechE 1976

simulation were selected to provide conditions identical to those used in the laboratory tests. The results in Figures 2 and 4 show that the model of the antilock system as synthesized on the digital computer performs similarly to the actual antilock hardware for the operating environment employed in the laboratory test.

16. The load on the axle simulated in the test shown in Figure 4 corresponds roughly to the average load on one of the rear axles of an empty truck during a rapid stop. The results shown in Figure 2 represent conditions similar to those occurring at the same axle with a fully loaded truck. Clearly, the operation of the antilock-vehicle system is highly dependent upon the axle loads involved.

17. In the unloaded rear-axle case, angular accelerations exceeding 100 rad/sec^2 are encountered while the brake pressure is decreasing, and angular decelerations exceeding 200 rad/sec^2 are developed while the brake pressure is increasing. Obviously, very large torque imbalances resulting in high magnitudes of wheel acceleration will be encountered on a lightly loaded axle. As will be discussed later, pitch and bounce motions of the sprung mass of the loaded truck can cause individual axle loads to become small, leading to large magnitudes of wheel acceleration and complicated patterns of antilock cycling not obtained in the laboratory tests for an axle with a heavy static load.

18. Furthermore, large wheel accelerations increase the importance of effects which cause small differences in timing, whether these timing differences are due to pitch and bounce dynamics or due to time lags in the antilock system. For example, at a wheel acceleration of 200 rad/s^2, the wheel speed will change 5 rad/s in 0.025 s. At high forward velocities, the free-rolling angular velocity of a typical truck tire is approximately 50 rad/s. Accordingly, a 5 rad/s change in wheel speed implies at least a 10% change in longitudinal slip (100% slip represents a locked wheel and 0% slip represents a free-rolling tire). This amount of change in slip can cause profound changes in tire force when the tire is operated in the vicinity of its maximum shear force capability (ref. 8).

19. The preceding simple considerations of the laboratory test results indicate the paramount difficulties which can be encountered in analyzing the braking performance of a truck equipped with an antilock system. All the factors which influence the torque balance at each wheel must be carefully represented in the simulation. These factors include brake effectiveness, fade, and side-to-side imbalance; peak and slide tire shear force properties; suspension kinematics, coulomb damping, and spring force characteristics; and vehicle inertial properties such as center of gravity location and pitch moment of inertia.

DETERMINATION OF PARAMETRIC DATA FOR THE SIMULATION

20. The objective of this activity is to obtain parametric values which describe the subject vehicle with sufficient accuracy for analyzing braking performance. Certain parameters such as weights and dimensions can be measured readily. Suspension spring rates and damping forces and inertial properties (for example, pitch moment of inertia and center of gravity height) can be obtained from laboratory tests if equipment for handling large weights is available (ref. 9). Ideally, it would be desirable to make precision measurements of brake, tire, and tandem-suspension characteristics in the laboratory. However, the lack of availability of suitable facilities for testing these components and the large burden in time and money involved in making component measurements precluded the use of a direct parameter measurement activity for determining brake, tire, and suspension characteristics.

21. Consequently, the subject vehicle was used as a test device for estimating pertinent component performance. Special vehicle stopping tests using only the brakes on individual axles at various levels of brake pressure were used to evaluate tire-road friction characteristics and brake effectivenesses. In addition, data on side-to-side brake imbalance and interaxle load transfer in the tandem suspension were obtained from braking tests of the vehicle.

22. Even though field tests for assessing component characteristics lack the precision of laboratory tests, they have important advantages. Clearly, field operating conditions which may be difficult to reproduce in the laboratory are available. Since there may be considerable variation in performance from one brake or tire sample to another, the testing of the actual components used on the subject vehicle is desirable. Furthermore, the process of dismantling the truck for laboratory testing of its components and subsequently reassembling it is costly and may alter the characteristics of the components involved.

DESCRIPTION OF THE TEST VEHICLE

23. A truck of standard design equipped with antilock braking systems on all three axles was used in this study. The vehicle weighed 69,400 N (15,600 lbs) without load and it carried an additional 102,300 N (23,300 lbs) when fully loaded. The vehicle had a wheelbase of 4.293 m (169 in). In the loaded test condition, the ratio of center of gravity height to wheelbase was 0.35 and the center of gravity of the sprung mass was located 3.332 m (131.2 in) behind the front axle. The rear suspension employed a "walking-beam" with leaf springs. Dual wheels were used on the rear axles. The truck had a solid front axle suspended on leaf springs.

24. The torque characteristics of the "S-cam" brakes installed on this truck were represented in the simulation by the following parametric data:

Front Axle Brakes (average values, not including imbalance)

Gauge Pressure P_a 10^5	"Unfaded Torque" T_i Nm
0.48	0
6.897 (100 psi)	13,560 (120,000 in lbs)

"fade factor" = 750°F

©IMechE 1976

Rear Brakes (identical on each axle of the tandem suspension)

Gauge Pressure P_a 10^5	"Unfaded Torque" T_i Nm
0.48	0
6.897	22,600 (200,000 in lbs)

"fade factor" = 750°F

25. Braking tests indicated that the right front brake was much more effective than the left front brake. The right front brake was capable of locking the right front wheel at $(4.14)10^5$ P_a (60 psi), but a pressure of $(6.21)10^5$ P_a (90 psi) was required to lock up the left front wheel. Thus the right front brake was treated as being approximately 1.5 times as effective as the left front brake.

26. The "fade factors" contained in the list of brake parameters were used to compute the applied brake torque, T, as follows:

$$T = T_i(1 - \theta/\theta_F) \qquad (8)$$

This equation for computing brake fade has been used in earlier studies (ref. 10). Nevertheless, this is an approximate empirical method and there is a continuing need to develop a better understanding of the phenomenon of brake fade. For this program the values of fade factor were estimated using average torque data from dynamometer tests conducted by the brake manufacturer on samples of these types of brakes.

27. The results from special vehicle tests designed to estimate tire characteristics indicated that the 11 x 20 highway rib front tires had a peak coefficient of friction of 0.79 and a slide coefficient of friction of 0.54 at a vertical load of 16,460 N (3,700 lbs) and a forward speed of 17.9 m/s (40 mph). The 10 x 20 rear tires had a peak coefficient of friction of 0.91 and a slide coefficient of friction of 0.63 at a vertical load of 4,450 N (1,000 lbs) and a forward speed of 17.9 m/s. (The load on one of the 8 rear tires during a rapid deceleration of the unloaded truck is approximately 4450 N.)

28. Tire forces at other speeds and loads were then extrapolated from these results, using the trends obtained from recent truck-tire testing programs (ref. 8).

29. Braking tests using the brakes on the two rear axles indicated that all four brakes were well balanced and that the torque rods attached to the rear axles were effective in reducing the amount of interaxle load transfer caused by brake torque. The test results indicated that approximately 5% of the total rear brake torque was reacted by a slight increase in vertical load on the tires of the center (front tandem) axle and a slight decrease in vertical load on the tires of the rear tandem axle.

30. Other factors which proved to be important for this truck were: (1) the front and rear air reservoirs lost approximately 21,000 P_a (3 psi) each time the corresponding antilock systems caused the brakes to cycle off; (2) the bump stop on the front axle was contacted on all high-deceleration stops and this factor had a significant influence on the magnitude of the pitch oscillations of the sprung mass; (3) the rear suspension had a significant amount of coulomb friction (approximately 26,700 N (6,000 lbs) for the entire tandem suspension), nearly enough to "freeze" the rear suspension, thus rendering the rear tire spring rates important in determining the amplitude and period of the pitch motions which were excited.

31. Figures 5 and 6 show results from stopping tests with the loaded and unloaded vehicle. Superimposed on these test results are simulated results. Depending upon one's point of view, the simulated and measured results may appear to be in reasonable agreement or they may be regarded as containing significant discrepancies. However, in view of the parameter sensitivity discussion presented in the next section, the computed results are remarkably similar to the test data for this vehicle. Furthermore, Figure 7 illustrates the differences in pressure and wheel-speed time histories which occur from one vehicle test to another test at nominally the same test conditions. Thus, the simulation results are believed to be representative of the measured performance. Finally, the measured and computed stopping distances from 26.8 m/s compared favorably as shown below:

	Fully Loaded	Unloaded
Measured	76.8 m	57.0 m
Computed	75.3 m	55.2 m

THE INFLUENCE OF TIRE, BRAKE, SUSPENSION, AND ANTILOCK SYSTEM PERFORMANCE ON BRAKING PERFORMANCE

32. Once the simulation is considered to be a valid representation of the vehicle under study, the model employed in the simulation can be used to (a) interpret the test results and (b) study the sensitivity of braking performance and wheel cycling characteristics to changes in components or operating conditions.

33. The following observations concerning the braking performance of the truck were developed using the computer model to account for the measured responses:

(1) The average front brake torque capability is approximately equal to the product of tire rolling radius times the maximum shear force capability of the front tires for the unloaded truck. Thus the antilock system does not release and reapply the front brakes very many times during a rapid stop. The tendency for the front brakes to cycle is enhanced by the large imbalance between the two front brakes and the fact that the antilock system operates on a worst-wheel option (i.e., the wheel with the stronger brake controls the system).

(2) Furthermore, the timing of front brake cycling is highly dependent upon the changes in vertical load caused by the pitching and bouncing of the vehicle. That is, when the front tires are unloaded, their maximum shear force capability is reduced sufficiently to allow the available brake torque to cause the antilock system to cycle.

(3) At the onset of braking for the fully-loaded truck, the front brakes may cycle before the maximum tire force is attained. This condition is due to the responsiveness of the air system. Upon

©IMechE 1976

rapid application of the treadle valve, the front brake pressure and consequently brake torque increases, rapidly creating a torque imbalance on the front wheels which is sufficient to cause the sensed wheel acceleration and velocity signals to meet the criteria for antilock cycling before the tire shear force has attained its maximum value and before enough load transfer has taken place to significantly increase the shear force capability of the front tires.

(4) For either the loaded or unloaded truck it is difficult to predict precisely how many times and when the front brakes will be released. Besides the influence of pitch and bounce dynamics, the rate and manner in which the brakes fade and the amount of air pressure lost during each cycle of the antilock system are important factors in determining whether another cycle will occur. A very accurate understanding of the amount and timing of brake fade is needed to explain precisely the observed variability in on/off cycling of the front brakes.

(5) For the unloaded truck, the rear wheels tend to operate at high values of longitudinal slip because the rise rate of brake pressure is fast, the available brake torque is large compared to the available tire shear force, and the tire shear force is limited by the light vertical load on the rear wheels. Furthermore, the antilock system allows larger wheel speed excursions towards zero angular velocity as the angular acceleration increases. This effect is represented in the simulation by the values of the quantities c and d in Inequality (2). In this case, the terms in Inequality (2) with coefficients c and d act as a delay in turning the brakes off, accentuating slip excursions for higher angular decelerations. (This delay may be due to hardware limitations or it may serve some useful purpose. Nevertheless, it exists in the system and it has an influence on the system's operation for the truck on a dry test surface.) In addition, the antilock system as modeled will turn the brakes on (increase pressure) whenever the wheel angular acceleration exceeds 108 rad/s. (See Inequality (6).) This level of angular acceleration is attained very shortly after the brake pressure starts decreasing. Consequently, the brakes come back on, prematurely forcing the rear wheel speed back towards or into lockup. For these reasons, the rear wheels on the unloaded truck have difficulty escaping from the high slip (above 40%) regime.

(6) The frequency of antilock system cycling is approximately 3 times per second. This frequency is close to the frequency of the "pitch mode" of the sprung mass of the loaded truck. (The frequency of the "bounce mode" of the sprung mass is approximately 1.6 Hz.) Consequently, increasing and decreasing of rear tire forces due to antilock cycling tends to sustain the pitch oscillations of the vehicle through the following mechanism: (a) increasing tire braking force causes the front of the vehicle to pitch down and the rear of the vehicle to pitch up; (b) when the rear of the vehicle pitches up, the vertical load on the rear tires is reduced, thereby decreasing the rear tire force, which causes the brakes to go off and also causes the rear of the vehicle to pitch back down; (c) when the vehicle pitches back down, the rear tires again become loaded, the brakes come on, and the tire braking force starts increasing again; (d) now the conditions of step (a) have been reattained and thus the oscillation is sustained.

(7) For this walking-beam, tandem suspension, the torque rods (brake reaction rods) are very effective in eliminating interaxle load transfer due to braking. Accordingly, the on/off cycling of brake pressures and the time histories of wheel speeds behave similarly for the two axles in the rear suspension.

34. In order to examine the effects of parameter variations on overall vehicle braking performance, a parameter sensitivity study was conducted from a baseline or reference set of parameters. The selected baseline parameters were identical to those used to simulate the test vehicle (Paragraph 23) with the exception of a small amount of brake imbalance on all three axles. Computer runs reflecting variations in tires, brakes, antilock logic, suspension, and loading configuration were performed and compared with the baseline results, with stopping distance serving as the principal numeric of braking performance. The baseline and each of the variations is listed in Table 1, along with corresponding stopping distances for both the loaded and unloaded configurations, except where noted.

Table 1

Baseline and parameter variations

Description	Stopping Distance Unloaded, m	Stopping Distance Loaded, m
Baseline	55.2	75.3
*Tire force capability increased by 10%		73.2
Peak tire force capability reduced by 10% (reflecting low inflation pressure tire test results)	59.7	77.4
Effectiveness of rear brakes reduced by 28%	57.0	74.4
*Brake hysteresis included		75.3
Front brake effectiveness increased to that of the rear brakes	56.7	71.0
Brake fade effects reduced by 33%	56.7	71.6
Brake imbalances removed from all axles	53.0	73.8
Antilock logic variation producing a faster prediction in pressure "OFF" command	57.3	77.4
Antilock logic time delay removed, resulting in greater wheel control particularly for lightly loaded configurations	54.6	75.9
*Rear suspension spring rate increased 50% and rear suspension coulomb friction decreased 50%		76.8

© IMechE 1976

Description	Stopping Distance Unloaded, m	Stopping Distance Loaded, m
*Front suspension spring rate increased 100%		75.0
*102,300 N payload lowered 0.456 m		72.8
*102,300 N payload moved forward 1.14 m		82.9
*Percent of rear suspension interaxle load transfer increased from 5% to 50%		70.4

*Loaded configuration only.

35. Each of the parameter variations was constrained to ranges judged as reasonable. It obviously would not be difficult to produce more dramatic sensitivity results by selecting unreasonable parameter variations. As an example, stiffening the rear suspension and greatly reducing its coulomb friction would produce severe bouncing motions of the rear suspension, resulting in a considerable loss of braking by the rear axles.

36. Several numerics in addition to stopping distance were examined as indicators of braking performance. These additional numerics were:

P_{10} = percent of time "best" or faster running wheel spent below 10% longitudinal slip for time interval from 26.8 m/s to 13.4 m/s (each axle).

P_{40} = percent of time "worst" or slower running wheel spent above 40% longitudinal slip for same time interval (each axle).

N_c = number of antilock cycles per stop for each axle.

L_p = average peak load on the front axle for the first three pitch oscillations (loaded configuration only).

L_m = average minimum load on the front axle for the first three pitch oscillations (loaded configuration only).

37. The baseline simulation runs for the loaded and unloaded vehicle produced the following numeric values:

		Loaded	Unloaded
Front Axle	P_{10}:	93%	29%
	P_{40}:	0%	0%
	N_c :	1	4
	L_p :	109,900 N	-------
	L_m :	43,600 N	-------
Front Tandem Axle	P_{10}:	51%	22%
	P_{40}:	17%	82%
	N_c :	13	15
Rear Tandem Axle	P_{10}:	52%	28%
	P_{40}:	21%	89%
	N_c :	13	15

As indicated, the front axle for the loaded baseline cycled only once, while the rear axles had no difficulty cycling but spent a large percentage of time in low slip regimes. In contrast, the rear axles of the unloaded baseline spent the greatest percentage of time in high slip regimes.

38. The same numeric calculations were performed for each of the parameter variation runs (see Appendix 2) with the following variations demonstrating a greater sensitivity in the loaded configuration:

1) front brake effectiveness/fade insensitivity
2) payload position
3) tire characteristics
4) rear suspension interaxle load transfer mechanism

The unloaded parameter variation runs indicated sensitivity primarily to tire characteristics and brake imbalance.

39. Much of the difference in braking performance between the loaded and unloaded configurations is directly attributable to the smaller front brakes which clearly contribute a much smaller percentage of braking effort during the loaded stops. In light of this, it is not surprising that front brake size and payload position (which greatly influences the amount of front brake contribution) appear as primary sensitivity parameters for this vehicle in the loaded configuration.

40. Equally interesting as the above sensitivities is the relative insensitivity of the braking performance to apparently reasonable variations in the antilock logic. Logic variations which resulted in wheel slip variations of about 3-4% at the time of prediction produced little significant increase or decrease in stopping distances. Possibly, if the relative effectivenesses of the front and rear brakes were matched to the prevailing load and surface conditions, then changes in the design of the antilock system would influence overall performance.

41. The interaxle load transfer variation, resulting in a greater amount of load on the front tandem axle, increased braking performance by retarding cycling on that axle. The combination of fade and increased load for the front tandem axle resulted in a loss of cycling after 1.6 seconds, causing the wheels to operate for the remainder of the stop in the 15-20% slip regime, thereby increasing considerably the average amount of braking performed by that axle.

CONCLUDING STATEMENTS

42. The braking performance of a new truck equipped with an antilock system for meeting U.S. Federal Motor Vehicle Safety Standard No. 121 (FMVSS 121) was analyzed in this paper. The study was limited to braking performance on a good, dry surface. From a safety standpoint, improvement of braking performance on slippery surfaces may prove to be more important in the long run than shorter stopping distances on dry pavement. Nevertheless, a very demanding aspect of FMVSS 121 is to be able

© IMechE 1976

to stop a fully-loaded vehicle in 277 feet (84.4 m) from 60 mph (26.8 m/s) on a surface with a 75 skid number.

43. Accordingly, the purpose of this paper has been to demonstrate a method which can be used by vehicle manufacturers to help ensure that their products will meet the requirements of the new braking standard and to aid them in developing an understanding of the important aspects of the braking performance of specific vehicles and antilock systems.

44. The method illustrated in this paper consisted of (1) laboratory testing of antilock systems, (2) measurement of vehicle parameters and braking characteristics, and (3) computer simulation. The results obtained for an example vehicle demonstrate that an indepth understanding of a vehicle's braking performance can be attained through this approach.

45. In this study a very detailed computer program was needed to calculate antilock system performance with reasonable accuracy. Nevertheless, the fidelity of the analysis of braking performance could be improved with (1) a more detailed knowledge of the control logic built into the antilock system, (2) a better understanding and representation of brake fade, and (3) further research into the pitch and bounce motions of the vehicle's sprung and unsprung masses.

46. The exemplary analysis contained herein was presented to illustrate a method for studying braking performance. The results for the exemplary vehicle are not meant to be extrapolated to other vehicle types or other antilock systems. In addition, many considerations go into the design of a given vehicle. For example, it may appear that more effective front brakes would improve the braking performance of the subject vehicle. However, the addition of larger front brakes may require redesign of the steering system and stronger front axles. Furthermore, improvements in antilock system design continue to be made. Thus it should be emphasized that the approach described here is intended as a design analysis tool which can be used to assess the influence of design changes, component variations, or vehicle condition on the braking performance of a given type of vehicle.

ACKNOWLEDGEMENTS

47. This work was carried out as part of a study of commercial vehicle braking and handling sponsored by the Motor Vehicle Manufacturers Association of America, Inc. The development of computer programs and parameter measurement techniques by the following members of the staff of the Highway Safety Research Institute at The University of Michigan is gratefully acknowledged: Messrs. J. Bernard, C. Winkler, R. Ervin, and T. Post. Special thanks are due to Mr. Rajiv Gupta for his work in conducting the vehicle tests and making computer runs.

REFERENCES

1. Federal Motor Vehicle Safety Standard No. 121, U.S. Motor Vehicle Code 49CFR571.121; S121-1 to S121-10.

2. TURVILL, P.A. FMVSS 121—air brake systems: what it means to the truck builder. Automotive Engineering, Vol. 81, No. 8, August 1973, pp. 27-37.

3. HICKNER, G.B. and HOWARD, D.W. Analog simulation as a design tool for advanced braking concepts. SAE Paper No. 700157, January 1970.

4. BLOSSER, R. Simulation of an air-brake antiskid system. Report on the Western Simulation Council Meeting, Simulation, January 1974, p. 20.

5. FANCHER, P.S. Prediction of braking and directional responses of commercial vehicles. Proceedings of a Symposium on Commercial Vehicle Braking and Handling, The Univ. of Michigan, Ann Arbor, Mich., May 5-7, 1975, pp. 439-462.

6. WINKLER, C., BERNÅRD, J., FANCHER, P., MACADAM, C., and POST, T. A computer program for predicting the braking performance of trucks and tractor-trailers. Phase III Technical Report, The Univ. of Michigan, Highway Safety Research Institute, Ann Arbor, Mich., June 15, 1975.

7. MACADAM, C.C. A general-purpose simulation for antiskid braking systems. Proceedings of a Symposium on Commercial Vehicle Braking and Handling, The Univ. of Michigan, May 5-7, 1975, pp. 119-158.

8. ERVIN, R.D., MACADAM, C.C., and FANCHER, P.S. The longitudinal traction characteristics of truck tires as measured on dry pavements. The Univ. of Michigan, Highway Safety Research Institute, Report No. UM-HSRI-PF-75-3, February 1975.

9. WINKLER, C.B. Measurement of inertial properties and suspension parameters of heavy highway vehicles. SAE Paper No. 730182, January 1973.

10. POST, T.M. Measurement and prediction of commercial vehicle brake torque. Proceedings of a Symposium on Commercial Vehicle Braking and Handling, The Univ. of Michigan, May 5-7, 1975, pp. 77-102.

APPENDIX 1

FEATURES OF THE DIGITAL SIMULATION

The principles used and the factors considered in the digital simulation are illustrated in Figure 8.

© IMechE 1976

APPENDIX 2 - TABULATION OF PARAMETER VARIATION RESULTS

Numeric

	Parameter Variation	Vehicle: Stopping Distance, m	Vehicle: Stopping Time s	Front Axle: P_{10}	Front Axle: P_{40}	Front Axle: N_c	Front Axle: $\frac{L_P}{10^3 N}$	Front Axle: $\frac{L_m}{10^3 N}$	Front Tandem Axle: P_{10}	Front Tandem Axle: P_{40}	Front Tandem Axle: N_c	Rear Tandem Axle: P_{10}	Rear Tandem Axle: P_{40}	Rear Tandem Axle: N_c
LOADED	Baseline	75.3	5.6	93	0	1	110	44	51	17	13	52	21	13
	Tire Force Increased 10%	73.2	5.3	94	0	1	112	43	59	13	11	59	16	12
	Peak Tire Force Decreased 10%	77.4	5.7	93	0	1	103	47	53	24	14	49	27	14
	Rear Brake Effectiveness Reduced 28%	74.4	5.3	88	0	1	116	38	59	0	3	62	0	4
	Brake Hysteresis Included	75.3	5.6	93	0	1	113	42	48	21	13	47	26	13
	Front Brake Effectiveness = Rear	71.0	5.3	55	6	5	108	47	46	22	12	46	27	12
	Brake Fade Effects Reduced by 33%	71.6	5.3	88	0	1	108	47	45	26	13	43	30	13
	Brake Imbalance Removed	73.8	5.5	90	0	1	110	45	42	15	13	41	16	13
	Antilock Variation -Faster Prediction	77.4	5.7	96	0	2	106	47	58	11	14	56	13	14
	Antilock Logic Time Delay Removed	75.9	5.6	91	0	0	107	47	50	4	13	60	6	13
	Rear Suspension Spring Rate Increased 50%; Coulomb Friction Decreased 50%	76.8	5.8	91	0	1	116	41	49	26	13	48	28	13
	Front Spring Rate Increased 100%	75.0	5.6	94	0	1	101	46	50	17	13	53	21	13
	102300 N. Payload Lowered 0.456 m	72.8	5.2	94	0	1	97	46	54	14	9	54	17	10
	102300 N. Payload Moved Forward 1.14 m	82.9	6.2	100	0	0	111	40	43	32	18	38	47	18
	% of Interaxle Load Transfer Increased x 10	70.4	5.1	94	0	1	103	47	37	5	4	40	49	13
UNLOADED	Baseline	55.2	4.0	29	0	4	-	-	22	82	15	28	89	15
	Peak Tire Force Decreased 10%	59.7	4.3	56	2	8	-	-	26	79	15	21	89	16
	Rear Brake Effectiveness Reduced 28%	57.0	4.1	53	0	5	-	-	29	50	13	26	70	13
	Front Brake Effectiveness = Rear	56.7	4.2	43	42	14	-	-	30	63	16	26	87	16
	Brake Fade Effects Reduced 33%	56.7	4.1	39	23	7	-	-	28	89	15	26	89	15
	Brake Imbalance Removed	53.0	3.8	22	0	2	-	-	9	75	13	9	77	14
	Antilock Variation -Faster Prediction	57.3	4.1	44	0	6	-	-	30	78	15	31	78	15
	Antilock Logic Time Delay Removed	54.6	3.9	30	0	4	-	-	31	36	16	27	35	16

© IMechE 1976

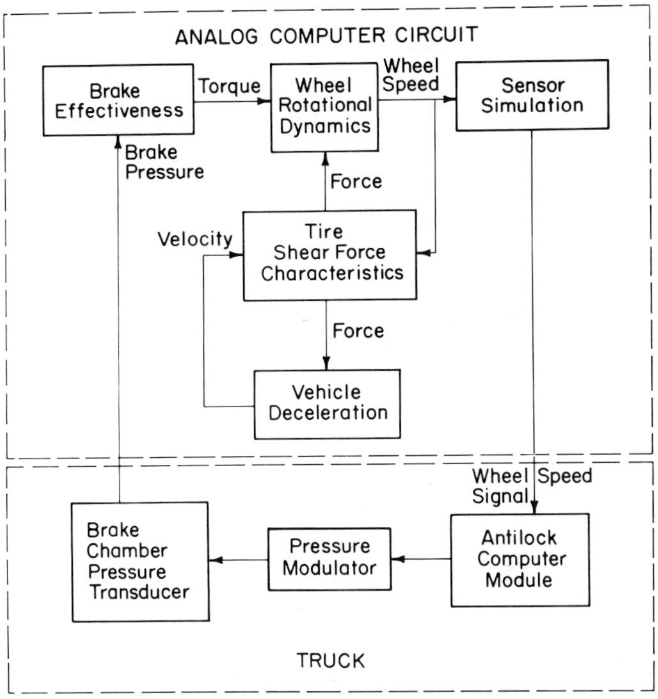

Fig. 1: Block diagram, antilock system testing

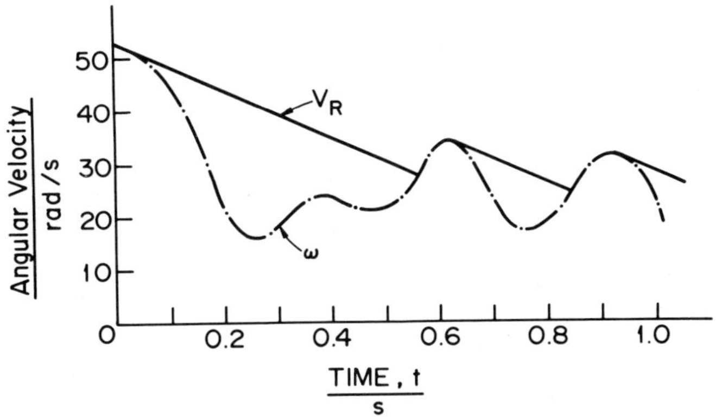

Fig. 3: Form of the reference velocity signal

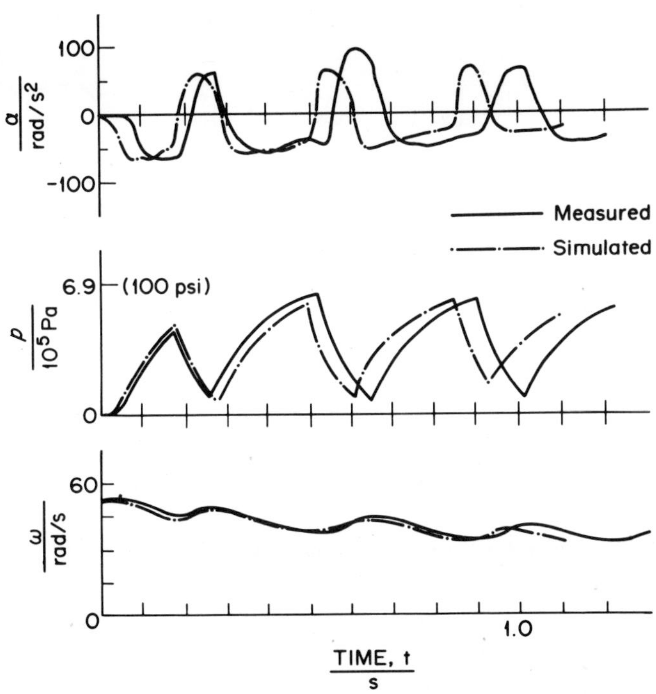

Fig. 2: Laboratory test for a rear axle on the fully-loaded truck

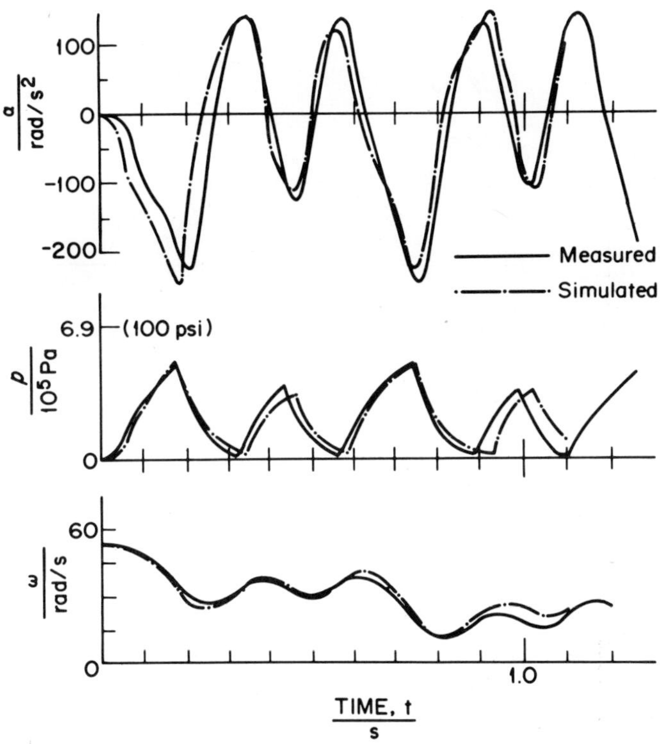

Fig. 4: Laboratory test for a rear axle on the unloaded truck

© IMechE 1976

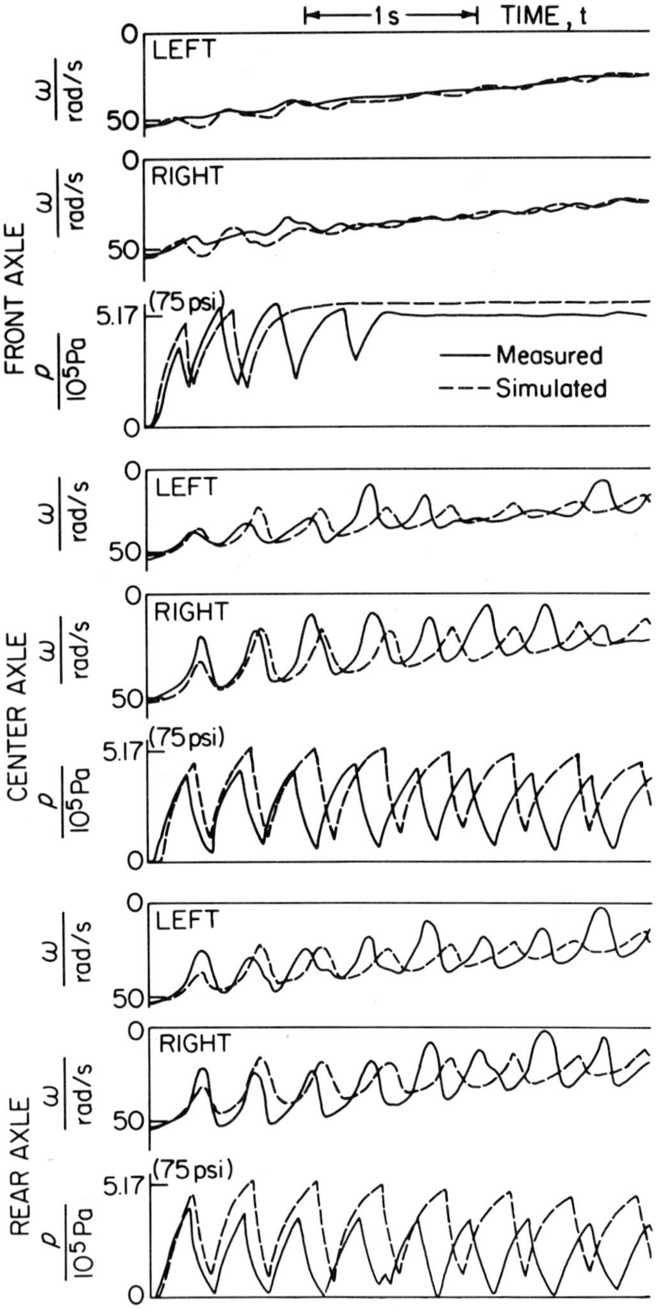

Fig. 5: Measured and simulated wheel speed control, loaded truck, initial velocity 26.82 m/s (60 mph)

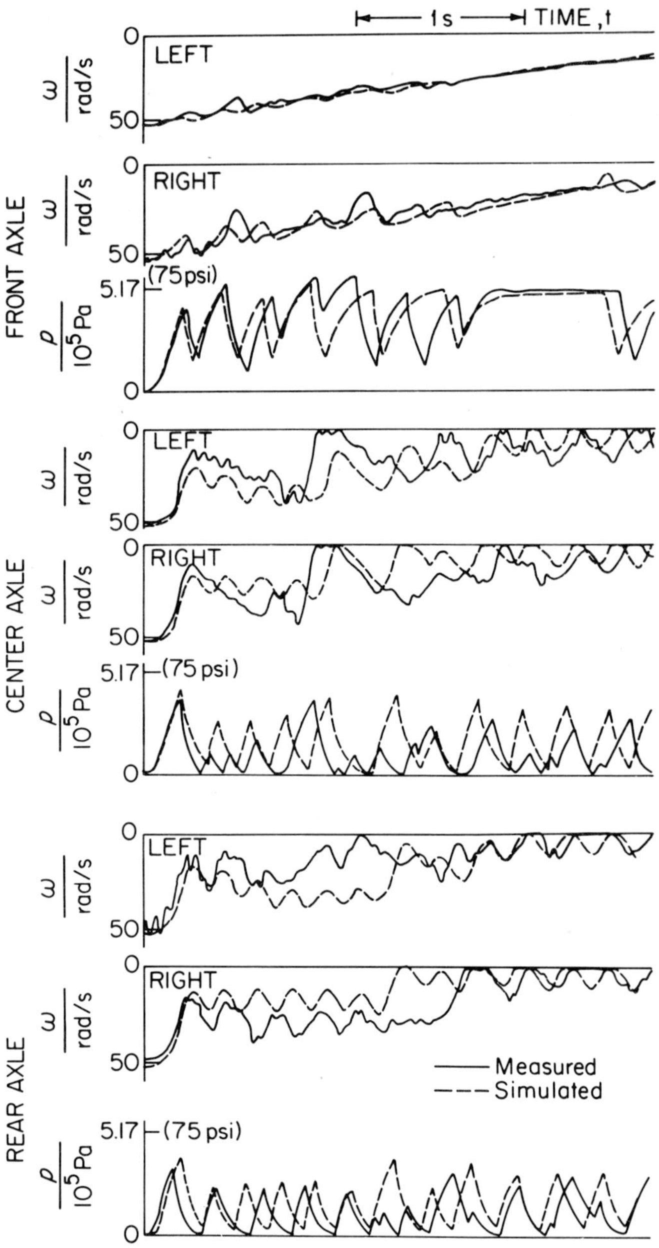

Fig. 6: Measured and simulated wheel speed control, unloaded truck, initial velocity 26.82 m/s (60 mph)

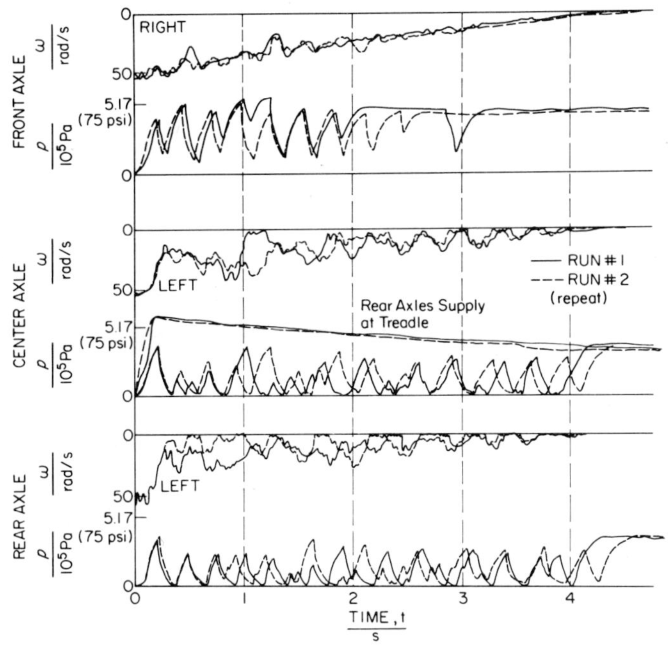

Fig. 7: **Variability of performance, unloaded truck**

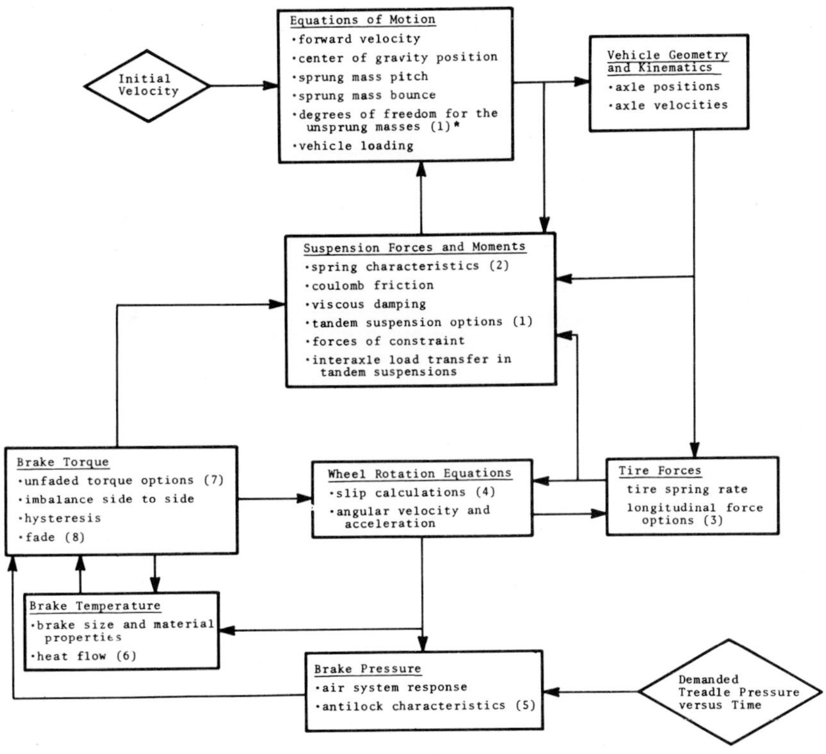

Notes:

(1) The computer program contains options for representing a variety of walking-beam, four-spring, or air suspensions. By setting "keys" in the input data, the mechanical properties of the specified type of suspension will be used in the calculations.

(2) Spring force characteristics may be represented by (a) linear spring rates or (b) tables of force versus deflection, thereby providing a means for including non-linearities such as bump-stop properties.

(3) The longitudinal tire forces can be represented by (a) tables of tire shear force as functions of slip, vertical load, and velocity or (b) a semi-empirical tire model.

(4) Changes in wheel slip during a time step of the digital calculation are computed using an analytical expression rather than numerical integration. This procedure provides a substantial savings in computer time without a significant loss in accuracy.

(5) The programming scheme for representing a variety of antilock systems is described in this paper.

(6) Drum surface temperature during a single stop is computed using a numerical approximation for solving the classical heat flow equation.

(7) Unfaded brake torque can be represented by (a) tables of torque versus brake pressure or (b) models of various types of brakes (e.g., S-cam, wedge, or disc).

(8) A simplified approximation for estimating brake fade is described in this paper. (See Equation 8.)

*Numbers in parentheses refer to the "Notes" above.

Fig. 8: **Concepts used in the digital simulation**

© IMechE 1976

© IMechE 1976

C33/76

PRACTICAL ASPECTS OF TESTING ANTI-LOCK SYSTEMS ON COMMERCIAL VEHICLES

W. J. P. ADAMS, CEng, MIMechE
and D. R. SPENCE, BSc(Eng), MSc
Girling Limited, Kings Road, Tyseley, Birmingham, B11 2AH

The MS of this paper was received at the Institution on 29 October 1975 and accepted for publication on 20 November 1975

SYNOPSIS This paper describes the test procedures and development test results obtained using anti-lock systems fitted to commercial vehicles. The limitations imposed by the potentially hazardous nature of the test sequences are dealt with together with the various procedures being proposed by legislation in Europe. Vehicle stopping distances and stability on various surfaces using current 'state of the art' equipment are tabulated.

1. Introduction

Since the early 1950's, Girling has been involved in the design and development of wheel slide protection (anti-lock) systems and various papers (1) and (2) have described the design philosophy and hardware which has been produced to meet this most challenging problem facing brake engineers. The early systems based on vacuum deboost principles were car orientated but since the late 1960's attention has focused on the problem of applying anti-lock to commercial vehicles. In the last five years experimental commercial vehicles using air/hydraulic or full air systems have been successfully equipped and tested with these systems and it is the practical aspect of testing such vehicles which is the subject of this paper.

2. The proving ground

All the test results in this paper have been obtained on a private proving ground in Warwickshire. This engineering facility covers 250 acres and has its own administration block, design office, instrumentation laboratory, machine shop, workshop facilities and control tower. The main circuit is 4 km (2.5 miles) long where high speed stops up to 195 km/h (120 m.p.h.) can be safely accomplished. To test anti-lock systems, however, a special area has been allocated where three different surfaces have been laid side by side. The entire area covering 152 m x 45 m (500' x 150') can be wetted by a sprinkler system located at each side of the area. During tests, the sprinkler is kept operating to ensure constant depth of water film to aid repeatability of results. The sprinkler heads rotate through 180° and the water is delivered at a rate of 45.5 litres/sec. (600 gall/min.) from a 91,000 litre (20,000 gall) reservoir.

2.1 Test surfaces

Stops on a dry high mue surface and at all speeds are carried out on the main circuit which has a Skid Number (SN) in the range 76-86 1/2 as measured in accordance with the A.S.T.M. practice E274-65T.

The skid pan area consists of three adjacent strips of different surfaces:-

(i) A fine textured asphalt strip 152 m x 12 m (500' x 40') having a skid number of 65 when wet.

(ii) A 152 m x 15 m (500' x 50') low mue surface consisting of an epoxy resin layer which, in turn, covers a specially prepared smooth concrete base. When wetted, this surface has a skid number of 8 at 64 km/h (40 m.p.h.). Durability of this surface has not been satisfactory due to climatic changes and also to the problem of tyres picking up small stones and heavily scoring the surface on locked wheel stops. As a precaution, all tyres have to be inspected and cleaned of foreign matter before testing and also the surface and approach road has to be swept at regular intervals.

An experimental strip of stainless steel sheet laid with a resin bond on smooth concrete gave a stable mue surface but again the climatic conditions broke the bond and the surface has since been removed. An elegant solution to the problem would be to lay a surface of polished marble as used by an Italian motor manufacturer.

(iii) The third strip 152 m x 18 m (500' x 60') consists of a polished bridport gravel surface laid to a detailed specification for aggregate and binder. When wet, this stable surface has a skid number of 32 at 40 m.p.h. which corresponds closely to the U.S.A. Federal Motor Vehicle Safety Standard (F.M.V.S.S.) No. 121 requirement of SN 30 for the 20 m.p.h. service brake stops.

The approach and exit to the skid area is of the same mue level as surface (i) above and hence vehicles can be checked for anti-lock performance on either a high to low coefficient transition stop or vice versa. In addition, because the three strips are laid side by side, the performance and stability of the vehicle, with or without anti-lock, can be checked when straddling a split coefficient surface viz. asphalt/epoxy or epoxy/bridport.

3. Safety precautions

The skid area is under direct observation from a control tower linked by radio to the admin-

© IMechE 1976

istration block and to the test vehicle. During tests, a cross country vehicle is on stand-by in the garage and can reach any part of the proving ground directed from the control tower. The vehicle is equipped with metal cutting, fire fighting and first aid equipment and workshop personnel and security officers are trained in their use.

Drivers and observers of test vehicles wear crash helmets and safety harness and the vehicles are fitted with roll over cages (Fig. 1), minimum trim and fuel capacity. Articulated vehicles are fitted with restraining chains to prevent damage on jack knifing or trailer swing.

The test procedures and safety precautions have been evolved over many years and have proved successful, there has been no injury to company personnel involved in anti-lock tests.

The hazards are not immediately obvious even to engineers within the industry and, on occasion, warnings and procedures have been ignored with consequent accidents and minor injuries to visitors. Other untoward hazards which do cause concern are the fact that two gliders and two light aircraft have made emergency landings on the test track during working hours.

4. Instrumentation

The performance analysis and development of anti-lock systems using static rigs and computer simulators require specialised laboratory techniques. The evaluation of vehicle performance, however, can be satisfactorily achieved using conventional instrumentation techniques as follows:-

(1) Thermocouples for measuring drum/disc temperatures are fitted to all brakes to ensure temperatures are between 60° - 80°C for each initial brake application.

(2) Vehicle speed and stopping distances are measured using a fifth wheel. Chalk gun stopping distances are sometimes used on high mue stops and always on stops involving pavé surfaces.

(3) A pedal effort gauge and decelometer are standard equipment as used in routine brake testing. Anti-lock 'on'/'off' stops are compared using a minimum pedal effort of 890N (200 lbf) thus simulating an emergency brake application which would result in locking wheels with a standard brake system.

(4) Pressure transducers are fitted to each brake chamber/wheel cylinder and fed to a U V recorder. The response of anti-lock valves can therefore be monitored together with wheel speed, vehicle speed, solenoid operation and air/hydraulic pressures at selected brakes.

(5) A 'motometer' deceleration trace is taken of stops involving transition braking on differing surfaces.

(6) Total vehicle angular deviation is measured on completion of the braking operation.

5. Test procedures

The object of the vehicle test procedure is to ensure the product meets the design specification which will include the capability of the system to:-

(1) Function on various road surfaces including transitional and split coefficient surfaces.

(2) Maintain vehicle stability and maneouverability on the above surfaces.

(3) Possess adequate energy resource to carry out sustained braking.

(4) Meet current and proposed legislation.

Vehicle tests can be split into three categories:-

5.1 Static tests

These tests entail checking reservoir capacity, compressor charging times at various engine speeds and system response times. An anti-lock system usually incorporates a logic checking circuit in the electrical system which will indicate to the driver particular modes of failure, e.g. a blown fuse or supply failure. This failure detection circuit is checked statically by operating the ignition system, applying brakes and noting the warning indicators.

5.2 Hill holding

The vehicle is held with minimum necessary service brake pressure with the engine switched off and in neutral gear on gradients from 1 in 6 to 1 in 4. A check is made on any tendency for the vehicle to roll back momentarily if:-

(1) the ignition is switched 'on' hence actuating the anti-lock system via the checking circuit.

(2) Starting the engine which may cause the anti-lock system to cycle via electrical interlocking of auxiliary equipment.

5.3 Braking on constant friction surfaces

Before any of the anti-lock test procedures are carried out, the vehicles brake system is checked out as regards bedding of linings, correct adjustment, basic brake performance at various speeds and functioning of all instrumentation.

Laden and unladen tests are then carried out as shown in Table 1 recording:-

(1) Surface conditions

(2) Anti-lock 'On' or 'Off'

(3) Vehicle speed and stopping distance

(4) Total angular deviation

(5) Reservoir pressures before and after each stop

(6) Unusual behaviour

5.4 Braking on split friction surfaces

Laden and unladen tests are conducted from 16 km/h (10 m.p.h.) and 32 km/h (20 m.p.h.) with the vehicle straddling two differing surfaces e.g.

© IMechE 1976

bridport gravel/epoxy and asphalt/epoxy. Four stops with anti-lock and four stops without are carried out alternately to reduce scatter. Total vehicle angular deviation and stopping distance are recorded.

5.5 Braking on transitional friction surfaces

These tests are to determine the speed of response of the anti-lock system when the vehicle is braked from 32 km/h (20 m.p.h.) and 48 km/h (30 m.p.h.) whilst traversing from one surface to another e.g. braking on asphalt onto epoxy. Two stops in each mode are made and note taken of angular deviation. Usually, observers at the side of the track check for unusual wheel lock condition or behaviour.

5.6 Lane change

Using the bridport gravel and epoxy resin surfaces, a chicane course is marked out using rubber cones to the scheme shown in Fig. (2). The vehicle is driven at increasing increments of speeds, without braking, to establish the maximum speed that the lane change can be made without hitting a marker cone. The test is repeated with anti-lock 'on' and with maximum braking effort applied at a distance of 15 m (50 ft) from a cone barrier in the approach lane. The steerability is determined from the ratio Vb (maximum braking speed) to the maximum drive through speed, Vd.

5.7 Radio interference

Checks are made that radio interference, wiper motors, etc. do not cause interference in the electronic circuit and thus lead to spurious operation of the anti-lock system.

5.8 Pavé test

Due to wheel deceleration and acceleration levels on this type of surface it is necessary to check that the electronic circuit can detect spurious levels of deceleration and inhibit them. Stopping distances on these surfaces are measured using a chalk gun triggered from the pedal.

6. Legislation

The braking requirements of commercial vehicles are comprehensively summarised in reference (3) and the two main sources of regulations which affect anti-lock braking test procedures are:-

6.1 U.S.A.

The published federal documents specifying braking performance are F.M.V.S.S. - 121 and F.M.V.S.S. 105-75 for air and hydraulic braked vehicles respectively. As written, there is no specific requirement for anti-lock systems but one of the stopping distances called for in F.M.V.S.S. 121 is 16.4 m (54 ft) from 32 km/h (20 m.p.h.) on a surface with a SN 30 rating. This has to be achieved laden and unladen without wheel lock-up, and the vehicle must remain inside a 3.6 m (12 ft) lane, a difficult target for a vehicle without some form of anti-lock. If an anti-lock system is fitted, both federal documents specify a warning signal in the event of total electrical failure of the anti-lock system. In F.M.V.S.S. 121, para S.5.5.2., the electrical power for an anti-lock system on a trailer must be obtained from the stop lamp circuit. The implications of this as regards failure indication to the driver are still being studied.

6.2 Europe

E.C.E. regulation 13, annex 10, states that vehicles equipped with anti-lock will be the subject of special requirements. What exactly the "special requirements" are is not yet published as the working party WP29/GRF is currently drafting a proposed annex for anti-locked equipped vehicles. It is understood, however, that two main requirements for anti-lock systems are being formulated in relation to energy consumption and adhesion utilisation.

6.2.1. Energy consumption

The object of the draft test procedures is to check that the reservoir capacity is not exhausted by the operation of the anti-lock system during a prolonged stop on a low adhesion surface. The current proposal is that the anti-lock system must be capable of operating for a time 't' secs where $t = \frac{Vf}{7}$ (Vf is the maximum speed of the vehicle in km/h with an upper limit of 160 km/h) on a surface with a coefficient of adhesion ≤ 0.3 and from an initial speed of 50 km/h. On coming to rest, five full brake applications are made, the fifth application must achieve emergency brake performance level.

For a typical commercial vehicle with a maximum design speed of 70 m.p.h. (112 km/h), the anti-lock operating time 't' will be 16 s thus representing the time required for one stop from 30 m.p.h. (50 km/h) on a surface with an adhesion level of approx. 0.1 (wet ice) and a stopping distance of 365 ft. (111 m). On the basis that this proposal provides adequate safeguard on energy capacity, the brake or vehicle manufacturer has two alternative methods of actually carrying out the test procedure viz;

(a) Carrying out a single stop for the specified time from 50 km/h on a very low adhesion surface e.g. 0.1 and, consequently, on a test track at least 400 ft (120 m) long. On coming to rest the five static brake applications are made. The disadvantage with this proposal is that test tracks of this length and adhesion are rare.

(b) The specified elapsed time can be made up in a series of stops on a higher adhesion surface, but not exceeding 0.3, thus permitting a shorter length of test track. This procedure results in complexity due to the need to isolate compressor feeds to the reservoirs between stages. As up to four stages may be required to complete the anti-lock operation time 't', the results are not easily repeatable as tyre/track conditions will vary between each stop.

From a brake manufacturers point of view, a less complex, viable and repeatable test would be to carry out one anti-lock stop from 50 km/h on an adhesion surface of approximately 0.3 followed by 'n' static full brake applications where the value of 'n' could be agreed between all parties for particular categories of vehicles and systems.

6.2.2. Adhesion utilisation (ε)

This parameter (ε) is intended to be a measure of an anti-lock systems capability of making use of the theoretical tyre/road adhesion (K) and is measured on two surfaces, one with a coefficient of adhesion ≤ 0.3 and the other of about 0.8. Laden and unladen tests from an initial speed of 50 km/h are proposed.

© IMechE 1976

Currently there are four proposals for evaluating ε (two of the proposals depend on measuring the factor K) and they can be summarised as follows:-

6.2.2.1. Adhesion method

Where all the wheels of the vehicle are controlled, the applicable formula is $\varepsilon = \frac{Z\ max}{K}$ where Z max. is the maximum braking rate of deceleration using the anti-lock system and K is the coefficient of adhesion. The anti-lock system shall be deemed satisfactory if $\varepsilon \geqq 0.75$. The value of K is derived from a test procedure which, using the brakes on one axle only, determines the maximum deceleration without locking the wheels from 50 km/h. This deceleration value, together with the physical dimensions of the vehicle e.g. wheelbase, height of c. of g. and rolling resistance are combined in a formula to give a value of K for a two axled vehicle as follows:-

$$K = \frac{Zm\ Pt - R}{Pi \pm \frac{h}{E} Zm\ Pt}$$

where:- Z_m = peak deceleration without locking the wheels

P_t = total weight of vehicle

P_i = static weight of axle i

R = rolling resistance of unbraked vehicle

h = height of c. of g.

E = wheelbase

6.2.2.2. Stopping distance method

The stopping distance 'S' in metres is measured with the anti-lock system in operation and for compliance, $S \leqq 0.15V + \frac{V^2}{190K}$ when both axles are controlled and V is the initial speed in km/h. Again the value of K has to be determined as previously.

6.2.2.3. Stopping efficiency method

This proposal compares the stopping distance of the vehicle when the anti-lock is in operation with the stopping distance of the vehicle without anti-lock and maximum brake force applied i.e. locked wheels. The anti-lock system will be deemed satisfactory if $\varepsilon \leqq 1.05$ where

$$\varepsilon = \frac{\text{stopping distance with anti-lock in operation}}{\text{stopping distance without anti-lock in operation}}$$

6.2.2.4. Mean braking rate

This method is based on the comparison between the minimum stopping distance so obtained by braking a single axle of the vehicle without locking the wheels (i.e. peak deceleration) to the stopping distance S_1 measured by braking the same axle with the anti-lock system in operation. The mean braking rates Zo and Z_1 of three successive stops are calculated and for compliance $Z_1 \geqq 0.75\ Z_o$.

Examining these four proposals from a practical aspect leads to the conclusion that proposals 6.2.2.1, 6.2.2.2 and 6.2.2.4 are not realistic for the following reasons:-

(a) Braking a vehicle on one axle only is not representative of service conditions because of the lower deceleration obtainable, the reduced weight transfer and the excessive energy absorption imposed on the brakes of the braked axle.

(b) The procedure only applies to two axled vehicles and is not suitable for vehicles equipped with different tyres front and rear.

(c) On low mue surfaces, attempting to brake at peak deceleration with unladen air braked vehicles is impractical due to the low level of air pressure required. Variations in brake pressure due to temperature, speed and driver skill can result in significant errors in measuring the surface coefficient.

(d) Measuring the peak deceleration of a single axle braked vehicle requires sophisticated instrumentation and the results can be influenced by driver skill.

From a brake manufacturers viewpoint, the only valid practical test is proposal 6.2.2.3. viz. comparing the locked wheel stopping distance with the stopping distance achieved by the anti-lock system. Further arguments in favour of this procedure are:

1. The test is suitable for all types of vehicles with single or multi-axle anti-lock equipment and equipped with different tyres front and rear.

2. The tests give repeatable results and do not depend on driving skill as maximum pedal effort is applied on each test.

3. The tests simulate real life situations with regard to deceleration level, weight transfer, energy absorption and also provide an accurate comparison of vehicle stability.

4. Manufacturers and government establishments have acknowledged the need for evaluating anti-lock systems and have independently arrived at the same conclusion that method 6.2.2.3 is a valid basis of comparison. In the U.S.A., the S.A.E. Wheel Slip Brake Control System Road Test Code J46 adopts this method together with the Swedish European Safety Vehicle programme and the Transport and Road Research Laboratory in the U.K.

From experience, the test method does have one expensive side effect in that locked wheel tests on the high mue surface results in rapid tyre wear. There is also the problem of testing articulated units with locked wheels which can result in hazardous situations.

7. Commercial vehicle anti-lock systems and typical test results

7.1 Air/hydraulic brake actuation

In parallel with the work being carried out on car anti-lock systems, reference (1), in the late 60's, a system for controlling the braking of rear wheels on commercial vehicles was also being developed. The first experimental vehicle was a TK Bedford of 12.7 tonnes G.V.W. equipped with 15 1/4" x 5" HLS/S front brakes and 15 1/2" x 6" H2LS rear brakes.

© IMechE 1976

The actuation was split 'vertically', with the rear brakes hydraulically operated from a 1 1/2" C.B. master cylinder acted on by a type 36 air chamber. Grafted on to the rear air system was the prototype anti-lock system as follows:-

(1) Wheel sensors:- Each rear wheel was fitted with a sensor device consisting of a 133.3 m.m. (5 1/4") diameter ring incorporating 100 teeth and two electro-magnetic pick-ups, one for the anti-lock signal and the other for instrumentation. The sensors gave a signal frequency of 9.5 Hz/km/h (15 Hz/mile/hour) the output being fed to the electronic module.

(2) The circuitry of the module was derived from experience on railway anti-lock systems and this prototype commercial vehicle circuit had additional adjustable settings for deceleration levels and inhibit timings such that the output could be trimmed during the development tests. A simple block diagram of this module is shown in Fig. (3) and the output of the module drove a solenoid in an air valve controlling the pressure in the type 36 rear brake chamber.

(3) The solenoid operated air valve dumped air from the air chamber on command from the module thus reducing the brake line pressure. Variable air proportioning valves and restrictors were fitted, again for trimming during development.

(4) The testing of this equipment provided a rapid course in the problems involved in commercial vehicle anti-lock systems. The system could cycle at 3-4 Hz and soon highlighted the problem of air consumption. The lowest practical threshold for detection of wheel deceleration was 1.8g, the limit being axle wind-up and wheel tramp. Typical stopping distances achieved are shown in Table 2 for the laden vehicle braking on the rear wheels only, the standard brake system locking the wheels.

The adhesion utilisation of this prototype system was such that it would comply with current parameters. The most significant result however, was that the anti-lock system retained stability on the surfaces, i.e. no angular deviation but that maximum use of the steering capability of the unbraked front wheels was required to keep the vehicle within 30° deviation on the medium and low mue surfaces when using the standard system. The significance of this aspect in relation to braking of articulated vehicles with the inherent problem of trailer swing and jack-knifing is discussed later.

A second air/hydraulic test vehicle has been equipped with an anti-lock system controlling both axles. This is a Ford D550 with a G.V.W. of 8.4 tonnes and equipped with 13" x 4" HLS/S front brakes and 13" x 5" H2LS rear brakes. The basic hydraulic actuation is provided by a 1 1/2" TV/CV tandem master cylinder operated by a dual air actuator, thus providing a 'vertical' split braking system i.e. one half of the master cylinder operates the front axle brakes and the other half the rear brakes. The anti-lock deboost valves are inserted in the front and rear hydraulic brakes lines and are controlled by solenoid operated air valves. The hydraulic deboost components are similar in design to that shown in reference (1).

On this test vehicle, a comparison has been made on the rear axle anti-lock system between

(a) fitting a single prop shaft sensor

(b) fitting wheel sensors to each rear wheel

Typical stopping distances (metres) from 48 km/hr (30 m.p.h.) on the wet bridport gravel surface are shown in Table 3. The comparable braking performance and ease of fitting of the prop shaft sensor has to be weighed against the cost of modifying the electronic module to cater for backlash in the rear axle drive. Furthermore, the stability of the vehicle when braking on split coefficient surfaces is inferior when fitted with a prop shaft sensor compared to rear wheel sensors, a conclusion also reached on evaluating car anti-lock systems.

7.2 Full air brake system

Anti-lock systems on an articulated vehicle have been developed on a combination consisting of a Leyland Super Comet tractor and a Dyson semi-trailer fitted with a water tank for ease of changing from the laden/unladen state. Details of this rig are shown in Table 4 and Fig.(1). The laden and unladen G.C.W. of the combination is 23.1 and 8.95 tonnes respectively.

The anti-lock system fitted to this vehicle comprises wheel sensors, electronic control module and air control valves as follows:-

(1) Wheel sensor

The signal for detection of wheel deceleration or acceleration is provided when a steel pressing exciter ring fixed to the rotating hub passes a fixed permanent magnet and coil attached to the brake backplate. The exciter ring has 120 slots arranged radially which determines the frequency of the signal.

As the exciter revolves, the ribs and slots interrupt the magnetic field and the coil produces an alternating voltage at a frequency proportional to wheel speed regardless of whether or not the brakes are applied. This self powered signal is an important feature of the failure detection circuit in the module. Atypical sensor installation in a hub is shown in Fig. (4).

(2) Control module

The electronic module is a solid state analogue computer using integrated circuitry and is hermetically sealed to provide protection against water and dirt.

The function of the module is to receive the signal generated by the wheel sensors and to interpret this signal in order to measure the rate of change of wheel speeds. By monitoring the deceleration of the wheel, the module detects when wheel lock is imminent and thus applies a signal to the air control valve via a solenoid thus reducing air pressure in the brake actuation. The reduction in brake torque prevents the wheel from locking up and due to the prevailing tyre/road adhesion, the wheel will subsequently accelerate up to vehicle speed. Again, depending upon the level of acceleration, the system logic will determine at what point the brakes should be reapplied. Various failsafe

© IMechE 1976

circuits are incorporated in the module e.g. failure of a sensor is detected and indicated to the driver by a warning light. Similarly, failure of the 24V supply, solenoid malfunction and earthing faults on any part of the wiring harness are also indicated to the driver.

(3) Air control valve

The solenoid operated air control valve Fig.(5) controls the braking effort on one wheel or one axle by being inserted into the line between the service brake control and the brake chambers. The construction of the valve is simple, consisting mainly of a solenoid, poppet valve and two independent diaphragm valves all contained in one housing. During a normal brake application, the solenoid is not energised and so the poppet valve remains open and the brake chambers are in communication with the air supply. On receiving a command signal from the module, the solenoid is energised and the poppet valve closes thus preventing air from the supply reaching the brake chamber. Simultaneously the upper diaphragm opens to allow the air from the brake chamber to be discharged to atmosphere. This cycle will continue until the electronic module changes its command or when vehicle speed falls below 3.2 km/h (2 m.p.h.). Should any malfunction of the valve occur, the brake system reverts to normal control. Also, if the solenoid is energised for more than 6 s the system causing the fault will be inhibited and again the brake system reverts to normal operation.

7.3 Development results - Leyland Comet and Dyson Trailer

7.3.1. Constant friction surfaces

Typical results of stopping distance (metres) on the medium and high mue surfaces shown in Table 5.

It should be noted that the greatest increase in stopping distance (5.4%) with the unladen trailer on the high mue surface represents a distance of 0.52 metre (20").

For comparison F.M.V.S.S. 121 specifies a stopping distance of 16.4 m (54') from 32 km/h (20 m.p.h.) a target which was met under all conditions of loading with the anti-lock and standard system. However, the other stipulation in this regulation is that the vehicle must stop in a 3.6 m (12') lane. This was met with the anti-lock system but jack knifing (especially in the unladen trailer condition) was severe with standard brakes for the combination and hence would have failed.

7.3.2. Split friction surfaces (see Table 6)

All stops with the anti-lock 'on' were inside a 3.6 m (12 ft) lane. In contrast, all stops with a standard brake system resulted in severe jack knifing of the combination (to the limit of the restraining chains). The tests on the tractor itself resulted in 90° spin. The stopping distances are measured in metres.

7.3.3. Failure simulation

One of the major problems in fitting anti-lock systems is to ensure that any failure in the mechanical or electronic devices does not result in a lowering of stability or braking standards, compared with the standard brake system, an aspect discussed in reference (2).

The current design of full air-anti-lock systems can achieve this level. The control module monitors various internal and external circuits continuously and any prohibited behaviour lights the warning lamp in the cab and, if necessary, blows the fuse supplying power to the module. Typical failures could be open circuit sensor leads, open or shorted solenoid, earthing of module unsatisfactory etc.

In all, a total of 130 tests have been carried out on the tractor/trailer combination checking out failure modes. In each case, the stability of the combination with an anti-lock failure was no worse than if the vehicle had been equipped with a standard brake system. For example, a failure in the trailer axle anti-lock components would result in trailer swing on braking but this would be no more than if the vehicle had standard brakes.

7.3.4. Lane change manoeuvre

A simplified version of the test procedure described in 5.6 has been adopted for demonstration purposes. Basically, the procedure for such a demonstration is to brake and attempt to steer round the barrier, with and without the anti-lock system on the bridport gravel surface. Using the standard brakes, the combination inevitably jack knifes and scatters the marker cones but with anti-lock the lane change manoeuvre can be accomplished even with inexperienced drivers.

7.4 Development results - Leyland Marathon Tractor and platform trailer

The driving axle only of this articulated unit has been equipped with an anti-lock system, this being a cost effective method of adopting such a system. A single air valve controls both brakes of the drive axle and the control module operates on the 'sense low' basis viz. the first wheel of the drive axle to approach lock point will actuate the module. A load conscious reducing valve was fitted to the tractor driving axle brakes and the trailer brakes and left in operation.

7.4.1. Tractor solo

Preliminary tests on the high mue surface at 60 m.p.h. indicated that 351.6 kN/m^2 (51 p.s.i.) air pressure produced heavy marking from the front tyres and thus no full pressure brake stops were attempted in case of overturning. The full test procedure on this high mue constant friction surface showed that stopping distances with the anti-lock system were never more than 3 1/2% longer compared with the standard system. The complete loss of control on the wet bridport with the standard braked tractor should be noted. Typical results are shown in Table 7.

7.4.2. Tractor and unladen trailer (G.C.W. 10.7 tonnes)

On the high mue surface, stopping distances with and without anti-lock were comparable i.e. within ± 2% at 64 km/h. However, the trailer wheels locked first in spite of the L.C.R. valve and thus induced trailer swing.

Table 8 shows typical results on the wet bridport gravel surface, the air pressure levels being low in view of the hazards involved. Several tests at higher brake pressures constantly locked the trailer wheels resulting in 20° deviation on the trailer.

7.4.3 Tractor and laden trailer (G.C.W. 32.3 tonnes)

Tests from 40 km/h, 64 km/h and 88 km/h on the high mue surface indicated an average reduction in stopping distance of 5% with the anti-lock system compared with standard brakes. There was a marked tendency for the front wheels to lock.

Similarly on the wet bridport surface, the front wheels locked first and also there was a tendency to jack knife with the standard system which was prevented when the anti-lock was in operation on the drive axle of the tractor unit. Typical results are shown in Table 9.

8. Conclusion

Current anti-lock systems, after many years of development, can provide an improvement in stability for articulated vehicles on all road surfaces with only a marginal increase in stopping distance on high mue surfaces. As most articulated vehicles are full air braked this has been the incentive for component manufacturers in the U.K. to market devices for controlling the brakes on one or more axles of such vehicles.

Experience in the U.S.A. of such systems has been published by N.H.T.S.A. in November 1974 where it is stated:- "millions of axle miles of anti-lock operation has shown a malfunction rate comparable to other equipment presently in service and no highway accidents attributable to the device". It is this aspect of anti-lock which is currently receiving most engineering attention in the U.K. The fail-safe modes and causes of failure are being vigorously investigated with the object to reduce the malfunction rate below that of current equipment in the standard brake system. The necessary service facilities, including diagnostic equipment for checking the electronic module, has an important bearing on the final cost of the vehicle anti-lock system.

9. Acknowledgements

The development of such equipment involves an engineering team made up from all departments and disciplines in an organisation. The authors wish to acknowledge this team at Girling and also to thank the Directors of Girling Limited and Lucas Industries Limited for permission to publish this paper.

References:-

(1) Design and Development of Hydraulic Powered Wheel Slide Protection Systems by R.A. Cochrane - S.A.E. Paper 720031 - January 1972

(2) Safer Braking Systems
by B. Ingram and P. Oppenheimer
Proc. Auto. Div., Instn. Mech.Eng. 1973 Vol. 187 paper 10/73

(3) Braking Regulations in Europe
by P. Oppenheimer - S.A.E. paper 740313, March 1974.

Figures:-

(1) Tractor/trailer test vehicle

(2) Lane change course

(3) Block diagram of prototype module

(4) Typical exciter/sensor installation

(5) Schematic layout of air control valve.

© IMechE 1976

Surface Mue Level	Speed Km/h			
	32 (20 m.p.h)	48 (30 m.p.h)	64 (40 m.p.h)	Max. safe speed
Main circuit	6 stops 'on' 6 stops 'off' Alternate	6 stops 'on' 6 stops 'off' Alternate	6 stops 'on' 6 stops 'off' Alternate	6 stops 'on'
Bridport gravel	"	"	"	6 stops 'on'
Epoxy resin	"	"	/	/

Table 1. Test schedule for braking on constant friction surfaces.

Speed km/h	Surface	Systems	Stopping distance metres
32 (20 mph)	Expoxy	Standard Anti-lock	35.9 34.9
48 (30 mph)	Bridport	Standard Anti-lock	48 46.2
48 (30 mph)	Dry tarmac	Standard Anti-lock	24.6 25.8

Table 2. Stopping distance comparison - prototype rear wheel anti-lock system.

Brake system	Front wheel sensors Rear wheel sensors		Front wheel sensors Prop. shaft sensor	
	Laden	Unladen	Laden	Unladen
Standard brakes	39.75	35.65	39.75	36.65
Anti-lock 'on'	37.84	34.01	38.22	33.13

Table 3. Stopping distance comparison - rear wheel sensors v prop shaft sensor.

	Tractor		Trailer
	Front axle	Rear axle	
Brake	15 1/2" x 4 1/2" 'S' Cam	15 1/2" x 7" 'S' Cam	15 1/2" x 6" 2LS
Air actuation	Type 16 D.D. 6 1/2" lever	Type 16 S.D. 7" lever	Type 16 S.D. Air cylinder
Anti-lock system	2 sensors 1 module 1 air valve (axle control)	2 sensors 2 modules 2 air valves (wheel control)	2 sensors 2 modules 2 air valves (wheel control)

Table 4. Brake equipment - experimental articulated vehicle.

© IMechE 1976

	Speed km/h	Medium mue (SN 30) Anti-lock 'ON'	Standard brakes	%	High mue Anti-lock 'ON'	Standard brakes	%
Tractor only	32	13.14	13.09	+.4	Testing too dangerous due to 'nose diving'		
	48	29.82	34.65	-13.8			
	64	57.52	67.54	-14.8			
Trailer unladen	32	12.80	13.04	-1.8	10.04	9.52	+5.4
	48	24.92	28.72	-13.2	20.87	20.18	+3.4
	64	56.16	69.54	-19.2	36.17	38.01	-5.1
Trailer laden	16	-	-	-	3.58	3.51	+2.0
	32	13.72	13.54	+1.3	9.87	10.0	-1.3
	48	30.95	34.4	-10.0	20.11	20.17	-0.3
	64	-	-	-	37.65	Dangerous	

Table 5. Stopping distance comparison - articulated vehicle - constant friction surfaces.

	Speed km/h	Bridport/epoxy Anti-lock 'ON'	Standard brakes	%	Asphalt/epoxy Anti-lock 'ON'	Standard brakes	%
Tractor only	16	5.27	4.89	+7.8	5.15	4.54	+13.4
	32	26.3	21.23	+23.8	24.21	18.19	+33.1
	48	66.7	Dangerous	-	62.6	Dangerous	-
Trailer unladen	16	5.29	4.05	+30.6	4.62	3.26	+41.0
	32	24.69	Dangerous		18.69	Dangerous	
	48	61.33	Dangerous		48.72	Dangerous	
Trailer laden	32	21.86	Dangerous		19.60	Dangerous	
	48	52.40	Dangerous		47.78	Dangerous	

Table 6. Stopping distance comparison - articulated vehicle - split friction surface.

Surface	Speed	Air pressure	Stopping Distance Std. brakes	Anti-lock on
Main circuit	96 km/h(60 mph)	351 kN/m^2(51 psi)	64.2 m(211 ft)	66.1m(218 ft)
Wet Bridport	64 km/h(40 mph)	689 kN/m^2(100psi)	76.7 m(251 ft) 180° spin	76.7m(252 ft)

Table 7. Stopping distance comparison - tractor solo.

Surface	Speed	Air pressure	Stopping Distance Std. brakes	Anti-lock 'ON'
Wet Bridport Gravel	40 km/h(25 mph)	96 kN/m^2(14 psi)	38.9 m(128 ft)	37.4m(123 ft)
	64 km/h(40 mph)	110 kN/m^2(16 psi)	92.2 m(303 ft) Trailer wheels locked	95.9m(315 ft)

Table 8. Stopping distance comparison - unladen case.

Surface	Speed	Air pressure	Stopping Distance Std. brakes	Anti-lock 'ON'
Wet Bridport Gravel	40 km/h(25 mph)	172 kN/m^2(25 psi)	41.2 m(135.5ft)	38.8m (127.7ft)
	64 km/h(40 mph)	172 kN/m^2(25 psi)	104.2 m(343 ft)	101 m (332.5ft)

Table 9. Stopping distance comparison - laden case.

© IMechE 1976

Fig. 1: Tractor/trailer test vehicle

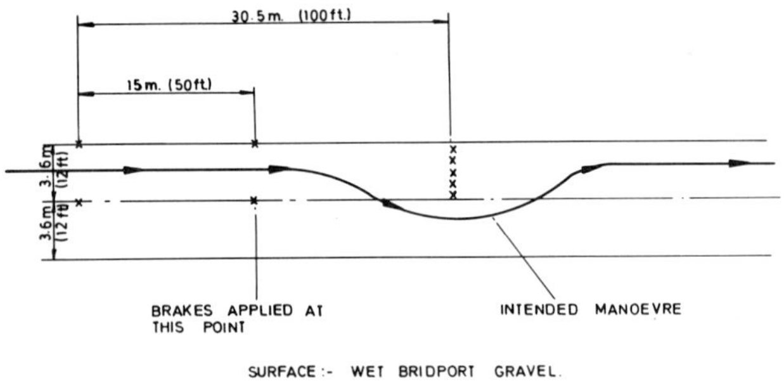

Fig. 2: Lane change course

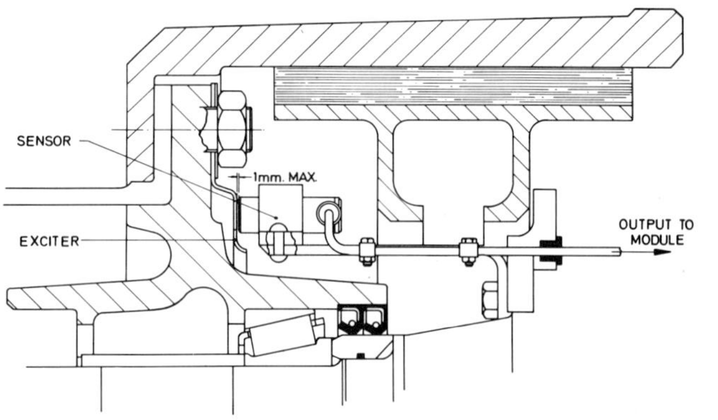

Fig. 4: Typical exciter/sensor installation

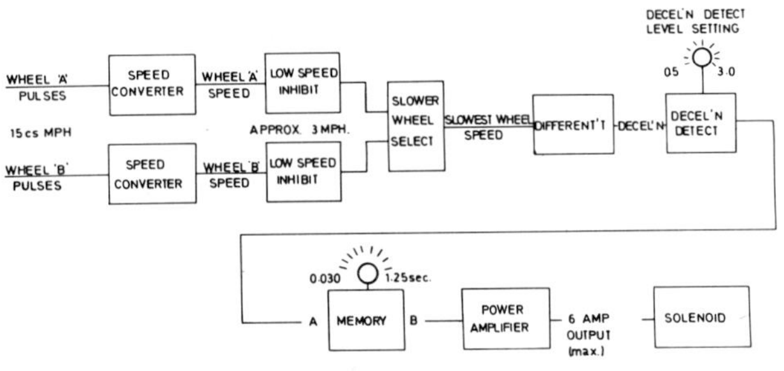

Fig. 3: Block diagram of prototype module

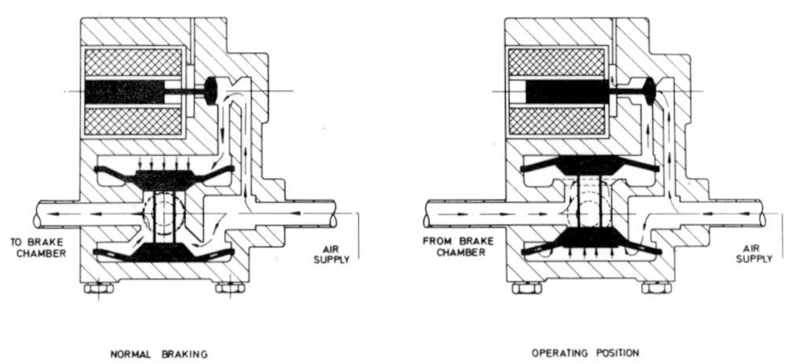

Fig. 5: Schematic layout of air control valve.

© IMechE 1976

C34/76

SOME SIGNIFICANT DEVELOPMENTS IN AIR BRAKE SYSTEM COMPONENTS

G. BURRIDGE, DipTech (Eng), CEng, MIMechE, MSAE
Bendix Westinghouse Ltd, Kingswood, Bristol

The MS of this paper was received at the Institution on 23 July 1975 and accepted for publication on 12 November 1975

SYNOPSIS Innovation and legislative changes have led to the development of present day brake equipment, from very simple beginnings. Divided systems and, subsequently, improved Service, Secondary and Parking performance have been achieved in a number of ways, some of which are described. More recently, means of regulating brake distribution, providing anti-skid and even supplying dry air have come to the fore.

INTRODUCTION

1. In any new field of work, some of the truly significant steps are the early ones and, in air brake technology, the work of George Westinghouse was fundamental. The direct air brake stems from his U.S. Patent No. 88929 in 1869 (re-issued as No. 5504 in 1873) (Ref. 1).

2. The Automotive Brake of 1872 and the Triple Valve of 1887 were early advances which turned the original concept into a practical proposition. By incorporating the latter device, brake application time on the 50th brake of a railway train was reduced from 20 seconds to 6 seconds. With further changes, this time was soon brought down to 2½ seconds and may now be appreciated as an essential step in overcoming the inherent weakness of the compressed air brake, that is the achievement of a fast response time with a low pressure compressible fluid. During this same period other subjects of topical interest today were given a great deal of attention. We may quote, for instance, Automatic Brake Adjusters, Self-Sealing Couplings and Electro-Pneumatic Braking Systems.

3. The foregoing concepts were applied, of course, to railway vehicles and the earliest evidence of automotive use of compressed air is in 1910, when it was employed in an air suspension system. It appears that air brakes were not used on road vehicles until some time later and we find them coming into general use in the early 1920's.

Automotive Air Brakes

4. The earliest brake systems comprised the very simple elements of compressor, storage reservoir, foot-operated control valve and actuators. Over the next 30 years, very little change occurred to the system and a comparatively minor updating of components and materials was evident. A mixture of brake cylinders and diaphragm chambers were employed to actuate the brakes and the latter were assembled by means of a ring of bolts.

5. The familiar brake chamber, comprising two pressings, a beaded diaphragm and a circular clamp ring, was conceived in the USA in 1950 and was first used in the UK in 1952 (Patent No. 711,635). This construction has subsequently been generally adopted as the standard form of service brake actuator on air-braked vehicles. Despite its rather low efficiency compared with a brake cylinder and the fact that its output varies somewhat with stroke, this unit provides an extremely "cost effective" means of converting pressure to force and is well suited to give reliable service when fitted beneath heavy trucks operating in a variety of climatic and environmental conditions.

Dual Brake Systems

6. In the mid-1950's, the heavy truck industry began to devote effort to the development of brake systems which were divided in such a way that any single failure would permit braking to be maintained on at least one axle. In terms of components, this created a need for suitable foot control valves. Early designs employed two conventional single valves mounted together and operated through a common mechanical linkage. A simplification of this concept, using an integral side by side arrangement, has been progressively developed over the years and remains in large scale use. (Fig. 1A). The diagram illustrates a straightforward design, giving a reliable, robust and easily understood mechanism. Apart from the input linkage, which can be made with an adequate factor of safety, all the operating components and the air system control are obviously separate and independent. Advantage may also be taken of the possibility of varying the relative operating lever lengths to bias the pressure to the front or rear brakes. This principle is best known in the full power hydraulic passenger car application, where movement of the pivot point on a similar valve is used to regulate the brake distribution, in accordance with load. In a less sophisticated manner, a valve fitted with a direct operating treadle may be set up to give increasing bias to the front brakes, by the movement of the treadle roller across the balance beam during its arc of travel.

7. Dual foot valves may also be of the tandem or concentric type and this generally provides a more compact design. (Fig. 1B). This valve uses the delivered air from one circuit to control the operating piston for the other circuit, so eliminating some of the variable factors which can influence the delivered pressure balance. However, this construction often leads to outputs which are biased in one direction and improvements to this characteristic can lead to problems with hysteresis

© IMechE 1976

and sensitivity. It is also necessary, according to interpretation of European legislation, to use double seals and a vented space in the centre of the valve, to ensure that a single failure between the circuits cannot go unnoticed.

8. These forms of foot control valve permitted the use of split braking systems on heavy commercial vehicles from the late 1950's and, in this respect, they were in advance of the majority of passenger cars. Certain vehicles in Europe have been produced with Triple Foot Control Valves, where one circuit is used for each of the prime mover axles and the third controls the trailer brakes. This allows the trailer brake application to be speeded up artificially with respect to the towing vehicle and, also enables the vehicle designer to alter the coupling pressure independently of the motor vehicle actuator pressure.

9. Alternative arrangements of duplicated control valve have been considered and designs have been postulated which include simultaneous upright (increasing pressure delivered as control is moved from "off" position) and inverse (decreasing pressure delivered as control is moved from "off" position) pressure delivery signals for use on spring-braked vehicles. Nevertheless, usage is concentrated on the straightforward dual types which, with the use of appropriate relay valves, actuators and foundation brakes, permit the vehicle designer a wide choice of systems, including vertical split (axle by axle separation), or, horizontal split (limited brake force at all axles from each circuit). Naturally, other forms of divided system, such as "diagonal", "three wheel tandem" and so on are possible, but have never, to the author's knowledge, been used for heavy commercial vehicles.

10. Although not specifically required by the "Construction & Use Regulations", these dual systems did provide one means of achieving the independent secondary brake, which became mandatory in the U.K. from 1968. More recently, European legislation has specified performance requirements for residual braking from the normal service control, in addition to the secondary system. Dual foot valves are, therefore, now becoming a standard fitting on all vehicles.

Secondary Brake Requirements

11. Legislation changes in the U.K. during the period 1964-1968 enabled goods vehicles to operate at higher gross weights than had been possible previously, but at the same time, more stringent requirements for brake performance were laid down. In particular, the independent secondary brake system, mentioned above, became mandatory. Engineering activity, therefore, moved from control valves to the brake actuators, where a need for multiple function was satisfied, first with triple, and then with the double diaphragm construction. (Fig. 2). These both met the same general requirement, in that independent operation from two circuits was possible and the integrity of the brake system could be protected on an adequate number of wheels, in the event of a single failure.

12. The two forms of actuators may be compared and it will be noticed that the central diaphragm of the triple arrangement merely acts as a separator and provides a vented space between the independent brake systems. Prevention of diaphragm scuffing when the rear circuit was pressurised, leading to premature failure, was an early problem and brought about the introduction of the large separating blocks, which contribute to the overall size. A significant innovation is shown in the double diaphragm construction, where protection of the front circuit, in the event of a rear diaphragm failure, is achieved by the means of a small sealing element positioned over the only possible air escape route. In addition to the obvious benefits of economy and ease of installation, the remaining diaphragm scuffing problems were alleviated by this design.

13. Widespread use of both forms of actuator commenced in 1964 but, on motor vehicles, we have seen a progressive change to other equipment and, more recently, the multiple diaphragm units have been largely confined to use on trailers. EEC legislation does not specify an independent secondary brake system for trailers and this use is, therefore, expected to diminish further.

Secondary and Parking Actuators

14. In 1968 the "Construction & Use" Regulations were revised and laid down specific parking performance for commercial vehicles. In order to preserve interchangeability between old and new, and between different makes of vehicle, the Industry agreed that, in all cases, a combination should be capable of being parked with the braking force generated on the motor vehicle. We were, therefore, faced with the prospect on many vehicles of generating parking effort on steered wheels. A small but significant change in the regulations was made at this time to remove the requirement for the parking brake to be maintained in operation by a direct mechanical action. These factors led to the evaluation of a large number of possible arrangements but only the Spring Brake and the Lock Actuator came into general use.

15. The operation of a typical Spring Brake is illustrated in Figure 3. and, although it introduced problems of inverse control pressure (sometimes combined with simultaneous upright control of trailer brakes) we can see, again, a simple and effective design. To the commercial vehicle user, a robust actuator capable of providing service, secondary and parking functions and which is easy to understand, is important. Maintenance of the vehicle designer's original braking performance throughout the vehicle life is also more certain. In the opinion of the author, the so-called "fail-safe" characteristics of the Spring Brake are of secondary importance. The advantages of preventing a vehicle from being moved without adequate air pressure are not questioned, but it is not considered desirable to construct the system in such a way that operation of the brakes can occur outside the control of the driver, except in the most unavoidable circumstance.

16. In addition to the type illustrated, many other forms of Spring Brake have been developed and the "remote" type, which does not include a service diaphragm portion, is extensively used for parking purposes on medium weight vehicles with air/hydraulic systems. Construction may be varied to include diaphragm operation of the spring portion but this tends to lead to a large unit, which is difficult to install, unless a low output spring or an expensive rolling diaphragm is used. A further alternative is to reverse the diaphragm

© IMechE 1976

and spring portions and this makes possible the construction of an actuator incorporating two diaphragms for the service and secondary brake functions, whilst retaining the spring for parking only. However, extensive experience has shown that early fears of short spring life and complications with control systems have not been realised and, generally, the spring portion can provide a very effective secondary brake, in addition to its primary parking role.

17. The Lock Actuator comprises a multi-diaphragm unit with a locking mechanism on the push rod. This has the advantage of upright control for the secondary and parking inputs and these are not limited to the power of a spring, which must be installed in a small envelope. Conversely, a third control line is needed for the lock mechanism and severe practical difficulties are encountered if a high parking force is locked into a hot brake, which subsequently cools and contracts. It therefore requires a sequenced control circuit to prevent locking in excessive forces and often needs an over-ride feature to generate high pressure when disengaging the locks. External, manual release of the brakes is also difficult to incorporate.

18. Installation of both types of units has presented the vehicle designer with considerable problems and has limited the package size permitted to the actuator designer. The constraints of overall vehicle width, turning circle, engine size, brake lining area and acceptable actuator durability all have to be weighed in the balance and a compromise found. Twin wedge brakes pose additional problems, by requiring two actuators diametrically opposite each other and mounted substantially perpendicular to the backplate. Fortunately, it is normally only necessary to use one Spring Brake per wheel, in conjunction with one single diaphragm brake chamber.

Hand Control Valves

19. Development of remotely controlled parking brake actuators which, in many cases, also provided secondary braking, led to the creation of new generations of hand-operated air control valves. The necessity for miniaturised valves, used in conjunction with relay valves, facilitating installation and pipework with tilt cabs, was quickly appreciated. For some time, however, the component manufacturers were ahead of the vehicle cab designers and many vehicles were fitted with conspicuous valves mounted in a variety of different positions.

20. A problem encountered with the small (Fig. 4B) and sometimes poorly located valves, was damage inflicted by the driver accustomed to a traditional mechanical handbrake lever (Fig. 4A). Goods vehicle drivers have adapted well to the sensible use of these valves but a more substantial construction is now used for off the road vehicles and on town buses (Fig. 4C).

21. Cab design and styling has, more recently, allowed for incorporation of the hand control valves and many examples of attractive and ergonomic design are now available. Nevertheless, there is still no standardised location and some vehicles still carry separate valves for the secondary and parking functions. The potential risk of such an arrangement is that the driver may instinctively operate the familiar parking control when a situation arises where it would be more appropriate to use the secondary brake. Since the parking valve is likely to be of the "on/off" variety and will often produce braking force on a limited number of wheels, the best interests of vehicle safety will not be served.

22. On a rigid vehicle, a graduating valve, operating in the inverse mode and with a handle latch to hold in the "air released" position, makes a convenient secondary and park control for a Spring Brake System. For combination vehicles, simultaneous upright control of the trailer brakes, with inverse control of the prime mover spring brakes, may be required. When parked, the spring brakes have to remain applied whilst the trailer brakes are exhausted. Early arrangements embodied comparatively simple hand valves, with an upright graduated output from one valve element and operation of an "on"off" second element for parking. A rather cumbersome inverse relay valve was then required downstream to control the spring brakes and to match their performance to the trailer diaphragm actuators. A most significant step forward in this field has been the development of a hand valve (Fig. 4D), capable of delivering simultaneous upright and inverse graduable signals to relay valves and with a unique arrangement to lift the operating cam when the "park" position is selected, and so to exhaust positively all air. Accurate matching of outputs, to ensure a good brake distribution and compatibility, is also more readily achieved with this arrangement, by careful attention to cam profiles.

Variable Load Valve

23. The requirement for regulating brake force, in accordance with weight distribution, has been well documented elsewhere (References 2 and 3), however, detailed progress of design has led to simplification of equipment and improved reliability.

24. Some early concepts utilised variable mechanical lever ratios, which were not capable of reacting to movement after the brakes were applied and, in practice, involved rather troublesome linkage arrangements. The valve illustrated in Figure 5 is a good example of the engineer's use of modern materials and incorporates a variable area diaphragm principle, made possible by the use of radially finned intermeshing plastic mouldings, so giving a simple and effective design which is capable of responding dynamically to weight transfer during a stop. The limitation now on reliability and accuracy of control of the brake force lies more with the hysteresis, production tolerances and settlement of the vehicle suspension system, which is used to generate the input control to the valve.

Electronic Braking Corrector

25. The general subject of skid control has been, and will be, covered in more detail elsewhere. However, it seems unlikely, until much more experience has been gained and reliability proven, that we shall see its widespread adoption. A realistic approach for the present is to concentrate attention on axles individually and, in particular, the drive axle of articulated vehicles, where skidding can have immediate and dramatic consequences. A limited introduction will also enable component manufacturers, vehicle builders and

© IMechE 1976

operators the opportunity to gain experience in this new field in a sensible and progressive manner.

26. Small-scale production and optional fitting may not permit the adoption of the best technical solution, which is likely to require individual wheel sensing and control, but a good practical arrangement is shown in Figure 6. This device uses propellor shaft sensing, together with a package comprising all the electronics and pneumatic control valves. It can, therefore, be fitted to standard vehicles on the production line, or subsequently added to vehicles in service, without change to the standard foundation brakes or actuators. The control philosophy is similar to that employed in other more complex systems and the results achieved are limited only by the "axle" sensing and control employed at this stage. It is expected that, in Europe, the increasing use of such equipment will be seen as a very necessary and significant step towards more comprehensive Anti-Skid Systems.

Air Dryer

27. Increased reliability, leading to reduction in maintenance time and enhanced safety, has become progressively more important as labour costs have risen and motorway mileage increased. In the case of compressed air brakes, some of the benefits of supplying the system with clean, cool and dry air are obvious, in particular, the elimination of freezing hazards and the ability to carry the intended reserves of compressed air, rather than reservoirs partly filled with liquid. For a number of years, many vehicles have been fitted with wet tanks and automatic drain valves, in order to trap and remove some of the contaminants. The efficiency of such devices is, however, limited and water condensate may still form in many parts of the system.

28. Economics and space prohibit the use of elaborate industrial type cleaning and drying devices on motor vehicles. The unit shown in Fig. 7 takes advantage of the short and relatively frequent compression/unload cycles and dries the air by adsorption in a micro-crystalline desiccant. At each compressor unload cycle, a small volume of dry air is expanded to atmospheric pressure and passed in the reverse direction through the desiccant to purge it for further use.

29. The major benefit anticipated was the prevention of freezing and reduction in the frequency of draining the main reservoirs. Field trials and subsequent service experience have, however, shown a remarkable improvement in the condition of system valves on vehicles fitted with Dryers, compared with control vehicles not so fitted, and a resultant benefit is that the overhaul periods of valves and actuators can be extended. Contrary to belief in some quarters that compressor oil is necessary in the system to maintain free movement of valve components, experience shows that, in fact, water deposited from the air tends to wash out the lubricant, replacing it with a mixture incorporating dirt and the inevitable pipe scale. Typical reservoir condensate also tends to be acidic, so promoting corrosion of metals and deterioration of elastomer materials. In addition to its drying role, the Air Dryer incorporates filters for the removal of carbon and oil and so has a beneficial effect in reducing problems with foreign materials of all kinds. Despite a high initial cost, such equipment may be expected to come into increasing use in the future.

Present Systems

30. In contrast with the basic direct air system, described earlier, the heavy vehicle produced today, in compliance with the EEC Directives on Braking and incorporating twin wedge brakes, could have a brake system as depicted in Fig. 8. This includes a horizontal split service brake system, together with spring brakes for secondary and parking purposes.

31. Examination of the system components would reveal the widespread use of die-cast alloy materials and plastics, in place of the iron and brass used in former times. The corrosion resisting properties and ease of manufacture of complex shapes has led to increased reliability and permitted more sophisticated designs. Weight saving, leading to payload advantages, is an additional benefit.

32. The use of thermoplastic tubing is also evident on several vehicles of recent design but colour coding of the systems is not generally used. Some work has been done in this respect but coding systems vary between Europe and the U.S.A. already, and it would be nice to hope that WP.29, or I.S.O. would issue a standard before widespread confusion is generated.

33. The system shown includes various special valves required by the EEC Directive. These include the single circuit protection valves at each reservoir, permitting continued replenishment of the remaining circuits, in the event of a single failure. (These may alternatively be combined into one multi-circuit protection device). Also illustrated, is a straightforward multi-relay valve arrangement for combining the various input signals for operation of the trailer brakes. It comprises a simple relay with a pair of changeover valves mounted integrally with the top cover. Finally, the Supply Dump Valve is now considered necessary with most two-line systems to cater for failure of the service line, by ensuring that a rapid trailer emergency brake application is initiated when the foot valve is fully applied.

Conclusions

34. In the relatively short period of 50 years, technical development, coupled with legislative changes, has led to more sophisticated systems containing many more items of equipment. Inevitably, the number of potential failure points has been increased but the benefits attendant on the increased number of circuits must be advantageous and produce safer braking systems. The author has, however, endeavoured to show that simple, robust and reliable design has been given a high priority and every engineer professionally engaged in this field must ensure that this continues to be the case, so that evermore complex equipment of unproven effectiveness is not incorporated on vehicles on the basis only of academic studies, or political convenience.

REFERENCES

1. PROUT, H.G. A Life of George Westinghouse. Benn Brothers, Ltd., London, 1922.

© IMechE 1976

2. ROUSE, J.A. The Distribution of Braking on Road Vehicles. I.Mech.E. Symposium on Braking & Cornering, 11th June, 1963.

3. SHILTON, B.R. Braking System Load Sensing Valves. I.Mech.E. Symposium on Auxiliary Power Services for Road Vehicles, 7th-8th July, 1969.

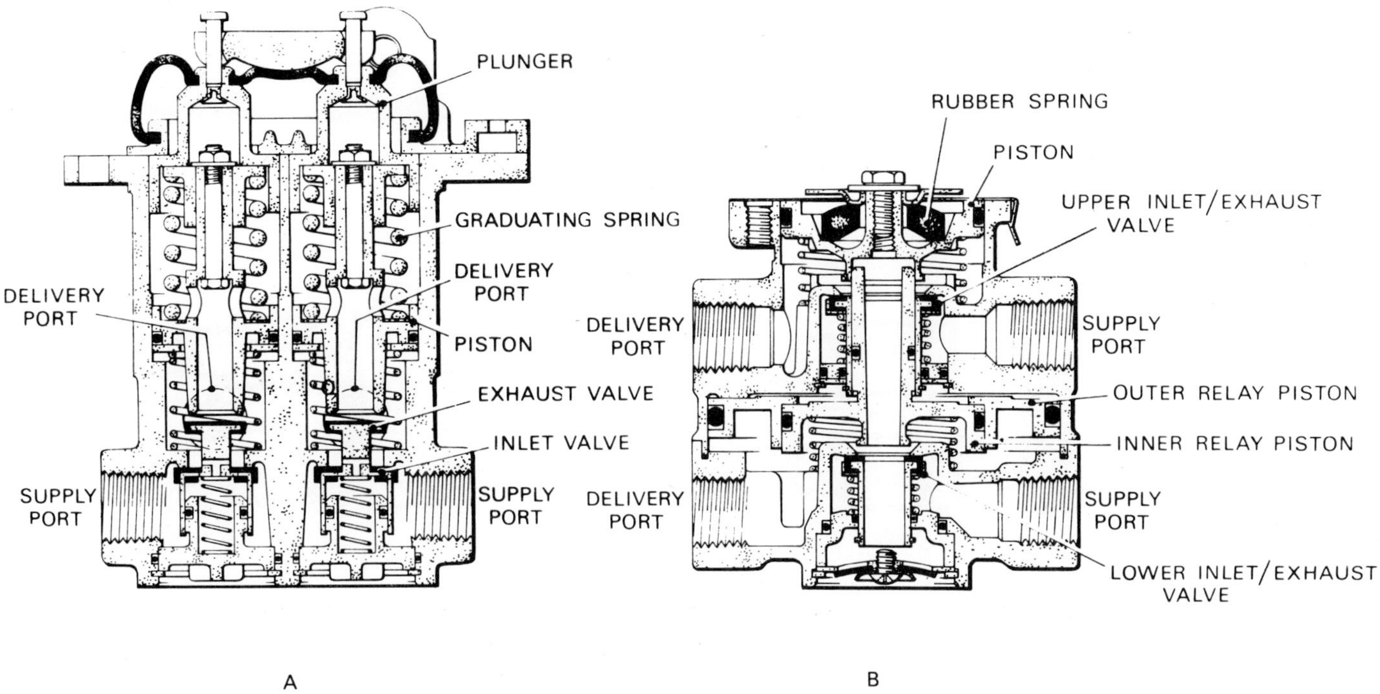

Fig. 1: Dual Foot Control Valves.

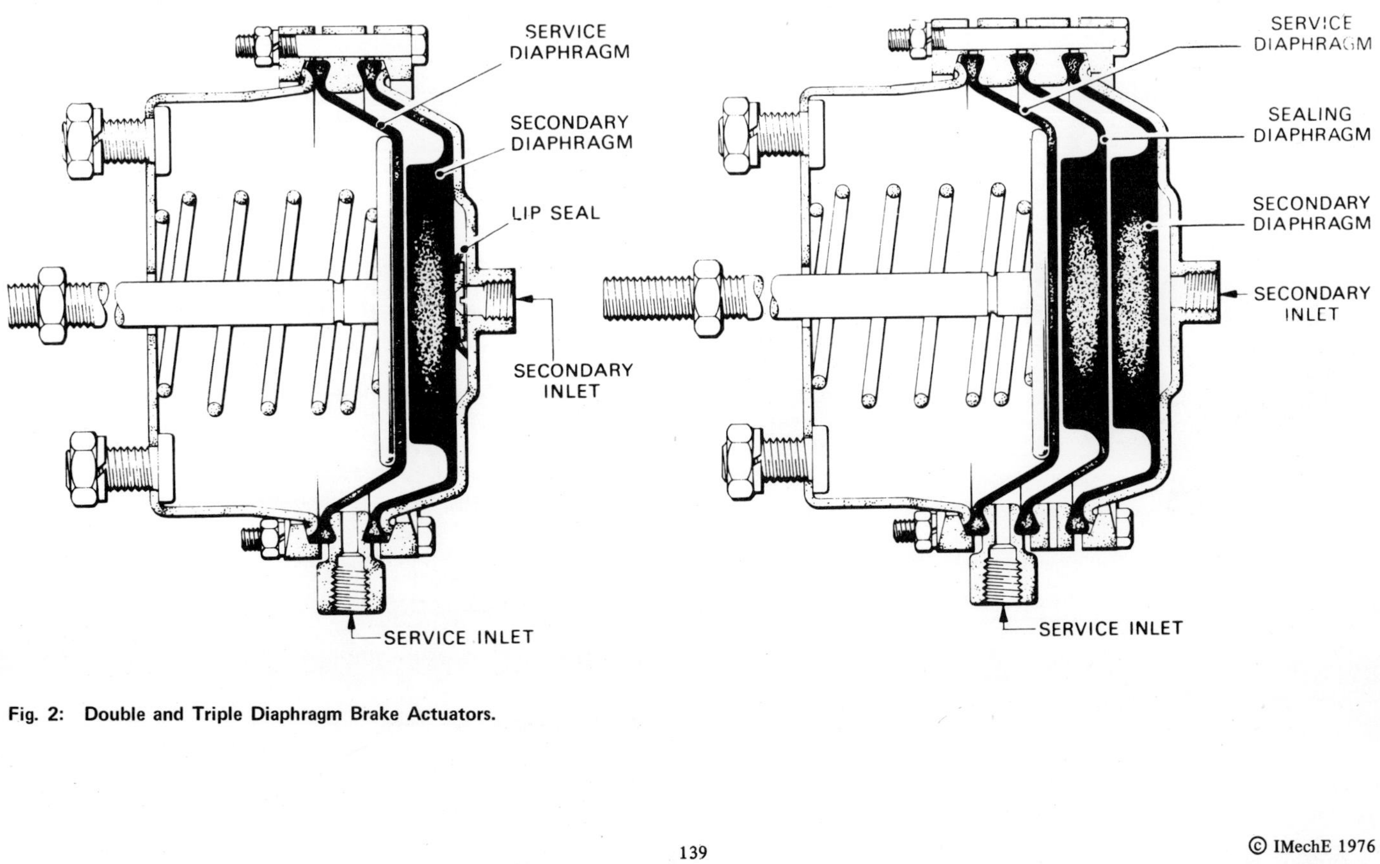

Fig. 2: Double and Triple Diaphragm Brake Actuators.

© IMechE 1976

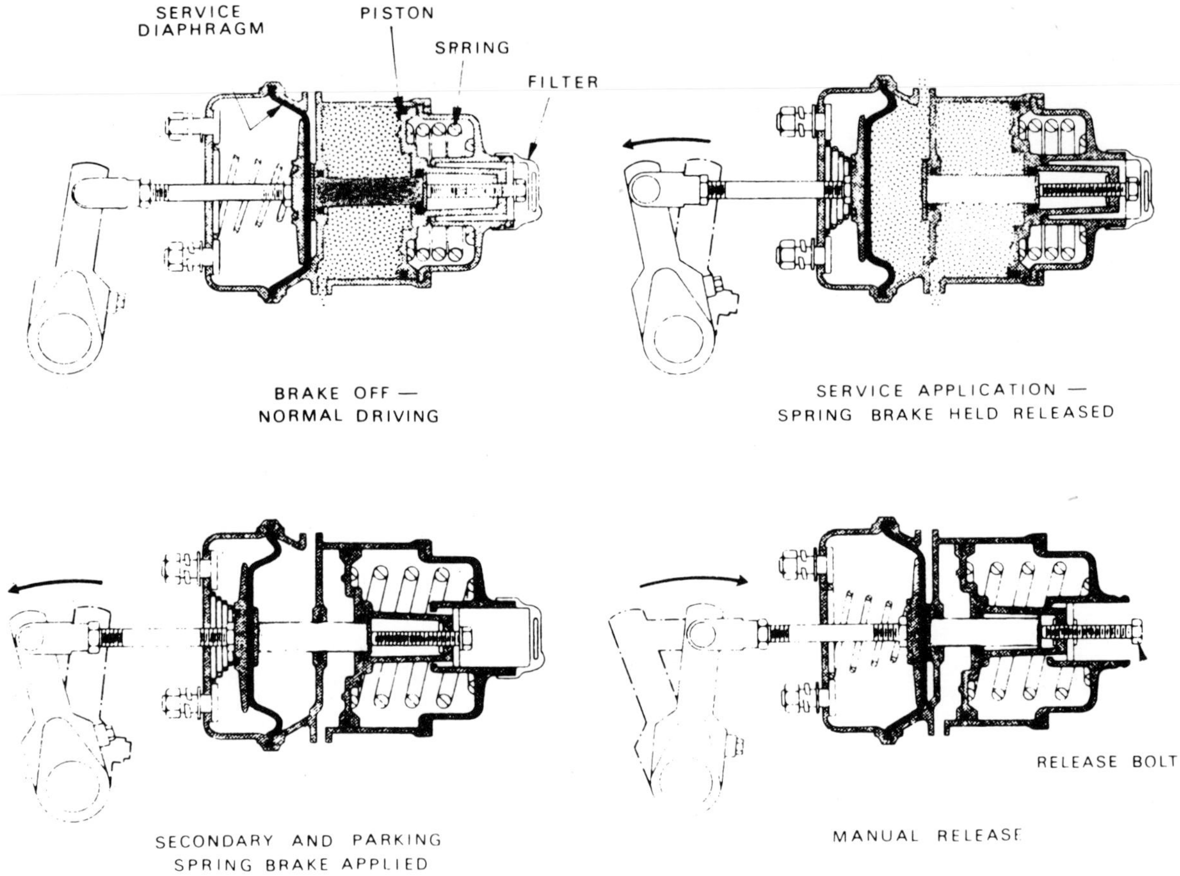

Fig. 3: Operation of the Spring Brake Actuator.

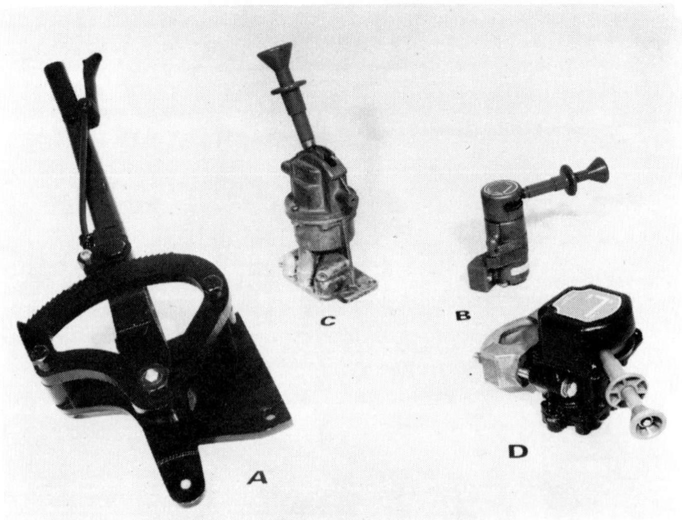

Fig. 4: Hand Controls.

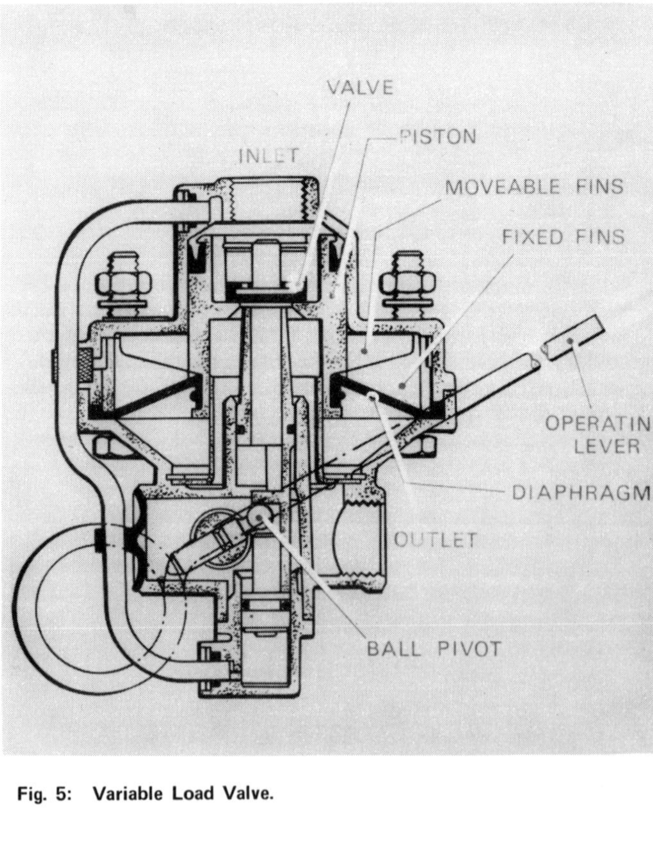

Fig. 5: Variable Load Valve.

© IMechE 1976

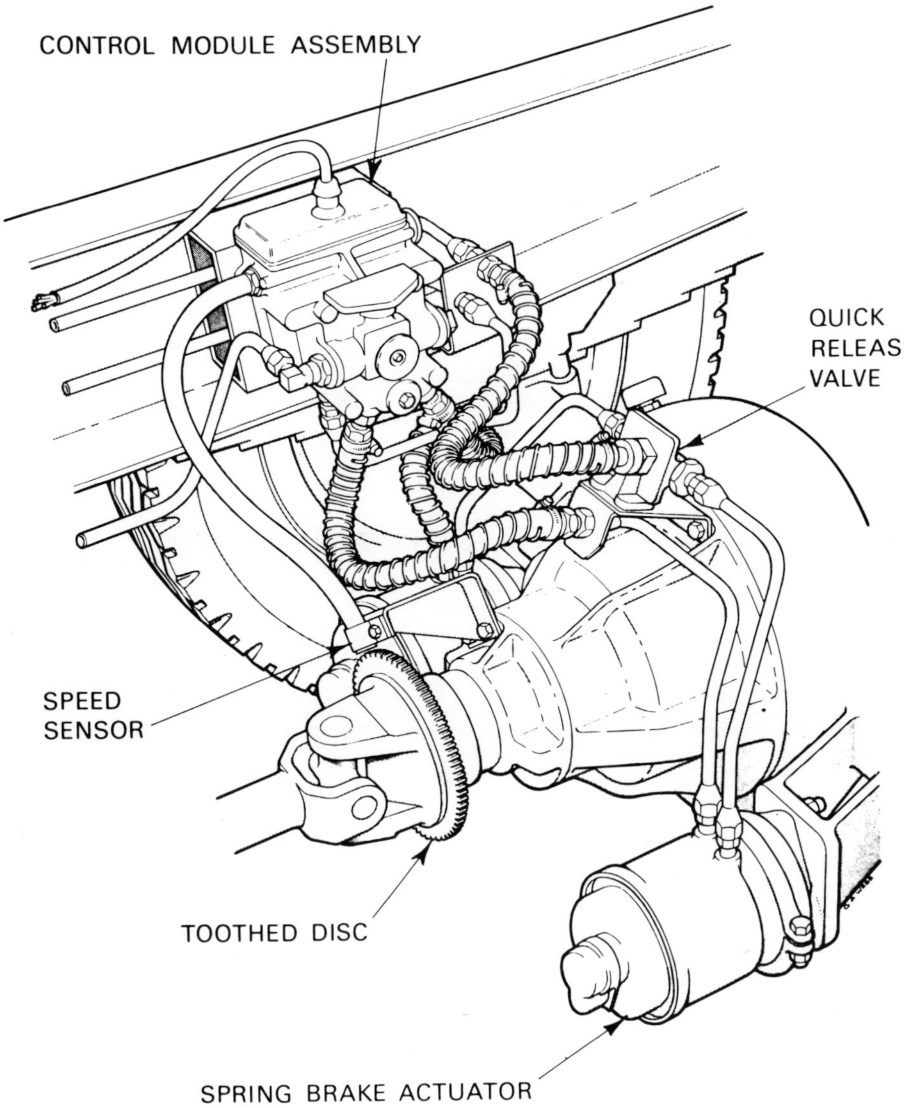

Fig. 6: Electronic Braking Corrector.

©IMechE 1976

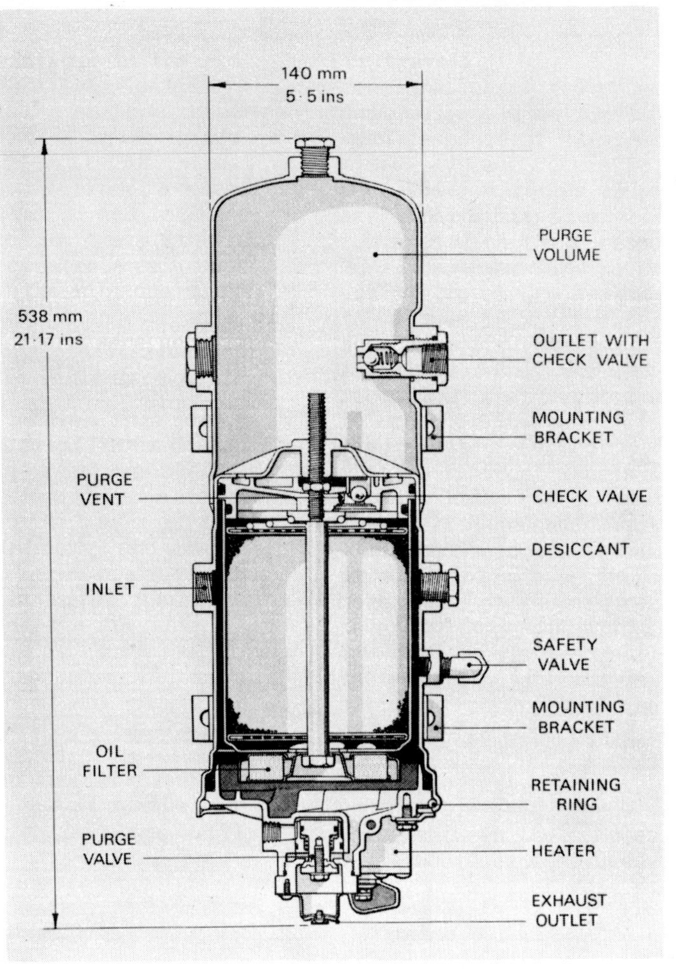

Fig. 7: Air Dryer.

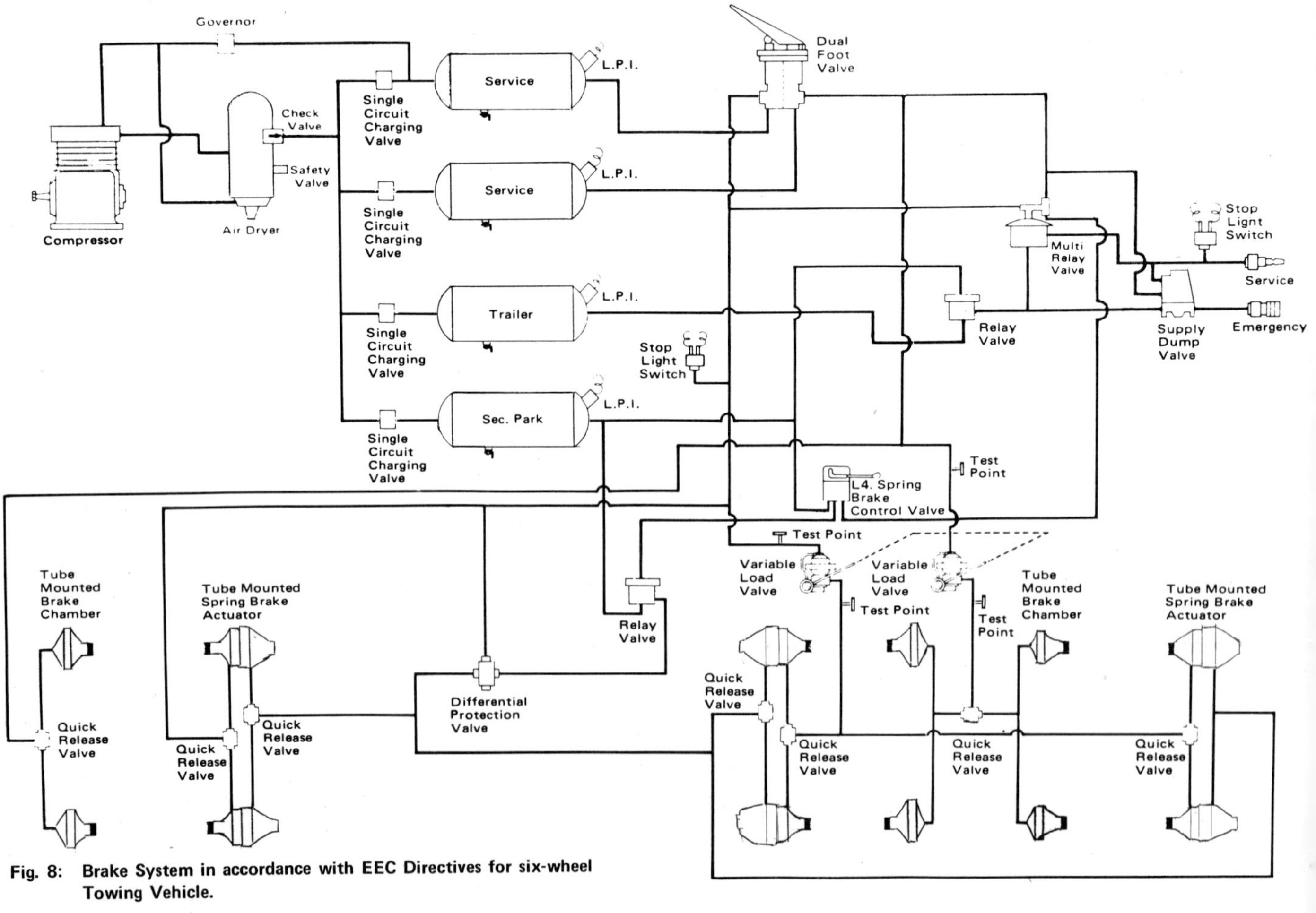

Fig. 8: Brake System in accordance with EEC Directives for six-wheel Towing Vehicle.

©IMechE 1976

C35/76

CAST IRON BRAKE ROTOR METALLURGY

B. J. CHAPMAN, BSc (H & SS) PhD and D. HATCH, BSc (Eng) PhD
Ferodo Ltd, Chapel-en-le-Frith, Derbyshire

The MS of this paper was received at the Institution on 17 September 1975 and accepted for publication on 10 November 1975

SYNOPSIS. Considerable differences occur in the material durability and friction characteristics of different brake assemblies which are to all appearances identical. The paper shows that these are due to differences between the cast iron brake rotors and that the amount of titanium present in the cast iron is responsible. Investigation of the mechanism whereby titanium affects performance has shown that other elements such as vanadium and niobium are also effective. These like titanium produce very small hard carbide particles in the cast iron. Ability to exercise control in a commercial foundry is demonstrated and the technological implications of the discovery are discussed.

INTRODUCTION

1. It has been known for many years that the characteristic friction level and wear rate of any given brake lining material varies according to the vehicle to which it is fitted. This has generally been attributed to mechanical differences between the models, i.e. the weight of the vehicle, the caliper design, the shielding and consequent cooling of the brakes, etc.

2. Because the friction materials are usually somewhat complex mixtures containing organic resins and asbestos, both of which degrade at high temperature, they were suspected of causing different performance characteristics when subjected to different usage on different vehicles.

3. The cast iron countermember was never considered likely to vary in performance since, unless very high temperatures are reached and heat spotting occurs, grey cast iron does not change in structure over the normal brake operating temperatures, remaining substantially the same combination of graphite and pearlite with a trace of ferrite and phosphide eutectic, throughout the life of the rotor.

4. However, it has been found that brake rotors, which are all nominally B.S. grade 17 grey cast iron, do in fact, vary greatly in the friction and wear characteristics they exhibit against brake lining materials. Using the same friction material, changing from a low to a high performance cast iron brake disc can increase friction by 50%, the material wear by a factor of 3 and the disc wear by a factor of 10. (Ref. 1.).

5. The variability in the performance of brake rotors was a serious problem to a research organisation relying on the results of sensitive dynamometer tests to determine the effects of formulation changes.

6. This paper describes investigations that led to the discovery of metallurgical differences which are responsible for variation in the performance of grey cast irons and describes how to control them.

HISTORY

7. A number of years ago, Ferodo Research Division decided to find which of the several differences between proprietary brake assemblies were responsible for their different friction levels when fitted with the same friction material on an inertia dynamometer.

8. Using various combinations of caliper and disc with pads of different shape and size, it soon became apparent that the discs made a considerable contribution to the range of friction obtained. The differences between discs remained even after they had all been reduced to the same thickness, reground on the same grinder and made geometrically identical.

9. The cause of the differences, originally attributed to the design of brake, was therefore shown to be due to differences in the metallurgy of discs used by different brake manufacturers.

10. During the late 1960's, attention was focused on improving the relevance, reliability and reproducibility of dynamometer test schedules. This resulted in much more sensitivity to real differences in performance and a great reduction in the scatter of both friction and wear measurements.

11. It soon became apparent that not only discs on different brakes, but also different discs for the one brake could behave differently, when one disc for a small popular saloon car, henceforward referred to as model A, gave 12% higher friction and 45% higher wear than another apparently identical disc, (Figure 1). It was clearly demonstrated that this was a real, reproducible phenomenon, which could not be altered by grinding and could only be due to the metallurgy of the discs.

12. Thus we now had differences in metallurgy leading to differences in performance between brakes on the same vehicle.

© IMechE 1976

13. In the early 1970's, numerous reports were received, of brakes pulling severely to one side or the other in service. At first, customers blamed this on variability of the friction material and referred the problem to us. Thorough investigation showed that the trouble lay almost invariably with the discs. Exchanging the near side and off side pads alleviated the problem temporarily, sometimes even switching the pull to the opposite side, which was what led service engineers to suspect the friction material, but after further running, the pull always returned to the original side.

14. Discs from numerous vehicles under complaint were tested on a dynamometer, usually with the original pads first and then after regrinding, and with a new set of pads.

15. It was generally found that the discs and pads as removed from the vehicle gave significantly different friction and wear characteristics. After regrinding, there was little difference between their friction levels initially, but during the equivalent of several hundred miles running, the friction on one disc gradually fell while that of the other remained relatively high. (Figure 2). This delay in the development of the imbalance accounts for the failure of previous investigations to determine the cause of the fault.

16. There are several reasons why differences in rotor metallurgy should become a serious problem at this time, when presumably cast iron has been just as subject to variability since brakes were invented without apparently causing any problems. First, until recently manufacturers of any particular vehicle tended to obtain their discs or drums from one supplier and therefore tended to get a fairly consistent product. In recent years there have been many mergers and takeovers followed by rationalisation programmes, which have resulted in components for the same vehicle being made in more than one foundry. With the advent of the Common Market and the development of multinational companies, the situation has become even more complicated.

17. Secondly, designers of motor cars favoured increasing amounts of off-set on the suspension system, which makes the steering much more sensitive to differences between the torque output from the two front brakes.

METALLURGICAL DIFFERENCES

18. There is a bewildering complexity of differences between any two samples of commercial grey cast iron when viewed under the microscope, in either a polished or etched condition, and every field of view on the same sample looks quite different from the previous one, even at X 100 magnification.

19. Substantially it consists of a pearlitic matrix of varying degrees of fineness interspersed with graphite flakes of assorted lengths, thicknesses and arrangements. A small amount of ferrite is sometimes present, usually situated in the centres of graphite rosette structures but occasionally along the larger graphite flakes. In addition, there is a variable but small amount of phosphide eutectic situated around the cell boundaries, and particles of manganese sulphide and titanium carbo-nitride are scattered throughout the whole matrix. Other trace elements present are concentrated at the cell boundaries around the phosphide eutectic lakes.

Graphite

20. When the differences in performance between the model A discs were discovered, a number of authorities on metallurgy were consulted and initially they could find no significant differences between these discs, both being normal grey cast iron of substantially similar structure and composition.

21. However metallographic examination indicated that the lower performance disc had a finer graphite flake structure than the other, and it was thought that this might be responsible for the difference in performance. Two sets of model A discs were prepared for testing, one with well-developed flake graphite and the other with a very fine undercooled graphite structure. These discs were found to behave very differently, the discs containing undercooled graphite giving lower friction and wear than any discs previously tested. (Figure 3).

22. Friction materials from a number of sources were run against these discs, and although there were minor variations to the pattern, all the materials tested showed much higher friction and wear on the flake graphite discs than on the undercooled graphite discs.

23. Examination of several pairs of discs under complaint in general supported the view that the larger graphite flake size, the higher the friction and wear would be. However, some discs could not be fitted into this overall pattern, particularly a set with very large graphite flakes which gave low wear and developed low friction. Therefore a series of discs was prepared with increasing graphite flake size from all undercooled, through mixtures of undercooled and fine flakes to all very large flakes in the hope that the whole range of performance characteristics would be covered. The range of graphite structures was achieved by making slight modifications to pouring temperature, cooling rate and composition : 0.065% and 0.15% titanium was added and phosphorus and chromium reduced to promote undercooling.

24. When tested, these discs gave either very high or low performance characteristics, none of them lay in the middle region where most commercially available discs belong.

25. The graphite structures of some thirty discs were examined and considered against their performance characteristics and it became clear that the answer was not here. The performance of some discs with very big flakes was very high and of others very low, and the performance of some discs with fairly small flakes was very high and of others very low. (Figure 4). Undercooled graphite however, had not given high performance on any disc examined.

© IMechE 1976

26. The possibility of the graphite itself being different was considered. It was abstracted from high and low performance discs by dissolving away the metals and silicon, and the residual graphites were examined by X-ray diffraction. However, no differences could be detected.

Matrix Structure

27. The matrix of all the commercial discs was substantially pearlitic with just a trace of ferrite in some of them. There was a tendency for the lower performance discs to contain more ferrite, and discs cast specially containing undercooled graphite contained much more ferrite and gave low performance. Thus there was a general correlation suggesting that increasing the amount of ferrite reduced performance.

28. In order to see whether the matrix structure was important, a number of fully pearlitic high performance discs were annealed, i.e. heat treated so that all the pearlite was converted to ferrite and others were normalised, i.e. heat treated so that most of the matrix reverted to pearlite but the areas around the graphite flakes remained ferritic. (Figure 5). These discs all gave the same high friction and wear characteristics after treatment as they had before.

29. Time and thought had been given to consideration of hardness variations, hardness distribution, the fineness of the pearlitic structure, eutectic cell size, grain size, the presence of impurities in solution etc. but no satisfactory correlations could be found and in the light of the results on annealed and normalised discs it seemed very unlikely that matrix structure could be of any importance in this context.

Inclusions

30. If neither the graphite nor the matrix was responsible for the performance differences it seemed that only inclusions were left.

31. A large number of pairs of discs which had been the subject of pulling complaints had been analysed for carbon, manganese, silicon, phosphorus, chromium, nickel and copper, but no element or combination of elements could be found to correlate with the direction of pull. Of the discs which had been tested on the dynamometer the only elements which gave any correlation were the titanium added to produce undercooling, and phosphorus.

32. At one stage there appeared to be a direct correlation between performance and the amount of phosphide eutectic seen in the etched sample. However, several proprietary discs were later found which did not fit the relationship. Chromium and some other carbide forming elements become concentrated around the phosphide eutectic lakes and etch darkly with picral. However, no correlation between darkly etching cell boundaries and performance could be found.

33. Manganese sulphide particles were scattered throughout the matrix of all the discs. In some they tended to be small and numerous and in others to be larger and fewer in number, but there appeared to be no correlation between either size or number and performance.

Titanium Carbo-nitride

34. Particles of titanium carbo-nitride were scattered throughout the matrix of all the discs, being particularly prevalent in the discs to which 0.065% or 0.15% titanium had been added to induce undercooling, and which gave very low performance characteristics. The number of particles per 20 fields of view under a microscope at X 400 mag was counted for all the discs available and there were significant correlations between the number of titanium carbo-nitride particles and both wear and the friction achieved. (Figure 6).

35. The particles are extremely small, being only 2 - 6 microns across and it seemed unlikely that the amounts present could be responsible, especially when the approximate percentage titanium present was substituted for the count scale in Figure 6. The curves suggested that in the presence of up to 0.01% titanium, friction and wear would be high, but in the presence of 0.03%, friction and wear would be low, and that increasing titanium up to 0.15% had little further effect. However, no disc could be found which did not fit into the overall picture.

36. The assumption that titanium is the factor responsible for performance would explain all the results, observations and correlations mentioned so far. It would account for the specially cast discs which contained only 0.005% titanium giving very high performance and those containing 0.065 and 0.15% developing very low performance. It would also account for annealing and normalising having no effect. Because titanium promotes the formation of undercooled graphite structure and increases the amount of ferrite present, the tendencies for discs giving low performance characteristics to have smaller graphite flakes and more ferrite than discs giving high performance would also be explained.

DISCS CAST WITH ADDITIONAL TITANIUM

37. With the help and co-operation of Mr. R. J. Edwards of the Department of Metallurgy at UMIST, a number of discs which gave very high performance characteristics and contained only 0.012% titanium were melted down in a laboratory induction furnace and recast. Some discs were cast without any addition and others with ferro-titanium added to bring the level up to 0.06 and 0.20%. Great difficulty was found in getting the titanium into solution as it tended to oxidise away in the slag leaving at most only 0.022% in the cast discs.

38. To overcome this problem, ferro-titanium and cast iron were heated together in a vacuum furnace to produce a master alloy, which was subsequently added to the charge in the induction furnace. In this way, discs containing 0.092% titanium were produced. The performance characteristics of discs produced without any addition were of exactly the same high friction and wear type as those of the original discs before melting, those of the discs containing 0.022% titanium were slightly, but quite significantly lower, and those of discs containing .092% titanium were as low as any discs so far tested. (Figure 7).

© IMechE 1976

39. These results confirmed that the amount of titanium present in the discs was indeed the critical factor and that the relationship between performance and titanium suggested by the curve in Figure 6 was probably valid.

RELEVANT MECHANISM OF TITANIUM

40. Titanium is present in the cast iron as discrete particles of titanium carbo-nitride which are very small, extremely hard and have a high melting point. It was difficult to believe that the presence of such very small amounts of a hard constituent could reduce friction and wear, as they could not carry any significant proportion of the load. Indeed it would be less surprising to find that the presence of hard particles increased performance, particularly wear rate.

41. Titanium is known to have a number of effects on cast iron ; it is a graphitising agent, with a tendency to produce undercooled graphite structures, which is why it had been added to the discs produced with undercooled graphite ; it removes gases from the melt such as oxygen, nitrogen and sulphur which in turn assists in the production of ferrite, strengthens the ferrite and increases the amount of manganese in solution. It seemed more likely that titanium was causing a reduction in performance by acting in one of these ways than by its physical presence as titanium carbo-nitride.

42. Therefore numerous discs were cast to which a variety of other elements which effect the cast iron structure were added. In general, 0.2% of each element was added initially, followed by a range of levels for those which appeared effective.

43. Two base materials were employed. Scrap castings containing 0.012% titanium were used for discs to which additions which might behave like titanium were to be made, and a pig iron containing 0.03% titanium was used for discs to which additions which might reduce the effect of titanium were made. Their analyses are given in Table 1.

Table 1

Analyses of the base metals used.

	Hot Blast Cupola Castings	Hematite Pig Iron
Carbon	3.20%	3.60%
Silicon	2.15%	2.00%
Manganese	0.62%	0.75%
Sulphur	0.06%	0.035%
Phosphorus	0.075%	0.010%
Titanium	0.012%	0.030%

44. Some of the additives were graphitisers like titanium ; some had the opposite effect being carbide stabilisers ; some were deoxidisers and others formed hard particles. In addition, nitrogen and sulphur, which titanium helped to remove, were reintroduced. The effects of the different additives are shown in Table 2.

Table 2

Effects of additives on cast iron performance.

	Graphitiser	Carb. Stab.	Deoxidiser	Hard Part.	Ferrite Prod.	Effect on performance
Ti	*		*	*	*	Reduced
Cr		*				None
Zr	*		*		*	None
V		*		*		Reduced
W		*				None
Mo		*				None
Al	*		*		*	None
Nb		*		*		Reduced
N						None
S		*				None

45. Most of these additives were fairly easy to get into the melt with the exception of zirconium which, like titanium, had to be added as a master alloy. The only elements other than titanium found to have any significant effect on performance were vanadium and niobium, both of which form very hard particles and which, so far as we know, have no other common effect with titanium. Thus it appears that it was the hard particles which were effective. Neither the introduction of sulphur nor any of the variations introduced appeared to reduce the effectiveness of titanium. Vanadium and titanium were found to be reasonably equivalent in their effectiveness and their effects to be additive.

46. In order to verify that no change in the iron was necessary other than the presence of these tiny hard particles, abrasive grit, size 1-7 microns was introduced into the surface of a high performance disc, in an amount equivalent to 0.10% titanium. The friction and wear were greatly reduced, being similar to those obtained with discs containing 0.09% titanium.

47. Thus we conclude that the presence of these very tiny, very hard particles in the cast iron is responsible for the variation in the friction and wear characteristics of brake rotors.

COMMERCIAL CASTINGS

Cupola Method

48. In foundries using conventional cupolas, mixtures of pig iron, self-returns and steel scrap are melted in intimate contact with coke. Samples of pig iron were found to contain between 0.03 and 0.11% titanium and samples of steel scrap all contained less than 0.01%, so if these were typical, the amount of titanium in the final product would be between 0.02 and 0.08% according to the source and amount of pig iron used.

49. However, it was found that castings made using pig iron with 0.103% titanium and steel scrap in ratio 3 : 2 plus three parts of self returned scrap contained only 0.04% instead of the calculated 0.06% titanium, indicating that some is lost

© IMechE 1976

during melting. These discs gave very low performance characteristics as predicted from the graph of performance versus % titanium in Figure 6.

50. Further discs were cast at the end of the cupola run using charges consisting of a pig iron containing 0.03% titanium, specially prepared returns containing only 0.01% titanium and steel scrap in the ratio 3 : 3 : 2. These discs contained 0.020% titanium, suggesting that there had been some contamination from previous charges in the cupola. However, they gave the intermediate performance characteristics in keeping with their titanium content.

51. We were unable to locate any pig iron containing less than 0.03% titanium, and in order to get very low levels, we tried a foundry making castings from all steel scrap using a hot blast cupola. Their scrap castings were found to contain only 0.012% titanium, and when recast in the laboratory, discs were found to give moderately high friction and wear characteristics. This material was used as the base to which additions were made to determine the effective mechanism of titanium.

Induction Melting

52. A foundry equipped with induction heated furnaces, melting all steel scrap, cast discs for us from their normal production melt and after adding 0.2% titanium as ferro-titanium to the ladle, and again after adding 0.1% vanadium as ferro-vanadium to the ladle. The discs with vanadium were cast to ensure the successful production of discs containing hard particles, in view of the difficulty experienced in getting titanium to go into solution, and the relative ease of getting the equally effective vanadium into solution.

53. When analysed and tested on a dynamometer, the results were as expected. The blank discs, found to contain 0.003% titanium, gave very high performance characteristics ; those to which titanium had been added and contained 0.018% titanium gave somewhat lower performance characteristics and those to which vanadium had been added, and contained 0.08% vanadium, gave extremely low performance characteristics. (Figure 8).

54. Second batches of blank discs and discs containing vanadium were produced and the results were very similar to those from the first batches. These results demonstrate that the performance characteristics of brake rotors produced in foundries using electric induction heated furnaces can be controlled very simply by adding the requisite amount of ferro-vanadium to the ladle and work is continuing to find the easiest way of adding titanium.

55. Six cars were fitted with various combinations of these discs and it was found that the brakes of those fitted with two high or two low performance members, even when each of the two discs was made from different batches, remained well-balanced, but those fitted with mixed high and low performance members pulled dangerously.

COMMERCIAL IMPLICATIONS

56. Although the work reported here was done on brake discs, it has been found that the metallurgy of brake drums affects the performance of drum brakes in a similar manner. The friction, wear, noise, fade, judder, crazing and speed sensitivity of a brake assembly all depend as much on the grey cast iron used for the brake disc or drum as on the formulation of the friction material. Currently the cast iron is not controlled in this respect and the performance characteristics of the brake rotor are left to chance.

57. We now know what factors affect the performance of cast iron rotors and that they can be controlled without adding greatly to the cost of the product, either simply by selecting the preferred raw materials or by making additions to the ladle.

58. If the cast iron was controlled, current problems due to its variability would be eliminated. Brakes on individual vehicles would be less subject to imbalance and so reduce warranty complaints of pulling. Vehicles of a particular model would show more consistency, for which there is increasing demand, and extremes which are the usual source of service complaints would occur less frequently. Vehicle manufacturers could fit rotors from different foundries without running into trouble, and the behaviour of production materials would agree with the results from prototype assessment tests.

59. The control of cast iron to a preferred level would enable much better properties to be obtained for a friction pair. Although it is not possible to improve all properties simultaneously, it is possible to improve the properties which are most important for any particular application. For example, by replacing rotors with high performance by those with low performance, material life and rotor durability can be greatly increased and noise much reduced, for some sacrifice of friction. Alternatively, thermal damage, crazing and judder can sometimes be reduced for some sacrifice of life. By suitably combining the friction material and the cast iron rotor, a much wider range of controlled performance characteristics can be obtained.

60. Although the behaviour of only one friction material has been described in this paper, the behaviour of a wide range of lining materials in general use has been examined, two of which are fully described in the patent literature. (Ref. 2).

61. All the materials gave substantially similar performance versus % titanium curves to those in figure 6, but the point of transition from high to low performance differed and some were more clearly sigmoidal in shape having flat regions at both low and high % titanium. Some materials which were known to give very variable performance were found to have their transition points at the level of titanium commonly encountered.

62. In order to achieve consistency in braking performance it is essential to control the cast iron so as to avoid the steep part of the curves.

© IMechE 1976

63. The importance of collaboration between the manufacturers of brake rotors and the producers of friction linings, and the necessity for reliable and reproducible test facilities was recognised by Wagenfurher who considered that optimisation of material pairings to be very complex. (Ref. 3). The authors agree entirely with these views, but believe that the problem has become much simpler since the factor which dominates the performance contribution of the rotor has been discovered, and that optimisation of performance characteristics can now be achieved.

REFERENCES

1.	Hatch, D.	Cast Iron Brake Discs J. Automotive Engineering October 1972 p.39
2.	Ferodo Limited	Control of Friction in Cast Iron - German Patent 2501 083
3.	Wagenfurher, H. et al	Development of friction-pair material by computer. ATZ Vol. 75, No. 7, July 1973 pp. 237 - 243 MIRA Translation No. 29/74

ACKNOWLEDGEMENTS

The authors wish to thank the Directors of Ferodo Ltd. for permission to publish this paper ; and to acknowledge the contributions of colleagues particularly Dr. A. M. Rizkallah who carried out many of the metallographic investigations and analyses, and Mr. M. W. Moore and his staff who did the initial work on dynamometers. The authors are also indebted to Mr. R. J. Edwards and Mr. R. Adams of the University of Manchester Institute of Science and Technology for their help and advice and the use of facilities at U.M.I.S.T ; and to Henry Wallwork and Co. Ltd. Ironfounders Manchester ; John Hawthorn and Co. Ltd. Engineers and Founders New Mills ; and Laycock Engineering Co. Ltd. Sheffield ; for supplying materials and casting discs to the authors specification.

© IMechE 1976

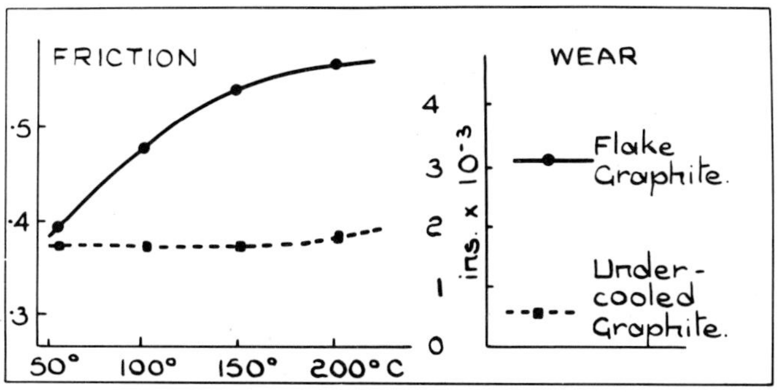

Fig. 1: Medium duty friction vs temperature, and wear rate of the friction material with 2 Model A discs.
——●—— used - -■- - - - reground

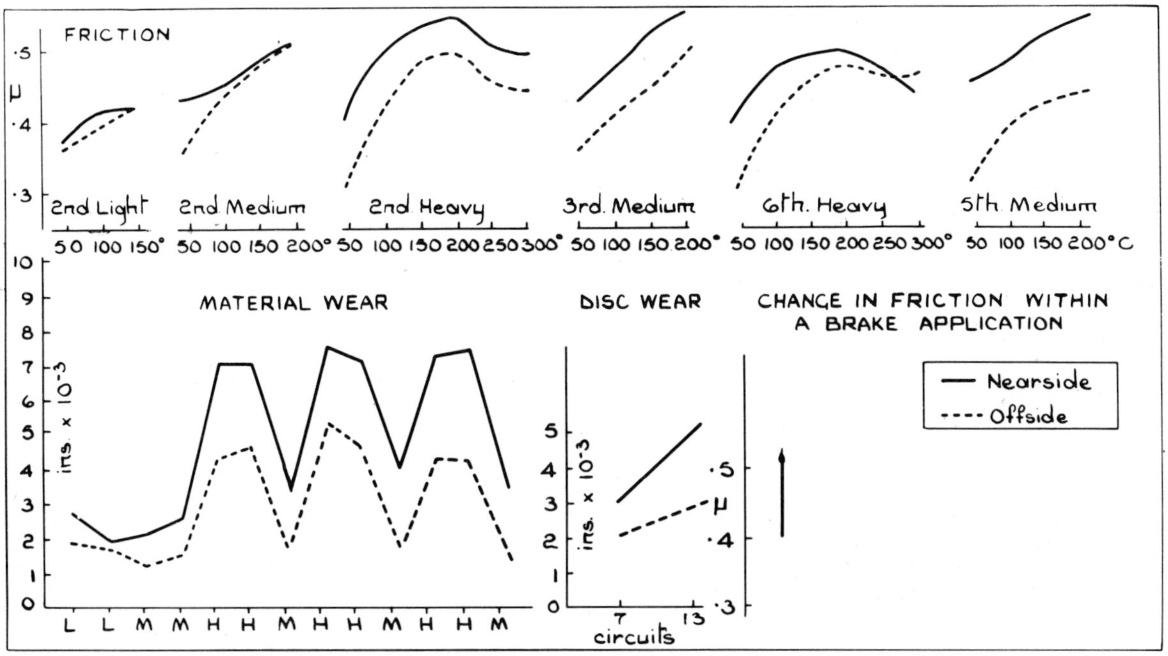

Fig. 2: Comparison of Friction and Wear characteristics developed during extended testing with a pair of discs which developed pull to the nearside.

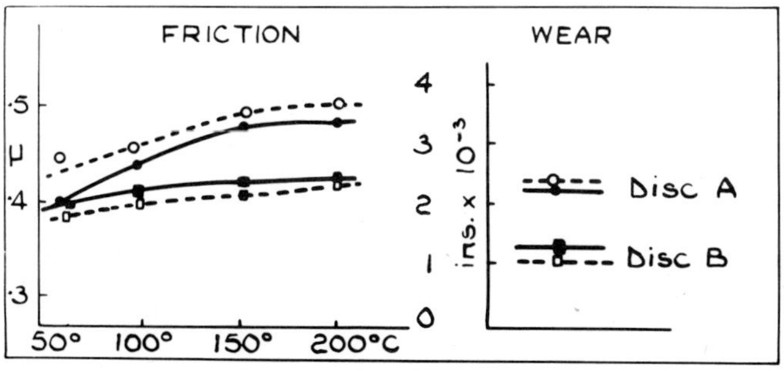

Fig. 3: Medium duty friction vs temperature, and wear rate of the friction material with specially cast Model A discs.

© IMechE 1976

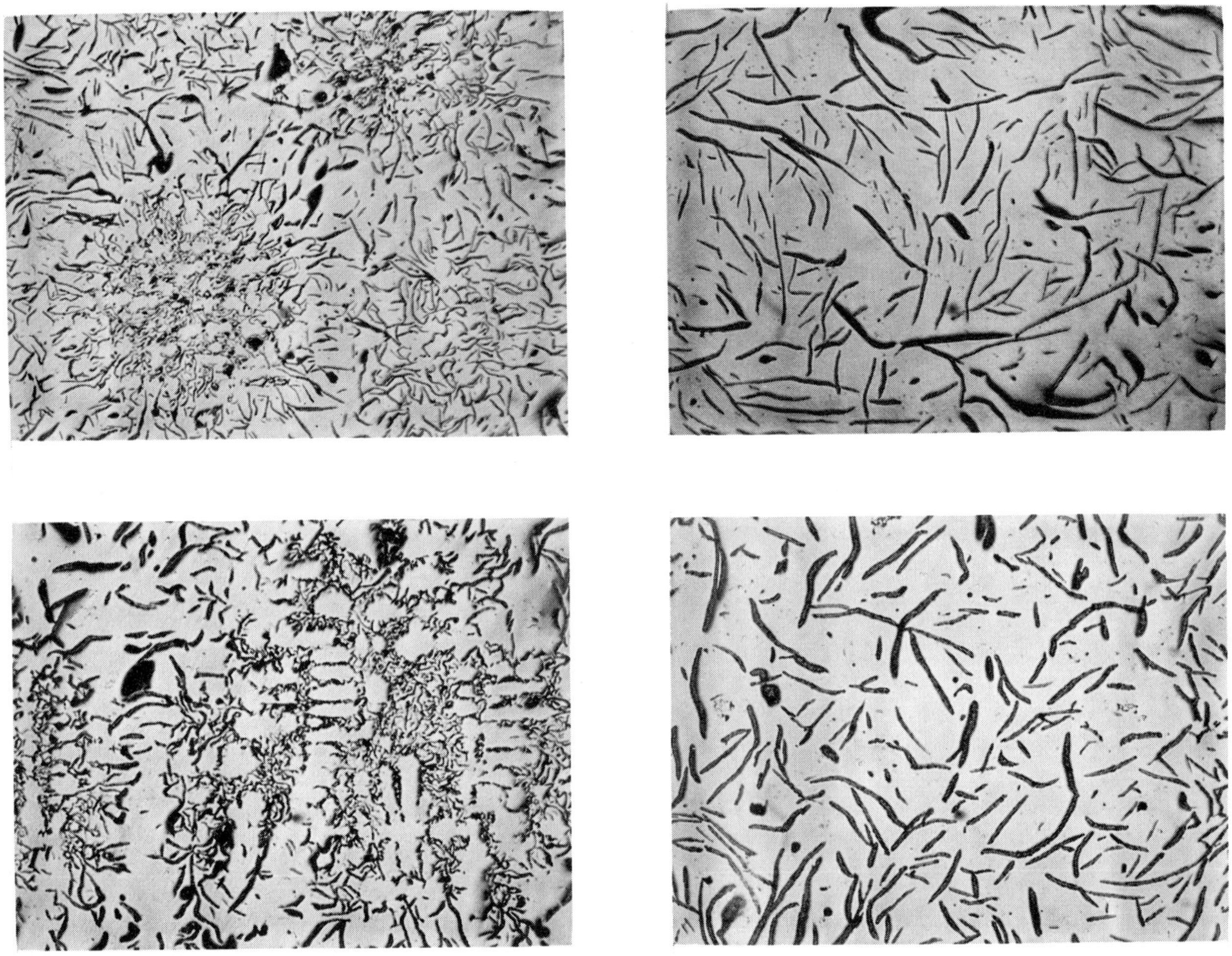

Fig. 4: Polished surfaces of brake discs at X 65 magnification. Upper pair both gave high performance characteristics. Lower pair both gave low performance characteristics.

© IMechE 1976

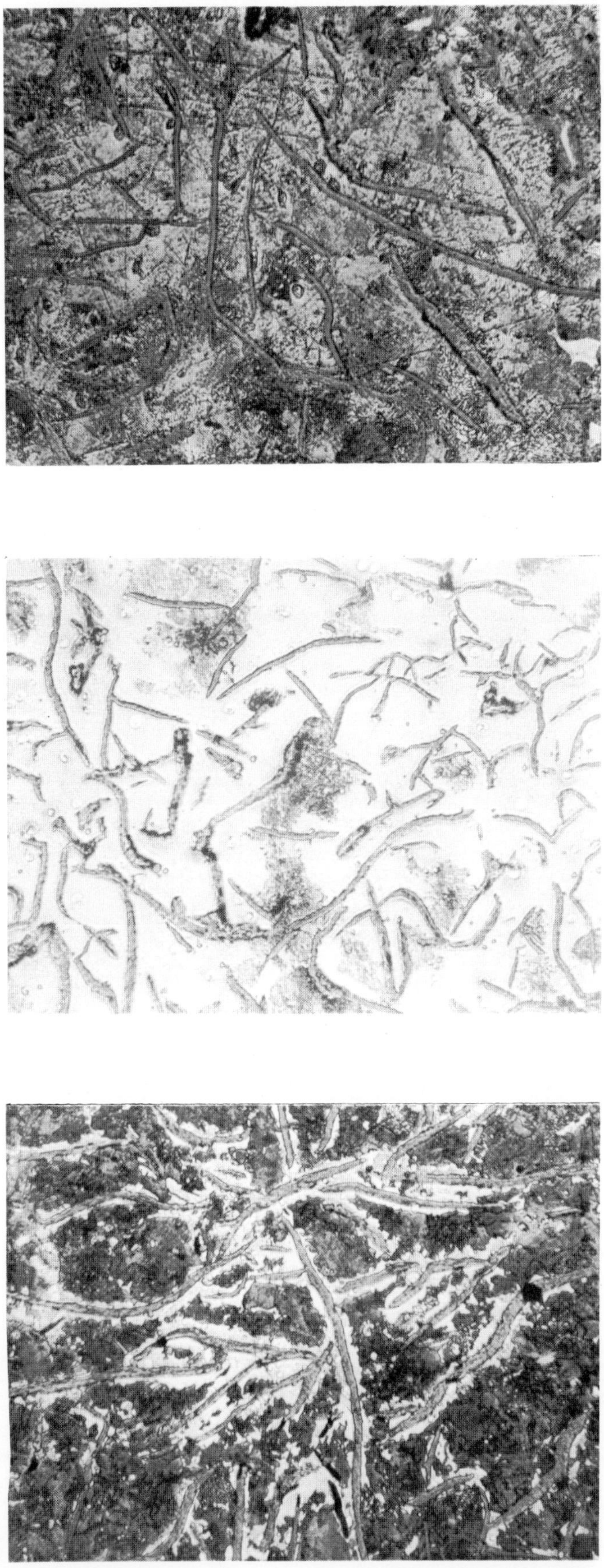

Fig. 5: Etched surfaces of four brake discs at X 120 magnification. Top is as cast, middle is after annealing, and bottom is after normalising.

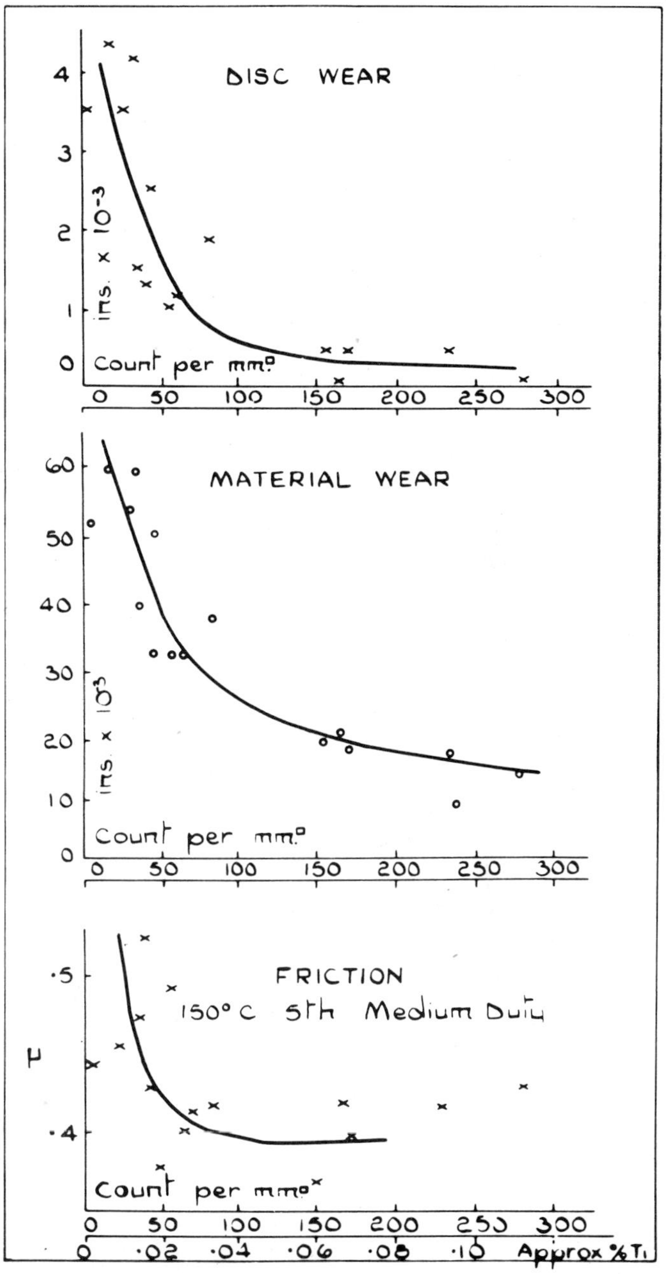

Fig. 6: Disc wear, material wear and friction vs number of titanium carbonitride particles per square millimeter of surface.

© IMechE 1976

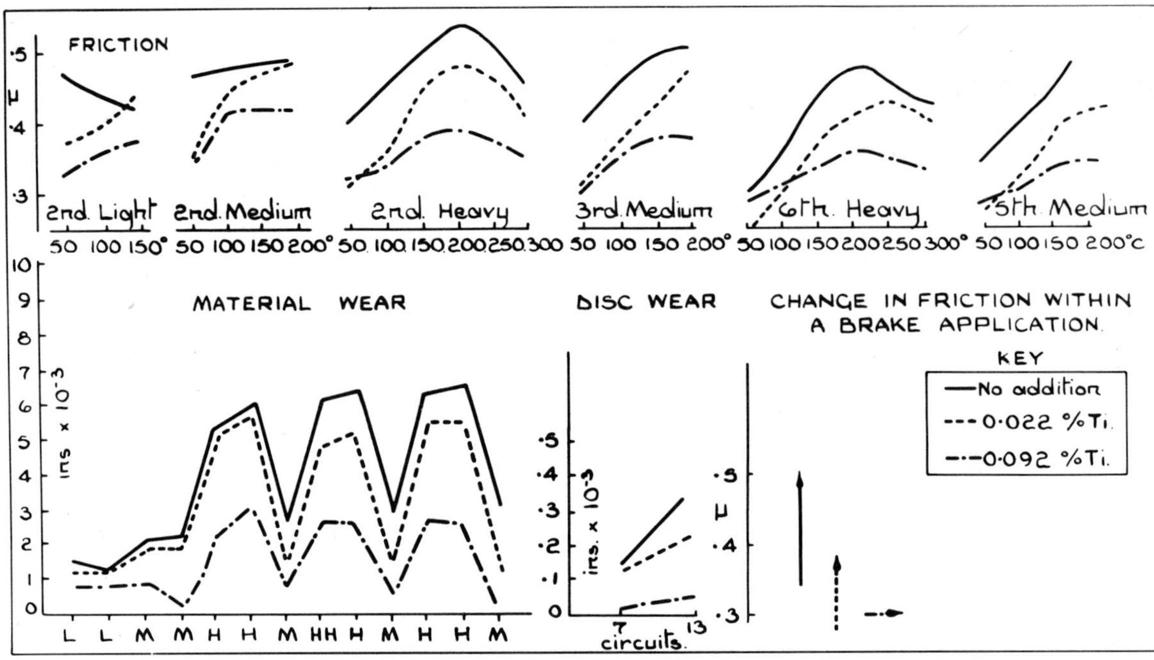

Fig. 7: Friction and wear characteristics exhibited by discs cast from laboratory furnace with and without added titanium.

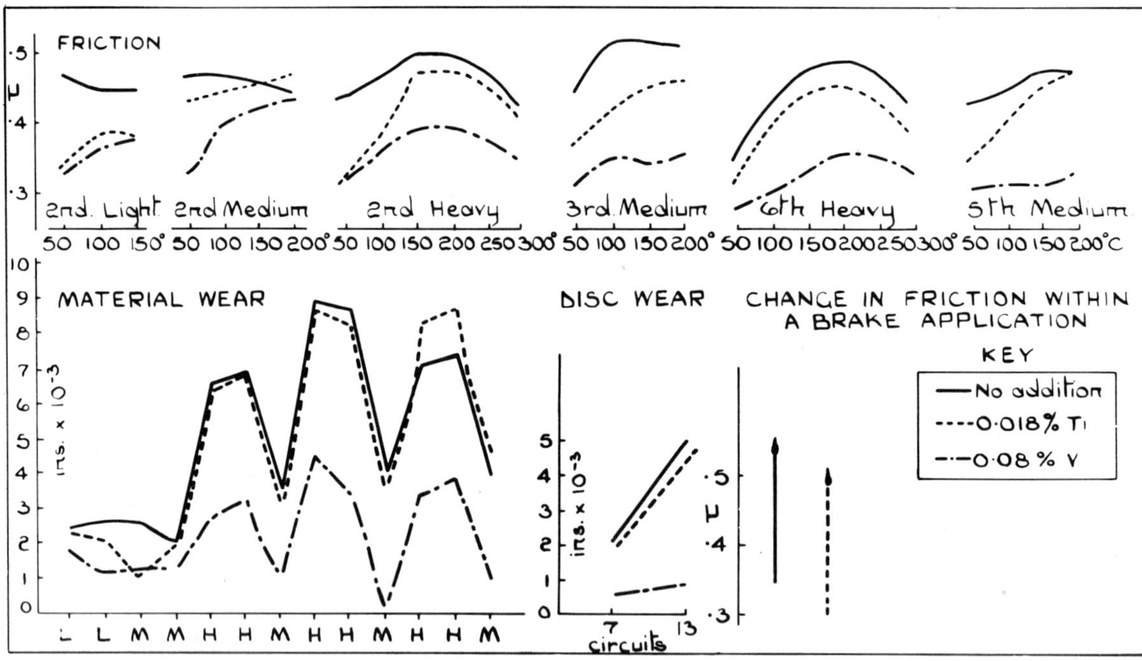

Fig. 8: Friction and wear characteristics exhibited by discs cast from induction heated furnace with and without added titanium and vanadium.

© IMechE 1976

C36/76

BRAKES – A REVIEW OF EXISTING DESIGN

W. R. NEWTON
A. C. W. WRIGHT BSc
Girling Limited, Kings Road, Tyseley, Birmingham

The MS of this paper was received at the Institution on 12 September 1975 and accepted for publication on 21 October 1975

SYNOPSIS The brakes which are fitted to the front and rear axles of passenger cars satisfy a range of functional purposes directed to the primary objective of safe, consistent stopping, low cost and low weight. International Regulations, changing patterns of use and increasing vehicle performance provide severe problems requiring compromise and this paper is concerned mainly with the many forms in which disc brakes now appear, describing the virtues and drawbacks of each. Sections are however included to explain adjusters and the drum brakes used for small vehicles and for the rear axles of many of the larger vehicles now in production.

INTRODUCTION

In this paper the design of automobile disc brake calipers and of drum brakes is shown to result from the interaction of numerous functional requirements with Government regulations and new vehicle concepts. Since the first application of the opposed piston disc brakes to passenger cars in the early 1950's, a large number of caliper designs have been produced throughout the world, each resulting from an effort to optimise the conversion of a vehicle's kinetic energy to heat, and its subsequent dissipation, whilst at the same time satisfying strict installation parameters and providing the maximum cost effectiveness ratio.

The factors involved in passenger car brake design are closely integrated with the vehicle characteristics since the brakes are part of the unsprung weight and need to keep pace with power, performance, weight, styling, tyre design, steering geometry and to give satisfactory pedal feel all at a low overall cost to the manufacturer.

FACTORS AFFECTING DISC BRAKE DESIGN

Energy dissipation

The prime object of the brake is to dissipate energy, usually by converting it to heat, and this must be achieved with controllability and stability, that is, there should be no tendency to grab or fade, and in a disc brake the geometrical factors which affect the sensitivity in drum brakes are substantially absent and this non servo action means that the brake reflects directly the friction performance of the lining materials. Disc temperatures have to be kept to a minimum, and the maximum surface area for heat radiation is desirable.

In heavy car installations, the energy dissipated at the disc/pad interface may be great enough to cause the propagation of surface cracks from the disc surface. Cracks are caused by the mechanical stress resulting from a very high temperature gradient at the disc surface and the consequent metallurgical changes occurring in the iron, they will produce judder when the brakes are subsequently applied and can eventually lead to complete failure of the disc. Again, to minimise the danger of disc cracking, as deep a braking path as possible is required.

Installation trends

Statistics have shown that braking action is involved in 50% of accidents, and the stability during braking is an essential factor contributing to vehicle safety. Instability during braking results from variations in lining friction level, and the tyre/road adhesion factor, which, in turn can be influenced by production tolerances, temperatures and cooling rate variations, and the effects of dirt, salt and water.

Significant improvements in these areas cannot be expected, but their effect may be minimised by the use of reduced offset steering to eliminate self steering characteristics brought about by unequal couples, produced by braking suspension and road forces about their respective swivels. The trend to lower steering offset has already shown itself in the majority of new European vehicles, being accelerated by the increase

© IMechE 1976

in popularity of front wheel drive cars.

Reduction of ground offset has to be achieved by moving the steering knuckle outwards from the centre of the car (see Figure 1) and generally the disc must follow this move so there is a corresponding decrease in space between the disc and wheel spider. This cannot be compensated by re-design of wheel-knave depth since this would lead to strength problems, and because the knave must remain protected from curb damage by the tyre form. This move also means that the disc outside diameter may become smaller as it is now limited by the wheel-knave, and not by the larger wheel rim.

Divided circuits

Most regulations are now demanding divided circuits in braking systems. Figure 2. This means that brakes must be capable of being operated by two separate hydraulic supplies, either tandem or binocular arrangements, for the more complicated 'LL' 'HI' or 'HH' split system arrangements. The simplest, most cost effective divided system is the front axle/rear axle 'V' arrangement. However, a typical small front wheel drive car would be unable to meet the failed system braking requirements when using only the rear axle system, so the 'X' arrangement is preferred on these cars. This allows calipers actuated by only one system to be used, but again demands reduced offset steering geometry for stability whilst braking in the failed condition.

Vibration affects

Calipers must be capable of withstanding the considerable accelerations transmitted to them from the axle, without failure or undue noise. Measurements on rough roads, pave, and at level crossings have shown forces equivalent to as much as 18g on occasions and having a mean value around 11g. This means that sliding or floating parts of caliper assemblies must be retained firmly in position by the addition of springs to avoid rattle and wear and these springs will inevitably increase the sliding loads on a moving part of the caliper.

Pad release

One of the most important features of the opposed piston caliper is the support provided for moving parts and after a brake application, the pads are readily knocked away from the disc, against the light friction loads produced by the pad anti-rattle springs, by disc run out and the effect of stub axle deflections.

Sliding calipers tend to have a massive sliding member which requires heavy springs to combat vibration noise and wear. The resulting sliding forces tend to create a clamping load differential across the disc, especially if aggravated by corrosion or dirt, and may result in a pad dragging on the disc in the brakes off condition. This will manifest itself as (i) excessive and unequal lining wear, (ii) twitter and squeal during cornering and (iii) in severe cases be the cause of uneven disc wear and judder.

Fluid vapourisation

More stringent safety regulations and the growth of tourism in Europe have led to the adoption of searching Alpine test procedures by major European car manufacturers. Test results have shown that many conventional caliper installations will not satisfy the full range of fast and slow Alpine descents which characterise the wide range of extreme duty cycles used to find the worst conditions, particularly if the brakes are not regularly maintained. Comparisons show that in some installations the fluid temperatures in 'sliding calipers' may be 30°C to 50°C lower than those of opposed piston calipers.

This results from:-

(1) the elimination of internal porting, particularly in the bridge near the disc:

(2) the fluid chamber being further from the braking path, since one piston is used:

(3) the longer piston providing a larger heat sink capacity between the fluid and disc:

(4) the component adjacent to the disc not containing the fluid chamber, i.e. (the carrier):

(5) the hydraulic elements are on the 'ventilated' side of the disc and not in the side shrouded by the wheel.

A caliper featuring these advantages is shown in Figure 1b and may be compared with the standard opposed piston caliper. Figure 1a.

DESIGN CONFIGURATIONS

Opposed piston calipers

The early successful caliper disc brakes were of the opposed piston type, incorporating the seal 'rollback' effect to give a small positive clearance between pad and disc.

This developed to the 'three piston arrangement' having one inboard and two outboard pistons giving an improved swept form and better installation into some wheel configurations.

The requirement for 'split systems', larger pads with better load distribution, and improved installability produced the '4 piston' caliper. This can now be seen having 4 equal bores, Figure 1a, or 2 large and 2 small bores satisfying the performance requirements of the 'HH' or 'HI' type split systems respectively.

© IMechE 1976

These types of calipers continue to give good service after many years because they use inexpensive manufacturing processes and materials, generally allow rapidpad changes, and are very suitable for use with divided systems. However, as they are difficult to install in many modern suspensions, may suffer vapourisation problems, and do not lend themselves easily to the addition of integral parking systems, and so a great deal of attention is now being focussed on sliding calipers.

Sliding calipers

The requirement for better installation in low offset suspension, improved performance, cost effectiveness, and the desire to incorporate simple integral parking brake arrangements have led to development of the 'sliding caliper'. This device has hydraulic cylinders on only one side of the disc which directly operates the inboard pad, whilst the reaction is transmitted to the outer pad by a sliding element.

Analysis of this type of caliper shows that the various designs fall into one of six basic arrangements depending upon:-

(1) whether the hydraulic actuator (a) remains stationary, or (b) moves, in order to compensate for pad wear.

(2) whether (a) both pads abut directly against an earth member, (b) one pad abuts an earth member whilst the other abuts an intermediate sliding member, or (c) both pads abut a sliding member which transfers the drag forces to earth.

The physical construction varies further, as discussed in later paragraphs and shown in Figures (1) and (3) by the use of:-

(1) Sealed, part sealed or exposed slides.

(2) Fist type, plate type or combined reaction members.

(3) Sliding, pivoting, swinging or stationary reaction members.

The hydraulic body

When using a fixed hydraulic body arrangement:

(1) Every seal must be regarded as a potential source of leakage and with two pistons, two seals and two boots, this arrangement is the same as the opposed caliper.

(2) It is likely to be longer than a floating body caliper because of the extra seal, boot and piston. These are disadvantages but:-
(3) It may be fed using standard bundy pipe as it does not move to accommodate pad wear or replacement.

(4) Because the hydraulic cylinder and mounting bracket are a composite casting a more efficient use of material allows grey cast iron to be used.

(5) No carrier between hydraulic body and bearing making the location on the axle easier.

The advantages of the sliding hydraulic body are:-
(1) Has only one seal, piston and boot.

(2) Gives a very compact unit.
but the disadvantages are:-
(3) Must be fed with a hose. (This may be the hose that accommodates suspension movement, but this is not always possible.)

(4) Has a shallower braking path depth, because of the relative positions of the mounting bracket and the hydraulic cylinder.

(5) Must use stronger, more expensive materials on the mounting bracket.

The pad abutment

If both pads are abutted directly against earth, Figure 1b, Figure 3a, the caliper will be less susceptible to noise problems. However, the drag loads of one pad may go to a floating member as in Figure 1c, d, Figure 3b, or both pads are reacted by a floating member which then transfers the forces to earth. The drag force may be reacted in such a way as to (i) break any corrosion forming on exposed sliding arrangements, (ii) create a twisting of the pad carrier which tends to reduce the separation of the pistons and therefore the fluid requirement.

This caliper may however tend to be very noisy in operation, because of the large forces and massive elements carried by the slides and such calipers employ a number of special measures to minimise noise,

(i) special lining materials, rivetted linings,

(ii) stiffer pad backplates,

(iii) use of a compliant material such as 'Wolverine' between piston and pad,

(iv) anti-rattle/squeal shim inserts,

In addition noise always tends to be associated with some high friction material which are otherwise very desirable for performance reasons, and there is usually a restriction of choice with this form of caliper.

The slide arrangements

If open slides are used they must have,

(1) Sufficient clearance to prevent simple rust

© IMechE 1976

seizure (normally this means not less than 0,15 mm) and this should be in combination with freedom in the normal direction.

(2) The transmission of brake torque loading and reversal to ensure the slides are 'worked'. The associated problem is the prevention of the now 'free' sliding member from rattle and subsequent slide face wear. This generally means the inclusion of a spring which in turn generates a slide friction load, and the possibility of outer pad drag.

The use of booted closed slides, normally achieved by a pin and hole device, enables the clearances to be eliminated, which, with the deletion of the anti-rattle spring, avoids drag and means that under 'knock back' conditions both pads can follow the disc to a greater degree without pushing fluid back to the master cylinder. The sealed slide structure is generally more costly in machining, parts and service complications and the slide sealing arrangement must be reliable.

The clamp member

The fist type (Figure 1b, 3a) is the most common in Europe at present. It has the advantages that:-

(1) When combined with a sliding hydraulic body it will give the minimum overall length sliding caliper.

(2) It gives a particularly compact installation outboard of the disc.

However, it does tend to deflect at high clamp loads so that the pads are loaded more at the periphery of the braking path than at the centre. This will give slightly increased brake torques for emergency braking but will put more energy into the edge of the disc and aggravate any disc heat cracking problems.

The plate type (Figure 1d, 3c) tends to require more overall length and circumferential length than claw arrangements and is particularly poor on installation outboard of the disc. However, it does allow very thin sections to be used over the disc, thus maximum disc sizes and the way in which the pads are clamped means that the whole disc face is loaded evenly.

A compromise solution is available, particularly for calipers having the outboard pad drag transferred to earth via the clamping member (e.g. Figure 1c). This again has the disadvantage of a requirement for a stiff outboard beam to support that pad, so a poor installation outboard of the disc, but because of the reduced bending moment on the sections over the disc it allows the possibility of low grade materials to be used without major sacrifices in disc size.

Pivoting calipers

Calipers which replace the slides by pivots have been in production for many years. The two types are shown in Figures 3e and f. Figure 3e has a sealed pivot, specially shaped pads, and incorporates a handbrake mechanism and adjuster. Figure 3f is a plate type construction, having an open pivot, shaped pads, and is designed to give a slight servo effect in the forward direction.

A survey of sliding caliper types

Figure 1a A standard opposed piston caliper having 4 equal bores suitable for 'HH' or 'LL' split systems or two unequal bores suitable for HI splits.

Figure 1b A fully sealed sliding caliper having both pad drag forces reacted directly to the earth member. This example has two equal cylinders suitable for 'HH' or 'LL' or two unequal bores suitable for HI type split systems, and shows the compactness resulting from a claw type single sided caliper design, allowing reduced offset suspension design.

Figure 1c A sealed sliding caliper having the outboard pad drag absorbed by the sliding member which transfers the load to the pin on which it slides. This makes use of grey cast iron bridge sections, and allows uncomplicated pad removal.

Figure 1d A plate type sliding caliper offering an in-expensive construction, large disc and simple pad removal, but having a stiff outboard beam.

Figure 3a A compact sliding claw type caliper having the pad drag forces reacted directly to earth, using exposed slides, and saving weight using aluminium for the hydraulic body.

Figure 3b A similar sliding claw type caliper having the outboard pad drag force transferred to earth via the sliding element.

Figure 3c A plate type caliper similar to Figure 1c, but having both pad drag forces reacted direct on the earth member, and the hydraulic cylinder sliding.

Figure 3d A sliding claw type caliper offering a sealed slide arrangement, large disc, and compact installation. The sliding element is supported on a sealed rubber bush inboard and on the outboard pad.

Figure 3e An example of a sealed pivoting caliper incorporating a handbrake mechanism.

Figure 3f An example of an unsealed pivoting caliper which develops a small amount of self servo action.

© IMechE 1976

Handbrake for disc brakes

The original handbrake calipers used a separate system of cable operated levers. This system must incorporate its own adjuster, uses small high μ linings, and being exposed to the elements is corrosion sensitive. Some vehicles have used a separate high factor duo servo drum brake (drum in disc system) acting on the disc bell. This system offers a good solution but the duo servo brake requires adjustment and the installation is often difficult.

The incorporation of handbrake systems into opposed piston calipers requires a separate automatic adjuster on both sides of the disc and uses an expander on only one side, thus relying on disc deflection, and wheel bearing clearance to transfer the clamp to both sides of the disc. The more recent caliper handbrakes concentrate on the incorporation of one expander and adjuster into a sliding caliper.

Several types of expander are currently in production:-

(1) Eccentric lever - Figure 4a

(2) Ball and ramp - Figure 4b

(3) Needle roller ramp - Figure 4c

(4) Toggle - Figure 4d

These devices can be designed to give variable ratio of application, initially coarse, around 10:1 rising to about 100:1 as the lever travel increases.

Adjusters for disc brakes

The requirement for disc brakes with integral parking systems has led to a great deal of effort developing automatic adjusters. These become necessary since to obtain sufficient parking clamp load in the caliper from the effort and stroke available from the driver, a very high mechanical advantage system is required, resulting in very small strokes being available for brake operation. This means that the pad/disc clearance must be kept to the very minimum compatible with free running in the brakes off condition, with no tendency to 'hang on' due to disc coning and expansion, or pad swelling and bending.

The most likely position for inclusion of an adjuster is in the service brake fluid, where it acts as a back stop to the piston, and carries only the parking brake loads. This has the advantages that:-

(1) The service brake travel is not adversely affected by the inclusion of adjuster components in the service brake load line.

(2) The adjuster can easily be placed in the fluid which will provide a suitable sealed, clean, lubricated environment.

(3) As the service and parking brake operate, potentially large forces 'work' the adjuster over its full stroke, giving very good reliability.

Designs of adjusters may be placed into groups depending on how many brake applications are required to eliminate the excess clearance. The linear ratchet adjuster (Figure 4a) will adjust all the excess clearance at one brake application. A similar result may be obtained using a fast thread strut, which is free to rotate during brake on condition, taking up the excess clearance, but being restrained from 'unwinding' during brakes off condition. These are known as 'one shot' adjusters.

Alternative devices eliminate the danger of over adjusting due to large caliper deflection on a very high duty stop by taking up clearances over a number of stops using a ratchet device (the incremental type). These have the disadvantage that during assembly to the vehicle or when fitting new pads a relatively large number of brake applications must be made to bring the caliper into adjustment.

A development of this is the Drive Ring 'exponential' adjuster Figure 4b and reference (1). This device reduces large clearances rapidly, but reduces small clearances relatively slowly, giving the installation advantages of the one shot device plus the safety features of the incremental device.

Adjusters that respond to the free travel of the caliper plus any deflections associated with the caliper operation are known as 'load sensitive' types. Those discussed so far fall into this category. When the caliper is undergoing a severe application these adjusters will sense relatively large movement due to pad compression and caliper deflection, and since the brake must not be left with a permanent clamp load the one shot adjuster clearance must be at least as large as this deflection plus an allowance for pad swell and running clearance. Incremental and exponential adjusters will not take up all this clearance in one go, and as stops of high severity are not common, the preset clearance of these adjusters may be less than for a one shot type. In order to set the brake at the minimum running clearance an adjuster should evidently be 'in sensitive' to caliper deflections. Devices of this type are shown in Figure 4c and Figure 4d and reference (1).

PART 2 DRUM BRAKES

DRUM BRAKES FOR PASSENGER CARS

Early development

Early drum brakes consisted of two shoes on common or adjacent pivots, and bearing upon an expanding device at the other ends of the shoes. The smaller sizes for passenger cars used a cam

© IMechE 1976

arrangement to expand the shoes, operated directly by cables or rods, or by roller wedge expanders.

The pivoted shoe arrangement was soon replaced by sliding shoes with the possibility of more uniform lining wear and the more stable behaviour due to more rapid bedding, (their construction also has the merit of requiring less accurate tolerance.)

The rod and cam expanders were replaced by a hydraulic system incorporating a simple wheel cylinder to expand the brakes, giving much better 'feel', equalisation between brakes and being far less susceptible to seizure or corrosion.

Basic configurations

The brake factor which is here defined as the drag force at the drum radius divided by the shoe tip input effort, increases in the order Leading/Trailing Shoe, Figure 5a, 2 leading shoes, Figure 5b, Duo-Servo Figure 5c. However, the stability of the brake or susceptibility to output change with lining friction change increases as the brake factor decreases. In Europe, cars fitted with drum brakes have two leading shoes or leading/trailing shoe brakes fitted to the front and leading/trailing shoe brakes on the rear wheels. The two leading shoe brake has the disadvantage that its good performance is lost when it becomes a twin trailing shoe brake in reverse, whilst the leading trailing shoe brake is more suitable for the rear (a) because it has a consistent brake factor in forward and reverse, (b) it lends itself more easily to the addition of adjusters and parking brake arrangements, because the two shoes can be operated by the same actuating force leading to simple and compact actuating arrangements, (c) because adjustment may be effected either using the parking linkage allowing the hydraulic system to follow or by adjusting the fixed shoe abutments.

In America and Australia where cars tend to be much heavier than in Europe, duo-servo brakes are favoured for the rear axle in order that the high factor can provide a high torque parking brake which would not require power assisted actuation.

Adjuster types

In recent years advancement in drum brake construction has mainly been confined to detail changes, for slight performance gains, manufacturing simplicity, or cost savings. However, some new adjuster arrangements have recently been introduced because the need to minimise servicing has become of great importance.

In the past automatic drum brake adjusters were often operated during application of the handbrake. In some parts of the world driving habits of leaving vehicles parked without handbrake application (to allow the car to be moved or to avoid the brakes seizing on in icy conditions) meant that this type of adjuster could not maintain the correct shoe to drum running clearance, and this has led to the use of adjusters sensing the action of the service brake.

Two basic types are available for leading/trailing shoe brakes, the 'one shot adjuster' which compensates for all the excessive clearance in the brake in one application, and the 'incremental adjuster' which eliminates excessive clearances during several brake operations.

Compared with the 'incremental adjuster' design the 'one shot' arrangement has the advantages that:-

(1) Brake setting during assembly is unnecessary.

(2) Only one brake operation is necessary to make the vehicle safe to drive when replacing shoes or fitting to a new vehicle. Incremental adjusters would require 20 or more applications.

(3) The adjuster device is not affected by maladjustment of the handbrake cable.

(4) Service brake shoe replacement is simplified.

(5) The handbrake lever/strut/adjuster is one sub-assembly.

(6) The number of components is reduced.

An example of the 'one shot adjuster' is shown in Figure 6 and reference (1).

The adjuster which fails to take account of elastic and temperature variations is far more limited in use with drum brakes than with the disc brake parking systems since the relative expansion of drum and shoes can be very large on such duties as hill descents. Thus the potential advantage in terms of pedal travel from an incremental adjuster are important for larger vehicles, and an example of this type of adjuster is shown in Figure 7 and reference (1).

The duo-servo brakes which have until recently been used widely for rear brake parking duty on heavy American and Australian cars have employed a linkage operated incremental adjuster which, by adjusting only when the adjuster wheel was unloaded, provided the load insensitive characteristic desirable when the absolute maximum output is necessary from the parking brake on these heavy vehicles.

Two leading shoe brakes are usually adjusted by hand operated 'snail cam' devices and since adjusters are unlikely to have to cater for handbrake operation they can be simple friction devices preventing the shoe return springs retracting the shoes too far from the drums. Each wheel cylinder piston then remains adjacent to its respective shoe tip.

Automatic versions of this one shot type of adjuster have been tried but in the past have proved difficult to engineer, since the built in clearance needs to be large enough to compensate for drum

© IMechE 1976

expansion, both elastic and thermal and for shoe deflection, and the effects of lining compression and swell.

This can result in rather poor pedal feel, as the disc brake overcomes many of these problems, a change to disc brakes would provide automatic adjustment, better pedal feel, and a more stable brake, at a slightly increased cost and places an upper limit to the weight of passenger axles to which drum brakes can satisfactorily be applied.

Acknowledgement:

The authors wish to thank the Directors of LUCAS INDUSTRIES LIMITED for permission to publish this paper.

References:

(1) Automatic Adjusters for Vehicle Brakes by A W Harrison - S.A.E. Paper 730199, Jan. 1973.

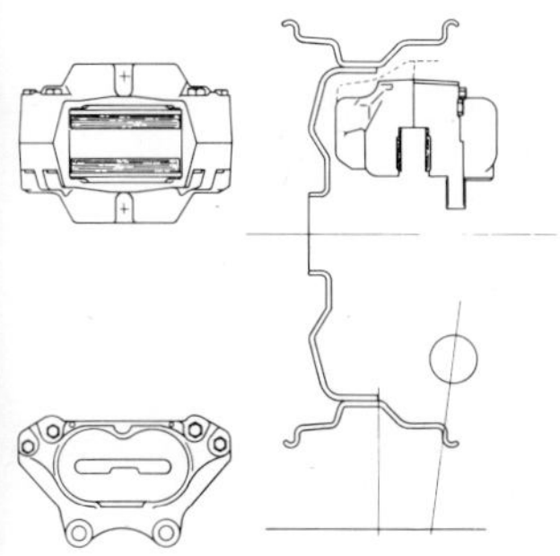

Fig. 1a: Opposed Piston Caliper

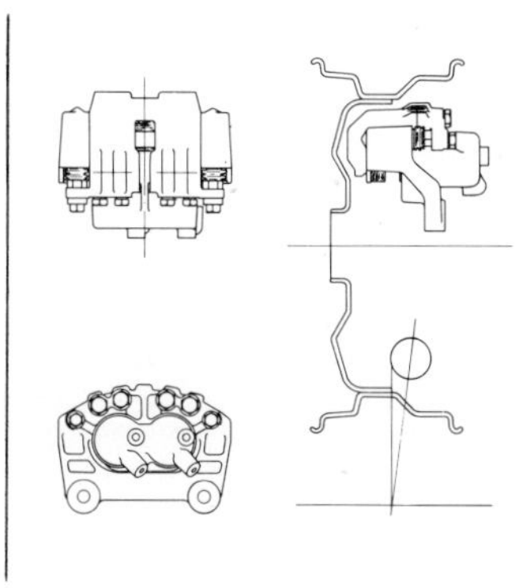

Fig. 1b: Sliding Caliper – 'Claw Type'

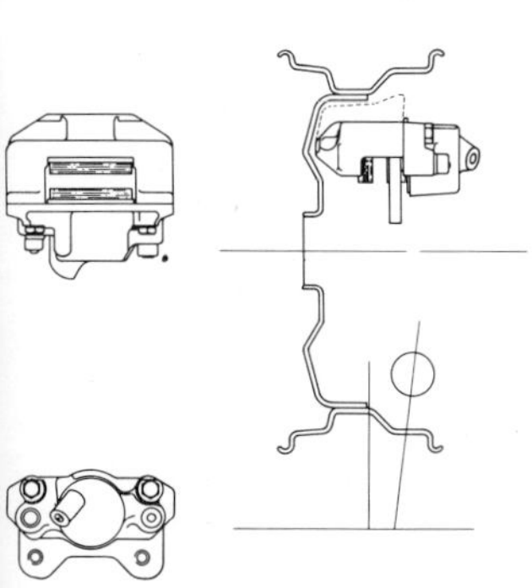

Fig. 1c: Sliding Caliper – 'Intermediate Type'

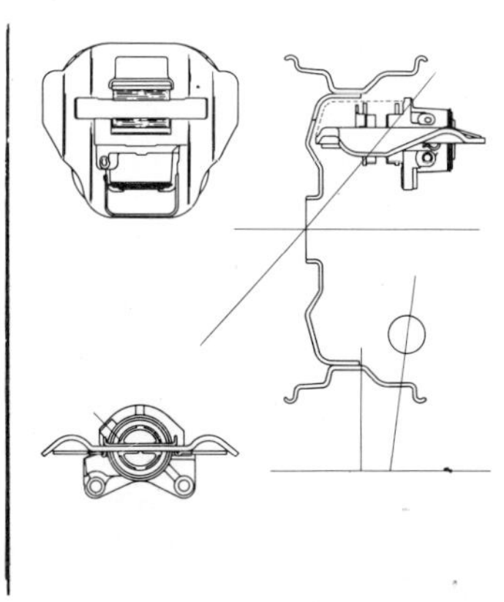

Fig. 1d: 'Sliding Caliper – 'Plate Type'

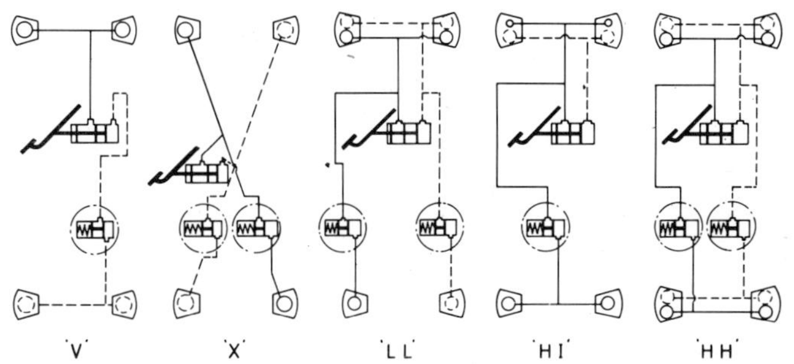

Fig. 2: Divided Circuits

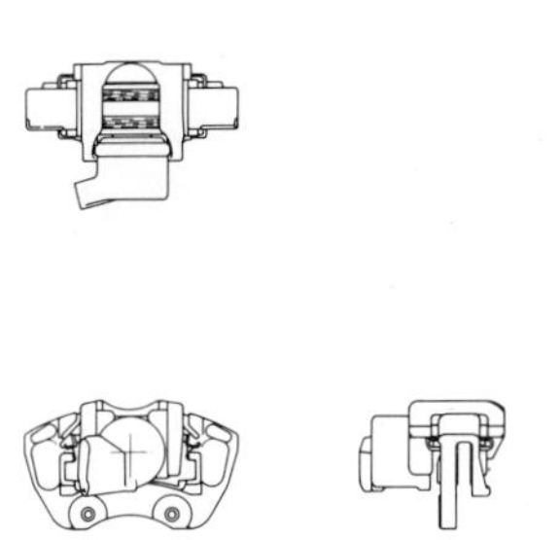

Fig. 3a.

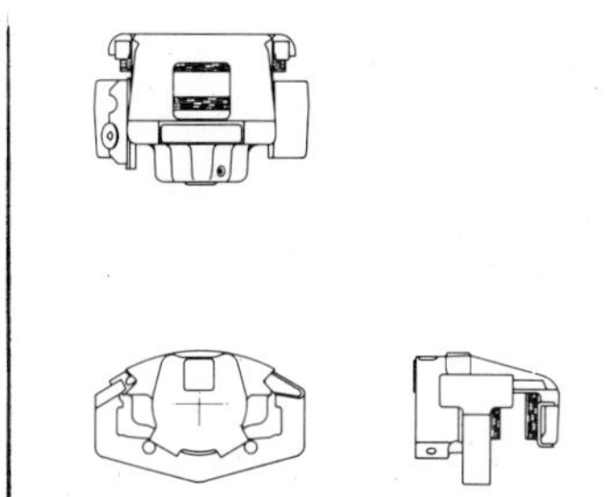

Fig. 3b.

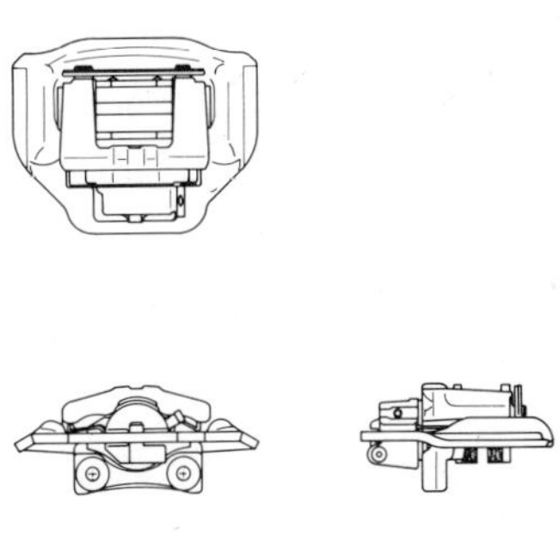

Fig. 3c.

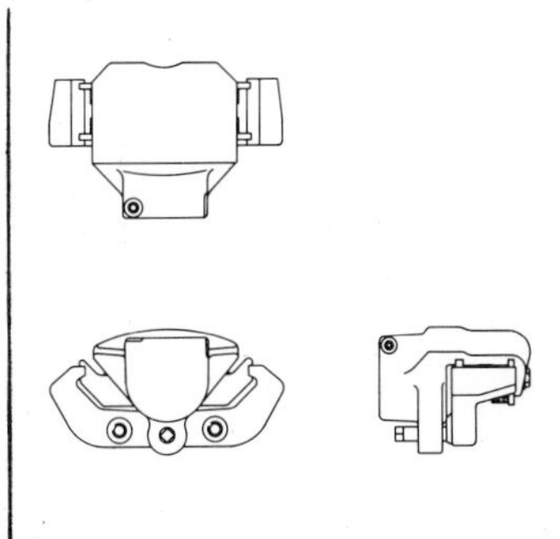

Fig. 3d.

Fig. 3a, b, c and d: Sliding Calipers

© IMechE 1976

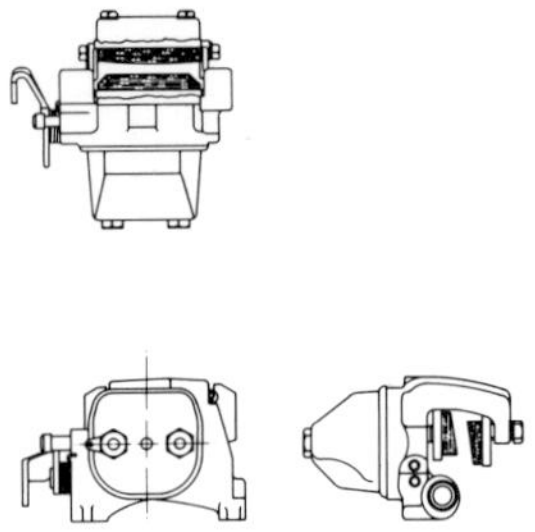

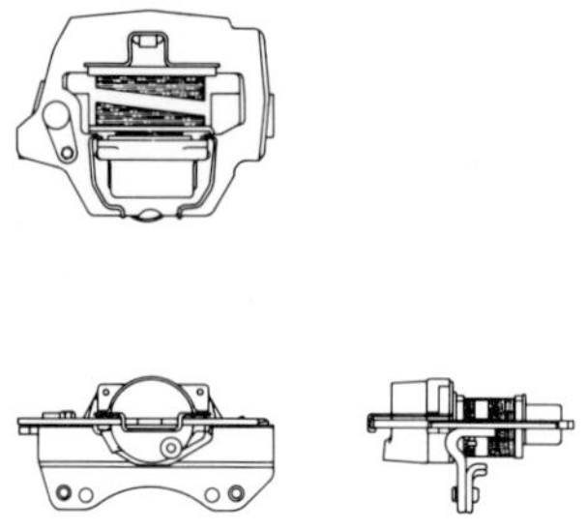

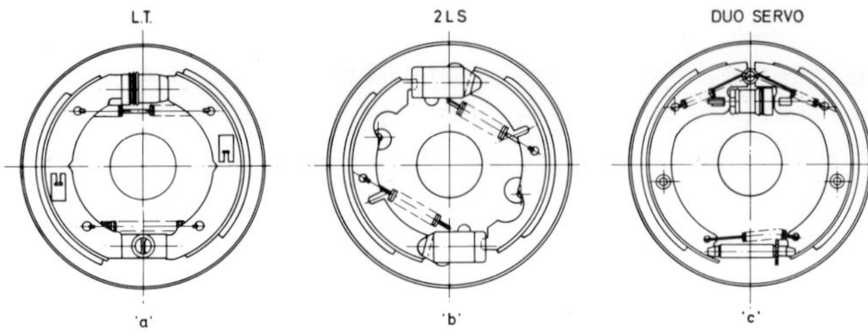

Fig. 5: The More Common Automobile Drum Brake Types

Fig. 3e: Swinging Caliper

Fig. 3f: Pivoting Calipers

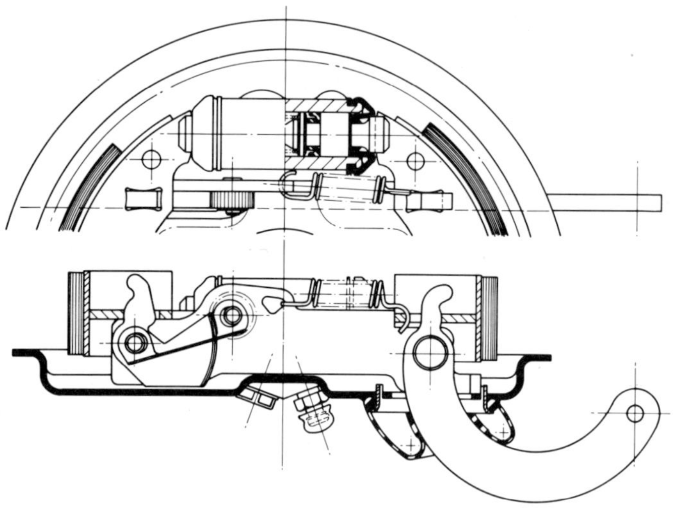

Fig. 6: A 'One Shot' Drum Brake Adjuster

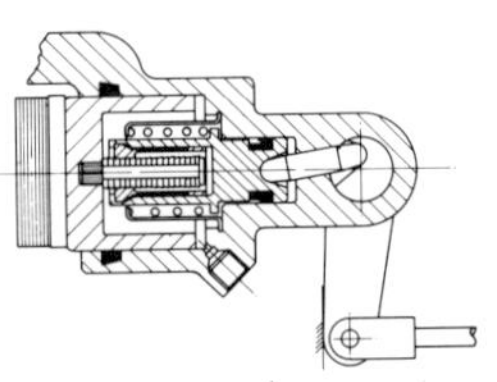

Fig. 4a.

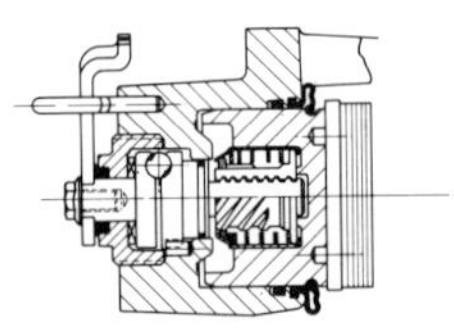

Fig. 4b.

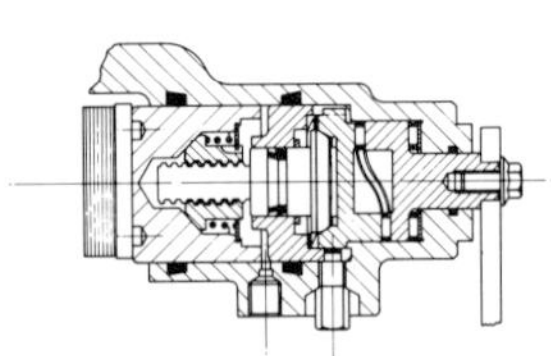

Fig. 4c.

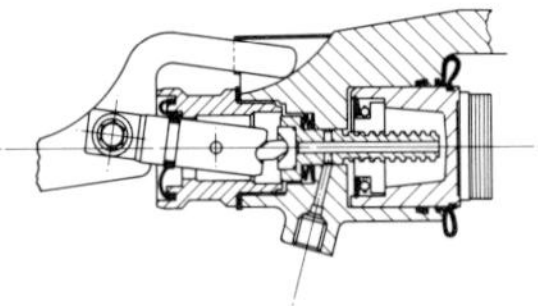

Fig. 4d.

Fig. 4a, b, c and d: Caliper Adjusters and Handbrake Mechanisms

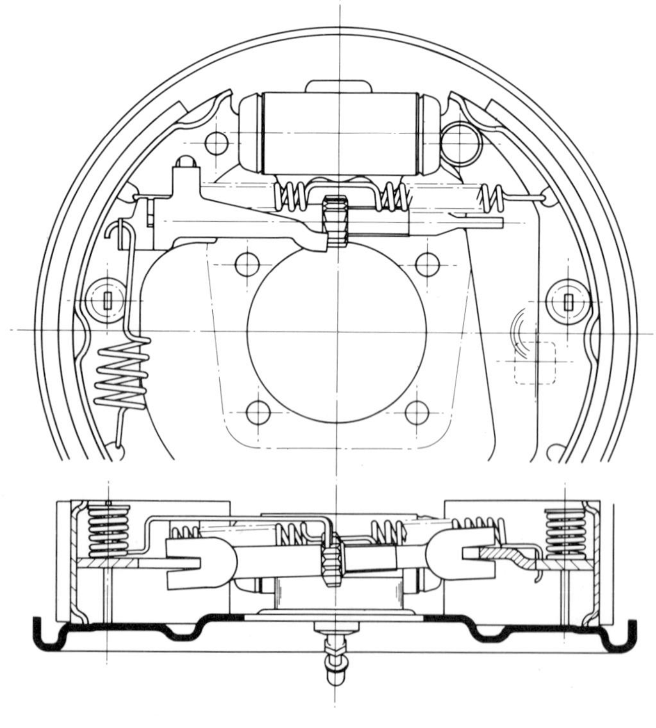

Fig. 7: An 'Incremental' Drum Brake Adjuster

© IMechE 1976

C37/76

RECENT PROGRESS IN BRAKING TESTS BY USE OF A CAR DYNAMOMETER

J. A. ODIER, FIMechE.
P. MOLINIER
J. THIRION de BRIEL, S.A.F. Ferodo, Paris

The MS of this paper was received at the Institution on 4 July 1975 and accepted for publication on 20 October 1975

SYNOPSIS After a short description of the apparatus used for the tests (indoor test bench with a car having the same behaviour as on the road) and the projection of the corresponding film, the progress made in the use of this bench for braking tests over the past few years is considered : application to the American braking test MVSS I05, analysis of the value of load sensitive pressure relief valves and proportionning valves, vibrations in braking, locking of wheels, several important and new studies (use of several types of adherences with their control ; systematic control and measurement (or record) of the braking ratio and of the pivoting torque on a car during braking ; road holding characteristics of a car when braking, etc....).
All the results come to the conclusion that with this new test bench, important progress has been made since it allows a more scientific approach to braking tests (progress in facility of use, in reproducibility, including several important new evaluations, etc...) as compared to road braking tests which lack precision, are expensive, and which, in practice, have not shown any progress for many years.

INTRODUCTION

I. In this paper, we are proposing to present to you a new test rig - known as a dynamometer - which allows the simulation in the laboratory of the road holding of a genuine car in view of its application to the development of braking tests.

2. This machine has already been covered in various publications either in relation to its design (I), its possible application to tyres (2) and to road holding (3), its limitations (4), its engineering (5), its application to braking (6), and primary safety (7), and more particularly, to the new American braking standard FM V SS I05 A (6) (8).

DESCRIPTION OF THE TEST BENCH

3. In the different publications and specially in publication (5) are described the main characteristics and advantages that can be expected from such a dynamometer.

USE OF THE TEST BENCH

4. Our Company has been using this test bench for the last few years (including the necessary period of development) with satisfactory results in the studies and testing carried out, and we have also rented it many times to a great number of other French, American, German and Italian Companies.

5. Generally speaking, the interest in this bench has been shown in the way that could be expected, namely :

(a) It allows, since the car is quasi-static, all the tests that are considered dangerous, inconvenient or inaccurate when they are carried out on the road, and particularly at high speed, at high deceleration (when braking) or at high centrifugal acceleration.

(b) It allows the simulation, particularly at high speed, of all the testing techniques possible (even if artificial), without the limits imposed by a given road circuit and thus in a reproducible manner by contrast with road testing which is influenced by various factors from one test to another (wind, rain, snow, temperature, air hygrometry, dust, etc...).

(c) It allows in the widest way possible, the taking of all kinds of different measurements (displacements, speed, acceleration, deceleration, stress and forces, torques, vibrations, filming of parts in motion, etc...) in a simpler, more accurate and more convenient manner than on the road, the quasi-static position of the car making the operations obviously much easier.

(d) It allows for the measuring or testing which is very difficult or impossible to make on the road (measurement of the reactions of the ground, pivoting torque, braking ratio, reduction or suppression of the factor "driver" if necessary, or of the factor "resistance of the air", etc...).

(e) It allows mechanical testing on a new type of vehicle, in all safety, instead of making them on the road (which often gives rise to some unsafe conditions).

6. Moreover, its price is much lower than that of a test track and the tests, which are much more accurate and reproducible, can be carried out at all seasons.

© IMechE 1976

BRAKING TESTS

7. We must now turn to the results obtained in braking tests and to the corresponding progress achieved (applicable also to other important applications apart from braking).

8. It is with the aim of improving the quality of braking tests - which, on the road, have made no important progress over the past thirty years - that the bench has been primarily studied, together with the objective of avoiding the very high cost of a braking test track.

9. Generally speaking, progress has been what was expected in view of the fact that a ventilating system servo-connected to the speed of the vehicle has enabled the same laws of heating of the brakes to be reproduced in the laboratory as on the road under the same conditions.

COMPARISON WITH THE ROAD AND REPRODUCIBILITY OF THE TESTS

10. An important condition in view of the interest raised by the bench is that the results obtained are identical to those obtained on the road (although if this similarity was not perfect, the bench would still be of general interest for comparative tests in braking, as is the case with the well known existing laboratory rigs) under the same working conditions, and the necessary studies carried out in this direction, despite the difficulty met in the comparison, have shown that this is quite the case (6) and that some of the criticisms that could be made on this matter are either unimportant, or unjustified (8).

11. This being said, one of the first advantageous characteristics of the bench that becomes evident is the fact that tests are reproducible.

12. Everyone knows that on the road, this is not possible and it is the reason why modern testing methods (refer later on to the American testing method FM V SS 105 A) impose special conditions - which are very expensive on the road (dry ground, no wind, etc...) in order to try to avoid these difficulties.
Appropriate parts of the method FM V SS 105 A are given in Appendix.

13. This car test bench allows for the fulfilment of these very important reproducible conditions and the group of tests that follows will show evidence of this.

14. Six tests have been carried out on a single vehicle (x) according to the method FM V SS 105 A. Three of them have been carried out on a test track, whereas the other three have been performed on the test bench. For each test, the brake pads have been replaced by a new set coming from the same manufacturing batch.

(x) - Total weight........... 1630 daN
Front weight........... 780 daN
Rear weight............ 850 daN

Four disc brakes

15. The stopping distance cannot be considered independently of the speed v and of the brake pressure ρ. For this reason, it has been found convenient to use the ratio $v^2/2\rho d$ for expressing the results.
With the following units :

speed.................... m/s
stopping distance........ meter
brake pressure........... bar

the average value for $v^2/2\rho d$ is about 110.

16. In the road test carried out, a fifth wheel was used in order to measure the speed and the stopping distance. On the bench, there is no need to use a fifth wheel since the speed and the stopping distance are directly measured on the rollers according to a very accurate system.

17. For each brake application in efficiency testing, each group of three tests gives three values of the ratio $v^2/2\rho d$ and the dispersion of these three results is evaluated in terms of their standard deviation.

18. Figure I shows the distribution of σ resulting from the tests. The average value of the population is 4,8 for tests on the bench.

19. For road test results, the dispersion is more important and the average value of the population is 7,3.

THE MOST IMPORTANT PROGRESS IN BRAKING TESTS

20. The most important progress, apart from the reproducibility of tests which has just been covered above, is what follows :

(a) <u>Knowledge of the braking ratio</u> i.e. distribution of braking between the two axles at each brake application and, during a given braking test, the drafting of the histogram for these distributions. Figure II gives an accurate and significant example of this on the same car as previously mentioned. As an abscissa is given the percentage of braking distribution on the front axle and as ordinates the number of brake applications. One can there distinguish two peaks (whereas only one is needed, with a very pointed histogram).

The explanation for this amazing distribution was found easily since the left part of the histogram around the first peak of 0,53 corresponded to the brake applications which preceded the fading test, and the other part around the second peak of 0,60 corresponded to the brake applications which followed the fading test.

Thus, the braking effectiveness of each axle therefore important to check at all times the braking ratio.

Such a change can make a car dangerous. It is therefore important to check at all times the braking ratio.

Such histograms, which cannot normally be obtained in road testing, are very important when calculating braking in the design office and for the development of car braking, whether the braking system is simple or whether it includes correcting devices.

© IMechE 1976

(b) Knowledge of the pivoting torque in braking, at each braking application, the importance of which is well known in vehicle dynamics : histogram of the value of this torque during a given test and study of its causes and of the maximum pivoting torque permissible.

Figure III shows the distribution of the pivoting torque around the yaw axis which is noted on a car (x) suffering from a small loss of hydraulic fluid at its left front brake. The average torque is about 50 m∧ daN.

When braking is normal, the form of this distribution is similar but the average value should be very close to zero. The width of the distribution makes it easy to estimate the reliability of the transversal equilibrium.

At this point, we have the problem of determining the maximum permissible torque for a given vehicle. The corresponding tests need an accurate measurement of the speed, braking forces, pivoting torque, lateral displacement, etc... On the test bench, this does not present any difficulty but what is more important is that danger is eliminated. Tests have been carried out on a few common cars, the brakes of which had been intentionnaly put out of balance, and the results obtained were interesting. For example, on one of the cars running at I20 km/h, it was found that a pivoting torque, caused by the front brakes being out of balance, became dangerous starting from 50 m∧daN. For most of the drivers, the highest acceptable limit was at 30 m∧daN. Obviously, these limits depend on the speed and the type of vehicle used.

On the other hand, the systematic recording of the pivoting torque enables the detection of differences in the delay responses of the left and right brakes. This is only a possibility and there are certain cars where the fugitive pivoting torque can reach about I00 m∧daN, under deceleration of 0,5 g.

(c) Accurate study of the swinging movement of the car during pitching at the early transient part of the braking during the so-called transient period : half a second for example (evolution of dynamic weight transfers on the front axle with overloading, loss of adherence, etc...) and, generally, an accurate knowledge of dynamic weight transfers at transient or steady-state conditions.

A comparative study has shown the position of such an accurate method of testing, with its advantages, and by comparison with the mathematical simulation - which is also of great interest but has limitations - used so far in such cases in place of tests on the road which are insufficient.

The pitching behaviour of a common car at the early transient part of the braking has been studied by means of a simulation using a digital computer. Then, the car (x) was tested on the bench and the results obtained were not in agreement.

There was no problem in finding the reason which was unexpected.

The program did not include conditions relating to wheel-locking. Such conditions had not seem to be necessary as the deceleration was not very high. However, the wheels did lock intermittently due to the oscillatory nature of the pitching movement. The locking of the wheels corresponded to the peak values of weight transfer ; the phenomenon was fugitive but it was sufficient to change the results.

(d) Accurate study of braking correcting devices of a car adhesion where a range adhesion values are explored.

It is well known that perfect proportionning valves allow, when braking the car sufficiently strongly but without brutality, the simultaneous (or quasi-simultaneous) locking of the four wheels within the whole range of adhesions that are normally encountered, from the lowest to the highest (9).

It is particularly easy to study braking proportionning valves on the bench. In effect, it is possible on the bench and for each brake application to measure simultaneously the braking forces and the weight on each wheel.

With these measurements, it is possible to draw very rapidly the curves of Figure IV which give, on the bench, the minimum adhesion which is necessary between the tyre and the ground in order not to lock any wheel at a given deceleration.

(This figure has been drawn for two axles, but it could have been drawn in the same way for the four wheels).

With ten brake applications at various decelerations, and using measurements of the braking and weight distribution of the car, it is possible to draw perfectly the curves of Figure V which show immediately and with accuracy which axle, during braking, will lock the first.

Therefore, it can be seen that without changing the rollers coefficient of friction (adhesion), it is quite possible to foresee the braking behaviour of the car, one wheel after the other, for all types of adhesions.

A rapid checking of the result can be made by changing the rollers coefficient of friction (adhesion).

Moreover, by using the curve of Figure IV, it is easy to find the curve of Figure VI which will give us the degree of "efficiency" of the car during braking.

(e) Accurate study of the road holding characteristics of a car during braking, the basic characteristics of a car for road stability being sometimes subject to important modifications according to the speed and the deceleration (for example, the car, which is primarily neutral in traction, tends to oversteer a lot during braking).

(x) - Car total weight.................. I370 daN
Car front weight.................. 740 daN
Car rear weight.................. 630 daN
Four disc brakes

(x) - Total weight..................... I2I5 daN
Front weight..................... 675 daN
Rear weight..................... 540 daN
Four disc brakes
Classical shock absorbers

© IMechE 1976

In Figure VII, you can see a car braking on a straight line whereas a lateral force of 80 daN is applied at the same time at its centre of gravity (weight of the car : I400 daN).

With braking at a deceleration of 0,3 g, the car is moving in the same direction as the force, without any change in the position of the longitudinal axle (yaw angle constant).

In case of braking at 0,6 g, the car is beginning to go in the same direction as the lateral force but at the same time pivoting around his yaw axle, which tends to make it rise against the force during braking.

(f) Accurate study of braking on a curve either by engine braking when relaxing the pressure on the accelerator during braking, or by braking with the wheel brakes in transient or steady-state conditions, with more and more increasing pedal pressure or in case of urgency (panic stop) which can go as far as locking one or several wheels. This is done with an appropriate analysis of the forces and torques which act on each wheel and on the whole car with its corresponding displacements.

In the example given in Figures VIII and IX, we show a classical type of vehicle (front engine, rear axle drive) fitted with radial tyres for a series of brakings on a curve. For all the tests, braking was performed with the engine in gear and the steering wheel holding.

Various radii of curves were successively used in order to reach an increasing lateral acceleration. For the type of braking illustrated in the Figures, the radius of the curve was I25 m, or an initial equivalent lateral acceleration of 0,5 g.

For all types of braking, the tendency was that :

(i) for braking at decelerations of 0,I to 0,4 g, the car tended to go inside the curve during braking.

(ii) for braking at higher decelerations, the car was more inclined to come out of the curve during braking.

When studying the curves more closely, it can be seen that in all cases, the car begins by moving slightly to the side where it will come out of the curve, although taking at the same time a yawing angle which will tend to make it go back inside the curve.

For braking at a deceleration of 0,I g to 0,4 g, it is the yawing angle which is the most significant and pushes the car inside the curve. With higher accelerations, the yawing angle, once the car has turned inside the curve, changes its direction and makes the car turn in the opposite way : the car then comes out of the curve.

In Figure IX, you can see an example of braking at a deceleration of 0,6 g with locking of the wheels inside the curve. It can be checked on the curves that once the wheels have locked, the lateral forces corresponding to the locked wheels cancel each other.

(g) Accurate study of vibration phenomena during braking such as juddering, which is facilitated by the quasi-static position of the car.

(h) Accurate study, on a car, of the locking of the wheels during braking. With the measurement of the components in the forces acting on the tyres, together with that of the speed of the vehicle, it is possible to study very accurately the locking of the wheels during braking.

Figures X and XI show the locking of a front wheel at 90 km/h, with a slackening of the locking at 45 km/h. On Figure X, we can see the evolution of the longitudinal force tangent to the wheel (braking force) and of the weight on that wheel (vertical force) in terms of time.

It is possible to see the braking force and the weight increase at the same time up to the locking, and then decrease in the same way. When the wheel unlocks, we can see the longitudinal force increase suddenly once the wheel has begun to rotate again.

Figure X shows the variations, still in terms of time, of the coefficient of adherence :

(ratio : $\dfrac{\text{longitudinal force}}{\text{vertical force}}$) and of the slipping of the wheel.

(ratio : $\dfrac{\text{angular roller speed - angular wheel speed}}{\text{angular roller speed}}$)

Figure XI is obtained from Figure X and gives the variations in adherence in respect of slipping before locking of the wheel as generally portrayed (IO) (II).

During other tests of the same type, it was possible to study the problem of locking - unlocking of the wheels during braking caused by the pitching oscillations of the car during braking, which modifie the weight on the wheel during their occurence.

(i) To this important technical progress, it is necessary to add the conveniency with which these tests can be carried out (for example, tests for going down mountains, whatever the season and without loss of time in reaching mountainous countries), and the fact that they are free from danger (possibility of violent testing at the limit of adherence and of braking tests on a curve until losing control of the car).

(j) A new practical application, which is particularly interesting, is the new American standard for homologation braking FM V SS I05 A (6) (8).

This testing technique, which is relatively demanding (flat ground, no wind, no rain, external temperature within a well defined bracket, etc...) is not very convenient and takes a long time (several weeks in general) when applied on the road).

The great number of tests carried out, in particular on French, American and German cars, have shown that not only this standard can be applied very successfully on the bench, but it also shows some very specific and important advantages as compared with road testing of a similar nature (6) (8) :

© IMechE 1976

(i) Shorter time (40 to 50 hours of testing as compared to several weeks on the road) with the corresponding economical advantages.

(ii) Reproducibility and precision of the tests.

(iii) Possibilities of higher control than on the road.

(iV) etc...

Some of the criticisms which had been made towards this method and which, in fact, called for a close investigation were found to have no real importance after the necessary studies had been carried out (8).

CONCLUSION

2I. In conclusion, it can be said that apart from the meteorological conditions - which, if necessary, could be considered - all the characteristics of car braking, as well as the corresponding dynamics, can be reproduced on this bench by means of accurate and reproducible tests.

22. The quality, the precision and the variety of braking tests can be very much increased by comparison with road tests, as shown by the results obtained over the past few years and supplied in this paper, and this confirms entirely the optimistic forecast which had been made before the building of this bench, at the design stage, and which was indicated at the Fisita Congress in Barcelona in 1968 (I).

23. We hope that the use of such dynamometers will increase - apart from their other applications to vehicle dynamics - in order to improve the quality of braking tests and the corresponding road safety.

24. Indeed, all the results given above come to the conclusion that with this new test bench, an important progress has been made as it allows a more scientific approach to braking tests by comparison with road testing (which lacks precision, is expensive and which, in practice, has not shown any progress for many years).

REFERENCES

I. ODIER Jean-Alfred - Conception et étude d'une nouvelle machine d'essai automobile simulant la tenue sur route.
Conférence prononcée au I2e Congrès International de la F.I.S.I.T.A., Barcelone, I9-25 mai 1968. Publiée dans le Journal de la Société des Ingénieurs de l'Automobile I0-68, 53I-38.

2. ODIER Jean-Alfred - Application aux pneumatiques d'une nouvelle machine d'essai permettant de simuler la tenue sur route d'une automobile. Publication dans la Revue Générale du Caoutchouc et des Plastiques, vol. 46, n°I0, 1969, p.II5I-53.

3. ODIER Jean-Alfred - State of the Art - Road Holding : Braking and Traction - I970 - International Automobile Safety Conference - Detroit - p. 204 à 209.

4. ODIER Jean-Alfred - Introduction to the Mechanical Problems Raised by the Simulation on the Bench of the Road Holding of a Car - 2nd International Conference on Vehicle Mechanics, Paris, I97I.

5. ODIER Jean-Alfred - A dynamometer on which the dynamic behaviour of a passenger car can be simulated. Proceedings I972, I. Mech. Eng., volume I86, 7/I2.

6. ODIER Jean-Alfred - Indoor Vehicle Braking Simulation : Correlation and Comparison with Road Testing - I4e Congrès F.I.S.I.T.A. I972 à Bruxelles et Detroit.

7. ODIER Jean-Alfred - Etude en laboratoire de la sécurité primaire - Le banc d'essai automobile à poste fixe - Publication dans le Journal de la Société des Ingénieurs de l'Automobile (S.I.A.), n°I, janvier I974, p. 53-57.

8. ODIER Jean-Alfred, MOLINIER Pierre, THIRION de BRIEL Jacques - Nouveaux développements du banc d'essai automobile à poste fixe - Conférence prononcée au I5e Congrès International de la F.I.S.I.T.A., Paris, I3-I7 mai I974.

9. ROUSE J.A. - The Distribution of Braking on Road Vehicles, I. Mech. E., I963.

IO. HOLMES and STONE - Tyre Forces as Functions of Cornering and Braking Slip on Wet Road Surfaces - I. Mech. E. I969.

II. GOUGH V.E. - Friction of Rubber - The Engineer 3I/IO and 7/II/I958.

APPENDIX

Main Parts of Motor Vehicle Safety Standards n° I05 a (FM V SS I05 A)
(Federal Register, Vol. 37, n°I72 - Saturday, September 2, I972).

S 6.6 - Ambient temperature. The ambient temperature is any temperature between 32° F. and I00° F.

S 6.7 - Wind velocity. The wind velocity is zero.

S 6.8 - Road surface. Road tests are conducted on a I2-foot-wide, level road-way having a skid number of 75.
Burnish stops are conducted on any surface. The parking brake test surface is clean, dry, smooth protland cement concrete.

S 7.3 - Service brake system - first (preburnish) effectiveness test.
Make six stops from 30 m.p.h. with a pedal force of I5 to I00 pounds.
Then make six stops from 60 m.p.h. with a pedal force of I5 to I20 pounds.

S 7.5 - Service brake system - second effectibeness test. Repeat S 7.3. Then, with a pedal force of 20 to I50 pounds, make four stops from 80 m.p.h. and four stops from the multiple of 5 m.p.h. that is 4 m.p.h. to 8 m.p.h. less than the speed attainable in 2 miles if that speed is 95 m.p.h. or greater.

© IMechE 1976

S 7.7 - Service Brake system - lightly loaded vehicle effectiveness test.
Make six stops from 60 m.p.h. with a pedal force of I5 to I20 pounds, with vehicle at lightly loaded vehicle weight.

S 7.I4 - Service brake system - fourth effectiveness test. Repeat S 7.5.

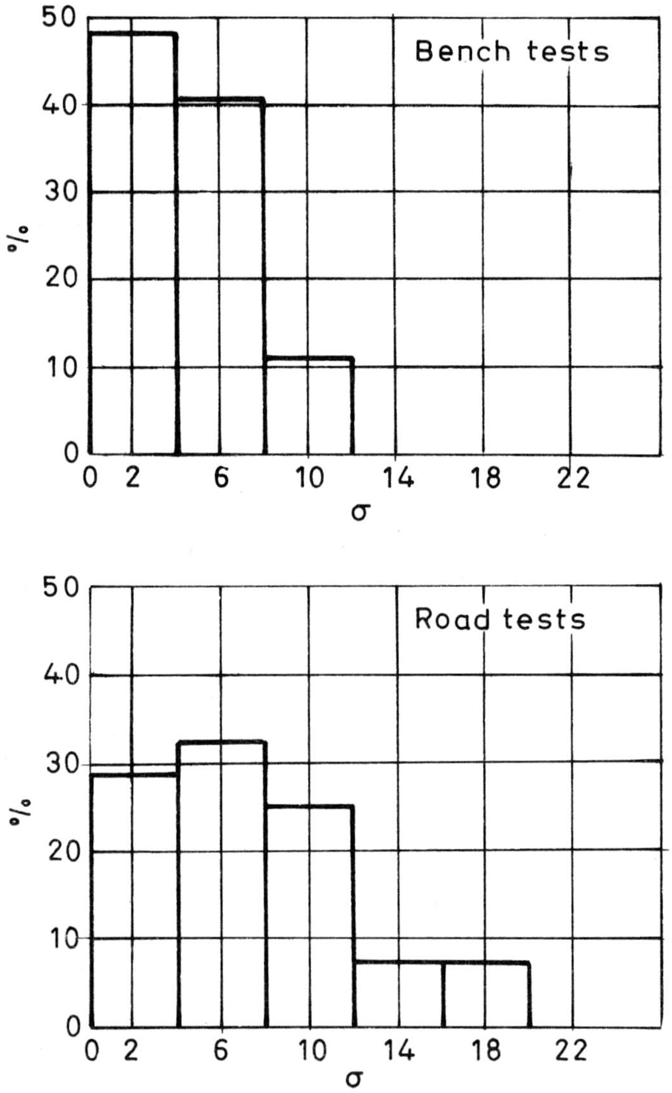

Fig. 1: FMVSS 105 A Effectiveness Tests Distribution of standard deviations

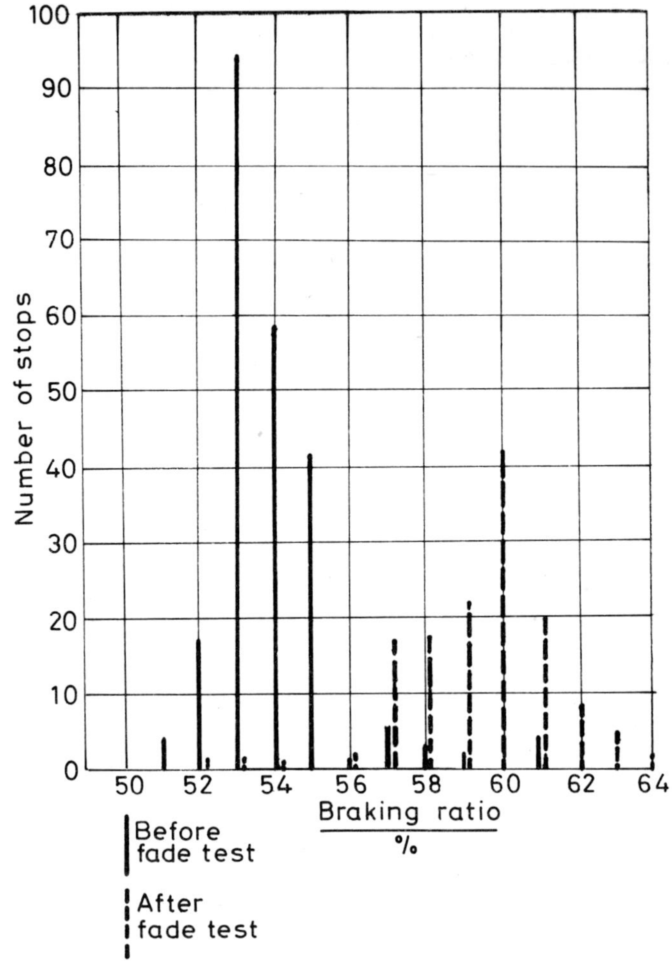

Fig. 2

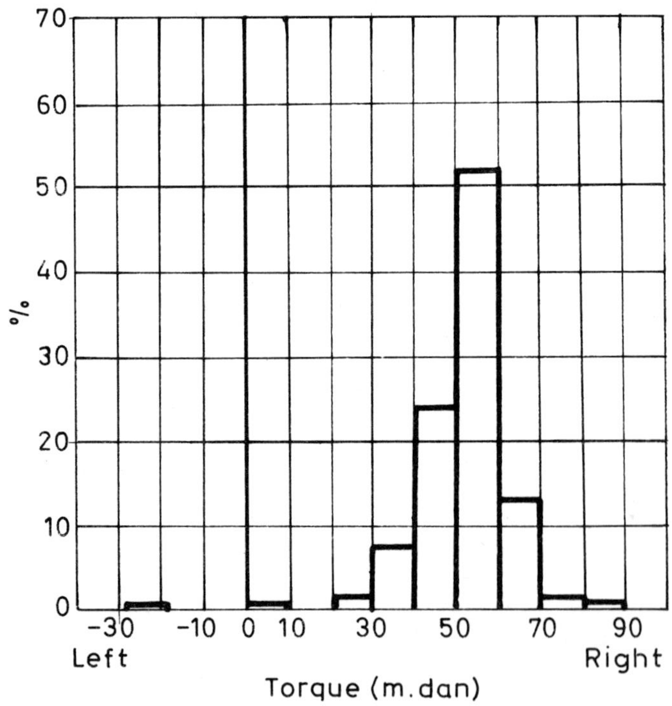

Fig. 3: Pivoting torque around the yaw axis

© IMechE 1976

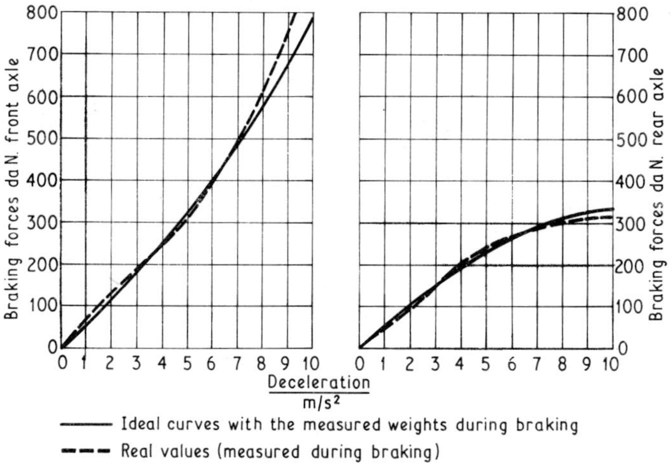

Fig. 4

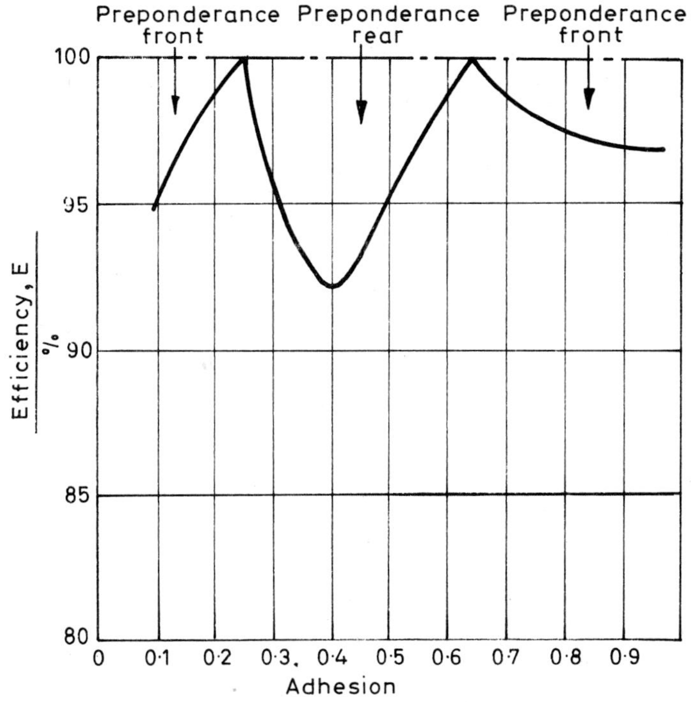

Fig. 6

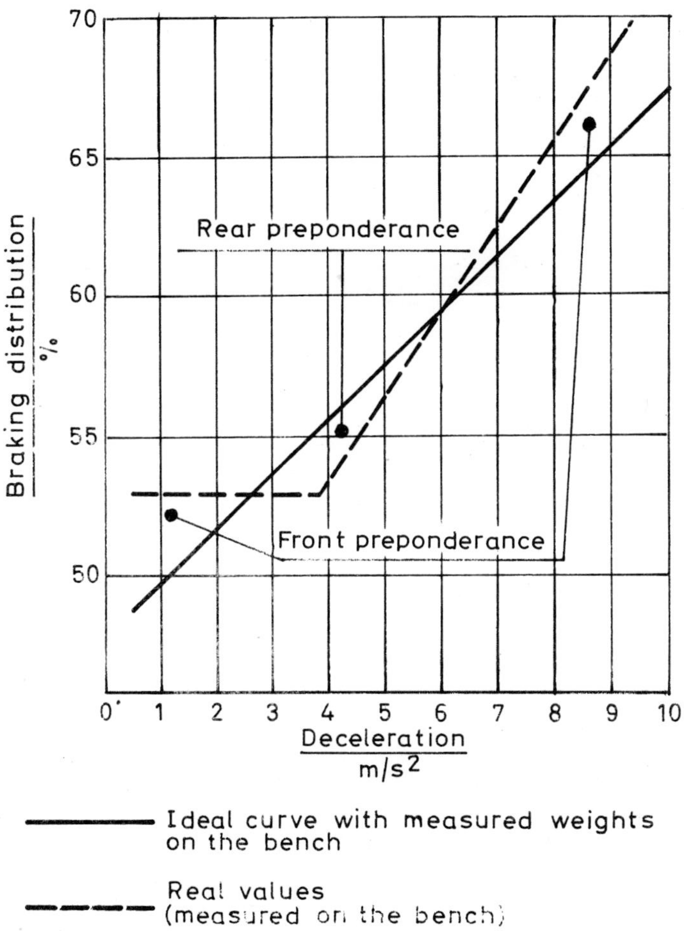

Fig. 5

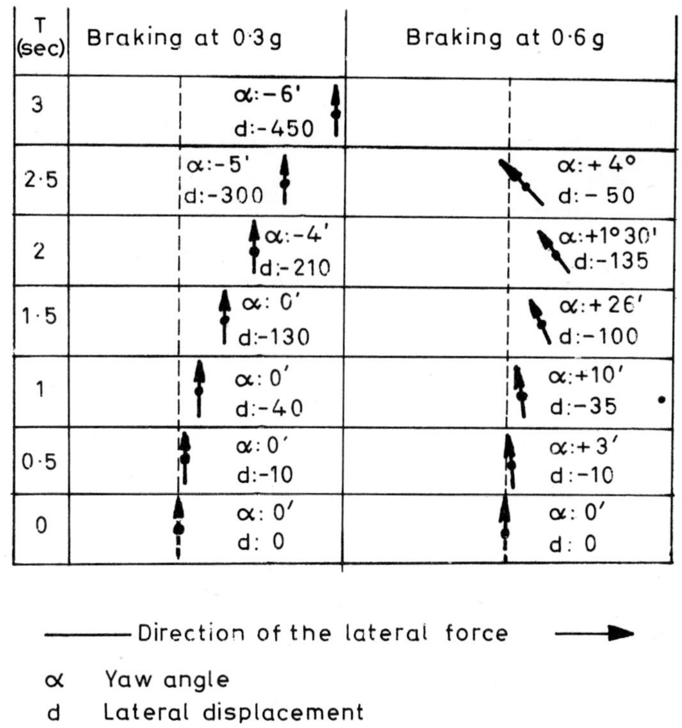

Fig. 7

© IMechE 1976

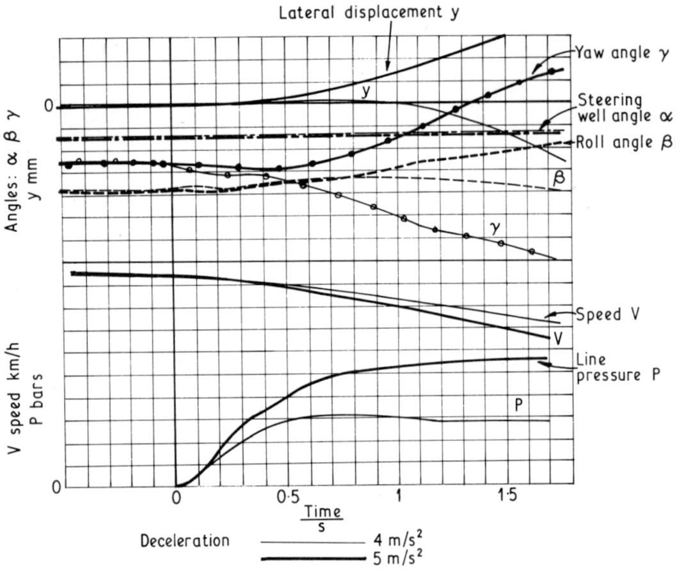

Fig. 8: braking in curve radius: 125 m initial speed: 90 km/h curve: to right

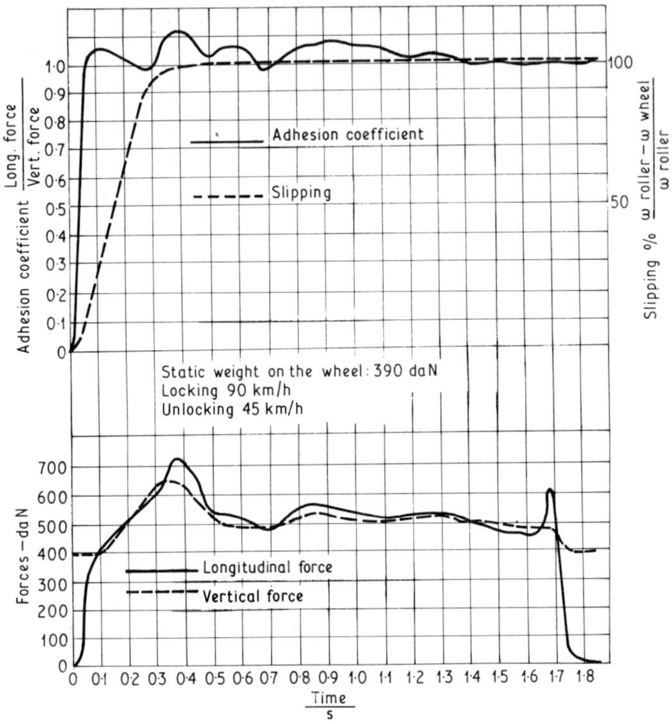

Fig. 10: locking - unlocking of one front wheel

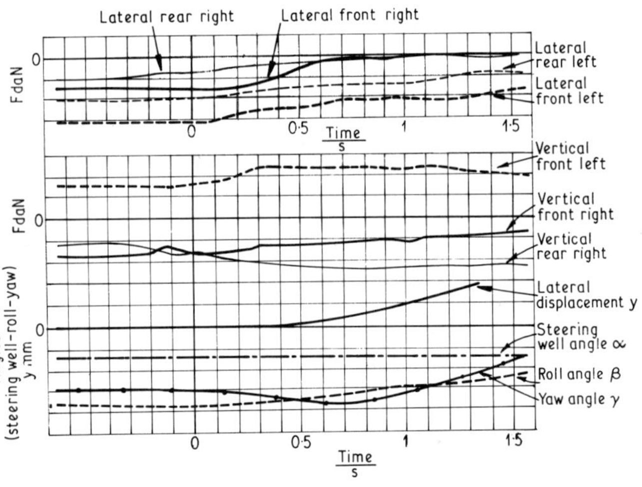

Fig. 9: Braking at 90 km/h at J = 6 m/s^2 on a radius of 125 m (curve to right)

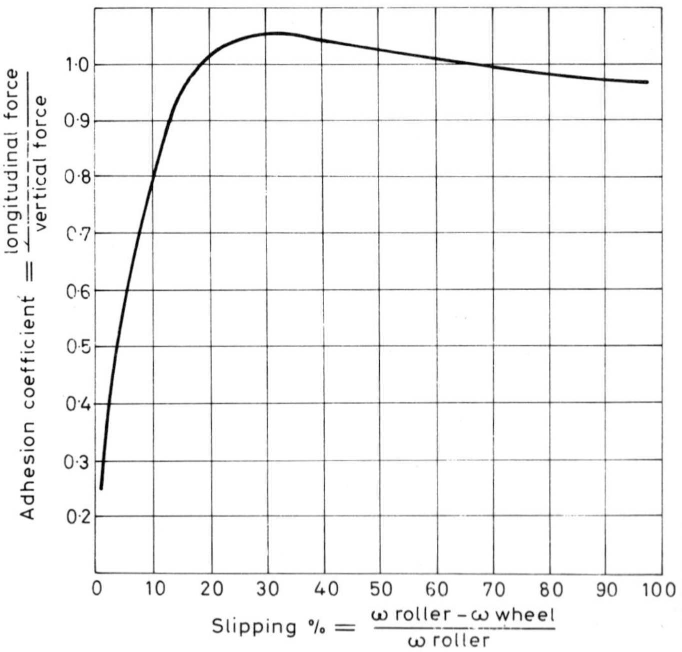

Fig. 11: Adhesion coefficient before wheels locking

©IMechE 1976

C38/76

DISC BRAKE SQUEAL

M. R. NORTH, MA, PhD, CEng, MIMechE
The Motor Industry Research Association, Nuneaton

The MS of this paper was received at the Institution on 11 August 1975 and accepted for publication on 22 October 1975

SUMMARY After a brief introduction and historical review, the paper examines in detail the four separate theories put forward since 1960 to explain the underlying mechanism of disc brake squeal. The salient features of the mathematics are reproduced and then the theories are compared, one with another, to bring out their essential features. A brief examination is then made to discover how the theories explain some of the practical observed facts of squeal.

It is concluded that there are two viable theories, with mathematical similarities, but physically quite different. A method of selecting the appropriate method for future exploration by any workers in the field is suggested.

INTRODUCTION

1. Friction brakes have always tended to suffer from unwanted vibrations; this has been true since the early days of train brakes, when brake block chatter was a problem, up to the present day disc brake in which squeal can cause considerable nuisance. Vibrations can occur at more or less any frequency, but squeal can be conveniently defined as a vibration within the range 1 kHz to 5 kHz and is characterised by the disc vibrating in one of its lower transverse modes such that the nodal diameters are spaced at approximately equal angles round the disc. The vibration is self excited, and is fugitive in the sense that it is extremely difficult to determine any repeatable test conditions in which a brake can reliably be expected to squeal. In addition, it is experimentally impossible to trace the source of the vibration since all vibration modes are coupled and all braking system components in the vicinity of the disc and caliper will vibrate.

2. The problem of squeal has usually been approached empirically in individual cases and, in the short term, minor modifications to system mass, stiffness or damping are often sufficient to eliminate the problem; longer term cures are extremely difficult to achieve and usually rest with some change in friction material properties or insertion of a shim between actuating piston and pad backing plate which can alter both the centre of pressure of the friction material and the stiffness between piston and backing plate. Such development programmes, which can sometimes occur during vehicle production runs when customer complaint reaches a high enough level, are expensive and time consuming and have led to a number of investigations of the squeal problem over the last forty years.

HISTORICAL REVIEW

3. The first concerted attack on brake squeal started in 1955 at MIRA, and led to a series of five reports, the last two of which were specifically concerned with disc brake squeal, References 1 and 2. This work was mainly experimental and produced an excellent fund of experimental data on disc brake squeal. Of particular interest were the frequencies and mode shapes measured on a squealing disc brake and details of the effects of systematic modifications to the disc and caliper assembly. Whilst no guaranteed cures were found, the conclusion of the final report was that a form of shim between the piston and pad back plate showed considerable promise as a means of alleviating squeal.

4. The earliest theory of squeal, which was current up to the early 1960's, was that of variable dynamic rubbing friction. This is a well known effect covered fully in the literature, for example Reference 3, in which an oscillatory instability can arise in a simple elastic rubbing system, Figure 1 (a), if the coefficient of friction decreases with increased rubbing velocity. Such a system is obviously similar in concept to a disc brake and pad, but the resulting oscillation is in the direction of the frictional force rather than transverse, and the frequency is controlled by the stiffness between the pad and some fixed point in space. Thus, while the theory may well account for certain brake system vibrations, it is not likely to account for squeal and has been superseded by other ideas.

5. The first attempt to move away from the variable friction model was in a paper published by Spurr in 1961, Reference 4. This considered the motion of a double cantilever system, Figure 1 (b), and showed experimentally that such a system did create stick-slip motion. The mechanism for the motion was explained qualitatively by considering O'P moving under the influence of the frictional force F until spragging occurs. Cantilever O'P is then effectively replaced by O"P which is at a smaller angle to the rubbing surface; F falls, allows the tip P to slip back, and oscillation occurs.

6. Squeal was also caused experimentally by the same author by loading a block of perspex against a rotating disc. Under these conditions the stress pattern consisted of a line running up at the

© IMechE 1976

spragging angle from the contact point to the specimen support.

7. This model excites transverse vibration of the disc and was the first to suggest that squeal might depend upon the actual value of μ rather than the slope of the μ/velocity graph.

8. Since the early 1960's there have been three major theoretical contributions to the squeal problem and these are now reviewed and compared in considerable detail. The mathematics quoted to illustrate the principles involved inevitably differs from that used in the original publication, but it is hoped that the salient points can be extracted by the use of simplified mathematics without too much distortion of the original work.

CANTILEVER - DISC MODEL

9. The first theory of disc brake squeal accompanied by significant analysis was that due to Jarvis, published in 1963, References 5 and 6. He chose a disc and cantilever model, Figure 2 (a), to represent a disc brake system and his mathematical analysis described the dynamics of the model.

10. The cantilever was clamped in a large, horizontally-pivoted block and the tip was loaded against the disc by means of a dead weight. The block was large enough to act as a seismic mass at high frequencies and the cantilever length, inclination and rubbing radius could all be altered.

11. The full mathematical analysis took account of the cantilever equations and disc equations, taking one mode only for each. In the simplified analysis presented here, which is due to Crisp and appears in the discussion to Reference 5, the motion of the tip of the cantilever is replaced by a mass M whose motion, u, is linear and perpendicular to the undeflected neutral axis of the cantilever. The mass M is constrained by stiffness, C_M, and damping, b_M, such that the frequency and damping of mass M are the same as for the cantilever tip. The direction of motion of M is at an angle γ to the disc surface.

12. The disc motion, w, is represented by a mass m with freedom to move in a direction parallel to the disc axis and this has associated stiffness, C_m and damping b_m, to give the same natural frequency and damping as the equivalent disc mode.

13. The two components are constrained to remain in contact and are coupled together by reaction force, R, and friction force $F = \mu R$, where μ is the coefficient of friction, as shown in Figure 2 (b). Accordingly, the equations of motion are

Cantilever $$M\ddot{u} + b_M\dot{u} + C_M u = R\,(\mathrm{Sin}\,\gamma - \mu\,\mathrm{Cos}\,\gamma) \quad (1)$$

Disc $$m\ddot{w} + b_m\dot{w} + C_m w = -R \quad (2)$$

Constraint $$w = u\,\mathrm{Sin}\,\gamma \quad (3)$$

These equations reduce to a single degree of freedom system.

$$k_1\ddot{q} + k_2\dot{q} + k_3 q = 0 \quad (4)$$

where q is the normal coordinate, u.

and $$k_1 = 1 - \Gamma \quad (5)$$

$$k_2 = {}^{b_M}/_M - \Gamma\,{}^{b_m}/_m \quad (6)$$

$$k_3 = {}^{C_M}/_M - \Gamma\,{}^{C_m}/_m \quad (7)$$

$$\Gamma = ({}^{m}/_{2M})\,\mathrm{Sin}\,2\gamma\,(\mu - \mathrm{Tan}\,\gamma) \quad (8)$$

14. Equation 4 appears as the governing equation of Reference 5, although, of course, in this simplified analysis, the expressions for k_1, k_2, and k_3 take rather different forms to the original expressions.

15. The requirement for oscillatory instability is that

$$k_2/k_1 < 0 \quad (9)$$

16. The original expression, k_1 was found to be always positive for the values of constraints in the model. In this instance the condition reduces to

$$1 > \Gamma > \frac{b_M}{b_m}\cdot\frac{m}{M} \quad \text{and} \quad \mathrm{Tan}^{-1}\mu > \gamma > 0 \quad (10)$$

17. There can, therefore, be a finite range for γ over which instability can occur for suitable values of $^{b_m}/b_M$ and $^m/M$. By the same token, such instabilities can be prevented by appropriate choice of μ, by appropriate damping and by control of the ratio, $^m/M$. It should be noted that, with such a model, added damping can, in fact, be destabilising.

18. Jarvis extended his analysis to include the second disc mode, although this is not essential in order to describe the mechanism of the instability; as he points out, two disc modes are required to describe the behaviour of the disc completely and the second mode is excited by the presence of the first in his model.

19. The basic cause of oscillatory instability in his model is the coupling together of a single disc mode to a single cantilever mode by the geometric constraint of having the two components remain in contact. This type of instability was called a kinematic constraint instability by Crisp and the mathematics in this instance reduce to the mathematics of a single degree of freedom.

20. The agreement of the experimental stability boundaries with the theoretically predicted boundaries was poor and attempts were made, both in the paper and in the discussion, to improve the agreement by examining the contact conditions between cantilever and disc. Minor improvements were achieved, but the correlation was never fully satisfactory.

PIN/DISC MODEL

21. Details of this model were published in 1971 by Earles and Soar, Reference 7. This model consisted of a disc, as in the Jarvis model, but his cantilever was replaced by a pin fixed to the end of a shaft and loaded against the disc; the pin had one translational freedom, x, along the axis of the supporting shaft and at an angle Θ to the disc surface, and one torsional freedom, λ, about an axis orthogonal to both the pin and supporting shaft axes, as shown in Figure 3. As in the case of the

© IMechE 1976

Jarvis model, the model was designed to be amenable to mathematical analysis and to allow good experimental control.

22. Although the model requires three degrees of freedom to describe it, the analysis was carried out by examining the coupling of the translational mode of the pin with the disc separately from the coupling of the torsional mode with the disc.

23. Referring to Figure 4 (a), it can be seen that the equations of motion of the translational pin mode coupled to the disc will be exactly as for the Jarvis model, and that the same kinematic constraint will apply. The equations derived to define the translational pin mode are therefore the same as for the cantilever-disc model.

24. Earles and Soar derive the same conditions for instability, but say that in their model, using their values, instability would require the damping, stiffness or mass ratios to be greater than 50 and that this is most improbable. They then go on to examine the torsional pin mode coupled to the disc.

25. As in the Jarvis model, the disc is represented by mass M, constrained by stiffness K_D and damping C_D. However, the pin, of length 1, is represented by moment of inertia I, constrained by rotary stiffness K'p and rotary damping C'p. The reaction force between pin and disc is given by N(t) and gives rise to frictional force $\mu N(t)$.

26. Referring to Figure 4(b), the disc and pin equations are, respectively

$$M\ddot{y} + C_D\dot{y} + K_D y = -N(t) \qquad (11)$$

$$I\ddot{\lambda} + C'p\dot{\lambda} + K'p\lambda = N(t)\ell \operatorname{Sin}(\theta + \lambda) - \mu N(t)\ell \operatorname{Cos}(\theta + \lambda) \qquad (12)$$

with the kinematic constraint

$$y = \lambda \ell \operatorname{Sin}(\theta + \lambda) \qquad (13)$$

27. Earles and Soar showed that, for small λ, Equations 11, 12 and 13 could be reduced to

$$\ddot{\lambda} + \left(\frac{2BMS}{I + MAS}\right)\dot{\lambda}^2 + \left(\frac{C'p + C_D AS}{I + MAS}\right)\dot{\lambda} + \left(\frac{K'p + K_D AS}{I + MAS}\right)\lambda = 0 \qquad (14)$$

where $A = \ell \operatorname{Sin}\phi$
$B = \ell \operatorname{Cos}\phi$
$S = \ell(\operatorname{Sin}\phi - \mu \operatorname{Cos}\phi)$
$T = \ell(\operatorname{Cos}\phi + \mu \operatorname{Sin}\phi)$
λs = Steady state value of λ
$\phi = \theta + \lambda s \doteqdot \theta$

28. A more accurate description of the kinematic constraint is, in fact,

$$\delta y = \delta\lambda\, \ell \operatorname{Sin}(\theta + \lambda) \qquad (15)$$

which gives $y = \int \ell \operatorname{Sin}(\theta + \lambda)\, d\lambda$ (16)

or $y = \ell(\operatorname{Cos}\theta - \operatorname{Cos}(\theta + \lambda))$
If $y = 0$ when $\lambda = 0$ (17)

which leads to a more linear relationship between y and λ.

This gives $\dot{y} = \ell \operatorname{Sin}(\theta + \lambda)\dot{\lambda}$ (18)

$$\ddot{y} = \ell \operatorname{Sin}(\theta + \lambda)\ddot{\lambda} + \ell \operatorname{Cos}(\theta + \lambda)\dot{\lambda}^2 \qquad (19)$$

For small λ $\ddot{y} = \ell\ddot{\lambda}(\operatorname{Sin}\theta + \lambda \operatorname{Cos}\theta) + \ell\dot{\lambda}^2(\operatorname{Cos}\theta - \lambda \operatorname{Sin}\theta)$ (20)

which gives $\dot{y} = \dot{\lambda}(A + B\lambda)$ (21)

$$\ddot{y} = \ddot{\lambda}(A + B\lambda) + \dot{\lambda}^2(B + A\lambda) \qquad (22)$$

29. This simplifies the derivation of Equation 14 and leads only to the alteration in the coefficient of $\dot{\lambda}^2$ to $(BMS/(I + MAS))$ once all the higher order terms are **ignored**.

30. The coefficients of $\dot{\lambda}$ and λ in Equation 14 are very similar to those obtained for the translational mode and Earles and Soar again consider that instability arising from a negative value of the coefficient of λ is most unlikely for the system under test.

31. Accordingly they examined the effect of the term in $\dot{\lambda}^2$ by putting the coefficient of $\dot{\lambda}$ in Equation 14 equal to zero and rewriting equation 14 as

$$\ddot{\lambda} + G\dot{\lambda}^2 + F\lambda = 0 \qquad (23)$$

32. For the initial conditions $t = 0$, $\lambda = -\lambda_o$, $\dot{\lambda} = 0$, it can apparently be shown that, for $G\lambda_o < 0.5$, the system described by equation 23 oscillates with stable limit cycles, but for $G\lambda_o \gtrsim 0.5$ the system is unstable. It is suggested that stable limit cycles will still occur for $G\lambda_o \gtrsim 0.5$ since the non-sinusoidal form of the vibration will cause harmonics to be generated which will increase the internal damping of the system. It is certainly an observed fact that brake squeal is typically accompanied by the generation of harmonics.

33. The instability of the torsional mode thus arises entirely from the non-linearity generated by the Sine and Cosine terms due to the effects of a linear freedom coupling to a torsional freedom; without these non-linear terms, the mathematics of the torsional mode are the same as the mathematics of the translational mode.

34. The experimental observations accompanying this work were that the translational mode did not induce squeal, whereas the torsional pin mode did; it was concluded that, if the resultant force on the pin tends to rotate it into the disc, then the pin system will have a dominantly torsional mode, resulting in squeal, whereas, if it tends to move the pin away from the disc, the mode will be dominantly translational with no squeal. The torsional mode, resulting in squeal, occurred in the range $\operatorname{Tan}^{-1}\mu > \theta > 0$.

35. This work has been extended in two further publications, References 8 and 9. The first of these examines the same pin-disc model as a multi-degree of freedom system from the point of view of receptances and characteristic frequencies. This

© IMechE 1976

form of analysis does not add particularly to the understanding of the nature of the self-excited oscillations, but adds to the general physical understanding of the system. Of more interest is the experimental work which was carried out using pins with a dome radius rather than a flat face and of differing length.

36. Experimentally, this reinforced the conclusions of their previous work, but, unfortunately, no mathematical analysis was carried out on the effect of the domed face of the pin.

37. Reference 9 is the most recent publication in the series of reports on the pin/disc model and examines a different pin arrangement with pin, P, supported at the end of a flexible cantilever, C, which is in turn attached to an arm A free to rotate in a horizontal plane about bearing, B, Figure 5(a) and 5(b).

38. The pin is represented by mass, m_1, and moment of inertia, I, and has three degrees of freedom; one freedom, ϕ, is torsional and two, X and Y, are translational, respectively normal and parallel to the disc. Freedoms X and Y are associated with translational stiffnesses k_1 and k_2 and freedom ϕ is associated with rotary stiffness k_3. The disc is represented by mass m_2 and translational freedom Xp: the associated stiffness is k_4. The tip of the pin is horizontal distance, l, and vertical distance, c, from the centre, O, of the pin axis system. The reaction force between pin and disc is given by N(t) from which the friction force $\mu N(t)$ arises. Equations of motion for the four freedoms, after linearising for small ϕ, are given by

$$m_1\ddot{X} + k_1 X = -N(t) \qquad (24)$$

$$m_1\ddot{Y} + k_2 Y = \mu N(t) \qquad (25)$$

$$m_2\ddot{X}p + k_4 (Xp - \ell) = N(t) \qquad (26)$$

$$I\ddot{\phi} + k_3\phi = N(t)(\mu\ell + c) \qquad (27)$$

After inclusion of the kinematic constraint between the pin and the disc, the characteristic equation is derived as:-

$$(m_1\lambda^2 + k_2)(a\lambda^4 + b\lambda^2 + d) = 0 \qquad (28)$$

where λ is the conventional symbol associated with the writing of characteristic equations and the values of λ satisfying the equation give the natural frequencies and damping of the normal modes.

Equation 28 indicates that the mode parallel to the disc is uncoupled from the other modes.

39. The conditions of stability of the other two coupled modes are $a>0$, $b>0$, $d>0$, $b^2>4ad$. From the detailed analysis, the magnitudes of the coefficients will depend upon the sign of c $(\mu\ell + c)$ for suitable values of the stiffnesses and inertias of the system; for negative values of c, or for disc rotation opposite to the direction of Figure 5(b), this expression leads to the same condition for instability as the previous model and the Jarvis model, that is

$$\mathrm{Tan}^{-1}\mu > \theta > 0 \qquad (29)$$

EIGHT DEGREE-OF-FREEDOM MODEL

40. This has been described in References 10 and 11 in some detail and the intention here is to reproduce sufficient of the mathematics to bring out the mechanism of the instability.

41. Figure 6 shows a much simplified, two degree-of-freedom version of the original model with the disc of thickness h sandwiched between layers of flexible friction material of total stiffness k' bounded by immoveable planes. The disc is represented by a beam of mass, M, and moment of inertia, I, and with two degrees of freedom, y and θ, these two freedoms being necessary to represent the generalised motion of the disc in the vicinity of the pads, as appreciated by Jarvis. The disc is restrained by translational stiffness k_t and rotational stiffness k_r; the pad material has the effect of additional translational stiffness k' and additional rotational stiffness $k'.\ell^2/3$ where 2ℓ is the pad length. No is the static preload between pad and disc and F_1, F_2 the frictional forces.

42. The equations of motion are given by:

$$M\ddot{y} + (k_t + k')y + (F_1 + F_2)\theta = 0 \qquad (30)$$

$$I\ddot{\theta} + (k_r + k'\frac{\ell^2}{3})\theta - (F_1 - F_2)h = 0 \qquad (31)$$

where $F_1 = \mu(k'y + No)$ (32)

$F_2 = \mu(-k'y + No)$ (33)

thus $F_1 + F_2 = 2\mu No = 2F$ (34)

and $F_1 - F_2 = 2\mu k'y$ (35)

Put $a = (k_t + k')/M \qquad b = {}^{2F}/_{M}$ (36)

$c = (k_r + k'\frac{\ell^2}{3})/I \qquad d = -2\mu hk'/I$ (37)

Equations 30 and 31 can be rewritten as

$$\ddot{y} + ay + b\theta = 0 \qquad (38)$$

$$\ddot{\theta} + c\theta + dy = 0 \qquad (39)$$

This gives a characteristic equation

$$\lambda^4 + \lambda^2(a + c) + (ac - bd) = 0 \qquad (40)$$

$$\lambda = \pm\sqrt{\frac{-(a + c) \pm \sqrt{(a + c)^2 - 4(ac - bd)}}{2}} \qquad (41)$$

The criterion for oscillatory instability is

$$(a + c)^2 < 4(ac - bd) \qquad (42)$$

which leads to stability boundaries at

$$k' = 0 \text{ and } k' = \frac{16\,M\,I\,F\,\mu h}{(I - M\ell^2/3)^2} \qquad (43)$$

43. These equations give, qualitatively, the same type of instability as the full eight degree-of-freedom model, but naturally the stability boundaries occur at different stiffness values. This type of instability is very similar to aircraft wing flutter and the mathematics describing flutter of a two degree-of-freedom system is known as binary flutter, Reference 12. The full eight degree-of-freedom

system contains the freedoms appropriate to the two pads and the caliper pressing as well; the stiffnesses between pads and caliper and ground are all represented and damping is included across all stiffness terms.

COMPARISON OF THEORIES

44. The comparison of these models indicates that there are four distinct theories as follows:

1. Two degree-of-freedom with kinematic constraint.

2. Non-linear self-excited oscillation due to the term in $\dot{\lambda}^2$

3. Three degree-of-freedom with kinematic constraint.

4. Binary flutter of two disc modes.

45. The differences and similarities between these theories make interesting comparisons.

46. Theory 1 assumed linear motion of the cantilever tip so that, with the kinematic constraint, a linear single degree of freedom equation is obtained. However, the cantilever tip executes non-linear motion and it is conceivable that Theory 1 could be transformed into a theory similar to Theory 2 by inclusion of non-linearities.

47. Theory 2 relies upon the non-linear motion of the pin tip to create energy input to the vibrating system, but the vibration amplitudes of a squealing disc are usually very small (typically 0.0001 inches, Reference 2) so that the angular rotation of a 1 inch pin would be typically $<10^{-3}$ radians for the angle of inclination, $\theta > 5$ deg. It seems, therefore, that the degree of non-linearity would normally be very mild and predictions on self-excited vibrations would need to take account of the degree of structural and other damping in the system.

48. In later investigations of the pin-disc system, the authors abandon Theory 2 and progress to Theory 3 which returns to a more general version of the kinematic constraint instability. The pin freedom parallel to the disc is found to be uncoupled from the other freedoms whereas the kinematic constraint couples the disc mode to the remaining pin modes, reducing the three degrees of freedom to two. The mathematics of these two degrees of freedom are simple if damping is ignored and the appropriate conditions for instability are easily obtained. From these conditions it is deduced that a necessary, but not sufficient condition, for oscillatory instability is

$$\mathrm{Tan}^{-1}\mu > \theta > 0 \qquad (44)$$

49. This condition is identical with one of the conditions predicted for Theory 1. The other condition for Theory 1 was concerned with the ratios of masses, stiffnesses and damping in the two modes, whereas the additional conditions for Theory 3 are evidently very much more complicated and have not been derived by the authors.

50. Mathematically the simplified version of Theory 4 is identical with Theory 3, since it is only concerned with the stability of a coupled, two degree-of-freedom system: the stability criterion is however necessary and sufficient and is much easier to derive. Physically the theory is quite different and relies only on the presence of the disc and the friction forces produced by pressure of elastic friction material. One significant difference between Theory 4 and the others is that the presence of two independent disc modes is vital to Theory 4 and incidental to Theories 1, 2 and 3. Theories 3 and 4 are similar only in the sense that they can both ultimately be reduced to two degrees of freedom coupled by the presence of a frictional force and normal force.

51. It is interesting to consider how such theories cope with the practical, observed facts of squeal.

1. Generation of harmonics.

2. Generation of squeal over a limited pressure range.

3. Generation of simultaneous disc modes.

4. Generation of vibrations over a wide frequency range.

52. All theories are linearisations of basically non-linear equations and could thus be expected, in practice, to produce oscillations with a range of harmonics present. There would also usually be some effect such as loss of contact between friction material and disc to generate the limit cycle conditions and simultaneously generate harmonics due to impact as contact is regained.

53. The generation of squeal over a limited brake line pressure range is explained satisfactorily by Theory 4, since the stability boundary is a function of pad material stiffness which, in turn, is a function of brake line pressure. The other theories do not explicitly take account of pad material stiffness and presumably explain this effect in terms of geometric variations in the brake system altering the effective pin or cantilever contact angles.

54. The generation of simultaneous disc modes is an explicit requirement of Theory 4. Other theories will only generate a second disc mode through internal disc coupling and this is not examined in detail except in Theory 1.

55. All theories cope satisfactorily with the generation of vibration over a wide frequency range since they only examine coupling between two selected modes out of a range of many possible disc modes and, where appropriate, cantilever, pin or pad modes. Coupling of other modes is clearly able to generate vibrations at other frequencies.

56. The question is then posed as to how one can select the theory which represents the conditions of brake squeal most accurately. The obvious area of approach is to examine whether the disc can be represented by one mode or must be represented by two. Experimentally it is likely that two modes will always be detected since the presence of a single mode will generally excite a second mode. However, the second disc mode of theories 1, 2 and 3 does not need to appear explicitly in the equations of motion of the system and will therefore exist as an uncoupled, free disc mode; the first disc mode, being coupled dynamically to the rest of the system, will occur at the characteristic frequencies of the system which will differ from the free disc modes.

© IMechE 1976

Thus, if the second mode is excited by virtue of the first, it will be excited at a frequency other than its natural frequency and will thus be in phase or anti-phase with the forcing mode at the low levels of damping present in an iron disc.

57. Theory 4 would normally require some intermediate phase angle between the disc modes, as demonstrated in Reference 10, and it is conceivable, therefore, that detailed measurement of the modes of a squealing disc, including time phase and circumferential phase, would indicate the degree of involvement of the second disc mode in the dynamics of a squealing disc.

DISCUSSION AND CONCLUSIONS

58. The two theories of squeal which have been pursued most recently are the pin-disc model with three degree-of-freedom representation, reducing to two by means of a kinematic constraint arising from pin-disc contact continuity, and the eight-degree-of-freedom representation. Mathematically these can be reduced to similar equations in order to demonstrate the instability mechanism, which arises in the one case from the kinematic constraint and in the other as a binary flutter type of instability.

59. These instability mechanisms are quite different and design procedures to alleviate squeal arising from extensions to the two theories are likely to be significantly different. It would therefore seem sensible to resolve the differences before more effort is invested in either direction without knowing which is the more suitable. The problem is not one of incorrect mathematics or invalid approximations, but of choosing a simple model which is amenable to analysis whilst still retaining the necessary features of disc brake squeal. In the case of a pin-disc model, the mathematics describing the pin-disc system are likely to be substantially correct, but there may be doubts as to how well the pin-disc system represents the typical disc brake.

60. In the case of the eight degree-of-freedom model, the mathematics represent a disc brake system sufficiently closely that the experimental work was carried out on the actual disc brake system modelled, although mounted on a rig rather than installed in a car. Measurements of brake system parameters such as pad and caliper stiffness and damping were difficult to achieve accurately and representation of the disc by an equivalent lumped mass was only an approximation anyway. Nevertheless, by adjusting the measured stiffness by small amounts, the mathematical model gave squeal frequencies in very close agreement with measured values; the associated mode shapes also gave fairly good agreement with measured mode shapes, giving broadly similar results for the amplitudes of all freedoms, and for the phases of the pad with the largest amplitudes. Additional evidence of satisfactory correlation was that the band of brake line pressure over which the model predicted squeal agreed closely with the measured band of line pressure.

61. To resolve the differences between the two models, it would appear necessary to measure the amplitudes and phases of the fundamental disc modes during squeal; these are difficult measurements to make, but the results could set future research on disc brake squeal along the right lines.

ACKNOWLEDGEMENTS

62. The author wishes to thank the Director of the Motor Industry Research Association for permission to publish this paper.

REFERENCES

1. Fosberry, R.A.C. and Holubecki, Z. 'Interim Report on Disc Brake Squeal'. MIRA Research Report No. 1959/4.

2. Fosberry, R.A.C. and Holubecki, Z. 'Disc Brake Squeal: Its Mechanism and Suppression'. MIRA Research Report No. 1961/2.

3. North, M.R. 'A Survey of Published Work on Vibrations in Braking Systems'. MIRA Bulletin No. 4, 1969.

4. Spurr, R.T. 'A Theory of Brake Squeal'. I Mech E Auto Div Proc No 1 1961/62 pp 33-40.

5. Jarvis, R.P. and Mills, B. 'Vibrations Induced by Dry Friction'. I Mech E Proc Vol 178, Pt 1, No. 32, 1963/64, pp 847-857, Discussion: pp 857-866.

6. Jarvis, R.P. 'Vibrations of a Disc Induced by Dry Friction'. PhD Thesis, Birmingham, 1962.

7. Earles, S.W.E. and Soar, G.B. 'Squeal Noise in Disc Brakes'. Paper C 101/71, I Mech E Symposium on Vibration and Noise in Motor Vehicles, July 1971.

8. Earles, S.W.E. and Soar, G.B. 'A Vibrational Analysis of a Pin-Disc System with Particular Reference to Squeal Noise in Disc Brakes'. Institute of Physics, Stress Analysis Group Annual Conference, 1974.

9. Earles, S.W.E. and Lee, C.K. 'Instabilities arising from the Frictional Interaction of a Pin-Disc System Resulting in Noise Generation'. Paper No. 75-DET-25. Design Engineering Technical Conference, Washington DC, September, 1975.

10. North, M.R. 'Frictionally Induced, Self Excited Vibrations in a Disc Brake System'. PhD Thesis Loughborough University of Technology, 1972.

11. North, M.R. 'Disc Brake Squeal - A Theoretical Model'. MIRA Research Report No. 1972/5.

12. Done, G.T.S. 'The Flutter and Stability of Undamped Systems'. Aeronautical Research Council R & M No. 3553, 1968.

© IMechE 1976

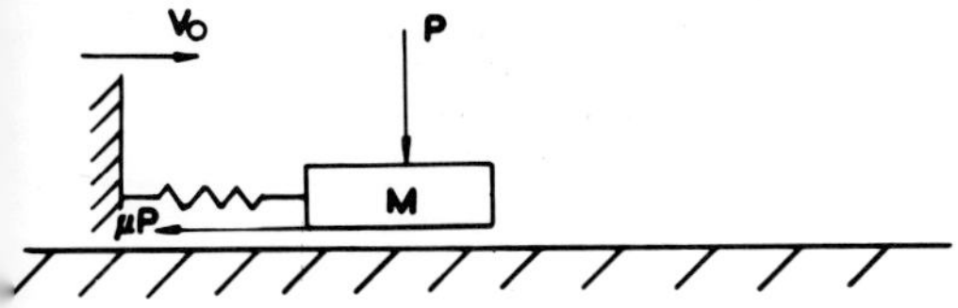

Fig. 1a: Simple elastic rubbing system

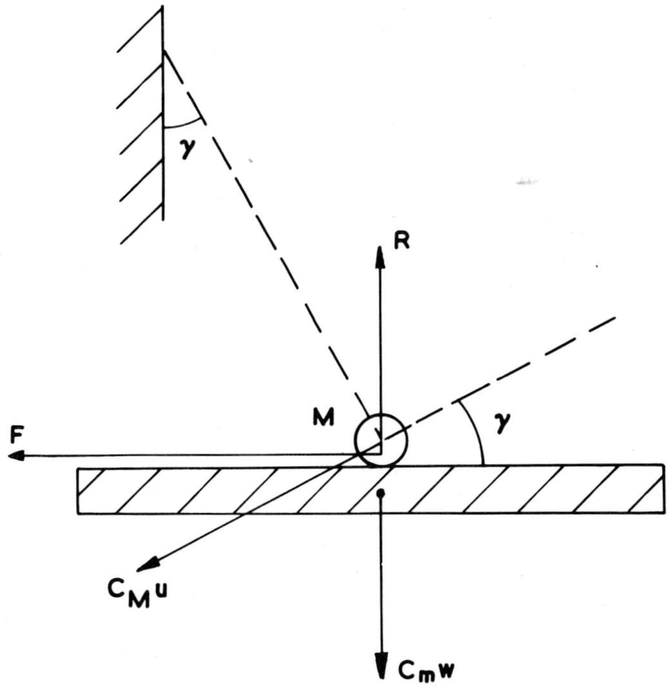

Fig. 2b: Simplified representation of the disc-cantilever model

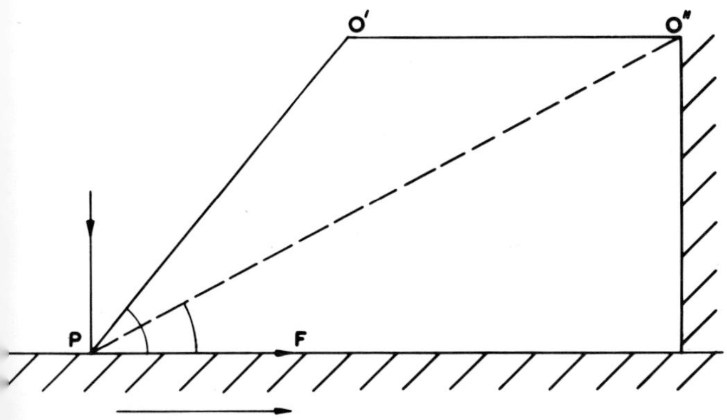

Fig. 1b: Double cantilever system

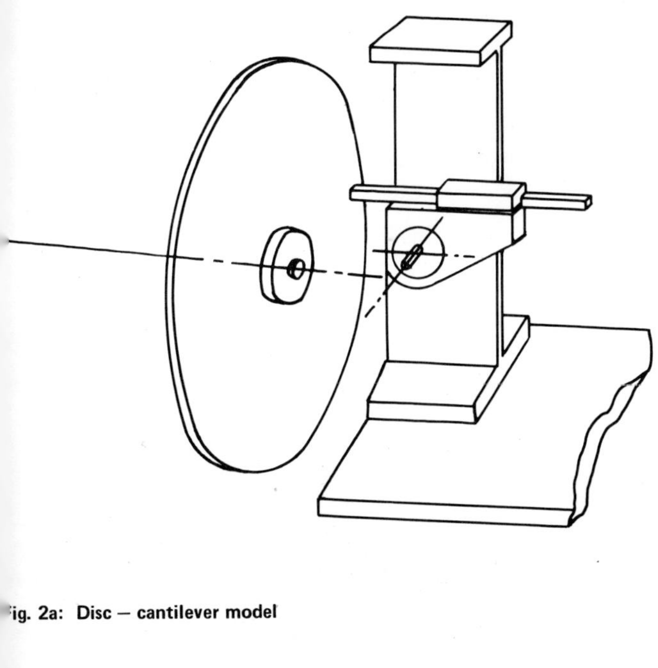

Fig. 2a: Disc – cantilever model

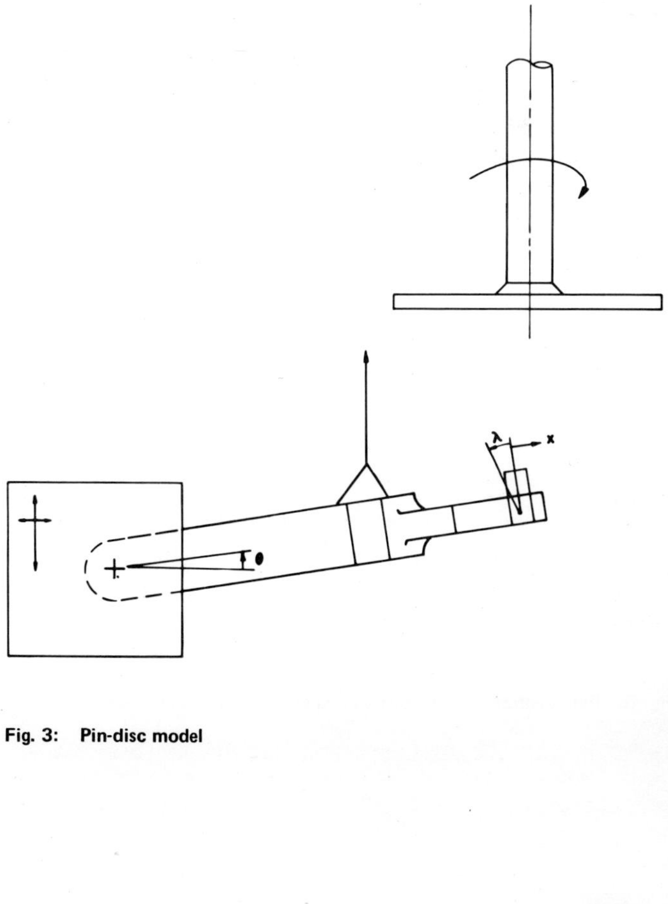

Fig. 3: Pin-disc model

© IMechE 1976

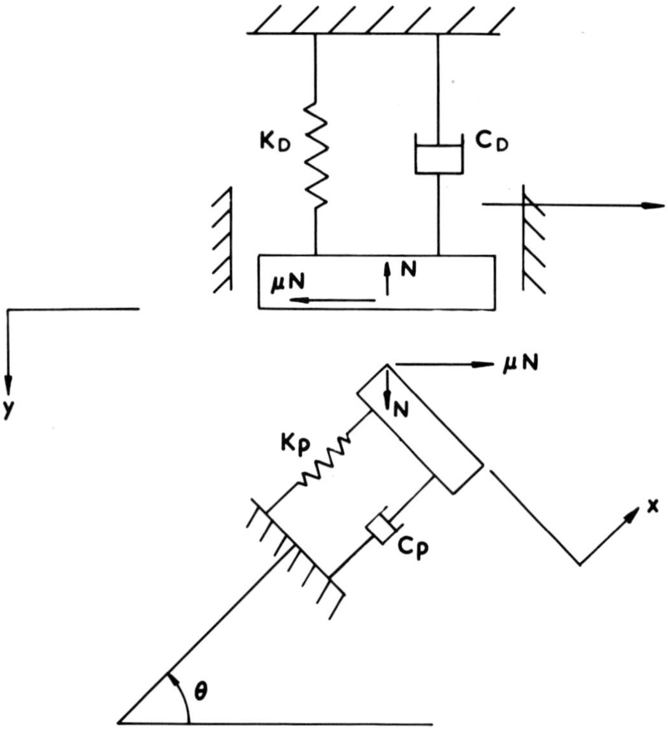

Fig. 4a: Representation of the translational mode of the pin-disc model

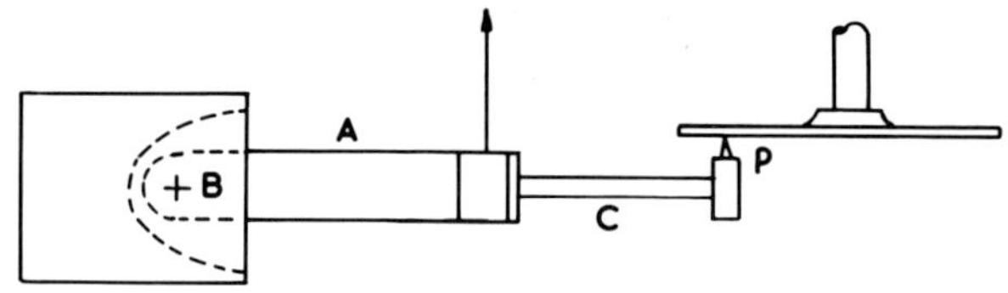

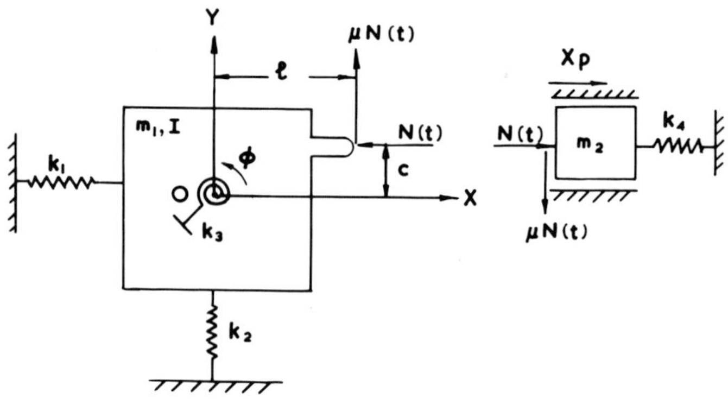

Fig. 5: Four degree-of-freedom pin-disc model

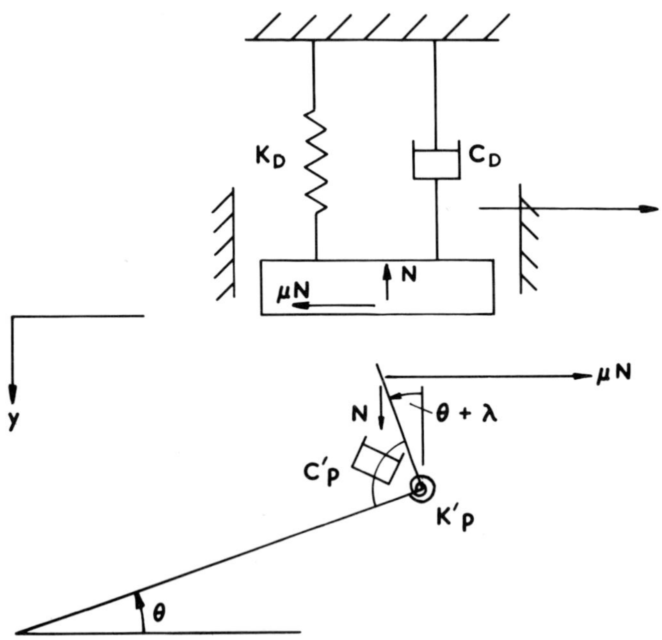

Fig. 4b: Representation of the torsional mode of the pin-disc model

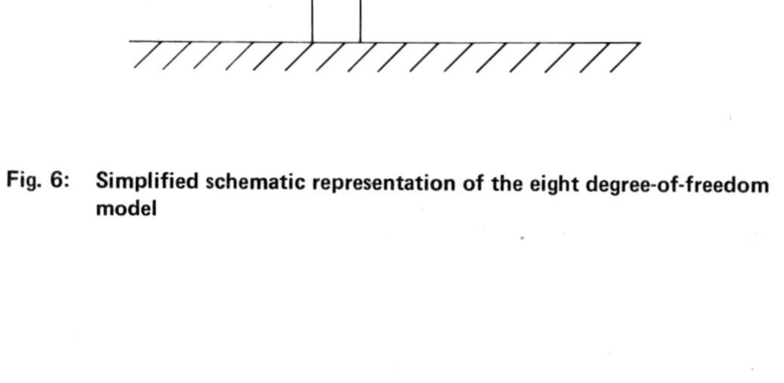

Fig. 6: Simplified schematic representation of the eight degree-of-freedom model

© IMechE 1976

C39/76

A THEORY OF DRUM BRAKE SQUEAL

N. MILLNER, BSc, MPhil
Ferodo Limited, Chapel-en-le-Frith, Derbyshire

The MS of this paper was received at the Institution on 17 September 1975 and accepted for publication on 10 November 1975

SYNOPSIS A theoretical analysis of drum brake squeal has been developed in which allowance is made for the vibrational modes and the contact geometry of the shoe and drum. Aspects of the theory have been verified experimentally.

INTRODUCTION

1. Brake squeal has been the subject of many experimental investigations. However no theory has been published that fully explains the mechanism of drum brake squeal or that suggests a permanent practical cure. Methods of alleviating drum squeal either give a reduced torque or use devices which may become ineffective after continual heating and cooling of the assembly. A better solution is to use shoes and drum which have mass, stiffness and damping values not conducive to squeal. The purpose of this work was to formulate a mathematical model for drum squeal which could be used to estimate such values.

NOTATION

a, b, c	Lagrange coefficients
a_1	constant in μ_d/V equation
E	Young's modulus
F, R	increase in the value of the friction force, normal force acting on an element.
k	difference between rubbing and lining radius
L	load
n	damping coefficient
P	brake shoe actuating force
q	normal co-ordinate
Q	normalised force
r	neutral axis radius
r_r	rubbing radius
t	time
T	lining thickness
u, v	radial, tangential deflection
V	rubbing speed
w	lining width
x, y	real, imaginary part of eigen value
$\gamma, \gamma_1, \gamma_2, \phi$	angles
ψ	normal function
κ	radius of gyration
λ	eigen value
ρ	density
σ	cross-sectional area
μ_d, μ_s	coefficient of dynamic, static friction

subscripts

s, d, l,	shoe, drum, lining
c, f, h.	crown, full, heel and toe contact

PREVIOUS WORK

2. Squeal is thought to be caused either by variations in the coefficient of friction of the lining or by a particular type of contact geometry between the lining and drum (Refs. 1-5). The theories of squeal use these criteria in simplified models to show how vibrational energy is generated. The simplest models employing a friction criteria simulate the shoe by a spring-mass-damper in which the mass is in rubbing contact with a friction plane representing the drum. Theoretically, oscillations based on the condition $\mu_s > \mu_d$ are favoured most by large values of V and L whilst models employing the equation

$$\mu_d = \mu_s - a_1 V \tag{1}$$

predict that the oscillation amplitude increases with a_1 and L.

3. In another model the shoe and drum are considered to be analogous to a strut held in contact with a moving plane (Ref. 5). Vibrations of the strut were explained by a sprag-slip action along the plane when the strut made an angle arc cot μ_d with the surface. This model showed that oscillations were possible for a constant value of μ_d provided the contact geometry was in a critical configuration. Similar ideas have also been applied to disc squeal theories (Refs. 6 - 8).

4. The fundamental objections to the present drum squeal models are that they are oversimplified in that only one mode of vibration is used for the shoe and none for the drum. Experimental evidence suggests that both shoe and drum vibrate with two degrees of freedom. The following theory attempts to formulate a more realistic model.

© IMechE 1976

THEORETICAL ANALYSIS

Basic assumptions

5. Jarvis and Mills, when investigating the action of a loaded cantilever against a rotating disc, found that the resulting vibrations were not primarily caused by a fall of μ_d with rubbing speed (Ref. 6). Using a constant value for μ_d, the oscillations were explained by assuming that there were two components in the disc vibration, both of the same mode order but with different arbitrary nodal positions. These ideas are applicable to and form the basis for the construction of the following drum squeal theory.

6. It is assumed that the shoe and drum both vibrate in one mode and that there are two components in the drum vibration. Their respective mode shapes are taken to be the same as those for the free vibrations of a thin curved elastic strip and a thin open-ended cylindrical shell (Refs. 9 - 10). This assumption enables the displacements to be expressed in terms of the products of the normal functions and the corresponding normal co-ordinates. The lining is assumed to be rigid in shear and the radial stress over the contact areas of the lining is assumed to be proportional to the radial compression, i.e. the difference between the gross displacements of the shoe and drum. The generalised forces are found in terms of the normal co-ordinates, the lining modulus, μ_d and the contact geometry. These expressions are then substituted into the Lagrange equations and the normal co-ordinates determined.

Mathematical details

7. A drum brake is shown diagramatically in Fig. 1. The shoe is represented by arc MN where, M and N are the actuating and abutment (or pivot) points respectively. The lining is considered to be of uniform thickness and extends over the arc AB. A′ B′ is a very small lining element in contact with the drum at angle ϕ, measured from ON.

8. Suppose the shoe and drum are disturbed from their equilibrium state. Lagrange's equations of motion (assuming the drum has two normal co-ordinates q_1 and q_2) are:-

$$a_s \ddot{q}_s + b_s \dot{q}_s + c_s q_s = Q_s \qquad (2)$$

$$a_d \ddot{q}_1 + b_d \dot{q}_1 + c_d q_1 = Q_1 \qquad (3)$$

$$a_d \ddot{q}_2 + b_d \dot{q}_2 + c_d q_2 = Q_2 \qquad (4)$$

The normalised forces are found by first deriving the shoe and drum deflections at A′ B′. Assuming that the elastic extension of the shoe and drum are small enough to be ignored their distortions will arise only from bending. This assumption makes

$$u = -\frac{\partial v}{\partial \phi} \qquad (5)$$

where

$$v = \Psi q \qquad (6)$$

These equations, on substituting for Ψ, and summing the contributions given by both normal co-ordinates of the drum, enable the radial and tangential deflections of the drum and shoe to be expressed as :-

$$u_d = q_1 \sin s(\phi - \epsilon_1) + q_2 \sin s(\phi - \epsilon_2) \qquad (7)$$

$$v_d = q_1 \frac{\cos s(\phi - \epsilon_1)}{s} + q_2 \frac{\cos s(\phi - \epsilon_2)}{s} \qquad (8)$$

$$u_s = -q_s \frac{d\Psi_s}{d\phi} \qquad (9)$$

$$v_s = q_s \Psi_s \qquad (10)$$

In equations (7,8), the drum mode orders are given by substituting $s = 2, 3, 4 \ldots\ldots$, and ϵ_1 and ϵ_2 are unequal arbitrary constants. An expression for Ψ_s may be found in the Appendix based on an analysis given in (Ref. 9). The respective positions of the drum nodes for co-ordinates q_1 and q_2 are chosen to be at $\phi = 0$ and $\phi = \pi/2s$ so that

$$\epsilon_1 = 0 \qquad (11)$$

and

$$\epsilon_2 = \frac{\pi}{2s} \qquad (12)$$

Substituting ϵ_1 and ϵ_2 into equations (7,8) gives

$$u_d = q_1 \sin s\phi - q_2 \cos s\phi \qquad (13)$$

$$v_d = q_1 \frac{\cos s\phi}{s} + q_2 \frac{\sin s\phi}{s} \qquad (14)$$

9. The increase in the values, R and F, of the normal and friction force acting on A′B′ are found by assuming that μ_d remains constant and that the effect of neighbouring elements on the elastic behaviour of A′B′ can be ignored. The coupling equations for the shoe and drum are therefore:

$$F = \mu_d R \qquad (15)$$

and

$$R = \frac{E_1 w r_r}{T}\left(-q_s \frac{d\Psi_s}{d\phi} - q_1 \sin s\phi + q_2 \cos s\phi\right) d\phi \qquad (16)$$

The contributions of A′ B′ to the generalised forces are then

$$\delta Q_s = R\frac{d\Psi_s}{d\phi} - F\Psi_s \qquad (17)$$

$$\delta Q_1 = R \sin s\phi + F \frac{\cos s\phi}{s} \qquad (18)$$

$$\delta Q_2 = -R \cos s\phi + F \frac{\sin s\phi}{s} \qquad (19)$$

Substituting R and F into equations (17, 18, 19) and integrating over the contacting arcs of the lining gives expressions for the generalised forces in terms of the normal co-ordinates and these are shown in equations (13 - 15) in the Appendix.

10. Using the expressions for Q_s, Q_1 and Q_2 in equations (2, 3, 4) makes

$$a_s \ddot{q}_s + b_s \dot{q}_s + c_s q_s = A_1 q_s + A_2 q_1 + A_3 q_2 \qquad (20)$$

$$a_d \ddot{q}_1 + b_d \dot{q}_1 + c_d q_1 = B_1 q_s + B_2 q_1 + B_3 q_2 \qquad (21)$$

© IMechE 1976

$$a_d\ddot{q}_2 + b_d\dot{q}_2 + c_d q_2 = C_1 q_s + C_2 q_1 + C_3 q_2 \qquad (22)$$

as the final equations of motion. Their solutions are

$$q_s = X_s e^{\lambda t} \qquad (23)$$

$$q_1 = X_1 e^{\lambda t} \qquad (24)$$

$$q_2 = X_2 e^{\lambda t} \qquad (25)$$

and were found by formulating the equations of motion into the matrix equation given in the Appendix. With the aid of a computer program it was possible to construct and solve the matrix equation for the eigen values, λ, and the eigen vector elements $\dot{q}_s$, $\dot{q}_1$, $\dot{q}_2$, q_s, q_1 and q_2.

THEORETICAL RESULTS

Interpretation

11. In the theoretical results obtained λ and vector $\underline{q}$ appear as complex conjugate pairs, i.e.,

$$\lambda_1 = x + jy \qquad (26)$$

$$\underline{q_1} = X + jY \qquad (27)$$

$$\lambda_2 = x - jy \qquad (28)$$

$$\underline{q_2} = X - jY \qquad (29)$$

Similar conjugate pairs are also found with λ_3 and λ_4 and with λ_5 and λ_6. Positive and negative values of x occur; these indicate either an increasing or decreasing oscillatory motion. An instability is therefore obtained when x is positive and the associated frequency is given by

$$\nu = \frac{y}{2\pi} \qquad (30)$$

12. In a brake assembly, physical constraints inhibit the continued growth of such oscillations and limit cycles exist. These may be explained using the proposed model by assuming that damping, stiffness and contact geometry vary with amplitude so that a critical condition is ultimately reached when the input energy and damping losses balance, at which point x falls to zero. Oscillations then continue without further increase in amplitude. The initial magnitude of x however is indicative of the net rate at which vibrational energy is generated at the onset of squeal and was therefore taken as a measure of squeal propensity.

13. Calculations were made to show how the brake actuating force, and lining, shoe and drum coefficients, affected the value of x for a leading shoe. Three shoe and five drum modes were investigated. Values for the Lagrange coefficients, a and c, were estimated from the mode energies and the relevant equations are given in the Appendix. The associated damping coefficient, b, was determined from the equation

$$b = 2\,na \qquad (31)$$

using an experimentally derived value for n. The range of integration in equations (16 - 24) in the Appendix was found from the contact arcs, calculated after allowing for the shoe and drum deflections (Ref. 11).

14. The following data for a proprietary brake assembly were used:-

$E_1 = 0.69$ GN m^{-2}, $E_s = 193$ GN m^{-2},

$E_d = 124$ GN m^{-2}, $\kappa_s = 9.0$ mm, $\kappa_d = 4.24$ mm,

$\rho_s = 7.83$ Mg m^{-3}, $\rho_d = 7.09$ Mg m^{-3},

$\sigma_s = 194$ mm^2, $\sigma_d = 419$ mm^2, $r_s = 89.2$ mm,

$r_d = 104$ mm, $n_s = 110$ s^{-1}, $n_d = 18.5$ s^{-1},

$w = 44.5$ mm, $T = 4.3$ mm, $r_r = 102$ mm,

$\gamma = 2.39$ rad, $\gamma_1 = 0.44$ rad, $\gamma_2 = 2.36$ rad.

15. Predicted squeal was found to be most prolific when the shoe and drum vibrated in their fundamental modes. This modal configuration gave the lowest frequency and generally the greatest squeal propensity. These values (except where stated) are plotted in Fig. 2 - 6. For simplicity, a squeal mode configuration is denoted by (S,D) where S and D are the respective shoe and drum mode orders.

Effect of lining coefficients

16. Fig. 2 shows the variation of squeal propensity and frequency with μ_d for crown contact (in the (2, 5) mode), (k = 0.25mm), full, and heel and toe contact (k = - 0.25mm). These conditions can occur in practice and represent the central, the whole, and two end portions of the lining contacting the drum. They are indicated in Fig. 2 by subscripts c, f and h attached to x and ν.

17. For full contact squeal comes in at $\mu_d \approx 0.29$ and x_f then increases progressively with μ_d. Over this range ν_f remains approximately constant at 3400 Hz. In the case of crown contact x_c is comparatively small and constant for values of μ_d up to 0.60, x_c then increases enormously. Heel and toe contact has no associated squeal when μ_d is less than 0.40; at a slightly higher value there is a large increase in x_h. The plots show that with non-uniform contact very large variations in squeal propensity are possible over a small μ_d range. However, in both cases the contact area increases with μ_d until the whole lining contacts the drum. Their predicted squeal behaviours are then identical with x_f and ν_f.

18. Minor instabilities (not shown) were also predicted in other modes. They were irregular in occurrence but generally the magnitude of x increased with μ_d. These instabilities occurred mainly in the (1, 3), (2, 2) and (2, 4) modes for crown contact and in the (2, 1), (2, 4) and (3, 1) modes for both full and heel and toe contact.

19. Fig. 3 shows the variation of x and ν with lining modulus for full contact. The magnitudes of x and ν increase with modulus presumably because the strength of the coupling between the shoe and drum increases. An implication of this result is that conformable linings are most suitable for reducing brake noise.

© IMechE 1976

Effect of actuating force

20. Fig. 4 shows the variation of x and ν with shoe actuating force for the different contact types. As above, the main squeal mode was (2, 5) for crown contact.

21. The theory shows that squeal behaviour is dependent on contact area but independent of contact pressure.
Therefore for full contact x_f and ν_f remain constant with P, as shown in Fig. 4.

22. In the case of non-uniform contact the contact area increases with P resulting in a variation in the squeal behaviour. For crown contact x_c and ν_c remain approximately constant in the range 0.2 - 1.0 kN despite the contact area increasing from 18% to 68% of the total lining area. At P = 1.0 kN the (2, 5) mode becomes stable and the (2, 4) mode is predominant resulting in the main squeal frequency falling from 9500 Hz to 6800 Hz. For heel and toe contact no squeal should occur for P less than 0.30 kN. Squeal was predicted initially in the (2, 2) mode with ν_h = 3200 Hz which disappeared when P increased and was replaced by the (1, 1) mode. When P = 0.79 kN the lining is in full contact with the drum and the behaviour is then identical with x_f and ν_f.

Effect of shoe and drum coefficients

23. Fig. 5 shows the variations of x and ν with densities for full contact. Subscripts s and d specify shoe and drum respectively. The squeal propensity falls steadily with increasing shoe mass whilst frequency remains approximately constant at 3400 Hz. Varying the drum mass has a pronounced effect on squeal behaviour, x_d increases sharply from zero to a maximum and then falls steadily. The frequency decreases monotonically with mass over this range. These results imply that the (1, 1) squeal mode may be eliminated by doubling the shoe mass or by reducing the drum mass to about a third of the datum value.

24. Fig. 6 shows the variation of x and ν with radii of gyration, κ, for full contact. Squeal propensity increases and decreases nearly linearly with κ_s and κ_d respectively whereas the frequency remains approximately constant. As stiffness is proportional to κ^2 it can be seen that the greatest reduction in squeal propensity results from an increase in shoe flexibility.

25. Varying the damping had no appreciable effect on squeal propensity or frequency. Increasing the shoe damping however increased the magnitude of x for the minor squeal modes whilst increasing drum damping had the opposite effect.

EXPERIMENTAL VERIFICATION

26. Experimental verification of the theory was obtained using a two leading shoe brake mounted on an inertia dynamometer. The squeal tests were made over practical ranges of speed, 200 - 800 rev/min, and actuating pressure, 1.3 - 5.6 MN m^{-2}, with a constant initial drum temperature of 100°C. The brake torques were measured and the squeal noise analysed for each application.

27. Tests were conducted using shoes and drums of different mass, radius of gyration and damping. The radius of gyration and damping coefficients were estimated from the free vibrations of a suspended shoe and of the drum mounted on the test rig. Relevant values are shown in Table 1.

Table 1 The drum and shoe coefficients

	DRUM		SHOE		
	Stand'd	Flex	Stand'd	Flex	Weighted
E GN m^{-2}	124.1	124.1	193.1	193.1	193.1
ρ Mg m^{-3}	7.09	7.09	7.83	7.83	12.66
σ mm^2	419.4	271.0	193.6	178.7	193.6
r mm	104.1	104.1	89.2	90.9	89.2
κ mm	4.24	2.64	8.99	7.19	8.99
γ rad	-	-	2.39	2.39	2.39
n s^{-1}	18.5	22.5	109.5	105.0	109.5

28. All possible shoe and drum combinations were investigated under conditions of full contact. The recorded torque values were used to construct performance lines (plots of torque against actuating pressure) from which the values of μ_d were derived. Good correlation between the experimental, ν_e, and calculated, ν, squeal frequencies was found, except for the flexible shoe - standard drum arrangement, as can be seen in Table 2.

Table 2 Comparison of experimental and calculated frequencies

SHOE	DRUM	μ_d	ν_e Hz	ν Hz	x
STANDARD	STANDARD	0.42	3800	3450	1484
STANDARD	FLEXIBLE	0.48	4400	4180	1975
WEIGHTED	STANDARD	0.47	3700	3404	1236
WEIGHTED	FLEXIBLE	0.50	4200	4159	100
FLEXIBLE	STANDARD	0.44	2600	3424	1530
FLEXIBLE	FLEXIBLE	0.52	4200	4152	1959

29. Subjectively the weighted shoes and flexible drum combination gave least squeal. This result is consistent with the comparatively small calculated value of x. Generally the trends in the observed squeal propensities were upheld by the values of x predicted by the theory.

30. The effect on squeal of the coefficient of friction was investigated in the range 0.32 - 0.44 by using different lining samples and the standard drum and shoes. Squeal occurred only for values of μ_d greater than 0.36, the subjective loudness increasing with increasing μ_d. The critical coefficient of friction contrasts with the calculated value of 0.29 estimated from Fig. 2. The discrepancy may have been caused by using coefficient values in the theoretical model measured when the assembly components were in a non-squealing state. Such values may alter when squeal occurs.

© IMechE 1976

31. The experimental results indicate that the occurrence of squeal is independent of speed and pressure as assumed in the theory for full contact. The brake torques often remained constant throughout squealing stops indicated that μ_d was independent of rubbing speed. This experimental result supports the present theory rather than those squeal theories based upon a friction/velocity criteria (Refs. 1 - 4).

32. Three drums with different internal diameters, viz 203.20mm, 203.84mm and 204.47mm were used to investigate the effect of contact geometry. The linings were bedded to the largest drum and then squeal tests were conducted on the smaller drums and so the degree of heel and toe contact varied. The test results showed that squeal under heel and toe, unlike full contact, was dependent on brake line pressure. As the amount of heel and toe increased the brake line pressure necessary to give squeal also increased. The theory also predicts such a trend. It can be seen in Fig. 4 that squeal is predicted to come in at 0.30 kN actuating force when k = - 0.25mm and μ_d = 0.45. This value compares favourably with the corresponding measured forced of 0.35 kN.

33. The agreement between measured and calculated frequencies at different brake line pressure for heel and toe contact was fairly good. There were however some discrepancies at the higher pressures but these were thought to be caused by undetermined variations in the contact geometry resulting in the excitation of a higher squeal mode. Sudden jumps in the squeal frequency were often observed. Predictably, the largest changes were brought about by exciting different drum modes.

34. Some measurements were also made under crown contact (k = 0.25mm) when μ_d = 0.40. No squeal was found and this result is consistent with the very small predicted squeal propensity shown in Fig. 2.

SUMMARY AND CONCLUSIONS

35. In this analysis of drum brake squeal a constant coefficient of friction is assumed and allowance is made for the vibrational modes and the contact geometry of the shoe and drum.

36. The theory was verified experimentally using a two leading-shoe brake. Trends in experimental results and theoretical predictions were in fairly good agreement, for example, predicted and measured squeal propensities increased with μ_d and increased from crown to heel and toe to full contact. Experimentally no squeal was found under crown contact and this result correlates with the compartively small value of squeal propensity predicted. In the case of heel and toe contact measured squeal only occurred above a critical shoe actuating force, the value of which agreed to within about 15% of the predicted value. When the brake coefficients e.g. stiffness, mass, were varied experimentally the occurrence of squeal correlated fairly well with the predicted squeal propensities. The calculated and measured squeal frequencies usually agreed to within 10%.

37. Some practical implications of the theory, for the type of brake considered, are that squeal propensity is reduced by increasing the mass and flexibility of the shoe, by reducing the drum mass (beyond a certain amount) and increasing the drum stiffness and by reducing the lining modulus. Increasing shoe and drum damping should have only a small effect on the squeal propensity of the main mode but it should respectively increase and decrease the propensity for the higher squeal modes.

38. The agreement between theoretical and experimental results suggests that the model could be useful for estimating values of the brake coefficients which are not conducive to squeal. However the theoretical results given here only apply to one particular brake assembly. In general the instability conditions are rather critical and so trends in the squeal propensities may not necessarily be the same for other assemblies.

ACKNOWLEDGEMENTS

The author would like to thank Dr. R.T. Spurr and M.R. Gunson for valuable assistance and criticism in the preparation of the paper and the Directors of Ferodo Limited for permission to publish.

REFERENCES

(1) TWISS, S.B. and BASFORD, P.R. 'Properties of Friction Materials - II, Theory of Vibration in Brakes,' A.S.M.E. 1957, paper no. 57-SA--97.

(2) SINCLAIR, D. 'Frictional Vibrations', J. Appl. Mech. 1955, 207/214.

(3) WAGENFUHRER, H. 'Bremsgerausche' (Noise from Brakes), A.T.S., August 1964, Vol. 66, No. 8, pp. 217 - 222. MIRA Translation No. 15/65.

(4) CHIKAMORI, S. 'Study on Brake Noise', Mitsubishi Tech. Rev., January 1968, Vol. 5, No. 1, pp. 35 - 45.

(5) SPURR, R.T. 'A Theory of Brake Squeal', I. Mech. E. Auto. Div. Proc., No. 1 1961/1962, pp. 33 - 40. Discussion : pp. 41 - 52.

(6) JARVIS, R.P. and MILLS, B. 'Vibrations Induced by Dry Friction', I. Mech. E. Proc., Vol. 178, Pt. 1, No. 32 1963/1964, pp. 847 - 857. Discussion : pp. 857 - 866.

(7) EARLES, S.W.E. and SOAR, G.B. 'Squeal Noise in Disc Brakes', Paper C101/71. I. Mech. E. Symposium on Vibration and Noise in Motor Vehicles, July 1971.

(8) NORTH, M.R. 'A Mechanism of Disc Brake Squeal,' Fourteenth FISITA Congr. Paper 1/9, June 25 - 30, 1972.

(9) LAMB, H. 'Flexure and the Vibrations of a Curved Bar,' London Mathematical Society, May 1888.

(10) LAMB, H. 'The Dynamical Theory of Sound' 2nd edition 1960 (Dover Publications, New York).

© IMechE 1976

REFERENCES

(11) MILLNER, N. and PARSONS, B. 'Effect of Contact Geometry and Elastic Deformations on the Torque Characteristics of a Drum Brake', Proc. Instn. Mech. Engrs. 1973 187, 317 - 331.

APPENDIX

The normal function of the shoe

The normal function is either odd or even with respect to the radial centre line of the shoe. To simplify the analysis given in reference (9) it is assumed that the ratio of the radius to the curved length of the shoe is large enough to make $\omega \approx 1$. This gives

(i) odd modes

$$\Psi_s = \sinh c_1(\phi - \gamma/2) + c_3 \sin c\,(\phi - \gamma/2) \quad (1)$$

$$c_1 = \frac{2}{\gamma}\left(\frac{1}{2}m_n + x_n - \frac{3\gamma^2}{8m_n}\right) \quad (2)$$

$$c_2 = \frac{2}{\gamma}\left(\frac{1}{2}m_n + x_n + \frac{3\gamma^2}{8m_n}\right) \quad (3)$$

$$c_3 = \frac{c_2\,(c_1^4 + c_1^2 - 2)\cosh c_1\gamma/2}{c_1\,(c_2^4 - c_2^2 - 2)\cos \frac{c_2\gamma}{2}} \quad (4)$$

$$x_n = -\frac{3\gamma^2}{4m_n^2}\left(1 + \frac{1}{2}m_n \tan \frac{1}{2}m_n\right)\tan \frac{1}{2}m_n \quad (5)$$

$$m_n = \frac{3\pi}{2}, \frac{7\pi}{2}, \frac{11\pi}{2} \text{ etc.} \quad (6)$$

(ii) even modes

$$\Psi_s = \cosh c_1(\phi - \gamma/2) + c_3 \cos c_2(\phi - \gamma/2) \quad (7)$$

$$c_1 = \frac{2}{\gamma}\left(\frac{1}{2}m_n + x_n - \frac{3\gamma^2}{8m_n}\right) \quad (8)$$

$$c_2 = \frac{2}{\gamma}\left(\frac{1}{2}m_n + x_n + \frac{3\gamma^2}{8m_n}\right) \quad (9)$$

$$c_3 = \frac{c_2(c_1^4 + c_1^2 - 2)\sinh \frac{c_1\gamma}{2}}{c_1(2 + c_2^2 - c_2^4)\sin \frac{c_2\gamma}{2}} \quad (10)$$

$$x_n = -\frac{3\gamma^2}{4m_n^2}\left(\frac{1}{2}m_n \cot \frac{1}{2}m_n - 1\right)\cot \frac{1}{2}m_n \quad (11)$$

$$m_n = \frac{5\pi}{2}, \frac{9\pi}{2}, \frac{13\pi}{2} \text{ etc.} \quad (12)$$

where γ is the angle of the shoe arc.

The generalised forces

The generalised forces for the shoe and drum are given by

$$Q_s = A_1 q_s + A_2 q_1 + A_3 q_2 \quad (13)$$

$$Q_1 = B_1 q_s + B_2 q_1 + B_3 q_2 \quad (14)$$

$$Q_2 = C_1 q_s + C_2 q_1 + C_3 q_2 \quad (15)$$

where

$$A_1 = -H\int_{\gamma_1}^{\gamma_2} (\Psi_{sr} + \mu_d \Psi_s)\,\Psi_{sr}\,d\phi \quad (16)$$

$$A_2 = H\int_{\gamma_1}^{\gamma_2} (\Psi_{sr} + \mu_d \Psi_s)\sin s\phi\,d\phi \quad (17)$$

$$A_3 = -H\int_{\gamma_1}^{\gamma_2} (\Psi_{sr} + \mu_d \Psi_s)\cos s\phi\,d\phi \quad (18)$$

$$B_1 = H\int_{\gamma_1}^{\gamma_2} \left(\sin s\phi + \mu_d \frac{\cos s\phi}{s}\right)\Psi_{sr}\,d\phi \quad (19)$$

$$B_2 = -H\int_{\gamma_1}^{\gamma_2} \left(\sin s\phi + \mu_d \frac{\cos s\phi}{s}\right)\sin s\phi\,d\phi \quad (20)$$

$$B_3 = H\int_{\gamma_1}^{\gamma_2} \left(\sin s\phi + \mu_d \frac{\cos s\phi}{s}\right)\cos s\phi\,d\phi \quad (21)$$

$$C_1 = H\int_{\gamma_1}^{\gamma_2} \left(\mu_d \frac{\sin s\phi}{s} - \cos s\phi\right)\Psi_{sr}\,d\phi \quad (22)$$

$$C_2 = -H\int_{\gamma_1}^{\gamma_2} \left(\mu_d \frac{\sin s\phi}{s} - \cos s\phi\right)\sin s\phi\,d\phi \quad (23)$$

$$C_3 = H\int_{\gamma_1}^{\gamma_2} \left(\mu_d \frac{\sin s\phi}{s} - \cos s\phi\right)\cos s\phi\,d\phi \quad (24)$$

$$H = E_1\,w r_r/T \quad (25)$$

$$\Psi_{sr} = -\frac{d\Psi_s}{d\phi} \quad (26)$$

The matrix equation

See page 7.

© IMechE 1976

$$\begin{bmatrix} \frac{-b_s}{a_s} & 0 & 0 & \frac{-(c_s - A_1)}{a_s} & \frac{A_2}{a_s} & \frac{A_3}{a_s} \\ 0 & \frac{-b_d}{a_d} & 0 & \frac{B_1}{a_d} & \frac{-(c_d - B_2)}{a_d} & \frac{B_3}{a_d} \\ 0 & 0 & \frac{-b_d}{a_d} & \frac{C_1}{a_d} & \frac{C_2}{a_d} & \frac{-(c_d - C_3)}{a_d} \\ 1 & 0 & 0 & 0 & 0 & 0 \\ 0 & 1 & 0 & 0 & 0 & 0 \\ 0 & 0 & 1 & 0 & 0 & 0 \end{bmatrix} \begin{bmatrix} \dot{q}_s \\ \dot{q}_1 \\ \dot{q}_2 \\ q_s \\ q_1 \\ q_2 \end{bmatrix} = \lambda \begin{bmatrix} \dot{q}_s \\ \dot{q}_1 \\ \dot{q}_2 \\ q_s \\ q_1 \\ q_2 \end{bmatrix} \qquad (27)$$

<u>The Lagrange energy coefficients</u>

Suppose u and v are the radial and tangential displacements, T the kinetic energy and V the potential energy of bending. The energy coefficients a and c are given by

$$T = \frac{r\rho\sigma}{2} \int (\dot{u}^2 + \dot{v}^2)\, d\phi = \frac{a}{2} \dot{q}^2 \qquad (28)$$

$$V = \frac{E\sigma\kappa^2}{2r^3} \int \left(u + \frac{d^2u}{d\phi^2}\right) d\phi = \frac{c}{2} q^2 \qquad (29)$$

where r is the neutral axis radius, ρ the density, σ the cross sectional area, κ the radius of gyration and E Youngs Modulus.

Substituting for u and v into the above equations and integrating between the appropriate limits gives the following expressions for the energy coefficients

<u>Drum</u>

$$a_d = \pi \rho_d \sigma_d r_d \frac{(s^2 + 1)}{s^2} \qquad (30)$$

$$c_d = \pi \frac{E_d \sigma_d \kappa_d^2}{r_d^3} (s^2 - 1)^2 \qquad (31)$$

<u>Shoe</u>

(i) <u>odd modes</u>

$$a_s = r_s \rho_s \sigma_s \int_0^{\gamma} \left\{ (c_1 \cosh c_1 \theta + c_2 c_3 \cos c_2 \theta)^2 + (\sinh c_1 \theta + c_3 \sin c_2 \theta)^2 \right\} d\phi \qquad (32)$$

$$c_s = \frac{E_s \sigma_s \kappa_s^2}{r_s^3} \int_0^{\gamma} \left\{ (c_1 + c_1^3) \cosh c_1 \theta + c_3 (c_2 - c_2^3) \cos c_2 \theta \right\}^2 d\phi \qquad (33)$$

(ii) <u>even modes</u>

$$a_s = r_s \rho_s \sigma_s \int_0^{\gamma} \left\{ (c_1 \sinh c_1 \theta - c_2 c_3 \sin c_2 \theta)^2 + (\cosh c_1 \theta + c_3 \cos c_2 \theta)^2 \right\} d\phi \qquad (34)$$

$$c_s = \frac{E_s \sigma_s \kappa_s^2}{r_s^3} \int_0^{\gamma} \left\{ (c_1 + c_1^3) \sinh c_1 \theta + c_3 (c_2 - c_2^3) \sin c_2 \theta \right\}^2 d\phi \qquad (35)$$

where $\theta = \phi - \gamma/2$.

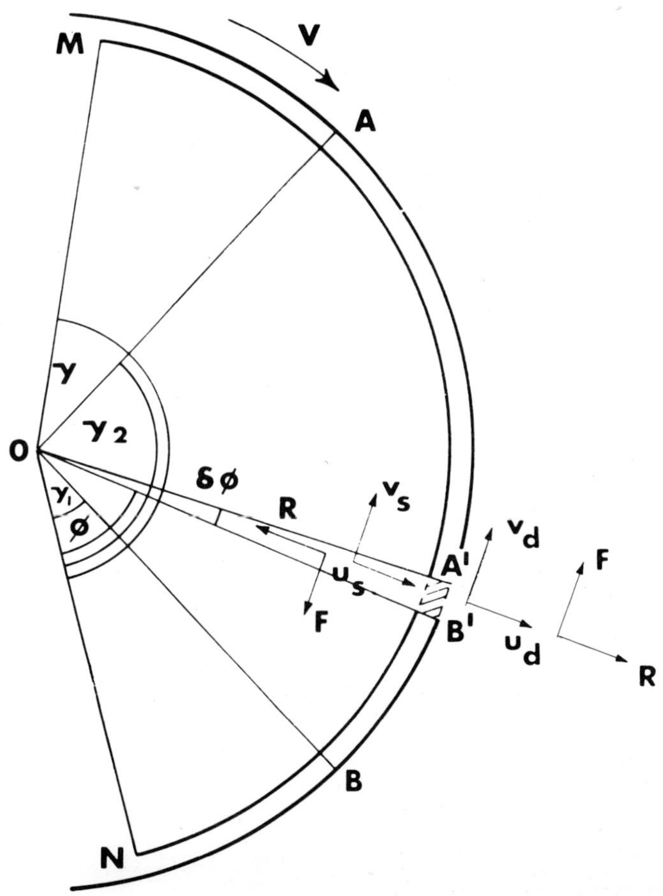

Fig. 1: Simplified force and displacement diagram for the drum brake.

© IMechE 1976

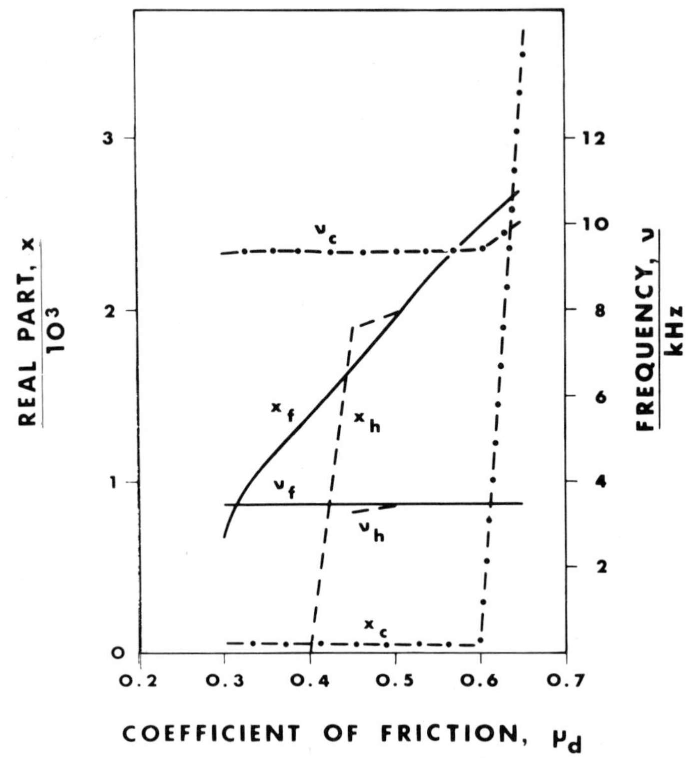

Fig. 2: Variation of x and ν with coefficient of friction for different contact geometries (P = 0.58 kN).

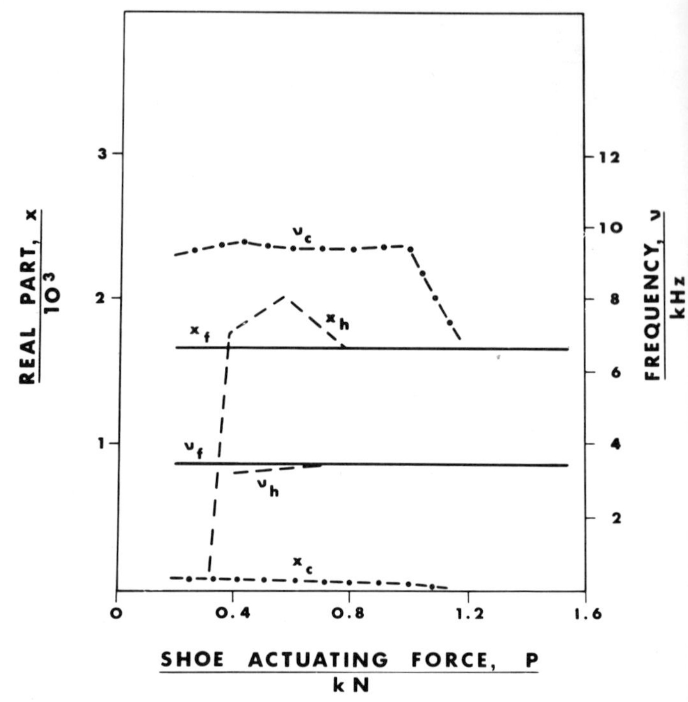

Fig. 4: Variation of x and ν with shoe actuating force for different contact geometries (μ_d = 0.45).

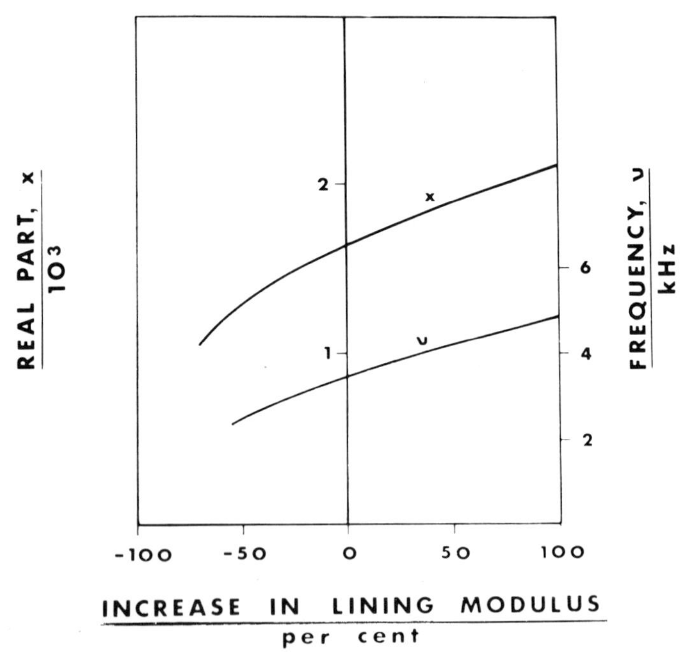

Fig. 3: Variation of x and ν with lining modulus (μ_d = 0.45).

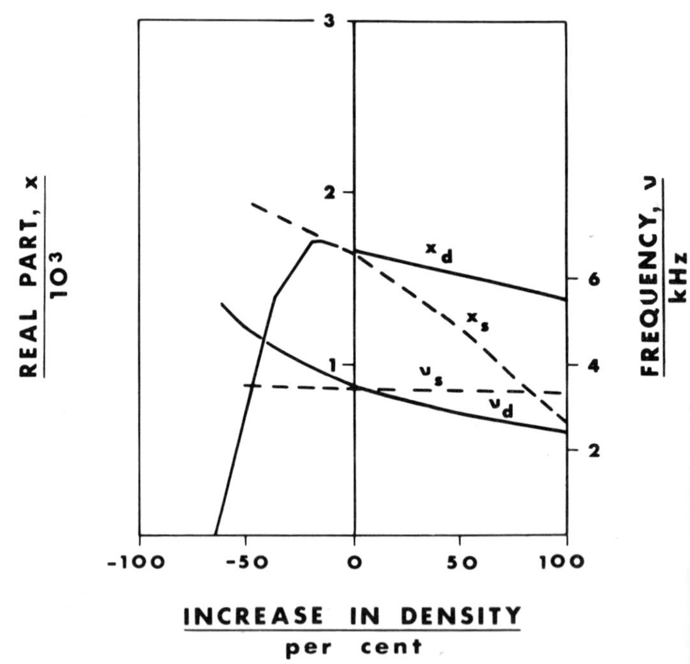

Fig. 5: Variation of x and ν with shoe and drum densities (μ_d = 0.45).

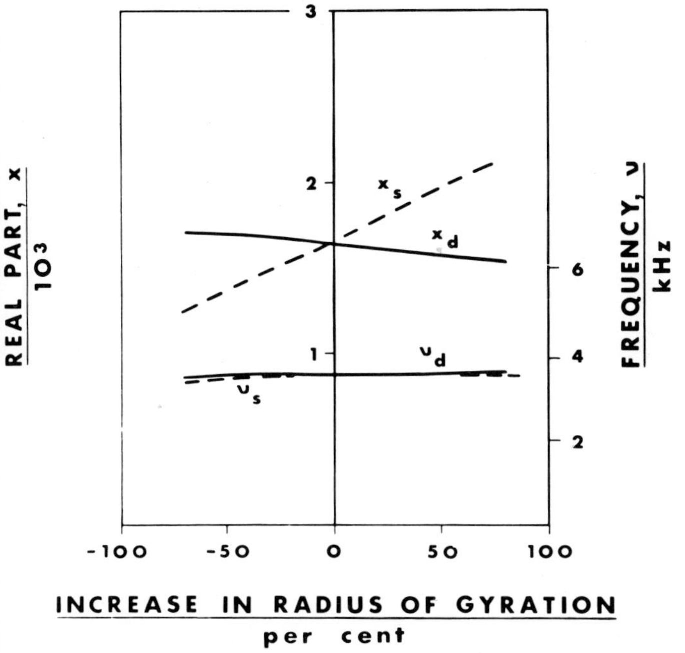

Fig. 6: Variation of x and ν with shoe and drum radii of gyration (μ_d = 0.45).

©IMechE 1976

©IMechE 1976

C40/76

IMPLICATIONS OF SOME CHARACTERISTICS OF DRIVERS FOR BRAKE SYSTEM PERFORMANCE

R. G. MORTIMER, PhD
Department of Health and Safety Education
University of Illinois
Champaign, Illinois 61820 USA

The MS of this paper was received at the Institution on 16 July 1975 and accepted for publication on 21 October 1975

SYNOPSIS Studies of the maximum foot pedal force of drivers, and the effects of some brake system design variables on driver-vehicle braking performance, were reviewed. A survey was made of knowledge of braking characteristics of vehicles among drivers and automobile service persons. It was concluded that the maximum brake pedal force should be not more than 340 N to achieve 5.8 m/sec^2 average deceleration from 97 km/h; brake systems should operate within an envelope of defined deceleration and pedal force values; anti-locking systems would aid in directional control and in reducing driver variability in braking; and training should emphasize basic knowledge, and testing on high coefficient surfaces.

1. INTRODUCTION

The regulations for the performance of vehicle braking systems in European and other countries and in the USA differ in various respects (1,2). But, they are phrased in such a manner that vehicle performance, in braking, is defined in terms which relate the application of force on the braking control to a measure of the deceleration or stopping distance which must be achieved by the vehicle. In the U.S., the braking system performance of passenger cars is governed by Federal Motor Vehicle Safety Standard No. 105 which is based on the SAE recommended practice J-937b.

The standard allows a fairly broad range of performance. For example, to achieve a deceleration of 6.1 m/s^2 from 97 km/h, the pedal force required may be 67-534 Newtons. Greater pedal forces are allowed in the inoperative power system test and recovery tests. The revision to this standard, FMVSS 105-75, effective from September 1975, allows a greater upper limit, the values being 67-688 N in the comparable effectiveness tests to achieve stated stopping distances.

There are two major considerations which affect braking performance of the driver-vehicle combination. Static considerations consist of anthropometric factors which impinge upon the ability of the driver to produce force upon the service brake pedal. Dynamic considerations involve the manner in which the vehicle responds to the application of force to the brake and the ability of the driver to maintain deceleration and directional control of the vehicle. The latter, therefore, is dependent on variables of brake system design which affect the manner in which braking torque is achieved at the wheels of the vehicle in interaction with the skills of the driver.

2. STATIC BRAKING CONSIDERATIONS

In order to operate the vehicle service brake the driver must place a foot on the brake pedal and apply a force sufficient to produce the desired braking torque at the wheels of the vehicle. This will also usually involve some displacement of the brake pedal, which will vary with certain brake system design characteristics. For example, vehicles with disc brakes generally require greater brake line pressures for the same braking torque than vehicles with drum brakes. These greater brake line pressures are usually achieved by the use of a power-assist and/or greater brake pedal displacement.

In order for a driver to apply force and displacement on the brake pedal with the foot, he must be suitably positioned so that the pedal can be reached and depressed through its operating range, and the pedal located such that the driver can exert the force needed to decelerate the vehicle.

A number of variables have been examined which have an affect upon the maximum brake pedal force which can be exerted by drivers.

2.1 Left vs right leg strength

Karim et al.(3) measured the maximum force that could be applied by 25 female pilots on the rudder of a light aircraft simulator with the left and right foot, and found that the mean values were almost identical, at 792 and 787 N, respectively. Using a sample of female and male drivers, Mortimer et al.(4) found that the maximum left and right foot forces correlated 0.93 to 0.96, depending on test conditions. It is, therefore, evident that left and right feet are able to exert similar maximum force levels on a foot control.

2.2 Sex

It is generally well known that females are the weaker sex when measured in terms of muscular exertion, at least when body build is taken into account. Asmussen and Heebol-Nielson (5) reported a study on the basis of which they concluded that women, on average, are about 77% as strong as men. In the study by Mortimer et al.(4) of the foot force capability of 276 females and 323 males, the females attained 5th and 50th percentile maximum forces of 311 N and 691 N, respectively, on the first trial. In the same test condition the

© IMechE 1976

sample of males attained 5th and 50th percentiles of 592 N and 1241 N, respectively. Thus, in that study the maximum foot force of females was about 55% that of males.

2.3 Body Weight

While body weight has been reported to be correlated with maximum strength, its relationship to maximum foot force is moderate. The study by Mortimer et al.(4) found a correlation between body weight and foot force of 0.24 while the study by Stoudt et al.(6) reported a correlation of 0.42.

2.4 Age

The study by Stoudt et al.(6) showed little correlation (0.09) between age and maximum foot force. For subjects seated in automobiles which were either static or moving Radlinski and Price (7) and Pierce et al.(8) found low negative correlations between age and maximum foot force. In an analysis of these data, Konz (9) suggested that there might be a decline of maximum force of about 0.9 N/year of age difference. This would imply that a person aged 65 would be able to exert about 26 N (6 pounds) less foot force than one of age 35.

2.5 Actuating Limb Angles

Important angles that are likely to affect maximum brake force are those between the thigh and the back and the angle produced at the knee. An experimental evaluation of variation in these angles by Aoki (10) showed that maximum foot force was developed when the thigh-back angle was about 73 degrees, and the knee angle 110 degrees. Dupuis et al.(11) reported a knee angle of approximately 150 degrees to be optimal for exertion of force on a foot pedal.

2.6 Other Variables

There are numerous other variables that can enter into the consideration of the maximum brake force that can be exerted by drivers. For example, the compliance of the seat, the effect of learning, fatigue or experience with the specific brake pedal configuration and vehicle, the motivational conditions in which the experiment is conducted, and the specific measurement taken, such as the peak or a value averaged over some time period, are all important considerations which have a bearing on the selection of a maximum pedal force value.

While it is pertinent to indicate the differences between the findings of studies of brake pedal force capabilities of drivers, such as was done by Hill and Lawrence (12) and Konz (9), Mortimer (13) has pointed to the overall similarity of the results of the studies notwithstanding the different procedures that were used.

2.7 Further evaluation of the effects of experience and motivation

Experience on the task and motivation have received considerable attention because test results indicate that they have a substantial effect upon the maximum force levels that have been attained.

Table 1 shows results of a pilot study by Mortimer et al.(4) who measured pedal forces of three groups of drivers. On the first day each group received the "standard" motivational instructions and they were able to view a gauge which indicated the force which they generated. On the second day the force gauge was visible to the males and the females in group A, but not to the females in group B. In addition, data were obtained on the left foot as well as the right foot. It will be noted that there was a minor increase in maximum pedal force for the right foot for the male group on the second day compared to the first. There was a substantial (34%) increase for the females who were able to see the force gauge on the second day. However, the females in group-B who did not see the force gauge on the second day, produced only a median of 13 N (3 lbs) more force on the second day compared to the first.

These data suggest that the males may be less timid initially in exerting force in a new situation, represented by the laboratory pedal-force measurement equipment, and produced only a slight increase on the second day compared to the first. On the other hand, the females of group-A produced a large increment, suggesting that their first trial may have been sub-maximal because of timidity or fear of harming themselves by maximum exertion on the task. The females of group-B produced a negligible increment on the second day, which can be attributed to the changed condition imposed by the removal of the force gauge on the second day's trial. This might be regarded as a case of negative transfer of training, since the initial trial was given with the force gauge visible.

Each group produced a further increment in the "induced" motivation condition, with the greatest relative increment being achieved by the females of group-B.

In summary, the pilot study showed that removal of a cue which may be either motivational or informative in character, as occurred for the females in group-B between day one and two, resulted in a reduction of maximum force exertion. Experience on the task, even such a simple one as being asked to exert maximum force on a foot control, without any other change in condition produced a significant increase in pedal force among females, with a negligible increase in males. Thirdly, creation of a more motivating environment, in this instance by the use of exhortation by the experimenter, produced additional increases in force exertion.

The effect of experience is also shown in the data of Stoudt et al.(6). The mean force of their female subjects increased from 730, 863, 912, 956, to 1001 N between the first and fifth trial that each subject received.

Similarly, the data of Radlinski and Price (7) show increases in the mean pedal forces achieved by their female subjects in the second of each pair of trials, in each of the three conditions in which they were used. Their subjects were seated in a car and depressed the brake pedal on two trials to achieve maximum force in each case. They were then given two additional trials in the same vehicle, but told beforehand that pedal travel would be somewhat increased, simulating a semi-failed braking system. This resulted in a reduction in the maximum force compared to the first two trials. While this may have been due to the approximately three inches of increased pedal travel, it may also have been partly attributable to the changed conditions from the first two trials, which produced a negative transfer effect. Support for the latter

© IMechE 1976

argument is implied by the fact that there was an increase in pedal force on the second of the two trials in the failed condition.

The study by Eaton and Dittmeier (14) placed subjects in an automobile which they drove around a track, occasionally braking to a stop in a sealed-off lane. These braking maneuvres were not severe. However, on the last trial the brake power assist booster was disabled, unknown to the driver, and the pedal force during 0.5 seconds was measured as she tried to stop the car. The pedal force values obtained were quite comparable to the static studies of Mortimer (4), Stoudt et al.(6) Radlinski and Price (7), and Pierce et al.(8).

Hill and Lawrence (12) involved twenty female drivers who drove a vehicle at 97 km/h (60 mph) and completed three high deceleration brake stops before a final braking maneuvre was carried out in which, unknown to the subject, the power-assist was failed. By comparison, the subjects of the Eaton and Dittmeier study did not have prior experience in severe braking, although they would have been motivated to bring the vehicle to a stop within the distance provided by the sealed-off lane. Therefore, there are two major differences between these two dynamic studies. One is that the GM subjects had voluntarily exposed themselves and successfully experienced severe decelerations of the vehicle from a high speed. This involved carrying out a severe maneuvre to keep the vehicle on proper path as well as exerting considerable force on the brake pedal. The subjects of Eaton and Dittmeier were exposed to normal deceleration levels from only 56 km/h. It is hypothesized that the subjects of the GM study were far more prepared to depress the brake pedal with considerable effort when the system was failed compared to the subjects used by Eaton and Dittmeier. Hence, the mean force exerted by the subjects in the GM study are expected to be greater than those in the other studies, particularly on the first trial, as was found.

The subjects in an extensive dynamic study of severe braking also had their maximum foot force levels measured in the laboratory equipment used by Mortimer et al.(4). It was found that the brake pedal force levels which they used to obtain maximum deceleration from the test vehicle on the dry pavement, when the braking system was configured to produce a low deceleration pedal-force gain, are correlated 0.78 with those obtained in the laboratory, and were generally greater than those in the static test.

Even more than in the GM study, the drivers used in this test had considerable prior exposure to severe braking as a part of the test procedure so that high brake force levels would be expected, compared to studies in which little prior experience in the test situation was provided.

While Hill and Lawrence (12) feel that their task was the most realistic, it is not necessarily true that it is also the most relevant. The review of these studies has tended to indicate that pedal force levels are likely to increase with familiarization on the task and if there is positive transfer from one test condition to another.

When viewed in this manner, it is possible to consider the roles of familiarity and transfer effects. Familiarity implies that there are frequent opportunities in driving for a driver to depress the brake pedal with high force levels. Secondly, positive transfer effects are intended to imply that the type of performance which precedes the test sample is similar and conducive to high force exertion.

Mortimer (15) has indicated that braking with high deceleration is more common than might be supposed. However, it is common in the sense that high deceleration levels are involved in the driving environment on a remarkably frequent basis; but, on a per-driver basis they are infrequent. In fact, it is estimated (15) that a driver may require to decelerate in an emergency situation at 4.9 m/sec^2 (0.5g) or greater approximately once in every 16,000 km, or once per year. Thus, drivers are infrequently exposed to high decelerations in driving and are unfamiliar with the sensation which accompanies application of a high level of force on the brake pedal and the subsequent violent response of the vehicle. It follows that, given a situation which demands high deceleration, the driver can derive little benefit from positive transfer that could be attributed to the previous thousands of brake applications made prior to the need for a severe brake application. Thus, braking skills of drivers are largely based on practice in achieving low or moderate decelerations and pedal force levels.

While it can be assumed that motivational effects will be high in an emergency braking situation, it is likely that any benefits of motivation for high force exertion on the brake pedal can be discounted due to the lack of positive transfer effects already mentioned. Therefore, for the purpose of setting an upper bound on the maximum pedal force required to achieve a high deceleration, it is reasonable to use data from studies where little practice was given drivers in developing high brake pedal forces.

In a recent review of four such studies of driver brake pedal-force, Mortimer (13) concluded that the fifth percentile female maximum brake-pedal force is about 400, and recommended that not more than this be required to attain close to the maximum braking capability (7.3 m/sec^2; 0.75 g) from an automobile. Ingram and Oppenheimer (2) feel that present European standards on maximum pedal force of 490 N for 5.9 m/sec^2 (110 lbs force for 0.6 g) are too high, and should be reduced to about the Australian requirement of 343 N. Such a value would be in reasonable agreement with that derived in our present and previous (4) analyses.

3. DRIVER KNOWLEDGE OF VEHICLE BRAKING CHARACTERISTICS

In order that a driver may exercise good control over the path taken by the vehicle during severe braking, it is probably necessary that he understands a few basic facts concerning the response of the vehicle under severe braking conditions. For this reason a small survey was made to try to discern the knowledge which drivers have of vehicle braking behavior. A total of 48 drivers were administered the questions which are shown in Table 2. There were 16 male and 13 female drivers who were selected from among office secretarial staff and randomly at a shopping centre, to represent the general driving population. A further 19 male subjects were obtained at service stations and garages, to represent a population of automobile repair men all of whom admitted to carrying out brake service as well as other tasks.

3.1 Results of the Survey

The results of the survey are shown in Table 3.

© IMechE 1976

Except for one male respondent out of the general driver sample, all subjects indicated that skidding during braking is more likely to occur on a wet surface than a dry one. The mean stopping distances from 97 km/h varied from 44-48 m between the samples with an overall mean of 47 m. There were no significant differences in these estimated mean stopping distances between the three groups of subjects. The corresponding mean stopping distances on a wet road were 69-73 m, with an overall mean of 72 m. Again, there were no differences between the three subject groups. But, they estimated that mean stopping distances were significantly greater on a wet road than a dry road, indicating that it would require 53% more distance to stop on the wet than the dry road from 97 km/h. It should be noted that the standard deviations of their estimates were 25% of the value of the mean distances for the dry road case, and almost 50% of the mean values in the wet road case. Thus, there was a large dispersion in the estimated stopping distances reported.

When asked whether the front or rear wheels are most likely to lock first in severe braking on a dry pavement, subjects' responses did not differ from chance nor between the three groups. For the case of a wet pavement the samples of males and females in the general driving population responded according to chance, but the service persons correctly responded significantly more frequently by indicating the front wheels would lock first. Similarly, on ice, 79% of the service persons indicated front wheel locking to occur first, but this was not different significantly from chance or the other groups.

When given the hypothetical situation that the front wheels only have locked during a severe braking maneuvre, subjects were asked to infer which of three vehicle responses will most likely occur, as shown in question 7. Responses of those in the general driving population did not differ significantly from chance. Sixty-three percent of the service persons correctly recognized that loss of steering would occur, which response differed significantly from chance among the three alternatives. In question 8 it was hypothesized that rear wheels only had locked, which would be expected to result in the vehicle spinning about the front wheels. The responses of each of the three groups of subjects differed from chance but both the male groups responded with the correct response as well as alternative (c), indicating that they felt that the vehicle could be controlled in direction about as well as when no wheels were locked. However, 69% of the females correctly identified the instability caused by rear wheel locking.

The remaining data indicated that the three groups had held driving licenses for about the same mean time period, but there were considerable differences in the distances they drove per year. The mean total distances driven in the previous two years by the males, females and male service persons, were reportedly 58,000, 17,000 and 112,000 km, respectively. While few of the respondents drove as an intrinsic part of their work, 13 of the 19 service persons had driven motorcycles for more than 500 miles.

3.2 Discussion of survey findings

The drivers tended to underestimate the stopping distance required from 97 km/h (60 mph) particularly on the wet pavement. At an average deceleration of 4.9 m/sec^2 (0.5 g), stopping distance would be 73 m which compares with the mean response of 47 m, which would require 7.7 m/sec^2 average deceleration. Assuming an average deceleration of 2.9 m/sec^2 (0.3 g) on a wet surface, a braking distance of 122 m is required, which compares with their response of a mean of 72 m. More important for the controlability of the vehicle in severe braking is an understanding of the manner in which wheel locking is likely to occur and its effects upon directional control stability. Only the sample of service persons recognized that front wheel locking is more likely to occur at low coefficient of friction between the tire and the pavement. This group also recognized that, with front wheels locked, directional control is lost. The effect of rear wheels locking was less well understood by the sample of subjects.

It had been hypothesized that individuals who have had experience in riding motorcycles would be more aware of the nature of vehicle behavior in severe braking situations, in terms of the order in which wheels are most likely to lock as well as the resulting effects on vehicle stability. Overall, the responses of those in the service trade differed from the other samples primarily in that more of the former were aware of front wheel locking on wet pavements and its effects on controlability. Otherwise there were no differences between the samples.

It would be concluded, based on this preliminary survey of a relatively small number of drivers and service persons, that their general knowledge of the responsiveness of vehicles in braking is poor. In the city of Urbana, in central Illinois, where the survey was conducted drivers will have encountered dry, wet and icy road conditions, so they will have had exposure to the types of situations about which they were asked to respond. It is not known whether drivers who obtain direct experience or, at least, knowledge of the behavior of vehicles in severe braking under various tire-pavement friction conditions, would be less involved in crashes. However, it is evidently a topic that should receive greater emphasis in driver training, education and improvement programs.

4. DYNAMIC BRAKING CONSIDERATIONS

4.1 Brake modulation in normal braking

The ability to exert force on the service brake is a prerequisite to achieving adequate deceleration from the vehicle. However, it is also essential that the driver be able to readily modulate the deceleration in a manner that is consistent with safe traffic flow as well as the prevailing tire-pavement friction level in avoiding a crash. It is important that drivers are able to obtain the needed levels of deceleration to avoid crashes, but it is also important for them to retain steering control of the vehicle during braking. Aside from the relative braking torque at front and rear wheels, the primary brake system variable which affects the performance of the driver in braking is the deceleration/pedal-force gain. A secondary variable is the brake pedal displacement. In controlling the deceleration of the vehicle with the brake pedal the driver is acting as one element in a servo-system. This implies that the driver ultimately modulates the pedal in order to obtain the desired deceleration or overall stopping distance dependent upon

© IMechE 1976

the conditions. The cues which he utilizes in order to accomplish this task are derived from the force and displacement applied to the brake pedal and from visual, auditory, kinesthetic, vestibular and proprioceptive stimuli resulting from the motions of the vehicle. It appears that the ability to estimate deceleration levels can be done based on kinesthetic, vestibular and proprioceptive cues, as found in a study by Spurr (16). Obviously, visual cues are essential for directional control of the vehicle or for maintaining a gap with respect to a vehicle being followed or an object which it is intended to avoid. In modulating the brake control the driver will generally seek modest levels of deceleration, below 2.9 m/sec^2 (15). (Lateral accelerations in cornering of 2.9 m/sec^2 (19) are also rarely exceeded.) This helps to maintain smooth traffic flow and shows an awareness that it will reduce the likelihood of being struck in the rear by a following vehicle, and provides driver and passenger comfort. In order to accomplish these moderate levels of deceleration it is necessary that the brake control can be suitably modulated by the driver.

4.2 Brake modulation in maximum braking

More related to crash avoidance is the need to allow drivers to attain maximum deceleration from the vehicle within the limits imposed by the prevailing tire-pavement friction coefficient. Thus, the driver must also be able to modulate the brake under severe deceleration conditions in order to avoid undue locking up of front or rear wheels so as to retain directional control over the vehicle. In such a situation, on a dry pavement, a high deceleration/pedal-force gain may be considered desirable in order to allow drivers to quickly reach the necessary brake pedal efforts required to obtain high deceleration levels. On the other hand, when the pavement is icy or wet, it is more desirable to have a brake system of lower deceleration/pedal-force gain in order to provide easier modulation and to reduce the likelihood of locking wheels. While it might be desirable to be able to modify the deceleration/pedal-force gain of vehicles in relation to the prevailing tire-pavement friction, this is not within present vehicle construction practice. Therefore, it is valuable to determine a range of deceleration/pedal-force gain levels which provide acceptable driver-vehicle braking performance in a variety of tire-road friction conditions.

4.3 Driver-vehicle braking tests

A study of the role of deceleration/pedal-force gain and pedal displacement was carried out as a master's degree thesis by Brigham (18) in 1968. He used four levels of deceleration/pedal-force gain between 0.02 to 0.04 m/sec^2/N (0.008 to 0.018 g/lb), and three levels of pedal travel between 3.8-14.0 cms (1.5-5.5 in). His eight subjects carried out both controlled stops from 80 km/h (50 mph) and emergency stops from 48 km/h (30 mph). The mean jerk and stopping distances were computed for the two types of tests. While few significant differences were found in this study on the driver-vehicle performance measures, the subjective evaluations made by the drivers indicated that they preferred the brake pedal with a low compliance and a high gain.

It should be noted that Brigham's study was conducted only on dry pavements and, secondly, the ratio of the highest to the lowest gain was about 2:1. Although the deceleration/pedal-force gain levels he used appeared to encompass the range found in his examination of 61 automobiles, the highest gain level was far below that which is frequently found in U.S. automobiles with power-assisted brake systems.

Therefore, a further study was carried out (4) (19), which utilized a wider range of deceleration/pedal-force gain levels from 0.009 to 0.142 m/sec^2/N (0.004 to 0.065 g/lb), and two levels of brake pedal displacement, 0 cms and 6.4 cms (2.5 in). The test was conducted on three road surfaces having sliding wheel coefficients of friction of about 0.82, 0.66 and 0.20 and rolling wheel coefficients of about 0.86, 0.71 and 0.40, measured at 48 km/h. A cross-section of 28 male and female drivers were used in the test.

The subjects' task was to bring the vehicle to a stop as soon as they detected a light signal which appeared along the track. Stopping distance was to be minimized while retaining control of the vehicle within the 3.1 m (10 ft) wide lane.

The results of the study showed that there was a highly significant effect upon braking performance attributable to the deceleration/pedal-force gain level (Figure 1). For example, on the dry pavement, gain levels of 0.026 to 0.081 m/sec^2/N (0.012 to 0.037 g/lb) provided significantly greater mean average decelerations than the other levels. However, on the wet-painted (Jennite) surface, gain levels of 0.081 m/sec^2/N and above produced significantly lower deceleration levels than lower gains.

It is also of interest to note that the mean decelerations achieved from 80 km/h and 48 km/h are not significantly different on the dry pavement, but are lower under wet pavement conditions. As Kummer and Meyer (2) noted, bulk water removal and the removal of the fluid film are processes which require time, and approach completion less and less as the speed of the vehicle increases, thereby causing a reduction in the effective friction level. In these tests the mean deceleration achieved at 80 km/h on the wet and wet-painted pavements was about 10 percent less than at 56 km/h. Effects of the pedal displacement variable were relatively minor, and primarily acted upon the frequency and duration of wheel lock-ups, which were somewhat greater for the non-displacing pedal.

The difficulty of the braking task for these drivers is exemplified by Table 4, which shows the percent of trials involving "loss of control" of the vehicle, defined by crossing the boundaries of the demarcated lane. At an initial speed of 80 km/h, 48 percent and 39 percent of the trials on the wet-painted surface resulted in loss of directional control with the non-displacing and displacing brake pedals, respectively, at the highest deceleration/pedal-force gain level used. As speed and the deceleration/pedal-force gain are reduced and pavement friction increases, there is a reduction in loss of control trials.

Subjective ratings of the "controlability" of the vehicle and the "pedal force" required with each of the gains used, correlated quite closely with the findings of the performance measures.

The results of the study were used to define a deceleration/pedal-force space within which are found those brake configurations which provided acceptable performance on each of the three pavement surfaces. While the results of the dynamic study were used to set the maximum and minimum deceleration/pedal-force gain levels, the results of tests of the forces exertable by

© IMechE 1976

females on foot controls were also incorporated to set a maximum force level commensurate with a high vehicle deceleration (Figure 2).

It is of some interest to note that the gain levels recommended as a result of this study include those which were investigated by Brigham in his dynamic test. By using a wider range of gain levels we were able to observe the reduction in performance which accompanies both greater and lower gain levels than he used, and which were representative of the vehicles that he surveyed.

Examination of the deceleration/pedal-force gains which are found on U.S. vehicles indicates that most of those with manual brakes have characteristics which fall within the envelope defined by this test, while those which fall outside the proposed boundaries are mostly vehicles with power-assisted braking systems having higher gains.

4.4 Variable ratio master cylinder

A master cylinder was recently described (21) having two output pressure/pedal force gains which are linked by a crossover region of operation. For example, such a master cylinder could be used to provide low deceleration/pedal-force gains for deceleration levels of up to about 2.9 m/sec^2 (0.3 g) to ensure good brake modulation for the bulk of brake applications and on low coefficient of friction surfaces. Beside providing a lower gain up to the crossover region, the higher gain beyond the crossover region could allow high deceleration levels to be attained at lower pedal efforts than with a conventional master cylinder, thereby reducing the need for high pedal efforts. Use of such a concept might be a more effective manner of utilizing the deceleration/pedal-force space recommended in Figure 2 than with a conventional master cylinder. The effect of the transitions into and out of the crossover region and the crossover region itself, where the deceleration/pedal-force gain is necessarily greatest, on driver acceptance and performance would need to be evaluated.

5. IMPLICATIONS FOR DRIVER TRAINING

The mean braking efficiency of the driver-vehicle combination, around optimum levels of deceleration/pedal-force gain, was about 65% on the dry surface, 57% on the wet surface and about 45% on the wet-painted surface. These results show that drivers are poor modulators of the brake system since they only achieve between 45 to 65% of the potential deceleration of which the vehicle was capable if it had been controlled by an ideal braking mechanism.

The results also show that braking efficiency was lowest on the surface with the minimum tire-pavement coefficient, which might suggest that it is on such pavements that greatest emphasis should be placed in driver training for emergency braking.

In fact, the emphasis in advanced driver training schools has been placed upon training in vehicle control both in steering and braking on surfaces providing very low tire-road friction levels. In northern states frozen lakes are often used.

A further examination of these data were made by comparing the performance of the best and worst subjects. Figure 3 shows the braking performance for the best and worst subject in terms of the mean deceleration achieved at each deceleration/pedal-force gain on the dry, wet and wet-painted pavements. On the dry surface the best driver achieved a mean deceleration of about 6.9 m/sec^2 (0.7 g), while the poorest subject achieved a mean deceleration of about 3.9 m/sec^2 (0.4 g). This represents about 60% of the mean deceleration of the best subject. On the wet-painted, Jennite, surface the best subject achieved a mean deceleration of about 1.9 m/sec^2 (0.19 g) while the poorest subject achieved about 1.6 m/sec^2 (0.16 g), which is about 85% of the performance of the best subject.

Therefore, the difference between the performance of the best and the worst subject on the lowest coefficient of friction surface is relatively small, and the poor performance of all subjects in this condition reflects the difficulty of the brake modulation task on this surface.

By contrast, it was noted that differences between these two drivers were substantially greater on the dry or wet asphalt pavements. This suggests that training in braking on these surfaces may be a means of improving overall driver braking performance by allowing the less able drivers to improve and attain braking skills more akin to that of the best drivers in the study. Indeed, since most driving is done under dry road conditions, with relatively little under extremely low friction levels such as found in the presence of ice, it would appear that the greatest benefit would accrue from improving braking performance on wet or dry pavements.

6. ANTI-LOCKING BRAKING SYSTEMS

One advantage of anti-locking brake systems would be to improve overall driver braking performance, by reducing the variability in skill levels in brake modulation. This reduction in driver variability would be expected to manifest itself mostly in braking performance on dry and wet pavements rather than on very low coefficient surfaces, as just described. In this respect, a study by Murphy and DiLorenzo (22) reported that on dry asphalt a four-wheel anti-locking brake system was only about 68% efficient, whereas on the wet Jennite the efficiency rose to about 95%. If this is representative of typical anti-locking brake system performance, then it would be expected that on dry surfaces such a system is about as efficient as the best driver used in the driver-vehicle test, thereby raising the performance of all drivers to this level. By comparison, the anti-locking system provided a very large reduction in stopping distances on the wet-painted surface compared with brake modulation by drivers.

However, the ratio of peak:lock-wheel tractive force on the Jennite surface was about 2:1, whereas the ratios were close to unity on the dry and wet asphalt. Thus, the anti-locking system had the greatest potential for reducing stopping distances on the wet-painted surface. Generalizations of braking performance by drivers with and without the anti-locking control to actual driving conditions on icy pavements are dependent on the extent to which the wet Jennite surface and ice exhibit similar peal:locked-wheel coefficients.

This consideration and the results of the driver test suggest that the major advantages of the anti-locking system are retention of steering control in maximum braking and in reducing variability in braking performance between drivers, rather than in significant reductions of stopping distances. Results of a Swedish study (23) on ice confirms that an anti-locking system allows drivers to brake and steer around an obstacle more often (81%) than with conventional brakes (49%), although about 90%

© IMechE 1976

of the drivers attempted to steer as well as brake with both brake systems.

7. CONCLUSIONS

(1) There are many variables that require investigation if braking systems are to be developed which provide improved driver-vehicle deceleration performance. However, it is believed that the experiments that have been done recently have provided new information which could be usefully employed for revising some aspects of current braking standards.

(2) Studies of the maximum pedal forces achievable by drivers have provided results that appear more homogeneous than heteregeneous, in spite of the fact that a variety of tasks have been used. Perhaps, it is precisely because of this variety of task situations that one can have a good degree of confidence in the generally systematic type of findings that have been obtained. Explanations for discrepancies have been provided earlier.

(3) The maximum brake pedal forces allowed in the U.S. and Europe, in present and forthcoming standards (i.e., MVSS 105-75) are considered to be more than should be permitted based on performance of typical 5th percentile females. These studies indicate that a maximum brake pedal force of about 400 N should be stipulated to achieve near to maximum deceleration from an automobile, or about 340 N to achieve a stopping distance of 62 m (204 ft), or 5.8 m/sec^2 (0.6 g) average deceleration, from 97 km/h.

(4) Deceleration/pedal-force gains employed in passenger vehicles and light trucks should adhere to the envelope defined by Figure 2 which was found by experimental study to provide effective braking over a wide range of tire-pavement friction conditions and driver characteristics.

(5) The results obtained in that study pertained to a vehicle which achieved a braking effectiveness of greater than 95%, by use of a proportioning valve and constant loading, on all three road surfaces. In practice, loading would affect the front-rear braking distribution. It would be useful to conduct some further study of the latter variable to assess its impact on the recommendations described by Figure 2.

(6) The use of anti-locking braking systems would appear to be primarily advantageous for directional control rather than stopping distance. Against this has to be balanced their cost and problems associated with maintainability (2). Their braking effectiveness needs to be improved on dry and wet pavements where they are most likely to be used and of value. While they will provide an advantage to drivers in those relatively rare situations where the pavement is icy, it is doubtful that they would be cost-effective in most countries on that basis alone. An added benefit will accrue in their use because of the reduction in variability in driver-vehicle braking performance that they will provide in all conditions.

(7) While training of drivers for emergency braking situations usually is done on very low tire-road coefficients, the variability in braking performance among drivers in those conditions appears to be relatively small. Therefore, such training should be done on wet or dry pavements of moderate to high coefficients where greater variability in driver skills was found.

(8) This would allow drivers to obtain experience in the violent maneuvres that result when decelerations of the order of 6.8 m/sec^2 (0.7 g) or greater are incurred, with impending loss of vehicle control. By providing such training under safe conditions, drivers may overcome the timidity that may be associated with attempting to carry out severe braking and steering responses.

(9) Whether anti-locking braking systems are adopted on a wide scale or not, the need for training in emergency braking will still exist, to ensure that drivers attempt to utilize the directional control which the anti-lock provides.

References

(1) Oppenheimer, P. Braking regulations in Europe. Society of Automotive Engineers, Report 740313, 1974.

(2) Ingram, B. and Oppenheimer, P. Safer braking systems, Proc. Instn. Mech. Engrs., 1973, 187, 87-97.

(3) Karim, B., Bergey, K.H., Chandler, R.F., Hasbrook, A.H., Purswell, J.L. and Snow, C.C. A Preliminary Study of Maximal Control Force Capability of Female Pilots, Federal Aviation Administration, Report FAA-AM-72-27, Department of Industrial Engineering, University of Oklahoma, 1972.

(4) Mortimer, R.G., Segel, L., Dugoff, H., Campbell, J.D., Jorgeson, C.M. and Murphy, R.W. Brake Force Requirement Study: Driver-Vehicle Braking Performance as a Function of Brake System Design Variables. Report No. PB-193-258, National Highway Traffic Safety Administration, Contract FH-11-6952, Highway Safety Research Institute, University of Michigan, 1970.

(5) Asmussen, E. and Heebol-Neilson, K. Isometric Muscle Strength in Relation to Age in Men and Women, Ergonomics, 1962, 5, 167-169.

(6) Stoudt, H.W., Crowley, T.J., Gruber, G. and McFarland, R.A. Vehicle Handling: Force Capabilities for Braking and Steering, Contract FH-11-6910 National Highway Traffic Safety Administration, Harvard School of Public Health, Harvard University, 1969.

(7) Radlinski, R.W. and Price, J.I. The Brake Pedal-Force Capability of Adult Females, National Bureau of Standards, Technical Note 557, 1970.

(8) Pierce, B.F., Woodson, W.E. and Selby, P.H. Human Force Considerations in the Failure of Power Assisted Devices, Report No. DOT HS 800 889, National Highway Traffic Safety Administration, Contract DOT-HS-230-2-396, Man Factors, Inc., 1973.

(9) Konz, S. Maximum Pedal Force Exertable by Weak Females, Society of Automotive Engineers Control Efforts Task Force (informal report), May 30, 1974.

(10) Aoki, K. Human Factors in Braking and Fade Phenomena for Heavy Applications-Problems to Improve Brake Performance. Bulletin of the Japanese Society of Mechanical Engineers, 1960, 3, No. 12, 587-594.

(11) Dupuis, H., Preuschen, R. and Schulte, B. Zweckmassige Gestaltung des Schlepper

© IMechE 1976

Fuhrer Standes, Max Planck Institutes fur Arbeitphysiologie, Dortmund, Germany, 1955.

(12) Hill, F.W. and Lawrence, A.P. A Review of Research on Maximum Human Pedal-Force Capability, Society of Automotive Engineers Control Efforts Task Force (informal report), Detroit, Michigan, February 15, 1973.

(13) Mortimer, R.G. Foot brake pedal force capability of drivers, Ergonomics, 1974, 17, No. 4, 509-513.

(14) Eaton, D.A. and Dittmeier, H.J. Braking and Steering Effort Capabilities of Drivers, Society of Automotive Engineers, Report 700363, 1970.

(15) Mortimer, R.G. Hard Braking Is More Common Than You Might Think, Automotive Engineering, August, 1971, 79, No. 8, 32.

(16) Spurr, R.T. Subjective Assessment of Brake Performance-Skilled Drivers Can Assess Deceleration on a Systematic and Reproducible Basis, Automotive Engineering, 1965, 55, 10.

(17) Ritchie, M., McCoy, W.K. and Welde, W. A Study of the relation between forward velocity and lateral acceleration in curves during normal driving, Human Factors, 1968, 10, No. 3, 255-258.

(18) Brigham, F.R. A Human Factors Study of Vehicle Braking Systems, University of Aston in Birmingham, M.S. Thesis, 1968.

(19) Segel, L. and Mortimer, R.G. Driver Braking Performance as a Function of Pedal-Force and Pedal-Displacement Levels, SAE/Fisita International Automobile Safety Conference, Paper No. 700364, 1970

(20) Kummer, H.W. and Meyer, W.E. Tentative Skid-Resistance Requirements for Main Rural Highways, National Cooperative Highway Research Program, Report No. 37, Highway Research Board, 1967.

(21) Nigg, R.L., Palmer, A.W. and Green, R.F. The variable master cylinder - a description of its function and operation. Society of Automotive Engineers, Report 750382, 1975.

(22) Murphy, R.W. and DiLorenzo, F.A. A procedure for evaluating braking performance, HIT-LAB REPORTS, University of Michigan, 1972, 2, No. 10, 7-12.

(23) Rundrvist, S. Swedish experimental safety vehicle program: steerability during emergency braking, Report 4-01, June, 1973.

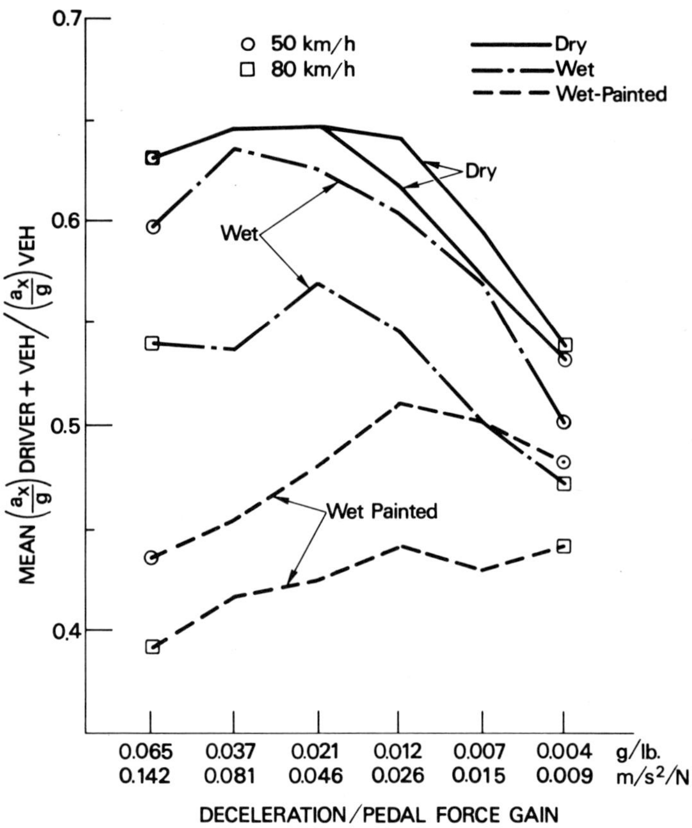

Fig. 1: Driver-vehicle braking efficiency as a function of deceleration/pedal force gain, speed and surface

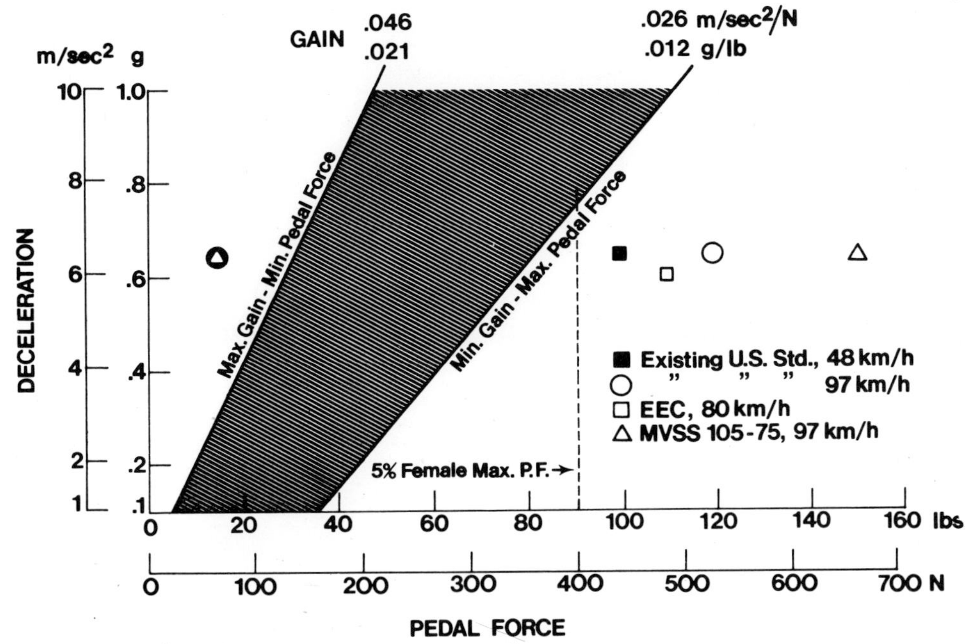

Fig. 2: The recommended deceleration/pedal force space

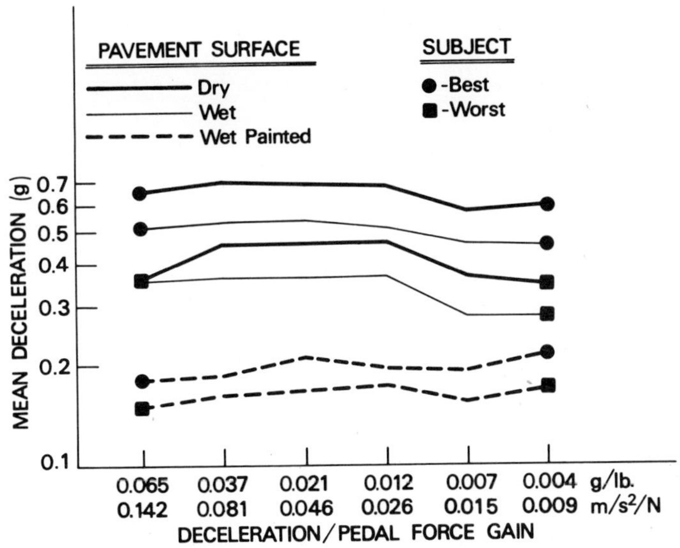

Fig. 3: Braking performance of the best and poorest subject

©IMechE 1976

C41/76

SIMULATION OF DRIVER BEHAVIOUR DURING BRAKING

D. McLEAN, PhD
T. P. NEWCOMB, DSc, CEng, FIMechE
Loughborough University of Technology
R. T. SPURR, DSc
Ferodo Ltd, Chapel-en-le-Frith

The MS of this paper was received at the Institution on 28 July 1975 and accepted for publication on 22 October 1975

SYNOPSIS The availability of reliable data on driver behaviour during braking has permitted the validation of a simple mathematical model which was used as the basis of the digital simulation described in the paper.

INTRODUCTION

1. A knowledge of what factors influence a driver and how they determine his behaviour in driving a vehicle is important. For, although humans are very adaptable in their behaviour, so that any driver can drive a vehicle moderately well even though its mechanical and also its ergonomic design depart from the ideal, there is a need to design the road environment and the vehicle to exploit the driver's capabilities and to minimize the effects of his limitations so that he can drive the vehicle safely over a wide range of conditions.

2. In the study described below only one aspect of driving, namely the braking manoeuvre, was considered. The work presented deals with some experimental studies carried out with a number of drivers and vehicles and with the attempt to provide a mathematical model, together with a digital simulation based on that model.

EXPERIMENTAL RESULTS OBTAINED FROM REAL BRAKING TASKS

3. It was considered important to establish experimentally the degree of accuracy in assessing the motion and relative position of his vehicle to be expected from a driver, for it is from such assessments that the braking manoeuvre is initiated. A series of experiments was made therefore to establish how the driver monitors the deceleration, velocity and position of his vehicle.

Deceleration

4. From earlier work (ref.1) it was evident that a driver did not use visual information to assist him in assessing deceleration. With the driver transferred to the passenger seat in those tests, it was established that his estimate of deceleration when blindfolded was indistinguishable from his estimate obtained when sighted and in the driving position. Further tests demonstrated that the assessment by the driver of his deceleration was not influenced by anything associated with the vehicle. For example, a driver made stops at 0.25g in a Ford Zephyr to familiarise himself with the brake and then endeavoured to make further 0.25g stops without viewing the vehicle instruments. He was transferred to a B.L.M.C. Mini and again endeavoured to make 0.25g stops. He obtained the same deceleration to within ± 0.03g in both vehicles. This result was typical of a group of skilled drivers used in the tests.

Velocity

5. These skilled drivers were also able to estimate with good accuracy the speed of the vehicle. This was shown by road tests in which the speedometer was obscured, the driver was instructed to travel at a nominated speed and the vehicle speed was recorded by an observer. Without giving the driver any information about his performance in that test he was asked to drive at another nominated speed, which was recorded, and so on. The group of drivers was able to estimate speed to within 2.25m/s. All had been permitted 15 minutes free driving for the purpose of familiarization.

Position

6. A subject stationed at about 75–100m from a stationary vehicle guided the driver of a moving vehicle to a nominated position. The stationary car, subject and mobile vehicle were all in a straight-line path so that the subject could not see both cars simultaneously and he was asked to position the second car so that both cars were the same distance from him. The estimates of position were accurate and without bias: the average individual values for a study group of fifty people all agreed to a close tolerance with the true distance.

7. From the braking tests (ref.2) it was also observed that the average deceleration

© IMechE 1976

of every driver increased with increased initial speed. Some drivers braked much later and therefore harder than others; in stops from 30m/s the maximum decelerations ranged between 0.25 to 0.65g (see Fig.1).

8. Unskilled drivers, who were also used in the tests, demonstrated erratic performance; they obviously monitored their performance continuously and made corrections which gave rise to an oscillatory response (see Fig.2). The skilled drivers appeared to have a firmly established repertoire of braking habits and once initiated, the braking task required very little further feedback information (see Fig.3).

9. To confirm that a skilled driver uses essentially a pre-programmed response which does not require continuous monitoring of the external situation, a series of tests was conducted to establish how a driver behaves when his vision is active only at periodic intervals during the braking run. By using spectacles with controlled opaque eye-pieces and blinkers a number of braking runs were carried out in which the driver was permitted to view the driving situation at every 0.2s initially. This was repeated until the sampling period was increased to 3s at which period the driver was discomforted. It was noted (ref.1) in the tests when the stops were made from low speeds that the deceleration was corrected after each glimpse only when the vehicle was within 10m approximately of the target.

THE MATHEMATICAL MODEL

10. The proposed mathematical model to describe the behaviour of an unskilled driver during braking is shown in schematic form in Fig.4. This particular form of model was selected after careful examination of the recordings obtained from the earlier experiments. There are four principal components of this model:

- Human Operator:
  - a motion detection and sensory characteristics
  - b limb dynamics and force characteristics
- c braking system
- d vehicle dynamics

A detailed description of each component is presented below.

Motion Detection and Sensory Characteristics

11. From the results presented in (3) it was established that every driver is able to detect position, speed and linear deceleration, and the braking task is performed using as visual clues either a combination of the three or any one of these motion signals. The detection of linear deceleration, however, is not primarily a visual skill; in the human operator linear deceleration is sensed via the utricles which are small sensory organs located in the inner ear. It has been shown (ref.3) that the relationship between subjective and vehicular decelerations may be expressed as:

$$\hat{a} + 0.67 \frac{d\hat{a}}{dt} = a(t - 0.3) \quad .. \quad .. \quad .. \qquad (1)$$

where $\hat{a}$ is the subjective deceleration and a is the vehicular deceleration from the utricles to the oculomotor system. However, there exists also some signal which has been termed "nystagmus cross feed" (ref.4). This signal results in some involuntary motion of the eye because the vestibular system of the eye has been excited. This visual, but secondary effect, due to linear deceleration, has not been included in the model. Some threshold has been shown to exist in the deceleration path (refs. 1 & 3); however, it should be noted that the value obtained in (ref.3), in respect of a flying task, was about one tenth the value of 0.1g determined in a braking task (ref.1). The threshold value of 0.1g was used in the model.

12. It has been assumed that the visual detection of motion occurs instantaneously and that the visual assessment of the current position of the vehicle with respect to the designated stopping point is instantaneous, precise and continuous. In (9) there was established the need to provide in the simulation (and by inference, in the mathematical model) an alternative position feedback path which is sampled, hence discontinuous. Initially each motion feedback path was given equal weighting: each was assigned unity gain. The effects of unequal weighting were assessed in the simulation. Any delay in motion detection has been incorporated, for convenience, in the description of the human operator.

13. A substantial amount of experimental and theoretical work (refs. 5,6,7,8) exists to provide appropriate describing functions for the tracking tasks of human operators. In such work, in addition to the describing function, it is usual to add a term called the remnant. It appears (ref.9) that the remnant, for full-attention, continuous-control operation, is of limited and secondary importance only, because over a wide range there is very little non-linear dependence of the describing function upon the amplitude of the forcing function. Consequently, because the braking task has been considered in this work to be a full-attention continuous-control action, the remnant term has been omitted from the mathematical description of the human operator. The omission results in the description of the human operation being entirely linear except for the decision elements, the operation of which have been regarded initially as discontinuous. It is possible therefore to consider the model as piecewise linear. One consequence of this is that the describing function for the sensory characteristic of the driver may be expressed as a linear differential equation the most appropriate

form for which, for this work, is that given below

$$n(t)+T_2\frac{dn(t)}{dt} = K_D\, e(t-\tau)+T_1\frac{de}{dt}(t-\tau) \qquad (2)$$

where n(t) is the brain signal which actuates the limb and e(t) is the composite output from the motion detection system and depends upon x, $\dot{x}$ and $\ddot{x}$. What is being implied by this representation is that, when the human acts as a sensor, his output (in this case, a signal to activate the limbs) is produced as a result of a prediction of the action required based upon an assessment of the viewed situation and the speed at which it is changing. From a literature study (refs. 3,4,5,7) the following values of parameters were used initially:

$$\left.\begin{aligned} K_D &= 25\ s^{-1} \\ T_1 &= 3.3\ s \\ T_2 &= 20\ s \end{aligned}\right\} \quad .. \quad .. \quad .. \quad .. \quad .. \qquad (3)$$

These were modified later in the simulation study (see para. 19). A choice of τ of 0.1s represents the value of the visual time delay measured for trained operators such as pilots. Experimental work showed that for drivers however the visual delay time was nearer 0.3s. Since a similar value is appropriate to the sensing of linear acceleration a block representing a mental processing time of 0.3s was inserted in the model of the human operator.

14. Although continuous visual monitoring by the driver is assumed to occur it is also assumed that the driver does not act until a particular distance from the target point, referred to as the braking threshold, is reached. Once the threshold has been crossed the operation of the limbs controlling the brake pedal has then to be considered. A representative description (ref.6) of the neuromuscular lag and limb dynamics was taken to be:

$$0.1\frac{d^3}{dt^3}\delta_A(t)+4.96\frac{d^2}{dt^2}\delta_A(t)+75.6\frac{d}{dt}\delta_A(t)$$

$$+360\,\delta_A(t)$$

$$= 360n(t)+46.0\frac{dn}{dt}(t) \quad .. \qquad (4)$$

where n(t) is the brain signal to the limb dynamics, as before, and $\delta_A(t)$ is the angular deflection of the ankle. Some feedback of ankle deflection is present in the model; although this limb motion feedback is regarded as inherent in the human operator a path has been included explicitly in the diagram of Figure 4 and is simulated directly.

The Braking System

15. Although braking torques are applied to each wheel of the vehicle, it was decided in the early stages of this work, at least, to regard the braking effect as a composite force, f_B, opposing the motion of the vehicle. A block diagram representing the braking system is shown in Fig.5. Several simplifications were made initially, the most important of which were

a) the brake torque was a linear function of pedal effort and independent of the vehicle speed, v;

b) the speed at which the brake pedal was actuated did not affect the resulting brake torque.

The reasons for choosing to make these restrictive assumptions were associated with the method of simulation: no function generation block was available in the version of the digital simulation language currently in use at the University. The values chosen for the parameters of the braking system are shown in Fig.5 and are based on the work described in (ref.10). In the block representing the brake dynamics the symbol D represents the operator $\frac{d}{dt}$.

Vehicle Dynamics

16. During braking the driving force f_D is assumed to be given by the relationship

$$f_D = ma - f_R - f_B \quad .. \quad .. \quad .. \quad .. \qquad (5)$$

where a is the vehicle deceleration,
m is the mass of the vehicle
and f_R is the retarding force

A suitable form for the retarding force (ref.11) is given by the equation

$$f_R = k + bv + cv^2 \quad .. \quad .. \quad .. \quad .. \qquad (6)$$

The omission of the aerodynamic drag term cv^2, does not affect significantly the results obtained, although it was included in the simulation. The parameters chosen for these simulation experiments were:

$$\left.\begin{aligned} k &= 20N \\ b &= 7.5Ns/m \\ c &= .025Ns^2/m \\ m &= 1000kg \end{aligned}\right\} \quad .. \quad .. \quad .. \quad .. \qquad (7)$$

DIGITAL SIMULATION AND RESULTS

17. By using the digital simulation language SLAM, which is a Fortran-based continuous system simulation language intended for use on the ICL 1900 series of digital computers, the simulation of the model of Fig. 4, which is based on the equations given in this paper, was performed digitally. The method of integration to be used in a SLAM program may be selected by the programmer from a group of five:

Simpson's
Trapezoidal
Adams - Moulton Predictor-Corrector
Runge-Kutta, 4th Order, Fixed Step
Runge-Kutta, 4th Order, Variable Step

The routine is selected by assigning an appropriate integer value to the variable

© IMechE 1976

z in the ALG statement in the INTINF block of the program, shown in the Appendix. From a number of experimental runs it was found that for this work the use of Trapezoidal method provided sufficient accuracy in the shortest time with a step length of 0.1sec.

18. The time delay associated with the human operator was generated numerically by using the block DELAY in which a value entered at integration step 0, for example, is not output until integration step 3. The delay can be specified by the programmer who may also specify the integration step length. The use of the blocks COMPAR and SWIN was to avoid any switching delay problems. For completeness, the aerodynamic drag term was included also.

Results and Discussion

19. Using the program shown in the Appendix a series of runs was made to verify the model. Several adjustments were required to the values of constants chosen initially. In particular, the braking force applied was made an inverse function of the composite error, named ERROR in the program. As the target point is approached too quickly with insufficient deceleration greater braking force is applied. To avoid a run error if the divisor is zero, there is added some auxiliary switching so that when ERROR is sufficiently small, the maximum braking force is applied with the divisor set at unity. The values of the equalisation characteristics chosen initially and given as eqn. (3) were changed to

$$\left.\begin{aligned} K_D &= 15.0s^{-1} \\ T_1 &= 1.0s \\ T_2 &= 10.0s \end{aligned}\right\} \quad \cdots \cdots \cdots \quad (8)$$

With these values more satisfactory correspondence between the simulation results and the experimental results were obtained. Such a substitution was to be expected since the original values of equation (3) were for highly trained operators (pilots) performing specific tracking tests (ref.3). Typical results for an unskilled driver for various initial velocities are shown as Figure 6. In each case the braking run was assumed to start at a distance of 200m for the target, although braking was not i initiated until the braking threshold value was reached. The braking threshold value was taken as a function of $v^{1.5}$, e.g. for initial speed of 10m/sec. the threshold value was 31.6m. Comparison of Figures 2 and 6 and 3 and 7 will indicate the effectiveness of the simulation: the character of the simulated braking responses is similar to the experimental responses although, in general, slightly higher levels of deceleration are maintained.

20. To assess further the effectiveness of the simulation a graph of normalized acceleration vs. non-dimensional time was plotted as Figure 8. On this same graph is plotted for comparison a curve representative of one of the drivers in the road tests (ref.2) and the similarity between these curves is plain.

21. Even without the inclusion of the remnant, which is usually added to describing functions for human operators, it has been demonstrated, from the results presented in Figures 7 and 8, that the deterministic mathematical model proposed in this paper provides a useful basis for simulating the behaviour of a driver in a braking task. It is noteworthy however that the use in the simulation of a fixed value for the operator gain K_D, produced unsatisfactory results at lower starting speeds. It became necessary to change K_D as a function of starting speed in order to ensure good matching of the simulated response with the observed experimental results. This indicates the need to extend the model to include the adaptive ability of the human operator which is clearly in evidence.

22. Although only private passenger vehicles were considered the simulation could be used for commercial vehicle studies by modifying slightly the braking system and supplying the appropriate data for the mass and the friction of the vehicle.

ACKNOWLEDGEMENTS

23. Dr. R.T. Spurr wishes to thank the Directors of Messrs. Ferodo Ltd., for permission to publish.

REFERENCES

1. SPURR, R.T. Driver behaviour during braking. Proc.Symp. on Psychological Aspects of Driver Behaviour, Holland 1972, Paper 1-2.

2. SPURR, R.T. Subjective aspects of braking. Auto.Engr. 1969, Vol.59, p.58.

3. STAPLEFORD, R.L., PETERS, R.A. and ALEX, F.R. Experiments and a model for pilot dynamics with visual and motion inputs. NASA CR-1325, 1969, May.

4. McRUER, D.T. and KRENDEL, E.S. Mathematical models of human pilot behaviour, Agardograph 1974, No. 188, Jan. p.45.

5. KRENDEL, E.S. and McRUER, D.T. A servomechanism approach to skill development. J.Frank.Inst.1960, Vol. 269, No.1, Jan. pp. 24-42.

6. McRUER, D.T., MAGADLENO, R.E. and MOORE, G.P. A neuromuscular actuation system model. Trans. IEE, 1968, Vol. MMS-9, No.3, Sept., pp. 61-71.

7. McRUER, D.T., GRAHAM, D. and KRENDEL, E.S. Manual control of single-loop systems. J.Frank.Inst., 1967, Vol. 238, No.1, Jan., pp. 1-29.

© IMechE 1976

8. ELKIND, J.I. Characteristics of simple manual control systems. Tech.Rept. No. 111, MIT,Lincoln Lab., 1956, 6th April.

9. KLEINMAN, D.L. and BARON, S. Manned vehicle system analysis by means of modern control theory. NASA CR-1753, 1971, June.

10. FISHER, D.K. Brake system component dynamic performance measurement and analysis. S.A.E. Transactions 1970, Vol. 79, Section 2, Paper No. 700373, pp. 1157-1180.

11. NEWCOMB, T.P. Work done by brakes. Auto.Engr., 1964, Vol. 54, p.98.

APPENDIX

Simulation Program

```
MASTER BRAKE
INTEGER Z, SERIAL
REAL TF, LFFB, KA, KD, KV, KP, KL, KE,
     KLF, KBF, M, LIMBS
*NER, KSLP, NATPER
LOGICAL U, COMPAR, S, ZD, START
ARRAY STORE (3)/3.0.0/,NUTS(2)/2*0.0/
     UTR(3)/3*0.0/
INPUT SERIAL, TF, Z
INPUT XD, KD, T1, T2, KE, BTH, KL, NATPER,
     DAMP, KSLP, VC,
*FRICN, VIS, STIC1, M, D, KA, KLF, KBF,
     KV, KP, XO,
TITLE 'HUMAN BRAKING PERFORMANCE SIMULAT-
     ION'
OUTHD 'TIME', 'ACCELERATION', 'VELOCITY',
     'POSITION',
*'BRAKE FORCE', 'ERROR DISTANCE'
ERROR   = SUM(XD, -VFB, -XFB, -AFB)
NER     = SUM(ERROR, -LFFB)
U       = COMPAR(BTH, ERROR)
SENS    = LEDLAG(NER, T1, T2, Ø.0)
SENS 2  = KD*SENS
ACTION  = DELAY(SENS 2,3,STORE)
Q       = SWIN(ACTION,O,U)
FORCE   = SECLAG(Q,NATPER,DAMP,O,O)
LIMBS   = KL*FORCE
FP      = LIMBS*2.0
FPED    = SIMLAG(FB,Ø.1,Ø.Ø)
PRES 1  = 150*ABS(FPED)
BSLP    = KSLP*FACT
TB      = PRES 1*BSLP
BRFOR   = SIMLAG (TB,0.02,0)
S       = COMPAR(V,O)
BRFOR 2 = SWIN(BRFOR, OM S)
STIC    = SWIN(STIC1, O, S)
RF      = STIC + VIS*V + FRICN*V*V
FACT    = 100*D
VREF 1  = SIMLAG(VC,Ø.1, 0)
VREF    = SWIN(0,VREF 1, 0)
START   = COMPAR (3, TIME)
TRUER   = XO - X
ZD      = COMPAR (TRUER, 1.0)
VAR     = SWIN (ERROR, 1.0, ZD)
BRFR 3  = DIV(BRFOR2,VAR)
TORK    = KE*VREF
TRUTQ   = SUM(TORK, -BRFR3, -RF)
ACCN    = TRVTQ/M
V       = INTGRL (ACCN, VC)
V1      = SWIN (O, V, START)
X       = INTGRL (V1, 0)
VFB     = KV*V
XFB     = KP*X
LFFB    = KLF*LIMBS
BFFB    = KBF*TB
AM      = SIMLAG(ACCN, 0.67, 0)
AMD     = DELAY (AM, 3, UTR)
AMD 2   = DEAD 2 (AMD, -0.1, 0.1)
AFB     = KA*AMD2
INTINF
ALG: Z = RKFS, RKVS, TRPZ, SIMP, ADMN
CI:  CI = 0.1
STEPS:  MIN = 20
ABSERR:  ABER = 0.0001
MONITOR:  I = 1
END
OUTECI TIME, ACCN, V, X, BRFR 3, TRUER
TERMINATE (TIME GE TF)
REPEAT (SERIAL LE 10)
END
```

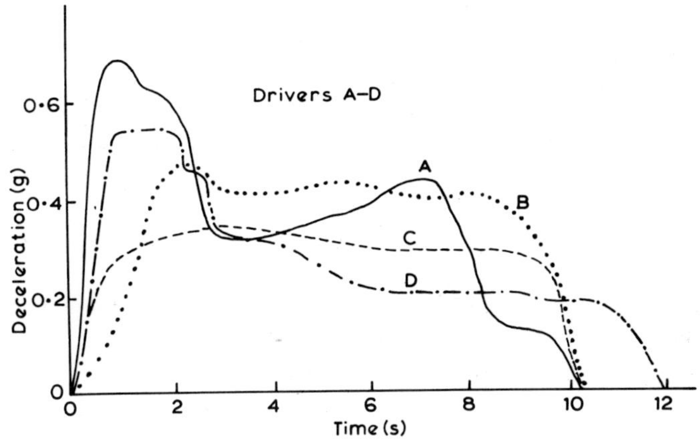

Fig. 1: Typical Braking Responses

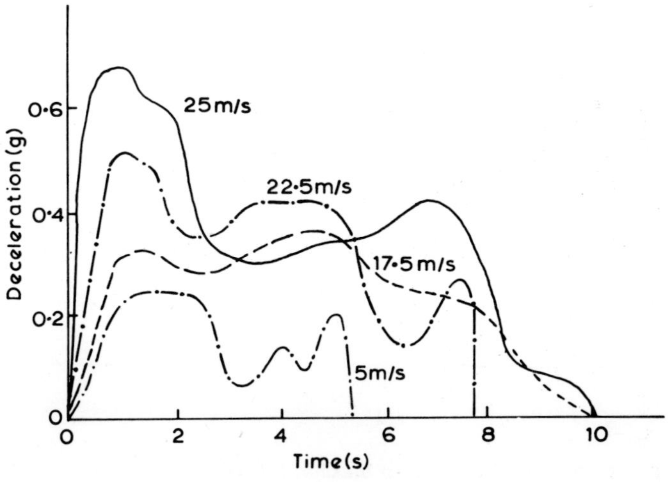

Fig. 2: Typical Braking Responses – Unskilled Drivers

© IMechE 1976

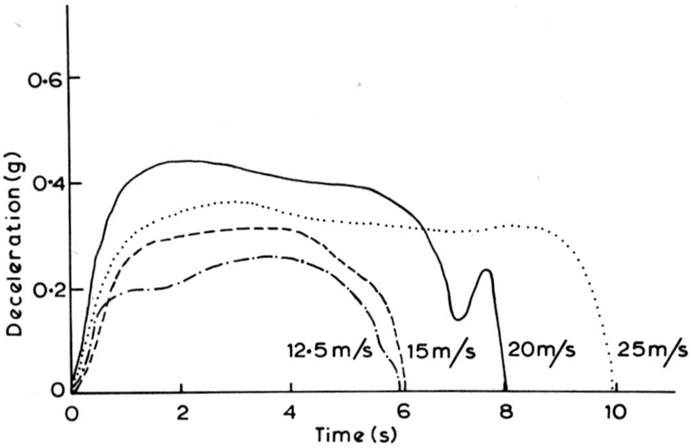

Fig. 3: Typical Braking Responses – Skilled Drivers

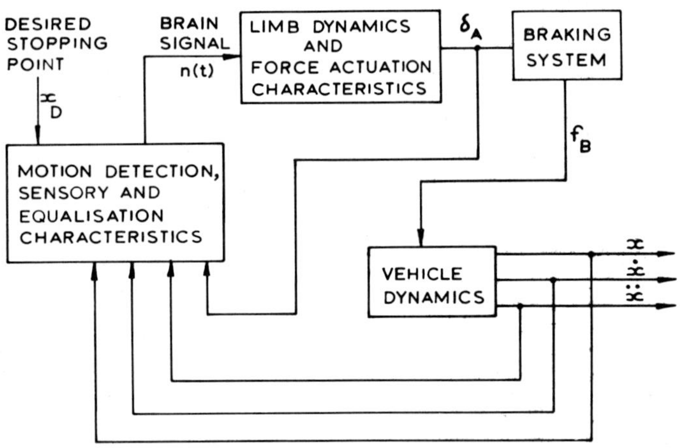

Fig. 4: Block Diagram of Mathematical Model

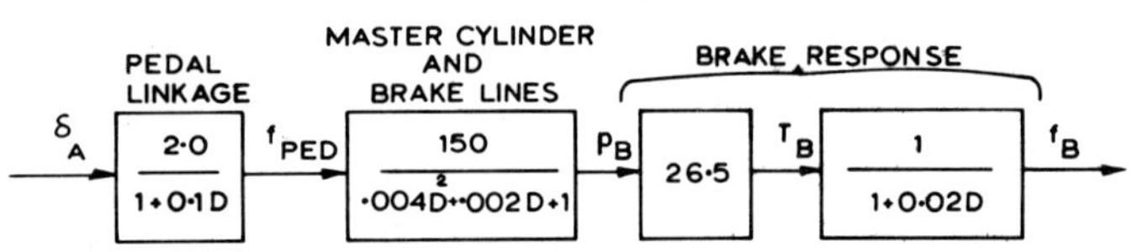

Fig. 5: Braking System

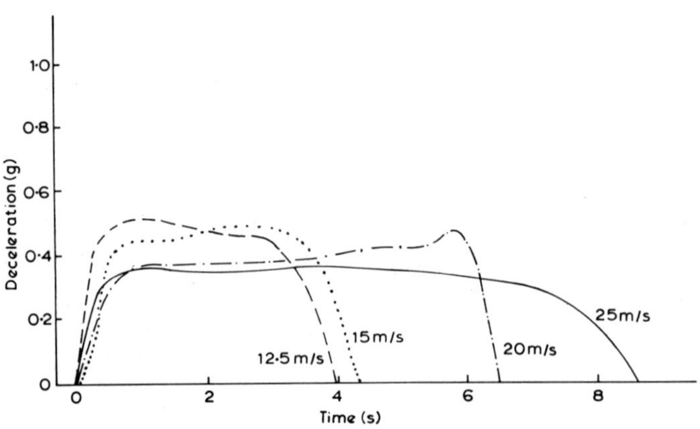

Fig. 6: Simulated Responses of Unskilled Drivers

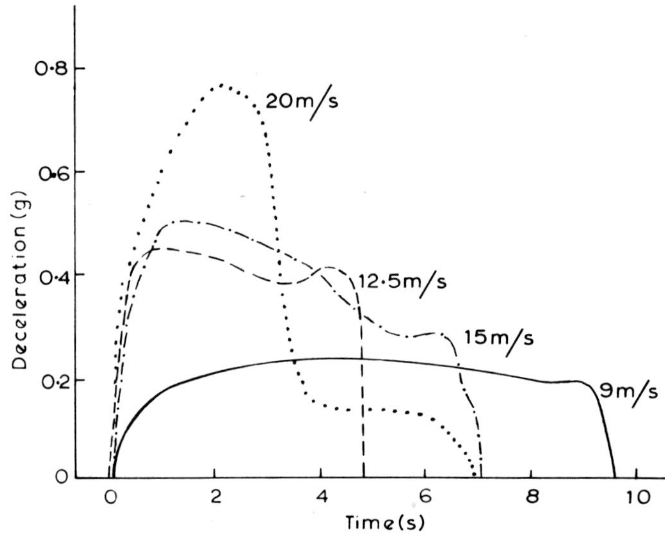

Fig. 7: Simulated Responses of Skilled Drivers

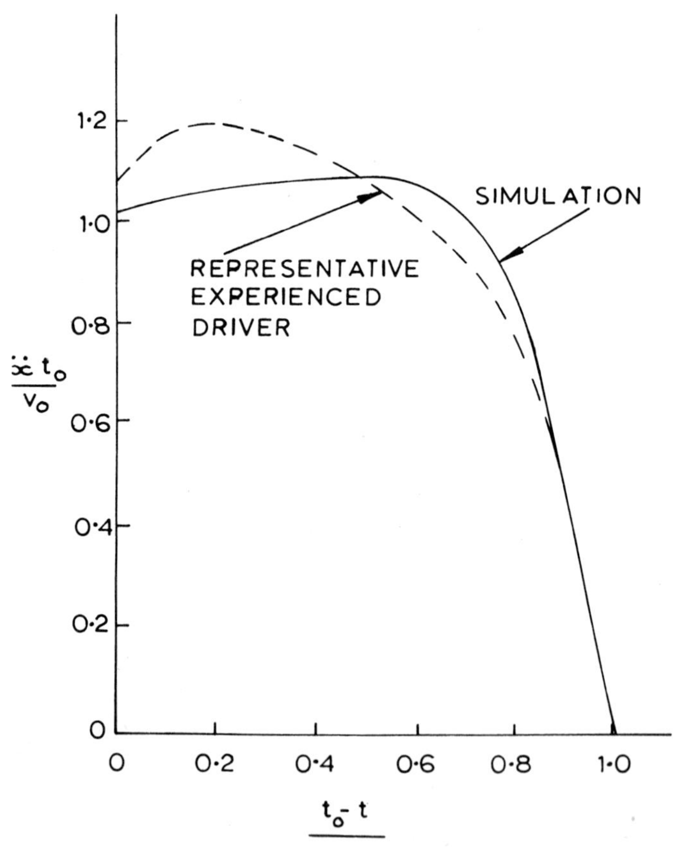

Fig. 8: Comparison of Normalised Responses

© IMechE 1976

C42/76

ACCIDENTS INVOLVING LOSS OF CONTROL WHEN BRAKING – A STUDY OF THE ON-THE-SPOT SURVEY DATA

I. D. NEILSON, MSc, DIC, AFRAeS
Transport and Road Research Laboratory
Department of the Environment, Crowthorne, Berkshire

The MS of this paper was received at the Institution on 28 July 1975 and accepted for publication on 21 October 1975

SYNOPSIS A detailed study of accidents has been carried out 'on-the-spot' by a team from the Transport and Road Research Laboratory with the primary aim of assessing the contribution of road, vehicle and road user factors to road accidents. This paper discusses 8 per cent of accidents including cars, in which control of a car was lost when its driver was braking. Leading characteristics of many of the braking loss-of-control accidents were as follows. The sites were divided almost equally into four categories:- left bends, right bends, intersections and other situations. Slippery surfaces were not necessarily present. Drivers had usually driven too fast either for the layout or for unexpected traffic hazards and were most at risk when driving for social reasons. Young drivers were particularly at risk on damp or wet roads outside urban areas. Braking loss of control ultimately resulted almost equally in spinning and drifting (ie side slipping). Car braking was generally well balanced though load and decleration sensing might be more used to increase the range of stable braking. Tyre or brake defects contributed to a small minority of accidents. The use of front and rear anti-locking brakes might be expected to halve the incidence of braking loss of control.

INTRODUCTION

1. It is appropriate at this conference on the braking of road vehicles, to report on the circumstances in which control of cars is lost whilst braking. The opportunity to do so arises with the completion of a four year study of 2,130 accidents in South East Berkshire by the TRRL 'on-the-spot' accident team. The resulting data have been tabulated and analysed and are now being reported; a published report (Ref. 1) gives brief details. The team concluded that control was lost when braking in about 8 per cent of these accidents. The actual proportion depends on the precise definitions of 'accident' and 'braking loss of control' which are discussed later. There were 2747 cars in 1832 accidents involving cars of which

- 140 control was lost when braking
- 429 control was lost in other circumstances
- 1262 were braked but control was not lost
- 916 were not braked and control was not lost.

The consideration of braking loss of control starts by discussing the background circumstances of where and when these accidents occur and which categories of drivers are involved. These accidents are then compared with other loss of control situations probably not involving braking and also with cases in which cars are braked but control is not lost. Consideration of why braking loss of control may occur is divided between the road, the driver and the car and some thought given to the action taken by the driver, so far as this is known. The various types of loss of control such as whether the car finally drifts or spins is also discussed. Finally, conclusions are drawn with particular attention to possible remedial measures.

2. The comments made in the paper are generally without the benefit of background information relating to the usage of cars in the same circumstances as those in which the accidents occurred. Also the sample of accidents is small and although the comments made are usually statistically significant, they may not apply nationwide. The comments are not always based solely on precise counts of the data recorded, but they also take into account other information recorded by the investigations.

DEFINITIONS OF 'ACCIDENTS' AND BRAKING LOSS OF CONTROL

3. The accidents reviewed in this study are those which the TRRL 'on-the-spot' team investigated, having been called to the scene of nearly all of them by the local Police as soon as they were notified. Thus they include injury accidents and those damage only accidents which Police attended because either they seemed to be serious or one of the parties concerned requested them or they impeded traffic flow.

4. Table 1 is a breakdown of the numbers of cars (1831) in the first three categories of accidents (1668) mentioned in the introduction. The subsequent Tables 3 to 8 are breakdowns of the numbers of accidents involving these cars. (In two tables a few accidents are missing because the data were incomplete).

5. 'Loss of control' is regarded as a circumstance rather than as a contributory factor to an accident. Contributory factors are regarded here as being related to either the vehicle, the road or the road user, whereas loss of control arises from the response of the vehicle to its driver's inputs. For the purpose of this study 'loss of control' is defined as occurring when a vehicle

© IMechE 1976

does not go where its driver intends, having reached a situation from which it is beyond the ability of the driver to regain control in time to avoid the incident. 'Braking loss of control' refers to those cases in which the driver is braking while the car is going out of control. Cases of braking but striking an obstacle in the path of a car are not included unless it is considered that the driver had had time to steer out of trouble by correct control of the car. One consequence of this definition is that it includes all cases in which drivers failed to maintain full control when braking by reason of tiredness, alcohol or illness. When investigating accidents these cases are not always distinguishable from loss of control by alert drivers and are therefore included in the same category. The team has divided loss of control situations into 'braking' and 'other' but the latter category includes some cases in which the driver did brake but this could not be ascertained from verbal evidence or from marks on the road surface. In one or two cases there is a subjective element in allocating accidents to the loss of control categories but this is thought not to greatly detract from the findings.

CIRCUMSTANCES IN WHICH BRAKING LOSS OF CONTROL OCCURS

6. Braking loss of control may be compared with 'other loss of control' and accidents in which cars braked but in which control was not lost. The remaining accidents involving cars are those in which the drivers did not lose control and did not brake or were not known to have done so. Of the 1832 accidents involving cars in this sample, approximately 8 per cent involved cars with 'braking loss of control' and 23 per cent 'other loss of control'.

7. Table 1 shows that the under 25 year old age group of drivers had a high involvement in both loss of control categories, together accounting for over 40 per cent of them. Since they drive less than a quarter of the total mileage they are at over twice the risk for older drivers per unit distance travelled.

8. Braking loss of control must be related to tyre road adhesion and Table 2 shows that when compared with other braking or loss of control situations it is relatively much less frequent for dry roads and higher for wet roads. The risk of braking loss of control seems to be about 4 times as high when the road is wet or damp compared with dry. On the other hand the problem of loss of control when braking would not be solved by eliminating all slippery road surfaces when wet; such a measure could at best prevent less than a half and probably would reduce them by a much smaller fraction. About twice as many braking loss of control accidents occurred in winter than in summer when the roads are more often dry. Three quarters of loss of control occurred on 70 mile/h roads in the local area, whereas braking without loss of control had only a half of its occurrences on such roads (Table 3). This implies that loss of control following braking is over 3 times more likely on 70 mile/h roads. There are no trends with class of road (Table 4).

9. Table 5 suggests that braking loss of control does not occur more frequently during the hours of darkness than during the day, and in this it differs from the other loss of control situations. It is of interest that of those cars which braked and did not avoid accidents, about a fifth lost control on damp and wet roads in daylight but only about a tenth did so on dry roads and on damp or wet roads at night.

10. Younger drivers (under 25) have proportionally many more accidents during the hours of darkness rather than in daylight than do older drivers and this is particularly true of 'other loss of control' situations but is not marked for braking loss of control (Table 5). It is thought that younger drivers cover a greater proportion of their driving at night than do older ones. A more detailed study of this situation is being carried out.

11. Younger drivers (under 25) had twice the proportion of dry to wet or damp loss of control accidents than that found for older drivers (Table 1). It appears that older drivers have a particularly high proportion of their 'braking loss of control' accidents on wet roads in daylight, though numbers are low to be certain; (Table 5). Unlit roads seem to present relatively the greatest problem for loss of control accidents at night in the area under study. On the other hand poorly lit roads seem to have had few accidents.

12. It appears from Table 6 that the average injury severity of 'other loss of control' accidents is somewhat greater than for the other two categories. Not all accidents in which a driver loses control of his car involves only his vehicle; about a third of 'other loss of control' accidents did involve other vehicles. Braking loss of control accidents often result from emergency situations with other vehicles and so it is not surprising that over a half involve more than one vehicle. Cars of which control is not lost when braking are listed in Table 7 as having only a quarter of their accidents involving no other moving vehicles. These include most pedestrian accidents and failing to avoid unexpected hazards such as stationary vehicles directly in the car's paths.

ROAD FEATURES CONTRIBUTORY TO THESE ACCIDENTS

13. Braking loss of control is primarily a failure of the interaction between driver and car, but it is worth considering what effect road features have. These accidents generally arise in two circumstances, either the driver finds he is driving too fast to cope with a road layout hazard or a traffic situation which has suddenly developed in his path. The road features may be corners or bends, intersections, slippery surfaces and confusing or misleading visual situations. Accidents associated with loss of control are much more often a consequence of incidents away from, rather than at junctions, particularly if no braking is involved (Table 8). Less than a quarter took place on very slippery wet roads or icy ones in this investigation, probably because slippery black spot sites had largely been dealt with before most of this survey took place and there were few icy roads, particularly with the mild winters and the liberal use of salt.

14. Accidents at left bends and at intersections each accounted for over a quarter of the braking loss of control accidents and right bends for under a quarter, the remainder being at sites with no such features. Drivers over 25 years of age seemed

© IMechE 1976

to have few accidents at bends when they had to brake and the roads were dry. 'B' roads had a higher proportion of braking loss of control accidents when compared with other accidents (Table 4). Unlit roads at night seem to be over-represented for the area of the survey. At many of the bends concerned, it would be important for a driver unfamiliar with the road to notice the bend signs and heed them. In a few cases general bend signs were ineffective for subsequent bends which were sharper than those just beyond the signs. Few road hazards contributed to this sample, apart from 28 accidents in which parked or temporarily stationary vehicles presented difficulty.

DRIVER FITNESS

15. Driver fitness can be said to be a matter of training, experience and health, including the influence of age, tiredness, alcohol and emotional state.

16. Men under 25 years totalled at least 44 out of 120 male drivers, while only 3 out of 20 women were under 25 years old. Lack of driving skill was apparent in 18 cases and 7 were unfamiliar with the cars driven. There was no unusual proportion of accidents at night (49 out of 139) but it was particularly noteworthy that a half of the drivers were using their cars for social purposes rather than to, from or during work. Interviewing of the drivers also showed that about 30 drivers had consumed alcohol in more than trivial amounts, many of these being in the twenties age group. This is to be reported separately in more detail.

17. It does not appear that braking loss of control is particularly common during the first year of driving but rather in the next few years, and may be associated with increasing confidence overtaking experience of how to cope with sudden emergencies when travelling somewhat fast for the conditions. These accidents seem to arise more particularly during casual driving for social purposes often when alcohol had been consumed. Occasional cases of prior emotional distress and of pre-occupation with extraneous matters have come to light.

VEHICLE FITNESS

18. The on-the-spot survey of vehicle factors has mostly been concerned with tyre and brake defects because other vehicle features either rarely prove to be contributory or cannot easily be discerned by the investigator. For braking loss of control, by definition the brakes must have worked to some degree, though detailed examinations showed some defects in many of the cars. Brake defects had two effects (Ref. 2), sometimes reducing the deceleration available to the driver so that he did not avoid an accident but also by loss of braking at particular wheels leading to loss of control by over-braking of the others. Numerically under a quarter of the braked loss of control cars had accidents for which defective brakes contributed and most of these appeared to have had reduced deceleration either resulting from seized cylinders or lack of adjustment and internal leaks leading to the need for pumping of the pedal. In few of these cases were the defective brakes clearly contributory, but there was one case in which only the rear brakes were effective, the rear tyres were smooth and the road was wet!

19. Badly worn or smooth tyres were noted in a few wet road accidents, not very different in number from that expected for cars in everyday use. The tyre pressure balance was wrong in a small number of cars but the contribution made by this was doubtful in most cases.

DRIVER ACTION

20. Few of the drivers who lost control when braking were considered to be driving too fast to cope with the road layout when left to themselves but a quarter were travelling too fast for it to be possible for them to handle suddenly arising traffic situations and a similar proportion found themselves travelling too fast to avoid braking loss of control with their level of driving ability. This leaves about a half who lost control when braking despite not travelling fast at the time and these include drivers trying to avoid other drivers who had made errors, those in difficult road conditions and in some cases those experiencing vehicle defects. No driving errors were particularly frequent apart from the taking of wrong braking action. Distraction and driving beyond the range of the car lights in use at night were noted.

TYPE OF LOSS OF CONTROL WHEN BRAKING

21. Four main situations are considered for braking loss of control; at a bend or along a straight road for either a drift or a spin. In a drift the car moves more or less in a straight line forwards, presumably with its front wheels locked. In a spin, the rear wheels lock or slip without the front ones doing so. Cases of violent snaking are included with spinning.

22. The results in Table 9 show that drifting and spinning when braking occurred in almost equal numbers and this was so for rear engined and front drive cars as well as for conventional cars. Drivers were alone in 83 out of 132 cars and this proportion is well above the national average of 45 out of 100. 60 per cent of the cars with only drivers spun rather than drifted. None of the 12 cars with rear passengers spun around on dry roads. This might be expected because the rearwards movement of the centre of gravity would reduce the probability of rear wheels locking. Cars appeared to be particularly likely to drift at bends on B roads but to spin when on straight sections. This latter trend did not appear for other roads. When the cars were divided into the two categories of having estimated speeds before braking of more than or less than 40 mile/h, it appeared that more of the faster cars spun rather than drifted and two-thirds of their accidents were on straight roads. However, two-thirds of the slower cars drifted rather than spun, more than a half being at bends.

23. In Table 10 the numbers of cars which drifted or spun on wet, damp or dry roads are tabulated. Wheel locking on dry, damp or wet road surfaces gives some indication of the grip available between road and tyre at the time of the accident. There is a clear indication that for cars of conventional layout most drifted on wet surfaces and more than a half spun on dry roads. This probably reflects the greater weight transfer to the front when brak-

© IMechE 1976

ing hard on a dry surface. This trend does not appear for front drive cars, for which a slightly more frequent condition seemed to be loss of control on damp or wet surfaces. Many car models are included in the braking loss of control sample and those for the first half of the survey may be compared with the roadside counts of their distribution in traffic (Table 11) carried out during daytime. This distribution is generally very similar to that found in this accident survey for 'victims', ie, drivers and cars thought not to have contributed to the accidents in which they were involved. This suggests that to the low degree of accuracy of this small survey, all car models have accidents in proportion to the distance travelled by them. Comparing the small number (89) of cars with braking loss of control with the traffic distribution, the only differences were that sports cars were over-represented and cars of conventional layout seemed to be under-represented. The front drive cars were high in the braking loss of control sample, but included several with inexperienced drivers. Much larger surveys would be needed to check the significance of these results.

POTENTIAL VALUE OF ANTI-LOCKING BRAKES

24. Anti-locking brakes are potentially of value in braking loss of control situations. The circumstances of each of the 138 braking loss of control accidents were studied to assess the probability on a subjective basis, that anti-locking brakes might have prevented the accident. They would not be expected to do so if a bend was being taken at a faster speed than a skilled driver could manage to cope with. Similarly, in some traffic situations there might be little chance of avoiding other vehicles even if control were maintained. It was concluded from a careful study of all the accidents that about a half of them might have been avoided if anti-locking brakes had been available and correctly maintained. This assumes that drivers of vehicles equipped with anti-lock brakes do not use up some of the extra safety margin by driving faster. Thus perhaps 4% of accidents involving cars might be prevented by anti-locking brakes at all wheels in this more rural than urban road locality. To this figure must be added for a complete estimate of the value of anti-locking brakes, the benefits in other situations. These would include incidents when drivers might have avoided obstacles ahead of them, because anti-locked front brakes would leave drivers with steering control and some incidents classified as 'other loss of control' because there had been no evidence of drivers braking, though they did so. The subjective assessment suggested that front and rear wheel anti-lock would have been equally beneficial in these loss of control accidents.

POSSIBLE REMEDIAL MEASURES

25. No single straightforward road safety action could eliminate the loss of control of cars when braking. The most effective possibility is the fitting of front and rear anti-locking brakes, which between them might halve the problem. This study shows that there is no serious fault in the braking balance of cars as a whole, whether they are front drive, rear engined or of conventional layout. More attention to the condition of brakes and tyres would make contributions. Greater ranges of safe braking conditions for individual car models might be achieved by the use of more sophisticated load and deceleration sensing valve layouts, if anti-lock is not to be introduced. It is to be remembered that the accidents under study form only a small sub-set of the total in this survey and the report on the total situation will be of more value for assessing road and driver remedial measures.

26. Road features which contribute to the problem include unexpected hazards such as bends on B or minor roads which are not clearly indicated or at which there are limited views of approaching vehicles, worn and slippery road surfaces and inadequate lighting of difficult sites. Eliminating road surfaces which are slippery when wet would probably reduce braking loss of control accidents by a relatively small fraction, certainly much less than a half.

27. Driver behaviour aspects might concentrate on driving too fast, not only for the road but also for the unexpected traffic situation, or driving after drinking which is a not unusual feature in these accidents. Driving beyond the range of headlights in use has also been noted. The parking of vehicles or the bringing of vehicles to rest in unforeseen situations can be important hazards for following drivers.

ACKNOWLEDGEMENTS

28. The work described in this paper forms part of the programme of the Transport and Road Research Laboratory and the paper is published by permission of the Director. The work of the 'on-the-spot' accident investigation team is particularly acknowledged as without their hard work of collecting and collating the data, this paper would not have been possible.

REFERENCES

1. SABEY, B.E. and G.C. STAUGHTON. Interacting roles of road environment, vehicle and road user in accidents. Fifth International Conference of the International Association for Accident and Traffic Medicine, London, September 1975.

2. HOBBS, C.A. Brake defects of cars in road accidents. Paper to Inst. Mech. Engrs Symposium 'Braking of Road Vehicles'. Loughborough, March 1976.

Crown copyright 1975. Any views expressed in this Paper are not necessarily those of the Department of the Environment. Extracts from the text may be reproduced, except for commercial purposes, provided the source is acknowledged.

© IMechE 1976

Table 1

Loss of control and braking of cars in accidents by road surface condition and age of driver

		Numbers of cars with:											
		Other loss of control				Braked loss of control				Braking no loss of control			
Road surface		Dry	Damp	Wet	Total	Dry	Damp	Wet	Total	Dry	Damp	Wet	Total
Age of car driver losing control etc.	Teen	35	11	11	57	3	3	8	14	50	14	20	84
	20-24	79	26	17	122	18	9	12	39	126	43	44	213
	25-64	115	50	45	210	22	28	33	83	527	158	229	914
	65-	2	3	3	8	0	0	2	2	20	4	11	35
Totals		231	90	76	397	43	40	55	138	723	219	304	1246
Incomplete data					32				2				16
					429				140				1262

From a sample of 2747 cars involved in 1832 accidents

Table 2

Road surface condition by daylight and darkness

	Percentage of time				
	Road surface:				Total hours
	dry	moist	wet	snowy icy	
Daylight	74	10	16	½	17454
Darkness	61	14	24	2	14418
Whole time	68	11½	19½	1	31872

Table 3

Loss of control and braking of cars on road of different speed limits by age of driver

		Numbers of accidents in which there were cars with:											
		Other loss of control				Braking loss of control				Braking, no loss of control			
Speed limit (mile/h)		30	40	50-70	T	30	40	50-70	T	30	40	50-70	T
Age of car driver losing control etc	Teen	18	4	35	57	8	0	6	14	38	7	36	81
	20-24	20	13	89	122	7	2	29	38	71	33	103	207
	25-64	40	18	151	209	13	1	69	83	297	111	403	811
	65-	1	1	6	8	0	0	2	2	12	2	21	35
Totals		79	36	281	396	28	3	106	137	418	153	563	1134

© IMechE 1976

Table 4

Loss of control and braking of cars on roads of different class by age of driver

		Numbers of accidents in which there were cars with:														
		Other loss of control					Braking loss of control					Braking, no loss of control				
Road class		M	A	B	Other	T	M	A	B	Other	T	M	A	B	Other	T
Age of car driver losing control etc	Teen	0	31	7	19	57	0	4	4	6	14	0	36	18	27	81
	20-24	0	82	16	24	122	1	18	10	10	39	0	125	32	50	207
	25-64	5	122	34	48	209	1	46	16	20	83	7	469	123	212	811
	65-	0	8	0	0	8	0	1	1	0	2	1	22	3	9	35
Totals		5	243	57	91	396	2	69	31	36	138	8	652	176	298	1134

Table 5

Loss of control and braking of cars by daylight, lighting and road surface conditions

		Numbers of accidents in which there were cars with:											
		Other loss of control				Braking loss of control				Braking, no loss of control			
Road surface		Dry	Damp	Wet	T	Dry	Damp	Wet	T	Dry	Damp	Wet	T
Car drivers losing control under 25	Daylight	33	12	7	52	8	5	13	26	85	27	28	140
	-unlit	44	20	8	72	7	3	2	12	38	12	10	60
	-lit (<20)	15	1	3	19	1	1	1	3	17	7	10	34
	-lit (<20)	21	4	10	35	5	3	3	11	28	10	16	54
Car drivers losing control over 25	Daylight	49	22	25	96	14	19	29	62	344	89	127	560
	-unlit	37	19	12	68	6	9	5	20	64	27	28	119
	-lit (<20)	7	6	4	17	0	0	0	0	32	12	17	61
	-lit (<20)	21	6	7	34	2	0	1	3	56	17	31	104

('Lit' refers to street lighting, of better quality on standards over 20' high, and of poorer under 20')

© IMechE 1976

Table 6

Accident severity by driver age in incidents involving loss of control or braking

Accident severity		Other loss of control: Fatal, severe or moderate injury	Other loss of control: Minor injury	Other loss of control: Damage only	Braking loss of control: Fatal, severe or moderate injury	Braking loss of control: Minor injury	Braking loss of control: Damage only	Braking, no loss of control: Fatal, severe or moderate injury	Braking, no loss of control: Minor injury	Braking, no loss of control: Damage only
Age of car	Teen	23	17	17	4	3	7	21	27	33
drivers	20-24	62	30	30	12	7	20	63	71	73
losing	25-64	77	68	64	29	24	30	242	214	355
control etc	65-	3	2	3	1	0	1	15	11	9
Totals		165	117	114	46	34	58	341	323	470

(Number of accidents in which there were cars with: Other loss of control / Braking loss of control / Braking, no loss of control)

Table 7

Single vehicle and other accidents involving loss of control or braking by age of driver

Number of vehicles involved		Other loss of control: One	Other loss of control: More than one	Braking loss of control: One	Braking loss of control: More than one	Braking, no loss of control: One	Braking, no loss of control: More than one
Age of car	Teen	46	11	7	7	21	60
drivers	20-24	82	40	22	17	62	145
losing	25-64	141	68	31	52	189	622
control etc	65-	5	3	0	2	9	26
Totals		274	122	60	78	281	853

(Numbers of accidents in which there were cars with: Other loss of control / Braking loss of control / Braking, no loss of control)

Table 8

Association of loss of control and braking accidents to cars with junctions* by age of driver

At or away from junction	Other loss of control: At	Other loss of control: Away from	Braking loss of control: At	Braking loss of control: Away from	Braking, no loss of control: At	Braking, no loss of control: Away from
Car drivers losing control etc.						
- under 25	30	149	18	35	163	125
- 25 or over	40	177	24	61	501	345

(Numbers of accidents in which there were cars with: Other loss of control / Braking loss of control / Braking, no loss of control)

*Junctions include private entrances.
Away from junctions includes at, but not associated with junctions.

© IMechE 1976

Table 9

Types of braking loss of control

	* Numbers of cars which lost control						
	at bend		on straight road		Totals		
	Drift	Spin	Drift	Spin	Drift	Spin	Drift or Spin
All cars	37	22	33	43	70	65	135
- Rear engined	2½	2½	4	3	6½	5½	12
- Front drive	9½	5½	8	14	17½	19½	37
- Conventional	25	14	21	26	46	40	86
Cars with							
- driver alone	19	14	17½	31½	36½	46½	83
- driver + front passenger(s) (no rear passengers)	8	6	6	7	14	13	27
- rear passenger(s)	9½	2½	7	3	16½	5½	22
Cars on road with							
50/70 mile/h limit	30	19	21	35	51	54	105
- other limit	6	3	12	8	18	11	29
Cars on -							
- A roads (incl. M)	15½	12½	19½	23½	35	36	71
- B roads	13½	5½	3	11	16½	16½	33
- other roads	6½	4½	10	8	16½	12½	29

* Entries split into halves when a car both drifts and spins during one accident

Table 10

Braking loss of control by type of car and condition of road surface

	* Number of cars which lost control on					
Road surfaces	Wet		Damp		Dry	
	Drift	Spin	Drift	Spin	Drift	Spin
Rear engined cars	2½	1½	1	1	3	3
Front drive cars	7	8	5	7	5½	4½
Conventional, front engined, rear drive cars	21	11	13	11	11	18
All cars	30½	20½	19	19	19½	25½

*Entries split into halves when a car both drifts and spins during one accident

© IMechE 1976

Table 11

Braking loss of control, by car model
(First half of survey)

Model Name	Braking loss of control Number	Braking loss of control %	Roadside Traffic Counts (1970) %
Mini	9	10	7½
1100/1300	9	10	7½
1800	2	2	2
A40	1	1	1½
Minor 1000	2	2	4
A50/60 series	3	3	4
Herald/Vitesse	4	4	3
Rover 2000	1	1	3
Anglia	0	0	2½
Escort	4	4	4
Cortina	5	6	12
Corsair	0	0	2½
Zephyr	5	6	3
Imp	2	2	2
Minx	2	2	about 5
Viva	4	4	4½
Victor	3	3	3½
Sports cars	10	11	about 2
Others	23	26	about 28
	89	100%	100%

© IMechE 1976

©IMechE 1976

C43/76

THE EFFECT OF SURGICAL OPERATION ON THE 'BRAKE-CLUTCH SIMULATOR'

C. WASTELL, I. WISE, J. F. COLIN
Surgical Unit, Westminster Hospital

The MS of this paper was received at the Institution on 22 August 1975 and accepted for publication on 23 October 1975

SYNOPSIS The ability to operate a brake and clutch assembly similar to that found in a car, has been measured in patients before and after left and right inguinal herniorrhaphy. The effect of the operation had disappeared by the 7th and 8th day after operation.

INTRODUCTION

A common question that is asked after surgical operation is; "when can I drive my car?". Driving is obviously a complex function. Since apparently almost no data exists to guide the surgeon we have attempted to reduce the variables to a controllable number and examine just two aspects of driving, namely, the operation of the brake and the clutch. We realise that such a system falls far short of simulation of car driving but it does have some points of similarity as well.

MATERIALS AND METHODS.

The brake-clutch simulator consists of an adjustable car seat fixed to a jig constructed of Dexion. Mounted at the forward end are a pair of pedals connected via conventional hydraulic links and cylinders to load springs which mimic the clutch and brake loads of an average car. As well as two gauges, provided to monitor, visually, the pressure in the brake and the clutch lines there is a transducer in the brake system for pressure recording. Micro switches are provided which, as well as standardising the position of the feet at the beginning of the test also indicate when either the left or the right feet leave the floor. Further switches indicate the moment of commencement of pedal depression and in the case of the clutch, when the clutch movement is complete. The stimulus to start the whole cycle of a simulated emergency stop is provided by a light operated by a button. A steering wheel is mounted but serves merely to standardise the position of the hands.

The time from when the light is switched on to the first movement of the feet is termed "reaction time" either right foot or left foot. The time from the movement of the feet to the beginning of pedal depression is termed "floor to pedal" time left or right foot. Satisfactory braking is achieved when a pressure of 600 lbs per square inch is developed in the brake line which is equivalent to a pedal force of 67 lbs, (30.5 kilograms). The time taken for this to be achieved is known as "pedal movement". In an average car braking system, this pedal force would produce an emergency stop of 0.87g deceleration from 30 mph. The time from the beginning of the movement of the left pedal until it is fully depressed has also been termed "pedal movement".

The various times from the switching on of the light are automatically measured, on-line, by an IBM 1800 computer. The times taken as the result are the mean of the three, consecutive, shortest times in a series of ten "emergency stops" and these are calculated automatically by the computer.

The subject sits in the seat which is correctly adjusted for his leg length and places his hands on the steering wheel. He is then told to watch the light and when it goes on, to carry out an "emergency stop" by depressing both clutch and brake pedals as quickly as possible. The light is switched on at random times with intervals between 30 and 120 seconds. All the subjects were men and all held a current driving licence. The first group of twelve patients were suffering from right inguinal herniae and the second group of ten patients from left inguinal herniae. The effect of anaesthetic was examined in 5 patients who received a general anaesthetic for a minor surgical procedure which did not involve either the trunk or the legs. The effect of "learning" was examined in 10 subjects who carried out repeated cycles of 10 "emergency stops" with suitable periods of rest between each cycle. Patients who underwent operation were tested before operation and then on post operative day 2, 4, 6 and in the case of left inguinal hernia day 8 and in the case of right inguinal hernia day 7.

RESULTS

1. Right inguinal hernia
   a. Right foot see table 1 and figure 1
   b. Left foot see table 2 and figure 2
2. Left inguinal hernia
   a. Right foot see table 3 and figure 3
   b. Left foot see table 4 and figure 4

© IMechE 1976

3. Controls

The mean time from signal to completion of pedal movement in the five male subjects of similar age to the patients with hernia before anaesthetic was .83 seconds for the right foot and .77 seconds for the left foot and after the anaesthetic was .79 seconds for the right foot and .76 seconds for the left foot. No significant difference could be shown between these mean times before and after anaesthetic.

4. The effect of learning.

The mean times from signal to completion of pedal movement for the right foot in seven cycles of ten stops each in 10 male subjects of similar age to the hernia patient were as follows: 0.63, 0.61, 0.63, 0.63, 0.66, 0.62, and 0.62 seconds. The mean times for the left foot were: 0.61, 0.61, 0.62, 0.60, 0.59, 0.60 and 0.62 seconds.

There would seem to be little or no effect therefore either of learning or of fatigue on the operation of the simulator

DISCUSSION

In the experimental situation that has been described, many factors may affect the operation of the simulator including sex, driver's status, age, intelligence, experience with the simulator, fatigue during a run, general anaesthetic, the surgical condition of the patient and the operation. Since each subject was acting as his own control, all were men and drivers the only variables to be taken into account are those concerned with the primary surgical condition, the operation, experience with the simulator and fatigue. It would appear that neither general anaesthetic by itself nor experience and fatigue affects performance and that therefore where times have been elongated after operation, this is due to the affects of the operation itself. In the case of the patients with the right inguinal hernias, R.foot, the mean time from completion of signal to pedal movement is significantly prolonged on the second postoperative day but thereafter no significant difference can be demonstrated from the pre-operative time. If the individual times for each foot are examined, it is seen that all the times on the second day after operation for the right foot are significantly prolonged and then only the reaction time on day 4 and again on day 6. The pattern of prolongation for the left foot is more variable, the reaction time being significantly prolonged on day 2 but significant difference cannot be shown from the preoperative level thereafter. The floor to pedal time is significantly longer in the left foot on post operative day 2 and day 6. All values for both left and right feet cannot be shown to be different from the preoperative levels by day 7. In the case of subjects with a left inguinal hernia the mean times from signal to completion of pedal movement are significantly prolonged on postoperative day 2 in both feet. On day 4 for the right foot but not the left and thereafter no significant difference can be shown with the preoperative time. If the individual times are examined, significant differences can only be shown on the second postoperative day in the floor to pedal movement for both left and right feet and for pedal movement in the left foot.

There was considerable individual variation, some patients apparently not being affected by operation at all. It may be that the preoperative values for hernia patients are depressed by the presence of the hernia. Some patients found, particularly when carrying out 10 emergency stops, that some discomfort was experienced in the region of the hernia but when their preoperative values are compared with the control, no difference could be shown to exist.

At first sight the results are hardly surprising in that we show that the movements required to operate brake and clutch pedals after operation take longer. It was however, something of a surprise to find that the reaction time for the right foot in those patients with a right inguinal hernia was significantly prolonged and we would speculate that this is due to the slower build up of neuronal action potentials consequent upon the patient learning to limit the velocity of all movements of the legs because of the experience of pain.

In conclusion, the changes are small and recovery to preoperative levels has occurred by the seventh postoperative day.

Acknowledgments;

We wish to record our gratitude to the Department of Medical Computing at Westminster Hospital for their considerable help to this project. In addition we would like to thank Girling Limited for designing and making the simulator. Finally, we thank Miss Althea Pudney for typing the manuscript.

RIGHT INGUINAL HERNIA - RIGHT FOOT

Mean times in seconds

	Reaction Time	Floor to Pedal	Pedal movement
Pre op.	·34 (·02)	·22 (·03)	·13 (·07)
Post op. Day 2	·42*(·02)	·33†(·04)	·37*(·08)
Post op. Day 4	·40*(·02)	·26 (·04)	·19 (·08)
Post op. Day 6	·40*(·02)	·30 (·03)	·20 (·07)
Post op. Day 7	·37 (·02)	·21 (·04)	·13 (·09)

Note.

1. * Indicates significant increase from pre op. P<·05
2. † Indicates significant increase from pre op. P<·02
3. The standard error of the mean is shown in brackets

Table 1

RIGHT INGUINAL HERNIA - LEFT FOOT

Mean times in seconds

	Reaction Time	Floor to Pedal	Pedal movement
Pre op.	·32 (·03)	·23 (·03)	·13 (·03)
Post op. Day 2	·47$^{\theta}$ (·04)	·33* (·04)	·14 (·04)
Post op. Day 4	·36 (·04)	·28 (·04)	·16 (·03)
Post op. Day 6	·38 (·03)	·32* (·04)	·15 (·03)
Post op. Day 7	·33 (·04)	·24 (·04)	·10 (·03)

Note.

1. * Indicates significant increase from pre op. P<·05
2. θ Indicates significant increase from pre op. P<·01
3. The standard error of the mean is shown in brackets

Table 2

LEFT INGUINAL HERNIA - RIGHT FOOT

Mean times in seconds

	Reaction Time	Floor to Pedal	Pedal movement
Pre op.	·38 (·02)	·17 (·01)	·09 (·02)
Post op. Day 2	·50 (·04)	·29*(·03)	·53 (·06)
Post op. Day 4	·42 (·03)	·20 (·02)	·14 (·03)
Post op. Day 6	·40 (·02)	·16 (·01)	·11 (·02)
Post op. Day 8	·44 (·04)	·15 (·02)	·08 (·04)

Note.

1. * Indicates significant increase from pre op. P<·05
2. The standard error of the mean is shown in brackets

Table 3

LEFT INGUINAL HERNIA - LEFT FOOT

Mean times in seconds

	Reaction Time	Floor to Pedal	Pedal movement
Pre op.	·32 (·02)	·23 (·02)	·10 (·03)
Post op. Day 2	·48 (·03)	·57$^{\theta}$ (·03)	·57† (·05)
Post op. Day 4	·37 (·02)	·25 (·02)	·13 (·03)
Post op. Day 6	·33 (·02)	·24 (·02)	·13 (·03)
Post op. Day 8	·33 (·03)	·24 (·03)	·08 (·04)

Note.

1. † Indicates significant increase from pre op. P<·02
2. θ Indicates significant increase from pre op. P<·01
3. The standard error of the mean is shown in brackets

Table 4

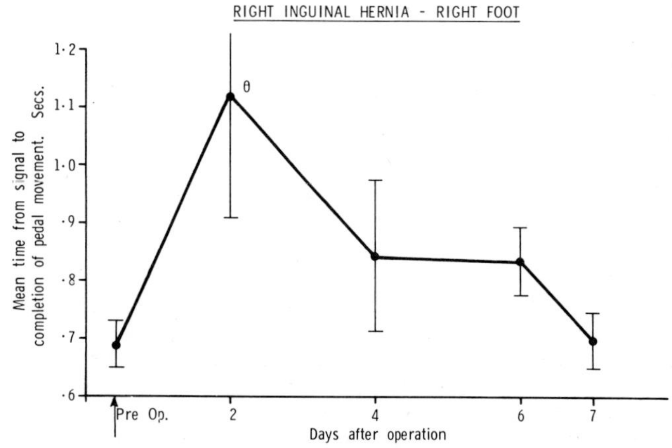

Note.

1. ± Standard error of the mean shown by vertical lines through points
2. θ Indicates significant increase from pre op. P<·01

Fig. 1

© IMechE 1976

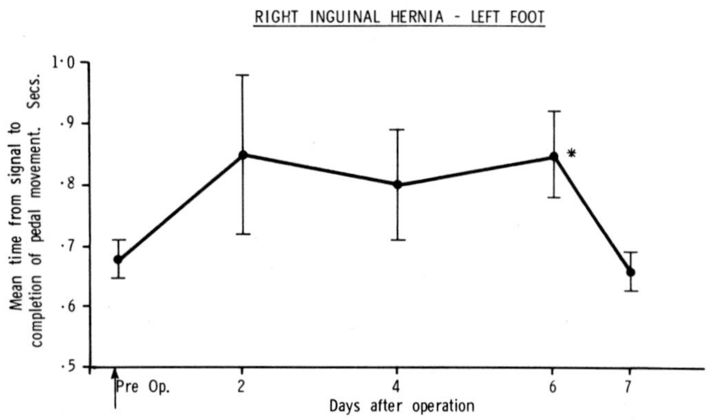

Note.

1. ± Standard error of the mean shown by vertical lines through points
2. ✱ Indicates significant increase from pre op. P<·05

Fig. 2

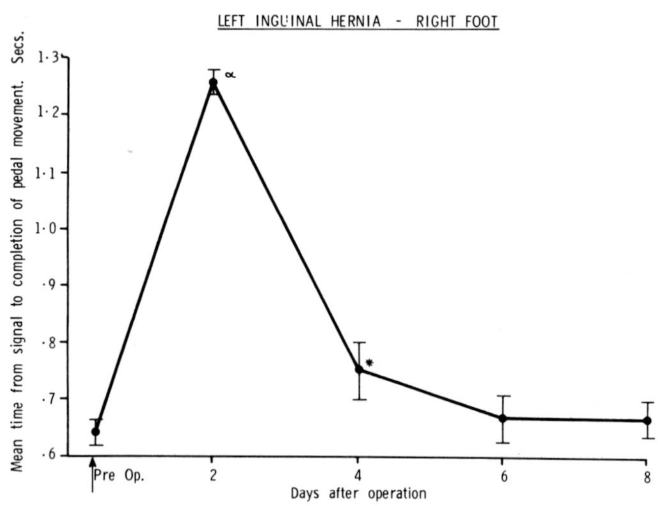

Note.

1. ± Standard error of the mean shown by vertical lines through points
2. ✱ Indicates significant increase from pre op. P<·05
3. ∝ Indicates significant increase from pre op. P<·001

Fig. 3

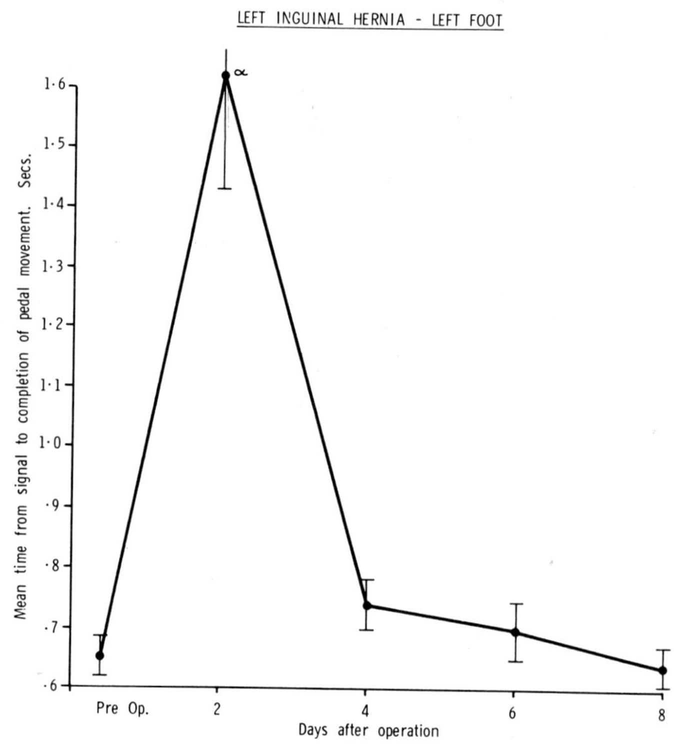

Note.

1. ± Standard error of the mean shown by vertical lines through points
2. ∝ Indicates significant increase from pre op. P<·001

Fig. 4

OPERATIONAL VIEWS AND REQUIREMENTS FOR MAINTENANCE, REPAIR AND EFFECTIVENESS OF BRAKES

P. H. WYKE SMITH FCIT, MIRTE
National Bus Company, London

The MS of this paper was received at the Institution on 14 August 1975 and accepted for publication on 12 December 1975

SYNOPSIS The introduction draws attention to the continuity of established brake design and emphasizes the changing environment with increasingly severe conditions. The reliability of the various brakes is discussed and it is argued that all too frequently production orientated design characteristics prove to be costly and temperamental in operation with consequential disaffection amongst engineering and driving personnel. The growing importance of accurate performance analysis and costing, which should be used to add a new dimension to the process of design and development to achieve higher levels of reliability, is discussed in some detail and concludes that manufacturers should no longer prove their products at the operators' expense.

INTRODUCTION

1. The contents of this paper concerns the environment in which brake systems on public service vehicles are operated and, to put the subject into perspective, a quotation from "Bus and Coach" annual review number of December 1962 is quoted: "Straight air pressure systems, actuating S-cam brakes of the leading-trailing shoe type continue to be the most popular braking equipment for public service vehicles. A few retain hydraulic actuation for which air pressure has now almost entirely superseded vacuum as the means of boosting pedal pressure. Vacuum operated brakes are rapidly becoming obsolete." This statement could well have been written to describe the situation in December 1975 as there has been no major shift in practice, although the recent period has witnessed various developments in concept and design to anticipate the technical standards of the European Community of which Directive No 71/320/EEC of 26 July 1971, concerning the approximation of the laws of the member states relating to braking devices, is an example.

Although the principles of design are relatively unchanged, all other related factors have been the subject of wide variation. In the last twenty years speeds have risen from 40 to 50 m.p.h (miles per hour) and the power to weight ratio from 9-12 b.h.p (brake horse-power) per ton to 11-14 b.h.p per ton. Design payloads have gone up but the vagaries of the passenger market have become pronounced with more and more journeys being undertaken with the ratio of empty to occupied seats escalating. Seating capacities have increased from 44-53 and 58-78 (single deck and double deck) and standing passengers from 8-23 and 8-20 (single deck and double deck) respectively.

If you ask a bus fleet engineer the question: "What is your major problem"? his answer is likely to be: "How can we improve brake imbalance"? Vehicle "down-time" has doubled because front brake liner life has an advantage over rear liner life to a factor of 4, 5 and even 6, requiring frequent intermediate brake relines. The bus fleet engineer is very aware of inherent defects in brake valves, and that the designer is encountering difficulty in calculating braking ratios to take account of the large proportion of bus miles that are burned off in average load conditions, and that during the last twenty years the weight of single deck buses has gone up by about 25% and the passenger carrying capacity by 30%, and where the corresponding double deck figures are 15% and about 25%.

The late 1950's and early 1960's did seem to provide an answer to the perennial imbalance symptoms in an application of continuous flow hydraulics to disc brakes on a Midland Red Bus Company's single deck chassis. This had an unladen weight of $5\frac{1}{2}$ tons and the front and rear axles were almost equally loaded, allowing single wheels on front and rear. The disc brake pads gave comparable life to the linings used for drum brakes, about 45,000 miles. But there was, at the time, no parallel to this special design in the open market where proprietary chassis were heavier with more heavily loaded rear axles having twinned wheels. Its advantage was that the brakes were lightly stressed and the energy generated during stopping was distributed in near equal proportions. Therefore, it can be said that near perfect brake balance was achieved at a point of time but in less severe operating conditions. The inescapable fact is that brake maintenance continues to be a problem area for the fleet engineer and revolves around considerations such as:

Is the quality of components satisfactory?

Is adequate attention given to protecting brake systems from contamination?

Have dual line systems reached the optimum stage of development?

Is valve design and engineering satisfactory in its application to brake installations?

Are twin cylinder reciprocating compressors adequate for compressed air systems of public

© IMechE 1976

service vehicles or should rotary units be adapted from other available designs?

The first can severely limit the ability of the system to introduce dry, uncontaminated air to the brake units and this feature will be developed elsewhere in the paper. But there are matters other than these miscellaneous comments closely related to the eventual reliability and maintainability of braking equipment. For example, what about the financial environment in which operator and manufacturer have to co-exist? The immediate past is no different from the present in that if no change can be discerned in the demand for established brake systems, a change for the worse in the availability, cost and design of vehicles and other operating hardware and, more importantly, the supply and cost of manpower resources and skills, is very obvious. For example, new vehicle deliveries in 1965 took one year, and in 1975 - two years. Financially, the commitment was around £6,500 per vehicle in 1965 and £20,000 in 1975. Clearly, these changes in the context of this paper have not been to the advantage of the operator.

These considerations are important because they have a bearing on the ability of the operating industry to cope with the maintenance of equipment which is often inaccessible, difficult to identify and inherently unreliable. There is, after all, one common denominator for correlating vehicle investment and maintenance and that is to measure the latter in money terms, but as the value of money continues to change to everyones disadvantage, it is questionable whether manufacturers, labouring under their own problems, are able to relate the effects of their design patterns to fleet engineers' maintenance problems allied as they are to restricted investment, and where there may be inadequate inspection pit and vehicle hoist accommodation coupled to shortages of staff.

Be that as it may, servicing requirements are entirely ordained by the design of the vehicle and maintenance becomes more efficient if ease of remedying defects is designed into the vehicle, which implies that manufacturer and maintainer should work in close partnership. However, the manufacturer is primarily concerned with producing a product most likely to minimize his costs, conferring near uniformity of each task, maximum throughput and the automation of his plant.

On the other hand, servicing a vehicle in running order can be closely regulated and repetitive, but the diagnosis and clearance of defects is not the uniform process which characterizes the manufacturing cycle. The other aspect of maintainability, the theme of this paper, is that of living with a shortage of spare parts. This is not a UK phenomenon. It is generally prevalent and implies that the best braking systems - the most reliable systems, are those requiring the minimum number of parts. Whether or not this aspect of maintainability should be taken into account during the process of selecting chassis specifications is arguable. If it is, then the operator who does not always enjoy full freedom of choice, may be required to purchase lower priced brake operating mechanisms containing more moving parts.

A study of manufacturers' maintenance manuals dealing with fixed anchor single leading and single trailing shoes on the one hand, and two leading shoe and self energising systems on the other, shows that the former attracts the shorter descriptive and instructional material and also has the briefest list of components. So there is some conflict here. Decisions flowing from these generalities depend largely on operating conditions and influence the maintainers' attitude towards the goal of minimizing vehicle "down-time".

BRAKE RELIABILITY

2. The mass of available information on the variety of brake systems to be found in the bus operating field presents a confusing picture, not because analysis of performance is lacking, but due more to the extremes of conditions in which the equipment is required to work. Provided there is recognition of the fact that for two-thirds of total miles operated, components are near to the upper limits of tolerance in terms of frequency of operation and temperature, the intensity of debate and opinion that exists concerning the choice of systems will be more readily understood. The extreme condition can best be described as the buses' peak hour environment when it has to contend with a minimum of five scheduled stops per mile, traffic delays, higher power to weight ratios, with drivers extending the equipment to the limit in maximum load conditions in the knowledge that most brake systems can provide very large operating forces at the brake assembly. Bus design has seen much change in the search for bigger payloads, lower floors, lower steps and, more recently, tighter specifications to reduce noise levels. Bus body design has not been conducive to securing good braking ventilation and the new features referred to have tended to be the opposite of progress. The rear-engined low floor double deck vehicle reveals a tendency to more costly rear axle maintenance for this reason. Inner rear tyre temperatures on the twin wheel assemblies are fairly critical and call for $22\frac{1}{2}$" rims to ensure good ventilation. In city operation with hilly terrain, rear inner rims are reaching temperatures above the level at which tyre beads can exist, leading to bead burning and, in extreme cases, blow-outs. Few of the older type vehicles (ie engine at the front) demonstrated this trouble: where they did it was partly due to poor braking distribution and partly due to large overall diameter brake linings, the drums for which practically filled the 20" wheels.

Due to the incidence of peak operation and other social reasons, it is relevant to repeat that drivers are less tolerant and drive to the limit of the vehicle's braking capacity. Here again, conflict of choice emerges. Drivers like two leading shoe brake assemblies, although they produce similar retardation figures to the conventional equipment. These brakes are keen, having high shoe factors and there is negligible delay in the system. These assemblies are usually mounted on light pressings and help to keep down the unsprung weight. The maintenance engineer has the choice of buying service from agents to sustain the equipment at peak performance or to employ skilled staff. By comparison, cam operated leading and trailing shoe assemblies linked with compressed air systems are found to have inherent delay factors to which the driver must adjust his reactions, and suffer from uneven wear on leading and trailing shoes due to the variable shoe factors; often to the extent of 2 : 1. However they can be serviced by semi-skilled staff at the

© IMechE 1976

brake assembly but have the disadvantage of being heavy castings, forgings, etc., which tend to maintain or increase unsprung weights.

When assessing the reliability of brakes in the knowledge that reliability is a function of the quality of production which, in turn, is a function of cost, it may be invidious to make direct comparisons of alternative braking systems which have been designed and produced at different periods and have thus encountered variable economic pressures and market demands on manufacturing. A study of the scale of design applied to vacuum brake units by comparison with currently used air brake system units illustrates the point, the former being generously proportioned but with the latter having characteristics and a complexity which require finer design tolerances, attracting the ongoing critique of the value engineer.

Triple servo vacuum brakes were generally regarded with affection by maintenance staff if not so much by drivers as they were not a prompt acting brake. These brakes were tolerant of road and workshop conditions, easily dismantled and re-assembled in the operating garage, and were self-repairing when small leaks developed, provided the exhauster kept pace. Brake adjustment spanners did, however, wear out fairly quickly owing to the frequency with which adjustments had to be made as the brake shoe liners were usually of woven material to obtain a high co-efficient of friction between the lining and the drum. So long as the exhauster, air hoses and adjustment received adequate attention, the system was reliable. Master servos, for example, in city operation, have been known to function satisfactorily up to 325,000 miles - about 6 years.

Hydraulic braking systems are not generally regarded as having a high reliability factor, particularly when comparing the heavier and lighter chassis. This may be due to the use of two leading shoe assemblies on the lighter chassis which rely on the friction co-efficient between drum and liner for their inherent self-servo action, and are, therefore, more prone to fade, whereas the heavier chassis are normally fitted with leading trailing brakes which, because of their lesser-servo action have better fade-resistance. But this assessment has little to do with the characteristics of the units in the hydraulic line.

Accumulator systems with twin accumulators and systems with tandem master cylinders are regarded more favourably, but require an adequate supply of trained, skilled labour at the operating garage.

Air over hydraulic systems are efficient in the sense that the advantages claimed for hydraulics are maintained, which is to say that efficiency is high, the components are small in volume and light in weight. The maintenance engineer says that ingress of air into the hydraulic system, making the brake feel spongy, is likely to occur when powered by compressed air, and that the whole system is more prone to leaks. This is not entirely correct because air leaks into the hydraulic system for a variety of reasons and from several sources, whether the brakes are continuous flow or air powered. Hydraulic fluid causes minimal corrosion of metal surfaces and deterioration of seals and the whole system, unlike compressed air brakes which tend to inhale dirt, is virtually sealed.

It is important, therefore, for the operating industry to pay particular attention to the discipline of cleanliness during dismantling and reconditioning operations in order not to detract from this valuable dirt occluding property. In the majority of applications to public service vehicles the brake is progressive and retains the confidence of the driver, and helps to reduce the cyclic demands for adjustment between normal periods of inspection and maintenance.

Adjustment tends to be planned on a blanket basis ie all vehicles at the same interval. However, there is a noticeable, gradual trend to shorten this interval.

The spring handbrake has also forced a more methodical approach - previously it was quite usual on non-urban work to adjust on drivers' and shunters' reports. In-so-far as these are a guide, vehicles with serious imbalance are reported more frequently.

The commentary on the reliability of compressed air systems could, however, be said to suffer from discontinuity from a diversity of causes. This is not to say that, given freedom of choice, the operating industry would opt for an alternative. Undoubtedly, compressed air is so widely used, due in part to an increase in the utilisation of compressed air systems with the introduction of air-operated change-speeds for gear boxes, compressed air powered passenger entrance and exit doors, windscreen wipers and brake stop lamps. Few would disagree with a typical opinion recently offered that "An air system, properly maintained and adequately drained, provides the best braking for our stage carriage operations."

Conversely, the development of the air brake has not been without its difficulties and frustrations and some of these are ongoing. Single line systems found general acceptance excepting the poor performance of twin cylinder 9-12 c.f.m (cubic foot minute) capacity compressors. Pipe layouts were intelligible and valves were robust. These were superseded by dual line systems to achieve improved safety margins, commendable and correct in principle, but involving more complex pipe layouts and more valves.

BRAKE SYSTEM MAINTAINABILITY

3. Many of these valves exhibit poor maintainability characteristics, partly because of their sensitivity to contamination by hot, carbon-deposited air. They have thereby delayed investigations to reduce brake imbalance between front and rear axle assemblies.

Why, it may be asked should there be any concern about air compressors? They appear to do the job; no doubt this can be proved - but - compressors capable of delivering 11 c.f.m of air at 1000 rev/min are known to have been working in an environment where no previous assessment of total air consumption had been made. If it had - capacity would have been increased. Pipe layouts have been so variable as to place unloader valves within three feet from the compressor in one instance and up to eight feet from the compressor in another.

It is universally agreed that nuisance costs rose sharply during the protracted period it took to increase compressor deliveries from 11 c.f.m to

© IMechE 1976

17 c.f.m at 1000 r.p.m and to introduce water cooled cylinder heads in order to reduce the necessity for premature changing of faulty unloader, brake and EP gear operating valves. Again, the total consumption of compressed air on the bus seems to have been overlooked and it could be that the value engineer was given too much rein in dictating cost/performance levels, and that the operating industry and manufacturer failed to agree a correct assessment of specified requirements. Oil blowby continues to be experienced with compressors generally, exceeding the accepted norm of 1cc (cubic centimetre) per cylinder per hour. Air compressors do not have separate oil systems and are pressure fed from the engine sump which results in accelerated wear rates. A life of one year (50,000 miles) has to be accepted.

It is good that dual line systems have been developed with time to spare for conformity with the EEC Directive No 71/320, but there remains a lot of work to be done to improve the brake valve performance. Dual concentric valves are not proving reliable. They are prone to valve flutter at low brake pedal pressures and to valve stem wear at point of contact with sealing rubbers and are bedevilled by variable spring lift. They seem incapable of supplying two outputs of equal pressure particularly at low pressures.

Some work with single valves to equalise airline pressure to the axles has been undertaken and the results, so far evaluated, have not provided a "cure". It is perhaps pertinent to suggest that two identical valves should be linked by a pivoted bar with the pedal force acting upon it. If the pedal force acts in the centre of the bar it will apply equal forces to the two valves, but this ratio could be varied by offsetting the point of action. This arrangement with an adjustable balance bar has not, so far as is known, been applied to public service vehicle chassis, using compressed air braking systems but might confer a means of obtaining the balance which eludes the dual concentric valve. It is appreciated that such adjustment would probably involve the use of a roller brake tester, at least in the early stages.

The criteria required to satisfy fleet engineers' expectations are:

(a) The brake must remain serviceable between adjustments at brake inspection cycles of not less than one week or 1000 miles.

(b) It should maintain its maximum performance over at least two-thirds of the adjustment period.

(c) It must be progressive and moderately sensitive and there should be a minimum of delay between the application of the pedal and the operation of the brakes.

(d) It should be tolerant of road and workshop conditions.

(e) It must be of adequate quality, easily dismantled and re-assembled with no possibility of incorrect assembly.

(f) The brake should reflect the success of the designer in providing a system which balances the braking effort at front and rear not because it depends on the system of operation but on the selection of the most suitable balancing ratios.

Fleet engineers are aware that the demand for improved brake performance is aggravating the inherent problems of accommodating the effect of weight transfer, when static loads carried by the front and rear wheels are not generally equal, and that during braking less load is carried on the rear and more on the front than when the vehicle is stationary.

However, conveying such problems to fitters and drivers and relating them to vehicle maintenance and driving characteristics is often difficult. The fact that some brake systems are complicated with many highly developed components fails to impress and if for one reason or another they demonstrate actual or alleged symptoms of unreliability they fail to achieve the operator's goal of maintainability, causing the relations between the engineering supervisor, driver and fitter, and between the fitter and driver to deteriorate and even to become emotive. Unless rigidly controlled the snowball effect on vehicle "down-time", particularly in public service vehicle operating centres, can have undesirable repercussions with adverse Trade Union re-action, loss of passenger confidence and the insistence by Vehicle Examiners that Statutory inspection cycles are supplemented by even more frequent intermediate checks.

Such comments on maintainability of brakes again point to the necessity for bringing nearer together the design and operating spheres. The reference to the complex and highly developed characteristics of brake systems can be illustrated by mentioning that on a typical heavy single deck chassis, currently in production, the operating industry's mechanical staff require to have a working knowledge of a compressed air installation comprised of some thirty separate valves, single diaphragm brake chambers, spring brake chambers, pressure protection valves, low pressure warning devices and so on. In these circumstances maintainability is a function of confidence in, and an understanding of, the design and operating principles by mechanical staffs. To achieve higher standards training has been intensified by the provision of instructors and technicians, scaling up the engineering "cover" to maintain serviceability whilst on-and-off-the-job courses are completed, the effect being to multiply the cost element of maintainability. In any event the Motor Vehicles (Construction and Use) Regulations 1973, more specific than previous regulations, render it imperative to secure a better interplay of operating and design experience. For example, the air spring parking brake was introduced at about the same time as the new regulations. Since it is generally installed with a dual system its function is legally to act as a parking brake on a 1 in 6 gradient with the vehicle unladen. It is, however, frequently checked against a 25% or 30% retardation standard. This brake marked a step forward in design but required a new brake testing technique, driver familiarization cadres and an ongoing flow of breed-improving modifications.

COST CONTROL

4. It is fundamental for the achievement of an acceptable level of reliability that a logical appraisal of the opposing claims of finance and cost control is applied to the standard of quality of those auxiliaries which are critical to the

© IMechE 1976

consequential behaviour of dependent mechanical devices in the overall layout and in which a malfunction will severely alter or negate the individual reliabilities. This criterion is applicable to any circuit or layout where malfunction at the power source will affect performance at every other point of mechanical activity. It is particularly pertinent when allied to safety and, especially, therefore, to brake systems. The essential element for the accurate appraisal of the relative importance of finance, design and production engineering considerations depends on the availability of accurate performance data. This must be fully representative of overall operating experience to obtain the confidence of those working on the analysis. By ensuring that it is wholly accurate and that those receiving the data know it can safely be used, firm decisions can be made as to which of the three considerations should take precedence in the circumstances.

CONCLUSION

5. There is room for improvement involving a more meaningful exchange of information from field to plant and vice versa. The process should be continuous and not at all identified with the popular but intermittent practice of complaining when things go wrong. There is scope for mutual benefit to be derived from an improved process of communication of what needs to be done to improve the flow of information along the channels which now exist between the manufacturer and the operating industry. In short, the operating side should be capable of identifying the functions where improved quality and additional cost are essential for the improvement of performance.

All of these points of contention can be expressed in the various maintenance costing systems already in existence or about to be introduced. The bus operating industry is in the process of updating established procedures to enable the engineer to adopt a uniform approach for an improved point of reference on all matters of principle and detail connected with bus engineering and to identify and relate costs to particular categories of maintenance according to vehicle make, model and year of manufacture. By this means the operating industry will provide added impetus and accuracy to the process of calculating vehicle whole life costing so as to determine with confidence and accuracy the information required for the uncompromising selection of the most suitable combinations of operating systems to guarantee reliability and maintainability. It is for this reason felt to be so important to make a plea for the free and informal exchange of views and to implement this exchange formally through the societies and representatives' bodies, indirectly contributing to the enhancement of the UK heavy commercial vehicle industry's reputation.

Reliable brake systems with minimal maintenance tasks providing the operator with measurable cost benefits continue to be elusive. A more effective formula by means of which manufacturers obtain rapid feed back on performance must be devised. Achievement of this aim will engender a better attitude in all personnel grades and foster that spirit of understanding and co-operation required to reduce the cost generated by unserviceable vehicles. It will divert the energies of Chief Engineers in the operating industry from unplanned demands on their time to deal with deficiencies, into the more objective and controlled engineering functions associated with fleet investment, specification and maintenance.

ACKNOWLEDGEMENTS

6. I have to thank Mr T P Newcomb and Mr R T Spurr for the opportunity to quote from their book "The Braking of Road Vehicles", Mr B T Hancock for information from I.M.E. paper C3/74 and Messrs A Gurley, K R Worrall, R J Hawkins and R A J Holding, Chief Engineers of NBC's subsidiary companies, for their assistance.

© IMechE 1976

C45/76

BRAKE DEFECTS IN CARS

C. A. HOBBS BSc, CEng, MIMechE
Transport and Road Research Laboratory
Department of the Environment, Crowthorne, Berkshire

The MS of this paper was received at the Institution on 3 December 1975 and accepted for publication on 19 December 1975

SYNOPSIS As a follow up to on-the-spot investigations, which indicated that brake defects made a major contribution to the occurrence of about 3 per cent of road accidents, a thorough examination of brake defects in two samples of cars has been undertaken.

In the course of a two year survey 436 accident involved vehicles and 341 vehicles belonging to staff of the Transport and Road Research Laboratory were examined, whether or not any defect was suspected. The condition of discs/drums, linings, hydraulic cylinders, adjusters, pipes, hoses and handbrake linkages were inspected. Where possible samples of brake fluid were taken from the master cylinder and a wheel cylinder for determination of water and dirt content. In the case of roadworthy vehicles, decelerometer readings were recorded for footbrake and handbrake.

The proportions of vehicles with defects were very similar for the two samples: 83% of the accident involved vehicles and 85% of the comparative sample vehicles had at least one defect. The number of defects per vehicle averaged 4½ in both samples and there were no significant differences in the number or types of defects found.

The relative importance of different types of defect is discussed. It is recommended that more regular checking is carried out as part of normal servicing. This would lead to improvements in brake maintenance and a reduction in the number of defects.

INTRODUCTION

1. Detailed examination of cars for brake defects have been carried out in two sample surveys, one of 436 accident involved vehicles, and the other of 341 vehicles belonging to staff at the Transport and Road Research Laboratory. This study, which has given an indication of the kinds of defects present, was a follow up to recent finding, from in-depth investigations of accidents, that brake defects made a major contribution to the occurrence of about 3 per cent of road accidents.

SURVEY DESCRIPTION

2. Over the two year period up to March 1974, out of a total of 1783 vehicles involved in 966 accidents covered by local on-the-spot investigations in East Berkshire, 436 vehicles were given detailed brake examinations. This sample of vehicles was biassed towards those more seriously damaged and thus immobilised in the accidents.

3. During the same period, 341 vehicles belonging to the members of staff of the Transport and Road Research Laboratory were examined for use a comparative sample, although this may not be wholly representative of cars in use in the area. The majority of vehicles examined in both samples were cars, with a few light goods vehicles and 3 wheelers.

4. Engineers employed by Automotive Products Ltd carried out the examination of vehicles under contract to TRRL. During the examination the fluid level in the reservoir and the state of adjustment of each brake assembly were checked. Each assembly was then dismantled and inspected in detail. The condition of the discs or drums, linings or pads, wheel cylinder seals and mechanical components were all checked. Master cylinder operation was observed and all pipes, hoses, joints and linkages were inspected.

5. Where possible brake fluid samples were taken for analysis from both the master cylinder reservoir and a wheel cylinder. These fluid samples were then analysed, under contract, by Burmah-Castrol Ltd. The analysis showed the percentage water content of the fluid and the percentage contamination with mineral oil and dirt.

6. The presence of accident damage often made it impossible to inspect every part of the brake system and in many cases it was not possible to road test the vehicle. Where it was possible to road test the vehicle, footbrake and handbrake decelerations were measured with a pendulum type decelerometer.

7. Analysis of all the data collected was aided by the use of a computer retrieval system.

RESULTS OF BRAKE EXAMINATIONS

8. Little difference was found in the numbers and types of defects in the two samples; 83% of the vehicles in the accident involved sample had brake defects compared with 85% of vehicles in the comparative sample (Table 1). In both samples there was an average of 4½ brake defects per vehicle.

11. Faulty adjustment was the most common defect found in the brake assemblies, that is the brake assembly was either too slackly or too tightly adjusted. It may be argued that faulty adjustment should not constitute a defect and in some cases this may be true. However, slack adjustment often results in an unbalanced braking system and excessive pedal travel, whereas a brake assembly that is

Table 1

Number of brake defects found in the vehicles examined

	Accident sample No. of vehicles	%		Comparative sample No. of vehicles	%	
No defects	76	(17)		51	(15)	
1 or 2 defects	91	(21)	(83)	68	(20)	(85)
3 - 5 defects	115	(26)		106	(31)	
6 - 10 defects	114	(26)		94	(28)	
More than 10 defects	40	(9)		22	(6)	
Total vehicles examined	436			341		

9. In many cases the defects recorded were not mutually exclusive; often one defect was the cause of one or more other defects. For example, a leaking wheel cylinder seal may have allowed brake fluid to contaminate the brake lining material. In this case two defects would be recorded, i.e. "leaking wheel cylinder" and "faulty lining". A summary of the defects is given in Table 2.

BRAKE ASSEMBLIES

10. One or more brake assemblies had defects on 312 (72%) of the accident vehicles. The types of defect are listed in Table 3.

too tightly adjusted may overheat. More than half the vehicles had one or more of its brakes out of adjustment and one in twenty vehicles had all its brakes out of adjustment. Significantly more rear brakes were found to be out of adjustment than front brakes. This may be accounted for by the larger number of front brakes being self-adjusting, often because of the presence of disc brakes.

12. Other more serious, but less frequent, defects were faulty linings, discs or drums (often due to brake fluid or oil contamination) leaking or seized wheel cylinders and worn linings.

Table 2

Summary of brake defects

	Accident involved vehicles		Comparative sample vehicles	
	No. of defects	No. of vehicles with defects	No. of defects	No. of vehicles with defects
Brake assembly	1314	312 (72%)	988	249 (73%)
Master cylinder	58	52 (12%)	43	39 (11%)
Brake pipe/hose	144	134 (31%)	116	103 (30%)
Other foot brake	161	125 (29%)	134	113 (33%)
Handbrake	164	139 (32%)	110	99 (29%)
Brake fluid	129	81 (19%)	101	64 (19%)
Total	1970	360 (83%)	1492	290 (85%)
Total vehicles examined		436		341
Average number of defect per vehicles		4.5		4.4

Percentages in ( ) are of total vehicles examined

Table 3

Brake Assembly Defects

	Accident involved vehicles		Comparative sample vehicles	
	No. of defects	No. of Vehicles	No. of defects	No. of Vehicles
Faulty adjustment	454 (28%)	222	361 (27%)	169
Worn linings	149 (9%)	94	100 (7%)	61
Faulty linings	184 (12%)	114	129 (10%)	81
Worn disc/drum	70 (4%)	51	64 (5%)	45
Faulty disc/drum	230 (14%)	127	185 (14%)	97
Seeping wheel cylinder	71 (4%)	63	66 (5%)	49
Leaking wheel cylinder	110 (7%)	72	46 (3%)	32
Seized corroded wheel cylinder	36 (2%)	32	24 (2%)	20
Seized-gummed wheel cylinder	10 (½%)	7	13 (1%)	8
Total defects	1314	312	988	249
Assemblies not examined	148		25	
Assemblies examined	1594		1339	
Total assemblies	1742 (436 vehs*)		1364 (341 vehs)	

Percentages in ( ) are of the assemblies examined
*Includes 2 three-wheelers

Table 4

Master cylinder, pipes and hoses and other foot brake defects

	Number of defects	
	Accident involved vehicles	Comparative sample vehicles
Master cylinder		
Fluid reservoir empty	3* (½%)	1 (¼%)
Fluid reservoir low	29* (7%)	16 (5%)
Reservoir breather blocked	9 (2%)	9 (3%)
Internal leak	10 (2%)	6 (2%)
External leak	0 (0%)	1 (¼%)
Seal sticking	3 (½%)	5 (1%)
Other master cylinder defect	5 (1%)	5 (1½%)
Total master cylinder defects	58 in 52 vehicles	43 in 39 vehicles
Brake pipes and hoses		
Brake pipe corroded	118 (27%)	87 (26%)
Other pipe defect	4 (1%)	3 (1%)
Hose perished/chafed	22 (5%)	26 (8%)
Total pipe and hose defects	144 in 134 vehicles	116 in 103 vehicles
Other foot brake defects		
Adjuster seized/faulty	83 (19%)	47 (14%)
Bleed nipple seized/blocked	21 (5%)	17 (5%)
Other mechanical parts faulty	52 (12%)	65 (19%)
Incorrect parts fitted	5 (1%)	5 (1%)
Total other footbrake defects	161 in 125 vehicles	134 in 113 vehicles
Total vehicles examined	436	341

Percentages in ( ) are of total vehicles examined

*A further 24 vehicles had empty or low reservoirs due to system leaks caused by the accident.

© IMechE 1976

MASTER CYLINDER, PIPES AND HOSES AND OTHER FOOT-BRAKE DEFECTS

13. Just over one in ten of the vehicles examined had a defect related to the master cylinder, although about half of these defects were lack of sufficient brake fluid in the reservoir (Table 4). The other most common master cylinder defects were blocked reservoir breathers and internal leaks. Apart from fluid level, visual inspection of the exterior of the master cylinder would not have revealed the defects. Internal leaks, however, could be detected by observing pedal creep at low pedal pressure.

14. In more than a quarter of the vehicles, brake pipes were found to be corroded though not necessarily to a dangerous extent (Table 4). A 200lbF pedal load burst test was passed by all the vehicles. However, the presence of corrosion does indicate the loss of the pipes' protective coating, leaving them vulnerable to more severe corrosion attack. Of the other defects found the majority were perished hoses.

15. About a third of the vehicles had some other mechanical defect affecting the footbrake operation, about one in six vehicles having at least one of its brake adjusters seized. A similar number of vehicles had other mechanical parts faulty due to siezure or wear.

HANDBRAKE DEFECTS

16. Some part of the handbrake mechanism was defective on about a third of the vehicles. Lack of lubrication of the linkage resulting in corrosion and wear were the most common faults. Nearly one in five vehicles were suffering from seized or faulty handbrake linkages and about 15% of the vehicles had poorly adjusted handbrakes.

BRAKE FLUID ANALYSIS

17. Most hydraulic fluids used in vehicle braking systems readily absorb water either directly or from the atmosphere. The presence of water in the fluid leads to corrosion of the braking system, and, when the brakes get hot, vapour locks may occur resulting in the brakes becoming spongy or possibly failing. A water content of 5% is usually considered the maximum safe level. In 90% of the vehicles the fluid samples taken from the master cylinder and wheel cylinder contained less than 5% of water (Table 6). In only about 1½% of the samples was the water content 10% or more. No significant differences were found between samples taken from the master cylinders compared to those from wheel cylinders. There were no known cases of vapour lock having occurred in the accident vehicles at the time of their accidents. In fact when interviewed after the accidents, none of the drivers of the accident involved vehicles complained of symptoms which could be related to vapour lock.

18. Any mineral oil present in brake fluid will tend to attack the rubber components in the braking system. Ninety-five percent of the samples were found to contain less than 0.1% of mineral oil and were considered satisfactory. There were 5 vehicles which had at least one fluid sample with 0.5% or more mineral oil. In two of these vehicles the samples taken from the master cylinders had over 1.0% of mineral oil. This was considered danger ous. No significant differences were found between the samples taken from the master cylinders and wheel cylinders.

19. Dirt present in the brake fluid will tend to wear the seals in the system and to block ports in the master cylinder. The dirt content was 0.5% or below in about 98% of the samples tested. Higher dirt contents were found in the wheel cylinder samples than in the master cylinder samples. By far the most common type of dirt present was rubber debris from seals and hoses. Some rubber was found in almost all the samples tested and a large proportion of the samples had non-ferrous metal present.

ROAD TEST PERFORMANCE

20. The Vehicle Regulations require that the footbrake and handbrake are able to decelerate the vehicle by 50%g and 25%g respectively. Pendulum type decelerometer readings were taken for footbrake deceleration from 25 mph and handbrake deceleration from 20 mph. It was possible to test only 22 of the accident involved vehicles in this way, but 191 were tested in the comparative sample.

21. All the vehicles tested in both samples achieved over 50%g in the footbrake tests. However, four vehicles in each sample failed to achieve 20%g deceleration with their handbrakes. Deceleration readings in excess of 90%g were achieved by 9 out of 10 of the vehicles in the footbrake test.

Table 5

Handbrake defects

	Number of defects	
	Accident involved vehicles	Comparative sample vehicles
Handbrake seized/faulty	82 (19%)	60 (18%)
Handbrake out of adjustment	65 (15%)	48 (14%)
Handbrake lock faulty	17 (4%)	2 (½%)
Total handbrake defects	164 in 139 vehicles	110 in 99 vehicles
Total vehicles examined	436	341

Percentages in ( ) are of total vehicles examined

©IMechE 1976

Table 6

Brake fluid analysis

	Number of defects			
	Accident involved vehicles		Comparative sample vehicles	
	Master cylinder fluid	Wheel cylinder fluid	Master cylinder fluid	Wheel cylinder fluid
Water 5.0 - 9.9% $^{v}/_{v}$	34 (9%)	31 (8%)	31 (10%)	25 (8%)
" 10% $^{v}/_{v}$ & over	6 (1½%)	5 (1½%)	1 (¼%)	6 (2%)
Mineral oil 0.1% $^{w}/_{w}$ & over	18 (5%)	18 (5%)	14 (4%)	13 (4%)
Dirt over 0.5% $^{v}/_{v}$	6 (2%)	11 (3%)	5 (2%)	6 (2%)
Total defects	64	65	51	50
Total brake fluid defects	129 in 81 vehicles		101 in 64 vehicles	
Total samples tested	394	366	319	311

DEFECTS RELATED TO ENGINE SIZE AND VEHICLE MILEAGE

Engine size

22. Vehicles with an engine size of less than 1 litre were found to have relatively more defects than those vehicles with larger engines (Table 7). One-third of the under 1 litre vehicles had the front brake assemblies out of adjustment: this is very significantly more than in the over 1 litre vehicles. The reason for this is most likely because physically smaller brakes are usually fitted to smaller vehicles.

Vehicle mileage

23. Vehicles with higher mileage readings had more defects than those with lower mileages (Table 8). Vehicles with mileage readings of 40,000 miles or over averaged nearly 6 defects per vehicle, whereas those with readings less than 20,000 averaged far fewer defects - less than 1. Certain defects which were related to corrosion such as: pipe corrosion, seized or faulty handbrake and faulty brake adjuster resulting in incorrectly adjusted rear brakes, occurred very significantly more frequently on higher mileage vehicles.

VEHICLES WITH BRAKE DEFECTS CONTRIBUTING TO THE ACCIDENT

24. From the detailed assessment of contributory factors in the whole four year on-the-spot survey brake defects were judged to have been main contributory factors in nearly 3% of accidents (Ref. 1). This includes 16 vehicles which had been given the detailed brake examination. It is likely that brake defects contributed in part to the occurrence of many other accidents without being identified as doing so in the on-the-spot survey. Of the 16 vehicles given detailed examination there were 11 cases where the brake defect had caused the vehicle to pull to one side or resulted in loss of control. In only 5 vehicles had the brake defect resulted in the vehicle simply not being able to stop. This indicates the importance of balance in a vehicle's braking system. Most of the 16 vehicles had additional brake defects which were also present but were not contributory to the accident. Ten vehicles had more than five contributory defects, one vehicle having 15 contributory defects and a further 5 non-contributory defects. (Table 9). The vehicles had an average of nearly 7 contributory defects each with an average of over 2½ other defects present.

Table 7

Brake defects related to engine size

	Accident involved vehicles		Comparative sample vehicles	
Engine capacity	< 1000cc	1000cc & over	< 1000cc	1000cc & over
No defect	11 (10%)	65 (20%)	4 (6%)	47 (17%)
1 - 5 defects	42 (40%)	164 (50%)	34 (52%)	140 (51%)
> 5 defects	53 (50%)	101 (30%)	28 (42%)	88 (32%)
Total vehicles	106	330	66	275

© IMechE 1976

Table 8

Brake defects related to mileage reading

	Accident involved vehicles			Comparative sample vehicles		
Mileage reading	<20,000	20,000-39,999	≥40,000	<20,000	<20,000-39,999	≥40,000
No defect	43 (57%)	18 (16%)	15 (6%)	21 (32%)	16 (20%)	14 (7%)
1 - 5 defects	32 (42%)	59 (52%)	113 (46%)	37 (56%)	53 (65%)	84 (44%)
> 5 defects	1 (1%)	36 (32%)	116 (48%)	8 (12%)	13 (16%)	94 (49%)
Average number of defects/vehicle	0.9	4.2	5.9	2.1	3.2	5.7
Total vehicles	76	113	244	66	82	192
Mileage reading not known		3			1	
Total vehicles examined		436			341	

DISCUSSION AND CONCLUSIONS

25. The examination of brake defects in a sample of 436 accident involved vehicles, mainly cars, has shown that a large proportion, 83% had one or more defects. A similar proportion, 85% of the comparative sample of 341 vehicles also had defects. In both samples there was an average of 4½ defects per vehicle, the two samples showing no significant differences in the numbers and types of defects found.

26. By far the largest proportion of the defects found in both samples were associated with brake assemblies, most of the contributory defects in accident occurrence also being related to the brake assemblies. The main areas requiring attention were adjustment of brakes (particularly in smaller cars i.e. less than 1000 cc), wear of linings and pads and leaking cylinders. All of these can be checked with very little effort.

27. Of the defects associated with master cylinders, about half were due to insufficient fluid in the reservoir. Thus, while defects in the operation of the master cylinder cannot be identified by visual inspection, more attention could readily be paid to checking fluid levels.

28. Corrosion of pipes and components occurred more frequently on higher mileage vehicles, particularly those having covered over 40,000 miles; if this is regarded as an acceptable service life, then the finding lends support to the recommendations given by manufacturers regarding replacement of pipes etc.

29. Brake fluid was not found to present a serious problem, relatively few vehicles having unsatisfactory water, oil or dirt content. No cases of vapour lock were known to have occurred in any of the accidents covered by the on-the-spot studies.

30. Despite the large number of defects observed, only 16 of the 436 accident involved vehicles examined were known to have brake defects which had contributed to their accidents. Overall in the whole on-the-spot survey brake defects were judged to be a main contributory factor in only 3% of accidents. However, it is likely that brake defects contributed in part to the occurrence of other accidents without this being identified. It may also be that more accidents do not occur due to such defects because drivers modify their driving behaviour to accommodate their inadequate brakes. Nevertheless, as has been indicated, there are a number of areas where more regular checking as part of the normal servicing of the vehicle would lead to improvements in brake maintenance and reduction in numbers of defects.

Table 9

Contributory brake defects

110 contributory brake defects in 16 vehicles in 16 accidents

	No. of defects	No. of vehicles
Faulty adjustment	34	14
Lining defects	29	9
Disc/drum defects	14	9
Wheel cylinder defects	15	8
Master cylinder defects	5	4
Other defects	13	6
Total	110	16

The vehicles had an additional 42 non-contributory defects

© IMechE 1976

ACKNOWLEDGEMENTS

The work described in this paper forms part of the programme of the Transport and Road Research Laboratory and the paper is published by permission of the Director.

The author wishes to thank the staff of Automotive Products Ltd, Burmah-Castrol Ltd and the various garages where the vehicles were examined, the Chief Constable of Thames Valley Police, the owners who allowed their vehicles to be examined and his colleagues in the Accident Investigation Division of the Transport and Road Research Laboratory who helped in carrying out the survey.

REFERENCES

1. SABEY, B.E. and G.C. STAUGHTON. Interacting roles of road environment vehicle and road user in accidents. Paper presented to the Vth International Conference of the IAATM, London, September 1975.

Crown copyright 1975. Any views expressed in this paper are not necessarily those of the Department of the Environment. Extracts from the text may be reproduced, except for commercial purposes, provided the source is acknowledged.

© IMechE 1976

©IMechE 1976

C46/76

DEFECTS IN ROAD VEHICLE BRAKING SYSTEMS – EXPERIENCE FROM STATUTORY INSPECTIONS

F. WHALLEY CEng, MIMechE
North Western Traffic Area, Department of the Environment

The MS of this paper was received at the Institution on 1 October 1975 and accepted for publication on 12 November 1975

SYNOPSIS This paper relates to the nature and incidence of defects found in automotive braking systems during statutory examinations of different classes of road vehicles. The methods of inspection are described along with the equipment used. Results are given which compare the incidence of brake defects at the various kinds of check and indicate the respective prevalence of individual categories of such defects found during some of the examinations.

INTRODUCTION

Information concerning the maintenance and reliability of various vehicle components including braking systems is collated by the Department of the Environment from the results of examinations carried out by its Vehicle Inspection Division at purpose built testing stations and elsewhere. Further information is available from the results of statutory tests on private cars and light goods vehicles undertaken by specially authorised commercial garages.

The types of examination carried out on the three main classes of road vehicle can be broadly categorised as follows:

(1) Heavy Goods Vehicles

- (a) Statutory annual tests at Government testing stations.
- (b) Roadside checks at specially selected sites.
- (c) Inspections in connection with the issue of licences to operate goods vehicles carried out either at Government testing stations or at premises made available by the vehicle owner.

(2) Light Goods Vehicles and Private Passenger Carrying Vehicles

- (a) Statutory annual test of vehicles over 3 years old carried out by commercial garages.
- (b) Roadside checks employing a portable test facility.

(3) Public Service Vehicles

- (a) Examinations for the initial grant, or renewal after 7 years or less, of a Certificate of Fitness. Such examinations are normally carried out at the manufacturer's works or the operator's premises respectively.
- (b) Periodic pre-notified inspections or spot checks, carried out usually at the operator's premises, at intervals not exceeding one year.

In addition to these routine examinations, 'in depth' investigations into road accidents and mechanical failures provide a further source of information concerning the performance and reliability of vehicles in service. The results of such investigations however, are not covered by this paper.

In all types of official examination the standards employed to assess the serviceability of vehicle parts and braking systems in particular, are based on the requirements of the Motor Vehicles (Construction and Use) Regulations. In the case of brakes, the Regulations require that a prescribed minimum braking force or rate of retardation shall be achieved; also that every part of the braking system, including its means of operation, shall be maintained in good and efficient working order and shall be properly adjusted. The prescribed minimum rates of retardation are generally:

Service brake	50% g
Secondary brake	25% g
Parking brake	16% g

Braking performance at Heavy Goods Vehicle Testing Stations is normally assessed by means of a power driven roller brake testing machine (Fig 1) which measures the reaction at the rollers to the braking force at each wheel. Total braking efficiency is then calculated from the expression

$$\text{Efficiency} = \frac{\text{summation of braking force at each wheel}}{\text{gross permissible weight of vehicle}} \times 100$$

© IMechE 1976

It should be noted that, whilst many vehicles are presented for test unladen, the braking performance is assessed in relation to the maximum permitted gross weight of the vehicle when laden. In the case of heavy goods vehicles the maximum permitted gross weight is displayed on a plate which the vehicle is required to carry, whilst for private cars and light commercial vehicles lists are maintained at each testing centre equipped with a roller brake tester, showing the appropriate weights for each model. When testing public service vehicles the gross weight is assumed to be the unladen weight plus the weight of fuel and passengers at 63.5 kg per person.

Due to the very high laden to unladen axle weight ratios on certain types of goods vehicles and the practical limits imposed by the coefficient of friction between the rollers of the brake testing machine and the tyres of the vehicle, it is often impossible to achieve even the minimum prescribed braking efficiency if the vehicle is tested unladen. Accordingly, a load simulator has been developed for use at Heavy Goods Testing Stations by means of which an imposed load of up to approximately 5 tonnes f can be applied to the axle undergoing test (Fig 2).

For roadside checks or vehicle inspections carried out at the operator's premises it is normally impracticable to conduct brake performance tests by means of a roller brake test machine. In such cases therefore, this method is replaced by a dynamic test using a conventional pendulum type decelerometer. Whilst the latter type of instrument has certain limitations, the results it provides are considered to be acceptable for this purpose when interpreted by experienced inspectors.

The extent of compliance with the general legal requirement, referred to previously, concerning the state of maintenance and adjustment of the brakes is assessed by a visual inspection accompanied where appropriate by handling of individual components during actuation of the braking system. The scope of, and standards of acceptance for the annual test of Heavy Goods Vehicles are contained in a Manual published by the Department of the Environment, a typical page from which is reproduced at Fig 3. Conformity with the standards laid down in this manual is assured by adequate staff training accompanied by a comprehensive system of physical and statistical standards control.

No such precise standards of acceptance as for the Heavy Goods Vehicle Testing Scheme are yet prescribed for inspections of other classes of road vehicle, or for examinations of goods vehicles for purposes other than the annual test. However references in this paper to the incidence of defects for all classes of vehicle are based on the categories and standards employed in the Heavy Goods Vehicle Tester's Manual. The relevant items in the manual are as follows:

- Item 34 Low air pressure or vacuum warning device
- Item 35 Rate of build-up of air pressure or vacuum
- Item 36 Hand levers controlling mechanical braking systems, including pawl mechanism
- Item 37 Service brake pedal
- Item 38 Service brake operation
- Item 39 Hand operated air pressure or vacuum brake control valves
- Item 59 Mechanical brake components including operating rods, levers etc and brake drum/shoe assemblies
- Item 60 Brake wheel actuator units
- Item 61 Brake pipes, master cylinders, reservoirs, valves and connections
- Item 70 Condition of trailer parking brake
- Item 71 Performance of service brake
- Item 72 Performance of secondary brake
- Item 73 Performance of parking brake

RESULTS FROM HEAVY GOODS VEHICLE TESTING STATIONS

Statistics obtained from a random sample of 32,500 tests carried out during 1975 indicate that approximately 36.5% of all defects which result in failure to receive a pass certificate are concerned with brakes. A more detailed analysis of the reasons for rejection in this sample is shown at Fig 4 from which it will be seen that low service and/or secondary brake performance is one of the most frequent causes of failure. Experience indicates that the majority of the defects under this heading are in the brake shoe/drum assemblies and it is evident that rejection in many cases is due to improper bedding-in of the linings and/or adjustment of the brake-shoes.

A further significant reason for failure concerns brake pipes, reservoirs and associated connections (category 61). Air leaks account for approximately 44% of all rejections in this category, the remainder being concerned principally with inadequate support for, and in some cases resultant chafing of the brake pipes. Only 5% of test failures under this heading concern hydraulic fluid leaks.

Defects in mechanical brake components (category 59) also figure prominently in the reasons for rejection at annual test. A random analysis of 100 tests in which defective brakes was one of the reasons for rejection indicated that 35% of all brake defects fall under this heading. Within this total 28% of the defects involved excessive wear in, or malfunctioning of, the compensator mechanism at the rear axle/s whilst a further 36% concerned improperly secured rod and/or cable connections.

The examination procedure at annual test does not facilitate checking the integrity of divided braking systems unless, by operating the brake pedal, the drop in air pressure in the respective parts of the system is apparent from gauges on the instrument panel. However, a special investigation of 100 vehicles equipped with air or air assisted brakes revealed 9 cases of defects in the non-return valves which should have isolated the secondary system air supply in the event of a failure in the service brake system. In only 3 of these 9 cases could the defective valve be detected by operating tha brake pedal. In the remaining 6 cases the

© IMechE 1976

defect could only be found by draining air from the service reservoir. The provision for draining the reservoirs in some instances could not be readily employed and it was considered that the difficulty of the task would not promote regular checking of this item during normal preventive maintenance inspections.

To obtain more detailed information on the condition of load sensing valves than is normally available from the annual test results a special survey was carried out to determine the serviceability of these components on 400 vehicles undergoing test. The results showed that 44% of all load sensing valves checked were defective in some respect. A closer investigation of the load sensing valves fitted on a proportion of these vehicles revealed that 30% of the valves were operating correctly, 50% were not functioning at all or only partially limiting the braking effort and 13% were potentially dangerous due to irregular limitation of air pressure from one brake application to another. The remaining 7% of load sensing devices had been deliberately disconnected before the test. Excessive wear, or failure or maladjustment of the operating linkage was the prime cause of defective valves on 60% of the vehicles in the proportional sample.

RESULTS FROM CHECKS ON HEAVY GOODS VEHICLES CARRIED OUT AT THE ROADSIDE AND OPERATOR'S PREMISES

Due to the limited equipment which can be employed to facilitate vehicle examinations at the roadside, such checks are of necessity confined to those safety critical items which can be readily checked without special facilities. So far as the braking system is concerned, all external parts are visually examined while the hand and foot controls are being operated as necessary, and a check is usually made on such items as compressor build-up times, maximum gauge pressures and the operation of low pressure warning devices. A brake performance test employing a decelerometer is then carried out where appropriate. The vehicles inspected are a random selection of those passing the check point and, as they have received no special preparation, their condition can be taken as representative of the condition of vehicles generally operating in that locality. The procedure for inspections carried out at operator's premises and commercial garages is similar to that for roadside checks except that a facility for under-vehicle inspection is usually available.

An analysis of inspections carried out as described either at the roadside or on private premises in a sample of 2,000 vehicles during 1975 revealed that 29% of all defects observed concerned some aspect of the braking system. It should be pointed out that the percentage of vehicles rendered defective due to brake faults was somewhat lower than this figure however, because some of the vehicles inspected were suffering from multiple brake defects.

Fig 5 shows a breakdown by age groups of the incidence in the sample of vehicles defective due to brake faults compared with the total incidence of defective vehicles. It will be evident from this analysis that the percentage of vehicles found to be defective as regards brakes increased at a fairly uniform rate for vehicles up to 4 years old but no further increase was apparent beyond this age. Whilst it is generally accepted that some increase in the level of defects may be expected as a vehicle gets older and that ultimately, with normal maintenance, a point must be reached when the incidence of defects levels off, it is not readily apparent why such a marked change should occur at the 1971 age point.

A more detailed analysis of the 2,000 vehicle sample relating categories of brake defect to the age of the vehicle is shown at Table I. It is apparent from this survey that defects in the brake component categories represented a considerably greater proportion of the total brake defects than in the results from Heavy Goods Testing Stations. Conversely defects in the brake performance categories in the sample represented a much smaller proportion of the total than at the annual test. The reversals in the order of these results can probably be explained in the case of the brake performance defects by the fact that the method of test used at the roadside or at operator's premises is less searching than in Heavy Goods Testing Stations. The reason for the greater incidence of brake component defects found at inspections other than the annual test is more obscure however, but it could indicate some variance in the standards of acceptance for these items or, more likely, careful attention to the rectification of any defects in these categories before the vehicle is submitted for annual test.

The practical limitations on the scope and depth of inspections carried out at the roadside and at most commercial premises preclude any reliable comparison between the condition of vehicles operating generally and those presented for annual test. However, such a comparison was made by diverting a sample of 960 vehicles selected at random from roadside checkpoints to Heavy Goods Testing Stations where they were examined in the manner and to the standards of the annual test. The results are summarised in Table II which shows that whilst vehicles in service were suffering from more defects generally, the total incidence of brake defects was less than on vehicles presented for annual test. It is interesting to note the significantly lower incidence of brake performance defects on vehicles in service which bears out the experience at Heavy Goods Stations referred to previously that many test failures are due to improper bedding-in and/or adjustment of the brake-shoes in the course of preparation for test.

RESULTS FROM STATUTORY EXAMINATIONS OF PRIVATE CARS AND LIGHT GOODS VEHICLES

Information available from the results of annual tests conducted by commercially operated testing stations is not sufficiently detailed to analyse the precise reason for failure in each case. However a random survey of 32,110 tests showed that 20.4% of the vehicles examined failed the test because of brake faults although some of these vehicles had other defects which themselves would have caused rejection.

More detailed information concerning the condition of private cars is available from a special check carried out by the Vehicle Inspection Division of the Department of the Environment in which 333 private cars were examined to the same standards as for the annual test for this class of vehicle. The results showed a total of 343 mechanical defects of which 39.7% concerned some part of the braking system. An analysis of the latter defects is given at Fig 6.

© IMechE 1976

RESULTS FROM THE EXAMINATION OF PUBLIC SERVICE VEHICLES

All buses and coaches are examined by Statutory Officers of the Department of the Environment before being granted a Certificate of Fitness to carry fare paying passengers. The inspection for the initial certificate is normally carried out at the chassis manufacturer's or the bodybuilder's works and the results of these examinations provide a commentary on the condition of this class of vehicle when new.

A survey of defects found on 1,081 new coaches incorporating several different makes of chassis revealed an average of approximately 6 mechanical defects per vehicle of which 21.25% concerned some part of the braking system. In isolated instances the brake defects were clearly as a result of work in connection with mounting the body. The majority, however, appeared to have been present when the chassis was delivered to the bodybuilders. A summary of the incidence of such defects under the various categories of brake components is shown at Fig 7. It will be noted that none of the vehicles failed to meet the statutory requirements concerning any aspect of braking performance. Faulty routing or security of brake pipes and leaks at brake valves or pipe connections (category 61) however were responsible for almost 10% of all chassis defects.

Inspections for the renewal of Certificates of Fitness at internals of up to 7 years and other routine pre-notified inspections at about one yearly intervals are supplemented by a system of random spot checking which covers approximately 20% of all public service vehicles each year. Generally, these inspections are carried out at the vehicle operator's premises without the aid of specialised equipment other than a means for under-vehicle examination and a pendulum type decelerometer.

Because many defects are rectified during the course of these examinations it is not always necessary to record all the faults located; as a result the statistics available for defects on operational public service vehicles do not accurately reflect the results of the examinations. However in a sample of 12,227 inspections of public service vehicles carried out in the North Western Traffic Area during 1974/75, 5.24% of the vehicles had to be suspended from service because of serious brake defects and the total of such defects accounted for 50.52% of all chassis defects. Fig 8 gives a breakdown of the brake faults located in this sample according to the categories in the Heavy Goods Vehicle Tester's Manual and Table III compares the results of these inspections with the results of the annual test of heavy goods vehicles. It will be noted from Table III that public service vehicles suffered a greater incidence of brake defects in almost every category than goods vehicles but the incidence of brake performance defects was similar for both classes of vehicle. In this connection however it should be borne in mind that braking tests on public service vehicles are usually conducted by decelerometer. If the tests were conducted by a roller brake tester it is likely that the incidence of brake performance defects recorded against public service vehicles would be much higher.

CONCLUSIONS

It is evident that brake defects constitute a large proportion of the total mechanical defects on all classes of road vehicle. The results of the annual test on heavy goods vehicles show that low braking efficiency is one of the prime reasons for rejection at the test although it seems likely that many defects in this category are due to poor preparation for the test rather than being typical of the heavy goods vehicle population at large. It is noteworthy that operational public service vehicles which usually are also specially prepared for examination are frequently found to be defective in the same respect.

Mechanical brake components along with brake pipes and reservoirs account for a large proportion of defects on all classes of vehicle. Whilst undoubtedly many of these defects could be prevented on vehicles in service by more diligent maintenance, it appears that greater attention by manufacturers to the detailed design and/or construction of such items as mechanical brake connections, joints in pneumatic systems and the support of brake pipes would do much to reduce the prevalence of defects in these categories. Similarly, better provision for checking the integrity of divided brake systems and improvements in the design of load sensing valves appear to be matters requiring more attention by manufacturers.

The paper indicates that statutory vehicle inspections particularly in respect of brakes play an important role in securing a standard of vehicle condition compatible with the requirements of safety and the law. It may be however, that even greater compliance with these requirements would be achieved if the scope and depth of all inspections were aligned more closely with the standards and procedures adopted in Heavy Goods Testing Stations.

ACKNOWLEDGEMENTS

The author is indebted to the Department of the Environment for permission to publish the information contained in this paper and to the Chief Mechanical Engineer and members of the Vehicle Inspection Division for their valuable assistance during the course of its preparation.

©Crown Copyright 1975

TESTER'S MANUAL ITEM	PERCENTAGE OF TOTAL BRAKE DEFECTS BY VEHICLE AGE GROUPS IN GOODS VEHICLE INSPECTIONS OTHER THAN AT HEAVY GOODS ANNUAL TEST								
	1975	1974	1973	1972	1971	1970	1969	1968	Pre 1968
Air/Vac Warning Device	-	-	-	-	-	4.17	3.45	8.33	8.33
Air/Vac Build-up	-	-	-	-	-	-	-	-	-
Mech Brake Hand Levers	-	20.0	27.27	-	15.0	4.17	10.34	33.3	-
Service Brake Pedal	-	-	-	-	-	4.17	3.45	-	16.67
Service Brake Operation	-	-	-	-	-	-	-	-	8.33
Air/Vac Hand Controls	-	20.0	18.18	15.79	2.5	4.17	-	8.33	-
Mech Brake Components	-	-	9.09	15.79	17.5	8.33	13.79	8.33	25.0
Brake Wheel Units	-	-	9.09	10.53	10.0	16.67	13.79	-	8.33
Brake Pipes and Reservoirs	-	60.0	27.27	31.58	52.5	54.17	44.83	25.0	16.67
Trailer Parking Brake	-	-	-	-	-	-	-	-	12.5
Service Brake Performance	-	-	-	-	-	4.17	6.9	8.33	-
Secondary Brake Performance	-	-	-	5.26	-	-	-	-	-
Parking Brake Performance	-	-	9.09	21.05	2.5	-	3.45	8.33	4.17

TABLE I

DEFECT CATEGORY	% DEFECTS AT ANNUAL TEST	% DEFECT IN NORMAL SERVICE
All defects	63.5	74.2
All brake defects	36.5	25.8
Air/vac warning	0.75	0.76
Build up of air/vac	0.21	0.18
Mech. brake hand levers	0.85	1.49
Service brake pedal	0.30	0.55
Service brake operation	0.14	0.18
Air/vac hand controls	0.57	0.40
Mech. brake components	4.05	3.2
Brake wheel units	1.14	1.02
Brake pipes reservoirs etc	6.93	4.65
Trailer parking brake	-	0.36
Service brake performance	8.91	4.87
Secondary brake performance	7.27	4.03
Parking brake performance	5.3	4.11

COMPARISON BETWEEN BRAKE DEFECTS AT ANNUAL TEST AND DURING NORMAL SERVICE

TABLE II

© IMechE 1976

BRAKE DEFECTS CATEGORY	% DEFECTS IN GOODS VEHICLES AT ANNUAL TEST	% DEFECTS IN PUBLIC SERVICE VEHICLES DURING ROUTINE EXAMINATIONS
All brake defects	36.5	50.52
Air/vac warning	0.75	1.43
Build-up of air/vac	0.21	0.80
Mech. brake hand levers	0.85	1.69
Service brake pedal	0.30	0.89
Service brake operation	0.14	0.27
Air/vac hand controls	0.57	0.71
Mech. brake components	4.05	7.84
Brake wheel units	1.14	5.26
Brake pipes reservoir etc	6.93	8.91
Service brake performance	8.91	8.11
Secondary brake performance	7.27	9.89
Parking brake performance	5.3	4.72

COMPARISON BETWEEN DEFECTS ON GOODS VEHICLES AT ANNUAL TEST AND PUBLIC SERVICE VEHICLES PRESENTED FOR ROUTINE EXAMINATION

TABLE III

Fig. 1: Power driven roller brake testing machine.

Fig. 2: Load Simulator.

Method of Inspection

Examine the mechanical components of the brakes which can be seen without dismantling, looking particularly for:

(a) Badly chafed rods or levers.
(b) Frayed cables.
(c) Corroded rods, levers or linkages.
(d) Wear in rod or cable ends.
(e) Wear in eyes of relay levers.
(f) Wear in clevis pins, bolts, stationary pins, pivots.
(g) Absence of locking devices on clevis pins.
(h) Thin brake linings.
(i) Fractures in brake drums.
(j) Any restriction of the free movement of the system.
(k) Any abnormal movement of levers indicating maladjustment.
(l) Security of brake back plates or discs (Including transmission brakes). ◀

Reasons for Rejection

1 Brake rods reduced in diameter by more than ⅓rd of original dimensions.
2 Cables so heavily corroded or with strands broken to such an extent that the strength is reduced significantly.
3 Serious reduction in strength of any component due to excessive wear.
4 The absence or insecurity of locking devices.
5 Brake linings less than $\frac{1}{16}$" thick at any one point.
6 Fracture in a brake drum.
7 Any restriction to the free movement of the system.
8 Brake actuating levers operating over centre.
9 Any abnormal movement of levers indicating maladjustment.
10 Insecure brake back plate. ◀
11 Insecure brake disc. ◀

59

Mechanical brake components

Motor Vehicles Part I No. 21
Trailers Part II No. 8

Fig. 3: Method of Inspection of Mechanical Brake Components.

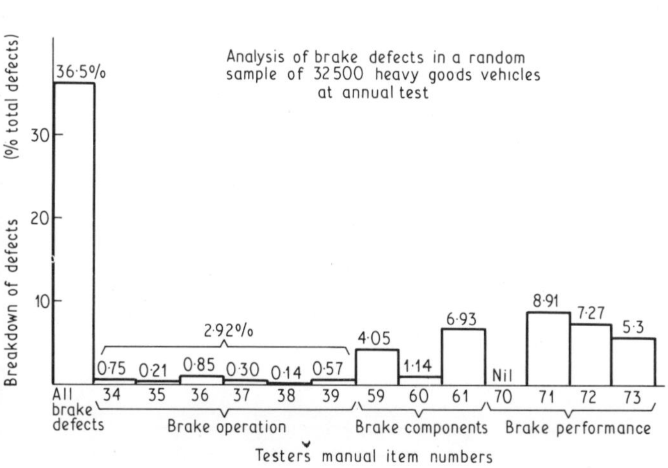

Fig. 4: Analysis of brake defects in a random sample of 32 500 heavy goods vehicles at annual test.

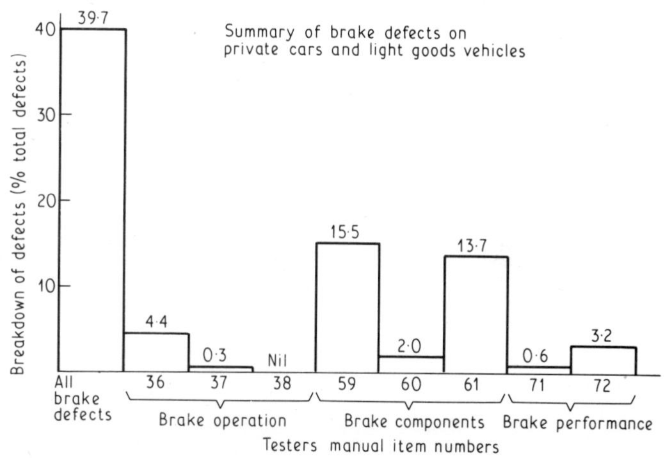

Fig. 6: Summary of brake defects on private cars and light goods vehicles.

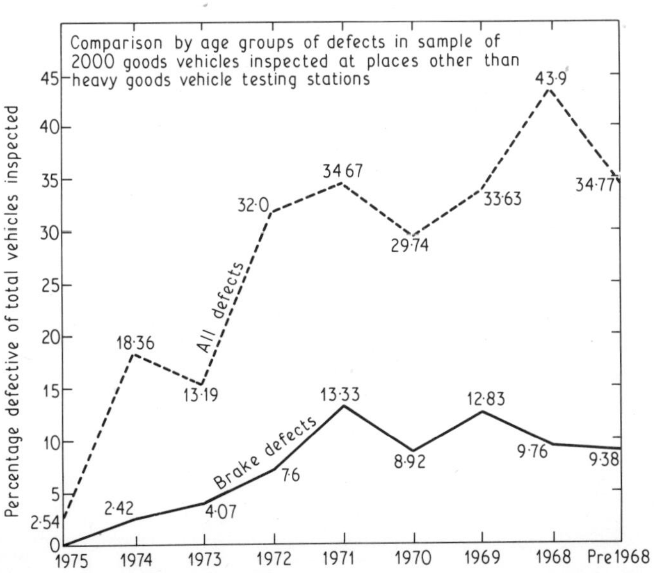

Fig. 5: Comparison by age groups of defects in sample of 2000 goods vehicles.

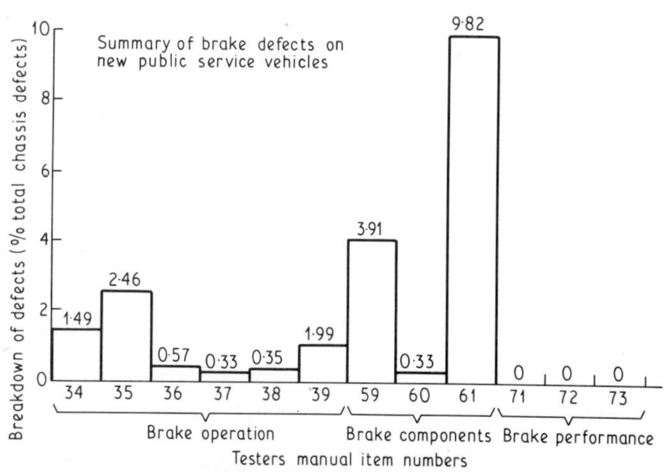

Fig. 7: Summary of brake defects on new Public Service Vehicles.

© IMechE 1976

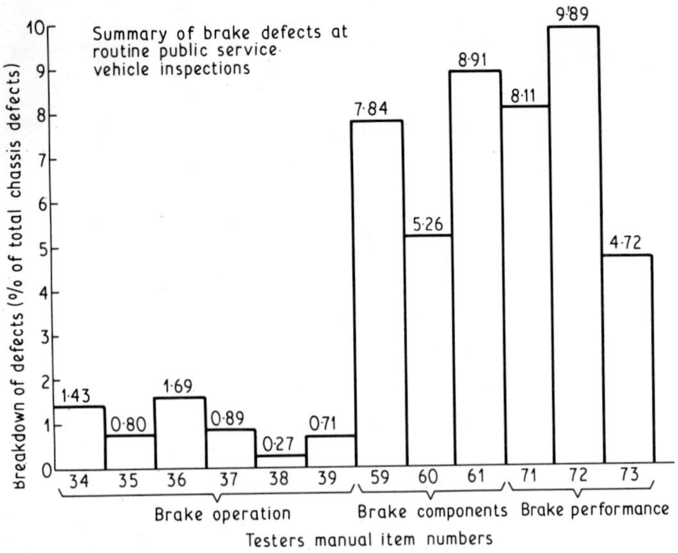

Fig. 8: Summary of brake defects at routine Public Service Vehicle Inspection.

© IMechE 1976

C47/76

THEORY OF RELIABILITY AND ITS APPLICATION TO VEHICLE BRAKES AND BRAKING SYSTEMS

Jovan TODOROVIĆ, dipl.ing.
Faculty of Mechanical Engineering, Belgrade University

The MS of this paper was received at the Institution on 22 July 1975 and accepted for publication on 22 October 1975

SYNOPSIS The application of Reliability Theory in the field of vehicle braking systems requires introduction of certain supplements in the basic theory in order to include their specific characteristics. The main problem is the derivation of adequate reliability function models which should be based on analyses of system construction and types of failures. Besides, owing to the system's complexity, it is highly advisable to analyze those elements only whose criticality is the greatest. The paper gives some recommendations about the way to solve these two problems. Some results obtained to this date in the investigation of braking system reliability are presented as well.

INTRODUCTION

1. Braking systems and car braking have long been a subject of comprehensive researches. The central point of these investigations is primarily connected with brake performances. Braking system reliability, however, received much less attention so far. One of the reasons is the lack of appropriate theoretical foundations. The development of Reliability Theory settles this basic problem to a great extent.

BASIC POSTULATES OF RELIABILITY THEORY

2. Theory of reliability is a science that investigates the regularity of failure phenomena in technical systems (Ref. 1 and 2). A start is made with the assumption that reliability is the property of a system to work without failure under specified conditions and over a specified period of time (Ref.3). Reliability is expressed through probability and its fundamental definition may be shown in the form

$$R(t) = P(t) = \frac{N_d(t)}{N_0} \qquad (1)$$

where $R(t)$ = reliability function,
$P(t)$ = probability of operation without failure,
$N_d(t)$ = number of faultless elements at instant t,
N_o = total number of elements observed (for t = 0).

The reliability function is graphically shown in Fig. 1a.

3. The basic definition of reliability makes it possible to determine other principles that define reliability in an indirect manner. In engineering practice, special importance is given to failure frequency and hazard rate functions, which are defined in the form,

$$f(t) = -\frac{dR(t)}{dt}\ ;\ \lambda(t) = \frac{f(t)}{R(t)} \qquad (2)$$

A graphic representation of the above functions is shown in Figs. 1b and 1c. As can be seen, these graphs have very characteristic shapes with three definite periods of time (I = early failures, II = normal usage, and III = weakening of materials).

4. Referring to the concept of failure, it is generally accepted that the term means such an event in the life of a system after which the system's output pass beyond some allowable limits. It is logical that this broad treatment of the concept of failure does not permit a sufficiently precise analysis of reliability. It would be helpful, therefore, to classify the possible failures of a system. How to classify failures in each individual case, that is, what failures are to be analyzed separately - depends on a series of circumstances and influences. The greatest importance is attached to the kind of system under observation and to its usage conditions.

5. By the term "system" Theory of Reliability means a group of elements united to perform a particular function. Connections between the elements may be series, parallel and combined ones (Ref.4).

6. Series connections (Fig. 2a) are characterized by the fact that failure of any one element results in failure of the system as a whole. Considering the reliability of element R_i, the system's reliability will have the value

$$R_s = (R_i)^n \qquad (3)$$

On the other hand, a system with parallel connections will not fail to operate even

© IMechE 1976

if a failure occurs in one of the elements in the parallel branch. The following is valid for the diagram in Fig. 2b:

$$R_S = 1-\prod_{i=1}^{3}(1-R_1)_i = 1-(1-R_1)^3 \qquad (4)$$

Combined, or parallel-series systems are obtained by combining the two basic techniques. The reliability of the system shown in Fig. 2c can be expressed in the form

$$R_S = R_1\left[1-(1-R_2)^2\right]\cdot R_3\cdots R_m = R_1^{n-1}\left[1-(1-R_2)^2\right] \qquad (5)$$

SPECIFIC CHARACTERISTICS OF BRAKING SYSTEMS

7. From the viewpoint of Reliability Theory, vehicle braking systems are characterized by numerous specific features. This particularly refers to the type of failure, system construction and differences in the significance of individual elements regarding the performances required (Ref. 5).

Braking system failures - element reliability

8. The failures of vehicle braking systems differ in type and character. In the first place, one should distinguish failures requiring replacement of elements from those that may be eliminated by adjustment or any other form of repair. In addition, it is highly important to consider not only failures resulting in breakdowns, but all kinds of malfunctioning as well. Namely, a decrease in braking system effectiveness may have a permanent nature, or may be of transient character. In the latter case, the failure will disappear all by itself, or spontaneously, after a while. This is well-known decrease in frictional properties or fading, which is due to high working temperatures, penetration of water, grit, and the like.

9. Therefore, any analysis of braking system reliability should take into account the following types of failures (Ref.6):

- According to character of failure: breakdowns and malfunctioning,
- According to manner of repair: failure that are eliminated by replacement of parts, by adjustment and transient failures.

10. To clarify this point, let us consider the possibility of defining the reliability function for one of the most important element of a braking system - the brake lining. It should be noted here that, above all, breakdowns and malfunctioning are mutually dependent events from the point of view of probability theory. This means that the brake lining reliability must be determined in relation to the two kinds of failures on the basis of conditional probabilities of events A and B, where A means that the lining is not inoperative and B denotes that the lining is not malfunctioning:

$$R_o(t) = P(t) = P(A\cdot B) = P(A)\cdot P(B/A) = R_k(t)\cdot R_p(t) \qquad (6)$$

In the expression (6) $R_o(t)$ indicates the lining reliability, P(A), indicates the probability that the lining has not failed completely, while P(B/A) shows the probability that the lining has not begun to malfunction, provided it has not failed completely by this moment. $R_k(t)$ and $R_p(t)$ represent conditional reliabities with regard to breakdowns and functional disturbances, respectively.

11. In an analogous way, a reliability function can be derived for the lining with relation to the differentiation of failures from the standpoint of their elimination. If, in this case, A denotes that the lining need not be replaced, B, that the lining requires no adjustment, and C, the absence of any transient failure. the reliability function can be written in the form,

$$R_o(t) = P(A)\cdot P(B/A)\cdot P(C/A\cdot B) = R_c(t)\cdot R_a(t)\cdot R_f(t) \qquad (7)$$

Element connections - general function of system reliability

12. The braking system is by its structure a complex technical system, composed of service, emergency and parking brakes (sometimes a retarder as well). Besides, there are various combinations in the design of service, emergency and parking brakes, especially when their mutual connections are concerned. Therefore, a braking system should be regarded as a set of several complex elements, interconnected in some specific ways.

13. It is known that the parking brake, besides its basic function, generally has the function of an emergency brake. This makes it necessary to have some parallel connections of service and emergency brake elements. Traffic legislation, however, allows the emergency brake performances to be lower than those of the service brake (Ref. 7). In accordance with the general definitions of failure and the basic principles of Reliability Theory, the emergency brake, therefore, has not the characteristics of a true reserve brake and such connections cannot be treated as parallel (Ref.4). However, these connections have a positive effect on vehicle braking, especially from the viewpoint of traffic safety. Therefore, these "quasi-parallel" connections (Ref.6) should not be neglected, but must be defined separately in a certain way.

14. This can be best explained by an example given in Fig. 3a. As it is seen, the entire braking system is divided into six elements, connected partly in a serial manner and partly in a quasi-parallel way. To determine the system's total reliability one must start from the fact that in case of failure on the service brake branch the emergency brake action still secures some deceleration, but on a lower level than the service brake. Let us indicate this reduction by the quantity K_{fp}, which has a value within the interval $0 < K_{fp} < 1$. In this way, and on the basis of random events algebra and the corresponding laws of proba-

© IMechE 1976

bility (Ref.8), the total reliability of a braking system may be determined as a probability of faultless operation, i.e. probability of realization of the favourable event E, that is $R_s = P(E)$. It should be emphasized here that E is a random event which represents a synonym for the good condition of the braking system. Event E can be realized by proper operation of either the service brake or the emergency brake, i.e., it comprises two random events E_1 and E_2. This dependence can be symbolically shown as a union of events E_1 and E_2 as follows:

$$E = E_1 \cup E_2 \tag{8}$$

15. On the other hand, the realization of E_i is conditioned by the accuracy of all series connected elements in the branch. Symbolically, this may be presented as a section of events E_{ii},

$$E_1 = E_{KS} \cap E_{TS} \cap E_{B1} \cap E_{B2} \qquad E_2 = E_{KP} \cap E_{TP} \cap E_{B2} \tag{9}$$

16. Then, one can write the following:

$$P(E) = P(E_1) + P(E_2) - P(E_1) \cdot P(E_2) \tag{10}$$

which is, on the basis of expression (9) and the given definition of coefficient K_{fp}, reduced to this:

$$P(E) = P(E_{KS}) \cdot P(E_{TS}) \cdot P(E_{B1}) \cdot P(E_{B2}) + P(E_{KP}) \cdot P(E_{TP}) \cdot P(E_{B2}) \cdot K_{fP} - P(E_{KS}) \cdot P(E_{TS}) \cdot P(E_{B1}) \cdot P(E_{B2}) \cdot P(E_{KP}) \cdot P(E_{TP}) \tag{11}$$

or

$$R_S = R_{B2} \left[1 - (1 - R_{KS} \cdot R_{TS} \cdot R_{B1})(1 - R_{KP} \cdot R_{TP}) - R_{KP} \cdot R_{TP}(1 - K_{fP}) \right]$$

The factor of performance decline K_{fp} can be most suitably determined as a ratio between the corresponding braking distances ($K_{fp} = s_e/s_s$, Ref. 9).

17. Braking systems of different construction can be analyzed in a similar manner. The scheme given in Fig. 3b and other typical system's construction are analyzed in Ref. 9.

Criticality of elements - simplified reliability function

18. To analyze the reliability of a braking system, the derived expressions for the general reliability function should be supplemented with corresponding reliability functions for each element. If this were done with all elements, that would result in very clumsy expressions. This would create considerable difficulty in practical work. We must only remember that comparatively simple braking systems of commercial vehicles contain over 1500 elements, 500 of which are different ones. It is therefore reasonable to include in the reliability function those elements only which have the greatest influence on the total reliability of the system, that is to say, the most critical elements.

19. The elements can be ranked by criticality taking into consideration different factors. These may be: an influence on the system's output (functional importance), suitability for bringing the element into good working order (accessibility for repair or replacement), the price of the elemet and its empirical level of reliability. All these factors, however, have different degrees of importance. In addition, each of them may present different relative levels, which are hardly comparable with one another. Theoretical treatment of these relationships is extremely hard, and even impossible.

20. Such being the case, the only solution of the problem is seen in a subjective, empirical ranking of elements. It must be mentioned here that this approach to the problem may have to stand varied criticism, but nevertheless it offers considerable advantages over other possible attempts. This method of attack has been elaborated in Ref. 10. According to the method, ranking is done by assigning points to the element, as follows:

- Functional importance F=10 - 30 points
- Empiric level of reliability R = 5 - 10
- Accessibility P = 1 - 5
- Repairability Z = 1 - 5
- Price of element C = 1 - 5 points.

The larger numeral indicates the less favourable condition which increases the element's criticality.

21. Considering the relationships among the individual factors, the developed method has introduced the following expression for criticality factor:

$$K = F \cdot R + (P + Z) \cdot C \tag{12}$$

It can be easily noticed that the criticality factor, defined in this way, contains the necessary logic.

22. The selection of elements that must be taken into account on the basis of criticality factors is best performed by use of the following expression:

$$K_c \geqslant K_m + \delta \tag{13}$$

where K_m = mean value of K for all the elements, and δ = standard square deviation.

23. As an example of this method, Fig. 4 shows part of a list of points for several elements having the highest and the lowest criticality factors. The example refers to a common drum brake, hydraulically operated, which contains 262 elements altogether (54 of which being different). After giving points to each element, the calculations produce K_m = 171 and δ = 57, so with $K_c \geqslant 228$ the total number of elements that should be treated as critical (and included in the reliability function) is only 24 (10 of which are different).

PURPOSE AND METHODS OF INVESTIGATING THE RELIABILITY OF BRAKING SYSTEMS

24. Reliability investigations are undertaken with various objects, but the general purpose will be to determine the failure distribution. In some cases, the point is to secure the greatest possible reliability, while in others the aim may be the

© IMechE 1976

most economical solution. In the given case of motor-vehicle braking systems the problems are far more complex. Reliability investigation must ensure a high level of reliability, i.e. higher road safety, and at the same time an economical design. In addition to increasing the reliability of the most critical elements, the investigation should therefore be directed towards developing a method of preventive replacement at the moment the element is still in working order but has reached a relatively high probability of failure at the following moment.

25. The basis of all reliability investigations methods is a precisely designed information system, which must provide all information on the system's operation, failures that take place (when, how, in what form), operating and enviromental conditions, etc. Where the existing systems are concerned, a well-organized investigation in service, with sufficiently large statistic sample (20 to 30 units for each characteristic operating condition) may produce information of great importance for making necessary decisions.

26. When the development of new systems or components is concerned, it is necessary to collect detailed information on working loads and conditions planned for the vehicle that is to be equipped with the new system. This can be accomplished by conducting appropriate tests of similar systems under corresponding operating conditions. Information obtained by these tests offers foundation for organizing laboratory reliability tests.

27. Laboratory tests of component's reliability is another major method of reliability investigation. These tests are also very delicate, particularly from the standpoint of simulating the actual working conditions, accelerating the testing procedure, and interpretating the results.

RESULTS OBTAINED TO THIS DATE IN THE INVESTIGATIONS OF BRAKING-SYSTEM RELIABILITY

28. As previously stated, no significant work has been done so far in the field of braking-system reliability, at least when the complex treatment of the system as a whole is concerned. However, the reliability of individual elements has been examined to a larger extent, the result being a broad fund of information and knowledge.

29. For the past few years much has been done in the area of organized investigations of motor-vehicle reliability in service (Ref. 11 and 12). So far, however, no detailed information on these investigations, especially considering braking systems, has been presented to the public. We only know that certain results were obtained in failure investigations with city buses in Madrid (Ref. 13), and with city buses and taxi-cabs in Belgrade (Ref. 14). It was shown in both cases that brake lining wear conformed with Weibull distribution. An illustration of results obtained in Ref. 14 is shown in Fig. 5 (Ref. 5). We should keep it in mind here that the parameters of Weibull distribution, and hence the reliability of lining, depend to a high degree both on working conditions and on lining quality. These results, therefore, cannot be generalized. The submission to Weibull distribution, however, may be estimated as a characteristic of the process of failure occurrence in this case.

30. On the other hand, braking systems have long been tested in actual usage conditions. Although these tests have not included all the parameters necessary to analyze braking system reliability, the results obtained are of great importance. The most significant contributions in this domain have been made by British researchers (primarily from the Ferodo firm, Ref. 15 to 18). Experimental determination of actual service loads on braking systems has also been conducted in the USA, USSR and some other countries (Ref. 19 to 21).

31. Comprehensive investigations of service loadings on the brakes of a 20-ton truck were carried out by the present author on carefully chosen road sections in the vicinity of Belgrade (Ref. 5). The road sections and other factors characterizing the operating conditions were chosen on the basis of previously conducted statistical investigations. On this basis, such information was obtained which has a general character and may be regarded as applicable to the braking systems of any similar commercial vehicle. This was made possible by introducing the concept of "standard road unit" (ref. 6). Table 1 and Figs. 6 to 8 present some characteristic information obtained in these investigations. It is essential to notice that all basic criteria of brake service loads prove to be in conformity with Weibull distribution, as shown in Table 2.

32. Most present-day laboratory reliability tests are based on working loads, established by testing under usage conditions. Criteria for evaluating the quality of individual elements are, however, mostly empirical. No adequate connections have been established yet between the actual life and the life obtained in laboratories. This is even better expressed with relation to the other forms of failures, especially transient failures.

33. At the Faculty of Mechanical Engineering in Belgrade a laboratory lining test procedure has been developed, based on the above-mentioned results of tests under usage conditions. Relations between the wear obtained in the laboratory and in the service were found out and defined, along with criteria for evaluating reliability from the viewpoint of transient failures due to thermal loads and penetration of water (Ref. 5 and 22). On the basis of such laboratory tests it was possible to evaluate the total reliability of a brake

© IMechE 1976

lining for definite working period under usage conditions. Although we still lack quite dependable information with which to appraise of reliability prediction based on such laboratory tests, the available information indicates that the developed method provides good conditions to attack this problem with relative success.

34. To evaluate the total reliability of braking system, organized investigations under usage conditions and corresponding laboratory tests should include the other elements, particularly those found to be most critical. So far, this has not been accomplished to a satisfactory degree. The developed methods of realibility analysis offer, however, good grounds for solving these questions as well.

CONCLUSIONS

35. The development of Reliability Theory gives a basis for a more dependable analysis of the total quality of braking systems and their components. To investigate the reliability of braking systems it is necessary to establish adequate reliability function models, based on careful analyses of construction methods, types of failures and other specific features of these systems. Besides, owing to the system's complexity and a large number of component parts, as well as to their different influences on the total function and the output of the system, it is highly advisable to analyze those elements only whose importance or criticality is the greatest. The paper gives some definite recommendations and ideas about the way to solve these two significant problems.

36. Once the adequately formulated reliability functions are on hand, reliability investigations of braking system elements should be turned in three basic directions: behaviour of the system in operation, actual working loads under usage conditions, and laboratory tests simulating the actual operating conditions. For each of these directions of research it will be necessary to develop such methods as will ensure most complete information, both in an absolute sense (for making direct decisions) and in a relative sense (regarding further use of the information throughout the reliability investigation project).

REFERENCES

1. BAZOVSKY, I. Reliability theory and practice, Prentice Hall, New Jersey
2. HAVILAND, R.P. Engineering reliability and long life design, Van Nostrand, Princeton, 1964
3. POLOVKO, A.M. The fundamentals of reliability theory (in Russian), Nauka, Moskva, 1964
4. LLOYD, D.K. LIPOW, M. Reliability: Management, methods and mathematics, Prentice Hall, New Jersey, 1964
5. TODOROVIĆ, J. Reliability of heavy vehicle brake linings under usage conditions (D.Sc. thesis, in Serbo-Croatian), Belgrade, 1972
6. TODOROVIĆ, J. An approach to braking system reliability, FISITA Congress, 1972
7. - Regulation ECE No.13: Uniform provision concerning the approval of vehicles with regard to braking, 1973
8. ROBERTS, N.H. Mathematical methods in reliability engineering, McGraw-Hill, New York, 1964
9. TODOROVIĆ, J. The braking system reliability function (in Serbo-Croatian), Conference "Science and Motor Vehicles 72", Belgrade, 1972
10. TODOROVIĆ, J., STANOJEVIĆ, T. A simplified method for analyzing the reliability of complex braking systems (in Serbo-Croatian), Conference "Science and Motor Vehicle 75", Belgrade, 1975
11. - Electronic data processing for a large or small fleet, SAE SP-295, 1967
12. INDIKT, A., KRIVENKO, I., SHERNJAININ, V.A., Reliability investigation in experimental transportation firms (in Russian), Avto.prom. 32/1966/12
13. ESTRADA SAIZ, S. Reliability in city transport vehicles. Int.Auto.Safety Conf., Detroit-Brussels, 1970
14. TODOROVIĆ, J., PAPIĆ, V., LENASI, J. Reliability investigation in large motor fleets (in Serbo-Croatian), Conference "Science and Motor Vehicles", Belgrade, 1971
15. CARPENTER, N. Some measurements of brake usage in a high-speed saloon-car, Proc. Inst.Mech.Engrs., Auto.Div., London, 1956
16. FROOD, A., MACKENZIE, D., NEWCOMB, P. Brake usage in a heavy vehicle, Proc.Inst. Mech. Engrs., Auto.Div., London, 1961
17. MACKENZIE, D., NEWCOMB, P., WATTON, B. European proving trial, Auto.Engr., No.1, 1966
18. SPURR, R., NEWCOMB, P. Testing of commercial vehicle brakes, FISITA Congress, 1968
19. ZECHEL, J., TURNER, L. Automobile brake evaluation under customer usage conditions, SAE SP-338, 1968
20. GUREVICH, V., GAPOAN, T., Energy loading on car braking systems (in Russian), Avto.prom. 36/1970/8
21. VALLIN, G. Legislative effect on brake design, SAE 680017, 1968
22. DEDOVIĆ, V., An investigation of brake lining transient failures relating to lining reliability (M.Sc. thesis, in Serbo-Croatian), Belgrade, 1975

© IMechE 1976

STANDARD ROAD SECTIONS	SIMBOL TYPE SURFACE QUALITY TRAFFIC CURVES DESCENTS		G SUBURB GOOD INTENSIVE MODERATE MODERATE	I HIGHWAY GOOD MODERATE MODERATE LOW	II FIRST RATE MODERATE INTENSIVE/MODERATE MODERATE MODERATE	III SECOND RATE MODERATE/POOR MODERATE/LOW MODERATE/SHARP STEEP
CHARACTERISTIC	MEAN SPEED DURING BRAKING $\left[kmh^{-1}\right]$	A*	12,08	42,80	26,05	31,02
		B*	8,76 ÷ 19,46	34,10 ÷ 55,30	22,49 ÷ 30,98	27,45 ÷ 35,68
	NUMBER OF BRAKE APPLICATIONS PER 1km $\left[km^{-1}\right]$	A	2,48	0,58	1,94	3,21
		B	1,46 ÷ 3,50	0,43 ÷ 0,73	1,43 ÷ 2,45	2,74 ÷ 3,68
	BRAKING TIME PER 1km $\left[s\,(km)^{-1}\right]$	A	9,94	1,78	9,38	16,80
		B	5,79 ÷ 14,09	1,36 ÷ 2,20	6,96 ÷ 11,80	14,36 ÷ 19,24
	BRAKING MOMENT PER UNIT WEIGHT $\left[kpm\,(Mp)^{-1}\right]$	A	63,00	39,10	46,30	37,30
		B	39,10 ÷ 86,90	30,07 ÷ 48,13	38,97 ÷ 53,63	32,44 ÷ 42,16
	RATE OF WORKING AT BRAKES PER UNIT WEIGHT $\left[HP\,(Mp)^{-1}\right]$	A	5,42	11,92	8,59	8,24
		B	4,27 ÷ 6,57	10,22 ÷ 13,62	7,51 ÷ 9,67	7,52 ÷ 8,96
	BRAKING ENERGY PER 1km AND PER UNIT WEIGHT $\left[kpm\,(km \cdot Mp)^{-1}\right]$	A	3885	1580	6050	9860
		B	2130 ÷ 5640	1164 ÷ 1996	4782 ÷ 7318	8550 ÷ 11170

A = MEAN MEASURED VALUE B = CONFINDENCE INTERVAL FOR P = 90%

NOTE: ALL TESTS ARE REALIZED WITHOUT EXHAUST BRAKE

Table 1

QUANTITY \ ROAD		G*	I*	II*	III*	"STANDARD ROAD UNIT"
RATE OF WORKING	β	1,25	1,45	1,35	1,75	1,55
	η	7,40	13,50	9,70	9,70	10,20
	m	6,90	12,25	8,90	8,62	9,18
	$F(P_i)$	$1-e^{-\left(\frac{P_i}{7,4}\right)^{1,25}}$	$1-e^{-\left(\frac{P_i}{13,50}\right)^{1,45}}$	$1-e^{-\left(\frac{P_i}{9,7}\right)^{1,35}}$	$1-e^{-\left(\frac{P_i}{9,7}\right)^{1,75}}$	$1-e^{-\left(\frac{P_i}{10,2}\right)^{1,55}}$
MEAN RATE OF WORKING	β	1,40	1,65	1,65	1,75	1,55
	η	5,9	10,50	8,20	7,60	8,0
	m	5,38	9,40	7,34	6,78	7,20
	$F(P_{sr})$	$1-e^{-\left(\frac{P_{sr}}{5,9}\right)^{1,40}}$	$1-e^{-\left(\frac{P_{sr}}{10,5}\right)^{1,65}}$	$1-e^{-\left(\frac{P_{sr}}{8,2}\right)^{1,65}}$	$1-e^{-\left(\frac{P_{sr}}{7,6}\right)^{1,75}}$	$1-e^{-\left(\frac{P_{sr}}{8}\right)^{1,55}}$
BRAKING ENERGY	β	1,00	1,12	1,10	1,05	1,10
	η	2,100	2,900	3,250	3,500	3,300
	m	2,100	2,800	3,140	3,430	3,200
	$F(A_k)$	$1-e^{-\left(\frac{A_k}{2,100}\right)^{1,00}}$	$1-e^{-\left(\frac{A_k}{2,900}\right)^{1,12}}$	$1-e^{-\left(\frac{A_k}{3,250}\right)^{1,10}}$	$1-e^{-\left(\frac{A_k}{3,500}\right)^{1,05}}$	$1-e^{-\left(\frac{A_k}{3,300}\right)^{1,10}}$
BRAKING TORQUE	β	2,15	2,00	1,85	1,75	1,80
	η	80	70	62	50	57
	m	72,50	53,50	62,00	43,50	50,70
	$F(M_k)$	$1-e^{-\left(\frac{M_k}{80}\right)^{2,15}}$	$1-e^{-\left(\frac{M_k}{70}\right)^{2,00}}$	$1-e^{-\left(\frac{M_k}{62}\right)^{1,85}}$	$1-e^{-\left(\frac{M_k}{50}\right)^{1,75}}$	$1-e^{-\left(\frac{M_k}{57}\right)^{1,80}}$

* FOR EXPLANATION SEE TABLE 1

Table 2

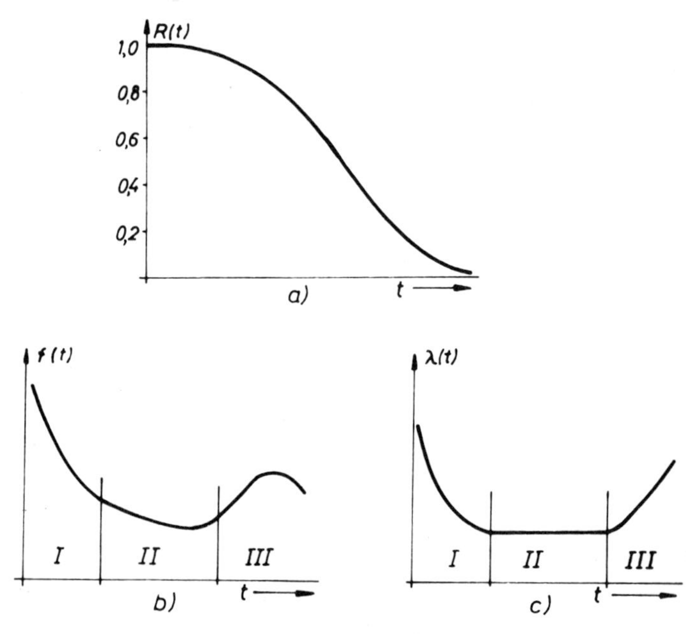

Fig. 1: Reliability Function

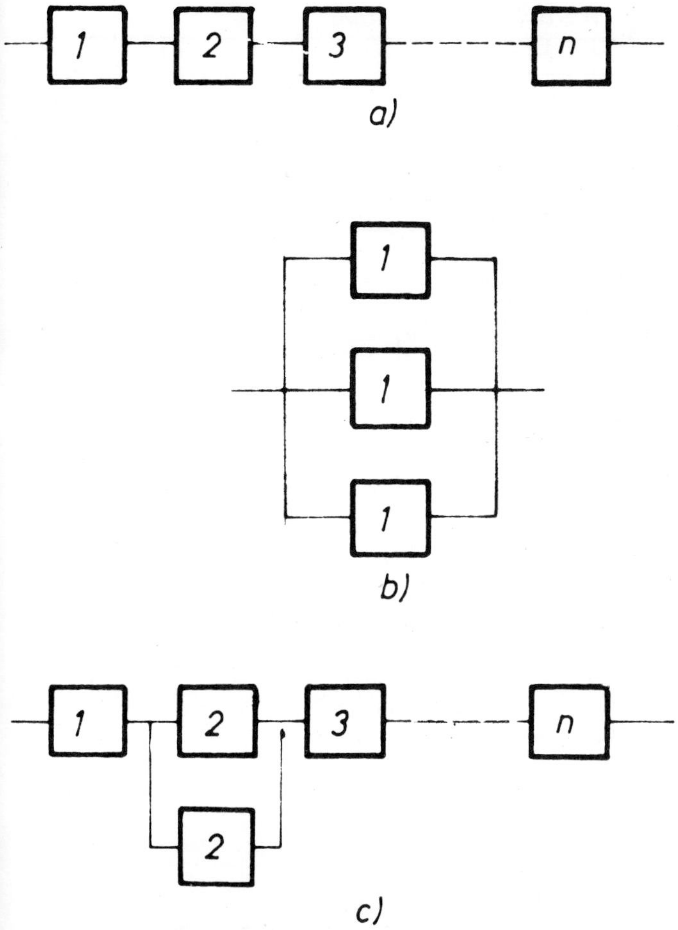

Fig. 2: Element Series Connections

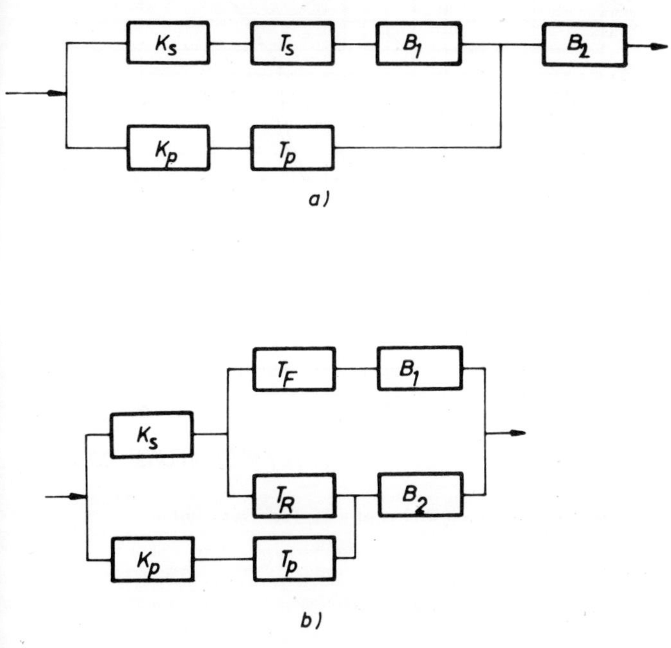

Fig. 3: Element Connections, Serial and Quasi-Parallel

DESCRIPTION	NO. REQD. PER BRAKE	FUNCTIONAL IMPORTANCE	RELIABILITY (EMPIRICAL)	ACCESSIBILITY	REPAIRABILITY	PRICE	CRITICALITY FACTOR
SHOE (LEADING)	1	30	7	4	4	4	242
LINING	1	30	10	4	5	2	318
DRUM	1	30	9	2	3	5	295
WHEEL CYLINDER	1	30	9	4	5	3	297
SEAL (PISTON)	2	30	10	4	5	1	309
PISTON	2	30	8	4	5	2	258
SCREW (BACKPLATE)	8	25	5	1	5	1	131
WASHER (SCREW)	8	20	5	1	3	1	104
RIVET (LINING)	28	20	6	4	5	1	129
SPLIT PIN	1	20	5	3	5	1	108
DUST CAP	1	10	9	1	5	1	96
BUSH (ADJUSTER)	2	20	6	4	3	1	127

Fig. 4: Criticality Factors

© IMechE 1976

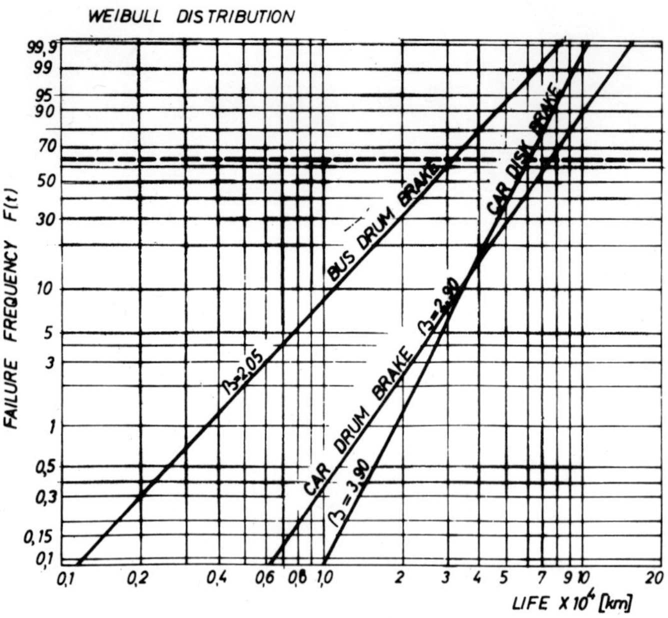

Fig. 5: Failure Investigations of City Buses and Taxi-Cabs in Belgrade

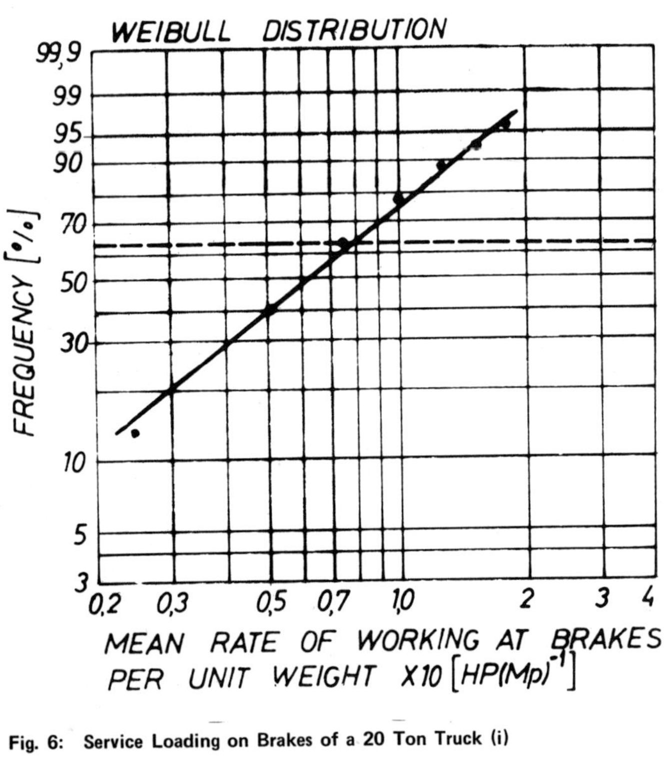

Fig. 6: Service Loading on Brakes of a 20 Ton Truck (i)

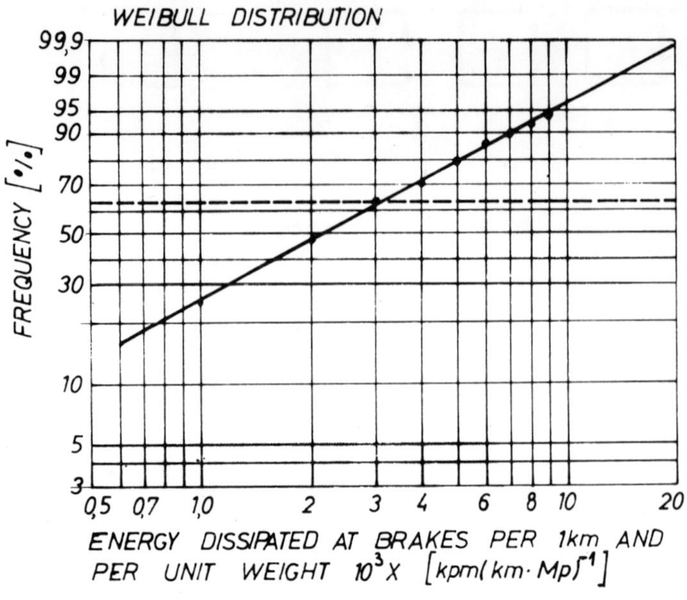

Fig. 7: Service Loading on Brakes of a 20 Ton Truck (ii)

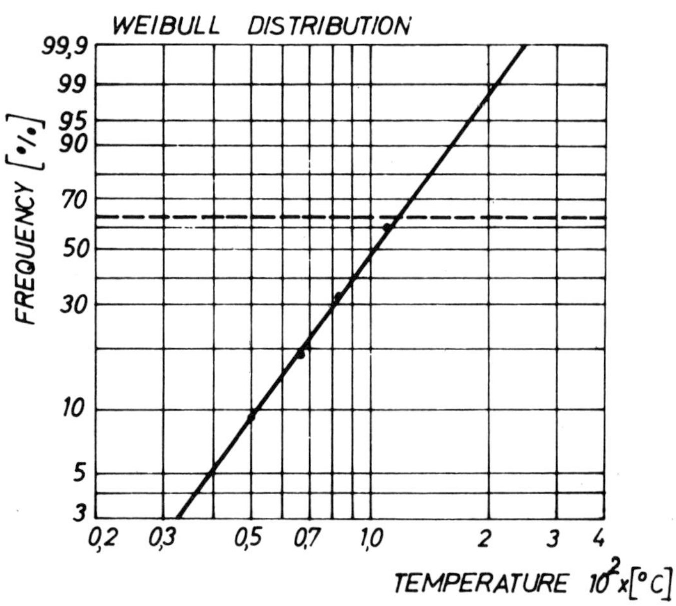

Fig. 8: Service Loading on Brakes of a 20 Ton Truck (iii)

© IMechE 1976

C48/76

RELIABILITY AND MAINTAINABILITY OF BRAKING SYSTEMS – MILITARY APPLICATIONS

J. D. W. PARSONS REME, CEng, MIMechE, AMIRTE
L. W. D. GODFREY REME, BSc
Royal Electrical and Mechanical Engineers

The MS of this paper was received at the Institution on 3 October 1975 and accepted for publication on 27 October 1975

SYNOPSIS The maintenance load imposed by vehicle braking systems will be explained in relation to the total maintenance load for a number of cargo vehicles in military use. The differing conditions of military usage, such as cross country driving, loading parameters, and storage time, will be explained and the consequent effects on braking performance assessed.

The characteristics of the different braking systems in use have different effects on the reliability of the system, particularly in regard to mud and dust contamination. The design requirements of military vehicles can sometimes result in brake component locations which are far from ideal, and the result this has on the reliability and maintainability is discussed. The possible ways of minimising these problems, such as the use of retarders, auto-adjusters, and "rolling road" testing methods, will be suggested.

INTRODUCTION

1. The Royal Electrical and Mechanical Engineers (REME) are responsible for the inspection, repair, recovery and modification for most of the Army's equipment.

2. Our facilities range from large well equipped workshops, through modern mobile units to a single mechanic with his tool box, working in the open.

3. Nearly half of REME's work is on wheeled cargo vehicles and this paper will concentrate on discussing the types with commercial equivalents.

MAINTENANCE LOAD OF VEHICLE BRAKING SYSTEM

4. As a result of scrutiny of computer printouts available to Senior Management, it was decided that there appeared to be a high repair load on vehicle brake repairs and adjustments. An investigation was put in hand by three officers on the Army's Officers Long Mechanical Engineering (Advanced) Course, as part of their project work.

5. Data gathered from these studies have been evaluated, and discussed with the R and D Establishment (Military Vehicles and Engineering Establishment) and with other responsible agencies, and compared with experience from large scale civilian usage.

6. Table 1 shows the percentage of the total repair load expended against braking systems for four vehicle types. The percentage varies from 7% to 13% and averages 10%.

Table 1

Vehicle Type	1972/75 Brakes % of Total
Austin 1800 RHD	7.0
Austin 1800 LHD	6.9
Land Rover $\frac{3}{4}$ Tonne RHD	10.3
Land Rover $\frac{3}{4}$ Tonne LHD	11.6
Bedford 4 Tonne RHD	12.2
Bedford 4 Tonne LHD	13.3
AEC 10 Tonne	11.4
Average	10.4

Table 2

Breakdown of Brake Work (By Man Hours) %

Vehicle type	Adjust & Bleed	Wear Items	Mech	Hyd	Vaccuum & Air
Austin RHD	23	53	2	13	9
Austin LHD	37	46	1	12	4
Land Rover RHD	39	4	7	10	-
Land Rover LHD	46	40	5	9	-
Bedford RHD	27	40	3	14	16
Bedford LHD	31	30	3	18	18
AEC	24	45	4	-	27
Averages	32.4	36.9	3.6	10.9	10.6

7. A further investigation was made to establish the breakdown of these brake repairs into the major areas of effort, and this is shown in table 2. Other general conclusions drawn from the data analyses were as follows.

Brake System Work as a proportion of Total Work

8. In the case of the Land Rover, Bedford and AEC, this was in the order of 12%. In view of the varied types of surface these vehicles are driven over, with the abrasive effects of cross country work, it is considered acceptable. In the case of the Austin 1800, a figure of 7% is consistent with the need for brake adjustment only on the rear wheels and the minimal cross country work.

Hydraulic System Work as a proportion of Total Brake Work (Austin, Land Rover, Bedford)

9. The frequency of repairs of hydraulic elements in the system was between 7% and 14%. This reflects reasonable reliability of the hydraulics.

© IMechE 1976

Air System Work as a proportion of Total Brake Work (Bedford, AEC)

10. Bedford 4 Tonne

Here at 17%, air system failure is proportionally more frequent than in the hydraulic system, showing apparently less reliability.

11. AEC 10 Tonne

At 34%, the frequency of repairs was considered high, as was 27% for the time expended. This appears to indicate the relatively higher complexity than with hydraulics and possibly a need for more in-depth training.

Adjust/Bleed Work as a proportion of Total Brake Work

12. In each case, this was unexpectedly high, but especially so in the Land Rover at about 60%. The indication of serious problems in this area was later confirmed during visits to repair workshops.

Theatre Comparison of Brake System Work

13. In general there was no significant difference in the proportion of work carried out on the various groups in the British Army of the Rhine (BAOR) and UK theatres. However, it appeared that in the case of the Austin 1800, more time was spent on adjustment in BAOR than in UK, and the reverse with respect to replacement of 'wear items'. This may be due to spares being more readily available in the UK.

Inspections

14. Approximately 10% of the total workload consists of inspections. Of the vehicles studied, the most frequent operational code, by frequency, was "in" and "out" inspections. In almost all cases, the vehicle brake systems will be inspected and this may well account for the high incidence rates of bleeding and adjusting.

Preventative Maintenance

15. Some investigation of data was carried out with a view to recommending that certain items might be removed at a predetermined mileage. The high reliability of the hydraulics and air components with respect to the life of the vehicle, however, precluded this in each case. In order to see how the replacement rate varies during the lifetime of the vehicle, we have presented the data at figure 1 in terms of the rate of brake shoe replacement actions per 100 vehicles per 5,000 miles. It can be seen from figure 1 that this particular component seems to show an almost constant failure rate after the initial period.

16. It is difficult to draw reliable conclusions from our various analyses, due to the influence of other factors affecting failure. A known defect on hub seals has, for example, led to an unusually high rate of lining replacement due to contamination. This effect is seen from perhaps 15,000 miles. Taking such effects into account, we concluded that in reasonable conditions, for Land Rovers, linings have an expected mean life of 30,000 miles and drums have an expected mean life of 70,000 miles (or about the life of the vehicle)

17. Further study of the printouts also showed that there was a high failure rate of some wheel hub oil seals, and a high replacement rate of some system components during the first 5,000 miles.

MILITARY USAGE CONDITIONS

18. We find that although many of the Army's conditions of use are not unknown in commercial use, the combination of these usage conditions often present unique problems. The result is usually greater workload, either in repair of failure, or in preventative inspection, and/or decreased reliability.

Cross Country

19. Many of the Army's General Service Vehicles are often required to travel in off-road conditions. Mud and dust contamination is the main problem, but allied to this is the time element, as it is sometimes not possible to stop and clean out as often as is needed. On occasion a guard provided specifically to keep out dust and dirt can be the cause of trapping loose stones against a rotating brake component and thereby cause damage. The requirements of cross country use normally mean that large suspension travel is required and this can mean more flexing of brake hoses and consequently shorter life.

Range of Loading/Usage Conditions

20. Much of the Army's equipment is designed to carry loads which are rarely carried in peacetime except on exercises. One would hope that this would result in longer lives from brake components but this is not always so because of usage factors. When vehicles are fully loaded such as on training, they are usually asked to perform cross country as well, thereby increasing the severity of the stress on the brake system.

21. Some cargo vehicles are used in peacetime much as they would be in war, but many vehicles, especially specialist types, are in use for their design purpose for only a small proportion of the year. The remainder of the time they can be used for static training or are not used at all. Since training is expensive it is usually concentrated into short periods of high usage. This means that a vehicle which may have hardly moved during the year is suddenly asked to perform intensive duties. This can, and does, cause unexpected failures on the way to the training ground! This high/low utilisation cycle can be a factor in giving misleading results from calculating "average" figures from analysis of repair activities.

Storage

22. It is a feature of the issuing system that vehicles can often spend large amounts of time in Vehicle Depots before issue to user units. This is often the case when a large production batch has been procured to meet a need which will extend over a considerable period. This can mean that a "new" vehicle issued to a user may well have stood largely inactive for a year or more, with consequent deterioration of components, and particularly of the brake fluid. Except for some specialised important equipment such storage does not take place under controlled temperature/humidity conditions, and it is this factor which is probably responsible for the relatively high work content shown in figure 1 for the early period of life.

© IMechE 1976

Climatic Requirements

23. One of the most difficult problems for designers to meet is to cope with a wide range of climatic conditions. It is relatively easy to design a brake fluid, say, for a Land Rover in Norway or for a Bedford in Nigeria. It is much more difficult to design a fluid for both vehicles in both countries. It is obviously sensible to avoid needing different fluids for different vehicles to ease supply problems. It is equally important that, should the "balloon go up" in a particular country, then it should not be necessary to change braking fluids or components before the vehicles can be sent. This factor can mean that for a particular vehicle in a particular climate, the brake fluid might be less than ideal. Whether it is always cost effective to design for the worst case is an old problem, of course.

Other Facilities

24. The military requirements for extra fitted equipment, such as stowage bins, shovel clips and so on, can often result in non standard routeing of pipes. A standard commercial chassis may be bought for later fitment of specialist equipment which can result in having to change locations of brake system components often to positions which are far from ideal. We also have to remember that the traditional heavy-booted driver is liable to put his foot on whatever is in the way, even if that something is a fluid reservoir! Provision of protecting guards to prevent such damage can reduce the accessibility of the item, consequently lengthening repair times, and can also sometimes discourage the inspector from looking at it.

PROBLEMS OF THE SYSTEMS IN MILITARY USE

25. The braking systems in use in the Army are many and varied, using almost all of the conventional systems in common use. They vary from the huge disc brakes of a Chieftain, through the smaller ones of Saladin Armoured Car to normal hydraulic and air hydraulic systems. Figures 2 to 5 show a selection of these systems.

26. This very wide range brings problems of its own. Since in any Army repair unit the braking system will be repaired by a vehicle mechanic, this places very great burdens on the training organisation who have to endeavour to teach skills which are quite different in scale and content for each system, and of course must compete with training time for the other aspects of the vehicle.

27. Contamination is the major problem with all these systems. The problem varies in seriousness according to the type of vehicle, oil contamination of the pads in the case of discs situated in the engine compartment, and continual water formation in cold damp climates (such as Northern Ireland).

28. It is not possible as a rule to find service stations at military convenience and so we have to take our brake fluid with us. For obvious reasons of economy this fluid is normally supplied to the Army in five gallon cans, which is rarely used all at once. This leads to serious contamination problems unless strict precautions are taken - and this is rather difficult in the desert. We feel certain that the use of smaller, sealed containers is preferable.

29. Because of the Army's role, much driver training takes place in off-road conditions, often in wet and sandy soils. This is a most harsh environment for the drum brakes found on most of our medium range vehicles and in one case resulted in a Land Rover having ground out a complete set of linings and drums in 60 miles.

30. As far as we are aware, brake manufacturers have not yet marketed a completely effective seal against the ingress of water and abrasives to the drum brake. Back plate and drum design can achieve reasonable success only under normal road conditions, denying ingress if possible but also allowing foreign matter to escape if it has found its way in. It is understood that there is a sealing system being developed by industry which claims to be completely effective. Such a device at the right price would be of great value to some of the Army's vehicles, especially those used for driver training, provided it was a simple bolt-on addition.

31. Disc brakes are not a major problem at present as long as the self adjusting action of the pistons maintains the contact between pad and disc. There are cases where the deflection plates are very good at retaining small pebbles if you get the vehicle deep into shingle beach. The resulting noise is usually sufficient to make a driver stop and remove them but if this is not done, severe damage can result to the discs and other components.

32. The problems of water contamination of conventional air brakes are well known, but are normally capable of being coped with if proper drain off procedures are used. However, in humid climates such as Singapore or Northern Ireland, there is sufficient water in the atmosphere to produce measurable quantities of water if the air reservoir is drained every hour. It is not always possible to do this every hour in an operational situation.

33. The various Construction and Use regulations that come into force from time to time affect us all; whereas it is possible to claim Crown Exemption for purely operational vehicles such as Tanks or Armoured Cars, this is seldom done for general load or passenger carrying vehicles, and the majority of our vehicles do comply with such regulations. This can and does cause considerable difficulties with conversion problems for our older vehicles (two to three line systems for example); with recovering or towing vehicles built to different C and U standards; and not least with using very old vehicles in a manner for which they are no longer suitable.

34. The last example above refers to the Humber 1 Tonne Armoured Car (usually referred to as the "Pig") which was rescued from oblivion by being the most suitable vehicle available for use in Northern Ireland. All of them are at least 19 years old and have had to have extensive rebuild. Much weight has been added and this has demanded a braking performance much in excess of that originally envisaged. One of the results of this has been that it has proved essential to have an improved servo unit fitted to give adequate braking performance in use. This has involved design, procurement and fitting of the new item in a very short time scale.

POSSIBLE IMPROVEMENTS

35. After the investigation into the maintenance workload had been evaluated, we were left with a basic question. Was this workload too high, and if so how should it be reduced? To this end it was decided to compare our results, albeit from very

© IMechE 1976

different conditions of use, with other major operators to see how relevant our results might be.

36. It was found that despite the great difference in use patterns that our results were of much the same order as other operators, and these are summarised in Table 3 below.

Table 3

Operator	% Brakes of Total Work
London Transport	15
Metropolitan Police	11.8
BRS	9.5
Army (average for 4 types of vehicle)	10.4

37. The London Transport figure is perhaps more comparable with our heavier vehicles, but our vehicles very rarely average the 800 miles per week of a London bus. Although repair and inspection policies vary within the organisations studied, it was concluded that military vehicle brake system's repair workloads are broadly comparable with other operators.

38. One item that the Police did confirm was that the Land Rover vehicle brake system requires a consistently higher manhour workload than other vehicles; this is due to the previously mentioned hub oil seal problem and to the difficulty of getting "a good pedal" on Land Rovers unless it is done exactly in accordance with the brake manufacturer's schedule.

39. A number of new techniques and developments have been studied to establish what benefits, if any, might apply if taken into military use. It is not claimed that this is an exclusive list, but it has included most of the well-known improvements that are currently available.

Retarders

40. The benefits that can be acheived from auxiliary braking systems can be exploited most by fleets that have a high utilisation. In general the Army does not have such a fleet, although there are some cases where cargo vehicles of the new logistic fleet are likely to be used in a similar manner to commercial concerns.

41. One big advantage of retarders is that they offer an additional degree of safety to heavy fast moving highway vehicles. Most military vehicles have not been designed for such use. However, the use of such systems can result in achieving acceptable braking performance within a wide spectrum of usage conditions, and fewer compromises are therefor necessary in the design of the friction brake. The friction brake has severe limitations in the endurance mode in terms of heat dissipation, and it could be that future designs could effectively utilise retarders in order to acheive smaller volume friction brakes, or make acceptable previously unsuitable locations of components. The use of wholly inboard friction brakes to prevent external contamination problems is usually difficult because of the heat dissipation required, but it is more easily possible if a non-friction device is taking care of the endurance braking mode.

42. It is therefore considered that the cost (around £700 for an AEC 10 Tonne) of retarders is sufficient disadvantage to prevent adoption of such devices on any but a limited scale at present. Should the widescale adoption of retarders bring down the unit cost, it would be worth re-examining both electric retarders and exhaust brakes for new vehicle requirements.

43. There are a few isolated cases where, in order to comply with legislation in EEC countries, we have bought some vehicles equipped with retarders. The effect of this is being closely studied and it is therefore hoped that we may be able to obtain direct comparative data.

Improved Brake Fluids

44. Moisture ingress, via the hygroscopic fluid, is worse during periods when the vehicle is out of use, and it is far from rare to find that a vehicle which has just been taken out of storage is suffering from leaking seals as a result of a contaminated braking system.

45. The least expensive way of combating water ingress during storage is to use a preservative fluid. As they are much thicker than conventional fluids they must be removed before the vehicle enters service, and they cost about 25% more than the normal fluid. For vehicles likely to suffer long term storage this seems an ideal arrangement. It is not currently done.

46. Silicone Based Brake Fluids

The most impressive characteristic of silicone brake fluid is its non-hygroscopicity.
Thus whatever the working conditions of humidity, the system can remain corrosive free. The very high vapour lock point also is attractive for heavy duty applications. However, being miscible with water, a leak into the system will still cause contamination, and drastically lower the vapour lock point. Other disadvantages are the lower lubricity and, of course, the cost (3 to 4 times as much as conventional fluids). If the British Army were to switch to silicone fluids this would result in an extra cost of some £35,000 per year. It is not yet known how much of this could be saved by reduced repair activities, but we will be studying this aspect.

Rolling Road Testers

47. The data analysis showed that over 10% of our total time spent on vehicle repair is for in/out inspections. In terms of annual expenditure this is over £300,000.

48. An investigation of inspection methods with regard to brakes, revealed that almost all inspectors considered an inspection incomplete without a brake test on the road. It seems that many vehicle inspectors would prefer to trust their own experience and "feel" instead of the anonymous readings of a Tapley meter or rolling road test apparatus. Admirable sentiments these may be, but this does lead to inconsistent standards, a degree of over inspection, and this is not particularly accurate.

49. Modern suspension design is quite capable of coping with imbalances of braking effort of as much as 30% in some cases, and this may not be detectable on the normal road test. The capital cost of a machine (say £2,000) which can not only detect such differences but also enable a reduction of inspection time by up to 20% is an

© IMechE 1976

attractive financial proposition. The more consistent test standards can also be worthwhile in cutting out unnecessary work due to particularly zealous inspectors.

Air Driers

50. There appears to be a good case for installing air driers on military vehicles with air brake systems. As many military vehicles are constantly subjected to cross country driving causing the potential ingress of dirt and moisture into the braking system, a device which will cool, filter, dry and purge the system with only two moving parts would be of great advantage.

51. The cost of such a device (say £70) seems relatively small in relation to the potential benefits, and could be recouped if it prevented the failure of just one major valve assembly per vehicle. An investigation showed that many components returned to manufacturers as defective, were in fact serviceable and could have been replaced on the vehicle after cleaning. If the air drier could prevent such unnecessary removals it would soon pay for itself.

52. It is considered that this is one of the most cost effective "bolt-on" modifications currently available.

Carbon Brakes

53. A carbon brake for commercial vehicles could be under half the weight of current drum brakes and could deal with braking requirements far beyond the capabilities of drum brakes. Presently there seems to be little military application, but once fully developed there could be a potential for the new range of heavy cargo vehicles. The long life and freedom from fade are obvious advantages, but it is probable that existing brake fluids would have to be improved accordingly to deal with the higher heat input. Detailed cost comparisons are not yet available, and development work continues.

Self Adjusters

54. The use of self adjusters is not new but as yet they are by no means universally accepted. The lack of confidence in them appears to result from some early designs which were prone to corrosion and jamming.

55. The use of self adjusters can reduce maintenance costs considerably but this should be balanced by the greater chance of damaged drums by the reduced frequency of inspection. "Fit and forget" is all very well up to a point, but in the Army we prefer to do our repairs at convenient rather than inconvenient times. It would be tempting to wait until scraping noises are heard before examining the lining wear, unless an indicating system is used in conjunction.

56. However, on some vehicles which have a high proportion of brake work spent on manual adjustment, this is a worthwhile risk. It is considered that the Land Rover fleet comes into this category (see paragraph 6).

CONCLUSIONS

57. The greatest factors affecting military vehicles compared with commercial use are; low utilisation, cross country work, and storage factors.

58. Considering such factors the reliability of the hydraulic and air system components of military vehicles was found to be generally satisfactory but could be improved cost effectively in several areas. Repair techniques need clarification in some cases, particularly in respect of the use of brake fluid from bulk containers. Work expended on hydraulic and air parts of the system, increase with system complexity which may indicate the need for more specific detailed training.

59. The braking system of a military vehicle has to be rugged, easily maintained, simple and reliable. This does not come about without considerable development effort and may need a far more detailed statement of the conditions of use, in order to acheive any improvements. It is considered that a proper balance of these factors must be struck in each case, in order to acheive the best available system effectiveness.

ACKNOWLEDGEMENTS

60. Grateful thanks are due to the members of the Officers Long Mechanical Engineering (Advanced) Course project team (Majors Bingham and Pierce and Captain Goodes) who carried out so much of the necessary fact gathering in preparation. The views and opinions expressed in this paper are those of the authors alone and do not necessarily represent REME or official Service policy.

©Crown Copyright 1975

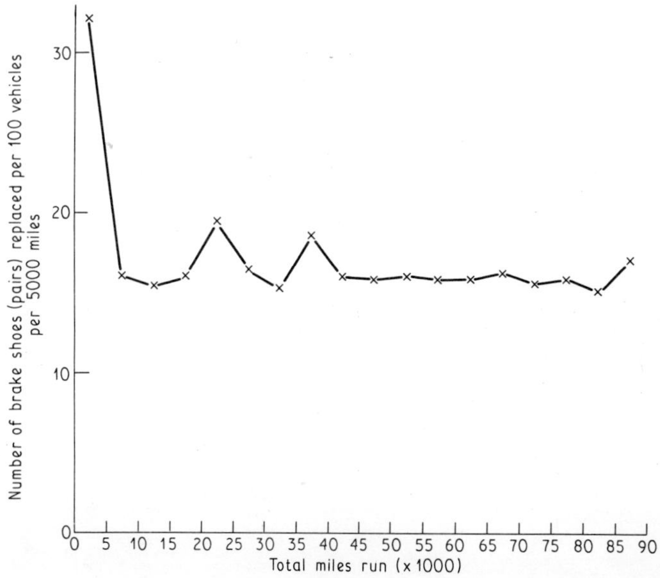

Fig. 1: Replacement rates of Landrover brake shoes

© IMechE 1976

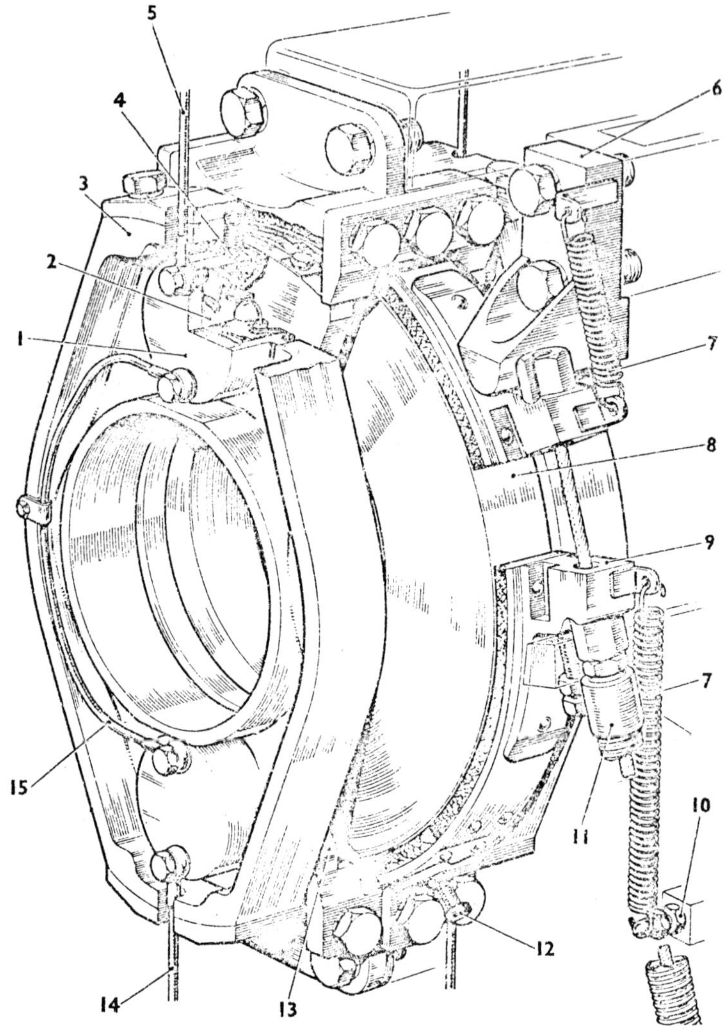

1 Cylinder
2 Piston
3 Carrier plate
4 Rubber boot
5 Fluid inlet pipe
6 Parking brake reaction bracket
7 Parking brake return spring
8 Brake disc
9 Parking brake band
10 Return spring anchor stud
11 Parking brake cable
12 Brake band centring screw
13 Reaction plate
14 FLuid outlet pipe
15 Interconnecting pipe

Fig. 2: Main and parking brake (Chieftain tank)

©IMechE 1976

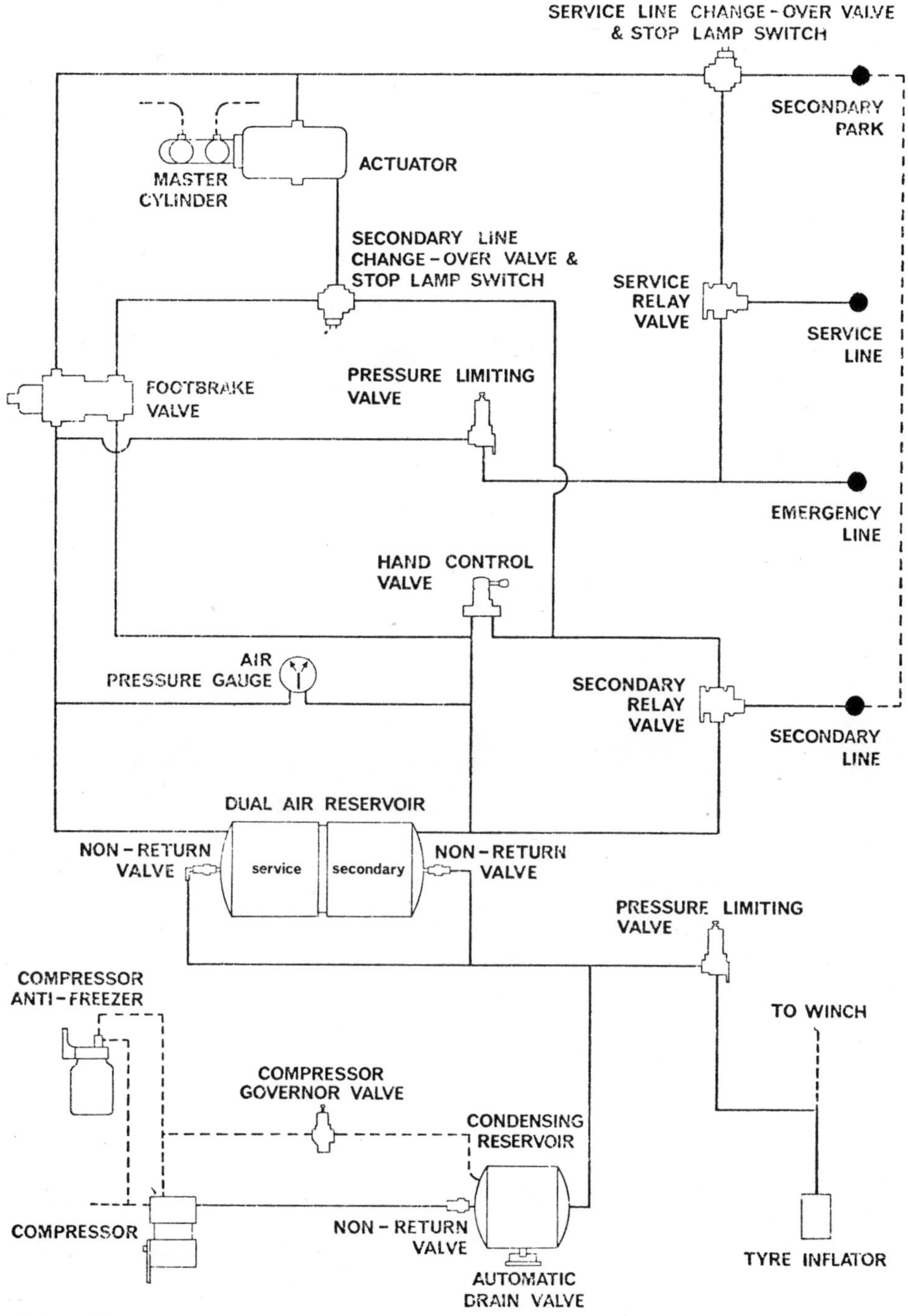

Fig. 3: Braking system — Bedford MK

1. Service line jumper coupling
2. Air line filter
3. EMPTY/LOAD control valve
4. Limiting valve
5. Emergency relay valve
6. Exhaust check valve
7. Brake air cylinder
8. Air pressure gauge
9. Air reservoir
10. Air reservoir drain cock
11. Air pressure gauge
12. Brake air cylinder

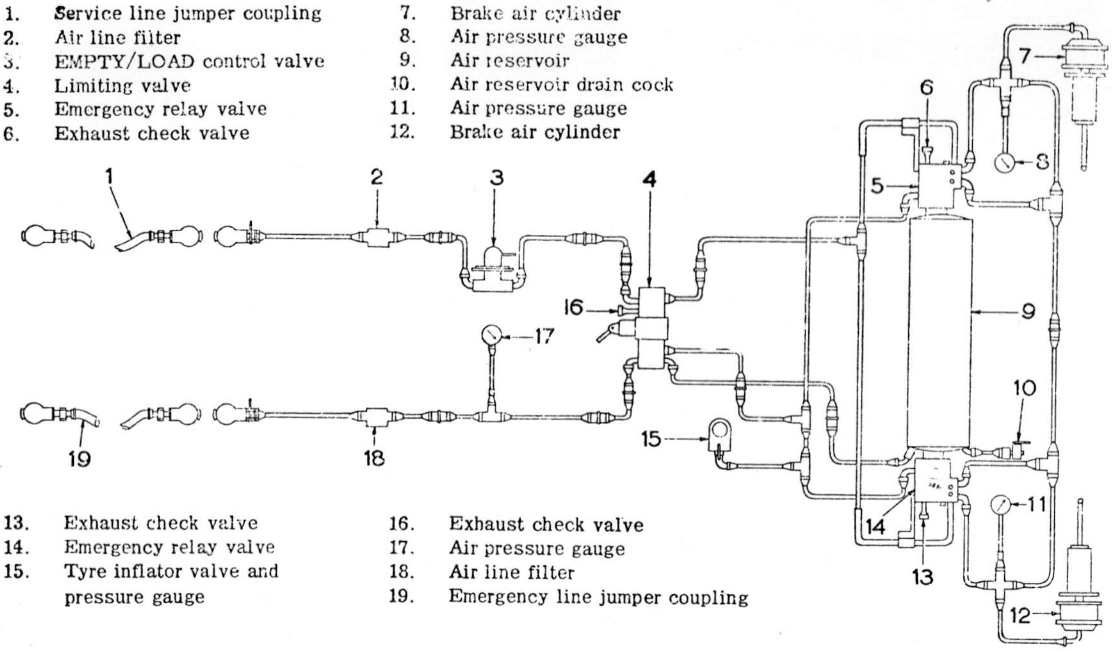

13. Exhaust check valve
14. Emergency relay valve
15. Tyre inflator valve and pressure gauge
16. Exhaust check valve
17. Air pressure gauge
18. Air line filter
19. Emergency line jumper coupling

Fig. 4: Trailer braking system

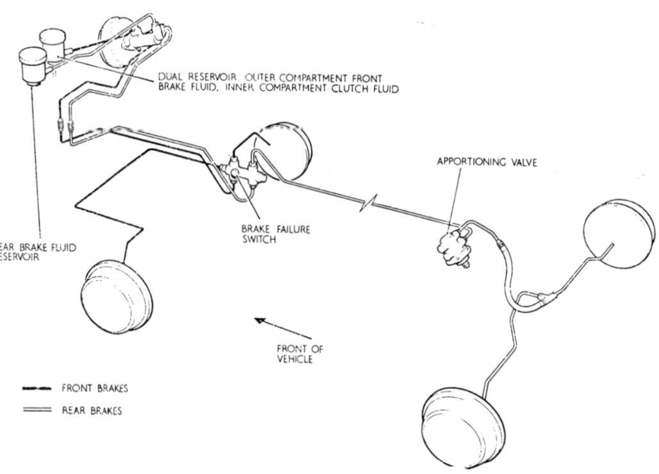

Fig. 5: Braking system – Rover 1 tonne

© IMechE 1976

C50/76

LEGISLATIVE CONTROL OF ARTICULATED VEHICLE BRAKING

P. D. LIVINGSTONE, CEng, MIMechE
Ford Motor Co, Laindon, Essex

The MS of this paper was received at the Institution on 6 November 1975 and accepted for publication on 8 December 1975

SYNOPSIS Matching the braking forces on the tractor and semi-trailer of an articulated vehicle to minimise situations leading to combination instability has been recognised and legislative control discussed for some period. Outlined are legislation developments from the early 1960's to the mid 1970's on articulated vehicle braking culminating in the EEC Directive 75/524/EEC. Insight is given into the way a technical regulation is developed by European Authorities.

CURRENT FRAGMENTED SCENE

1. The articulated vehicle has progressively increased its impact on modern society during the last fifty years. It has been developed into a highly efficient and sophisticated transport medium with payload capabilities ranging from 1000 Kg to 30,000 Kg or even greater for specialised types. The commonisation of articulated vehicle design and operational features on such aspects as gross combination weights, length, coupling heights, overall vehicle heights, cornering swept paths and braking systems has been a major objective on the international legislative scene for over ten years. Many of these features are still in the melting pot of environment versus transport need versus removal of non tarrif barriers to international trade and established national requirements, however, combination braking performance and compatibility has been resolved at least within the European Economic Community (EEC) and hopefully will provide a catalyst for even greater harmonisation in the future.

2. The UK has been in the forefront of combination braking performance and compatibility control for many years and together with industry developed a code of practice in the mid 1960's, proving a major step towards increased vehicle safety. This 'Code' based on early international discussion within the United Nations Economic Commission for Europe (ECE), led the way to the present harmonised European requirements.
The UK Code of Practice introduced from 1968 rear axle brake proportioning as mandatory on tractive units and stipulated a laden tractor and semi trailer braking relationship according to trailer control line air pressure and system responsed and release times. Three line upright trailer to tractor braking systems were required either full air or air/mechanical providing service, secondary (failed service) and emergency (supply and trailer breakaway protection) performance. Initial teething problems with load apportioning valve moisture ingress and corrosion clouded the initial picture, however, with valve improvements and some uplift in maximum combination weight the dramatic rise in articulated vehicle volume and acceptance in the market place was ensured.

3. France had tended to parallel the UK situation with three line systems (two line if trailer is fitted with spring brake actuators) but biased towards a design rather than performance standpoint, approved systems being recognised dependdent on tractive unit weight. Switzerland requires an inverted two line system (service line pressurised during normal running and vacated to apply trailer brakes) giving similar safeguards to the three line upright system. The majority of other territories have been split into the 'German influence group' on largely inverted single line and others using upright two line. Sweden and Norway led the adoption of preliminary ECE and EEC combination braking requirements, with Sweden introducing for 1971 model year vehicles laden retardation against control line pressure for tractive unit and their trailers/semi trailers, air pressure build up times at the coupling, maximum supply pressure in the trailer feed line and tractor and trailer system failure warning provisions. Norway introduced similar provisions effective 1 January 1972.

4. Rear axle load apportioning valves for tractive units over 7500 Kg gross vehicle (not combination) weight were required by W Germany in 1963 and Norway 1972, for units over 8000 Kg by Switzerland in 1972, over 3500 Kg by Sweden from 1971 and Denmark September 1972 and all tractive units in Czechoslovakia 1970. Sweden in December 1973 rescinded mandatory fitment owing to function and reliability problems.

5. France and Holland currently require trailer or semi-trailer brakes to be actuated at the same time as or before the towing vehicles brakes where as Norway and Sweden state an initial brake application pressure range (0.2 to 1.0 Kg/cm^2) and a faster pressure build up time for the towed vehicle compared to the tractor.

INTERNATIONAL LEGISLATIVE MACHINERY

6. Highly competent papers have already been written (1) (2) (3)* on braking legislation and upon European law making procedures, influences and inputs and further elaboration is unnecessary apart from the following brief outline.

© IMechE 1976

7. United Nations Economic Commission for Europe (ECE), centered in Geneva, established a working party (WP29) within their inland transport committee in 1958 to harmonise vehicle and automobile equipment construction legislation throughout Europe removing unnecessary barriers to international trade. Participation in ECE-WP29 activities, is now world wide with twenty four government and ten non-government (manufacturer, operator, standards) agencies involved in producing recommended legislation for national enforcement. Fifteen countries currently are signatories (empowered to test and certify) to the 1958 contractural agreement. Eight sub groups report to ECE-WP29 covering all aspects of vehicle construction, that concerned with braking is the Groupe des Rapporteurs Freinage(GRF). Voting rights within WP29 and its reporting group are restricted to government agencies (from respective Ministries of Transport). Manufacturers are represented directly through the Bureau Permanent International des Constructeurs d'Automobiles (BPICA) and certain more specialised trade associations and indirectly through standards organizations. The BPICA Technical Committee sub group on brakes is the Group des Experts Freinage (GEF).

8. The European Economic Community (EEC) with nine member territories develops legislation under the auspices of its Commission. Within the EEC Commission are twenty Directorate Generals (DG) of which Internal Markets (DG X I) - concerned with the removal of technical barriers to trade, assisted by DG V111 - Transport, has primary responsibility for automobile construction legislation. Plenary committee meetings establish inputs from fringe interest such as DGV - Social Affairs and DG 111 - Industrial, Technological and Scientific Affairs. The major legislative committees established by DG X I concerned with automobiles are Groupe des Entraves aux Exchanges resultant des Dispositions d'Ordre Technique - Vehicules a Moteur (GDE) (Working Group on the Elimination of Technical Barriers to Trade - Motor Vehicles) responsible for new legislative topics and Comite pour l'Adaptation au Progres Technique des Directives (CPA) (Committee for the Adaptation of the Directives to Technical Progress) responsible for revisions to existing requirements. In general the same government delegates (within the nine) attend both ECE and EEC. Industry is represented by the Comite des Liaison des Constructeurs Automobiles (CLCA), by Comite des Liaison des Constructeurs d'Equipements et des Pieces d'Automobiles (CLEPA) and to a minor extent on commercial vehicles by the Committee of Common Market Constructers (CCMC) (restricted to European based manufacturers) again with no voting rights. New legislation is referred to the European Parliament and Economic and Social Committee before final submission to the EEC Council for ratification and official publication. In general EEC vehicle construction legislation is issued as approval** building blocks (called Directives) to the base vehicle type approval Directive 70/156/ EEC largely dealing with documentation.

* Numbers in parentheses designate References at end of paper.

**Type approval is the process of submitting a component or vehicle for compliance assessment against a legal requirement, this being accepted by the Authorities as representative of future series production and certified by manufacturers on that basis.

9. ISO are involved in a peripheral sense with ECE and EEC legislation development through their Technical Committee (TC 22) - Automobiles, and respective sub-committees and are increasingly gearing their workload and priorities to this end. Sub-committee 2 and its associated working groups is largely responsible for braking matters.

10. The emergence of the EEC as a regulatory body on vehicle construction was not apparent until the issue of the base type approval document 70/156/ EEC in 6 February 1970. From 1970 on the pressure on ECE discussions increased considerably and the brake Directive 71/320/EEC was no exception. Eventually joint working programmes were established in 1973 to minimise duplication of effort and present a co-ordinated front. The principle of direct carry across of legislation between ECE and EEC is paramount and this two edged sword, an extremely powerful regulatory tool, is kept on the rails by common (within the restricted sense of the common market) Governmental and Industry representation.

INTERNATIONAL BRAKE LEGISLATION

11. ECE-WP29 spent most of the 1960's developing Regulation 13 - 'Uniform Provisions concerning the Approval of Vehicles with regard to Braking' which was finally issued in 29 May 1969. In its early finalised form it covered general brake equipment specification and function and dynamic performance requirements. Holland and Italy were the two sponsors required by ECE procedure willing to accept the Regulation and submit it to the UN in New York for ratification. Further acceptance of ECE Regulation 13 by other territories did not materialise and events were finally overtaken by publication of the EEC Directive 71/320 'On the Approximation of the Laws of the Member States relating to the Braking Devices of Certain Categories of Motor Vehicles and their Trailers' published in the Official Journal of the European Communities 6 September 1971. The EEC Directive encompassed many of the annexes to ECE Regulation 13 before final agreement within ECE-WP29 and over two years were to elapse before some of the annexes to 71/320/EEC were covered by appropriate amendments to ECE Regulation 13.

ECE AND EEC ARTICULATED VEHICLE BRAKING

12. ECE Regulation 13 when it was issued 29 May 1969 included the fundamentals of the system design and performance requirements for articulated vehicle tractors and semi trailers notably:

a. On tractive units over 12,000 Kg gross vehicle (not combination) weight dual line brakes are mandatory unless the system split was such that the semi-trailer braking was independent of the tractor.

b. Tractor unit system protection in the event of semi-trailer system failure.

c. Semi-trailers over 3500 Kg total axle weight must have powered service brake systems either continuous (for example direct air) or semi-continuous (for example an hydraulic or mechanical tractive unit system which triggers a trailer brake control valve).

d. Progressive and graduated service and secondary (service brake failure) semi-trailer braking.

© IMechE 1976

e. Mandatory operation of the semi trailer's brakes is required if the tractors secondary brake system is actuated and if the tractive unit has a dual line system each part must be capable of applying graduated braking to the semi trailer. This has led to the generation of a new type of trailer control valve having two upright trailer trigger lines from each circuit of the service control and one inverted line connected to a secondary spring braked system (one where the actuator is held off by air pressure and applied mechanically by spring force).

f. If the semi-trailer control line fractures (or its equivalent) then it must still be possible to actuate the brakes of the semi-trailer by the tractor service brakes, secondary brakes or a separate control.

g. Semi-trailer automatic braking coupling failure breakaway protection is required with all axle weights over 1500 Kg.

h. All wheels of the semi-trailer must be braked by the service brake system.

i. Semi-trailer must have its own mechanical parking brake system which can be released and applied from the ground (except for articulated buses).

j. No 'redundant' members may be used in a system which are normally in a position of rest until brake system failure. This outlaws the general concept of the UK 3 line air system where the third line is purely an emergency brake and prohibits the use in many instances of the double check valve giving a dual feed to various types of valves or actuators.

k. System failure on tractive units must still permit the feeding of energy (air) to the other system(s) required to achieve secondary brake performance. This continual supply requirement has considerably increased the complexity of system protection valves leading to modulised triple and quadruple versions incorporating one way valves, to prevent system feedback.

l. Full power brake systems must have at least two independent energy reserves (reservoirs) however only one power source (compressor/pump) is required. Low air pressure warning must occur at or above 65% of normal system pressure.

m. Cold dynamic service/secondary brake performance for tractive units between 3500 Kg and 12,000 Kg total weight must achieve a stopping distance of 29/51 m from 50 kph and over 12,000 Kg achieve 20/34m from 40 kph (4.4/2.2 mm/s^2 developed retardation) on a road offering 'good adhesion' and without wheel lock.

n. Semi-trailers must achieve a force at the tyre equivalent to 45% of the static maximum axle weight.

o. Hot dynamic performance is split into repeated braking (snub) and continuous application tests. For semi-trailers the continuous application test has created significant problems for manufacturers and an alternative test procedure more compatible with tractive unit testing has been promoted vigorously by the UK Industry and Government. ECE-WP29 June 1975 rejected the UK proposals owing primarily to German resistance to back off from a continuous application test and claims of no problems from other European Governments (although previously a common front had been established by Industry-BPICA). German operators and industry in May 1975, however, were discussing semi-trailer braking instability problems resultant from the types of brake linings and their usage coincident with the introduction of two line trailer systems from April 1974, and thus it is probable the issue will be resurrected within ECE or EEC at a later date.

13. The EEC Directive 71/320/EEC when it was approved by the 'Council' 26 July 1971 had additional design constraints over ECE Regulation 13 for articulated vehicle tractors and semi-trailers covering for example:

a. Dual line brakes on all vehicles as from 1 October 1974 (excluding trailers & semi-trailers).

b. Compressor performance 0-65% and 65-100% of normal system pressure within 3 and 6 minutes respectively for solo units and 6 and 9 minutes for solo units plus a simulated trailer/semi-trailer.

c. Air reservoir capacity to enable eight full service brake applications whilst still maintaining secondary (failed service) performance, pressure drop being additionally limited to 50% for towing units.

d. Spring brake and lock actuator system provisions covering air supply requirements, depletion capacity, hold off pressure and system control features.

e. System response times of 0-10% and 0-75% asymptotic pressure in 0.2 and 0.4 seconds respectively for towing units with a full control actuation of 0.2 seconds measured at the coupling head, this compares with a 0.6 second requirement to achieve prescribed dynamic service performance pressure on solo units. Trailer/semi-trailer response time from 10-75% asymptotic pressure is 0.4 second. This balance of response time requirement of solo unit to towing vehicle to trailer/semi-trailer is intended to ensure operation of the towed vehicle's brakes at the same time as, or before, the tractors.

f. Two line (or more) trailer braking system.

g. System pressure in trailer supply line must be between 6,5 and 8,0 Kg/cm^2, control line pressure being 7,5 Kg/cm^2 maximum allowing 0,5 Kg/cm^2 for system fill depletion.

14. To convert the prescriptions within 71/320/EEC into a workable document and to establish a test procedure from the defined performance requirements was essential to ensure a uniform basis for vehicle type approval. Many regulations contain subjective, rather than objective, requirements leading to confusion on the part of the manufacture in being unable to design to ensure compliance and non-uniformity of interpretation by Governmental Approval Inspectors both nationally and on an international basis. Without uniformity of interpretation obviously the viability of harmonised legislation becomes suspect and non-tariff barriers to trade are created in direct conflict to the initial objectives of the law

© IMechE 1976

makers. With the UK entry into the EEC on 1 January 1973 it was essential to provide EEC vehicle type approval facilities to overcome the need for UK manufacturers to submit their products to each individual territory of the nine for approval. The UK had to recognise EEC legislation as an alternative to, or in place of, existing national law, a time limit of 18 months from issue of the Directive is generally stipulated, however, this was extended in certain instances during the period of accession of UK, Eire and Denmark. The brake directive 71/320/EEC became effective in the UK as from 1 July 1973. Throughout 1972 and in the early part of 1973 the UK Government (Department of the Environment-DOE) and Industry (Society of Motor Manufacturers and Traders-SMMT) were involved in establishing 'Interpretations' and 'Test Procedures' against 71/320/EEC and many other finalised Directives.

15. Uniform 'Interpretation' within the EEC is implicit in national discussions, however, during UK Government and Industry negotiation it was apparent agreement had already been reached within W Germany, France and Italy on certain fundamental aspects within the Directive. Tractive unit brake system redesign was required in some instances by UK manufacturers at a very late stage in approval negotiations to ensure acceptance of the 'approved' unit within other members of the EEC. The major design change was associated in meeting a qualification on the time from trailer/semi-trailer brake operation, via its emergency relay valve, with a control line coupling failure (see item 12(f)). The limit of 2 seconds to deplete the trailer supply line to 1.5 bar required additional isolation of the tractive unit system to minimise direct feed to the trailer.

VEHICLE COMBINATION BRAKING COMPATIBILITY

16. ECE-WP29 consolidated its early work on vehicle combination coupling and braking compatibility during the late 1960s with the draft W/TRANS/WP29/307 dated 15 April 1969. This draft, initially intended to cover combinations over 24000 Kg, stipulated:

a. Laden tractor and trailer/semi-trailer braking rates (braking force at the tyre divided by vehicle /axle weight) against trailer control line pressure. For a braking rate of 0,45 for the tractor and trailer/semi-trailer the pressure at the control line must be respectively 4,0 to 5,0 Kg/cm^2 and 4,5 to 5,5 Kg/cm^2 braking rate giving inadequate trailer braking and overbraking of semi trailers was not supported and led to a request from ECE-WP29 to ISO-TC22 and BPICA for their proposals.

b. Brake linings on both parts of the combination must approach the drums at a control line pressure between 0,2 and 1,0 Kg/cm^2.

c. Location of air and electrical couplings on the tractor and towed vehicle.

d. Mechanical coupling design, location and articulation provisions.

17. Detailed Industry discussion resulted in the publication of state of the art papers by both the W. German (VDA) and French (CSCA) manufacturer organisations during 1971 on both combination compatibility and tyre adhesion used under braking. The linking of these two topics became essential to ensure desired wheel locking sequence (tractor front, semi-trailer then tractor rear) with balanced braking between tractor and semi-trailer although this philosophy became clouded over the following months.

18. A decisive BPICA-GEF meeting of some thirty delegates from six countries and eighteen manufacturers was held in London February 1972 followed by the twenty second meeting of ECE-GRF in Rome March 1972. The first ISO-TC22 meeting on compatibility had been held January 1972. Broad agreement was achieved on the following fundamental aspects:

a. Equal braking on tractors and draw bar trailers

b. Compatibility to be assessed against static axle loadings.

c. Tractive unit braking rates less than 0.1 would be disregarded from compatibility restrictions.

d. An empirical 20% factor was taken for increased semi-trailer coupling loading at a braking rate of 0.4 owing to weight transfere. This corresponded well with the UK Code of Practice differential between tractors and semi-trailers albeit higher factors than 20% were recommended for specialised semi-trailer types.

e. Laden vehicle centre of gravity would not be established by a relationship against vehicle weight or other factor but would be that specified by the vehicle manufacturer.

f. Tyre adhesion under braking would be assessed primarily by calculated performance predictions supported as necessary by practical testing and the provisions are to be such that brake load apportioning valves become mandatory on the majority of commercial vehicle types.

19. An adhoc committee was created by ECE-GRF comprising France, Italy, Spain, W. Germany, UK and BPICA to formulate braking adhesion utilisation legislation. France had also submitted proposals to EEC late 1971 to accelerate progress on this matter. Detailed industry comment and Government discussion occurred at an unprecedented pace both nationally and internationally. An outline of the initial proposals concerning articulated vehicles is given below:

a. Only solo tractive units and fully laden tractive units were considered for tyre adhesion under braking assessment.

b. Tractor unit rear axle lock before front axle lock was not permitted for braking rates between 0.15 and 0.30.

c. Loading state of the vehicle shall not influence trailer supply or control line pressure.

d. Fully laden tractive unit dynamic axle weight was established assuming a 'typical' semi-trailer.

e. Tandem axles if less than 2,0 m apart were to be considered as one axle for performance calculations.

f. Semi-trailer wheel lock would be established from a braking rate to control line pressure relationship for laden, half laden and unladen conditions allowing a 20% tolerance either side of the norm to be established for semi-trailer

© IMechE 1976

compatibility with towing units.

20. The initial proposals on braking distribution and adhesion utilisation were modifed during the remainder of 1972 and testing programmes initiated by both Government and Industry to validate the major contentious issue concerning articulated vehicles and their semi-trailers. Computer exercizes were conducted to validate the proposed dynamic weight transfere factor (revised from the initial French proposal mid 1972) of a standard semi-trailer on tractive unit axle loadings against actual axle loadings established for a variety of laden tractor and semi-trailer configurations. At the end of 1972 general agreement had been reached on the content of the draft ECE Regulation, subsequently to be referred to as Annex 10 of ECE Regulation 13, on all major issues apart from that of the unladen semi-trailer.

21. Activity by ISO-TC22 and BPICA-GEF on braking rate to trailer/semi-trailer control line pressure also progressed to an agreed position for laden combinations during 1972. ISO-TC22 in March 1972 over ran itself by issuing a proposal including the dynamic weight transfere factor established for articulated vehicle tractive units in braking distribution discussions on top of braking rate criteria which had already been amended to cater for weight transfere, needless to say this proposal was abandoned. Initial proposals by France for a delay in towing vehicle braking of between 0,1 and 0,4 seconds (in addition to the biased response build up times mentioned in item 13(e)) was not supported.

22. For convenience the compatibility curve established for laden semi-trailer braking rate to control line pressure was extended to the unladen condition in July 1972, albeit other moves from practical test to tyre adhesion utilisation were proposed and disregarded until the end of 1972. ECE-GRF consolidated this approach at their meeting January 1973 by extending all the laden compatibility provisions for towing vehicles and their trailers/semi-trailers to cover the unladen conditions, these provisions were subsequently referred to as Annex 13 to ECE Regulation 13.

23. Activity within EEC commenced in earnest the beginning of 1973 with the Annex 10 to ECE Regulation 13 being issued as Annex 10 to 71/320/EEC under the auspices of the EEC Comite d'Adaptation (CPA). ISO-TC22 activity, sponsored largely by the UK, for a test alternative to calculations flowered mid 1973 but ended as an abortive exercise, testing however was still maintained within Annex 10 to be conducted if felt desireable by individual Authorities. A conflict on the desirability of such an approach existed between passenger car and commercial vehicle sectors of industry owing to the vast complexity of commercial vehicle options and variants which potentially would require test assessment.

24. Problems with the situation that had developed on vehicle combination provisions, particularly with respect to articulated vehicles following the ECE-GRF meeting in January 1973 endorsed by ECE-WP29 June 1973 were readily apparent to UK Industry and preliminary meetings were held with the UK Government May 1973. Major issues were:

a. Braking adhesion utilisation and compatibility provisions had become divorced into Annex 10 and and Annex 13 and the emphasis within EEC was primarily on Annex 10. Coupling an Annex 10 tractor to an Annex 10 semi-trailer or trailer would not offer any guarantee of combination braking stability.

b. The drastic reduction in unladen semi-trailer braking required by Annex 10 (or Annex 13) compared to existing vehicles would result in a dangerously unstable combination if a post-Annex 10 semi-trailer was coupled to an 'old' tractive unit owing to semi-trailer over run leading to premature tractor unit rear axle lock - the classic jack-knife situation.

c. Extending the laden provisions of Annex 13 to unladen articulated combinations infered that the degree of weight transfere from the semi-trailer to the tractor was the same laden and unladen, a premise which was obviously incorrect resulting again in under braking of the unladen semi-trailer.

d. The differential between laden and unladen braking force required for a common braking rate to control line pressure relationship dictated not only tractive unit rear axle and trailer/semi-trailer brake load apportioning valves of unprecedently high rates (in the region of 6:1 laden to unladen) but also load apportioning valves on the tractive unit front axle(s). The practicalities of maintaining such valve settings in practice was suspect.

e. Complex failure sensitive load apportioning valves would be required in many situations to offset the high unladen braking reduction to meet the dual line system circuit failure dynamic performance requirement within ECE Regulation 13 and 71/320/EEC.

25. A brochure of over 50 pages of technical argument, illustration and revised proposals was formulated by Ford Motor Company (subsequently endorsed and adopted by the SMMT and BPICA) and lobbying conducted with each member territory of the EEC and with the Commission at the end of 1973 and early 1974.
Political pressure, however, in France forced the EEC-CPA to endorse the deficient version of 'Annex 10' at their meeting 19 December 1973 and eventually a new Directive 74/132/EEC was issued 11 February 1974. The French Government recognised that a need for change may be required and were willing to make further amendments if discussion within ECE-GRF suggested this was desireable.

26. ECE-GRF March 1974 recognised the problem within what was now the EEC Directive 74/132/EEC and directed BPICA to establish a co-ordinated technical solution to the articulated vehicle situation.
Detailed technical discussion ensued between largely Mercedes, Fiat, Ford and the French Trailer Industry Association (Chambre Syndicale Nationale des Constructeurs de Remorques Semi-Remorques et Grands Containers). Eleven semi-trailer test cases were established with the UK trailer industry to cover the broad spectrum of types in operation and sophisticated computer programmes established by Ford to permit a rapid and accurate assessment of various alternative proposals from France, UK and W Germany. The sifting was conducted on a wheel locking sequence basis against overall combination retardation

© IMechE 1976

potential without wheel lock and considering various semi-trailer loading situations.

27. The prime essential of the exercise was to adjust the degree of semi-trailer braking by a constant dependent upon its physical characteristics to make it compatible with the tractor designed for a 'standard' semi-trailer and its dynamic weight transfere characteristics. Any imbalance of the tractor and semi-trailer braking relationship automatically shifts the optimum braking characteristic of the combination leading to premature axle locking.
It became apparent early 1974 that the unladen combination (rather than the solo tractive unit) should be considered for the tractive unit, that both laden and unladen semi-trailer braking rate to control line pressure relationship should be altered dependent upon its physical characteristics and that a separate laden and unladen tractive unit braking rate to control line pressure relation-ship was both a viable proposition and minimised the hazards of mismatching pre and post regulation units. It was also apparent that a simple formula for semi-trailer braking rate correction constant would not adequately cater for all operational situations and a more correct complex formula was established.

28. EEC issued their Directive 75/524/EEC 25 July 1975 embracing the proposals established in conjunction with BPICA on articulated vehicle compatibility, wheel locking sequence and minimum retardation without wheel locking across the entire road tyre adhesion spectrum ranging from wet ice to the best dry road conditions. This Directive will provide a major step forward to vehicle safety, however, care will be required by both manufacturer and operator to ensure:

a. Premature front axle locking does not occur on poor adhesion surfaces. An artificial maximum front axle loading is considered for adhesion calculations and manufacturer's laden vehicle centre of gravity height would probably be biased towards a high load condition. This situation is less serious on the articulated vehicle owing to the factors involved.

b. Consistent braking performance. Many factors are involved for example: air actuator and linkage to brake travel and efficiency, control valves and their setting and tolerance, load apportioning valve settings (particularly with limited rear suspension travels) and setting variation owing to road spring settling and general component wear, base brake design and mechanical ratio (more stable designs may be required in the longer term to give added safeguards on compatibility) and brake lining friction consistency both in production and in service (minimising brake to drum bedding period, lining work and speed sensitivity). The ECE/EEC documentation requires lining manufacturer and type to be specified, however, in service implications have still to be resolved and brake lining classification (at present being studied by ISO at the request of ECE-GRF) has proven a highly contentious topic for some years.

c. Adequate matching of pre and post EEC/ECE level tractors and semi-trailers. Performance compatibility is unlikely to be an issue in the UK, however, certain European territories (Item 12(o)) are already experiencing problems with 71/320/EEC and these will be aggravated by 75/524/EEC. Mixing single line and two line and two line and three line vehicles will largely be a matter for national control and in the UK the SMMT have drafted a British Standard on recommended approaches, operational difficulties will be a hurdle for many years to come.

d. Vehicle types must not become too restricted in their operation. This will require a careful balance on the part of the manufacturer to cover the broadest spectrum of operating conditions and loading whilst recognising that more specialised types (for example concrete mixers or low loaders) may require revision to the vehicle's base brake specification.

FUTURE DEVELOPMENTS

29. Brake wheel lock sensing systems originally gave an exemption from what is now 75/524/EEC however this was deleted before ratification. Such an exemption could have minimised the problems on certain vehicles with minimal suspension movement, this may be permitted on a national basis. More detailed legislation on vehicles equipped with wheel lock sensing systems is currently under consideration within ECE-GRF.
ECE have still to endorse the provisions of 75/524/EEC as an Annex to Regulation 13, this will be discussed during their 26th Session 1976.

30. Articulated vehicles towing draw bar trailers are popular in Sweden, Australia and the USA and are now permitted in the UK under special types authorisation. 75/524/EEC adequately covers this situation for the laden vehicle (albeit overall performance variation is increased) however owing to the differing approaches between articulated vehicles and vehicles with draw bar trailers for the unladen condition a 20% mean performance differential exists between the articulated vehicle and a draw bar trailer. Unladen draw bar trailers for this type of usage would require an uplift of 20% in their braking rate compared to that specified within 75/524/EEC. The alternative of deflating the articulated vehicle unladen performance, although offering a superior performance match with the draw bar trailer in the one vehicle laden one vehicle unladen or part laden situation, would not be a palatable solution for the reasons outlined in Item 24.

REFERENCES

1. K. Aoki 'A Comparison of World Braking Standards with Reference to the Development of Japanese Braking Standards' SAE Transactions Vol 81 (1972), paper 720030.

2. P. Oppenheimer 'Braking Regulations in Europe' SAE Automotive Engineering Congress 1974, paper 740313.

3. E.J. Cutting and K. Teesdale 'The Legislative Maze' I.Mech.E/SMMT 1974 Motor Show Conference.

4. EEC Directive 70/156/EEC 'On the Harmonsiation of Member States' Legislation on the Approval of Motor Vehicles and their Trailers' Council Directives of 6 February 1970.

5. ECE Draft Regulation 'Uniform Provisions Concerning the Approval of Vehicles with Regard to Compatibility between Drawing Vehicles and their Trailers or Semi-Trailers' W/TRANS/WP29/307 dated 15 April 1969 updated by TRANS/SC1/WP29/R12 dated 27 March 1974.

© IMechE 1976

C51/76

BRAKING REGULATIONS IN EUROPE

PAUL OPPENHEIMER, BSc (Eng), MSc, CEng, MIMechE
Girling Limited, Birmingham, England

This paper was originally published by the Society of Automotive Engineers as SAE Paper 740313 and accepted for publication in this Conference on 4 July 1975

SYNOPSIS Important new braking regulations for motor vehicles and trailers have recently been introduced by Sweden, the Economic Commission for Europe (Geneva), and the European Economic Community (Brussels).

This paper describes the relevant rulemaking procedures and the international organisations which provide for industry participation. The technical content of these regulations is summarised and specific examples of difficult, interesting, or unusual demands are highlighted. Some comparisons with the appropriate United States federal standards have been included and the European method of type approval is explained against the background of self-certification in the United States.

Several new European proposals for tractor/trailer compatibility, brake apportioning, and antiskid systems are reviewed to illustrate the current status of legislative progress in Europe.

1. The FMVSS (Federal Motor Vehicle Safety Standards) have shown the important effects that legislation can have on the design, performance, and cost of motor vehicles. Similar regulatory activities ongoing in Europe are becoming of increasing interest to the multinational manufacturers outside Europe and to the engineering profession throughout the world.

2. The purpose of this paper is to clarify some of the elaborate rulemaking procedures in Europe and to explain the principle of type approval. The technical contents of the most important European braking regulations are summarised and compared with the corresponding federal standards in the United States. By highlighting the major legislative differences between Europe and the United States, this paper also aims to contribute toward the eventual establishment of unified regulations throughout the world*.

BACKGROUND

3. Almost every country in Europe has its own braking regulations. These regulations have evolved over the years with some regard for vehicle performance, traffic conditions, and road safety, but with little regard for legislation in other countries. The ensuing variety of brake performance requirements has been reported by Aoki (1)**, illustrating the diversity of legislation in this field - in fact, there are often different requirements for passenger cars, trucks, buses, trailers, motorcycles etc. (The legislative requirements for motorcycles, agricultural tractors, and other off-highway machinery have been omitted from this presentation). Nevertheless, this paper aims to show that some commonisation is now under way.

4. For more than 15 years, the United Nations' Economic Commission for Europe (ECE) has progressively "harmonised" the individual European regulations for vehicles and equipment, despite numerous national, political, and social obstacles and enormous differences in attitudes and languages existing throughout Europe. There are also different engineering philosophies within the European motor industry which have been incorporated into production and which are sometimes difficult to reconcile. The recent establishment of the Common Market European Economic Community (EEC) has accelerated the adoption of uniform vehicle regulations within the Community. Consequently, it now becomes possible to confine the discussion of braking regulations in Europe to the ECE, EEC, and Swedish requirements, because these regulations include all the most difficult technical specifications and, at the same time, they cover most of the important motoring territories in Europe.

RULEMAKING PROCEDURES

5. Two supranational organisations dominate the rule-making scene in Europe today: the ECE and the EEC. Despite many apparent similarities between these organisations, it is vital to recognise their fundamental differences, as well as their respective territorial boundaries, in developing and implementing new vehicle regulations.

ECE

6. The ECE (Fig. 1) is a United Nations organisation with headquarters in Geneva; one of its prime

*Note: The explanations, interpretations, and summaries in this paper are presented in good faith but without authority, and should never be applied without reference to the latest official text of the corresponding regulations.

**Numbers in parentheses designate References at end of paper.

© IMechE 1976

objectives is to promote free trade throughout Europe by eliminating trade barriers. Accordingly, the ECE aims to harmonise the various European national regulations, which currently constitute an invisible tariff barrier, and all European countries are potentially involved.

7. The ECE procedure for harmonising vehicle and equipment regulations is governed by the 1958 Geneva Agreement "Concerning the adoption of uniform conditions of approval and reciprocal recognotion of approval for motor vehicle equipment and parts". This provides for the development of draft regulations by WP29 (Working Party 29) or the "group of experts on the construction of vehicles", which become available for voluntary adoption by any of the contracting parties to the Geneva Agreement. When two such countries agree to adopt a draft regulation, the latter is submitted to the United Nations in New York for ratification and then becomes an ECE regulation.

8. The two sponsoring countries thereby agree to approve equipment which complies with the ECE regulation and to accept equipment which has been approved to the same ECE regulation elsewhere. An appropriate E-mark is assigned to the equipment by the relevant national authority to denote approval. Other countries may subsequently adopt the same ECE regulation on identical terms. The voluntary nature of this sechme is evident from the selective adoption by various contracting parties of the available ECE regulations (Fig. 2).

9. Additionally, many other European and extra-European countries are basing their domestic regulations and approvals on ECE regulations. However, ECE Regulation 13 - "Uniform provisions concerning the approval of vehicles with regard to braking" (2) - has only been adopted by the two sponsoring countries: Italy and the Netherlands. The reasons for this apparent lack of interest by other countries will be explained later.

10. Amendments and annexes to existing ECE regulations are also discussed and approved by WP29, whose three or four meetings per year in Geneva usually last a whole week and are conducted in two official languages, French and English. Membership of WP29 is limited to official government delegates, usually from each country's Ministry of Transport, but it is not confined to the contracting parties. In fact, 22 governments currently particpate in this rulemaking action, including representatives from countries outside Europe such as Australia, Canada, Japan, and the United States (Table 1.). Observers from various international organisations are also invited to attend and represent the manufacturers and users of the relevant vehicles and equipment. This representation is naturally limited and very few manufacturers actually attend the WP29 meetings.

11. Nevertheless, the manufacturers' comments, objections and proposals may be voiced at WP29 meetings via one or more of the following communication channels:

1. Via the national government delegate.
2. Via the national manufacturers association and the national government delegate.
3. Via the national manufacturers association and the worldwide industry association Bureau Permanent International des Constructeurs d'Automobiles (BPICA).
4. Via the national standards organisation and the worldwide International Organisation for Standardisation (ISO).

12. Fig. 3 symbolises the central, autonomous government committee structure and the limited industry participation from the "outside" via the BPICA and ISO; Appendix A lists the corresponding abbreviations. Fig. 3 also explains how WP29 has delegated the detailed preparation and discussion of technical proposals and amendments to several "groups of rapporteurs" of which the Groupe des Rapporteurs-Freinage (GRF) deals with braking problems. GRF meetings are again composed of government representatives, but from only 12 countries, and with more specialised knowledge of the subject. Usually, one permanent representative from ISO and three or four selected delegates from the BPICA are allowed to present the manufacturers' viewpoint to the GRF.

13. For this purpose, the BPICA has established "groups of experts" from the industry to correspond with the "groups of rapporteurs" from the governments, and most of the important European and Japanese manufacturers are represented at the Groupe des Expert-Freinage (GEF) meetings, which usually precede the GRF meetings by 24 h and deal with the same agenda, thus attempting to establish a common manufacturers' opinion for presentation to the GRF by the selected BPICA delegation.

14. The ISO links with WP29 afford another opportunity for the manufacturers to explain their arguments; indeed, ISO are often charged with the development of specific test procedures or component standards for existing or proposed ECE regulations.

15. The relevant ISO structure comprises TC22 (Technical Committee 22, Automobiles) and its SC2 (subcommittee 2, Braking and Mechanical Couplings), whose current work programme has been assigned to six working groups, each dealing with a specific subject enumerated in Fig. 3. SC4 (Subcommittee 4, Caravans and Light Trailers) also covers some braking topics, such as vacuum and electric braking systems.

16. The ISO working group procedure involves nominated experts from various countries who participate in the development of appropriate draft standards which are later approved, amended, or rejected by the official national delegation at subcommittee level. The next stage is a postal ballot of the ISO membership, eventually culminating in a new ISO standard. It is important to note that ISO standards are entirely voluntary and no country or national standards institute is obliged to adopt an ISO standard. However, the draft may also be communicated to the sponsoring ECE committee and then be adopted as a basis for legislation, as well as standardisation, thus making it vital for the appropriate manufacturers to participate in the ISO deliberations.

EEC

17. The EEC, or Common Market, was created by the Treaty of Rome in 1957 and originally comprised six member states; three further countries, including the United Kingdom (UK), joined on Jan. 1, 1973 (Fig. 4). The administrative headquarters are in Brussels, where the EEC Commission staff of over 6000 people is responsible for implementing the Community policies.

© IMechE 1976

18. The Treaty of Rome includes provisions for the creation of a single market, involving the removal of all trade barriers and thus leading to the harmonisation, or "approximation", of vehicle regulations within the Common Market territory. The implementation of a type approval procedure for motor vehicles and trailers will enable any vehicle which complies with Community regulations approved by the authorities of a member state to be sold and used throughout the territory of the Common Market. Two technical committess of the Commission are concerned with vehicle legislation:

1. The Comité pour l'adaptation (CPA), or Committee for the Adaptation of the Directives to technical progress.

2. The Groupe des entraves (GDE), or Working Group on the Elimination of technical barriers to trade - motor vehicles.

19. Both are staffed by "Eurocrats" from the EEC Commission, whose allegiance is confined to the EEC Community and who act without favour to any of the member states. Nevertheless, the predominant use of the French language in all aspects of the committee work is particularly noticeable, often resulting in subsequent differences of interpretation. Government representatives from all the Common Market countries are invited to participate in the committee work.

20. Representation from the motor industry was originally confined to a single delegate from the Comité de Liaison des Constructeurs Automobiles (CLCA), comprising the national manufacturers' associations within the Common Market. The equivalent equipment manufacturers' association Comité de Liaison des Constructeurs d'Equipement et des Pièces d'Automobiles (CLEPA), and the European Standards Organisation Comité Européen de Normes (CEN) have also been privileged to attend more recent meetings in Brussels, which are held irregularly and often at short notice.

21. Because of the importance of these meetings, the Committee of Common Market Constructors (CCMC) has recently appointed two permanent representatives in Brussels to follow the proceedings. The CCMC, whose existence was announced in 1972, includes all the major European vehicle manufacturers and aims to ensure that government regulations on vehicle safety and environmental problems are identical throughout the Commmon Market.

22. The EEC technical requirements for vehicles have to be approved by the Council of Ministers and are issued in the form of directives to all member states. The intent of such directives must be introduced into the national legislation within a specified time (usually 18 months) and vehicles complying with the directive cannot be refused EEC or national approval after that date in respect of the equipment legislated in the directive. The corresponding national regulations may be retained for a certain period of time, during which vehicle manufacturers have the option of choosing to meet either the EEC directive or the relevant national regulations of all the EEC countries in which they market their vehicles.

23. The essential difference between EEC regulations and EEC directives is that the former are voluntary and the latter are compulsory within the Common Market countries. Consequently, progress in Brussels has been relatively slow, because the approval of a new directive or annex commits the nine member states to the introduction of appropriate national legislation within a specified time limit and to the subsequent admission of (foreign) vehicles type approved to the new directive, whereas approval of an ECE regulation involves no such positive commitment.

24. The committee structure and technical representation in Brussels is less elaborate and, in consequence, less challenging than in Geneva, and because all the EEC governments (except Luxembourg) also participate in WP29, it is not surprising that most of the current EEC directives (Table 2) are virtual copies of the corresponding ECE regulations.

25. EEC Directive 71/320 "On the approximation of the laws of the member states relating to the braking devices of certain categories of motor vehicles and of their trailers" (3) for example, is almost identical to ECE Regulation 13, except that certain editorial changes have altered the numbering of the paragraphs and annexes. Because this EEC directive on "Braking" is now valid within the Common Market countries, there is no further incentive to adopt ECE Regulation 13. In fact, France has already announced that the EEC directive on "Braking" will become compulsory for new vehicles registered after Oct. 1, 1975 (12 months later for cars and light trucks) and the national "Code de la Route" will be abolished. Other member states are expected to make similar announcements.

26. Amendments and new annexes to existing directives are handled by the GDE and CPA, who have also established a variable time scale for the introduction dates of such amendments and annexes, and for the dates after which vehicles can no longer be approved to the former requirements.

27. The similarity between ECE and EEC procedures also extends to the approval marking scheme: the EEC use a small "e" instead of the ECE's large "E" to denote type approval of the relevant equipment.

Sweden

28. The Swedish government has pioneered a variety of road safety regulations, imposing more rigorous vehicle/equipment specifications than ECE, EEC, or any other individual country. This is particularly true in the case of vehicle braking, making it imperative to consider Swedish requirements in any review of braking regulations.

29. The Swedish rulemaking procedure does not involve other governments and is, therefore, relatively straightforward. New vehicle regulations are prepared by the Swedish National Road Safety Board (NRSB) and issued under government authority, as are amendments and interpretations. As a member of WP29, Sweden has adopted many ECE regulations (refer back to Fig. 2), while several other Swedish national regulations have been based on the corresponding ECE regulations, but with modifications to suit the local conditions or their special requirements.

© IMechE 1976

30. Consultation with vehicle manufacturers prior to new legislation may be direct or via the Swedish Association of Automobile Manufacturers and Wholesalers. Occasionally, joint working groups may be established and BPICA viewpoints invited. Exceptionally, a formal meeting between the NRSB and the manufacturers may be arranged, such as the June 1971 conference in Stockholm to discuss the proposed introduction of antiskid legislation in Sweden by 1975.

TYPE APPROVAL

31. All of the important new European vehicle regulations (including those emanating from ECE,EEC and Sweden) are based on the principle of type approval. This makes the manufacturer responsible for complying with all the legal construction and performance requirements in respect of every one of his vehicles in the new condition; previously, some regulations such as the UK "Construction and Use" requirements, had placed the onus for compliance on the user or owner of the vehicle. At the same time, type approval also commits the authorities, who are obliged to confirm that their interpretation of the regulations has been satisifed.

32. The type approval procedure entails submission of a representative sample vehicle to the designated national type approval authority for compliance testing and formal approval. The subsequent type approval certificate allows the vehicle to be sold without restriction in the appropriate territory; in the case of an EEC approval, throughout the Common Market and with the corresponding e-mark.

33. The EEC type approval scheme also compels manufacturers to certify "conformity of production" and to issue a certificate of conformity with each vehicle sold. The authorities reserve the right to make verifications and check that every vehicle produced is identical to the sample submitted in respect of the legislated equipment and performance; this may infer that adequate production records of product quality and performance should be maintained.

34. The EEC type approval is carried out by the individual national approval authorities. In the UK, vehicle tests will be performed at the Motor Industry Research Association (MIRA) under supervision from the Department of the Environment (DOE) or at approved manufacturers premises, but again under supervision of an inspector from the DOE. The actual brake test procedure is still under development, but includes both static and dynamic tests using some sophisticated equipment for the continuous monitoring and recording of stopping distance, average deceleration, and maximum ("fully developed") retardation.

35. The brake type approval test in Sweden is handled by the NRSB in Stockholm and includes road testing on a nearby military airfield, whose main runway is maintained at a Skid Number between 75 and 85; this is particularly important for the checking of the wheel-locking sequence. The parking brake test is performend on an inclinable platform with a special surface to simulate the stipulated 0.6 coefficient of friction in a dry condition. The overall test sequence has now been clearly defined and made available to the manufacturers. It begins with the water soaking and recovery test, which is rather unusual and may lead unsuspecting manufacturers to fail the subsequent fade and effectiveness tests. Occasionally, and especially during the long Swedish winter, these approval tests may be carried out at the manufacturer's test track under supervision from the NRSB.

36. Neither the ECE/EEC nor the Swedish brake type approval test procedures include any brake lining burnishing. It is accepted that the manufacturer will prepare his sample vehicle in accordance with his own schedule, but this raises an interesting question as to how much "preparation" is permitted. Some manufacturers may wish to "prefade" or "condition" their brake linings to provide a better fade resistance during the approval tests; others may wish to optimise the parking brake adjustment, etc. There is no official guide available, but it is hoped that these items will be covered in the development of a standard brake test procedure in the ISO/TC22/SC2 Working Group 6.

BRAKING REGULATIONS

ECE/EEC

37. The similarity between ECE Regulation 13 and EEC Directive 71/320 has already been explained. The only real differences lie in the amendments and new annexes, some of which have been approved by the ECE but not yet introduced by the EEC. In general, therefore, these two regulations can be discussed together.

38. Fig. 5 summarises the ECE/EEC brake performance requirements* and Appendix B summarises the brake system requirements; both are applicable for all types of braking systems including air and hydraulic actuation. The variety of vehicle categories is immediately apparent: M for passenger-carrying vehicles, N for goods vehicles, and O for trailers. (The ECE also includes L for motorcycles)

39. There are three or four further weight classifications within each vehicle category, based upon the maximum weight permitted by the manufacturer. Some dual-purpose vehicles, which can carry either goods or passengers, may be difficult to classify appropriately. This classification philosophy allows towing vehicles and trailers to be approved independently, but then necessitates certain compatibility requirements which have not all been finalised.

40. The prescribed braking effectiveness is based on the stopping distance, and there are tests scheduled in gear and in neutral, fully laden and lightly laden, and from various speeds up to 80% of the maximum vehicle speed.

41. Motor vehicles are subject to a fade test by repeated "snub" braking applications, while trailers undergo a continuous drag test. The heavier buses (M3), trucks (N3) and trailers (O4) must also pass the Type II drag test, which is particularly severe for the trailers because they have no benefit of engine braking. In fact, there is some opposition to this 12 min drag test, which may not be representative of service conditions and which may necessitate special brake linings and "pre-

*The units of measure in this paper correspond to the units used in the regulations.

© IMechE 1976

conditioning", thus defeating some of the tractor-trailer compatibility objectives.

42. The emergency brake performance must be achieved after any single failure in the service brake system, either by the surviving service brake circuit or by the parking brake; but, in the latter case, the "failed" service brake must still guarantee a certain, lower performance level. "Any single failure" includes a power source failure or inoperative power-assist unit. An elegant corollary prohibits the use of "redundant" members, that is, components which are intended to be automatically activated by a failure, but whose effectiveness is not checked in normal operation.

43. Most of the annexes are concerned with air brake systems: spring brakes, lock actuators, energy sources and reservoir capacities, and measurement of the brake system reaction times. The latter annex describes the use of a simulator to measure the reaction time of trailer brakes.

44. Another lengthy annex prescribes the design criteria and test procedures for overrun (or "inertia") brakes for caravans and light trailers up to 3500 kg maximum weight. This simple type of brake actuation is very widely used throughout Europe; recently a further simplification has been introduced by eliminating the reversing catch in conjunction with a very low brake factor in the reverse direction of service brake operation.

45. The final annex describes the type approval certificate, and considerable importance is being attached to the inclusion of the brake lining material "make and type"; by inference, the identical replacement lining must be used in service.

Sweden

46. The Swedish braking regulations F.18 (4) were published in 1969 and became effective for 1972 models; several amendments and interpretations have been issued subsequently. These regulations have already made a considerable impact in Europe, because they contain a number of novel features and in toto they represented the most stringent braking regulations that were effective in the world at that time. The reasons for these severe requirements are probably twofold:

1. Sweden is exceptionally safety-conscious and there are strong pressures from trade unions for public safety.

2. Sweden has long periods of snow and ice, and subsequent salt on the roads, leading to skidding accidents and corrosion.

47. Fig. 6 summarises the brake performance requirements for the designated vehicle categories and Appendix C summarises the brake system requirements. The following more onerous requirements deserve particular attention:

48. (1) Wheel-locking sequence (only applicable for motor vehicles up to 3500 kg). "On a road surface coefficient of friction of 0.8 and with any permitted vehicle loading, no wheels may lock at decelerations up to 5.8 m/s^2 (0.59 g) and the rear wheels may not lock before the front wheels at decelerations between 5.8 and 8.0 m/s^2 (0.81 g)."

This requirement has led to the application of rear-axle brake apportioning valves on the majority of new vehicles in Sweden, with consequent improvements in stability during braking.

The concurrent requirement for a deceleration of 2.9 m/s^2 (0.3 g) after a circuit failure has raised further interest in the use of alternative divided-circuit solutions instead of the common front-to-rear split braking systems in use today (5).

49. (2) Automatic lining wear adjustment. "Brake lining wear must be automatically compensated by operation of the service brake or a light must warn the driver when adjustment is necessary to guarantee the emergency brake performance."

Divided braking systems can only provide continued assurance of the emergency brake performance if a requirement of this kind is included; yet no other braking regulation contains this provision and the reservoir capacity specification in FMVSS 105a does not cover this situation.

European brake manufacturers did not have such rear drum brakes with service-brake autoadjusters available, and this particular requirement was subsequently postponed for 12 months (and until Jan. 1, 1974 upon application by the vehicle manufacturer); but a variety of such brakes have now been developed and their use is spreading outside Sweden.

50. (3) Brake pipe corrosion. "Brake pipes shall have a corrosion resistance which is at least equal to that of a steel pipe covered with 0.025 mm of zinc."

The background to this requirement may be found in the statistics published by the Swedish Government Testing Stations, which carry out the compulsory annual inspections on vehicles in-use; brake pipe corrosion was found to be a major deficiency on a wide range of vehicle models. The zinc-coated brake pipes for Sweden have also been voluntarily introduced for other territories by some manufacturers, but there is no conclusive evidence yet that the zinc coating has improved road safety.

51. There are several other unusual requirements compared with the ECE/EEC regulations: the brake system must be capable of withstanding a pedal effort of 100 kg, there are specifications for brake release (as well as application) times, and the parking brake performance (16% grade) must be achieved on a road surface coefficient of friction of 0.6. The Swedish braking regulations F.18 also specify that used vehicles, at random checks or annual inspections, must achieve 90% of the type approval cold effectiveness requirements.

52. One unfortunate exclusion from both the Swedish and ECE/EEC braking regulations is the pressure vessel specification for compressed air reservoirs (and for full-power hydraulic accumulators). Sweden, Germany, and France have particularly stringent national requirements with respect to the materials and manufacturing processes for automotive air reservoirs and there is strong pressure at the BPICA to include an appropriate "harmonised" annex into the ECE/EEC and Swedish braking regulations thus including the air reservoirs in the type approval procedure for the braking system.

© IMechE 1976

PROPOSED AMENDMENTS AND ANNEXES

53. Most proposals for new European braking regulations derive from the ECE's WP29/GRF committees. The following three annexes are typical of the subjects under investigation.

Distribution of braking between axles (Annex 10)

54. This controversial addition to ECE Regulation 13 has also been accepted by the EEC authorities, but some important alterations are envisaged. Annex 10 (6) applies to all vehicle categories, except vehicles fitted with antiskid braking systems. It requires manufacturers to submit curves of "adhesion utilised (f)" plotted against "rate of braking (z)" for every axle and for both the fully laden and lightly laden conditions (Fig. 7). There are two basic requirements relating to vehicle stability and to adhesion utilisation, respectively:

55. (1) The front axle curves must lie above the corresponding rear axle curves between $z = 0.15$ and 0.8 for passenger cars, and between $z = 0.15$ and 0.3 for all other vehicles except semitrailers.

56. (2) For values of k between 0.2 and 0.8, the rate of braking must not be less than $0.1 + 0.85 (k - 0.2)$, where k = theoretical tyre-to-road adhesion coefficient.

57. There are different requirements for semitrailers, based on the compatibility bands developed for Annex 13.

58. In all cases where an apportioning device is used to meet the Annex 10 requirements, the failure of the device or its control must not preclude the emergency brake performance; and the stroke and exit pressure range of load-sensing valves must be indelibly marked (except for passenger cars).

59. Finally, the type approval authorities may carry out pratical checks to verify compliance with this annex. This latter requirement has caused particular concern to the manufacturers, because there is no specified procedure for carrying out these practical tests, which may conceivably cause the vehicle to fail its type approval. Consequently, there is an urgent need to develop an acceptable practical test which verifies the intent of Annex 10.

60. This development is hampered by the difficulties encountered in measuring the theoretical tyre-to-road adhesion coefficient k in the formula. Conventional road surface friction measurements, such as the ASTM Skid Number, tend to measure the sliding or locked-wheel braking force coefficient, which is not appropriate to k in the Annex 10 formula. It would appear necessary to evaluate the peak braking force coefficient, which is difficult to measure accurately and consistently, and which may vary significantly with speed, load, tyre parameters, etc. The provision of a range of road surfaces corresponding to $k = 0.15 - 0.8$ would present enormous practical difficulties.

61. There is also some opposition to the use of design calculations as a basis for legislation, because regulations should be based on performance standards. In this case, the variability of the brake lining friction coefficients may be cited as having considerable effect on the wheel-locking sequence; this variability is not taken into account in the Annex 10 curves; similarly, the effects of engine braking are ignored, yet they may well cause premature rear-wheel sliding, especially on powerful cars with automatic transmission. For these reasons, there is further support to develop an equivalent practical road test as a substitute for the design calculations in Annex 10.

Compatibility between towing vehicles and trailers (Annex 13)

62. Because towing vehicles and trailers are approved independently, there is an obvious need for a regulation to describe the compatibility specifications with respect to the mechanical coupling devices, electric plugs, brake pipe connections, and the braking performance for the parts of any vehicle combination.

63. These latter braking compatibility requirements are contained in Annex 13 (7) to ECE Regulation 13, which was approved by WP29 in June 1973. The detailed specifications were worked out jointly by the BPICA and ISO braking experts during 1972, culminating in the three graphs of "braking force/static axle weight" plotted against "air line pressure" for trucks and trailers, tractors (for semitrailers), and semitrailers, respectively (Fig. 8). The permissible relationships are contained within the hatched areas and must be respected under all conditions of loading. This final condition is being strongly opposed by the vehicle manufacturerers, because the original graphs were based only on the "fully laden" condition - and the extension to "all conditions of loading" would necessitate some form of braking pressure modulation on every axle of all new drawing vehicles and trailers.

64. This important alteration was introduced by WP29 without consultation with industry, and many arguments are being advanced against extending the use of load-sensing brake apportioning valves, especially with higher input/output pressure ratios exceeding 6:1 for rear axles. In the case of trailers, there is only a limited available spring deflection and the same applies to front (steering) axles of drawing vehicles, where there are also installation difficulties for load-sensing valves.

Wheel-lock preventers (antiskid braking systems)

65. The ECE study programme includes the subject of "wheel-lock preventers", and early in 1973 a special ad hoc GRF study group was established to prepare appropriate recommendations which may ultimately form the basis of a new annex to ECE Regulation 13 and the corresponding EEC Directive 71/320.

66. The European manufacturers consider such legislative initiatives to be premature, preferring to restrict the international activities on this subject to the development of uniform test procedures and the standardisation of the associated electric plug and socket for vehicle combinations equipped with antiskid braking systems; both these topics are on the ISO/TC22/SC2 agenda.

© IMechE 1976

67. The manufacturers feel that vehicles with antiskid braking systems should be subject to the same regulations as all other vehicles, and the progressive manufacturer who voluntarily adopts an antiskid braking system should not be penalised by having to meet additional regulations; in fact, such additional regulations might well inhibit the voluntary adoption of antiskid braking systems.

68. Alternatively, the manufacturers prefer to retain the option of meeting more stringent future regulations by judicious brake selection, by fitting apportioning devices, or by adopting antiskid braking systems.

69. The legislators argue that the use of antiskid braking systems could lead to additional safety hazards on the road, thus demanding extra safeguards. For example, a German government proposal suggests that the antiskid system energy reserve should be capable of sustaining a full stop from maximum speed on a 0.2 adhesion surface and that the braking stability should not be impaired while crossing from a high- to a low-adhesion surface. The French government representative has suggested the following parameters as a basis for antiskid legislation:

70. (1) Rules regarding performance.

(a) Avoidance of wheel-locking over a range of speeds and adhesions.

(b) Minimum effectiveness.

(c) Rapid system response to variations in the prevailing adhesion.

(d) Stability and maneouvrability.

(e) Energy consumption during prolonged braking.

71. (2) Rules regarding failures.

(a) Immediate warning to the driver.

(b) Partial failure effectiveness.

(c) Appropriate logic for antiskid failure on one wheel or one axle out of two.

72. (3) Rules regarding vehicle combinations.

(a) Compatibility of combinations comprising equipped and non-equipped vehicles.

(b) Compatibility of different antiskid systems.

73. Further meetings of the ad hoc GRF study group will examine all the above-mentioned topics and supporting demonstrations of the associated test procedures may soon lead to the adoption of a common proposal.

74. The Swedish legislators have also intimated their interest in antiskid braking systems. Their current regulations state that "later on, brake systems shall be so constructed that all available friction between the wheels and the road surface can be utilised for braking; possibly in about 1975". As a result of the joint conference with industry in Stockholm in June 1971, this date will almost certainly be postponed.

75. Meanwhile, a joint project between the NRSB and the Swedish car manufacturers has been initiated under the Experimental Safety Vehicle (ESV) programme to evaluate "steerability during emergency braking" by accident investigations, simulation experiments, field testing, and cost-effectiveness studies (8). Successful conclusions from this project might include recommended test procedures and suggested performance standards for antiskid braking systems as a basis for Swedish legislation.

COMPARISONS BETWEEN EUROPE AND THE UNITED STATES

76. The analysis of the rulemaking procedures, type approval and the major braking regulations in Europe enables comparisons to be made with the corresponding situation in the United States.

Rulemaking procedures

77. Rulemaking in Europe has been shown to be very different from the federal procedures established in the United States. First, the European rulemaking action is confidential within the government and industry participants, and there is no opportunity for the public, the press, or the consumer to take a direct part in rulemaking. Furthermore, there are no open docket files to keep the public informed of any proposals or their progress.

78. Second, industry participation in supranational rulemaking is also strictly limited and there is no obligation on the legislator's part to respond to any objections; in fact, some regulations and amendments are approved without even an opportunity for industry comments. Nevertheless, there is usually a good understanding and mutual respect between the various European legislators and their national industry, with effective consultation making up for formal procedural deficiencies.

Type approval versus self-certification

79. The federal self-certification procedures concentrate all responsibility for meeting legislation on the manufacturers; compliance may be verified on any randomly selected vehicle and penalties may be imposed for failing to meet the federal standards.

80. The European type approval procedure would appear to offer important advantages for the manufacturers. Foremost, the manufacturer is able to select and prepare his sample vehicle which is to be tested by the authorities for type approval. There must inevitably be less chance of "failure", but even a failure to meet the regulations would incur no penalties from the authorities. The "conformity of production" certification is unlikely to involve statistical deviations to the same extent as the federal procedures.

81. There are also psychological and political advantages in involving the authorities in accepting the vehicle model in principle. By granting the type approval certificate, they are confirming that the regulations have been met and thus remove a heavy burden of responsibility from the manufacturer. At the same time, the authorities are committed to "interpreting" the regulations and defining the test procedures. The official publication of such interpretations and procedures is

© IMechE 1976

invaluable to the manufacturers, because the regulations themselves are often ill-defined in many respects. In the case of the EEC directives, it is further intended to "harmonise" these interpretations and procedures in Brussels, so that all member states will approve and reject vehicles on an equal basis.

82. But type approval may also bring disadvantages for the manufacturers; for example, any change in the regulations such as the addition of a new annex to a directive would require manufacturers to resubmit all current vehicle models to the authorities for verification that the new requirements had been met; this may or may not involve changes in the vehicle equipment. Conversely, the manufacturer might choose to introduce a change in the vehicle equipment, for example, due to sudden supply shortages. Any significant change in the equipment specification would have to be resubmitted for type approval, thus causing costly delays in production and marketing.

83. In any case, the type approval of a new model range with all the available options may prove extremely time-consuming, and because type approval must be obtained before vehicles can be sold, it often becomes necessary to submit preproduction or even prototype vehicles for type approval.

European regulations versus federal standards

84. The current ECE/EEC and Swedish braking regulations have been compared with the corresponding FMVSS 105a, Hydraulic Brake Systems (9), and FMVSS 121, Air Brake Systems (10), which are due to become effective on Jan. 1, 1976 and Jan. 1, 1975, respectively.

85. The first notable difference between the European and United States braking regulations concerns the method of vehicle classification (Fig. 9). The elaborate ECE/EEC categories easily embrace the other classifications, except that the United States has also chosen to differentiate between air and hydraulic brake actuation systems. Such classifications recognise the inferior vehicle performance of trucks and trailers, and accordingly prescribe reduced hot and cold brake effectiveness requirements. But the imposition of speed limits on many highways and the increasing power-to-weight ratio of commercial vehicles questions the need for classifications in the future; in the interests of road safety, all vehicles on the same roads and travelling at the same speeds should possess the same stopping capabilities.

86. Meanwhile, the differing stopping capabilities of cars and trucks are implicit in the comparisons of Figs. 10 and 11 which show the minimum legal brake effectiveness requirements in Europe and the United States, plotted in terms of maximum permitted stopping distances. The single most important feature of any braking system is the service brake stopping distance, and on this basis the federal standards are significantly more stringent than the European equivalents-by an extraordinary margin of about 20%. This trend is not repeated for the emergency (or secondary) brake, where the European regulations are generally more demanding than the federal standards. For trucks, Sweden has imposed more stringent effectiveness requirements than the ECE/EEC, but surprisingly the Swedish effectiveness is measured in terms of "fully developed retardation" rather than the more meaningful "stopping distance".

87. There are also many supplementary specifications concerning the effectiveness tests in each of the regulations: the permitted range of pedal efforts, the allowable course deviations during the tests, the road test surface specification, and the vehicle test weights are examples of additional parameters which affect the potential stopping distance.

88. The parking brake specifications are summarised in Fig. 12 and the difficult FMVSS 105a requirement of 30% grade holding is immediately apparent, especially when associated with the maximum permitted hand lever effort of only 90 lb. Vehicles with automatic transmission may opt for the 20% grade hold, which is more in line with the other international parking brake requirements. The apparently less stringent Swedish performance (16%) is associated with a Skid Number of 60, which can prove troublesome in the "unladen" condition as tested on the tilting platform; but this is a vehicle weight distribution problem rather than a braking problem, and may lead to the adoption of front wheel parking brakes.

89. Figs. 13 and 14 express the "average power absorbed per cycle" and the "total energy input per fade test" per kilogramme of the maximum vehicle weight. Only the second fade test has been included for FMVSS 105a, because there are recovery and reburnish procedures between the two fade tests. The average power absorption in FMVSS 105a depends on the acceleration capability of the vehicle on test, because the cycles are repeated at 0.4 mile intervals, rather than at fixed time intervals. The main feature of these two figures is the high power absorption during FMVSS 105a for vehicles up to 10,000 lb, yet the total energy input is less than in Europe for the light vehicle categories.

90. There are surprisingly few differences between the heavy truck fade tests in Europe and the United States. But the heat energy input is only a part of the fade test; even more important are the residual and recovery performance requirements, as expressed in Fig. 15 for passenger cars. There are no recovery stops in the European procedures and up to 40% of fade may be permitted during "hot" effectiveness tests; there is no limit on over-recovery. By contrast, FMVSS 105a includes five recovery stops and prescribes permitted positive and negative pedal effort variations. Compared with actual vehicle experience during mountain descents, the use of target decelerations in FMVSS 105a is more representative than constant pedal effort used in the European fade tests. Furthermore the higher decelerations during the FMVSS 105a heating cycles are likely to constitute a more searching test.

91. There is no water recovery test in the ECE/EEC procedure, but the Swedish test might appear more severe than FMVSS 105a (Fig. 16), partly because there is less energy input per stop to dry the brake linings, but mainly because the permitted increase in pedal effort is considerably less than in FMVSS 105a. However, the increased number of recovery stops may alleviate these problems.

© IMechE 1976

92. The final comparison figures are principally concerned with air brake systems. Both the brake application and release time requirements in FMVSS 121 would appear to be more stringent than in Europe (Fig. 17); similarly, the energy reservoir capacity specification and compressor pump-up time in FMVSS 121 appear to exceed the corresponding European specifications (Fig. 18). There is some speculation why the requirements should be so different. In the case of reservoir capacity, the ECE/EEC requirement is a true performance standard, whereas the Swedish and FMVSS 121 requirements are design specifications.

93. Perhaps even more important than the above comparisons may be those legal requirements which are NOT repeated on both sides of the Atlantic. Tables 3 and 4 summarise the respective brake test procedures and brake system requirements, clearly illustrating the subjects which are unique specifications in either ECE/EEC, Sweden or the United States.

94. Most noteworthy tests (Table 3) are the ECE/EEC effectiveness tests "in gear" at up to 80% of the maximum vehicle speed, the Swedish wheel-locking sequence specifications, the FMVSS 105a preburnish effectiveness tests and spike stops, and the FMVSS121 wet road stops on a surface of Skid Number 30 and dynamometer brake tests.

95. In terms of system requirements (Table 4), there are no unique specifications in the ECE/EEC. The Swedish system strength test (100 kg pedal effort) bears some resemblance to the FMVSS 105a spike stops, but the static execution precludes validation of the dynamic torque loadings. The brake pipe corrosion resistance specification and automatic lining wear adjustment by operation of the service brake are unique Swedish requirements which have been explained in a previous section of this paper. The Swedish demands for brake fluids and hoses refer to the corresponding SAE specifications. The important unique features in the federal standards are the anti-lock failure indicator, which officially recognises that such systems exist - an important omission in the current European regulations; and the associated FMVSS 106 and 116 on brake hoses and brake fluids, respectively, which impose stringent performance and labeling requirements. The parking brake indicator and fluid reservoir capacity/labeling requirements are additional safeguards without legal counterparts in Europe.

96. Of course, there are other brake system requirements in the United States apart from the federal standards. For example, there are state laws on disc/drum machining marking and on brake lining classification; such laws are aimed at the aftermarket. There are no equivalent laws in Europe, except that type approval implies that all critical replacement parts must be identical to the original equipment specification.

97. Further comparisons may be based on "in service" brake performance requirements. Sweden has already implemented such specifications and the ECE/EEC are expected to prescribe similar effectiveness demands at 90% of the type approval requirement level. The federal recommendations for "in service" braking standards appear to be less demanding than in Europe, being primarily aimed at checking the correct functioning of the various system components, rather than imposing high levels of overall vehicle braking performance.

98. Any comparison of braking regulations between Europe and the United States must acknowledge the spectacular pace of the federal legislation: the comparatively recent inception of the National Highway Traffic Safety Administration (NHTSA) and the introduction of the first federal standards in 1968 have been matched by corresponding practical improvements in the braking systems of American vehicles, while the forthcoming application of FMVSS 121 and 105a may anticipate further significant advances in the art of vehicle braking.

SUMMARY

99. Three European braking regulations are of overriding importance:

(a) ECE Regulation 13, not because it has been widely adopted, but because the originating WP29/GRF intergovernmental committee structure provides a unique worldwide forum for formulating new braking regulations.

(b) EEC Directive 71/320, because its provisions are becoming mandatory within the nine member states.

(c) Swedish Regulation F.18 (effective since 1972) because it contains numerous novel and stringent technical requirements not found elsewhere.

The technical content of these three braking regulations has been summarised and the status of several controversial new proposals has been explained.

100. The rulemaking procedures in Europe are essentially confidential, the effective date of new regulations is often unclear and unique interpretations are difficult to obtain; the variety of European languages serves to accentuate these problems. In contrast, the federal procedures are available to everyone, with clearcut rules about effective dates and interpretations provided by NHTSA - and there are no linguistic problems in the United States.

101. The European vehicle regulations are based on type approval and certificates of conformity, whereas the FMVSS rely on self-certification; there are associated advantages and disadvantages for the manufacturers.

102. The extensive comparisons between the European braking regulations and the corresponding federal standards confirm some of the exceptional requirements in FMVSS 105a and 121. But despite significant differences in system requirements, test procedures, and acceptance levels, there are no fundamental reasons why the technical requirements in Europe and the United States cannot be reconciled, thus leading to the eventual adoption of uniform worldwide braking regulations.

© IMechE 1976

ACKNOWLEDGEMENTS

103. The author gratefully acknowledges the support from colleagues at Girling and the exchange of information within the motor industry. Permission to publish this paper has kindly been granted by the directors of Girling Ltd. and of Lucas Industries Ltd. The S.A.E. has granted permission to reprint this paper for the 1976 I. Mech. E. Conference on "Braking of Road Vehicles".

REFERENCES

1. K. Aoki, "A Comparison of World Braking Standards with Reference to the Development of Japanese Braking Standards". SAE Transactions, Vol. 81 (1972), paper 720030.

2. "Concerning the Adoption of Uniform Conditions of Approval and Reciprocal Recognition of Approval for Motor Vehicle Equipment and Parts". Addendum 12: Regulation No. 13, "Uniform Provisions concerning the Approval of Vehicles with regard to Braking". United Nations E/ECE/TRANS 505/Rev.1/ Add. 12, May 29, 1969.

3. "On the Approximation of the Laws of the Member States relating to the Braking Devices of certain Categories of Motor Vehicles and of their Trailers" (71/320/EEC). Council Directive of July 26, 1971, published in the Official Journal of the European Communities, No. 202, Sept. 6, 1971.

4. "Regulations for Braking Systems on Motor Vehicles and Trailers which are coupled to Vehicles" (F.18-1971) by the Swedish National Road Safety Board. Translated and published by Intereurope Ltd., London, England.

5. B. Ingram and P. Oppenheimer, "Safer Braking Systems". Proc. Auto. Div. Inst. Mech. Engrs., Vol. 187 (1973), paper 10/73, March 1973.

6. "Annex 10: Distribution of Braking among the Axles of Vehicles". United Nations ECE/WP29 Fortieth Session: W/TRANS/WP29/493/Add. 1, May 18, 1973.

7. "Annex 13: Requirements for Compatibility between Drawing Vehicle and Trailer". United Nations ECE/WP29 Fortieth Session: W/TRANS/WP29/ 493/Add. 1, May 18, 1973.

8. S. Rundkvist, "Steerability during Emergency Braking". Paper presented at 4th International ESV Conference in Kyoto, Japan, March 1973 (Swedish experimental safety vehicle programme).

9. Federal Motor Vehicle Safety Standard No. 105a, Hydraulic Brake Systems. Docket No. 70-27; Notice 8, published in Federal Register, Vol. 38, No. 96 (May 18, 1973).

10. Federal Motor Vehicle Safety Standard No. 121, Air Brake Systems. Docket No. 73-13; Notice 1, published in Federal Register, Vol. 38, No. 109 (June 7, 1973) - and referenced previous notices.

© IMechE 1976

Appendix A

List of Abbreviations

Government organisations and committees

ECE = Economic Commission for Europe (United Nations) - Geneva

WP29 = Working Party 29 (Group of Experts on the Construction of Vehicles)

GRF = Groupe des Rapporteurs-Freinage (Group of Reporters on Braking)

EEC = European Economic Community (Common Market) - Brussels

CPA = Comité pour l'Adaptation au Progrés Technique des Directives (Committee for the Adaptation of the Directives to Technical Progress)

GDE = Groupe des Entraves aux Echanges résultant des Dispositions d'Ordre Technique - Véhicules à Moteur (Working Group on the Elimination of Technical Barriers to Trade - Motor Vehicles)

NRSB = National Road Safety Board (Sweden)

DOE = Department of the Environment (UK)

DOT = Department of Transportation (U.S.A.)

NHTSA = National Highway Traffic Safety Administration (U.S.A.).

Manufacturers associationsand committees

ANFIA = Associazione Nazionale Fra Industrie Automobilistiche (Italy)

CSCA = Chambre Syndicale des Constructeurs d'Automobiles (France)

SMM&T = Society of Motor Manufacturers & Traders (UK)

VDA = Verband der Automobilindustrie (Germany)

AMA = Automobile Manufacturers Association (U.S.A.)

MVMA = Motor Vehicle Manufacturers Association (U.S.A.)

JAMA = Japan Automobile Manufacturers Association

BPICA = Bureau Permanent International des Constructeurs d'Automobiles (Paris)

TC = Technical Committee of the BPICA

GEF = Groupe des Experts - Freinage (Group of Experts on Braking)

CLCA = Comité de Liaison des Constructeurs Automobiles (Liaison Committee of the Vehicle Manufacturers Associationsin the EEC)

CLEPA = Comité de Liaison des Constructeurs d'Equipements et des Pièces d'Automobiles (Liaison Committee of the Equipment Manufacturers in the EEC)

CCMC = Committee of Common Market (Vehicle) Constructors

MIRA = Motor Industry ResearchAssociation (UK)

Standards organisations and committees

AFNOR = Association Francaise de Normalisation (France)

BNA = Bureau des Normes de l'Automobile (France)

BSI = British Standards Institute (UK)

DNA = Deutscher Normenausschuss (Germany)

FAKRA = Fachnormenausschuss Kraftfahrzeug-industrie (Germany)

UNI = Ente Nazionale Italiano di Unificazione (Italy)

CUNA = Commissione Tecnica di Unificazione nell'Autoveicolo (Italy)

ANSI = American National Standards Institute (U.S.A.)

JIS = Japan Industrial Standards

JASO = Japan Automobile Standards Organisation

ISO = International Organisation for Standardisation (Geneva)

TC22 = Technical Committee 22 (Automobiles) of the ISO

SC2 = Subcommittee 2 (Braking and Mechanical Couplings) of TC22

SC4 = Subcommittee 4 (Caravans and Light Trailers) of TC22

CEN = Comité Européen de Normes (Standards Committee of the EEC)

Appendix B

Summary of Brake System Requirements - ECE/EEC

1. Road test surface must have good adhesion.
2. No wheel lockup or deviation from path during testing.
3. Type approval certificate includes brake lining "make and type".
4. Divided service brake circuits (after Oct. 1, 1974 for M1-M2-N1-N2).
5. Manual or automatic lining wear adjustment.
6. Adequate reserve travels when brakes are hot or partly worn.
7. Transparent fluid reservoirs or low fluid level warning light.
8. Hydraulic failure warning light (red).
9. Low energy level warning (optical or acoustical) at 65% of normal pressure (in addition to pressure gauge).
10. No "redundant" members normally at rest and only activated by failure.
11. Brake system application time to prescribed performance at least favoured axle not to exceed 0.6 s. (N.B. Additional requirements for towing vehicles and trailers with air brakes).
12. Auxiliary equipment must not deplete service brake energy reservoirs below 65% of normal pressure.

© IMechE 1976

13. Full-power systems must have two independent reservoirs and circuits (each with low pressure warning device), but may use single pump.

14. Energy reservoir capacity must retain emergency brake performance after eight full applications.

15. System pump-up time at maximum engine rpm to 65 and 100% of the service brake pressure must be within 3 and 6 min, respectively, (6 and 9 min, for combinations).

16. Failure of parking brake power-assistance must enable another energy source to be used (also for release).

17. Failure of coupling or trailer brakes must not prejudice tractor emergency brake performance.

18. Trailer air brakes must use at least two lines.

19. Failure of coupling must automatically stop trailer (except single-axle trailers of less than 1.5 t using chain or cable).

20. Normal maximum trailer feed pressure to be between 6.5 and 8.0 bars.

Further specifications are contained in the following annexes:

1. Measuring the reaction time.
2. Energy sources and reservoirs.
3. Spring brakes.
4. Lock actuators.
5. Overrun brakes.

Future annexes under study by the GRF include:

1. Distribution of braking between the axles of vehicles.
2. Compatability between towing vehicles and trailers.
3. Wheel-lock preventers (antiskid systems).
4. Methods and equipment to be used for braking tests.
5. Brake linings.
6. Electromagnetic brakes.
7. Hydrodynamic brakes.
8. Full-power hydraulic braking systems.
9. Test requirements for trailer brakes.
10. Use of spring brakes as service brakes.
11. Parking brakes for combinations of vehicles.
12. Test method for mechanical coupling components.

Appendix C

Summary of Brake System Requirements - Sweden

1. Road tests on 0.8 friction coefficient (interpreted as Skid Number 80).

2. No uncontrolled deviations during testing.

3. Divided service brake circuits.

4. System must resist 100 kg pedal effort.

5. Brake pipe corrosion resistance equivalent to 0.025 mm zinc on steel.

6. Brake fluid to SAE 70R3.

7. Service brake automatic lining wear adjustment or warning indicator when lining wear precludes emergency brake performance.

8. Adequate reserve travels when brakes are hot and partly worn.

9. Transparent fluid reservoirs or low fluid level warning light.

10. Hydraulic failure warning light (red) and activation pressure.

11. Low energy level warning (red light or acoustical) at 65% of estimated pressure (in addition to pressure gauge).

12. No "redundant" members normally at rest and only activated by failure.

13. Brake system application (0-75%) and release (75%-10%) times not to exceed 0.6 s (0.4 s at coupling head of towing vehicles; 0.8 s at trailers in combinations).

14. Pressure tapping (M16 x 1.5) at least favoured air brake chamber.

15. Air reservoirs in accordance with Swedish Pressure Vessel Commission.

16. Auxiliary equipment must not deplete service brake energy reservoirs below 60% of estimated pressure.

17. Service brake energy reservoir capacity not less than 12X combined brake chamber volume at 2/3 maximum piston movement (8-12 for trailers).

18. System pump-up time at maximum engine rpm to 65% of estimated pressure within 4 min (6 min for combinations).

19. Towing vehicle-trailer compatibility relationships for control pressure versus deceleration.

20. Failure of coupling or trailer brakes must not preclude service brake function on towing vehicle.

21. Trailer air brakes must use two lines and predetermined coupling locations.

22. Maximum trailer feed pressure to be between 6.5 and 8.0 bars.

23. Trailers over 3500 kg must have automatic lining wear compensation.

24. Trailer axles whose laden-to-unladen weight ratio exceeds 4:3 must have automatic brake force regulators.

25. Failure of coupling or supply pressure must automatically stop trailer.

© IMechE 1976

Table 1 - ECE/EEC Membership Review

	Member of WP29 (ECE)	Contracting Party (ECE)	Member State (EEC)
Australia	X		
Austria	X	X	
Belgium	X	X	X
Bulgaria	X		
Canada	X		
Czechoslovakia	X	X	
Denmark	X		X
France	X	X	X
Germany	X	X	X
Hungary	X	X	
Ireland	X		X
Italy	X	X	X
Japan	X		
Luxembourg		X	X
Netherlands	X	X	X
Poland	X		
Spain	X	X	
Sweden	X	X	
Switzerland	X	X	
United Kingdom	X	X	X
United States	X		
Union of Soviet Socialist Republics			
Yugoslavia	X	X	

Table 2 - EEC Directives

No.	Subject	Date
70/156	Type approval of motor vehicles and trailers	Feb. 6, 1970
70/157	Noise levels and exhaust systems	Feb. 6, 1970
70/220	Air pollution (gasoline engines)	March 20, 1970
70/221	Fuel tanks and rear protection	March 20, 1970
70/222	Fitting of rear number plates	March 20, 1970
70/311	Steering systems	June 8, 1970
70/387	Doors, locks and hinges	July 27, 1970
70/388	Audible warning devices	July 27, 1970
71/127	Rear view mirrors	March 1, 1971
71/320	Braking equipment	July 26, 1971
72/245	Ignition suppression	June 20, 1972
72/306	Air pollution (diesel engines)	Aug. 2, 1972

© IMechE 1976

Table 3 - Brake Test Procedures in ECE/EEC, Sweden and U.S.A.

Brake Test Procedures	ECE/EEC	Sweden	USA 105a	USA 121
Instrumentation check	X	X	X	X
Preburnish effectiveness			X	
Burnish procedure			X	X
Cold effectiveness (in neutral)	X	X	X	X
Cold effectiveness (in gear)	X			
High-speed stops (80% V_{max})	X	X		
Wheel-locking sequence		X		
Wet road (SN30) effectiveness				X
Partial failure (emergency brake)	X	X	X	X
Inoperative power (assist) units	X	X	X	X
Fade and recovery	X	X	X	X
Water recovery		X	X	
Spike stops			X	
Parking brake (grade-hold)	X	X	X	X
Parking brake (dynamic tests)	X	X		X
Dynamometer brake tests				X

Table 4 - Brake System Requirements in ECE/EEC, Sweden and U.S.A.

Brake System Requirements	ECE/EEC	Sweden	USA 105a	USA 121
Divided circuits	X	X	X	
System strength		X		
Adequate pedal travel reserve	X	X		
No "redundant" members	X	X		
Brake application time	X	X		X
Brake release time		X		X
Warning indicators for:				
Pressure loss	X*	X*	X	X
Low fluid level	X*	X*	X	
Antilock system failure			X	X
Parking brake "applied"			X	
Component specifications for:				
Hydraulic brake hoses		X	X	
Hydraulic brake fluid		X	X	
Brake pipe corrosion resistance		X		
Automatic lining wear adjustment		X		
Transparent fluid reservoir	X*	X*		
Fluid reservoir capacity			X	
Fluid reservoir labeling			X	
Lining material approval	X	X		
Air and vacuum hoses		X	X	X
Energy reservoir capacity	X	X		X
Energy source capacity	X	X		X

*Alternatives.

© IMechE 1976

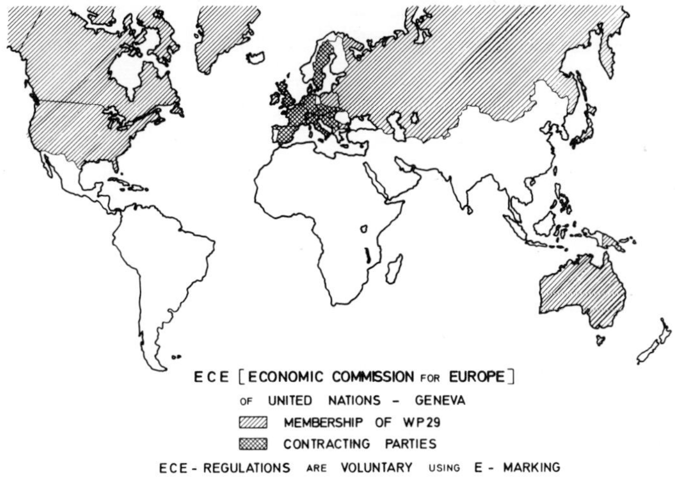

Fig. 1: ECE – Membership of WP29 and Contracting Parties

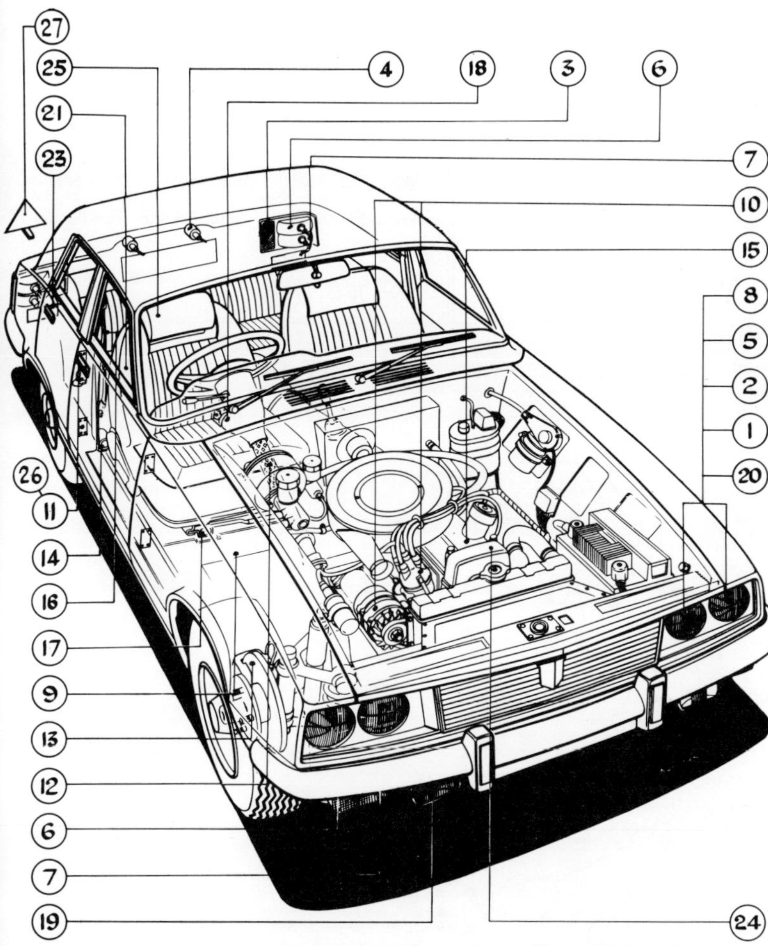

Nº	'E' REGULATIONS SUBJECT	GERMANY (W) E1	FRANCE E2	ITALY E3	NETHERLANDS E4	SWEDEN E5	BELGIUM E6	HUNGARY E7	CZECHOSLOVAKIA E8	SPAIN E9	YUGOSLAVIA E10	UNITED KINGDOM E11	AUSTRIA E12	LUXEMBOURG E13	SWITZERLAND E14	POLAND	DENMARK
1	HEADLAMP UNITS	X	X	X	X	X	X	X	X	X	X	X	X		O		
2	HEADLAMP BULBS	X	X	X	X	X	X	X	X	X	X	X	X		O		
3	REFLEX REFLECTORS	X	X	X	X	X	X	X	X	X	X	X	X		O		
4	NUMBER PLATE LIGHTING	X	X	X	X	X	X	X	X	X	X	X	X		O		
5	SEALED BEAM HEADLAMPS	X		X	X	X	X		X	X	X	X	X		O		
6	DIRECTION INDICATORS	X	X	X	X	X	X		X	X	X	X	X		O		
7	STOP, TAIL & SIDE LAMPS	X	X	X	X	X	X		X	X	X	X	X		O		
8	HALOGEN MAIN & DIPPED UNITS. H1.	X	X		X	X	X		X	X	X	X	X		O		
9	NOISE			X					X	X	X						
10	RADIO SUPPRESSION	X	X			X			X	X	X	X	X		O		
11	DOOR LATCHES & HINGES	X	X		X	X	X		X			X					
12	STEERING COLUMN COLLAPSE	X	X		X	X	X		X			X					
* 13	BRAKING			X	X												
14	SEAT BELT ANCHORAGE	X	X		X		X		X								
15	POLLUTION (PETROL ENGINES)	X	X	X	X		X		X	X		X			X		
16	SEAT BELTS	X	X		X		X		X	X					O		
17	SEATS AND THEIR ANCHORAGE	X	X		X	X			X			X					
18	ANTI THEFT	X	X		X		X		X	X		X					O
19	FOGLAMPS	X	X	X	X	X	X		X	X	X	X	X		O		
20	HALOGEN MAIN & DIPPED UNITS. H4.	X	X	X	X	X	X		X		X	X	X		O		
21	INTERIOR FITTINGS	X	X		/	X	X		X			X					
22	PROTECTIVE HELMETS (M/C's)				X	X	X										
23	REVERSING LIGHTS	X	X	X	X	X	X		X	X		X					
24	POLLUTION (DIESEL ENGINES)	X	X		/				X	X							
25	HEAD RESTS	X	X		X		/		X			X					
26	EXTERNAL PROJECTIONS	/	X			X	X		X			X					
27	ADVANCE WARNING TRIANGLES	/	X		X	X	X										
28	AUDIBLE WARNING	/	X	X		X				X							
	WINDSCREEN, DEFROST & DEMIST.																
	FIELDS OF VISION																
	COMPATIBILITY, VEHICLE & TRAILER																
	FIRE RISKS																
	VEHICLE STRUCTURE IN ACCIDENTS																
	CARAVANS																
	IMPACT PROTECTION (W'ON & O'TURN)																
	PNEUMATIC TYRES																
	ILLUMINATING & SIGNALLING LAMPS																
	REAR VIEW MIRRORS																
	ACCESSIBILITY OF CONTROLS																
	IMPACT PROTECTION (REAR & SIDE)																
	IMPACT ABSORPTION (FRONT)																
	IMPACT ABSORPTION (REAR)																
	POWER/WEIGHT RATIO (C.V.)																
	SUN VISORS																
	WINDSCREEN WIPERS																
	WINDSCREEN WASHERS																
	ACCELERATION (C.V.)																
	CLIMBING SPEED (C.V.)																

X ACCEPTING TERRITORIES
/ ACCEPTANCE PENDING
O 'E' APPROVAL IS ACCEPTED AS COMPLIANCE WITH NATIONAL REGULATIONS.

Fig. 2: ECE – List of Regulations and Adopting Countries

© IMechE 1976

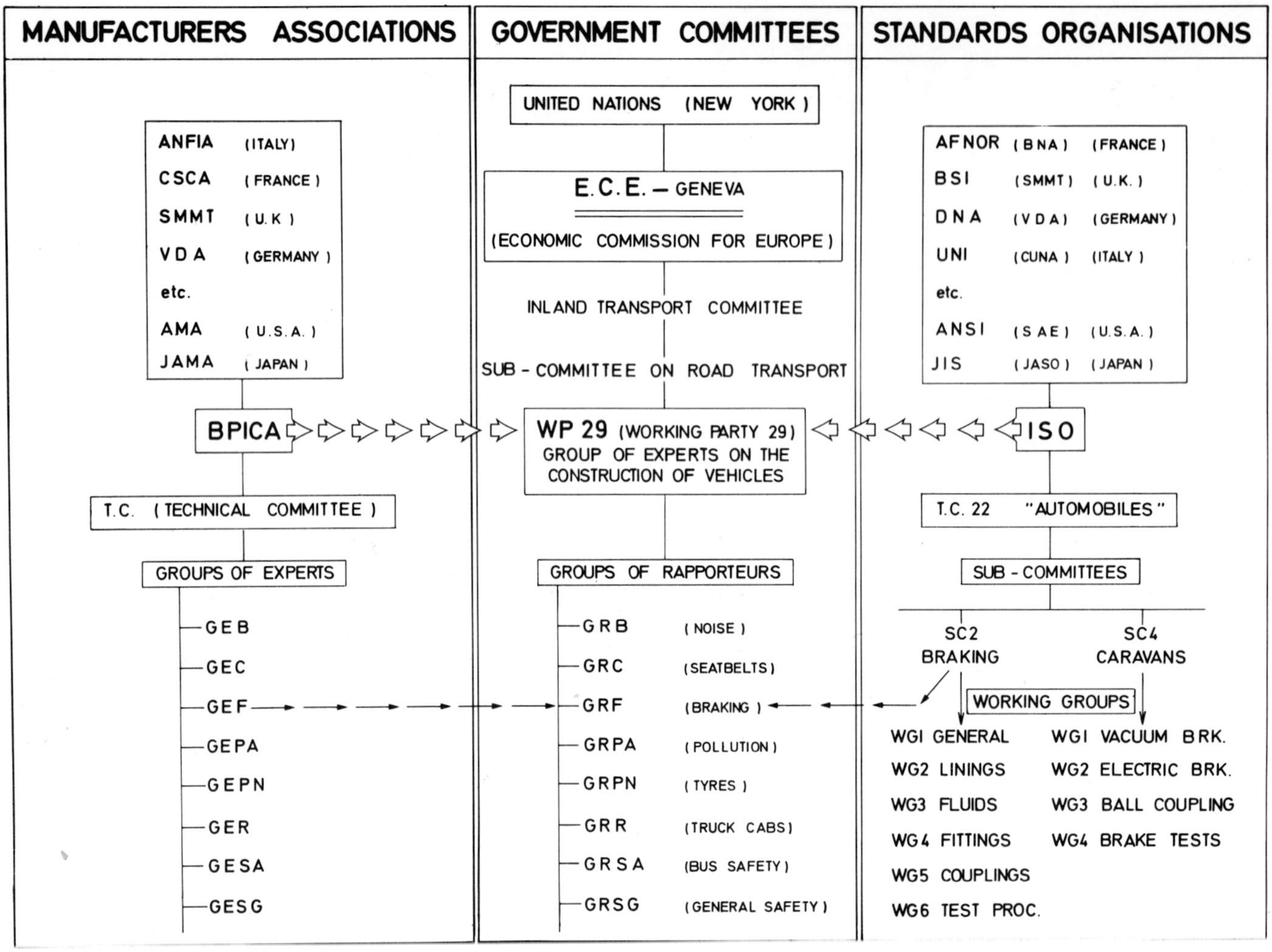

Fig. 3: ECE – WP29 Committee Structure and Associated Organisations

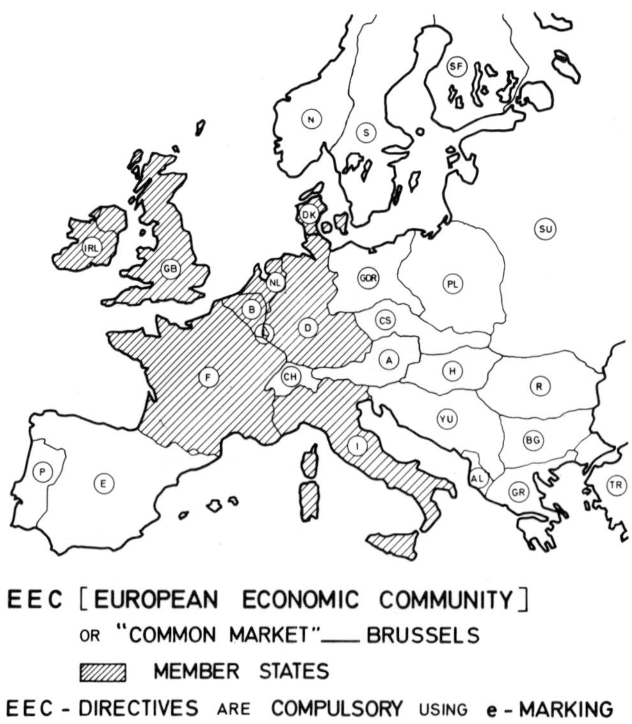

Fig. 4: EEC – Member States of the Common Market

© IMechE 1976

SUMMARY

ECE REGULATION 13 & EEC DIRECTIVE 71/320

BRAKE PERFORMANCE TESTS

FL = FULLY LADEN
AL = ANY LOADING

				CARS & BUSES M1	M2	M3	TRUCKS N1	N2	N3	TRAILERS O1	O2	O3	O4
		VEHICLE CLASSIFICATION		M1	M2	M3	N1	N2	N3	O1	O2	O3	O4
		No. OF PASSENGER SEATS (+DRIVER)		≤ 8	> 8	> 8							
		MAX. VEHICLE WEIGHT (TONNES)			≤ 5	> 5	≤ 3½	3½ – 12	> 12	≤ ¾	¾ – 3½	3½ – 10	> 10
	SERVICE BRAKE – ACTING ON ALL WHEELS, PROPERLY DISTRIBUTED, SYMMETRIC A/C VEHICLE												
1	EFFECTIVENESS TEST (TYPE 0 TEST) NEUTRAL – COLD – AL	INITIAL SPEED	KPH	80	60	60	70	50	40	SERVICE BRAKE IS OPTINAL	OVERRUN BRAKES PERMITTED EXCEPT FOR SEMI-TRAILER	TYRE DRAG ≥ 0.45 OF TRAILER MAX. WEIGHT (USING ≤ 6.5 BARS)	
		STOPPING DISTANCE	M	$0.1V + V^2/150$	$0.15V + V^2/130$		$0.15V + V^2/115$						
		MIN. DECELERATION	M/SEC^2	5.8	5.0	5.0	4.4	4.4	4.4				
		PEDAL EFFORT	KG	50	70	70	70	70	70				
2	HIGH SPEED TEST (TYPE 0 TEST) IN GEAR – COLD – AL			EFFECTIVENESS TESTS VARIOUS SPEEDS FROM 30% TO 80% OF V MAX.									
3	FADE TEST BY REPEATED BRAKING (TYPE I TEST) IN GEAR – HOT – FL. FIRST SNUB AT 3.0 $M/SECS^2$, REMAINING SNUBS AT SAME PEDAL EFFORT	V1 = 80% V MAX. BUT, ≤; V2 = ½ V1	KPH	120	100	60	120	60	60				
		TIME INTERVAL	SECS.	45	55	60	55	60	60				
		No. OF CYCLES		15	15	20	15	20	20				
4	FADE TEST BY CONTINUOUS BRAKING (TYPE I TEST) HOT – FL									DRAG TEST AT 40 KPH FOR 1,7 KM ON 7% SLOPE.			
5	RESIDUAL EFFECTIVENESS AFTER TYPE I TESTS NEUTRAL – HOT – FL			PERFORMANCE TO BE ≥ 80% OF TEST 1 REQUIREMENT AND ≥ 60% OF TEST 1 ACHIEVEMENT.									
6	DRAG TEST (TYPE II TEST) IN GEAR – HOT – FL					DRAG TEST AT 30 KPH FOR 6 KM ON 6% SLOPE (OR 0.5 M/sec^2 BY ENGINE ALONE)							
6a	DRAG TEST (TYPE II BIS TEST) IN GEAR – HOT – FL (FOR P.S.V. WITH > 8 PASS. SEATS AND MAX. WEIGHT > 10T.)					DRAG TEST AT 30 KPH FOR 6 KM ON 7% SLOPE (OR 0.6 M/sec^2 BY ENGINE ALONE) – WITHOUT USING SERVICE, EMERGENCY OR PARKING BRAKE.							
7	RESIDUAL EFFECTIVENESS AFTER TYPE II TESTS NEUTRAL – HOT – FL. N.B. TYPE II TESTS ONLY APPLY TO M3 – N3 – O4					PERFORMANCE TO BE ≥ 75% OF TEST 1 REQUIREMENT.							
	EMERGENCY BRAKE – MAY BE OPERATED BY SERVICE OR PARKING BRAKE CONTROL												
8	SERVICE BRAKE FAILURE (TYPE 0 TEST) NEUTRAL – COLD – AL	INITIAL SPEED	KPH	80	60	60	70	50	40				
		STOPPING DISTANCE	M	$0.1V + 2V^2/150$	$0.15V + 2V^2/130$		$0.15V + 2V^2/115$						
		MIN. DECELERATION	M/SEC^2	2.9	2.5	2.5	2.2	2.2	2.2				
		PEDAL EFFORT	KG	50	70	70	70	70	70				
		HAND LEVER EFFORT	KG	40	60	60	60	60	60				
8a	POWER ASSIST FAILURE – AS 8 BY SERVICE BRAKE OR AS 10 IF PB = EB			NB EMERGENCY BRAKES (AND DIVIDED – CIRCUITS) OF TRACTORS FOR O_3 – O_4 TRAILERS MUST ALSO ACTIVATE TRAILER BRAKES GRADUALLY.									
	PARKING BRAKE – CONTROL MUST BE INDEPENDENT OF SERVICE BRAKE CONTROL												
9	HILL HOLDING TEST NEUTRAL – COLD – FL	GRADIENT	%	18	18	18	18	18	18	PARKING BRAKE IS OPTIONAL	18	18	18
		HAND-LEVER EFFORT	KG	40	60	60	60	60	60		60	60	60
		FOOT-OPERATED EFFORT	KG	50	70	70	70	70	70				
	N.B. ALL PARKING BRAKES MUST BE CAPABLE OF APPLICATION "IN MOTION"			N B PARKING BRAKE OF TRACTORS MUST HOLD COMBINATIONS ON 12% SLOPE.						NB OPERABLE FROM OUTSIDE (AND FROM WITHIN ON P.S.V.) – WHEN SEPARATED FROM TRACTOR			
	DIVIDED CIRCUITS – OPERATED BY SERVICE BRAKE CONTROL ON SUFFICIENT WHEELS												
10	CIRCUIT FAILURE NEUTRAL – COLD – AL	MIN. DECELERATION (M/SEC^2) WITH 70 KG PEDAL EFFORT	FL	1.74	1.50	1.50	1.32	1.32	1.32				
			LL	1.45	1.25	1.50	1.10	1.10	1.32				
	N.B. NOT APPLICABLE TO TRACTORS FOR SEMI-TRAILERS WITH INDEPENDENT SERVICE BRAKES			AFTER 1 OCT. 1974			AFTER 1 OCT. 1974						

Fig. 5: Summary of ECE/EEC Brake Performance Requirements

© IMechE 1976

SWEDISH BRAKING REGULATIONS F.18

SUMMARY

BRAKE PERFORMANCE TESTS

FL = FULLY LADEN, AL = ANY LOADING

Test	Parameter	KG	Motor Vehicles <3500	Motor Vehicles >3500	Trailers <3500	Trailers >3500
SERVICE BRAKE — ACTING ON ALL WHEELS PROPERTY DISTRIBUTED SYMMETRIC A/C VEHICLE						
1 EFFECTIVENESS TEST, NEUTRAL–COLD–FL.	INITIAL SPEED	KPH	80	60	60	60
	MIN. DECELERATION	M/SEC²	5.8	5.0	5.8	5.0
	PEDAL EFFORT	KG	50	70		
2 WHEEL LOCKING TEST, ON 0.8 TYRE-TO-ROAD μ, NEUTRAL–COLD–AL.	NO WHEEL-LOCK ⩽	M/SEC²	5.8	5.0		
	NO REAR-LOCK BEFORE FRONT-LOCK ⩽	M/SEC²	8.0			
3 HIGH SPEED TEST, NEUTRAL–COLD–FL.	INITIAL SPEED	KPH	80%VMAX			
	MIN. DECELERATION	M/SEC²	4.0			
	PEDAL EFFORT	KG	30			
4 FADE TEST, BY REPEATED BRAKING, IN GEAR–HOT–FL.	VI = 80% V MAX BUT ⩽	KPH	120	60	(100) BUSES ⩽ 5000 KG.	
	V2 = ½ VI					
	TIME INTERVAL	SECS	45	60	(55) BUSES ⩽ 5000 KG.	
	No. OF CYCLES	–	15	20	(15) BUSES ⩽ 5000 KG.	
FIRST SNUB AT 3.0 M/SECS² REMAINING SNUBS AT SAME PEDAL EFFORT	MIN. DECELERATION	M/SEC²	1.8	1.8		
5 FADE TEST, BY CONTINUOUS BRAKING, –HOT–FL.					DRAG TEST AT 40 KPH FOR 1.7 KM ON 7% SLOPE	
6 RESIDUAL EFFECTIVENESS, AFTER TEST 4 OR 5, NEUTRAL–HOT–FL.			PERFORMANCE TO BE > 80% OF TEST I REQUIREMENT & > 60% OF TEST I ACHIEVEMENT			
7 WATER RECOVERY, AFTER 2 MINS. SOAK, NEUTRAL–COLD–FL.	INITIAL SPEED	KPH	40	40		
	DECELERATION	M/SEC²	2.5	2.5		
	MAX. PEDAL EFFORT	KG	50	70		
	No. OF CYCLES	–	10/15	15		
8 RECOVERY EFFECTIVENESS, AFTER TEST 7	PEDAL EFFORT AT–	[V. BASE PE]	+20LB			
	STOP No. 10		TO–40%			
	STOP No. 15		+20% TO –40%			
EMERGENCY BRAKE — TEST 9 MAY BE OPERATED BY SERVICE OR PARKING BRAKE CONTROL						
9 CIRCUIT FAILURE, NEUTRAL–COLD–AL.	INITIAL SPEED	KPH	80	60		
	MIN. DECELERATION	M/SEC²	2.9	2.5		
	PEDAL EFFORT	KG	50	70		
	HAND LEVER EFFORT	KG	40	60		
10 POWER ASSIST FAILURE, NEUTRAL–COLD–FL.	INITIAL SPEED	KPH	80	60		
	MIN. DECELERATION	M/SEC²	2.9	2.5		
	PEDAL EFFORT	KG	50	70		
PARKING BRAKE — CONTROL MUST BE INDEPENDENT OF SERVICE BRAKE CONTROL						
11 DYNAMIC TEST	INITIAL SPEED	KPH	20	20	20	20
12 HILL-HOLDING TEST, ON 0.6 TYRE-TO-ROAD μ, NEUTRAL–COLD–AL.	GRADIENT	%	16	16	16	16
	HAND-LEVER EFFORT	KG	40	60	60	60
	FOOT-OPERATED EFFORT	KG	50	70		

Fig. 6: Summary of Swedish Brake Performance Requirements

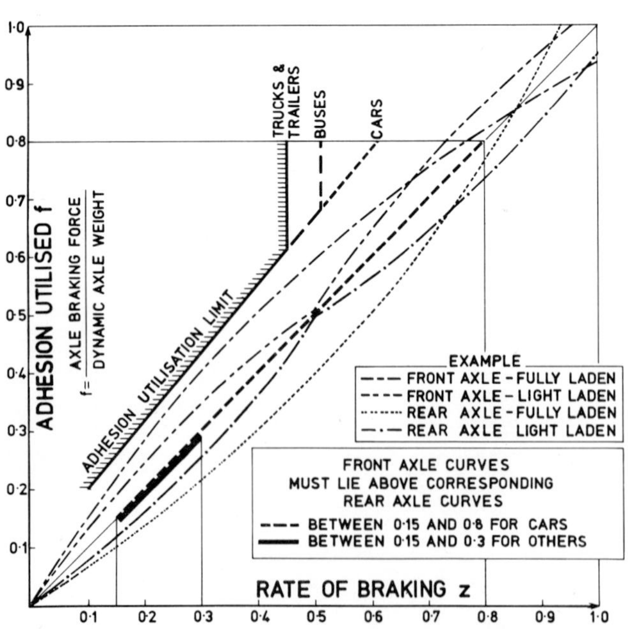

Fig. 7: ECE Annex 10 – Distribution of Braking between Axles

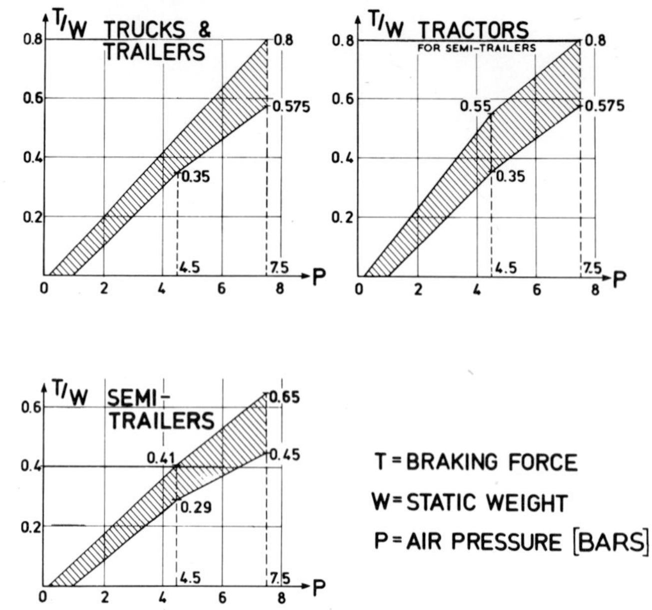

Fig. 8: ECE Annex 13 – Compatibility between Towing Vehicles and Trailers

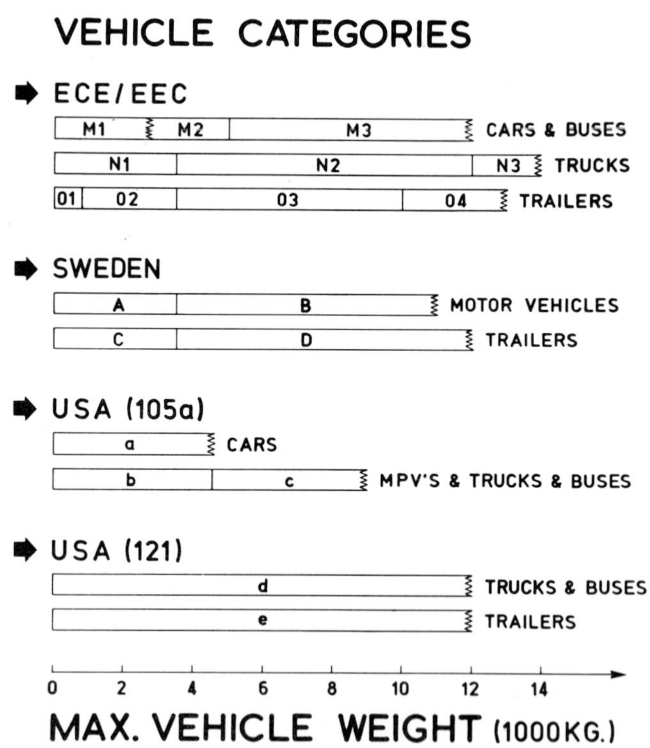

Fig. 9: Comparison of Vehicle Categories in ECE/EEC, Sweden and USA

© IMechE 1976

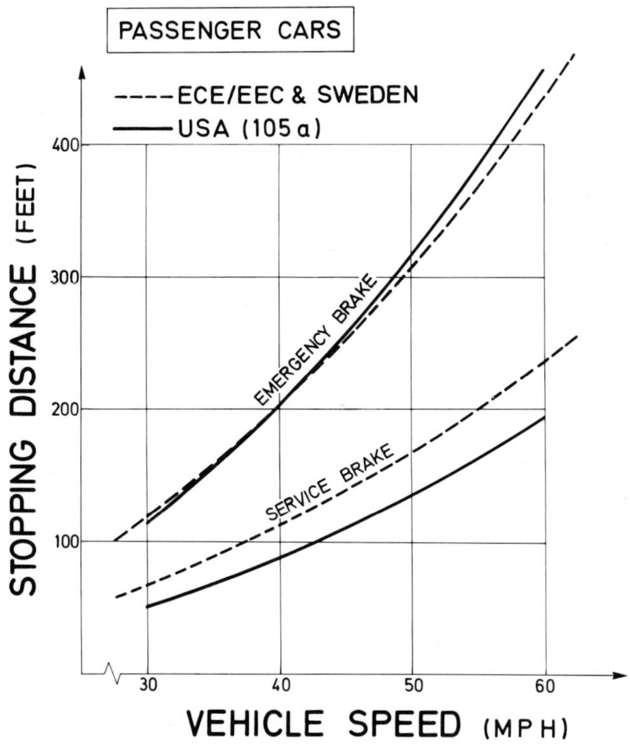

Fig. 10: Comparison of Stopping Distances for Passenger Cars

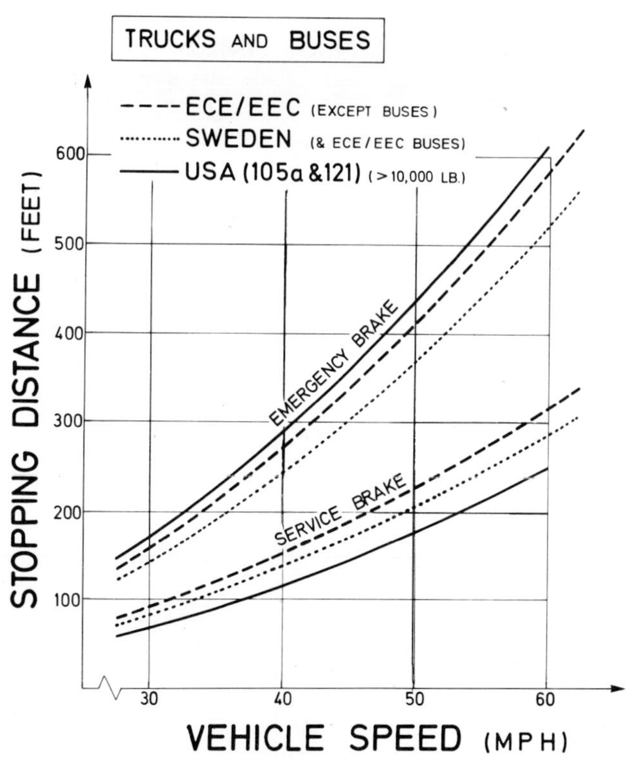

Fig. 11: Comparison of Stopping Distances for Heavy Trucks and Buses

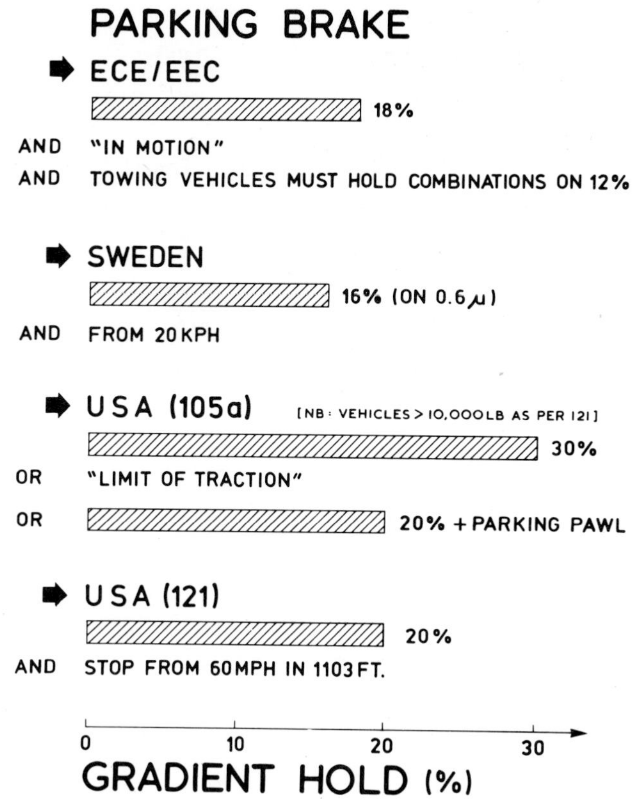

Fig. 12: Comparison of Parking Brake Requirements

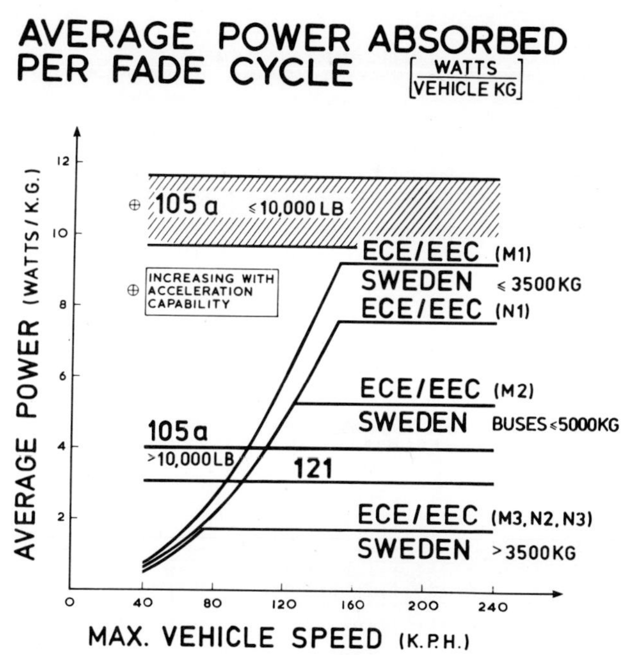

Fig. 13: Comparison of Fade Tests by Average Power Absorption

© IMechE 1976

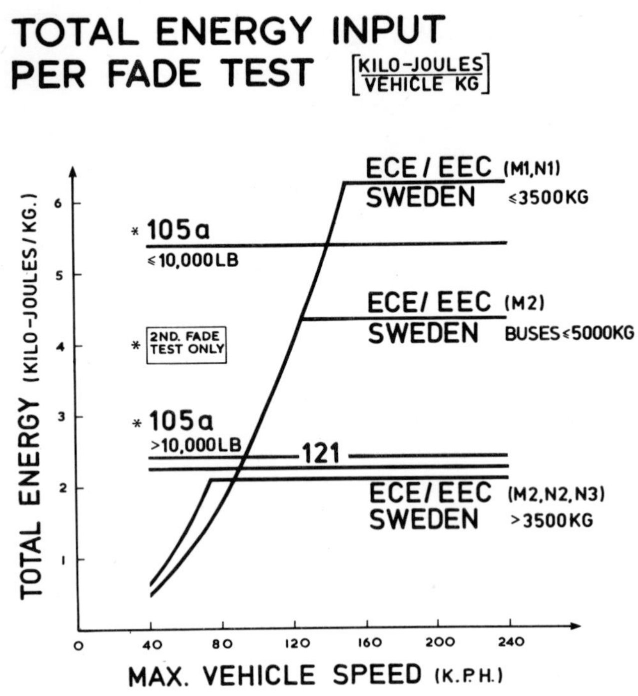

Fig. 14: Comparison of Fade Tests by Total Energy Input

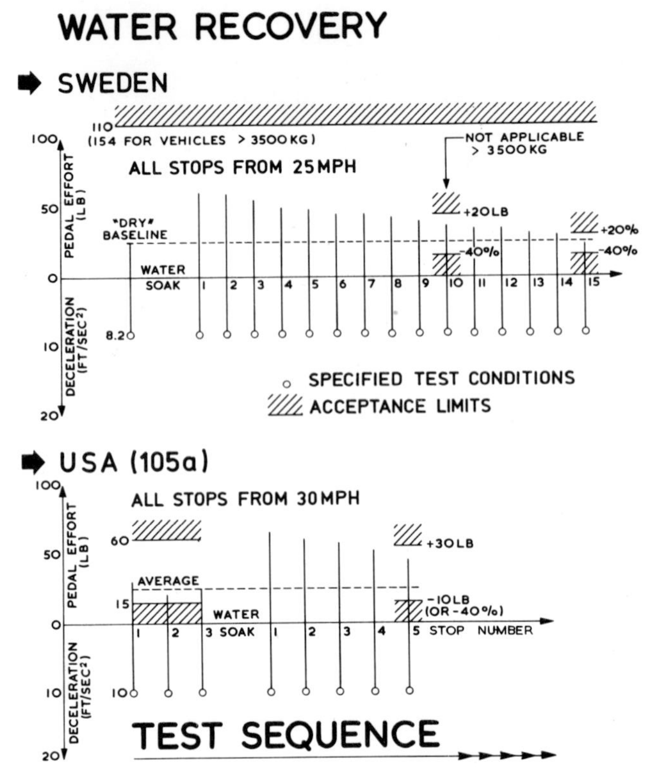

Fig. 16: Comparison of Water Recovery Requirements

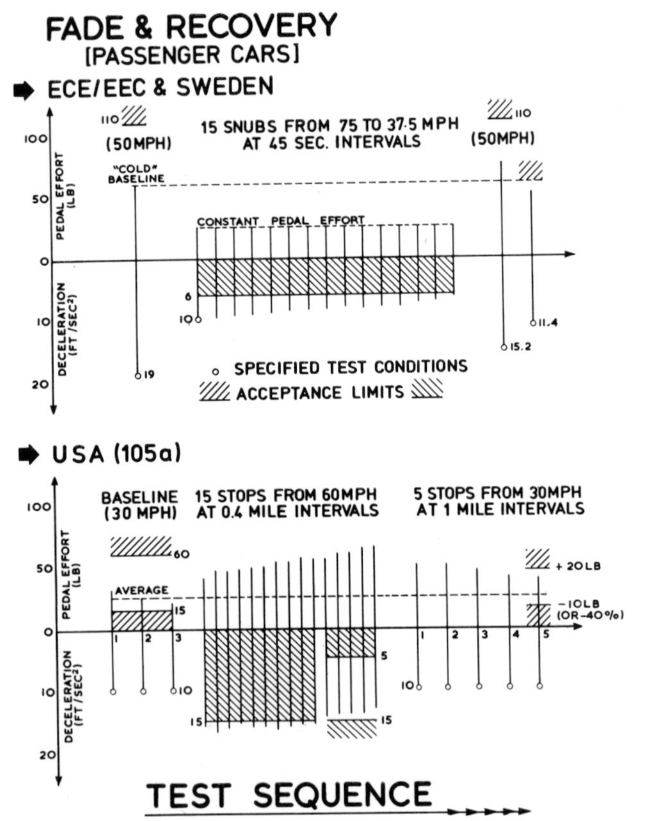

Fig. 15: Comparison of Fade and Recovery Requirements (Passenger Cars)

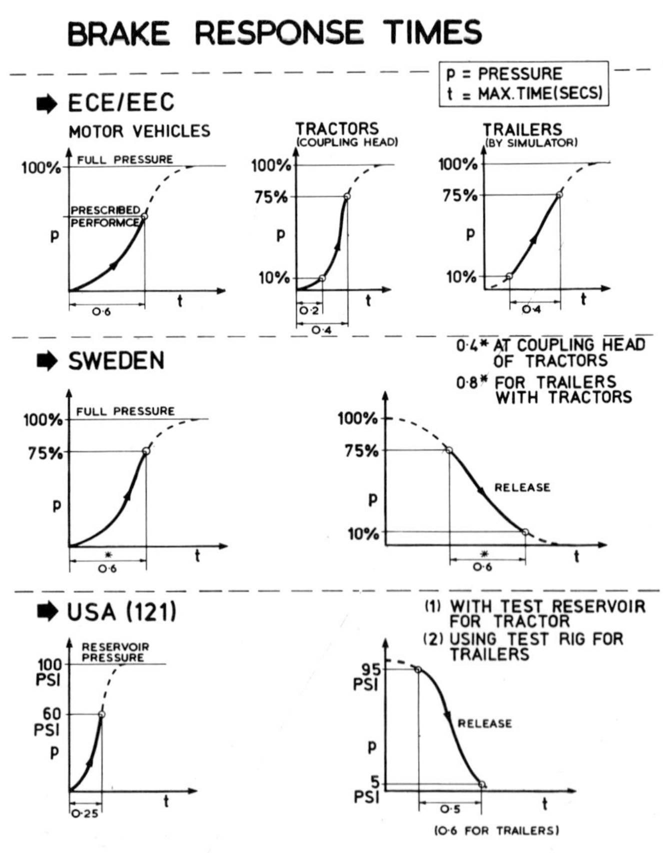

Fig. 17: Comparison of Brake Response Times Requirements

© IMechE 1976

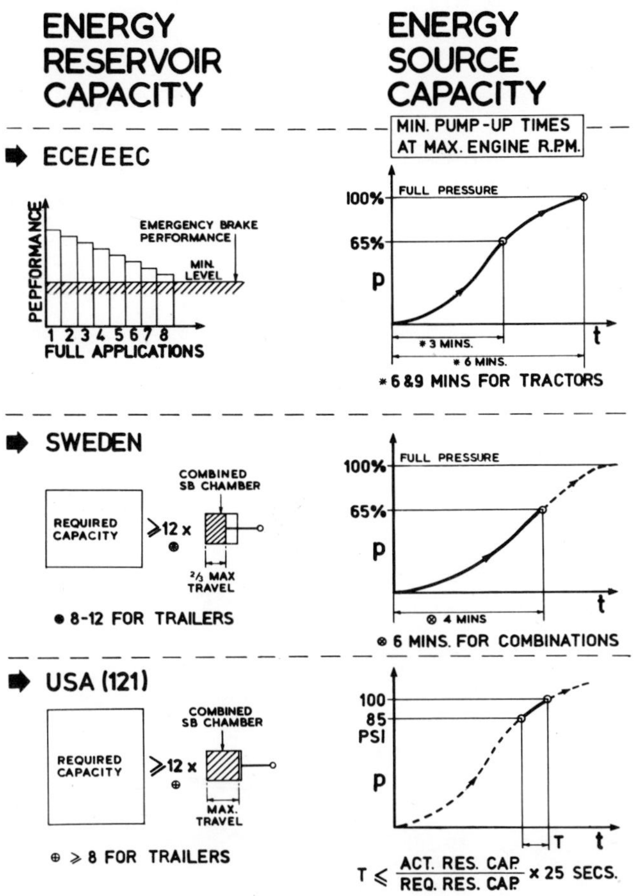

Fig. 18: Comparison of Energy Capacity Requirements
left – Energy Reservoir Capacity
right – Energy Source Capacity

© IMechE 1976

C52/76

REVIEW OF BRAKING LEGISLATION IN BRITAIN DURING THE PAST DECADE

W. L. BAXTER, CEng, MIMechE
Department of the Environment, Southwark Street, London SE1

The MS of this paper was received at the Institution on 29 September 1975 and accepted for publication on 23 October 1975

SYNOPSIS The main changes in braking legislation for road vehicles are reviewed and indications of possible future changes are given.

BRIEF HISTORY

1. The principal changes in braking legislation in Britain during the past 10 years or so occurred in 1964 (ref.1) and 1966 (ref.2) and primarily concerned commercial vehicles. The earlier regulations increased the permitted maximum weight on a single axle from 9 tons to 10 tons and allowed the gross weights of vehicles to be increased as follows:-

Rigid Vehicles

2 axles from 14 tons to 16 tons
3 axles from 20 tons to 22 tons
4 axles from 24 tons to 28 tons

Articulated Vehicles

3 axles from 2o tons to 22 tons
4 axles from 24 tons to 32 tons (or more)

The heavier weights were subject to minimum axle spacing requirements in order to safeguard roads and bridges.

2. the 1964 Regulations applied to new vehicles which took advantage of the heavier weights but only if specific braking performance standards were satisfied. Additionally the vehicles had to be 'plated' by the vehicle manufacturer to show the designed axle, gross and train weights.

3. The 1966 Regulations extended the 'plating' and braking requirements to virtually all new vehicles and in particular to those which did not take advantage of the heavier weight concessions. Light weight goods vehicles under 1525kgs unladen, cars and motor cycles were already required to conform to specified braking performance requirements. For trucks and other motor vehicles, the requirements which earlier specified the secondary braking system to be operated by a hand control (usually the parking brake control) were changed to allow, as an alternative, a 'split' or 'divided' system under the foot pedal control to be deemed the secondary braking system. Additionally it became permissible to fit spring brakes or lock actuators to vehicles to provide the parking brake facility. This concession was coupled with a requirement that the parking facility should be maintained without the use of stored energy and must hold on a gradient of 1 in 6.25 under all conditions of loading. A multi-pull type handbrake was acceptable as a parking brake but not as a secondary brake unless it became fully effective at the first pull. Vehicles mostly affected were rigid goods vehicles in the 3 to 14 ton gross weight category and articulated vehicles up to 24 tons gross weight.

4. Thus by 1966 specific minimum braking requirements were established for all new vehicles as follows:-

service brake	.5g
secondary brake	.25g
parking brake	must hold on a 1 in 6.25 (16%) gradient.

The parking and secondary brake could be combined and spring brakes and lock actuators were permissible. Vehicles had to be marked with their maximum design weights.

5. Prior to the introduction of the 1964 and 1966 Regulations the general legal requirements for most commercial vehicles were that they must have 'brakes sufficient under the most adverse condition to bring the vehicle to rest within a reasonable distance'. The parking brake had to be capable of holding the vehicle at rest and had to be actuated by direct mechanical linkage. Spring brakes and lock actuators for parking were not permissible. Multi-pull hand brakes were permitted for either the secondary or parking brake.

6. Before the 1964 Regulations were introduced discussions had been taking place in the ECE forum • on a draft international regulation for vehicle braking. The regulation had not been finalised but the then Ministry of Transport had contributed to the discussions and the main recommendations. These were taken into account when the 1964 and 1966 British regulations were introduced but there were differences. For example the ECE proposals were that braking performance should be measured by stopping distance from given speeds whereas the British regulations specified maximum deceleration. In effect however the results obtained are not too dissimilar and it was decided that it would be more appropriate in the context of the British requirements to adopt maximum deceleration since this could be more readily checked by operators and enforcement officers in garages or at the roadside. Stopping distance tests, which necessitate the use of test track facilities, would have to wait for the introduction of mandatory vehicle type approval.

7. Although the 1964 Regulations were sufficiently specific as regards braking performance and were not incompatible with the then draft ECE proposals on braking it was possible to adopt different types

© IMechE 1976

of braking systems which would meet the 1964 Regulations or ECE proposals but not both. For example with articulated vehicles the ECE proposals accepted either two-line or three-line braking systems for semi-trailers. Also, proposals in the ECE meant that reservoir sizes on British vehicles would have to be considerably increased. Discussions between the Ministry of Transport and Industry led to the formulation of a Code of Practice setting out agreed standards for braking systems.

8. The Code covered many aspects of braking for commercial vehicles and applies particularly to air braked systems which were undoubtedly encouraged as a result of the heavier weight vehicles and the specified braking performance standards. The important features of the Code are:-

a. for design purposes the following braking performance standards are recommended for new vehicles to allow for slight deteriorations in service.

service brake	- .6g
secondary brake	- .3g
parking brake	- hold on a 1 in 5.26 gradient (19%)

b. compatibility of service braking efforts on towing vehicle and semi-trailer at 85 lbs per sq. inch (5.9 bar) when the vehicle is decelerating at .6g and semi-trailer braking effort is .5W; W being the maximum weight imposed on the road by the trailer axles.

c. under all conditions of vehicle loading the braking distribution between axles to be arranged such that the rear wheels of the tractive unit of an articulated combination do not lock during braking.

d. for semi-trailers a three-line air pressure braking system (2 lines for service and a single line for the secondary brake) is recommended.

9. The 1966 regulations also introduced braking performance requirements for 'existing' commercial vehicles. These vehicles were those which existed before 1 January 1968 and which did not take advantage of the higher weights granted in the 1964 Regulations. The performance standards were:-

2 axled vehicles

Service brake	.45g
Secondary/Parking brake	.2g

3 or more axled vehicles

Service brake	.4g
Secondary/Parking brake	.15g

Proposals for up-rating these braking standards to .5g and .25g were circulated later but were never implemented.

10. In 1968 the 'Plating and Testing' Regulations were introduced and these required all goods vehicles over 1525 kgs unladen weight to be submitted to Government Testing Stations for checks on roadworthiness. A brake test related to each vehicles maximum axle or gross weights was carried out. These tests are normally made on roller brake testers. These checks were mandatory and continue on an annual basis.

11. In July 1973 the EEC Directive 71/320 concerning vehicle braking became effective in Britain (ref. 3) as optional requirements to the national rules. These requirements are very detailed and apply to most types of road vehicles but not to motor cycles.

EFFECTS OF LEGISLATION 1965-1972

12. Legislation in Britain has had a marked effect on the design of commercial vehicles during the past decade. Prior to 1964 the regulations gave little encouragement to articulated vehicles as regards carrying capacity. Tractive units were often short wheel-based tipper chassis to which a coupling had been fitted by the trailer manufacturer or operator. The braking systems of tractive units were seldom compatible with those of other makes and types of tractive unit and yet interchangeability of semi-trailers was usually possible due to standardised coupling attachments. As a result the braking distribution between various axles of complete articulated vehicles was such that vehicle instability during braking was common. Despite these deficiencies the numbers of articulated vehicles increased for sound operational and economic reasons. Overloading of goods vehicles was common and resulted in many vehicles being under braked. Vehicles were sold with emphasis on maximum payload capacity without due regard to the maximum permissible axle and gross weights intended to safeguard the vehicles or the roads on which they were used. There was virtually no enforcement action to combat vehicle overloading. The average speeds of vehicles were less than those of present day vehicles.

13. The 1964 Regulations permitted heavier weights which greatly favoured the articulated vehicle and resulted in motor vehicle manufacturers providing tractive units specially designed for articulated vehicle operation. In order to meet the braking performance requirements of the regulations and in particular those for the secondary and parking brakes new types of braking systems were introduced. The requirements of the Code of Practice agreed by Industry considerably helped to standardise these systems. Air pressure braking systems rapidly replaced the older vacuum types. Triple and double air brake actuators were introduced. Load apportioning valves to adjust the braking effort on the tractive unit drive axle according to the load carried on that axle were fitted virtually as standard equipment on British vehicles by about 1967. Secondary braking systems operated by a separate control to that of the normal parking brake and which actuated brakes on the front axle of the tractive unit and the brakes of the semi-trailer became common. These systems gave improved braking performances and provided very stable braking. Additionally they largely replaced the separate control of the trailer brakes. This latter control which was sometimes wrongly but appropriately called the "dead man's handle" was very common on early designs of articulated vehicle and was helpful in steadying the vehicle on a downhill descent. Unfortunately the same control, if used harshly in an emergency when the combination was lightly laden, caused the trailer wheels to lock and promoted trailer swing accidents. The parking brake control operating on the drive axle alone was not favoured for secondary braking performance because of the inherent danger of wheel locking which could lead to 'jack-knifing' in the case of articulated vehicles.

© IMechE 1976

14. The permitted use of spring brakes and lock actuators coupled with the agreed aim of vehicle manufacturers to obtain a parking capability for laden vehicles on a gradient of 19% led to these devices being fitted to both front and rear axles of tractive units and to many rigid vehicles. In the case of articulated vehicles this meant that all the parking requirements could be obtained from the tractive unit thus avoiding complicated braking systems for operating the trailer parking brake from the tractor cab. In the case of spring brakes these were also often used as part of the secondary braking system. The earlier type of parking brakes which operated by direct mechanical linkage from the control to the brakes, as in most cars, was virtually abandoned for heavier vehicles.

For medium weight vehicles the use of dual air-over-hydraulic braking systems replaced either the then current vacuum-over-hydraulic or single air-over-hydraulic systems.

15. The disadvantages of all these changes were, of course, the added complexity of the new systems as compared with the earlier types and the need to inform vehicle mechanics and maintenance engineers of the new arrangements.

16. The legislative changes of the 1964-1966 also encouraged the fitting of 'split' or 'divided' braking systems for rigid goods vehicles and for public service vehicles. Today such systems are virtually standard on public service vehicles. Other developments during the period under review but not due to legislation were the increasing use of exhaust brakes and electric retarders. These developments were considered at an I. Mech. E. Conference in January 1974 and are not discussed in detail here but clearly such devices can be regarded as additional safety items particularly for vehicles operating in hilly districts.

17. At this stage it might be pertinent to comment on the general accident situation. There is no doubt that during the early and mid-1960s there was much public concern about accidents involving commercial vehicles particularly articulated vehicles. Vehicle run-away accidents were not uncommon. 'Jack-knifing' accidents - whether or not the 'jack-knife' was the cause or result of the accident - usually got wide publicity and were frequent occurrences. Unfortunately, such accidents still occur but they are less frequent despite the increase in numbers of vehicles on the roads and the increase in vehicle mileages. From figure 1 it will be seen that the number of articulated vehicles in use has increased from 83,000 in 1967/68 to about 96,000 in 1973, or an approximate increase of 3% per annum.

It will also be noted that the number of vehicles involved in jack-knifing incidents fell from 254 in 1969 to 174 in 1973, or a reduction of between 7.8 and 8% per annum. The most significant drop of incidents occurred in the period from 1970 to 1972 where the annual rate fell from 0.28% to 0.185% of accidents per articulated vehicle in service.

18. To what extent the 1964-1966 legislation and the requirements of the Code of Practice affected these decreases in accident rates is not possible to apportion but it is believed that they were important factors. Equally important was the 1968 legislation which introduced compulsory annual inspection of heavier goods vehicles and Ministry 'plating'. These factors coupled with the braking performance standards of the 1966 Regulations led to many older vehicles being either scrapped or provided with modified braking systems. A further factor in the reduction of accidents must have been the greater attention now being given by drivers and operators in avoiding vehicle overloading due no doubt to 'plating' and the increased activities of weight enforcement agencies. New roads and road improvements will also have assisted in easing traffic hazards.

EFFECTS OF LEGISLATION 1973-75

19. As stated above the EEC Directive on vehicle braking became effective in Britain in July 1973 as an optional requirement to the national rules. The Department of the Environment was ready to evaluate vehicles and carry out the necessary tests to the Directive in January 1973. A considerable number of British Manufactured vehicles including cars, trucks and tractive units have been given approval to the EEC requirements. Also a number of approvals to trailers fitted with overrun brakes have been granted. Few approvals have so far been given for semi-trailers. It is believed that where approvals have been granted the vehicle types are mostly for export purposes. It is too soon to forecast changes in braking system design for vehicles in use on roads in Britain.

20. It may be of interest to consider some of the essential features of the EEC and current British requirements.

a. Like the British regulations of 1964 and 1966 and the Code of Practice for commercial vehicles the EEC Directive takes account of the ECE Braking Regulation. In fact, the EEC wording is identical to ECE in most respects.

b. The Directive applies to most motor vehicles and trailers which are separately defined into categories. Different standards of braking performance apply to different categories of vehicles when tested under ideal conditions on a test track. This is unlike the British requirements at present which in general calls for minimum performance standards which are applicable to the vehicles as soon as they are used on the roads in service and failure to meet those requirements may be an offence by the operator.

c. Approval to the Directive requires the manufacturer to submit technical data and a sample of the vehicle model or components to the Department of the Environment for testing. If the vehicle meets the requirements then the manufacturer will be given a Type Approval Certificate in respect of that vehicle type. The Certificate will be valid in all EEC Member States without the need for further testing. Occasional checks have to be made by the DOE to ensure that the vehicle manufacturer continues to produce vehicles which conform with the sample tested. Failure to conform to the type could mean that the approval certificate may be revoked and production must legally cease.

d. Braking performance standards are in terms of stopping distance from prescribed speeds whereas the British requirements are in terms of maximum deceleration.

e. The EEC deceleration requirements for the

© IMechE 1976

service brake of cars are more severe than the current British regulations. However, in practice most British cars already meet the EEC requirements.

f. EEC approved vehicle combinations fitted with air pressure braking systems may have two or more pipe lines. British practice at present requires three pipe lines.

g. Overrun braking requirements are prescribed in great detail in the EEC Directive. No such requirements exist in British regulations.

h. Brake fade requirements are explicit in the EEC Directive. In the British regulations no such details are prescribed, but it is a legal requirement to maintain prescribed braking efficiencies at all times.

i. An amendment to the EEC Directive issued in 1975 (ref. 4) intended to improve the distribution of braking between axles makes it virtually essential to fit load apportioning devices to many motor vehicles and trailers. The EEC Directive does not permit anti-locking brakes to be utilised as an alternative to load apportioning devices at present but these are likely to be allowed when further regulations have been finalised. So far as British articulated vehicles are concerned the tractive units in current production appear mostly capable of satisfying the EEC requirements. Semi-trailers will have to be fitted with load apportioning devices to take advantage of the EEC Directive. It seems however that for semi-trailers the use of anti-locking braking systems may be preferable since these devices do not rely on limited spring deflections for their effectiveness and could be more reliable in service. Under British rules anti-locking brakes are permitted for motor vehicles or trailers and this practice is likely to continue. Many such devices are already fitted.

j. With the EEC Directive there is a requirement that the energy supply for the braking system shall be maintained to each remaining brake reservoir in the event of a failure in a braking circuit. This means that the energy supply to the reservoirs may have to be through a special non-return/distribution valve. Current British requirements do not specify such devices and provided a warning of an impending failure is given to the driver this is deemed to suffice. It would then be expected that the driver would stop his vehicle utilising the remaining unfailed system and arrange for the defective system to be rectified. Continued disregard of the failure and use of the brakes could result in the energy supply system to the unfailed system being lost also. With the EEC system the failed braking circuit would not adversely affect the continued supply of energy reaching the remaining system. However, the practice of drivers' continuing to use a vehicle with a known defect in the braking system is open to criticism on safety grounds.

k. The EEC Directive requires residual service braking eg 'split' or 'divided' systems, in the event of a single failure. British legislation does not at present specify this requirement but many current production vehicles are so fitted.

21. The foregoing differences between the current British braking requirements and those of the EEC Directive are not comprehensive. The differences are not major in technical content but the question of whether there should be two or three pipe lines for vehicle combinations is controversial. Vehicle operators now have the possibility of chosing either whereas under British regulations three lines were required. It is suggested that for commercial vehicles the changes in braking systems made during 1964-1966 had already prepared British manufacturers for the transition to the ECE/EEC requirements.

POSSIBLE FUTURE LEGISLATION

22. There is little doubt that the EEC Directive on vehicle braking, together with amendments which will be introduced from time to time as technology progresses, will be the main basis for future British legislation. In the foreseeable future the legislative changes are likely to result in the following for new vehicles:-

a. cars and light commercial vehicles will be required to have residual braking performances in the event of a single failure in the service brake, and this must be under the foot pedal control.
b. brake fade requirements will be specified.

c. anti-locking brakes will continue to be permitted but technical performance standards for such devices will be prescribed.

d. air pressure braking systems will be required to have suitable test points for checking load apportioning valves and other parts of the braking system.

e. technical requirements for power hydraulic systems will be prescribed.

f. means to readily check the state of brake adjustment will be required.

g. construction standards for air pressure reservoirs could be prescribed.

h. brake liners to be classified according to their performance ability.

i. anti-corrosion requirements for brake pipes will be prescribed.

23. In addition to the changes mentioned above the adoption of a British National Type Approval System in which manufacturers will be required to seek approval for their vehicles before sale will be introduced. Cars are likely to be subject to Type Approval requirements at first but other classes of vehicles will be covered in due course.

CONCLUSIONS

24. Significant changes in the design and construction of commercial vehicle braking systems have occurred during the past 10 years. These changes primarily involved articulated vehicles but other goods vehicles and public service vehicles were affected. The application of the secondary and parking brakes have been improved both as regards performance and stability during braking. Service braking systems have been improved through better braking distribution and load apportioning devices. Anti-locking brakes have been encouraged and are fitted to a significant number of goods vehicles. Spring brakes and lock actuators together with double and triple diaphragm actuators were introduced. Braking systems have become more complex as compared with earlier systems.

© IMechE 1976

25. During the period under review the maintenance of heavy goods vehicles has been improved due primarily to the mandatory annual checks at Government Testing Stations introduced in 1968.

26. Means for combating the overloading of goods vehicles through 'plating' were introduced in 1966 and 1968. This action, together with increased activity at the roadside by weight enforcement officers, has caused vehicle manufacturers and operators to give greater attention to the axle and gross weight capacities of their vehicles. This in turn has helped to minimise vehicle overloading and brake overheating problems.

27. In the period being considered reductions in vehicle accident rates have occurred. It is believed that the vehicle legislation requirements introduced in 1964, 1966 and 1968 were important factors in the reductions affecting goods vehicles.

28. The changes in brake design which have occurred as a result of the 1964-66 legislation have considerably reduced the need for changes which otherwise would now be necessary to align with the EEC requirements.

REFERENCES

1. Motor Vehicles (Construction & Use) Regulations 1964 - 1169

2. Motor Vehicles (Construction & Use) Regulations 1966 - 1288

3. Motor Vehicles (Type Approval) Regulations 1973 - 1199

4. O.J. E.E.C. Vol. 18 No L.236 (75/524/EEC)

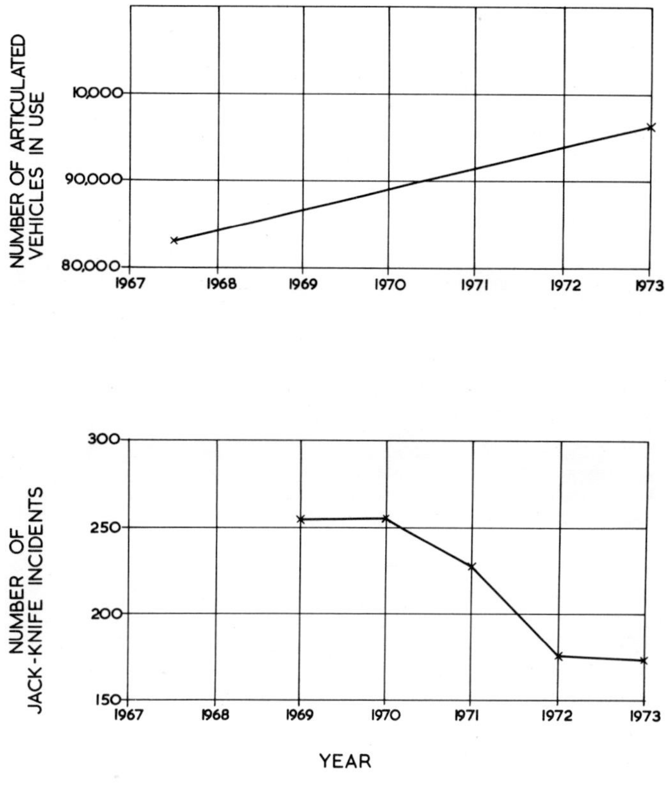

Fig. 1

© IMechE 1976

C76/76

AIR BRAKING SYSTEMS WITH INCREASED PRESSURE

OTTO EHRICKE, Dipl.-Ing.
WABCO Westinghouse GmbH, Hannover, Germany

The MS of this paper was received at the Institution on 12 November 1975 and accepted for publication on 12 January 1976

SYNOPSIS Increased pressure is a means to reduce reservoir capacity and space requirements of the equipment in air braking systems. The technically possible high pressure level is determined by the properties of the components. It is found by examining the various parts of a braking system. Different methods of applying the increased pressure level to vehicles are described. Their main characteristics are explained and compared.

1. INTRODUCTION

Modern traffic and legal requirements (for instance EEC-Directive "Braking Systems") imply for the traditional pressure levels considerable reservoir volumes. The increasing number of auxiliaries contributes to this effect. So reservoir volumes of 80 l in a medium truck, 100 l in a heavy truck and 120 l in a bus for inter-urban service are today quite common. The mounting of such volumes on the vehicles creates engineering difficulties which were the reason for studying the possibilities of reducing the reservoir capacities by increasing the pressure level.

The problem required to be approached from two sides. On one side the most practicable pressure level had to be found, on the other side the appropriate method for its application in vehicle braking systems had to be searched for. According to these two questions the following report will be organized.

2. NOTATION

HP = High Pressure
LP = Low Pressure
HPR = High Pressure Governor - or Unloader - Controlled
HPS = High Pressure self-balanced
P = Symbol for Proportional Reduction of Trailer Supply Pressure (in HP/HP-P-System)
d = Diameter
p = Pressure
S = Surface

3. The Best Pressure Level

The most practicable high pressure level is determined by the properties of the various components of a vehicle braking system. Consequently each kind of components will be investigated in this chapter for relevant characteristics.

3.1 Compressor

The delivery volume of a compressor decreases with increasing delivery pressure until it becomes 0. This fact is due to the re-expansion of the compressed medium contained in the dead space. The pressure level at which the delivery volume becomes 0 is the "self-balancing pressure". The self-balancing pressure of a given compressor depends on the size of its dead space, meaning the self-balancing pressure will increase as the dead space decreases.

These relations can be seen in Fig. 1. This shows in idealised form for a given speed the corrected delivery volume as a function of delivery pressure of a 300 cm^3-1-cylinder-compressor with three different dead space heights, namely:

0.28 mm in line 1
0.96 mm in line 2
3.60 mm in line 3

A compressor can now be operated either (as is traditional) below its self-balancing pressure by means of a pressure governor (HPR-principle) or by using the self-balancing effect for limiting the pressure (HPS-principle). In an HPR-compressor the dead space is as small as possible being limited by technical and production considerations, while in an HPS-compressor it has a rated value (for instance characterized by its height).

The practicable pressure level is limited on one hand by thermal conditions, on the other hand by drive problems.

The thermal limits of a compressor are determined by the development of oil carbon within the compression chamber. Oil carbon will settle on the piston head and the clearance volume surface and will thus reduce the dead space, with the consequence that the self-balancing pressure grows. In HPS-compressors, this creates increased drive torque oscillations and is therefore especially critical.

From a certain thickness, the carbon layer will become unstable and peel off, the pieces will contaminate the devices of the braking system, thus impairing their functions and service life.

The occurrence of oil carbon is determined by the temperatures of the internal compressor surfaces and the compression end temperature. These can be held under control by proper cooling up to approximately 20 bars for HPS-compressors and 15 to 18 bars for HPR-compressors, the difference being based on the generally lower temperature level per-

© IMechE 1976

taining to the former. The efforts to provide this cooling depend greatly on the rotary speed and the cylinder volume. Cylinders with displacements up to about 160 cm^3 require no special cooling provisions. Compressors, however, with higher cylinder displacements - for instance those with 225 and 300 cm^3 customary for heavy vehicles - necessitate particularly intensive cooling. This can be provided by conducting the cooling medium - be it air or water - as close as possible around the valves and as close as possible to the surface of the compression chamber, for instance by including the valve carrier in the cooling circuit. Thus the temperatures of valve carrier, valves, piston head and cylinder wall are reduced to such extent that the compression end temperatures, too, go down.

The drive problems of high-pressure-compressors are more serious. They have their source in the pulsation of the driving torque. The driving torque varies during one crankshaft revolution between positive (energy-absorbing) and negative (energy-generating) phases. The drive problems depend on the peak torque in case of belt drives as well as on the value of the negative torque in cases of gear drives. In the modern gearwheel drives the change of torque causes an intermittent loading of the teeth surfaces inducing noise and capable of overloading and fracturing the gear. A negative torque of not more than 150 to 200 cmkp can produce such difficulties. In this respect, the elasticity of the driving components plays an important role, a more elastic drive will be less susceptible to torque oscillations.

Fig. 2 shows for several HPR-units of different sizes how their maximum positive and negative torques grow with the governed pressure. Fig. 3 shows the positive and negative peak torques of several HPS-compressors as functions of the self-balancing pressure. The graphs indicate that the torque values increase proportionally approximately with the pressure and they show the difference between the two types: at a given pressure the HPR-principle requires a drive torque level about twice that of the HPS-principle; the negative peak torques are considerably less than the positive torques (about one fourth) in the HPR-case and approximately equal to the positive peak torques in the HPS-case. At a certain pressure the torque varies with the compressor speed, also, this influence is, however, at the usual speed level small and negligible for this subject matter.

According to these different properties, the possibilities of mastering the drive problems differ also. HPR-compressors with small displacements up to approximately 160 cm^3 are easily feasible with gearwheel drives while such compressors with big displacements require elastic and damping members in such drives, for instance a gearwheel containing a rubber ring similar to a vibration damper. For belt drives HPR-compressors necessitate, because of their higher positive peak torques, additional belts or bigger pulley diameters. Among the HPS-compressors only those with small displacements permit gearwheel drives, but with special provisions for elasticity and damping as described above. Belt drives for HPS-compressors can be easily adapted to the positive torque; for bigger axle distances these need, however, a tensioner to prevent belt shimmy due to torque oscillations.

A substantial reduction of peak torques and torque oscillations and hence of the drive problems can be achieved by splitting the cylinder displacement into two cylinders. This is illustrated by the characteristics of the 2-cylinder-inline-1-stage-300 cm^3-compressors in Figs. 2 and 3. Another approach to this problem could be a rotary piston compressor. We are working on and testing one design. As such a new design will have to match with the high reliability standards of the traditional piston compressors, there is still considerable development work to be done, so that this principle will not be a short-term solution.

Another important feature for the practical applicability of one principle or the other is the delivery volume. As Fig. 1 indicates, the delivery in the vicinity of the self-stabilizing point is considerably less than that of a governed compressor at the same pressure. How far this affects the practical value of the two principles will be explored in a later chapter.

3.2 Reservoirs

From the point of view of reservoirs, any pressure increase is technically possible. The stored energy remains constant if the reservoir size is modified inversely proportional to the pressure.

In practice, the admissible pressure will be limited by reservoir weight and approval regulations. For instance in Germany and some other European countries, it is recommendable that the stored energy fulfills the condition:

$$p \times V < 1000$$

in which case a type approval for the reservoirs is sufficient.

3.3 Pipes and Screw Couplings

The maximum pressure for pipes and fittings customary today is 40 bars.The application of plastic pipes, however, is according to their diameter limited to between 16 and 21 bars.

With traditional pipe diameters the timing characteristics of a braking system improve as the pressure increases.

On the other hand, the pipe diameters may be reduced if the delay and build-up time can remain unchanged. This aspect has, however, little importance. Trials have shown that in this case the pipe diameter obeys the formula:

$$d_{HP} = d_{LP} \times \sqrt[5]{p_{LP} / p_{HP}}$$

which means that the diameter decreases very slowly as the pressure grows.

3.4 Braking Equipment

Under the condition that the forces remain unchanged, it can be said for all air-actuated devices that the overall dimensions decrease with growing pressure level. Pressure-operated surfaces scale down inversely proportional to the pressure-quotient:

$$S_{LP} / S_{HP} = p_{HP} / p_{LP}$$

© IMechE 1976

The pressure is limited to 20 bars by the sealing properties of O-rings. Higher pressures would require the application of other sealing components, involving higher costs.

As for friction of the sealing elements and hence the hysteresis of the complete devices, trials have shown that no difficulties should be expected.

4. The Best Method of Applying The Principle of Increased Pressure to Vehicle Braking Systems

For trailer air braking systems the statutory maximum pressure is 7.5 bars (see EEC-Directive "Braking"). This means that a pure high-pressure vehicle system is practically not possible because air braking systems for motor vehicles are generally designed for supplying and controlling trailers. The necessary combination of pressure levels can be based on different principles which will be compared in the following sections. For this purpose each principle will be projected into two sample vehicle combinations.

4.1 Description of The Different Methods

An HP/HP-system is an installation in which the whole motor vehicle braking system is operated by high pressure. The trailer is supplied via a pressure-limiting regulator valve and controlled by a pressure-reducing trailer control valve.

An HP/HP-P-installation is a system with a constant proportion pressure-limiting valve for the trailer supply instead of a customary regulator type. Fig. 4 shows a basic piping diagram for such installations.

An HP/LP-installation is shown in Fig. 5. In this case, the high pressure generated by the compressor and stored in the reservoirs is reduced for the drawing vehicle's cylinders as well as for the trailer.

The high pressure is assumed in all cases to be 15 bars, and the pressure for the trailer as well as in the latter system for the drawing vehicle cylinders to be 7.5 bars.

As fourth system, a traditional low-pressure (LP-) installation is included in our investigations which will serve as the basis for all comparisons.

4.2 Characteristics of The Sample Vehicles

The first truck/trailer-combination consists of a 16 tons 2-axle truck with a 22 tons 3-axle trailer. The second combination is a 22 tons 3-axle truck with a 16 tons 2-axle trailer.

The relevant data of the sample vehicle combinations are to be found in table 1. The actuator diameters of the HP/HP-systems have been found with the above formula and with an addition for reduced efficiency. The actuator displacement is inserted at 2/3 of the total stroke to account for the elasticity of the braking mechanisms. Thus the results will have a considerable reserve. The dead spaces of the pipes and valves are estimated figures.

4.3 Criteria for Valuation and Results

4.3.1 Determination of Minimum Reservoir Sizes

The EEC-Directive "Braking Systems" sets forth in Annex IV, paragraphs 1.2 and 1.3, requirements for the brake performance and the pressure level in the trailer control line after eight full-stroke brake applications, without re-feeding of the system and with auxiliaries disconnected.

Under the condition that both parts of a vehicle combination comply with these requirements separately, the relevant characteristics of the principles in question for the calculations below are the following:

In a vehicle combination with conventional LP-braking system no overflow from the truck reservoirs to the trailer will take place.

In a vehicle combination with an HP/LP-truck as well as with an HP/HP-truck, air from the truck reservoirs will replace the consumption in the trailer as long as the reservoir pressure in the truck exceeds the cut-off pressure of the valve in the trailer supply line. In another respect, these systems differ, however, considerably. In the HP/LP-installation, the reservoir-pressure may drop to minimum legal trailer control-line pressure, whilst in the HP/HP-system the limit is much higher (in the calculated cases double), due to the conversion ratio in the trailer control valve.

In a vehicle combination with an HP/HP-P-truck, again no overflow will occur because the constant proportion valve in the trailer supply line remains closed.

Based on these rules and characteristics, the necessary reservoir capacities of the four installation systems for each sample vehicle combination were calculated.

The calculations were made on the assumptions:

- isothermal change of state
- no leakages and no pressure losses in the system
- effective pressure identical with line pressure

The results are shown in Fig. 6 in absolute and relative units. In the HP/LP- and HP/HP-P-installations, in the order of 30 to 45 per cent of the reservoir capacities of a conventional braking system can be spared. The HP/HP-installation, however, brings for one sample combination an increased, for the other an only limited reduction of reservoir capacities and is therefore not interesting. This negative result is due to the characteristics described above.

4.3.2 Reservoir Sizes For Constant System Output

With the outputs of traditional low-pressure systems with reservoir capacities of 60 or 80 l as starting point, the reservoir sizes of the 3 high-pressure systems necessary to maintain the output were calculated.

The output is in this connection defined as "the number of full-stroke brake applications possible without re-feeding of the system until the pressure in the trailer control line has dropped to the level necessary for secondary braking effect". The output figures for the traditional installation are 14 for the sample combi-

© IMechE 1976

nation with the 2-axle truck and 12.5 for the combination with the 3-axle truck.

Fig. 7 shows the results of this examination. Again the mixed HP/LP- and the HP/HP-P-systems would permit to cut the reservoir capacities in the order as above. The HP/HP-system has improved but is still unfavourable.

At this point a special feature and interesting advantage of the mixed HP/LP-system is worth mentioning. Fig. 8 shows for the sample combination with the 2-axle truck the braking ratio versus the number of brake applications. In the HP/LP-system, during the first four brake applications, the achievable deceleration remains constant. This is due to the difference between storage and operating pressure. The actuators are fully pressurized before the reservoir pressure has dropped to cut-off pressure, (which is rated actuator pressure). This is a typical effect of this system independent of failure of supply and therefore advantageous also in normal operation.

4.3.3 Capacity of Pressure Generators

In this chapter the two compressor principles will be explored separately to find out whether or not a limitation of the applicability is existing. The investigation is based on two different criteria with the target to receive results as practice-related as possible.

4.3.3.1 Capacity According to EEC-Regulations

Statutory filling times are defined in the EEC-Directive "Braking", Annex IV, paragraph 2. These requirements are hardly appropriate for high pressure systems. For instance, in HP/LP and HP/HP-installations, at 65 per cent of the rated pressure in the motor vehicle reservoirs, the trailer reservoirs are already filled to 100 per cent. This means that such HP-systems and traditional pressure systems are hardly comparable under this criterion.

Nevertheless, the filling times of the various systems have been calculated on the basis of the reservoir sizes for constant system output as per chapter 4.3.2 with the assumption of trailer reservoir sizes as per the EEC-Directive. Without going into details it can be said that under these conditions with a 300 cm^3 compressor all systems with HPR-compressor as well as with HPS-compressor remained far below the statutory limit, namely within the first third of the same. Obviously this criterion does not permit any conclusions in respect of the applicability of the compressor principles.

4.3.3.2 Recovery Capability

Owing to the fact that the compressor capacity requirements according to EEC-Regulations are not a real challenge another criterion must be found. The author proposes to take for such criterion:

> "the number of possible brake applications per minute with secondary braking ratio with the compressor running at 1/3 of its rated speed and with the brake applications uniformly distributed on the interval of one minute."

Following this definition the number of possible brake applications has been calculated for the sample vehicle combinations on the assumptions of 300 cm^3 HPR- and HPS-compressors with a rated speed of 3000 min^{-1} and isothermal conditions.

The calculation data and the results are shown in table 2. The number of possible brake applications is between 5 and 6 for the LP-system and all HPR-systems. With HPS-compressors the numbers are around 4 for the mixed HP/LP- and about 3 for the full HP/HP-system. As had to be expected, these figures show a decreasing tendency for HPS-compressors. The problem is now to define a practice-related limit for the number of brakings. For a final decision this will have to be further explored. Without anticipating such decision the HPS-principle in an HP/LP-system seems to be an alternative deserving a serious investigation. It might be a good solution in single motor vehicles and will reach its boundary of application earliest in vehicle combinations with important auxiliary air consumers.

4.4 Drainage

An increased pressure level has no disadvantage with regard to water condensation. Saturated air at atmospheric pressure and a temperature of 20°C contains 15.2 g water per kg dry air. After isothermal compression to 8.5 bars absolute pressure, the water content is 1.8 g per kg in saturated state, 13.4 g per kg have condensed. A further compression results in only little more condensate. At 16 bars absolute pressure, only another 0.85 g per kg condenses.

The HP/LP-system is in this respect advantageous. An absolute reservoir pressure of 16 bars and an absolute working pressure of 8.5 bars be assumed. The air in the reservoirs will be saturated and contain at 20°C 0.95 g water per kg air. This water content remains unchanged during isothermal expansion to the working pressure of 8.5 bars. At this pressure and the said temperature, the air requires for saturation a water content of 1.8 g per kg. As only 0.95 g per kg are available, the air at this pressure level is far from being saturated and there is no danger of condensation in the braking equipment. In partial brake applications where the storage air is further expanded this effect is more noticeable.

5. Comparison of The Different Systems

As the HP/HP-installation implies an increase of reservoir capacity, this system is not included in this valuation.

Table 3 shows the comparison of the HP/LP-system and the HP/HP-P-system with the traditional LP-system and with each other. We see that, compared with the traditional system, the HP-systems offer important advantages as well as considerable disadvantages. It is therefore not possible to grant a general preference to one system or the other. For the particular case all features have to be evaluated and the individual decision so to be found. Among the HP-systems, the mixed HP/LP-system is superior. Only this system is available with relatively short lead time, and it offers plain advantages. These will be counterbalanced only in special cases by the reduced space requirements for valves and actuators - their availability provided - of an HP/HP-system.

© IMechE 1976

6. CONCLUSIONS

The limit for the high pressure level is set by the compressor. Its cooling capacity and drive problems resulting from high drive torques and torque oscillations limit the currently acceptable pressure level to 15 to 20 bars. The compressor can be operated with pressure regulation or with self-balancing control. The latter principle contains a delivery volume reduction. It has to be further investigated whether and by what extent this effect restricts the applicability of this principle.

For commercial vehicles high pressure can be applied in full high pressure or in mixed installations. The various high-pressure systems, when compared with each other and with conventional installations, have advantages and disadvantages that must be weighed in the particular case.

7. ACKNOWLEDGEMENTS

This paper reflects the actual status of research and experience of Wabco Westinghouse. The author is grateful to the directors of the company for their permission to publish this knowledge and to his colleagues for the support he received in collecting and preparing the material and the illustrations.

© IMechE 1976

	System	16 tons	22 tons	22 tons	16 tons
Brake Actuator Type	LP	20 16/24	30 16/24	20 16/24	30 24
Actuator Vol. at 2/3 stroke	LP,HP/LP	2.54	5.16	4.24	3.96
	HP/HP	1.43	5.16	2.39	3.96
Dead Volume	LP,HP/LP	1.00	1.10	1.20	0.90
(Pipes & Valves) (1)	HP/HP	0.80	1.10	1.00	0.90
Specified Pressure as per EEC-Directive (bars)	LP	7.5	7.5	7.5	7.5
	HP/LP	15.0	7.5	15.0	7.5
	HP/HP	15.0	7.5	15.0	7.5
Initial Braking Ratio of Combination at 7.5 bars in Tr. Contr. Line	LP	0.65		0.65	
	HP/LP	0.65		0.65	
	HP/HP				
Pressure in Trailer Contr. Line for Sec. Braking Ratio (bars)	LP				
	HP/LP	2.6		2.6	
	HP/HP				

Table 1 Specifications of sample vehicle combinations.

		System	16 tons	22 tons	22 tons	16 tons
	Delivery Pressure = Prot. Value Closing Pressure (bars)	LP	6.0	–	6.0	–
		HP/LP	6.0	–	6.0	–
		HP/HP	9.0	–	9.0	–
	Sec. Braking Press. in Actuators (bars)	LP,HP/LP	2.6	2.6	2.6	2.6
		HP/HP	5.2	2.6	5.2	2.6
	Corr. Air Consumption at Sec. Br. Pr. per Application (1)	LP,HP/LP	11.92	21.75	18.66	16.84
		HP/HP	13.25	21.75	20.36	16.84
HPR	Compr. Corr. Del. Vol. at 1/3 Speed & Above Clos. Press. (1/min)	LP	210.0	–	210.0	–
		HP/LP	210.0	–	210.0	–
		HP/HP	195.0	–	195.0	–
	Poss. Br. Applic. with Sec. Br. Ratio per Minute	LP				
		HP/LP	6.2		5.9	
		HP/HP	6.0		5.6	
HPS	Compr. Corr. Del. Vol. at 1/3 Speed & Above Clos. Press. (1/min)	LP	210.0		210.0	
		HP/LP	140.0		140.0	
		HP/HP	110.0		110.0	
	Poss. Br. Applic. with Sec. Br. Ratio per Minute	LP	6.2		5.9	
		HP/LP	4.2		3.9	
		HP/HP	3.1		3.0	

Table 2 Parameters and results of recovery capability.

© IMechE 1976

Table 3 Comparison of LP-, HP/LP- and HP/HP-P-systems.

Feature	Advantages & Disadvantages Compared With LP-System		Favourable	
	HP/LP-System	HP/HP-P-System	HP/LP	HP/HP-P
Compressor	Disadvantage: more work of compression.	Disadvantage: more work of compression (also against HP/LP).	+	
Reservoir Capacity	Advantage: 30 to 45% reduction.	Advantage: 40 to 45% reduction.		+
Space Requirement	Except reservoir like LP.	Advantage: reduction		+
Pipes	Like LP.	Disadvantage: restricted use of plastic pipes.	+	
Brake Equipment	Normal LP-equipment.	Disadvantage: completely new program of devices necessary.	+	
No. of Valves	Disadvantage: two to four additional pressure limiting valves.	Disadvantage: One or two additional pressure limiting valves.		+
Filling Time	Disadvantage: a little longer.	Disadvantage: a little longer.		
Recovery Capability	Disadvantage: less possible brakings per minute (but more than HP/HP-P).	Disadvantage: less possible brakings per minute.	+	
Drainage	Advantage: due to expansion no condensate in equipment.	Appr. like LP.	+	
Braking Ratio	Advantage: constant cylinder pressure, constant braking ratio	Like LP: dropping cylinder pressure diminishing braking ratio.	+	
Braking Ratio, Compr. Defective	Advantage: several applications with full cylinder pressure.	Like LP: immediate cylinder pressure drop with braking ratio reduction.	+	
Braking Ratio Circuit Defective	Like LP.	Like LP.		

© IMechE 1976

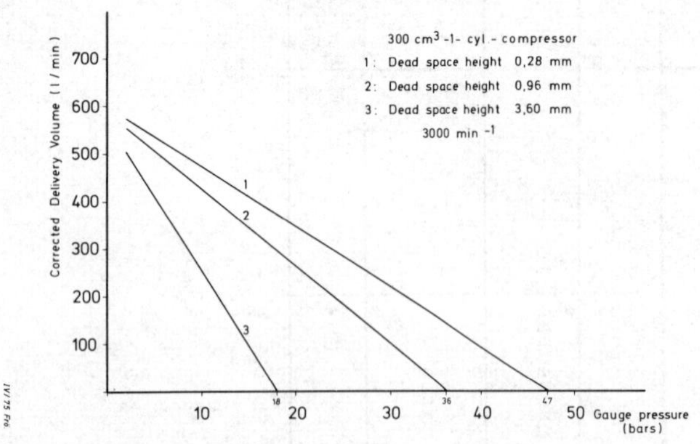

Fig. 1: **Compressor delivery versus delivery pressure with different dead dead spaces.**

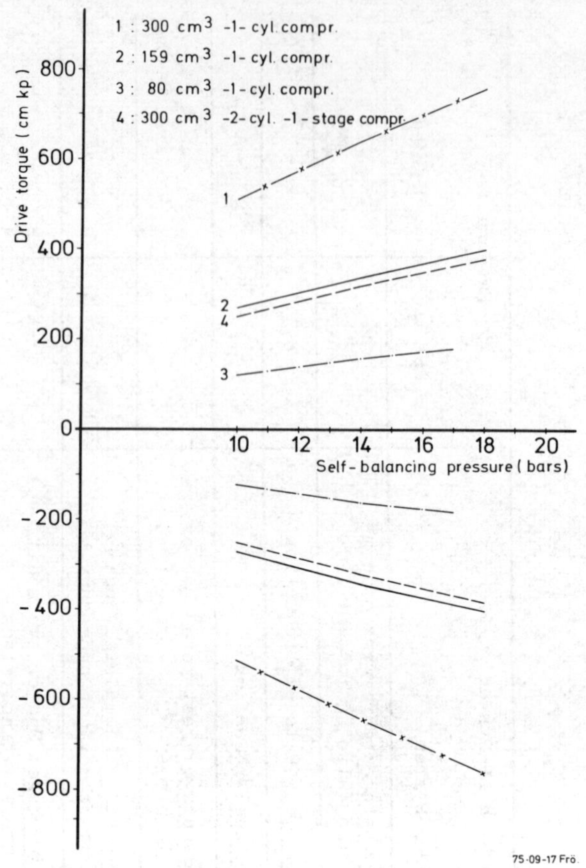

Fig. 3: **Drive torque versus self-stabilizing pressure of HPS-compressors.**

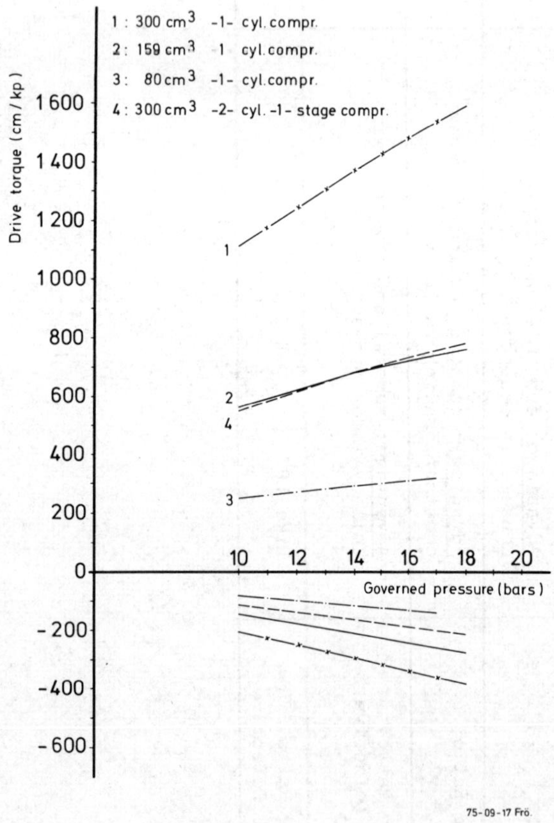

Fig. 2: **Drive torque versus governed pressure of HPR-compressors.**

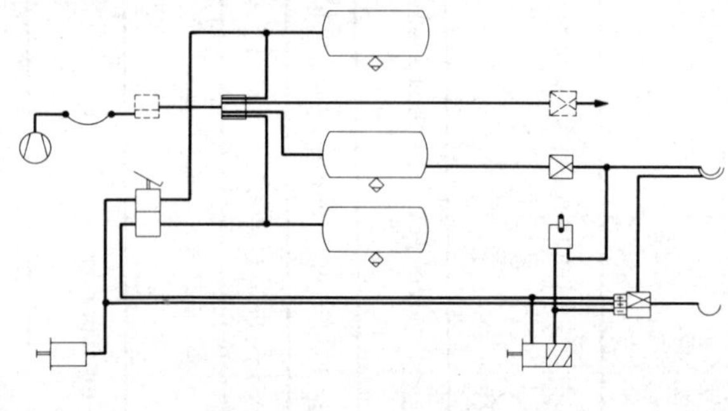

Fig. 4: **Schematic of HP/HP braking-system.**

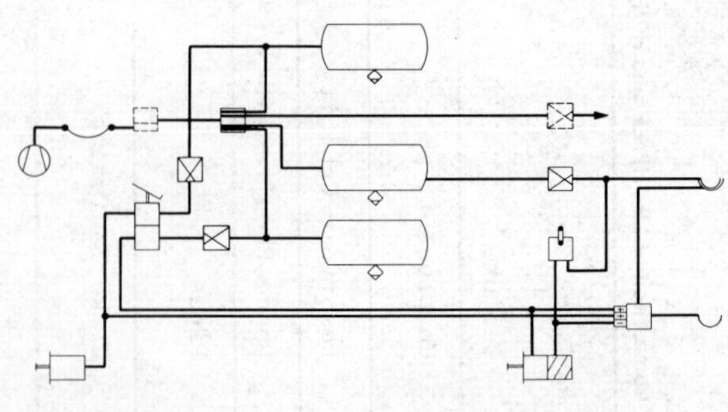

Fig. 5. **Schematic of HP/LP braking-system.**

© IMechE 1976

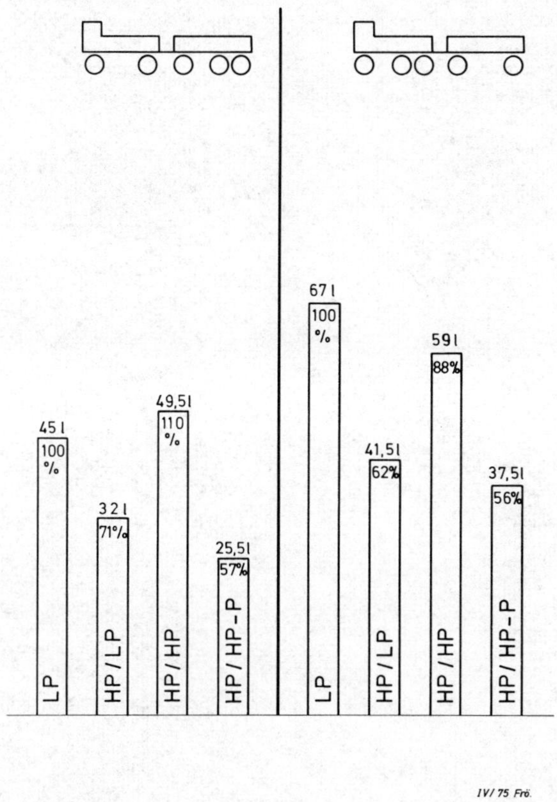

Fig. 6: Minimum reservoir sizes for compliance with EEC-Directive

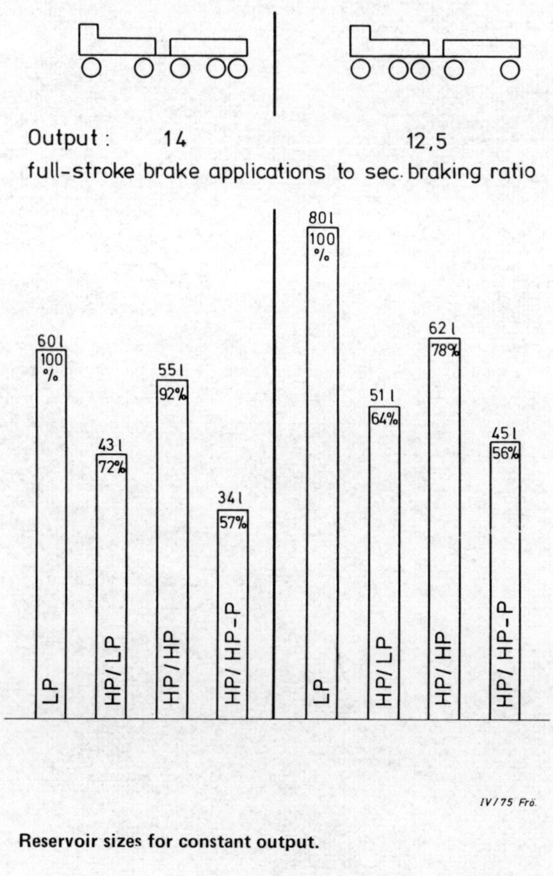

Fig. 7: Reservoir sizes for constant output.

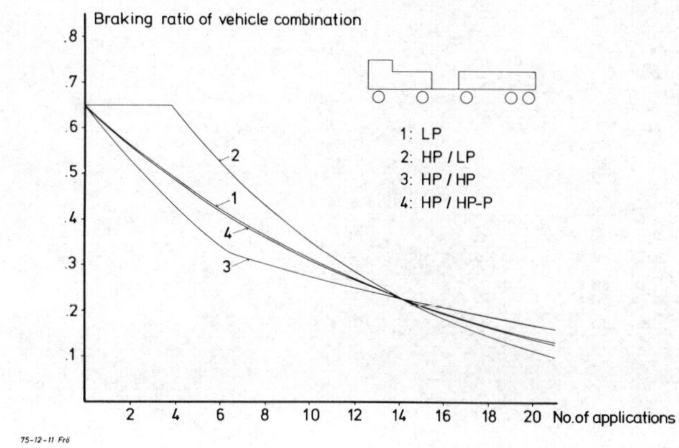

Fig. 8: Braking ratio as function of No. of applications.

© IMechE 1976

C79/76

A REVIEW OF COMMERCIAL VEHICLE BRAKES

I. R. SLACK, BSc
British Leyland UK Ltd, Truck & Bus Group, Engineering & Development Division, Leyland

The MS of this paper was received at the Institution on 15 December 1975 and accepted for publication on 19 January 1976

SYNOPSIS The paper is in three parts, commencing with a review of the foundation brakes currently in use on commercial vehicles. The scope of commercial vehicles, in this context, extends from delivery vans (excluding car based types) to maximum capacity road haulage vehicles. Public service vehicles are also included.

The review indicates the way in which currently available brakes are deployed by the British commercial vehicle industry and points to changes foreseeable in the future. Such changes may arise from legislation, newly available brake types and changes in vehicle usage.

Section II deals with the selection of brakes against particular applications. Some ways of rating brakes are indicated, along with factors involved in selecting friction materials and typical methods of approving the brakes selected.

In Section III, some developments in S-cam brakes are discussed. These include specific examples of improvement in mechanical efficiency, liner utilisation and freedom from vibration. Ways of improving liner and drum life are also discussed, with other reliability considerations.

The paper concludes with some indication of the future requirements of the commercial vehicle industry for changes in foundation brakes and possible ways in which the brake manufacturers may meet these requirements.

INTRODUCTION

1 The approach to brakes by the British Commercial Vehicle industry has moved along the same general lines as the industry's approach to other major components of the vehicle. Originally the British CV industry consisted of a large number of relatively small companies each making their own components, in accordance with their own particular experience and requirements.

In the last fifteen years mergers and consequent rationalisation have completely changed that pattern.

There has been an inevitable move to the use of proprietary components and as small firms have been absorbed by the larger groups, many of their traditional units have disappeared. In some cases the units used by such small firms have been adopted as standards by the larger group and this has inevitably placed compatibility restraints on the way in which units are mixed on a given vehicle.

In addition to the effects of rationalisation which have been developing over this longer period, there are more recent influences which are forcing quicker change. The more acute of the two major influences is legislation and in the UK where a significant amount of production is even now subject to EEC directives, this is having a major effect.

The other longer term major influence is the effect of continually rising vehicle performance. Axle weight and GVW are both substantially higher than even ten years ago and power to weight ratios have more than doubled in many cases.

This paper deals specifically with the foundation brakes and the way in which traditional approaches have led into today's situation. There follows from this an outline of a way of selecting brakes for the future and the requirements of future brakes to comply with all the constraints now placed upon vehicle design.

A REVIEW OF COMMERCIAL VEHICLE BRAKES IN CURRENT USE

2 The term "Commercial Vehicle" in this case includes goods vehicles from 2 TONNE GVW to maximum capacity artics and passenger vehicles from the light weight "truck-based" psv to maximum capacity double deckers. There is therefore a tremendous range of types and sizes of brakes involved.

It is possible however, to categorise brakes into groups and sub-groups on the basis of mating surface (drum or disc), the type of shoe anchorage (pivoted, sliding, etc) and the type of actuation (direct hydraulic, wedge, cam, etc).

Fortunately not all possible combinations of the major variables are used and current CV brakes are limited to about ten distinct types. (Table 1). Different manufacturers have their own particular version of some of these types but the fundamental principles are the same.

Distribution of brake types

2.1 The different types of truck brake tend to be distributed according to the GVW of the vehicle. The lighter vehicles are generally more cost and

© IMechE 1976

weight conscious. A basic, internal wheel cylinder, hydraulic brake is therefore used in this case. As weight increases a servo is added to this system until full air becomes standard at the heavy end. Full power hydraulics are used in one or two cases but are not yet widely applied. Fig 1a indicates the weight ranges in which different types of actuation apply.

Hydraulic brakes are not generally sensitive to the type of hydraulic system used, except that vehicles without servo tend to use duo-servo brakes in order to give adequate secondary performance. In the middle weight category there is often a tendency to use external actuation at the rear brakes in order to get efficient hand brake operation.

Higher up the middle weight range a two-leading shoe air wedge brake is often used. This leads into the heavy range where cam brakes have been used almost exclusively.

Increasing interest is now being shown in dual actuation, two leading shoe, air wedge brakes. Major attractions of this brake are the 'cleaning up' of the axles by removing the brake chamber mountings, etc and the extra torque capacity. These brakes also incorporate automatic adjustment as a rule. An easy switch can be made to hydraulic actuators giving a high rated hydraulic brake particularly suited to power hydraulics.

Originally the main attraction of these was the facility for horizontal split. Now, as the need is recognised for duplicate load sensing and anti-lock valves, the desirability of horizontal split becomes debatable. Consequently other attributes of this brake assume greater relative importance.

Brakes on the lighter PSVs tend to be similar to those on trucks of similar GVW. Heavy coaches and double deckers still use cam brakes, usually differing according to the plant of manufacture. PSV brakes have for a long time used 3/4" thick linings whereas most heavy goods vehicles have used 1/2" or 5/8" thick liners.

<u>The range of brake sizes in current use in relation to prevailing influences</u>

2.2 Fig.1 shows the way in which total brake lining area relates to GVW, for some typical 1974 European trucks.

Certain features are shown by the distribution. The heavier bands show the main concentration of areas, while the broader more lightly shaded band indicates the spread within the weight range. The figures are based on about 16 different makes of vehicle generally available for European operation.

In general, British made vehicles tend to occupy the top region of the scatter band, showing rather higher averages for lining area/tonne, and marks indicate where some typical British Leyland vehicles fall in the band.

The effect can be seen of overlap between certain weight ranges. Scatter is increased, particularly in the lower weight range, by variations due to axle fitment policy.

The inset to Fig 1 shows how drum sizes increase between 3.6 tonnes GVW and 14 tonnes GVW. From 14 tonnes upwards there is a move to 394 mm drums. This tends to coincide with the change to air brakes. In the range below this, the drum diameter is more restricted by wheel size and increases from 254 mm to 393 mm in steps of about 30 mm each. British trucks do not generally use drums larger than 394 mm but some British PSVs and trailers use 406 and 419 mm drums. This size is also more prevalent on some Scandinavian trucks and obviously allows a higher rated brake for a given width.

Fig 1b shows for some British Leyland trucks, the relationship between GVW, static axle load, specific liner drag pressure and specific power dissipation rate per unit drum area. Fig 1c shows similar relationships for some examples of Leyland PSVs.

The lines across the plots for the two brake rating parameters indicate how vehicles generally fall into the ranges recommended in Section 3.2.

The lining area relationships follow a similar pattern to that shown in Fig 1a but the way can be seen in which axle load influences specific drag pressures and power dissipation rates.

The ratings for 28 ton Mammoth Major rear bogie are treated axle by axle since a limiting valve is used to minimise the effect of load transfer within the bogie. Section 3 indicates in more detail how the relationship between brake size and axle loading is governed.

1968 was a "watershed" in the development of C V brakes because of the effect of revision of the C & U regulations. These revisions quantified the statutory requirements for parking, service and secondary brake performance.

The regulations' main effect was on brake torque, often resulting in increased lining area. The secondary brake requirement was one of the main influences in the selection of duo-servo rear brakes for the lighter vehicles.

In a similar way 1975 and 1976 will be a critical Time for brake development as EEC Type Approval begins to take effect. A conspicuous result of EEC Directives is a significant increase in many front brake sizes, due to brake ratio changes.

In some cases to meet EEC Directive 75/524 (71-320 plus stability and compatibility) the changes in front/rear brake ratio can be as high as 45/55 becoming 55/45. These changes in ratio result from the stability criteria contained in the Annexes to the original 71-320.

Vehicles exemplified in Figs 1b and 1c are not modified in accordance with 75/524 and for certain vehicles the change in brake ratio will necessitate an increase in front brake size, to maintain rating parameters within acceptable limits. For example in Leyland Reiver, ratio has changed from 29/71 to 43/57. Consequently front brakes must increase from 6" to 7" wide.

The increase in rear axle weight from 11 tonnes to 13 tonnes, for French requirements in particular, has increased the rear brake torque requirements in roughly the same proportion. A general increase in torque requirements is one of the reasons for the move to two leading shoe brakes in some heavies. However, a two leading shoe brake has a higher mu sensitivity (Ref 1) hence, to maintain the same fade resistance and stability a better brake lining is necessary.

© IMechE 1976

THE SELECTION OF BRAKES AGAINST PARTICULAR APPLICATIONS

3 Section 2 indicates the wide range of brake sizes and types available.

In this section some of the requirements of particular types of operation are considered in terms of their effect on brake rating parameters.

Special Requirements of different applications

3.1

3.1.1 Short Haul and Delivery vans

This application gives frequent brake applications at lower speeds. In some cases higher performance in this weight category, coupled with a lack of professionalism among drivers, can lead to higher duties.

A liner with good cold life is desirable for this application. High-mu liners and high factor brakes can be used to give reasonable pedal loads without power assistance.

This type of vehicle usually employs two leading shoe front brakes and duo-servo rears.

3.1.2 Mixers & Tippers - Mid weight range

Here, relatively high power/weight ratios and again a relatively unrestrained style of driving can lead to heavy and frequent brake usage. Speeds are not particularly high but high decelerations are used. This coupled with a high laden C G position tends to work the front brakes quite hard.

This type of vehicle is often working in "off-road" situations so dirt and mud contamination must be prevented or tolerated. Water recovery is important.

Two leading shoe front and leading trailing shoe rear brakes are usual for the lighter vehicles. Many of the heavier ones use cam brakes all round. Front brakes should be generously sized.

3.1.3 Haulage Vehicles - Mid weight

These are vehicles on medium haulage operations. Driving standards are higher but the type of operation is often arduous. They are four wheelers with high axle loads, often working in hilly areas, on farm delivery and collection but a lso doing a lot of town work.
Frequent transition from hilly routes to fast motorway running can create thermal cycling problems.
Drum crazing can be a problem, as can fade and liner wear.
Full air brakes are usual, either cam or wedge but the higher factor brakes need more stable linings. Water recovery can also be critical on this type of vehicle.

3.1.4 Long distance haulage and tankers

Certain aspects of this operation tend to neutralise each other. Although this type of vehicle spends a much greater part of its time fully laden, most journeys take place on level roads. Average speeds, and mileages are much higher but the general standard of driving and maintenance are also higher. The number of brake applications per mile is relatively low, but when applications do take place they are from high speeds with cold drums and thermal stresses can be high.

Drum crazing can easily become the major brake problem on this class of vehicle. Ample drum capacity should be provided. Liners with low speed sensitivity should be selected and if any significant amount of hill work is to be undertaken an auxiliary brake is a desirable feature.

3.1.5 Touring Coaches

There is much in common between the touring coaches and the mid-weight haulage vehicles. The main difference is that brake applications tend to be lighter.

Quite large differences can exist in the energy dissipation between drivers, some drivers dissipating up to 25% more energy than others (Ref 2). Drum crazing can again be a problem and it is desirable to balance 'drag point' pressures to ensure simultaneous contact of front and rear liners to drums.

Retarders are often fitted to touring coaches mainly for economic reasons due to their effect on liner life.

3.1.6. Stage Carriage Buses

Cooling is the biggest problem in this application. High temperatures can be achieved resulting in poor life and often brake squeal. Buses and coaches represent the most sensitive area where squeal and vibration are concerned and public reaction is often swift and vociferous.

Line pressures are often low and balance between front and rear brakes is sometimes erratic (Ref 3).

GVWs of buses have increased and so has performance. Rear brakes are often as big as can be practically accommodated, and manufacturers are having to resort to other devices to increase cooling.

Buses are usually fitted with automatic slack adjusters. The type of adjustment to be used needs consideration in the brake specification.

Some adjusters contain clutches which are sensitive to camshaft torque, and it is desirable to keep this torque within certain limits.

All the above considerations are important in the selection of brakes and can substantially modify the parameters given in Section 3.2.

A basis for commercial vehicle brake selection

3.2 Fig 2 shows a standard method of selecting brakes against a particular vehicle application.

Once the brake is defined the actual brake torques and lining areas are known and the two main parameters can be computed; viz tangential liner drag/unit liner area ('specific drag') at $6 m/sec^2$ deceleration and specific power dissipation/unit drum area at $5 m/sec^2$ and 96 Kph.

The power dissipation parameter gives an indication of sensitivity to drum crazing and can be considered in conjunction with drag/unit area to indicate liner life and fade.

The values for 'specific drag' have been shown by experience to give a reasonably trouble-free brake if they do not exceed $485 KN/mm^2$ at $6 m/sec^2$.

For stage carriage PSVs drag pressure should be no

© IMechE 1976

more than 450 KN/mm^2 because of the high temperatures achievable in this type of operation. When drag pressures are below about 345 KN/mm^2 and $6m/sec^2$ glazing can be a problem if low duty running is occurring regularly.

Tests have now shown that for certain brake linings in some applications, a drag pressure of 585 KN/mm^2 can be tolerated. Clearly, in this case, some sacrifice has to be made and liner life is noticeably reduced. Other problems such as noise and vibration are also more noticeable.

The specific power dissipation rates are based on tests conducted by liner manufacturers, to compare drum crazing properties of different friction pairs. It has been shown that, if power dissipation rate exceeds $1.8W/mm^2$ of drum swept area, in a stop from 96 Kph at $5m/sec^2$ this will lead to early drum crazing.

Some vehicles are not capable of 96 Kph in any case and on this basis, stage carriage PSVs can tolerate up to $1.9W/mm^2$.

For touring coaches and motorway trunkers a value of $1.6W/mm^2$ is desirable because of the proportion of high speed stops on cold drums.

The experience gained with these empirical values tends to justify their retention as a basis for initial selection and Figs 1b and c show some examples of current Leyland vehicles in this context.

Brake liner selection

3.3 Brake liners are influenced by the same considerations. Applications are important and certain liners are better in some features than others.

Most modern liners have a reasonable level of fade resistance and if a liner is such as to change the brake factor by no more than 30% at 300°C then it will normally pass all the current legislative fade tests, if brakes are adequately sized.

Cold Friction level (mu) is often regarded as the fundamental characteristic of brake liners, but this is only the starting point. Hot mu is equally important and so is the way in which the transition is made from cold mu to hot mu.

Ideally a liner will show a decreasing friction level with increasing temperature. Some liners will retain their friction up to a certain temperature and then suddenly begin to lose it. Others will drop their friction steadily with increasing temperature. As long as the drop is not too rapid the former situation can be advantageous as it warns the driver that he is overheating his brakes.

A liner which shows friction rising with temperature is not desirable. Whereas the opposite relationship has a self limiting effect on cross axle temperature differential, the rising mu/temperature characteristic will give a divergent condition which can develop into pull.

In a similar way, a liner which does not fade can be run up to destructive temperatures and then fail suddenly due to loss of mechanical strength. This is more dangerous than a steady loss of performance.

Delayed fade is quite unacceptable because of its insidious nature.

Over recovery can be an embarrassment, as it can cause pull if cross axle cooling rates are not identical. The resulting high mu levels often result in noise and vibration.

The other source of performance change is contamination and the water recovery test is an important part of a liner approval programme.

The main problem areas involving modern brake linings are not so much in the fundamental areas of performance and fade (although speed sensitivity can be an embarrassment, particularly with high mu liners) but rather are they problems concerned with the broader aspects of use, such as liner life, drum crazing, noise and vibration.

A major influence in the selection of brake linings should be power/weight ratio and vehicle performance.

A heavy truck with a high power/weight ratio (say 12 bhp/ton) can often climb a long gradient in half the time taken by a traditional vehicle. In consequence the descent can be started with brake temperatures between 50 and 100°C hotter than the slower vehicle. The rise during the descent will be similar for both vehicles if engine braking and GVW are similar. In many cases however, engine braking is less, due to the use of turbo-charged engines, with relatively low swept volume and care should be taken to allow for this.

Because of these factors, modern heavy vehicles on hilly routes can potentially achieve temperatures between 150 and 200°C higher than the traditionals. The consequent effect on liner life is tremendous.

Performance and fade characteristics are usually determined by specific tests. Ultimately however, only service experience can prove the suitability of the liners and brakes selected.

BRAKES FOR THE FUTURE

4 Changes in modern vehicle operating conditions, and increase in vehicle performance have combined with the effect of legal requirements (Ref 4) to necessitate a major reappraisal of some of the traditional ideas on braking. This section considers some of the foundation brake changes which could conceivably occur during the next decade of the Commercial Vehicle industry. The type of changes foreseeable in the future are those which will increase brake performance, reduce stopping distance and improve drivers' control. Performance is adequate to specified standards but there is always a need to improve durability, serviceability and reliability.

Further refinement is needed to free brakes from squeal, judder and other undesirable vibrations.

4.1 Performance

The maximum brake torque which is available is governed by the available actuation energy, the brake factor and the energy dissipation capability of the brake.

For an S-cam brake the maximum input torque is about 18 KNm (Type 30 x 6.5 bars x 175 mm lever). With an effective cam radius of about 15 mm and liners of cold mu about 0.37, this brake can

© IMechE 1976

generate about 165 KNm of torque, using a 394 mm diameter drum.

To increase torque outputs above this value either higher input torques or higher brake factors are needed.

Increased torques require more energy to be supplied by the actuating system, so greater storage or generating capacity is necessitated.

This can be achieved either by the use of a power hydraulic system or the use of higher air pressures. Power hydraulic systems are available and in many ways preferable for rigid vehicles. Problems of trailer compatibility can arise with both systems in the case of articulated vehicles.

The other means of increasing torque capacity is by increasing the overall brake factor. This can beexpedited by increasing the drum diameter, by increasing mean shoe factor or by increasing mechanical efficiency. A reduction in cam radius would be similar in effect to increasing lever length and will be disregarded.

An increase in drum diameter is attractive as it also creates greater lining area without increasing width. Drum sizes are, however, limited by wheel size and a reasonable air gap must be maintained between wheel rim and drum. A standard B7.5 x 20 rim limits truck drums to 420mm but for PSVs on stage carriage operation 394 mm drums are preferable unless effective heat shields are provided.

Mean shoe factors can be increased either by raising liner mu levels, by increasing arc length or by changing the brake configuration to two leading shoe. Whichever method is used the higher shoe factors will cause increased mu sensitivity and more stable liners are necessary.

Mechanical efficiency can be improved by using roller followers with S-cams or by using a strut-cam arrangement. High stiffness of camshafts, housings, shoes and drums is also necessary.

By means of higher strength/weight ratio of shoe assemblies and reduced friction at cam and pivots, pull-off spring loads can be reduced. Automatic adjustment can ensure that operation always takes place within the most efficient range of actuator stroke.

To achieve large changes in torque capacity fundamental changes are needed and it may be necessary to forsake cam brakes in some cases, in favour of two leading shoe wedge brakes.

Alternatively, system changes could give more energy at the brake actuation or improvements in response would allow acceptable stopping distance, with no significant torque increase.

Ultimately, stopping distance is the main performance criterion and if suitable values can be achieved with low braking torque this is desirable.

Durability

4.2 The main factor affecting durability is stress. In a brake, stress can be thermal as well as mechanical in origin. As the torque increases so do the stresses in the brake and the simple approach would increase physical sizes in suitable proportion. However, this is not practicable, particularly for thermal stress, because of space limitations.

In practice, either working temperatures must be reduced or components must be used which will survive temperatures in the region of 500°C.

To reduce working temperature, without extra space requirements, drums can be designed which will lose heat more effectively.

Composite drums may well be developed, shaped to give better heat transfer to air, and of low density material giving a good ratio of exposed area to mass.

Forced cooling by blown air can be applied, although this does complicate the vehicle.

Finally, if a multi-plate disc brake is used it can be fully enclosed and circulated liquid cooling is possible.

Other improvements in drum design are also desirable to reduce drum crazing in high power applications. Such solutions are likely to be in the metallurgy of the drum.

The other course of action is to use a brake which can survive the higher temperatures. The part of the brake most sensitive to working temperature is the lining, although hydraulic actuators need careful siting. For a heavy duty drum brake with a hydraulic system, external actuation is desirable.

Sintered brake linings are under development and have been used for some time in heavy duty clutches. These liners are potentially capable of operating at 500°C but they are prone to speed sensitivity and can shorten drum life.

The most likely brake to operate reliably at high temperatures is an open disc brake. Mechanical actuation is preferable to direct hydraulic under these conditions.

The surrounding environment must be suitably protected if high brake temperatures are to be the norm and temperature critical components should be kept away or shielded from radiant heat.

Noise and Vibration

4.3 This is a problem which has been attacked, in the main, by palliative methods. The dynamics of a brake are so complex that analytical approaches have yielded relatively little and most improvements have been empirical in origin.

Squeal can be dealt with in two ways. One is by stopping the friction pair from generating vibrations. The other is to damp out the vibrations before they develop in amplitude.

Combination liners have been used with some success and these fall into the first category. Although their effect is not fully understood they were originally proposed as a means of equalising the load between leading and trailing shoes.

High temperature operation influences squeal in drum brakes by causing heel and toe contact when the drum cools. In some cases squeal has been reduced simply by lowering the maximum working temperatures.

© IMechE 1976

Liner solutions to the problem may work through either the relationship between hot and cold wear rates, the relationship between mu and rubbing speed or even by the type of debris created by wear.

Palliatives include drum mouth dampers, damper bands, and damper pads applied at the shoe web. Resilient bushes at the fulcrum can also have an effect.

The other major vibrational problem is judder and in PSVs a serious manifestation of this occurs when reversing. This problem has existed for some years on S-cam brakes and consists of a high amplitude, low frequency vibration at certain speeds and pressures when reversing.

The problem is basically that the brake is part of a feedback loop. Small fluctuations in torque are inherent to the brake's operation. In reverse these fluctuations are fed back by their effect on the inertia of the operating components (pushrod, slack adjuster, etc) (See Fig 3). It has been shown that about 30N of friction applied to the push rod is sufficient to prevent judder from occurring. A full analysis is complicated by effects at the cam, and alternative feedback loops.

Vehicles with roller followers are modified by fitting a moulded urethane friction bush to the brake chamber push rod. This bush is made of an elastomer with very good wear and abrasion resistance and is an interference fit on the push rod. It also gives protection by sealing the brake chamber against dirt ingress. (See Fig 4).

These are some of the ways in which existing and anticipated problems can be dealt with in the near future.

In some cases pure analytical solutions may be found. In others, simple empirical approaches must be used but ultimately the performance in service is the only true indication of the brake engineers' success.

CONCLUSION

5 Brake problems first became significant when the 20 mph speed limit was removed from heavy vehicles. Since then weights and power have increased and legislation has specified higher performance standards. Over this period, the space available for brake installation has, if anything, reduced.

Improvements in temperature can be achieved by forced convection and by improved drum design to increase dissipation. However, the rate of input of power can also create problems and drum brakes should be considered very carefully for high energy applications.

Disc brakes do not offer advantages in the same proportions as when fitted to motor cars. For a given energy rating, relative to a drum brake the space required by a disc brake is not much different. Working temperatures are generally higher with consequent effects on pad life and immediate environment.

Multi-plate disc stacks are now a feasible proposition, particularly with carbon or sintered metal facings. Dissipation of heat to air, by forced convection, is difficult in this case and one of the brakes of the future may be a liquid cooled multi-plate stack. This is actually being tried at present by an American manufacturer, in a PSV application.

Such a brake can substantially increase both torque and energy ratings but piping and heat exchangers complicate the vehicle.

Auxiliary brakes and retarders can accept a considerable part of the total braking energy, but such devices are not without their problems. Not least of these is the vehicle stability problem resulting from the application of up to 10% of the vehicle's plated gross weight, as braking force, at one axle.

In the case of articulated vehicles on low mu surfaces, this can cause serious adhesion loss and suitable load or retardation sensitive modulation is required.

Current brake technology can still cope with current axle loads and performance. However, unless brakes are developed which can allow greater capacity in the available space, serious consideration should be given to increasing the number of axles on a vehicle if its performance is to increase still further.

The vehicle industry is once again at a critical point in brake development, but now the limit of traditional brakes is nearer, and within the next decade some fundamental changes can be expected in commercial vehicle brake equipment.

ACKNOWLEDGEMENT

The author greatly appreciates the various forms of help from staff of British Leyland Truck & Bus Engineering & Development Division, Leyland. Particular thanks are due to Mr D R Parkinson for his valuable advice and encouragement and to Mrs M Smith without whose patience and co-operation the task would have been much more difficult.

Acknowledgement is also due to the Directors of BL Truck & Bus Group for permission to publish.

© IMechE 1976

TABLE 1

RANGE OF VARIABLES BETWEEN BRAKE TYPES. (MAIN COMBINATION IN CURRENT USE)

Mating Surface	Actuation	Configuration	Shoe Anchorage	Adjustment
Discs	Hydraulic Single cyl Multi-cyl	Caliper & disc	-	Usually self adjusting if hydraulic
	Mechanical cam (ball and ramp type)	Multi-plate wet or dry	-	
Drums	Hydraulic	2LS	(Sliding abutment ((	Snail cam or peg cam
		LT	(Incline or (parallel	Flat faced cone and screw
		DS	Anchor pin	Turn buckle and strut
	Mechanical Cam and Strut	LT	Sliding	(Flat faced cone (Expanding strut
	Mechanical S-cam	LT	Pivoted	Worm and Wheel in actuator lever
	Single Wedge	LT) 2LS)	Sliding	Flat faced cone or expanding strut
	Dual Wedge	2LS	Sliding	Auto by expanding strut

NB 1 Basic types of adjustment given. Other change in detail between manufacturers.

2 All mechanical brakes shown usually operated by air and often by hydraulics in case of wedge brakes.

3 Not many disc brakes currently apply to CVs. Table shows also types which may be applied in future.

4 2LS = Two Leading Shoe

LT = Leading/Trailing Shoe

DS = Duo Servo

© IMechE 1976

TABLE SHOWING, IN BROAD TERMS, THE DISTRIBUTION OF BRAKES THROUGHOUT THE WEIGHT RANGE

TABLE II

GVW Weight-tonnes	Brake Type Front	Brake Type Rear	System Service & Secondary'	System Parking
TRUCKS				
2-4.5	*2LS - Int Cyl Occasional Discs (not general rule)	DS - Int Cyl	Direct Hydraulic Split for some markets. Parking brake used as secondary in UK	Rod or Cable
5-8.5	2LS - Int Cyl	DS - Int Cyl More LT - ext cyl	Servo Hydraulic Some Full Power Hydraulic	Rod or Cable Often remote spring brakes
9 - 11	2LS - Int Cyl	LT or 2LS Int Cyl or Ext Cyl	Air assisted or Air over Hydraulic	Rod or Cable Air assisted. Some remote springs
11 - 15	2LS - Int Cyl - Wedge	2LS - Int Cyl - Wedge	Some full air. **Some HP air on continent. Still predominantly hydraulic	As 9 - 11 More direct springs
16 - 17	2LS - Wedge LT - S-cam or Strut cam and sliding shoe	As Front	Full air Vertical split and spring brakes	Direct springs Some lock actuators. Very few air assisted
22 - 28 Rigid	As 16 - 17 tonne Rigids. Almost exclusively air. More complex parking and secondary systems			
28 - 44 Artics				
PSVs				
10 - 12 seats	As 2 - 4.5 tonne trucks			
26 - 36 seats	As 5 - 8.5 tonne trucks			
41 - 56 seats	As 11 - 15 tonne trucks			
Single Stage Carriage, Double Deck	Broadly as 16 - 17 tonne trucks S-cams or Strut cams Auto Adjusters with S-cam		Some full power hydraulics. Mainly air. Vertical Split. Often spring secondary	Air assisted or springs

* 2LS = Two Leading Shoe
LT = Leading/Trailing
DS = Duo Servo
** HP = High Pressure Storage approx 15 bars
Int Cyl = Internal Wheel Cylinder
Ext Cyl = External Wheel Cylinder

© IMechE 1976

APPENDIX

References

Ref 1 Newcomb, T P & Spurr, R T. Braking of Road Vehicles, PP 116–120

Ref 2 Froode, A D M, Mackenzie, D K & Newcomb, T P. Brake Usage in a heavy vehicle, I Mech E Paper AD 18.60

Ref 3 Baker, A K & Collinson, A. Beating Bus Brake Burn-Out, Commercial Motor, Jan 26 1973, Vol 137, No 3498, P40–43

Ref 4 Oppenheimer, Paul. Braking Regulations in Europe, SAE Paper 740313. Automotive Engineering Congress, Detroit, February 25 – March 1, 1974

Bibliography

Wilson, A J. Brakes & Brake System – Chapter 7, Commercial Vehicles – Engineering Operation, I Mech E 1967

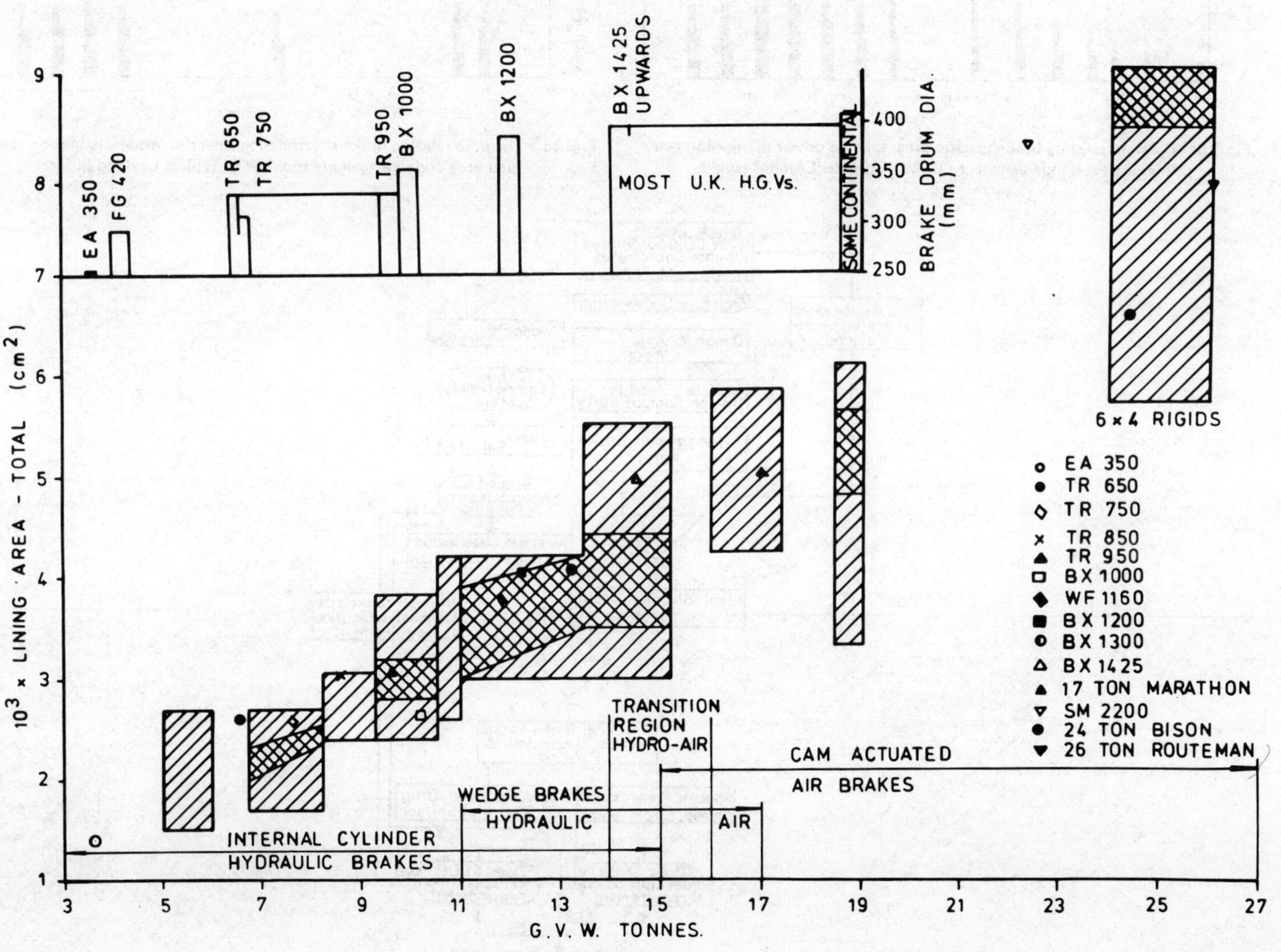

Fig. 1a: Relationship between GVW and Brake Liner Area. Rigid Trucks available in Europe 1974. 5–19 tonnes GVW – 2 axle 24–26 tonnes GVW – 3 axle

© IMechE 1976

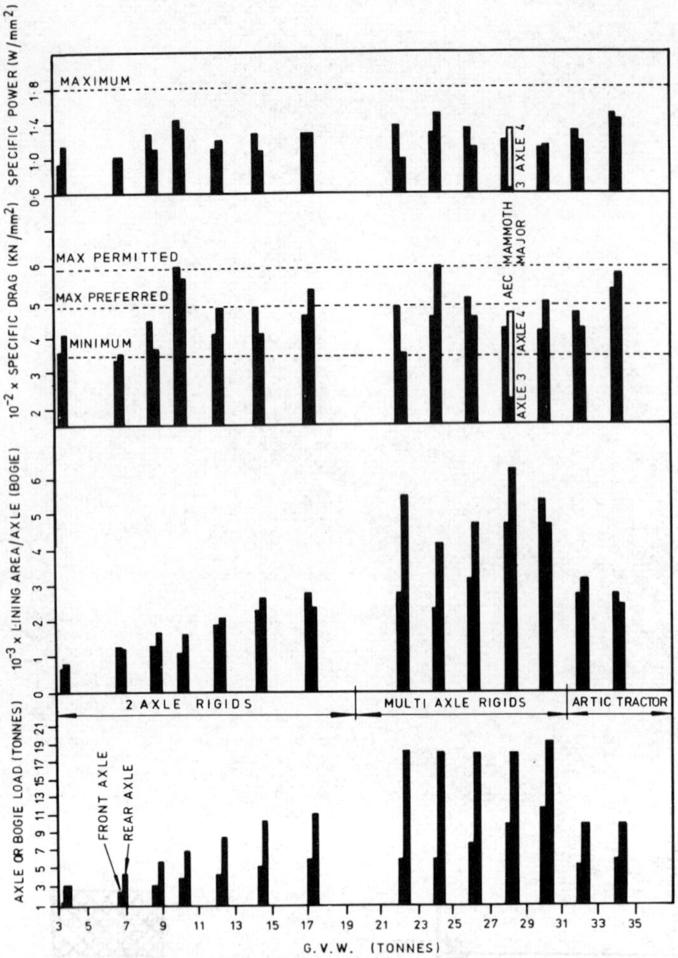

Fig. 1b: Diagram relating brake specific drag, specific power dissipation rate, liner area and axle weights to GVW of typical Leyland trucks.

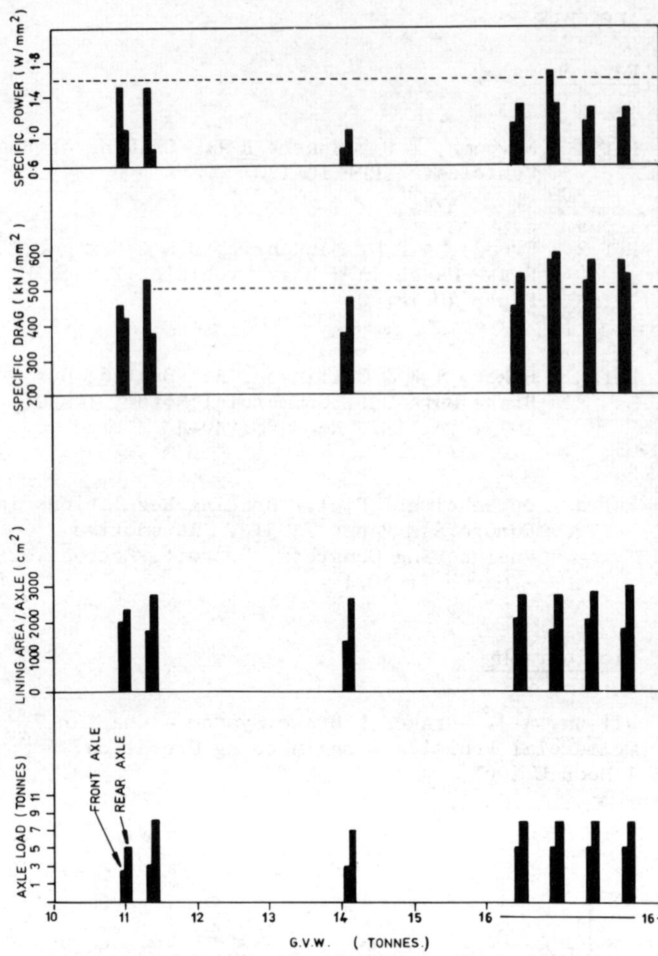

Fig. 1c: Diagram relating brake specific drag, specific power dissipation rates liner area and axle weights to GVW of typical Leyland buses.

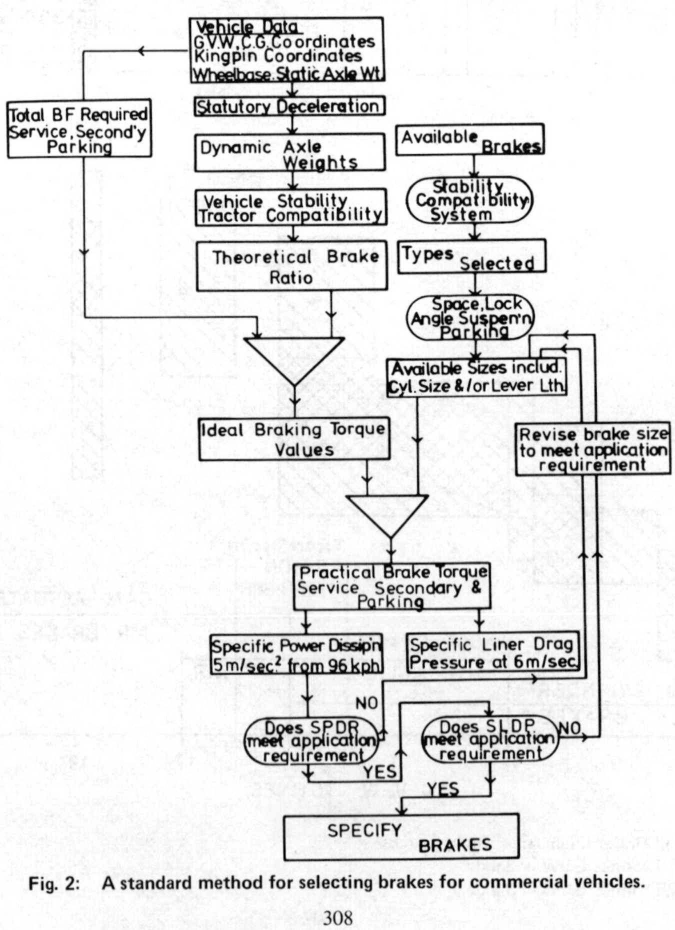

Fig. 2: A standard method for selecting brakes for commercial vehicles.

© IMechE 1976

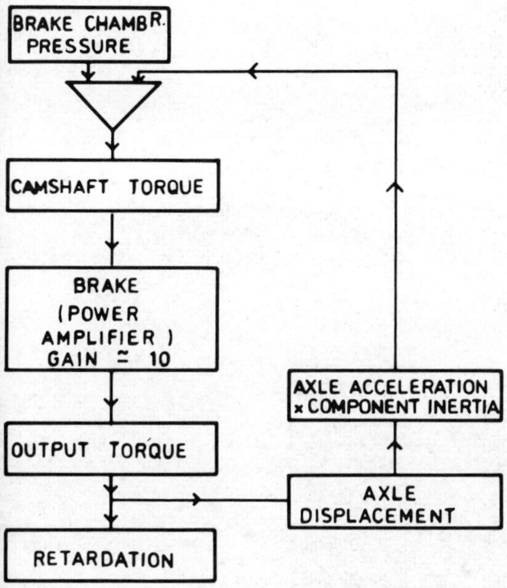

BLOCK DIAGRAM OF FEED BACK LOOP

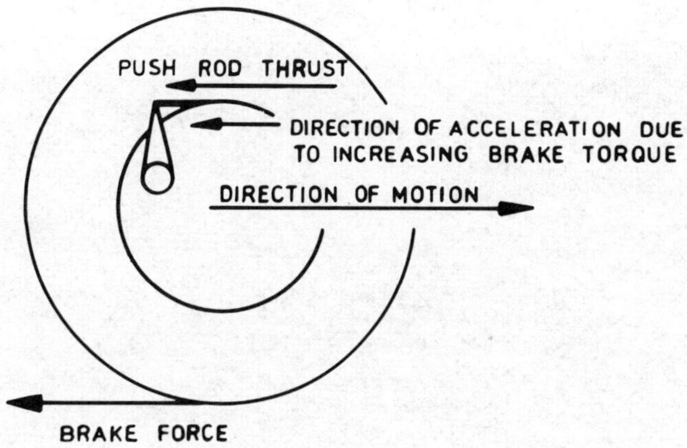

DIRECTIONS OF FORCES WHEN REVERSING

Fig. 3: Simplified diagram of reverse judder feedback loop.

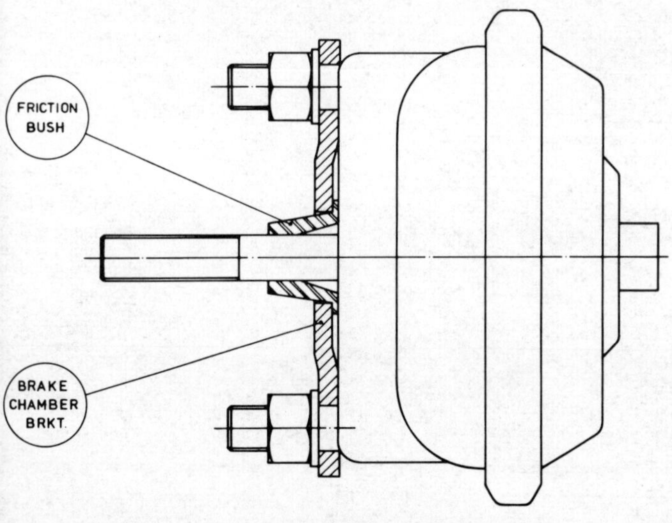

Fig. 4: Installation of friction bush to eliminate reverse judder.

© IMechE 1976

C71/76

DISC BRAKES FOR COMMERCIAL VEHICLES

HERBERT SCHMIDT,
Alfred Teves, GmbH, Frankfurt/Main

INTRODUCTION

Since, for many years, the disc brake has become standard equipment in the majority of passenger cars due to its improved fading behaviour, its thermal superiority and its easier service compared to drum brakes, there are now many signs which indicate a future application of this brake type in commercial vehicles.

Compared to the drum brake, the braking operation is improved considerably by means of disc brakes since the response of the brake is more sensitive due to the omission of the self-energising effect. Furthermore the vehicle maintains its direction when braking, since the brake forces on the wheels of one axle are effective to almost the same degree.

By introducing the disc brake into buses and commercial vehicles, the load-sensitive control of the brake forces used today and the future brake force control by means of an anti-skid system would be improved considerably, as the disc brake allows for a better brake force control because of its almost hysteresis-free actuation and its low degree of sensitivity.

In view of these advantages, disc brakes for commercial vehicles would be a considerable contribution to increased traffic safety.

It was in the mid-sixties that American and British producers of construction vehicles began to provide for the installation of disc brakes in their vehicles (fig. 1). The reason was that they had realised that disc brakes, for specific conditions of use, were of advantage to construction vehicles compared with drum brakes. The main advantages are the relative insensitivity of the disc brake to water and mud, improved quality with regard to thermal load (heat fading) and, in particular, high cost savings in the pad replacement and the reconditioning of the disc brake, as this work can be carried out on the spot.

Within the last few years, brake manufacturers in the United States and Europe have begun to intensively push on the development of disc brakes for the commercial vehicle sector. In England, the Midland Red Bus Company introduced disc brakes for their buses in the early sixties. In the United States the new regulations of the brake standard 121, with its strict requirements, contributed to the enforcement of this development, since, due to the required stops from high speeds (96 km/h) and to the so-called "failed system stops", drum brakes were no longer considered as being adequate.

FADING BEHAVIOUR OF DRUM AND DISC BRAKES

One of the accepted advantages of the disc brake as compared to all types of drum brakes consists, as stated above, in the improved fading behaviour which led to the introduction of disc brakes into passenger cars in the late fifties. As is known, the fading behaviour of a friction brake is identical to the sensitivity of the brake characteristic value C^+ to variations of the friction coefficient. Since drum brakes, due to their self-energising effect, have a greater dependency of their brake torque on the friction coefficient (μ) of the brake pad than disc brakes, the disc brake must automatically be superior with regard to the stability of the brake forces, e.g. the fading behaviour of the drum brake. In terms of mathematics, the brake characteristic value is a function of the friction coefficient

$$c^+ = f(\mu)$$

The characteristic value is defined as the ratio of the sum of the circumferential forces (U) divided by the clamping load (S) of a brake

$$C^+ = U/S$$

The degree of sensitivity ε of a friction brake is determined by forming the first differential quotient, i.e.

$$\varepsilon_B = \frac{d\,C^+}{d\,\mu}$$

Fig. 2 shows characteristic values and sensitivities of various drum brakes presently used in commercial vehicles compared to those of the disc brake.

It is shown that even the cam-shaft actuated Simplex brake, with its relatively low characteristic values, and μ being 0.38, still has a sensitivity which is approx-

© IMechE 1976

imately 4 times higher than that of the disc brake. This means that a commercial vehicle which is equipped with disc brakes has a considerably more stable behaviour during the braking operation and thus offers greater traffic safety compared to those vehicles which are equipped with drum brakes.

Fig. 3 shows the fading behaviour of a Duo-Duplex wedge brake 410 Ø x 160 as against a disc brake with a disc diameter of 368 mm and a width of 47 mm in ventilated design after 20 consecutive stops at one-minute intervals and with constant actuation pressure. The temperature rise during the 20 stops is shown in the diagram of fig. 3; during these stops, the energy which had to be absorbed for each stop was 52,500 mkp for the Duo-Duplex brake and 78,700 mkp for the disc brake. The increased temperature level can also be attributed to this 50% higher energy consumption of the disc brake. Although the disc brake had to absorb 50% more energy for each stop, the brake torque drop for all stops was only 12.5% compared with 39.5% in the case of the Duo-Duplex brake (wedge brake).

PERFORMANCE RATING AS A CRITERION OF THE THERMAL BEHAVIOUR OF DRUM AND DISC BRAKES

The thermal loads of commercial vehicle brakes are highest during duration braking and consecutive stops.

Definition of Performance Rating

The permissible thermal load which can be absorbed by a brake within a specific time is dependent on the friction pads used and on the counter-material, namely the drum and/or brake disc and is defined as performance rating.

Duration Braking

As long as there is no temperature stability, the heat volume

$$Q_1 = c.\gamma.v.d\Delta\vartheta \quad (1)$$

is accumulated in the brake disc and/or drum for each time unit and is dissipated by convection.

$$Q_2 = \alpha.F_K.\Delta\vartheta \quad (2)$$

This means that the friction brake absorbs the heat volume

$$Q_t.dt = (Q_1+Q_2).dt = c.\gamma.V.d\vartheta + \alpha.F_k.\vartheta.dt \quad (3)$$

This equation which is part of Newton's cooling law

$$d^2 Q = \alpha(\vartheta-\vartheta_o).d_F.dt \quad (4)$$

can be used for duration braking after respective conversion and integration and reads as follows:

$$Q_D = \frac{\Delta\vartheta_{max}.\alpha.F_K}{1-e^{-\frac{\alpha.F_K.t}{c.\gamma.V}}} \quad (5)$$

Consecutive Stops

As far as consecutive stops are concerned, it is assumed that the heat volume q of the first braking operation is accumulated loss-free. The temperature of the disc body then amounts to:

$$\vartheta_o = \frac{q}{c.\gamma.V} \quad (6$$

The equation

$$\gamma.v.c.\frac{d\vartheta}{dt} = -\alpha.\vartheta.F_K \quad (7$$

is applied to the cooling process after the first braking operation.

Upon integration of this equation and formation of the corresponding geometrical progression, the equation can be obtained for the temperature achieved after the performance of "n" stops.

$$\vartheta_n = \frac{q}{c.\gamma.V}\cdot\frac{1-e^{-n\frac{\alpha.F_K.t_s}{c.\gamma.V}}}{1-e^{-\frac{\alpha.F.t_s}{c.\gamma.V}}} \quad (8$$

These two equations can only be sufficiently evaluated in the case of relatively thin brake discs since they are based on the assumption that the entire disc body involved in the heat absorption process has the same temperature at a chosen time "t" and that the heat transfer coefficient α is constant. Brake discs with a more sophisticated ventilated design can thus only be determined in the form of a thermal balance. However, both equations have a tendency to show the relative influences of the main parameters of a brake design in relation to the absorbable performance rating Q_t. They also illustrate the effect of changes, e.g. in the disc ventilation.

Heat Flow Distribution

According to the law of the heat flow distribution, the heat volume generated in the friction layer is distributed to the brake pad and to the counter-friction area, namely the drum and/or disc.

In this connection it is interesting to examine which amount of the heat volume produced can be additionally carried off through the pad and the brake pad carrier.

To determine the heat distribution on the brake pad/counter material, the equation

$$\frac{Q_S}{Q_b} = \frac{\sqrt{\mu_s.\lambda_s.c_s}}{\sqrt{\gamma_B.\lambda_B.c_B}} \quad (9)$$

can be used with certain reservations with regard to exactness. The term $\sqrt{\gamma.\lambda.c}$ is a mere material value and characterizes the heat penetration coefficient.

When using organic pads with a thermal conduction coefficient of $\lambda = 0.8$ kcal/mh°C, the thermal share at the pad side is approximately 3 per cent, whereas, when using sintered pads with relatively high

© IMechE 1976

thermal conduction coefficients of $\lambda = 20$, this share can amount up to 18 per cent (fig. 6). Besides, due to the increased share in the heat accumulator volume,

$$Q_B = c \cdot \gamma \cdot V \cdot \vartheta_B$$

the performance rating Q_t of a disc brake can be improved considerably when using sintered pads. However, what may cause some trouble is the heating of the hydraulic actuation because of the good heat conductivity of the sintered pad.

Whilst, in the case of the drum brake, the temperature difference between the temperature produced in the friction area and the surrounding air due to the thermal conductivity through the drum wall is relatively low (fig. 4), it is a maximum value for the disc brake, as the friction area is directly surrounded by the air.

With the heat accumulator volume

$$Q_s = c \cdot \gamma \cdot V \cdot \vartheta \qquad (10)$$

and the cooling area F_K being the same, the performance rating Q_t of a disc brake is better than that of the drum brake since the heat dissipation shares caused by convection and radiation

$$Q = \alpha \cdot F_K \cdot \Delta\vartheta \qquad (11)$$

are by far greater due to the higher temperature difference.

$$\vartheta_D = \vartheta - \vartheta_L \qquad (12)$$

In the case of the floating caliper brake 2 F 90 developed by ATE, the ventilated brake disc was dimensioned in such a way that, in its cooling area, it is superior by approximately 24% to a drum brake with the dimensions 410 ∅ x 180 and by 37% to a drum brake with the dimensions 410 ∅ x 160.

BRAKE DISC MATERIALS

In case of stops with high energy consumption, a temperature difference is produced in the brake disc cross-section (fig. 5) which results in surface tensions, since the cooler layers which are at a larger distance from the friction area have less expansion. Due to the shearing stresses caused by thermal expansion and consecutive shrinkage, the disc brake material is fatigued which results in surface cracks.

The thermal expansion is defined as

$$\varepsilon = \bar{\alpha} \cdot \vartheta \qquad (13)$$

Based on the permissible expansion caused by tension according to the law of Hocke,

$$\varepsilon = \frac{\sigma}{E} \qquad (14)$$

the permissible surface temperature can be determined as follows :

$$\vartheta = \frac{\sigma}{\bar{\alpha} \cdot E} \qquad (15)$$

This equation shows that the best disc materials are those which have the highest permissible tensions possible and the lowest thermal expansion and elasticity modulus E possible.

In fig. 5 the initial inclination of the temperature curve is given by the equation

$$\frac{\Delta\vartheta}{\delta} = \frac{q}{\lambda} \qquad (16)$$

the value of q (Kcal) being the thermal volume absorbed for each stop.

The equation demonstrates that the gradient $\frac{\Delta\vartheta}{\delta}$ of the temperature curve changes depending on λ. A large λ value therefore results in a more even temperature curve, thus reducing the danger of crack formation due to surface tensions (curve 2, fig. 5) and improving the performance rating Q_t.

In the case of the grey cast-iron discs, sensitivity to thermal cracks already begins to exist with a heat density of $Q = 400$ kcal/m^{2}s. Steel cast-iron which is superior to grey cast-iron with regard to heat resistance, can hardly be considered because of its high price, its bad machinability and its unfavourable friction pairing with organic pads. A further decisive disadvantage of steel cast-iron is the fact that its specific heat "c" is approximately 65% lower than that of grey cast-iron, thus reducing the obtainable performance rating Q_t according to equation (5) for steel cast-iron discs by approximately 28%.

In view of this situation, it had to be examined whether an improvement of the so-called thermo-shock resistance of grey cast-iron discs is possible up to a thermal flow density of approximately $q = 600$ kcal/m^{2}s, which, for example, can be attained by fast-driving coaches. After extensive tests carried out by ATE, a new grey cast-iron specification was developed in which the thermo-shock resistance of grey cast-iron was improved to the extent that practically no brake disc damages occur under all kinds of stresses.

DISC BRAKE DESIGN

With regard to thermal resistance, the ventilated brake disc is superior to the full disc due to its bigger total surface. The surface increase of a disc brake with a 368 mm diameter with radial blading, for instance, is approximately 90%, whereas, in the case of curved blades and optimum design, the surface gained as against the full disc can be up to 130%.

Furthermore the curved blade allows for an increase in the flow speed and thus in the heat transfer coefficent α because there are no shock losses at the channel inlet.

According to the results of investigations made by Bode and Dorner, the performance rating Q_t can be increased by 19% by improving the inner ventilation of the brake disc when using curved blades. The disadvantages of this brake disc design, however, are the manufacturing costs as well as the demand for a left-handed and right-handed brake disc design.

© IMechE 1976

DEVELOPMENT OF DISC BRAKES FOR COMMERCIAL VEHICLES

In the early sixties, after having concluded the development of a disc brake system (fig. 7) for a heavy track-laying vehicle, ATE began the development of the disc brake for commercial vehicles. For the first time, extremely high thermal requirements had to be met by the brake pad and the disc, because, for a total vehicle weight of 42 t, only two disc brakes could be provided with minimum installation space available and as extremely high temperatures occur in the friction areas in case of stops from speeds up to 65 km/h. After extensive tests on the dynamometer and on the vehicle, it became evident that the dimensioning of the ventilated brake disc with a 462 diameter x 90 width was sufficent for all thermal loads. It was decided to use alloyed steel cast-iron as brake disc material, since grey cast-iron could not be considered because of insufficient thermal strength and the friction pairing for the high-wear-resistant ceramic sintered pads. The development of a parking brake was a further problem, since, due to military requirements, the vehicle weight of 42 t had to be held on a 60 per cent slope without any additional power assistance, i.e. with mere muscle power.

The development of disc brakes for heavy vehicles was followed by the development of the first production disc brake system for commercial vehicles with total weights from 6.5 to 10 tons. The design of the disc brakes provided for this purpose was stongly influenced by the portal axles which were under discussion at that time so that the disc brake finally had to be based on the conventional fixed caliper. As the customer wanted to have disc brakes on both axles of the vehicle and as a separate arrangement of the parking brake at the gearbox was out of the question for various reasons, it became necessary to combine the fixed caliper disc brake with an additional mechanical brake actuation.

The combination of service brake and parking brake on one fixed caliper was only possible after an automatic adjusting device had been found which had a high degree of reliability and allowed for the clearance of brake pad and brake disc on both pads to be kept constant and small and which functioned as a return stop for the piston when actuating the parking brake. In case of a parking brake actuation which is effected through spring accumulators, a reaction force is built up at the piston side opposite the mechanical actuation side after having overcome the clearance by slightly deforming the brake disc in the vertical plane in order to obtain the required brake torque on the one hand and to avoid too high a bending stress of the brake disc on the other. Fig. 8 shows the combined 2-cylinder fixed caliper disc brake for the 6 t DB Unimog, whilst the combined 4-cylinder fixed caliper for the 10 t Unimog vehicle is illustrated in fig. 9.

When developing these two disc brake systems, the limited installation space caused considerable problems regarding the strength to be achieved and the permissible deflection of the combined fixed calipers, since, in this brake design, the actuating forces can be superimposed by the service and the parking brake. The axial force is 11 t for the combined fixed caliper of the 6.5 t vehicle, whereas, in the case of 10 t vehicles, the axial force of 16.5 t must be transmitted through the bolt connections. In view of the specific use of the two vehicles, the parking brake actuation was designed to allow for a stop in a 60% slope. The experience gained in the meantime with both vehicles has shown that the disc brake has a clear superiority as against the drum brake with regard to thermal aspects, fading behaviour and employment in muddy terrain. Owing to the introduction of the disc brake, the time spent on pad replacement could be reduced from 12 hours to 1 hour.

THE ATE FLOATING FIST-TYPE CALIPER CONCEPT

Whilst the fist-type caliper design had to be used for the portal axle of the Unimog vehicles with 6.5 tons and 10 tons due to the installation space predetermined by the backgear and considering the fact that, for the reasons mentioned below, there was no choice of using another design for these vehicle weights, a new disc brake concept had to be developed for heavy commercial vehicles with a total weight up to 16 tons.

The reasons why fixed calipers cannot be used for heavy vehicles with weights ranging from 10 to 16 t, are as follows :

a) lack of installation space for a sufficient dimensioning of brake pad and brake disc thickness (fig. 12)
b) limited size of brake pad surface
c) the problem of how to install a parking brake actuation which is integrated into the caliper
d) the unfavourable utilisation of the abrasive pad volume by the adjusting wedge-shaped wear
e) insufficient thermal load resistance

The result of an extensive study of all functional and design parameters under consideration was that the floating fist-type caliper concept is the optimal solution of all problems arising in connection with a disc brake design for heavy vehicles.

The hydraulically actuated ATE floating fist-type caliper mainly consists of 2 parts, namely the caliper and the brake carrier (fig. 10). The floating caliper is guided on two pins via exchangeable bushes, these pins being mounted on the brake carrier. The caliper part transmits only the clamping forces for the service and parking brake actuation, whereas the circumferential forces are transmitted onto the brake carrier via the pad back plate. By means of this arrangement it can be achieved that the caliper guide is free from circumferential forces. It is only

© IMechE 1976

stressed by the inertial forces of the caliper. A further important advantage of this pad arrangement consists in the fact that the circumferential forces are absorbed at the leading side of the disc and that the wedge-shaped wear of the brake pads which we know from the other disc brake designs can be avoided. (Fig. 13)

Fig. 12 illustrates the space requirements of a fixed caliper in comparison with the floating fist-type caliper. The comparison shows that, with the same installation space available, a fixed-type caliper can only be installed when reducing the pad and brake disc thickness.

Brake Pad Replacement

As shown in fig. 11, the brake pad replacement is very easy since, after the removal of a guide pin, the caliper can be tilted up, thus having direct access to the brake pads to be replaced.

The brake pads of a commercial vehicle can be replaced within 2½ hours including the removal of the wheels. This is a considerable cost reduction compared with the replacement of drum brake linings particularly as there is no excessive turning of the pads; furthermore the utilisation loss of the vehicle which is caused by the pad replacement can be reduced considerably.

Caliper Guide

When designing the ATE floating fist-type caliper, special attention was paid to the caliper guide since, in the last analysis, it is of decisive importance to the service life of the brake pads and to the safe functioning of the brake. A requirement in this connection was that the caliper guide had to be free from dirt and corrosion in order to ensure a clearance between disc and lining upon completion of a braking operation. Since this requirement cannot be met with the known groove-guided floating calipers (fig. 14), ATE has developed a guiding element which allows for a sliding guidance of the disc brake caliper at the cylinder side, while, at the rim side, this guiding element has been modified in such a way that an active resetting of the caliper is obtained. The clearance between pad and disc at the cylinder side is ensured by the roll-back effect of the piston seal.

Two-Circuit Brake Actuation

Due to the twin cylinder actuation of the caliper, the real dual circuitry of the brake system can be easily realised. Thus, in case of a failure in one brake circuit, the brake force distribution remains unchanged and the stability of the vehicle is not negatively influenced during the braking operation, in particular with low friction coefficients between tyre and road surface. In addition to this, the emergency brake condition according to the new EC regulations are complied with.

Parking Brake Operation

The parking brake actuation was designed in such a way that the arrangement of the spring accumulator actuation which is used today for cam-actuated drum brakes, could be maintained. In order to ensure a high mechanical efficiency, it was envisaged to transmit the power by means of a maintainance-free recirculating ball and thread device co-operating with an automatic adjustment device integrated into the piston space and keeping the actuation angle at a constant level. (fig. 15)

The adjusting device which is fully protected by the surrounding brake fluid adjusts the wear travel of the brake pads which occur when actuating the service brake and thus corresponds to the new legal requirements.

Due to the superimposition of service and parking brake, axial forces up to 32 t can occur on the caliper causing problems regarding the fatigue strength and the required stiffness with regard to the fact that the weight of this component should be as low as possible. Due to the use of a high-quality nodular iron, it was finally achieved that the weight of the total disc brake system is even lower than that of a corresponding drum brake system for a 16 t vehicle.

SUMMARY

By using the disc brake for commercial vehicles, a further contribution can be made to increase traffic safety due to the more stable behaviour of the brake forces of the disc brake, the improved thermal characteristics and the realisation of a real dual circuitry of the brake system. Last but not least, the disc brake can noticeably reduce the maintenance costs incurred by the vehicle owner, as the pad replacement and the out-of-service time of the vehicle are reduced to a small portion of those relating to the drum brake.

References :

1) "Archiv für Eisenbahntechnik", Folge 15 Röhrig Verlag, Darmstadt, 1961

2) O.Bode, H.Dorner
Konstruktion von Kraftfahrzeugbremsen hoher Dauerleistung
Deutsche Kraftfahrtforschung 121/1958

3) H.Hasselgruber, Berechnung der Warmespannungen in Bremstrommeln.
ATZ 2/1954

4) F.W.Bielecke, Warmetechnische Nachrechnung von Kraftfahrzeugreibungsbremsen.
ATZ 9/1956

5) Newcomb, Journal Mechanical Engineering Science, Vol. 2 No. 3, 1960

© IMechE 1976

List of Symbols Used

C^+ = brake characteristic value

U = circumferential force

S = clamping force

μ = friction coefficient

ε_B = sensitivity of brake

Q_1 = accumulator heat

Q_2 = heat volume dissipated by convection

Q_t = performance rating/heat volume

Q_D = performance rating/duration braking

q = heat volume for stop

c = specific heat

γ = specific weight

V = volume of disc body

F_K = cooling area

α = heat transfer coefficient

λ = thermal conduction coefficient

t_s = time between 2 stops

t = time of duration braking

$\Delta\vartheta$ = temperature/disc body

ϑ_o = temperature/disc body, stop

ϑ_H = temperature/disc body, consecutive stop

ϑ_L = temperature of surrounding air

ϑ = temperature of friction area

ϑ_D = temperature difference

ε = permissible expansion

$\bar{\alpha}$ = thermal expansion coefficient

E = elasticity modulus

σ = permissible strength

δ = thickness of layer

© IMechE 1976

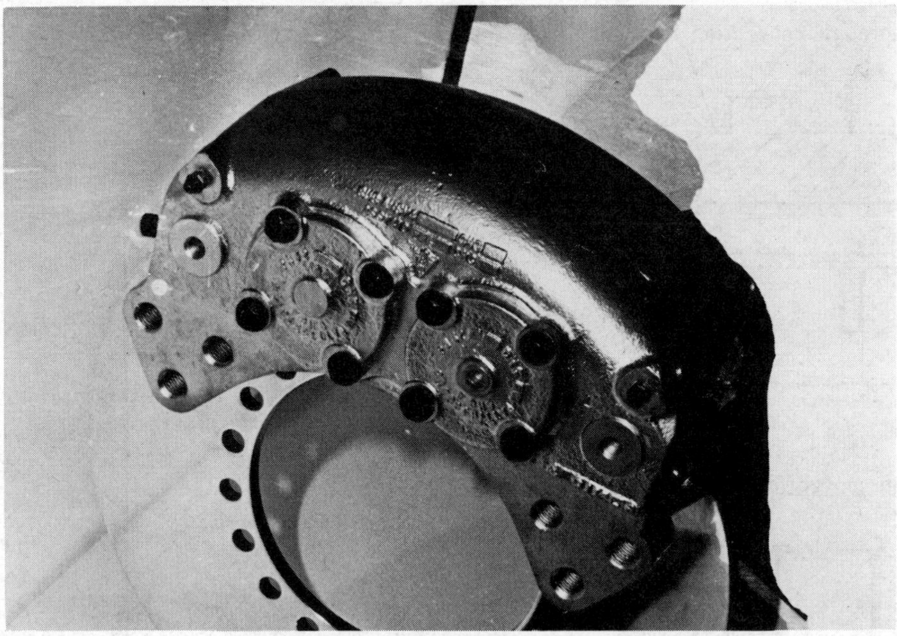

Fig. 1: Disc Brake for Construction Vehicles used in U.S.A.

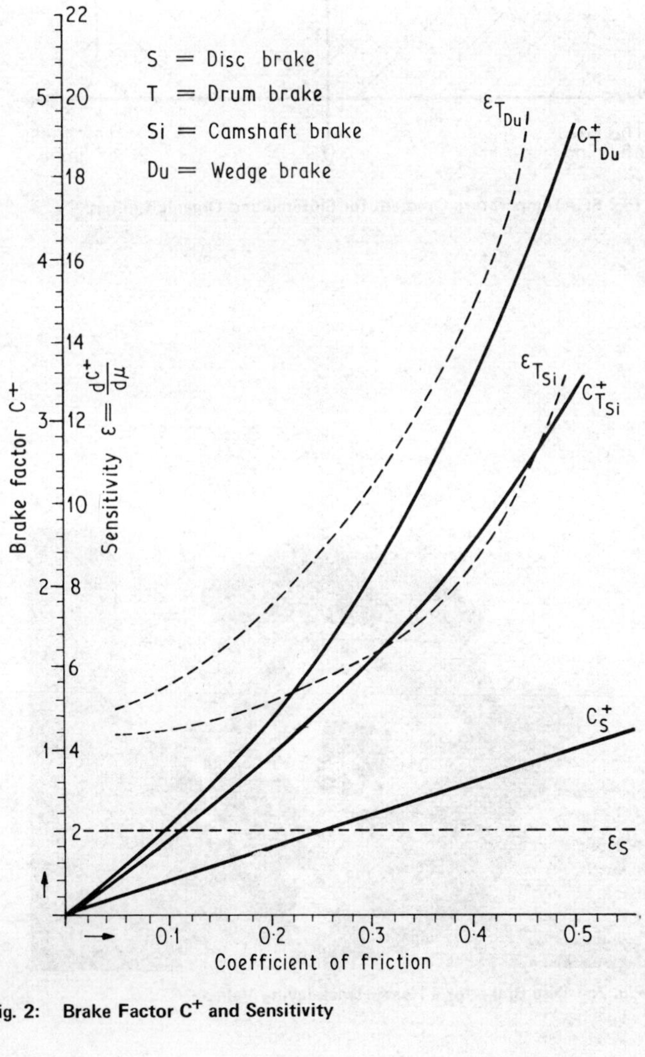

Fig. 2: Brake Factor C^+ and Sensitivity

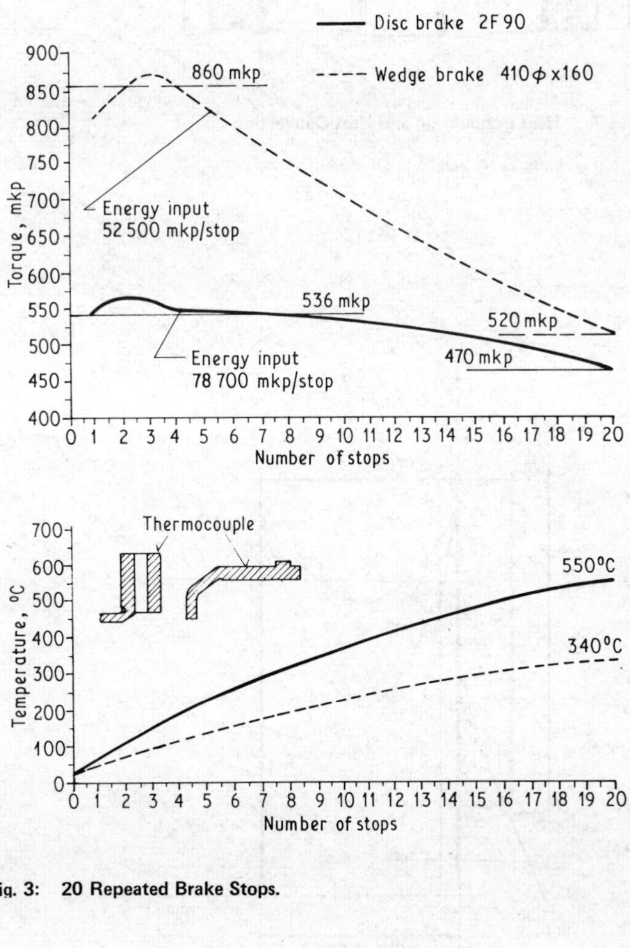

Fig. 3: 20 Repeated Brake Stops.

© IMechE 1976

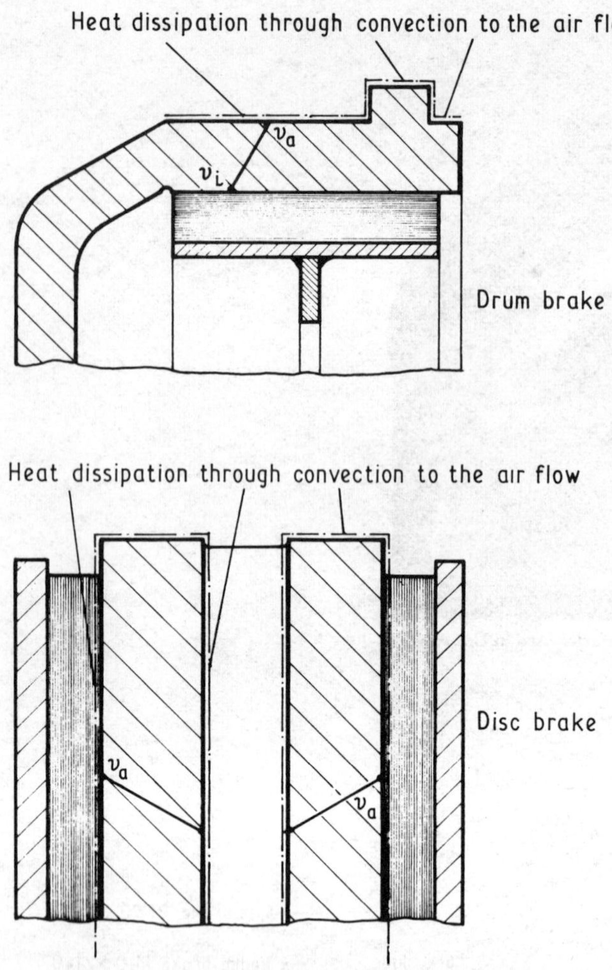

Fig. 4: **Heat Conduction and Heat Convection**

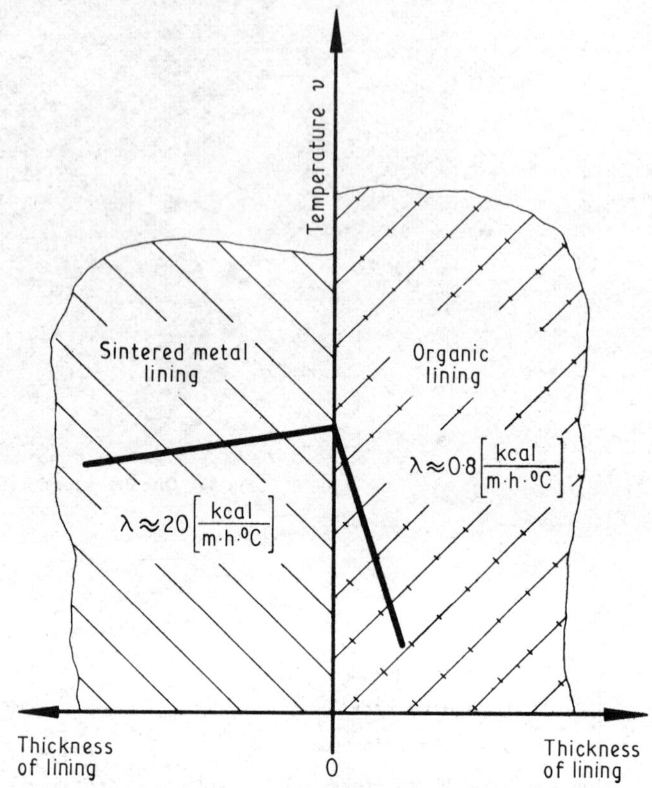

Fig. 6: **Temperature Gradient for Sintered and Organic Linings**

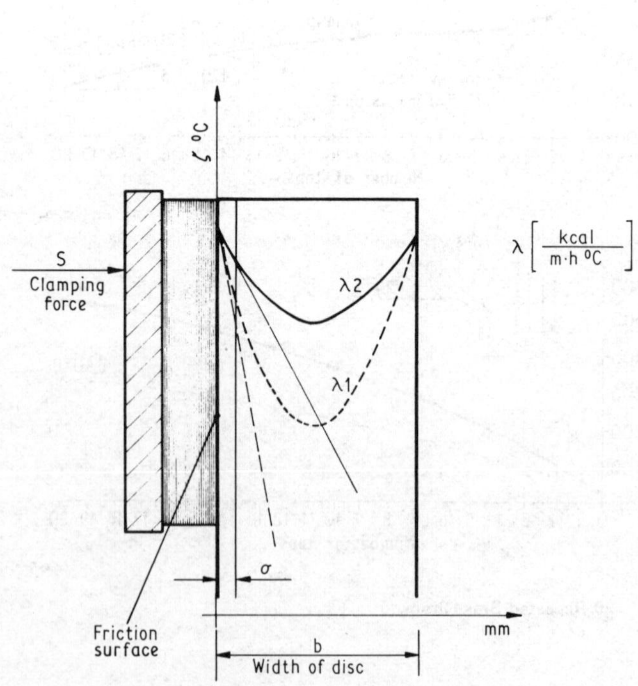

Fig. 5: **Heat Conduction within the Brake Disc**

Fig. 7: **Disc Brake for a Heavy Track-laying Vehicle**

© IMechE 1976

Fig. 8: **ATE Combined 2-Cylinder Fixed Caliper for the 6.5 ton D.B. Unimog**

Fig. 11: **ATE Floating Fist-Type Caliper. Brake Pad Replacement**

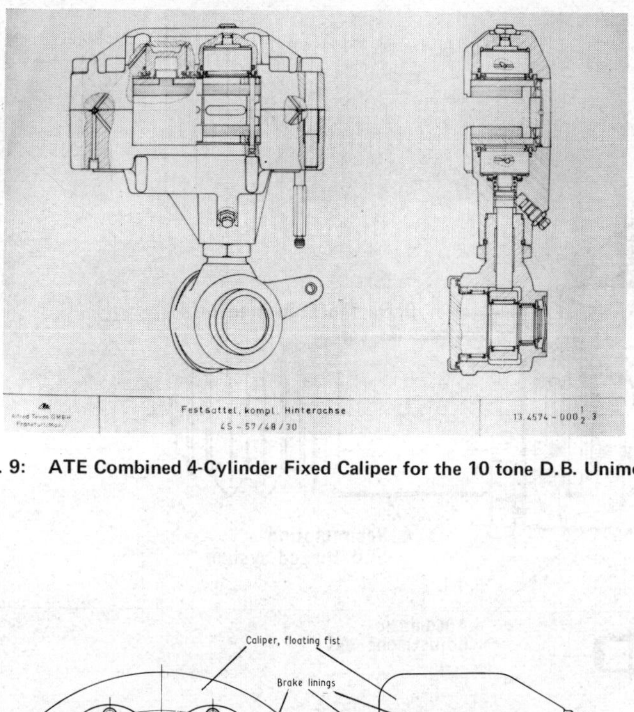

Fig. 9: **ATE Combined 4-Cylinder Fixed Caliper for the 10 tone D.B. Unimog**

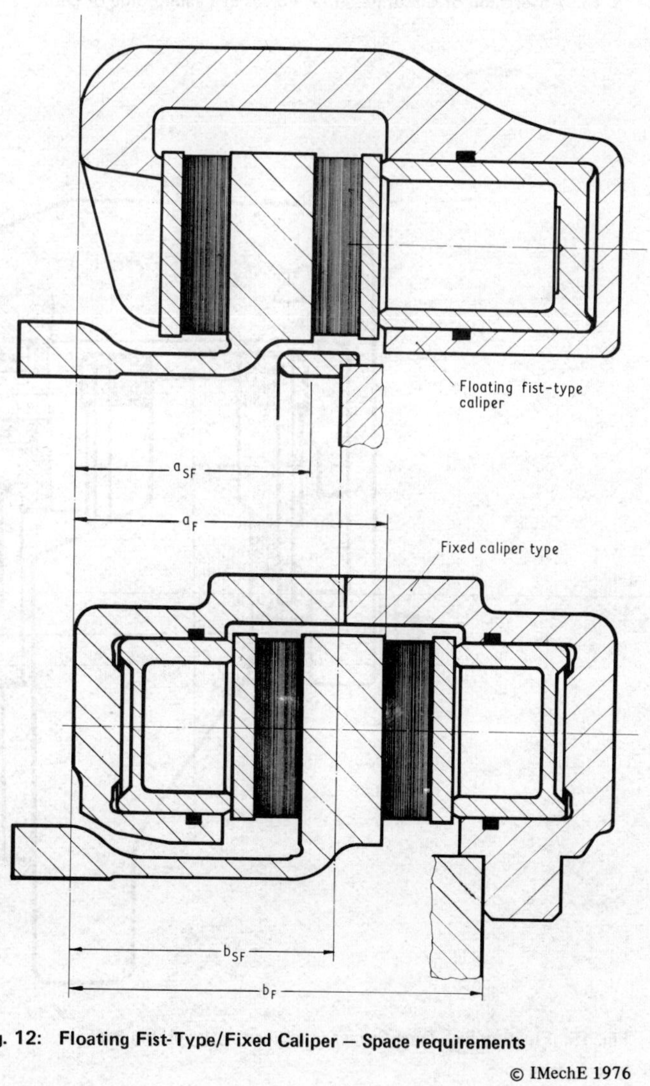

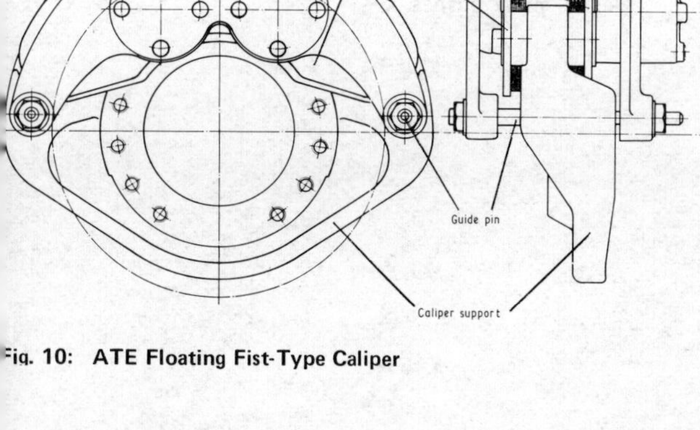

Fig. 10: **ATE Floating Fist-Type Caliper**

Fig. 12: **Floating Fist-Type/Fixed Caliper – Space requirements**

© IMechE 1976

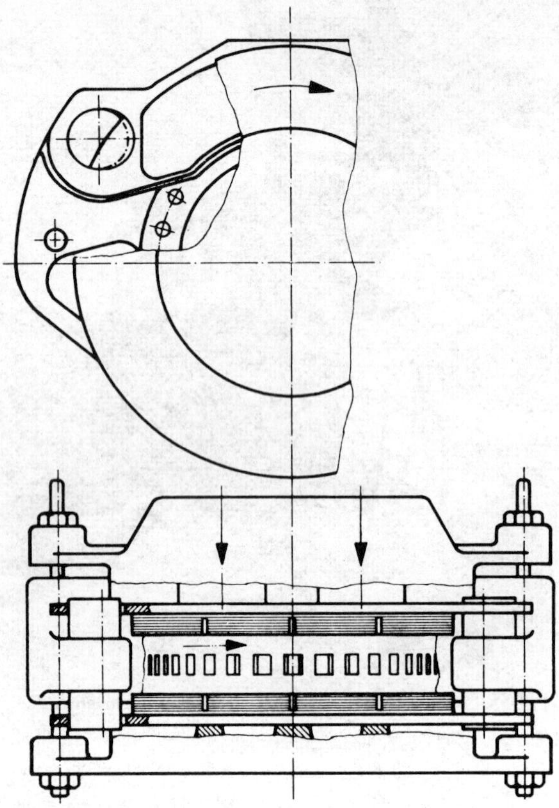

Fig. 13: Absorption of Circumferential Forces at Leading side of Disc

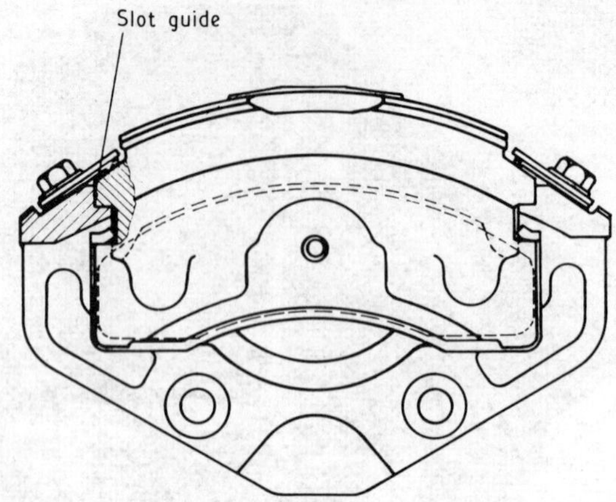

Fig. 14: Slot-Guided Caliper

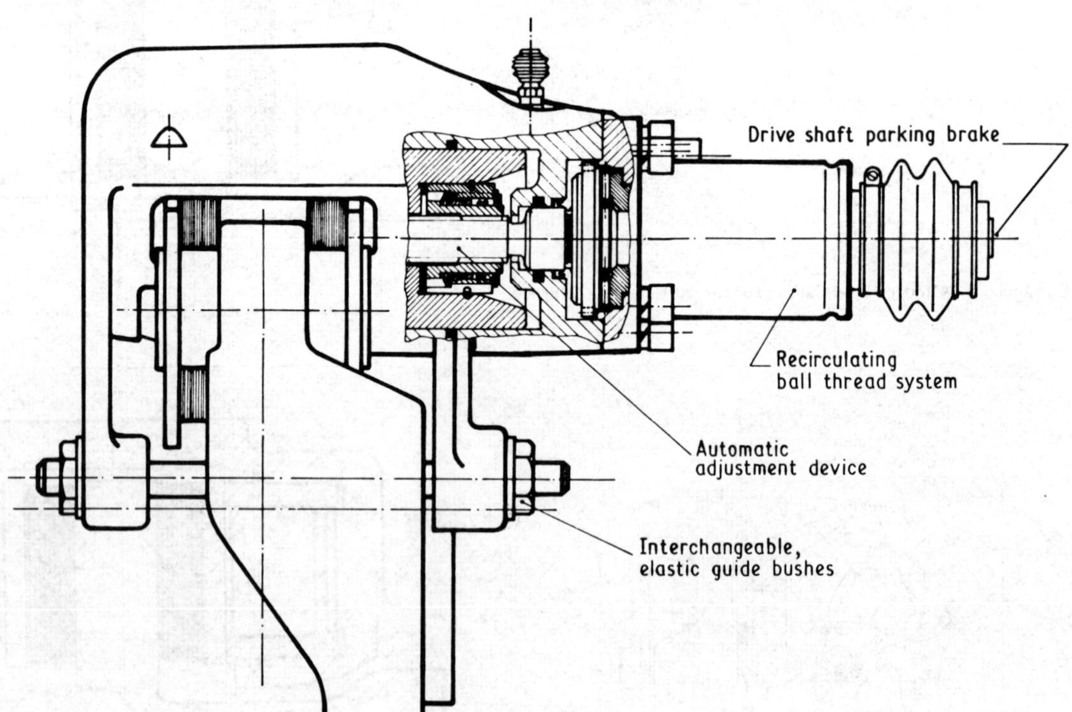

Fig. 15: Floating Fist-Type Caliper combined with Parking Brake

© IMechE 1976

C78/76

BRAKE PERFORMANCE AND COSTS – COMMERCIAL VEHICLES

ROGER DENNISS, C.Eng, MIMechE, MIRTE
Bass Production Ltd, Burton-on-Trent

The MS of this paper was received at the Institution on 22 August 1975 and accepted for publication on 19 January 1976

SYNOPSIS

The 1968 Road Traffic Act brought about a significant improvement in commercial vehicle brake performance. Currently brake maintenance costs account for up to 12% of total maintenance costs. Failure of oil seals is a major contributing factor to brake repair costs. Load sensing valves regularly adjusted regulate jack-knifing. Accidents due to brake failures have become a rarity. Air brake noise is an environmental nuisance. The exhausting of compressed air from brake systems, quick release valves, etc. and brake squeal is still an environmental nuisance.

INTRODUCTION

2. The prime objective of a Fleet Engineer is to provide a fleet of safe, reliable, vehicles to the required specification at the lowest cost and at a predictable cost, bearing in mind the need to provide a good fleet public image.
The influence of braking systems on the achieving of this objective has in the past been under-estimated by many fleet operators.
This paper sets out to review current typical commer cial vehicle fleet braking performance and costs.

3. Methods of Monitoring Performance

3.1 The 1968 Road Traffic Act introduced an "Operators Quality Licence" which permitted companies to operate commercial vehicles providing that they could satisfy the Department of Environment that they were capable of maintaining their vehicles in a safe and roadworthy condition.

3.2 This Act initiated an annual eighty item inspection to defined standards for commercial vehicles at purpose-built test stations throughout the country.

3.3 There can be no question that this proved to be one of the most significant strides towards improving commercial vehicle maintenance standards.

3.4 A major spin-off was that operators who had not in the main attempted to record and analyse cost and performance were presented with a basis of meaningful system, and many were forced into noting the cost of maintaining those components that did not automatically achieve the required Department of Environment minimum standards. A test failure meant that a vehicle was off the road, which cut into profits.

3.5 The 'O' licence or 'Quality' licence also encouraged operators to introduce regular, periodic inspections to prescribed permissible wear tolerances by independent examiners.
The frequency of those inspections being varied, depending upon the content of the reports thrown up, but usually ranging from monthly to three monthly intervals.
The data so gathered has been analysed in some cases using computers. Indeed, there are now a number of flourishing concerns providing vehicle maintenance analysis computer packages.

3.6 Another source of component reliability data is the statutory drivers daily defect report. Whilst the information so presented will certainly vary with the day of the week and weather conditions, the report presented by the average driver contains valuable information that does not appear to be used by either manufacturer or operator to maximum advantage. The success of modification to components can be easily monitored through the Drivers Defect Report Sheet.

4. Initial Specification

4.1 A fleet operator today has little choice in the vehicles brake component specification, and has to accept what the manufacturer offers. A fleet engineer may well determine through empirical work that brakes with larger lining areas may be cost effective. However, the only method of achieving this would probably be by "going up" to the next chassis dimension, which overall would not be economic.
The fitting of over-size axles would create a hybrid vehicle and consequent spares problems.

4.2 The normal brake options available are as follows:

4.2.1 Load Proportioning Valves
4.2.2 Anti wheel lock devices
4.2.3 Air brake anti-freeze equipment
4.2.4 Automatic adjusters
4.2.5 Automatic drain valves
4.2.6 Air drying tanks
4.2.7 Retarders

© IMechE 1976

4.3 Weight. Currently the annual Road Fund Tax is £72 per ton. If the restrictions authorised by the Dykes Act are implemented whereby local authorities have the power to totally prohibit vehicles above 3 or 5 tons unladen weight, weight is of even greater importance.
In certain cases a 450 grams weight reduction is valued at £2 per year.

4.4 The typical unsprung weight of brake components are shown below:

Vehicle Type	Weight of Brake components as a % of total wt.	Cost of brake components as a % of total cost
32 ton tractor	11.5	6.00
40' semi-trlr.	4.0	6.00
8 ton rigid	8.0	6.00

4.5 Optional Extras

Component	Add.Cost £	Appr.add. weight.	Operational cost saving per annum
Load Sensing Valve	25	2 kilos	-
Anti lock device	300	15 "	-
Antifreeze equip.	20	Negligible	unknown
Auto.adjusters	165	2 kilos	£65-£130
Auto.drain valves	8	500 grams	
Air drying tank	100	10 kilos	unknown
El.mag.retarders	800	322 kilos	£400

4.6 Antifreeze Equipment

Methanol drip antifreeze systems appear to be simple, cheap and effective, providing the reservoir is topped up.
Air drying tanks have proved trouble and maintenance free, effective, but relatively expensive. In times when economies top the list of priorities it is difficult to justify the air dryer, and, as with anti-skid devices, one looks towards legislation to force the issue.

4.7 The brake drum diameter does influence road wheel size and the overall platform height.
Lower deck heights mean more stable vehicle operation. The possible introduction of 38/44 ton vehicles in the 1980's with little increase in the permissible overall vehicle length, will tend to increase load heights, thus the small diameter compact lightweight multi disc brake would appear to offer a number of advantages, particularly if the maintenance requirement is completely eliminated.

5. Performance in Service

5.1 In 1968 at the introduction of annual testing the test station failure rate due to substandard braking performance was 30%.
It will be recalled that initially certain older vehicles did not have to meet current standards of .5g service brake and .25g. secondary brake. The current failure rate at test stations is 16.03%.

5.2 Typical service brake performance efficiencies are shown at Fig. 1 for 32 ton two axle tractor and at Fig. 2 for a 27/29 ton gross tandem axle trailer.
The conclusions one draws is that given regular inspection and adjustment there should be no problem in achieving the minimum braking efficiencies required by the Ministry.

5.3 Whilst the specialist vehicle and PSV operator has not been able to justify more frequent renewal of his fleet, the higher standards of maintenance demanded by the DoE through their test stations has encouraged own account operators and road haulage contractors to renew fleets more frequently.
It has been my experience with commercial vehicles that those made since 1970 have had little difficulty in meeting braking efficiency standards, and that in the main heat crazing of drums has been eliminated.
Failure of oil seals necessitating relining is a common occurrence and not confined to my Company, or for that matter, type of operation, although I understand this is not a PSV problem.
Load sensing valve/brake apportioning valves do require regular calibration.

5.4 The machining of new lining facing to the drum diameter, preferably in situ, centred on the hub, is a quick and economic method of attaining instant target efficiencies.
Drum surface condition is considered to be very important - a turned finished being preferred to ground.

5.5 Whilst I am no brake shoe facing expert, I must respect the experts opinion that the choice of lining is very much a case of 'horses for courses' although I do feel that the manufacturers original choice in this respect in the past could have been very much improved to give wider satisfaction.

5.6 The period between relining of trunker vehicle brakes obviously varies with operating conditions, nevertheless, 100,000 - 150,000 mile life is commonplace. Oil seal failures appear to promote as many relinings as does fair wear and tear.

5.7 The low mileage retail vehicle employed on stop-start work could be involved in as many as six relines in a 100,000 mile vehicle life.

5.8. Brake Defects. Analysing five hundred driver daily vehicle defect reports showed that over 33% of the reports on trunker vehicles involved brake deficiencies - the most common being minor air leaks, quickly rectified by tightening the tapered thread.
It has been suggested that reversion to the parallel thread and anti luce washer might solve this problem.

5.9 Recurring defects. It is noteworthy that there has been a significant reduction in the recurring defects noted over the last five years. So significant that the author feels that a complete revision of the brake component replacement interval recommended by the manufacturer is required (Refer to paragraph 10).
The failure of oil seals has already been mentioned. Seizure of brake cam shafts and shoes on anchor pins was once a problem, overcome by the sparing use of a boundary lubricant. Periodic lubrication of cam shafts is considered to be a retrograde step, in that there is always the possibility of the lubricant entering the braking drum, and the overall aim must be to eliminate every conceivable maintenance duty.

© IMechE 1976

5.10
Proportion of jobs involving brake reps./adjustment.

Code No.			Brake Repairs / Total Repairs x 100
A.1. *	11.2%	-	low mileage 32 ton tractor
A.2 *	9.7%	-	trailer
B.1. *	17.9%	-	high mileage 32 ton tractor
B.2. *	25.8%	-	flat bed two axle trailer. Brakes represent by far the largest maintenance potential.

5.11 Brake Maintenance Cause

Code	Scheduled.	Unscheduled.	Break-down	% fair wear and tear
A.1.*	52%	46%	2%	98%
A.2.*	59%	40%	1%	99%
B.1.*	69%	30%	1%	93%
B.2.*	79%	18%	3%	96%

6. Brake Costs

Brake costs vary overall from 5% to above 12% of total maintenance costs.

Code	Brake Costs			
	Cost per vehicle p.a. £	% of total maintenance	pence per mile	parts labour
A.1.*	48	6.4	0.24p	1.17
A.2.*	24	6.4	0.18p	4.00
B.1.*	120	9.0	0.10p	0.9
B.2.*	31	9.1	0.06p	0.8
C.1.*	120	12.4	0.61p	
C.2.*	40	6.4	0.21p	

7. Braking Noise Levels

Manufacturers and the Department of Environment appear to turn a deaf ear to brake noise emission frequently over 100 dB(A). This is very much an environmental nuisance. However, our air exhausts can be reduced by as much as 16 dB(A) by fitting a cheap air silencer. Our tests over three years have not indicated that there is a significant increase in air exhaust time. The silencer has shown itself to be self-cleansing (See Fig.3). Brake Squeal. Whilst apparently a diminishing problem, isolated instances do occur that prove time consuming and costly to rectify.

8. Accidents

A study of our own trunker fleet accident statistics reveals that not one incident that could be attributed to braking inefficiencies has been experienced in approximately six million miles. Certainly load sensing valves regularly checked and adjusted have contributed to the elimination of brake induced jack-knifing, which very much confirms the conclusions drawn by H.A. Wilkins in his TRRL Report LR 474 'Field trials of load sensing valves fitted to brake systems of articulated vehicles.'

9. Binding Brakes

The correct adjustment of commercial vehicle drum brakes cannot be over emphasised.
Insufficient lining clearance can cause over heating with resultant brake inefficiencies and premature failures of components.
Heat cracking of drum working surface becomes progressive, lining debris being forced into the fracture preventing it from self-healing.
A binding brake has been demonstrated to absorb 3 hp per axle, which could represent £52 per year fuel cost per axle or approximately £200 per vehicle per year if allowed to persist.

10. Scheduled Replacement and/or Overhaul of Brake Components.

10.1 It has already been stated that brake repair costs of high mileage trunker tractors can represent 12% of the total maintenance costs or £120 per annum. However, if the manufacturers recommended major brake component replacement schedule was followed there would be an additional cost for a trunker vehicle covering 100,000 miles a year or more of some £300.

10.2. The recommended replacement life of major components are generally as follows:

Component	Replacement/Service Period
Hand control valve	100,000
Unloader valve	50,000
Foot valve	100,000
Pressure Regulating Valve	100,000
Quick Release Valve	100,000
Spring Brake DIAPHRAGMS	50,000
Renew Spring	100,000
Inverse Spring Valve	100,000
Stop light Switch	100,000
Low pressure indicators	100,000
Relay Valve	100,000
Load sensing valve	100,000
Automatic Drain valve	100,000

11. Conclusions

1. There is a need for motor manufacturers to re-assess the recommended brake component replacement intervals.

2. The introduction of a compact multi disc brake offers many advantages. There is a need to provide automatic brake adjusters which would eliminate frictional losses and allied fuel costs.

3. The 1968 Road Traffic Act involving annual brake roller testing has been successful in improving manufacturers and commercial vehicle braking standards.

4. Given regular inspection and systematic preventative maintenance it is neither difficult nor expensive to continually achieve the statutory braking standards on British commercial vehicles manufactured over the last five years.

5. Vehicles operating in high density traffic record as high a brake maintenance cost as does the high mileage trunker vehicle.

---

* A.1. - low mileage 32 ton tractor
A.2. - trailer
B.1. - high mileage 32 ton tractor
B.2. - flat bed two axle trailer.
C.1. -) urban delivery vehicles.
C.2.)

© IMechE 1976

6. Oil seal failures are responsible for increasing b rake maintenance costs, and there is an urgent requirement for commercial vehicle manufacturers to take positive action to alleviate this problem.

7. Over a wide spectrum of commercial vehicles 40% of brake defects were found to be minor air joint leaks, emphasising the possible need to improve manufacturing tolerances.

8. Overall, 10/20% of all repair orders are concerned with brake repairs. 6/12% of overall maintenance costs involve brake repairs.

9. There is a requirement for reducing air brake noise levels and brake drum squeal.

10. Systematically checked and adjusted load sensing valves will eliminate brake induced jack-knifing, however, better design of equipment is urgently required.

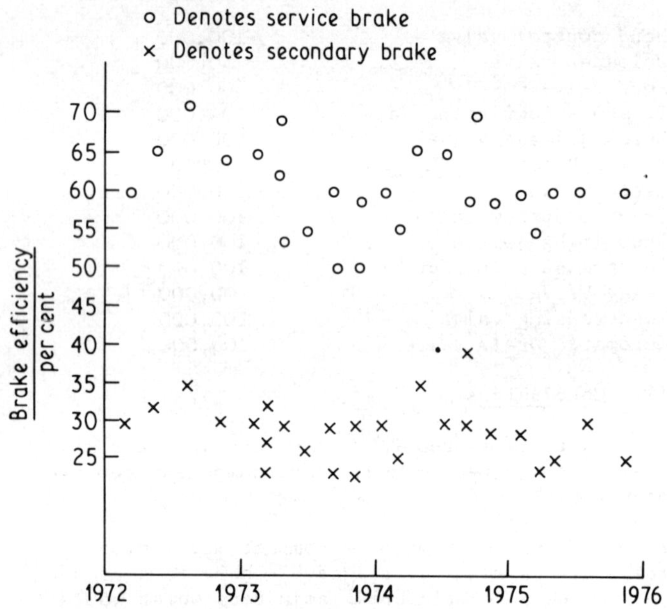

Fig. 1: 32 Ton Gross Two Axle Tractor

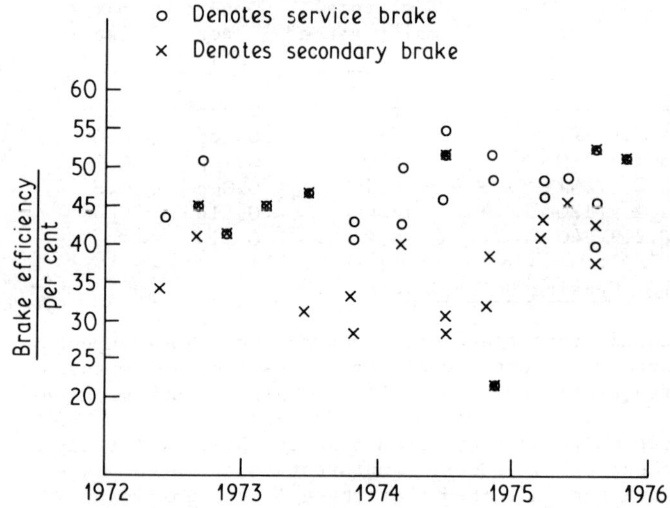

Fig. 2: 27/29 Ton Gross Tandem Axle Trailer

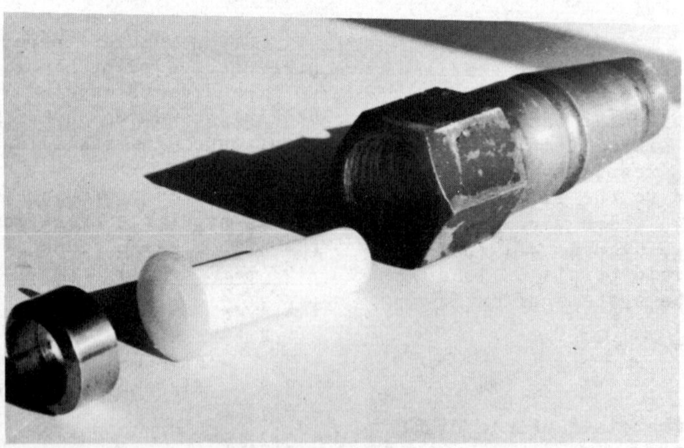

Fig. 3: The New Internal Emergency Line Silencer

C72/76

AIR BRAKING SYSTEMS FOR COMMERCIAL VEHICLES IN COMPLIANCE WITH THE EEC-DIRECTIVE 71/320 BRAKING

OTTO EHRICKE, Dipl.-Ing.
WABCO Westinghouse GmbH, Hannover, Germany

SYNOPSIS The EEC-Directive including some important official interpretations is analysed for requirements that have initiated special technical developments in the air braking equipment industry. Equipment having resulted from such developments is represented, its functions and outstanding characteristics are described. A method facilitating the design of braking force controls for compliance with the relatively new adhesion and compatibility requirements for a certain kind of control devices is introduced.

1. INTRODUCTION

The statements which follow are based on the present state of the EEC legislation on vehicle braking i.e. EEC-Directive 71/320 as amended by Directives 74/132 of 11th February 1974 and 75/524 of July 25th 1975. The complete legislation comprises the construction and performance requirements for all vehicles, but for the purpose of this presentation only commercial vehicles falling in the categories N_2, N_3, O_3 and O_4 will be considered.

2. NOTATION

P = Weight or axle load
h = Height of centre of gravity above ground
E = Wheel base
T = Braking force
FBR = Front braking ratio $T_1/(T_1 + T_2)$
i = Reduction ratio
z = Braking ratio
p_m = Pressure in trailer control line coupl. head

Suffixes

1,2 = Front ..., rear ..., or specification of vehicle category
3,4 = Specification of vehicle category
L = Laden
U = Unladen

3. Construction Requirements

In summarized form, the construction requirements say:
For motor vehicles:

a) The total braking system must contain three functions, namely service braking system, secondary braking system and park braking system. The performance of the first two must be graduable. The parking brake may be non-graduable, it must however, be applied with mechanical means only.

b) Service brake, secondary brake and parking brake are allowed to have common parts. There must, however, be two independent controls from the driver's seat, and the control of the service brake must be separated from that of the parking brake.

c) The service braking system must consist of two circuits which are completely independent from the air reservoirs (including these) onwards. Each of these circuits can serve as secondary braking system on condition that it provides the required performance. The only exception is semi-trailer tractors which need not be equipped with a dual circuit service braking system providing that the circuits of the tractor and semi-trailer are fully independent and separated.

d) Where separate controls for service brake and secondary brake are used simultaneous actuation must not result in a failure of service braking or secondary braking system.

e) The energy supply of the secondary braking system must continue in the event of a failure in the service braking system.

f) The brakes must be easily adjustable either manually or automatically and must have sufficient travel reserve to compensate for thermal effects and wear.

g) Each circuit of the service braking system must have an optical or acoustical warning device showing that the energy level has dropped to 65 per cent or less of its rated value. This is also the limit for energy loss caused by use of or possible defects in energy-consuming auxiliary equipment.

h) When a motor vehicle is designed for towing a heavier trailer - category O_3 or O_4 - the following special regulations have to be observed:

The service brake must be of the continuous or semi-continuous type. Motor vehicle and trailer brake must be connected by a dual line or multiple line system of which the first mentioned is the preferred solution.

Should the trailer brake fail or should the energy connection between the vehicles break then the motor vehicle braking system above must provide the secondary braking performance. Furthermore, in the event of a rupture of or a leakage in the connections, the driver must be able to actuate the trailer brakes if they do not work automatically. The latter demand not being sufficiently definite, it

© IMechE 1976

has been interpreted by a performance requirement which will be discussed in a later paragraph.

In case of a partial break-down of the service brake of the motor vehicle, the remaining part must be able to actuate gradubly the trailer braking system with full or partial performance. Finally the secondary braking system of the motor vehicle must also be able to operate the trailer brake with a graduable action.

For trailers:

i) Trailers of the categories under consideration must be equipped with a continuous or a semi-continuous service braking system acting on all wheels of the trailer. The braking force distribution on the wheels of an axle must be symmetrical.

k) As in motor vehicles, the brakes must be easily adjustable either manually or automatically, they must have sufficient travel reserve for compensation of thermal effects and wear.

l) In the event of a breakage in the coupling device the trailer brakes must be applied automatically. If a device to pneumatically release the trailer brakes is fitted it must return to the normal service position when the supply line is recharged.

m) A trailer in the categories referred to must have a mechanical parking system able to hold the trailer separately. It must be operable by a person standing beside the trailer.

Other requirements refer to both vehicle classes:

n) For spring brakes:
These shall not be used as service brakes. Their feed circuit should have its own reservoir unless they can be held off by at least 2 independent systems.

The pressure at which the spring begins to decompress shall not be higher than 80 per cent of the minimum value of the normally available pressure, and an optical or audible warning device shall indicate when such pressure is arrived at.

On motor vehicles with spring brakes equipped for operation with trailers of the heavier categories, an automatic application of the spring brakes must also cause the trailer brakes to apply simultaneously.

o) For lock actuators:
There must be an acoustical or optical warning device in the locking circuit operating when the pressure has dropped to a level close to locking pressure. A braking cylinder with locking device must be operable by two separated energy reserves. It must not be possible to disengage the locking device unless the brake system is operational. Finally, there must be an auxiliary unlocking installation.

3.1 Vehicle Installations

Figs. 1, 2 and 3 show schematics of motor vehicle braking systems according to the EEC legislation. Fig. 1 is a system with normal pressure level and spring cylinders which may serve, depending on the properties of the vehicle, as secondary braking system or merely as parking braking system. Fig. 2 is the schematic of an air over hydraulic system with increased pressure storage. Fig. 3 is a system with traditional pressure level and lock actuators.

Figs. 4, 5 and 6 show schematics of EEC-trailer braking systems. Fig. 5 represents the installation of a 3-axle trailer with long wheel base and correspondingly long loading area for which an additional braking force regulator is recommended.

3.2 Equipment Initiated by Construction Requirements

Among the construction requirements, those referring to separation of circuits, continuation of supply and feeding of spring cylinders as well as those concerning trailer control and mechanical parking braking were a great challenge to the brake equipment manufacturer, the results of which are the quadruple protection valve, the two-line trailer control valve and a special parking brake valve.

3.2.1 Quadruple Protection Valve

As can be seen in the schematics, Figs. 1 - 3, the quadruple protection valve can be considered the heart of an EEC-braking system. This device separates the two circuits of the service braking system. It ensures that the secondary braking system - be this one of the service braking circuits or the parking braking system - continues to be fed in case of a failure in one of the service braking circuits. It enables the spring cylinders to be fed by each of the service braking circuits, thus making an extra reservoir unnecessary in such cases where the service braking reservoirs are large enough and the closing pressures of the charge valves low enough. Finally, the quadruple protection valve insulates the braking system from the circuit of the auxiliary devices. Alternative circuitry is also possible, for example such that the valve separates the motor vehicle braking system from that of the trailer.

Fig. 7 shows a quadruple protection valve and explains its way of functioning. The version shown comprises two pairs of charging valves with limited return flow. The valves in each pair supplying ports 21 and 22 and ports 23 and 24 respectively are in parallel with each other, whilst the two pairs are in series. The two service braking circuits of the motor vehicle are connected with the ports 21 resp. 22, trailer, parking braking system, and auxiliaries are connected with the ports 23 resp. 24. In case of a defect in one of the circuits, the pressure in the other 3 circuits is maintained at the level of the opening pressure of the charge valve in the defective circuit.

An air-consumption, let us say, in circuit 23, will cause a flow from circuit 21 and 22 as well as circuit 24 to this port until the closing pressure of the charge valve in this circuit is reached with the other circuits losing insofar pressure.With sufficient reservoir capacities in the service braking circuits and proper choice of the closing pressure in the charging valves, this effect can be utilized for compliance with the requirements regarding spring cylinders and number of trailer brake applications without an additional air reservoir.

© IMechE 1976

This quadruple protection valve principle is rather versatile. The check valves can be arranged so that circuits 23 and 24 are totally separated from each other, the charge valves can be replaced by charging valves without return flow, additional check valves and by-passes can be installed. Thus this valve can be adapted to practically any possible braking system characteristics.

3.2.2 Two Circuit Trailer Control Valve

Fig. 8 shows a trailer control valve designed in accordance with the EEC legislation. This valve provides a graduable control of the trailer service braking system separately by either of the motor vehicle service braking circuits or the spring cylinder circuit. Motor vehicle service braking circuit 1 is connected with port 41, service braking circuit 2 with port 42 and the spring brake circuit with port 43; the supply is connected with port 13 and port 2 is the outlet for the trailer control line. In normal operation, the trailer is controlled by the motor vehicle service braking circuit 1 via port 41. The piston fed from this port operates the valve and governs the trailer service brake pressure. Service braking circuit 2 is neutralised by the spring brake pressure above the diaphragm between ports 42 and 43 and by the trailer control pressure. In the event of a defect in motor vehicle service braking circuit 1, the pressure from service braking circuit 2 below the diaphragm together with the supply pressure pushes the valve upwards against the upper piston, thus closing the exhaust and permitting the passage between supply and port 2 to open; the trailer control pressure level is now governed by the piston between the ports 13 and 2. The function of the valve by the spring cylinder circuit is achieved by reducing the pressure above the diaphragm in which case the supply pressure pushes the valve upwards opening the passage until the system is balanced by the trailer control pressure.

The illustration shows a version of this trailer control valve in which a trailer supply cut-off device necessary in some cases because of the interpretation mentioned above in h) is incorporated. This point will be reverted to below.

This trailer control valve is also available with pressure pre-dominance in the trailer control line.

3.2.3 Parking Brake Valve with Check Position

The call for a merely mechanically applied parking brake in combination with that for a halting of vehicle combinations alone by the motor vehicle parking braking system above results in an unsatisfactory situation for vehicle combinations. This means in other words that the complete vehicle combination has to be held in parked position by the brakes of only one or the motor vehicle axles. On a gradient with unfavourable friction properties this condition cannot be achieved, the combination can be held only with the assistance of the trailer service braking system. Because of this situation, the vehicle inspection authorities of some EEC-member countries have agreed upon an interpretation according to which a simultaneous application of the trailer service braking system with the motor vehicle parking brake can be approved under the condition that the driver has the possibility of checking whether the motor vehicle parking brakes alone are able to hold the vehicle combination.

This interpretation has initiated the development of a parking brake valve with check position. Fig. 9 shows the cam plate and the pressure characteristic of a hand-operated braking valve for spring cylinders with such feature. As the characteristic shows the valve is graduable within a certain hand lever travel; with the lever moved on, the valve is locked in parking position in which in addition to the spring cylinders in the motor vehicle the trailer service brakes are also applied. Beyond parking position is a test position to which the lever can be temporarily pushed and in which the trailer brakes only are disengaged and from which the lever returns automatically to parking position.

4. Performance Requirements

In the performance requirements are prescribed:

a) Minimum braking performances for cold and hot brakes, for motor vehicles in terms of stopping distances, for trailers in terms of braking forces in relation to weight. Further the fading properties are determined as well as, for motor vehicles, the minimum capability of the secondary braking system and of the residual service braking circuit with the other circuit defective.

b) Minimum gradients on which vehicles and vehicle combinations must be able to be parked.

c) Maximum initial delay and build-up times and procedures for measuring them.

d) Minimum reservoir sizes, defined in terms of residual pressure at the coupling head of the trailer control line after 8 brake applications with a differentiation between motor vehicles and trailers.

e) Minimum compressor sizes determined by reservoir pump up times.

f) Relationships for the braking force distribution between the axles of a vehicle and for the compatibility of motor vehicles and trailers with the call for an automatic function of a braking force regulator contingently required and a provision for residual braking performances in the event that such device fails.

g) Finally, by an interpretation by the vehicle inspection authorities of France, Italy and Germany, which has in the meantime been introduced into ECE-regulation 13 and which has been adopted by the DOE, precise conditions for the build-up of the automatic braking of the trailer in case of pressure loss in the trailer connections.

Of these requirements, the last two items deserve special attention within the scope of this article.

4.1 Build-Up of the Automatic Trailer Braking

The constructional requirement prescribing the automatic braking of the trailer in case of defects in the trailer air connections was considered insufficiently precise for practical technical vehicle inspection by the authorities of the countries mentioned above. Therefore they have agreed upon the interpretation that this qualitive requirement should be taken as complied with if

- upon full actuation of the service brake control of the drawing vehicle the pressure in the feed

© IMechE 1976

line falls to 1.5 bars within the next two seconds,

- the relay emergency valve operates when the pressure in the feed line has dropped to 2 bars.

The second condition is not a problem. The first condition, relating to motor vehicles, calls for a minimum volume of the trailer feed line and the flow into the same via the protection valve to be as small as possible. On the other hand, a certain minimum pressure level in the trailer control line after 8 brake applications is necessary which calls for a trailer supply line volume and the flow from the protection valve to be as big as possible. To master this contradiction, many motor vehicles require an additional installation which is shown as a separate device in the diagram Fig. 1 and integrated with the trailer control valve in Fig. 8.

The trailer supply flows through the lower cylindrical space of this additional valve. The piston of the same is fed on its upper side from a service braking circuit and is on its lower side held under pressure from the trailer control line and the trailer supply. In normal operation, the piston is held in upper position and the trailer supply has free passage. When the trailer control pressure falls to a level 4.5 bars below service braking pressure, then the piston is pushed downwards, its lower pin enters into the supply passage choking the supply thus ensuring the statutory evacuation speed for the trailer supply line.

4.2 Braking Force Distribution

In vehicles of the categories previously specified semi-trailers excluded, the braking forces must in all loading conditions be distributed on the axles so that for braking ratios between 0.15 and 0.30 the adhesion utilized on the front axle is higher than that on the rear axle and that for braking ratios between 0.1 and 0.61 the adhesion on any axle is inferior or equal to $k = (z + 0.07)/0.85$.

The braking force distribution of vehicles with more than two axles is considered in compliance with the above if for braking ratios between 0.15 and 0.30 the adhesion required on one front axle is higher than that required on one of the rear axles.

For 2-axle vehicles there is the alternative that the adhesion on the front axle must not be higher than that on the rear axle if for braking ratios between 0.15 and 0.30 the adhesions lie between $k = (z + 0.08)$ and $k = (z - 0.08)$ and if for braking ratios greater than 0.3 the adhesion on the rear axle is less than or equal to $k = (z - 0.02)/0.74$.

In EEC-Directive 75/524, these requirements are drawn in a graph.

Normally, verifying a given or a projected vehicle for compliance with these rules is a rather complicated job with a lot of bothersome calculation work. Therefore, Isernhagen (1) has developed a method of finding the necessary braking force distribution by means of only a few basic vehicle parameters, the required data being each for laden and unladen vehicle:

Front axle load

Total weight

Height of centre of gravity above the ground

Wheel base

The nomograph Fig. 10 shows the interdependence of these data and the front braking ratio $FBR = T_1/(T_1 + T_2)$. Originally, this nomograph covered only the basic legal requirement. As presented here, it has been extended and also includes the alternative for 2-axle vehicles. The nomograph gives for each set of vehicle data a range for the permissible relative front axle braking forces. Above the upper boundary of this range, the front axle adhesion would supersede the limit line $k = (z + 0.07)/0.85$. The lower boundary of this range is determined by two conditions. An FBR under the line left of the knee point would mean an intersection of the front axle adhesion curve with the line $k = (z - 0.08)$ below $z = 0.3$. An FBR below the line right of the knee point would mean that the rear axle adhesion curve would intersect the boundary line below $z = 0.61$. Additionally, the nomograph bears in dash-and-dot-line the old lower boundary still effective for multi-axle vehicles.

For further clarification a sample vehicle with the following data will be examined:

Front axle load laden: $P_{1L} = 3600$ kp

unladen: $P_{1U} = 2930$ kp

Total weight laden: $P_L = 10000$ kp

unladen: $P_U = 4545$ kp

Height centre of gravity laden: $h_L = 1504$ mm

unladen: $H_U = 896$ mm

Wheel base: $E = 3200$ mm

Hence:

For laden vehicle: $P_{1L}/P_L = 0.36$; $h/E = 0.47$

For unladen vehicle: $P_{1U}/P_U = 0.645$; $h/E = 0.28$

Carrying over these quotients into Fig. 10 (thin vertical and horizontal lines) shows that the relative front axle braking force can be:

For the laden vehicle: $0.52 \leq FBR_L \leq 0.71$

For the unladen vehicle: $0.76 \leq FBR_U \leq 1.0$

(higher values omitted as senseless).

There is no common FBR-range for the laden and the unladen vehicle. This means that with the braking specification of the laden vehicle the unladen vehicle will not have a statutory braking force distribution. Therefore a braking force control device is necessary for the unladen vehicle.

The next question refers to the qualitative and quantitative properties of the required braking force controller. The front axle braking force ratio for the laden and unladen vehicle can be generally described as follows:

$$\frac{\text{Braking force laden}}{\text{Braking force unladen}} = i_1 = P_{1L}/P_{1U} \quad (1)$$

© IMechE 1976

Similarly for the rear axle:

$$\frac{\text{Braking force laden}}{\text{Braking force unladen}} = i_2 = P_{2L}/P_{2U} \qquad (2)$$

A quotient of these ratios can be formed which reads after the introduction of T_1 = FBR $(T_1 + T_2)$ resp. $T_2 = (1\text{-FBR}) \cdot (T_1 + T_2)$ and some transformations:

$$\frac{i_2}{i_1} = \frac{1/FBR_L - 1}{1/FBR_U - 1} \qquad (3)$$

These interdependences are drawn in the nomograph Fig. 11. For the unregulated front axle, i.e. $i_1 = 1$, this nomograph gives directly the rear axle control ratio. Let us continue the investigation of the above sample vehicle. We assume its front braking ratio laden to be $FBR_L = 0.55$. The desired relative front braking ratio unladen is assumed to be the minimum $FBR_U = 0.76$. Only the rear axle will be regulated. As the thin lines in Fig. 11 indicate, under these conditions a rear axle control ratio of $i_2 = 2.75$ is necessary.

It has to be pointed out that these nomographs are applicable only to 2-axle vehicles or such multi-axle vehicles that can be mathematically represented by a 2-axle system. Furthermore, these nomographs do not cover semi-trailer tractors with laden semi-trailer. The braking force control characteristic is defined definitely by these nomographs only in case of control devices with constant ratio for constant load. These restrictions however do not devaluate the method represented. Even with these restrictions, it covers by far the majority of vehicles and has therefore a great practical importance.

4.3 Compatibility of Motor Vehicle and Trailer

To ensure their compatibility, trailers with air braking systems and motor vehicles equipped for drawing such trailers must correspond to certain correlations of braking ratio and pressure in the coupling head of the trailer control line. These correlations are defined in the EEC-Directive 75/524 in the form of graphs showing tolerance ranges for the braking ratio as a function of the coupling head pressure $z = f(p_m)$. There exist all in all three graphs: one for trucks and trailers, one for semi-trailer tractors and a third one for semi-trailers.

While the graphs for trucks and trailers and for semi-trailer tractors contain different tolerance ranges for the laden and the unladen vehicle, there is only one range for semi-trailers. However, this is only a basic range. It serves for developing the laden/unladen ranges of an individual semi-trailer with the vehicle characteristics by means of an additional nomograph.

The problem is now to check whether the braking force control found for proper braking force distribution also ensures compliance with these graphs and to assess what alterations have to be made. For this purpose, also, Isernhagen has introduced in his article already referred to a method facilitating the matter considerably. This method again delivers definite specifications for control devices with constant ratio for given load. Figs. 12 and 13 show the nomographs pertaining to this method. Fig. 12 represents the ratio of the braking forces of laden and unladen vehicle according to the formula

$$i = \frac{T_L}{T_U} = \frac{z_L}{z_U} \cdot \frac{P_L}{P_U} \qquad (4)$$

Therein z_L/z_U is the quotient of the braking ratios of laden and unladen vehicles for which the correlation graphs say approx.

$$0.54 \leq z_L/z_U \leq 1.18 \text{ for trucks and trailers}$$

and $$0.44 \leq z_L/z_U \leq 1.10 \text{ for semi-trailer tractors}$$

Thus the family of lines in Fig. 13 defines a range within which the data of a vehicle have to lie to ensure compliance with the correlation graph.

Fig. 12 represents the formula

$$\frac{i_2}{i_1} = \frac{1/FBR_L - 1}{\frac{1}{(i/i_1) \cdot FBR_L} - 1}$$

These nomographs apply to trucks and trailers as well as to semi-trailer tractors. Only the special location of the z_L/z_U-range for semi-trailer tractors (dash-dotted in Fig. 13) has to be borne in mind.

For further clarification, we revert to our sample vehicle for which we have found a rear axle control ratio of $i_2 = 2.75$. It has to be assumed that the laden vehicle lies within the correlation range. The nomograph Fig. 12 gives for the rear axle ratio $i_2 = 2.75$ (when $i_1 = 1$) and the relative front braking ratio of the laden vehicle $FBR_L = 0.55$ a value of $i/i_1 = 1.40$. As the front axle is not regulated, this value is identical to i. By this i and the weight ratio of the laden and unladen vehicle $P_L/P_U = 2.2$ a point is defined in the nomograph Fig. 13 located within the statutory z_L/z_U-range. This means that the sample vehicle with the said rear axle ratio is in compliance with the correlation requirements. However, there is another condition to be observed. The position in the z_L/z_U-range as found by Fig. 13 can be realised only if the laden vehicle has an appropriate situation within the correlation range. A low z_L/z_U-ratio means that the laden vehicle lies at a lower boundary of the correlation range, a high z_L/z_U-value means that the laden vehicle lies at the upper boundary of the range. Thus the general z_L/z_U-range of Fig. 13 is for the individual vehicle confined by the location of the laden vehicle in the correlation range.

The thin lines in Figs. 12 and 13 represent the situation of the sample vehicle.

The sample vehicle is a motor vehicle with, from the point of view of braking force distribution and braking ratio / coupling head pressure requirements, very unfavourable characteristics. Nevertheless this vehicle could be brought to compliance with a braking force controller only on the rear axle and a control ratio of not more than 2.75. Therefore the general conclusion seems to be permitted that for motor vehicles a control device is required only on the rear axle and that the beam characteristic load apportioning valves on the market with a maximum reduction ratio of approx. 4 are sufficient.

For trailers, the situation is different. In these normally the weight distribution on the axles varies

© IMechE 1976

little with the loading conditions. This means that with a view to braking force distribution these vehicles need not be regulated. However, in nomograph Fig. 13 unregulated trailers will normally lie outside of the statutory z_L/z_U-range which means that a braking force control device is enforced by the compatibility requirement. The reduction ratio can be found by the nomograph Fig. 13. First the location of the laden trailer in the correlation range has to be determined. Thus the individual z_L/z_U-range of the trailer in Fig. 13 is specified. Then in Fig. 13 the weight ratio of the trailer together with the minimum and maximum individual z_L/z_U-lines define the minimum and maximum ratios of the braking forces for the laden and the unladen vehicle. Other reflections show that for constant front braking ratios for laden and unladen vehicles the front axle ratio must be $i_1 = i$ and thus $i/i_1 = 1$. According to the nomograph Fig. 12, this requires $i_2/i_1 = 1$, i.e. $i_2 = i_1$. Consequently, both trailer axles have to be controlled with the same ratio.

The control ratio of trailers is determined by their weight ratio and their position in the correlation range. An analysis of usual trailers with regard to these factors leads to the conclusion that also for trailers the control devices on the market with a ratio of about 4 will suffice.

In practice, generally the braking forces on trailers are controlled by one single device which, for installation reasons, is mounted on and piloted by the front axle controlling the pressure for the brake cylinders of both axles or axle groups. However, for trailers with long load areas involving a considerable risk of non-uniform load distribution separate control devices on and for front axle and rear axle or rear axle group, as shown in the schematic Fig. 5 have turned out to be more favourable. Of course in such case the ratios of the control devices are identical.

Finally, some equipment required for meeting the legal demands described in the previous chapter and in this chapter are introduced.

Fig. 14 shows an emergency relay valve with pressure pre-dominance in the cylinder ports. This pre-dominance is necessary to compensate the threshold pressure of the foundation brakes. The pre-dominance is individually adjustable from outside of the valve by means of the screw modifying the spring tension. The characteristic shown is just an example. The pre-dominance could be placed in the trailer control valve of the motor vehicle as well. It has, however, been found that for compatibility purposes an individually tuned pre-dominance in the trailer is more favourable.

Fig. 15 shows a normal mechanically controlled load apportioning valve and its characteristic. Fig. 16 shows the same valve, however, with a cam operated valve assembly. This is a special development for applications in which the call for secondary braking effect in case of failure of the braking force control device is difficult to be coped with. If the mechanical linkage system of this valve should break then the cam disc would go into a rest position where the pressure reduction of the valve is roughly 2 : 1. This feature is also recognizable in the valve characteristic.

Fig. 17 shows a mechanically operated load apportioning valve especially suitable for installation on trailer front axles. This valve responds only to static vehicle spring deflections and it governs the cylinder pressure only from a certain value upwards, thus contributing to a rapid overcoming of the threshold pressure of the foundation brakes.

Fig. 18 shows a load sensing valve piloted by the pressure in air suspension.

Last but not least Figs. 19 and 20 show hydraulic governors for installation in hydraulic or air over hydraulic braking systems. Fig. 19 is a cut-off type governor with the special feature that upon failure of the front axle circuit the rear axle circuit is fully pressurized. Fig. 20 is a governor allowing free passage up to a certain pressure depending on the force on the lever and then reducing it at a constant rate. The big advantage of this type of governor is its uncomplicated design principle involving low cost and the possibility to approximate the braking force distribution to the optimum. The actual state of the EEC-legislation with its extended correlation ranges has improved the applicability of such governors for vehicles for trailer towing.

5. CONCLUSIONS

The new demands imposed by the EEC-legislation can be met by a series of newly developed equipment.

A universal method for defining the braking force control requirements and the lay-out of regulators for motor vehicles including to a certain extent semi-trailer tractors and trailers implied by the statutory braking force distribution and compatibility is described. By means of a sample vehicle the application of this method in practical lay-out work is demonstrated and the conclusion is made that no higher governor ratios than 4 : 1 and only rear axle controls seem to be necessary.

6. ACKNOWLEDGEMENTS

The author gratefully acknowledges the support he has received whilst preparing this paper from the directors and his colleagues at WABCO Westinghouse. He especially thanks Mr. F. Isernhagen for his technical assistance and Mr. R. Dore of WABCO United Kingdom for revising the language of this paper.

REFERENCES

(1) Isernhagen, R. "Application of the EEC-Directive Braking in Brake Calculations" (in German), in atue, Wabco Westinghouse Journal, 2/75, 10 - 15.

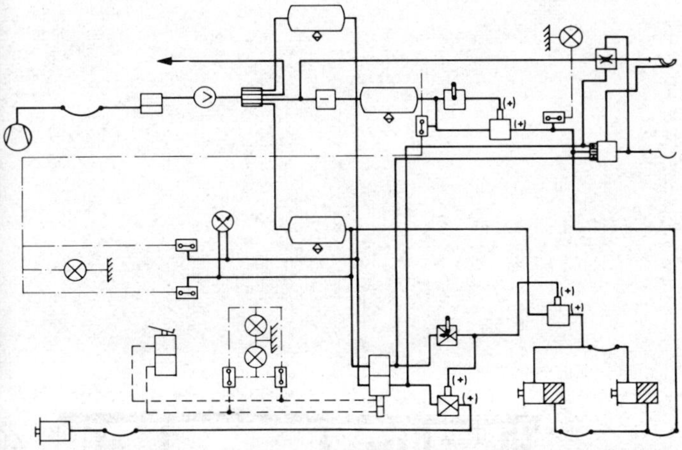

Fig. 1: **EEC Motor Vehicle Truck Braking System Schematic with Hydraulic Remote Control Dual Brake Valve.**

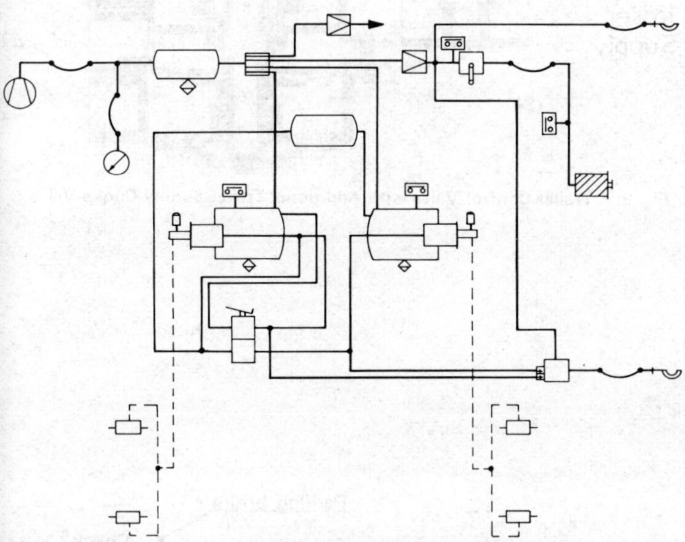

Fig. 2: **EEC Air Over Hydraulic Motor Vehicle (Semi-Trailer Tractor) Schematic with High Pressure Generation and Storage and Normal Pressure Operation.**

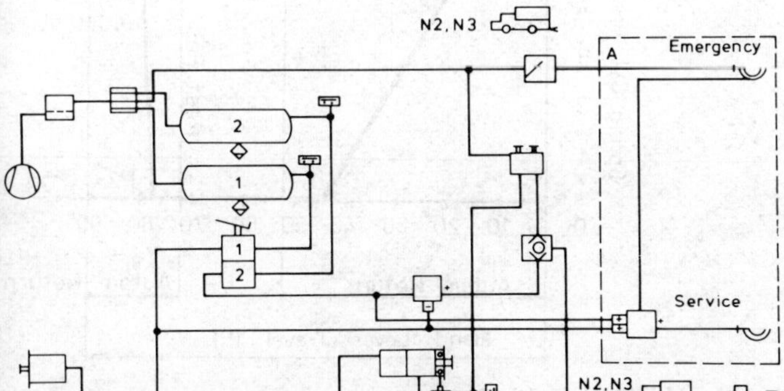

schema Nr. 3 30.6.75

Fig. 3: **EEC Motor Vehicle Braking System Schematic with Lock Cylinders**

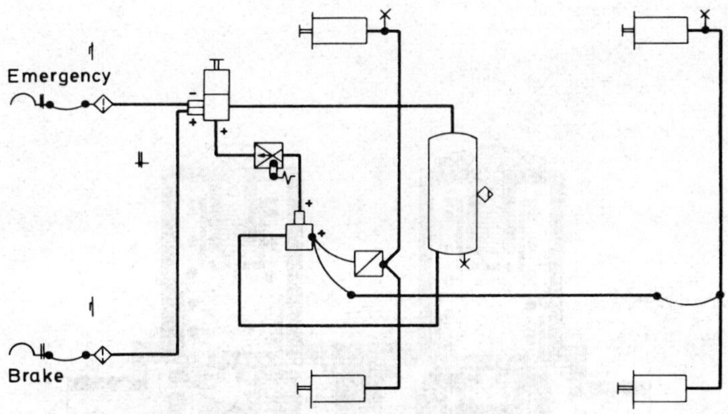

Fig. 4: **EEC Braking System Schematic of 2-Axle Trailer.**

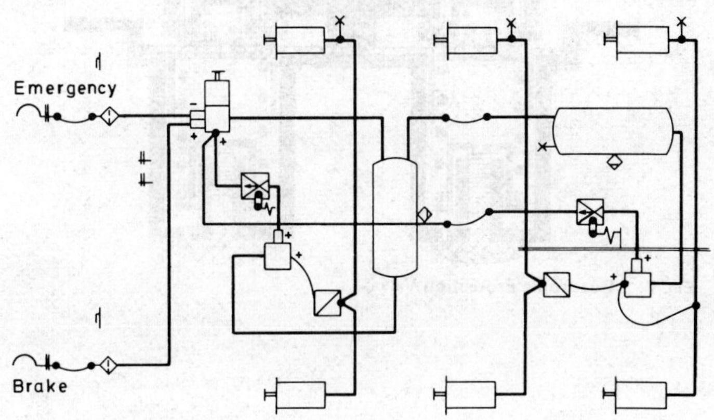

Fig. 5: **EEC Braking System Schematic of 3-Axle Trailer with Separate Load Apportioning Valves.**

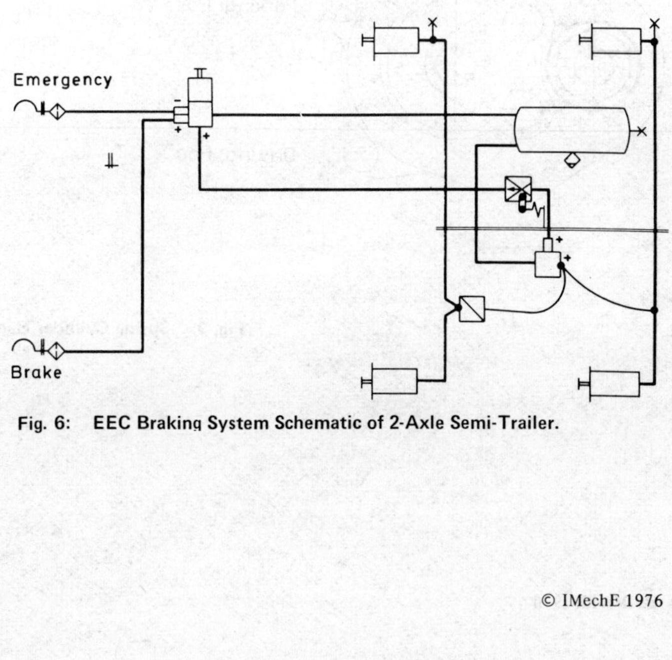

Fig. 6: **EEC Braking System Schematic of 2-Axle Semi-Trailer.**

© IMechE 1976

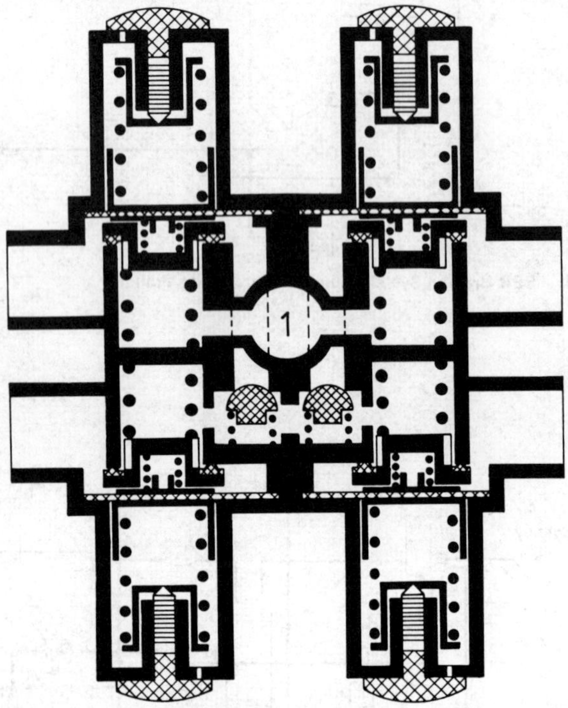

Fig. 7: **Quadruple Protection Valve.**

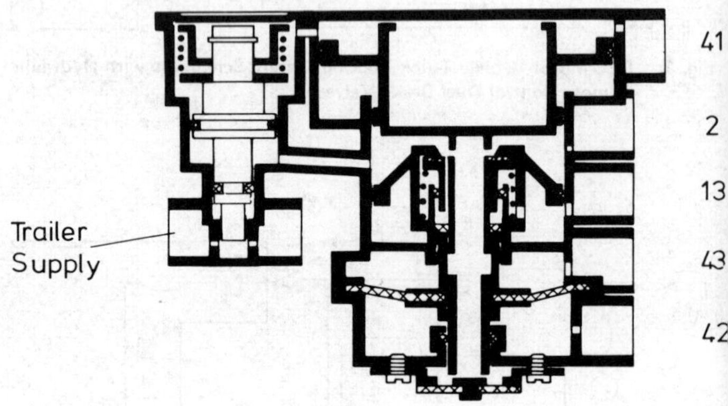

Fig. 8: **Trailer Control Valve with Additional Trailer Supply Choke Valve.**

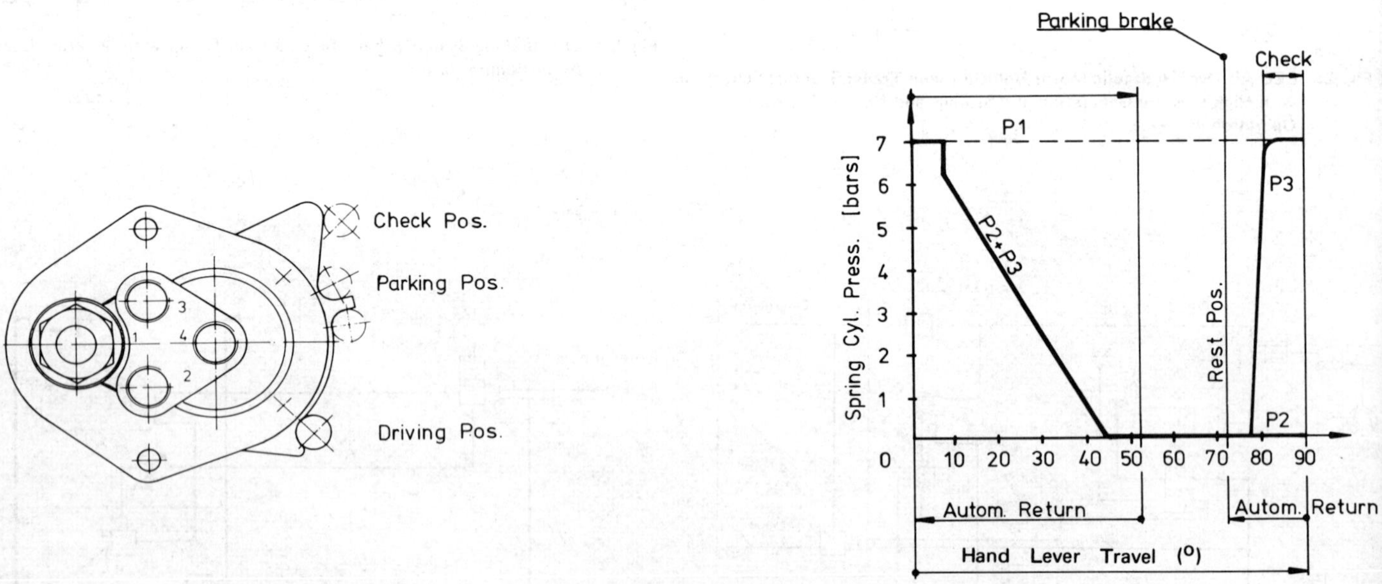

Fig. 9: **Spring Cylinder Handbrake Valve with Check Position.**

© IMechE 1976

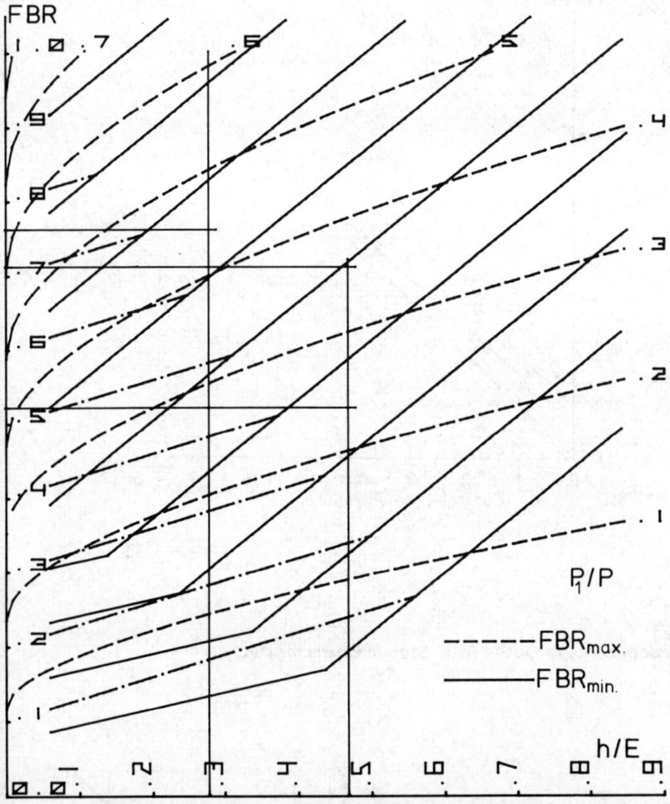

Fig. 10: Nomograph for Determination of Front Braking Ratio with Lines of Sample Vehicle.

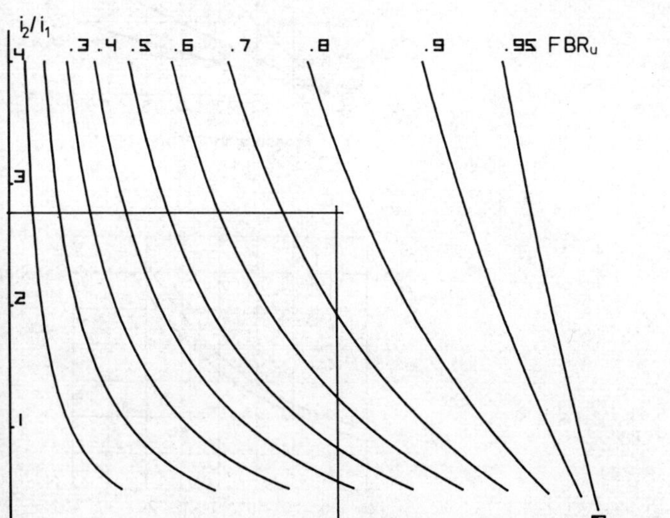

Fig. 11: Nomograph for Determination of Control Ratio (for Statutory Braking Force Distribution) with Lines of Sample Vehicle.

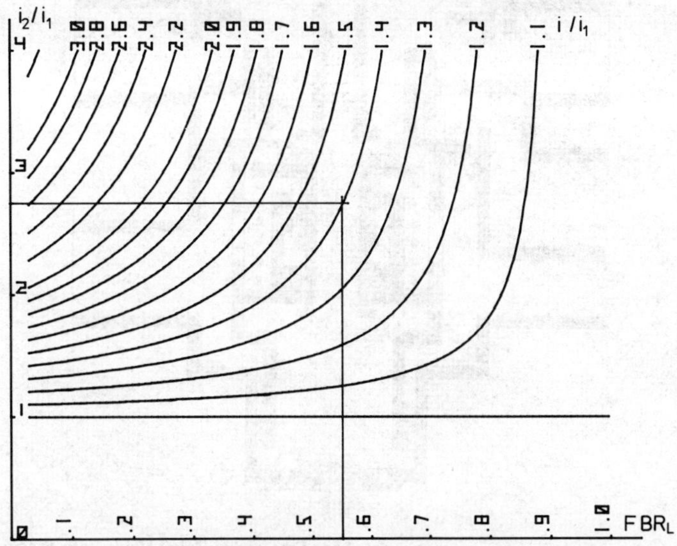

Fig. 12: Nomograph for Determination of Control Ratio (for Statutory Compatibility) with Lines of Sample Vehicle.

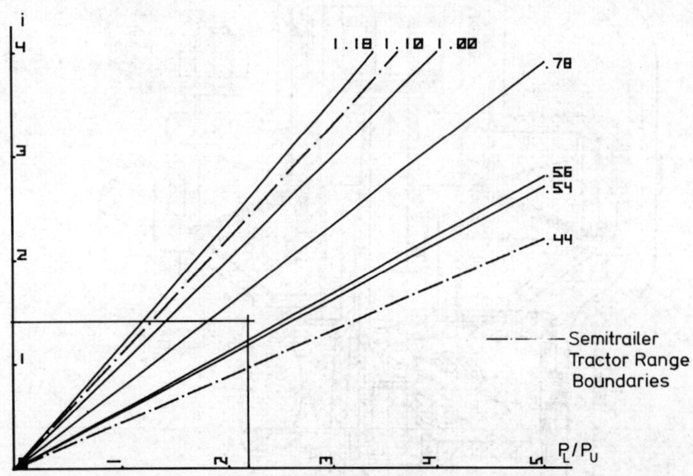

Fig. 13: Additional Nomograph to Nomograph Fig. 12 with Lines of Sample Vehicle.

© IMechE 1976

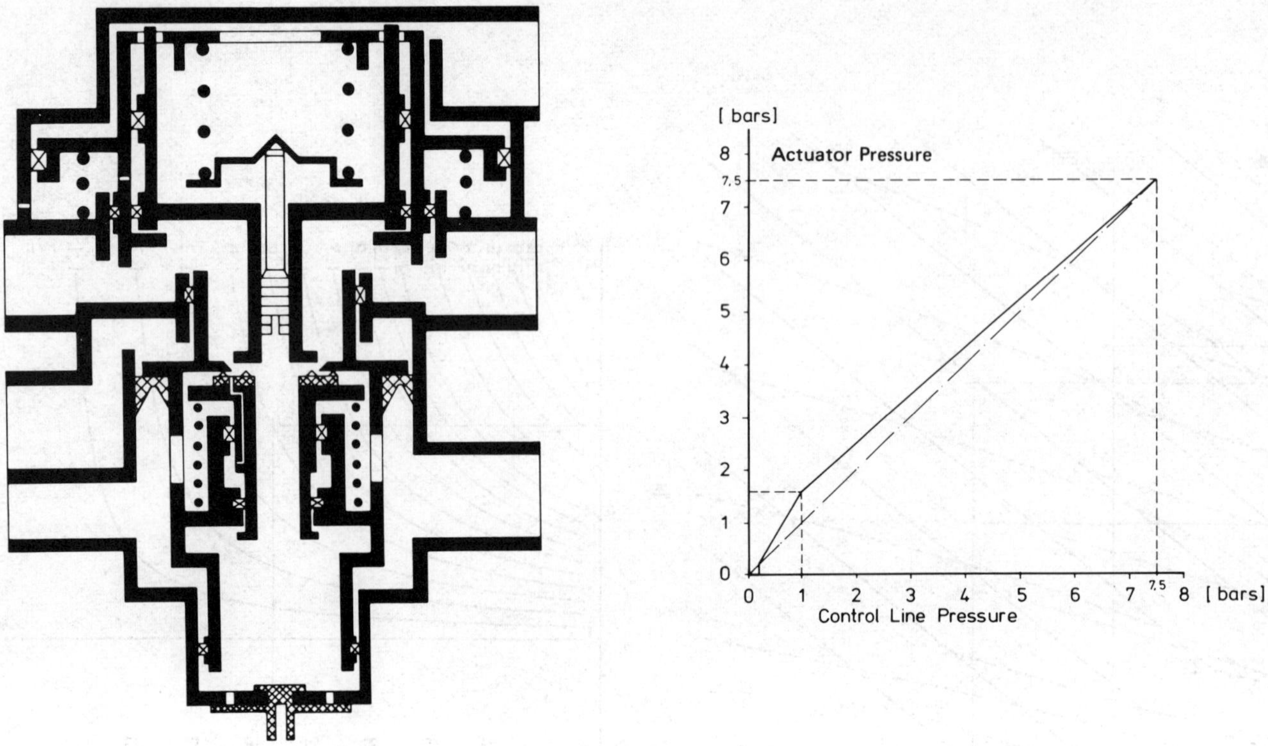

Fig. 14: Emergency Relay Valve with Pressure Predominance in Outlet with Sample Characteristic.

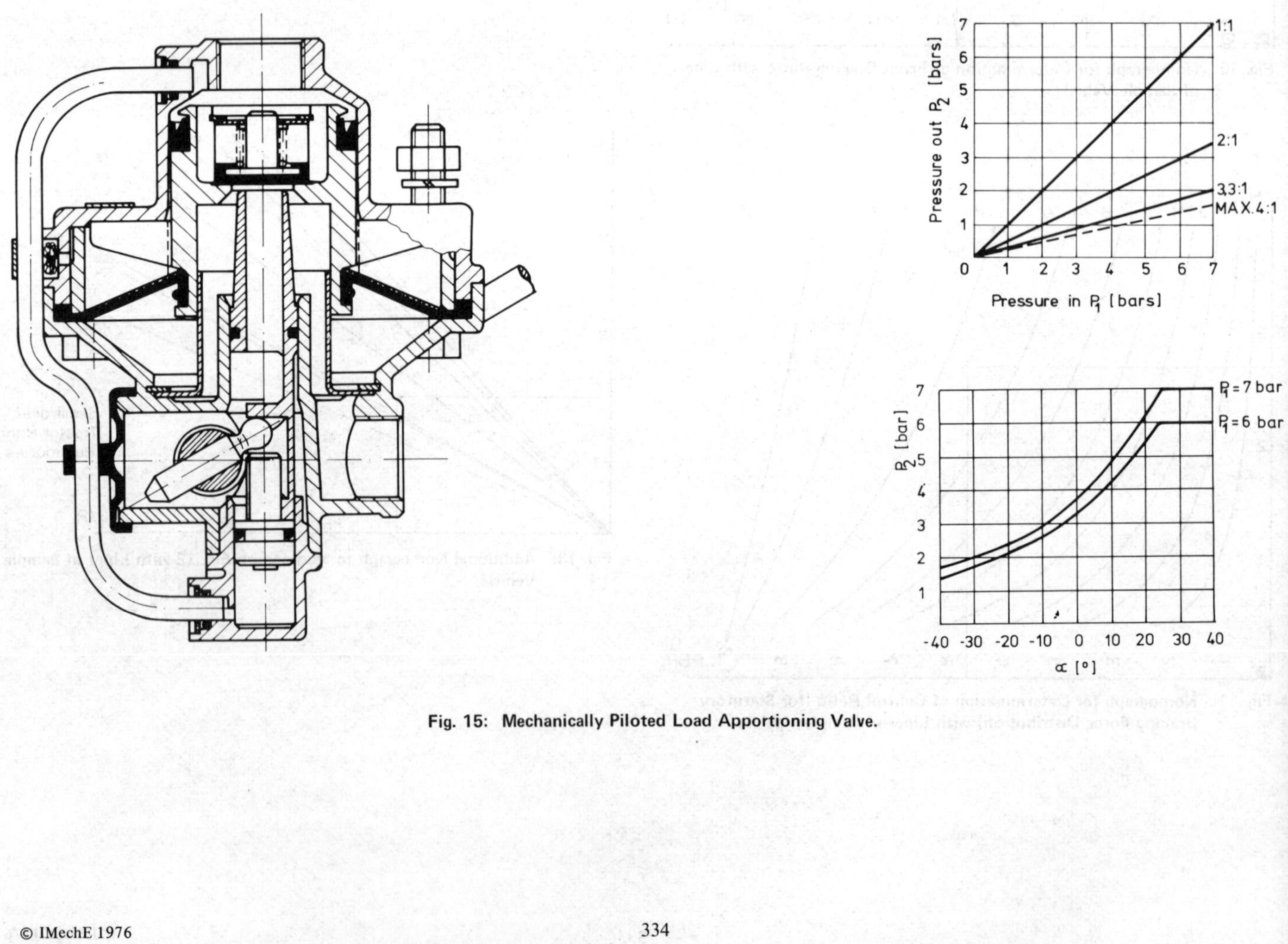

Fig. 15: Mechanically Piloted Load Apportioning Valve.

© IMechE 1976

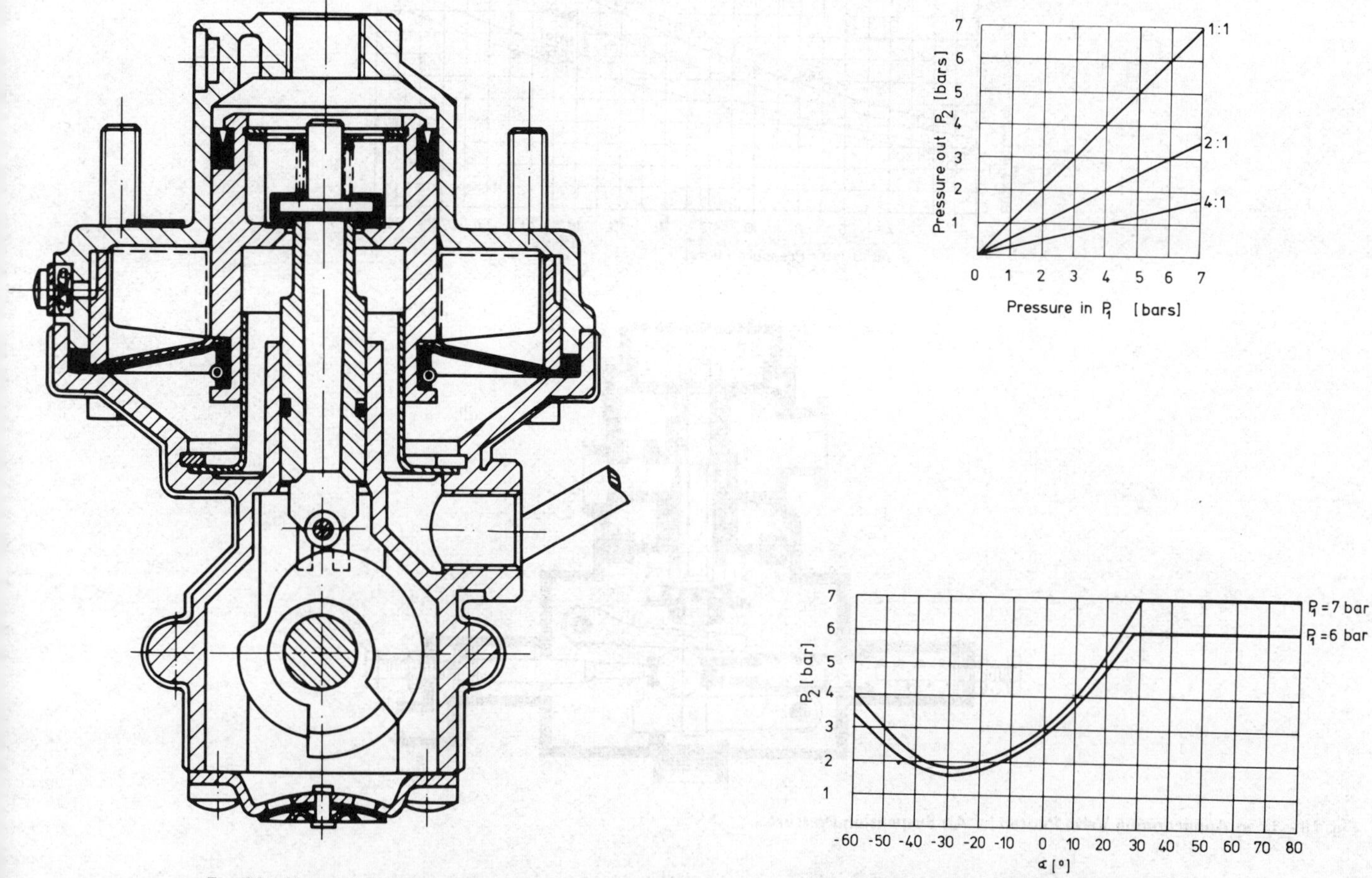

Fig. 16: Mechanically Piloted Load Apportioning Valve with Special End Position for Case of Linkage Breakage.

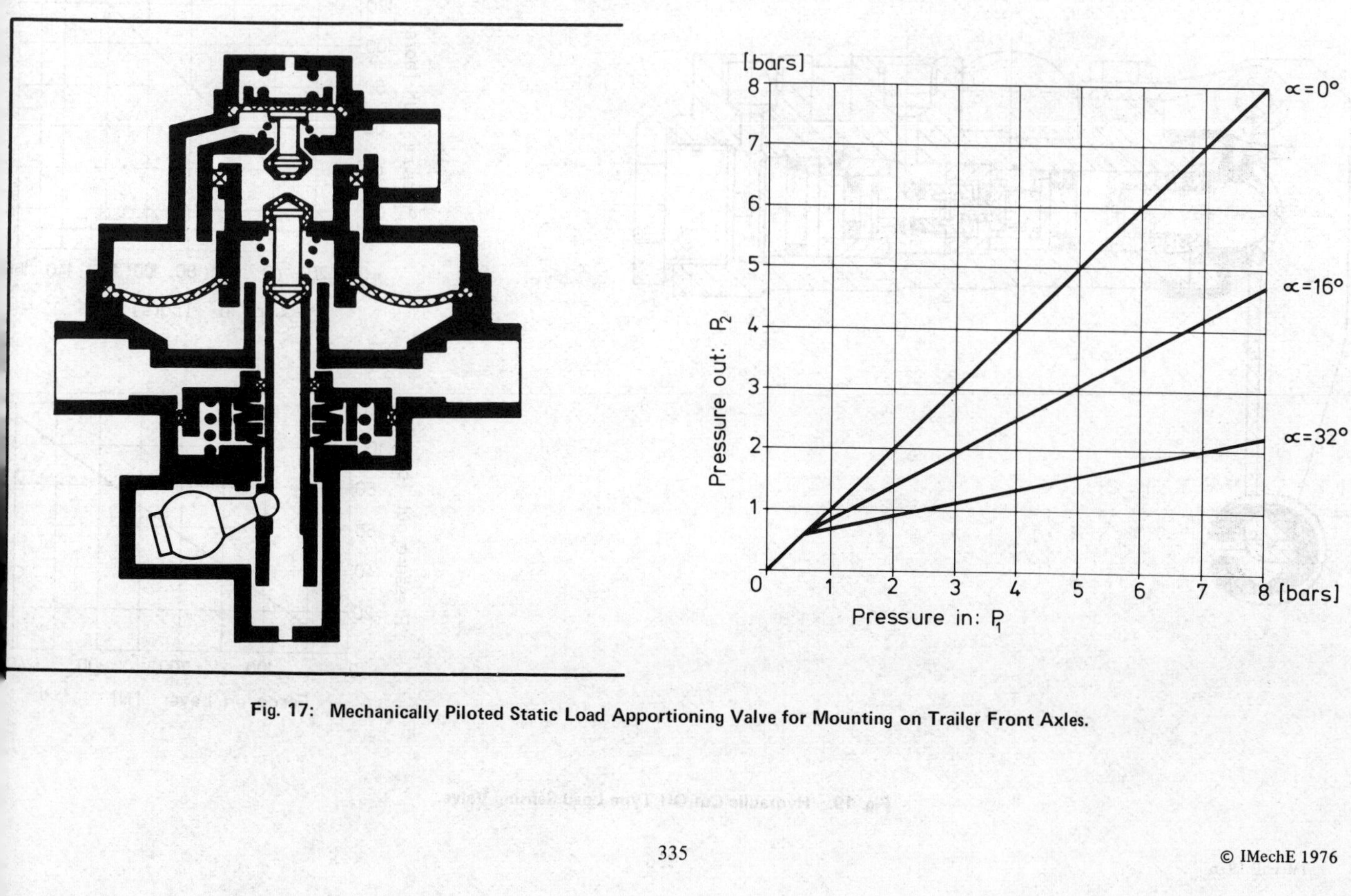

Fig. 17: Mechanically Piloted Static Load Apportioning Valve for Mounting on Trailer Front Axles.

© IMechE 1976

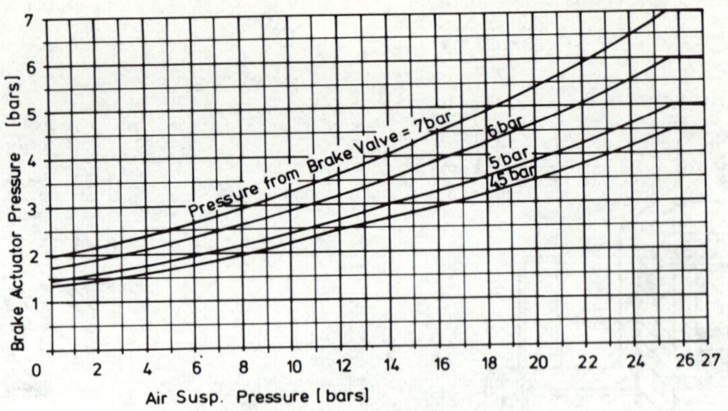

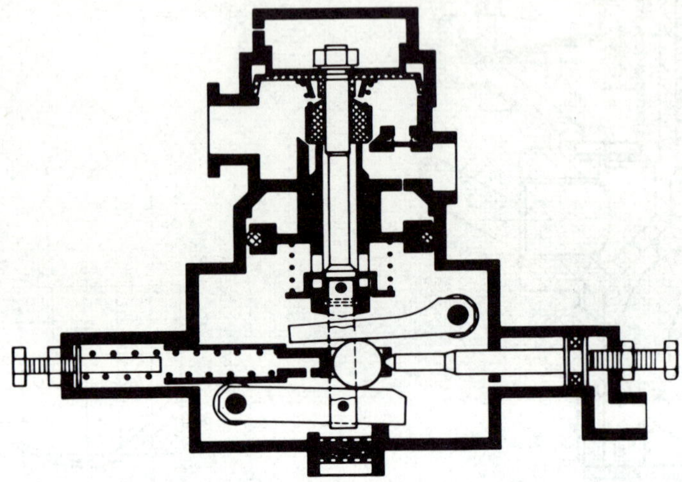

Fig. 18: Load Apportioning Valve Piloted by Air Suspension Pressure.

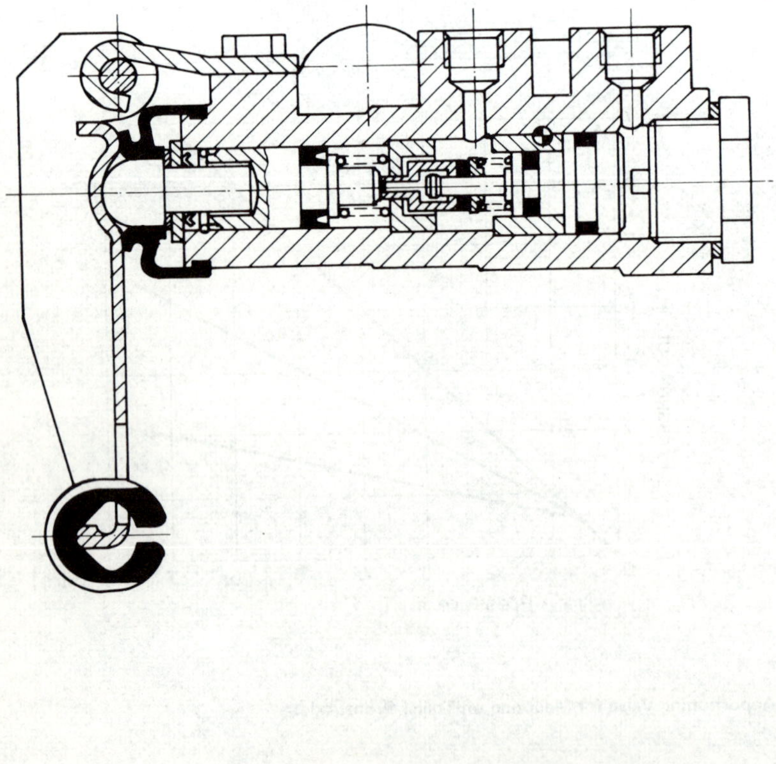

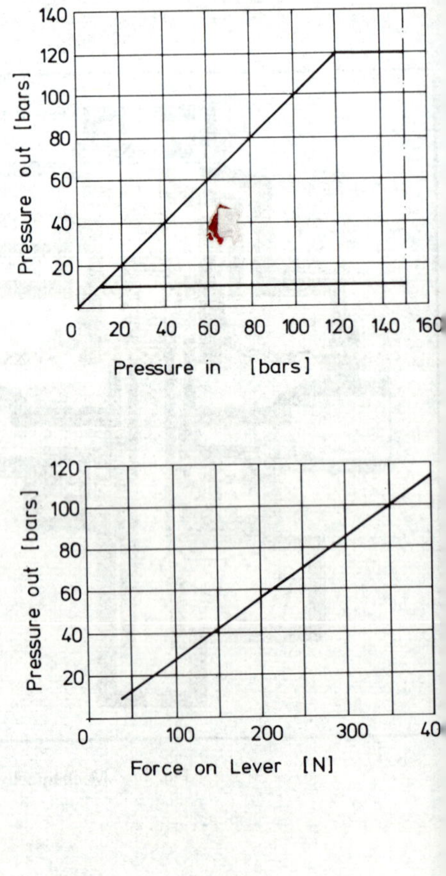

Fig. 19: Hydraulic Cut-Off Type Load Sensing Valve.

© IMechE 1976

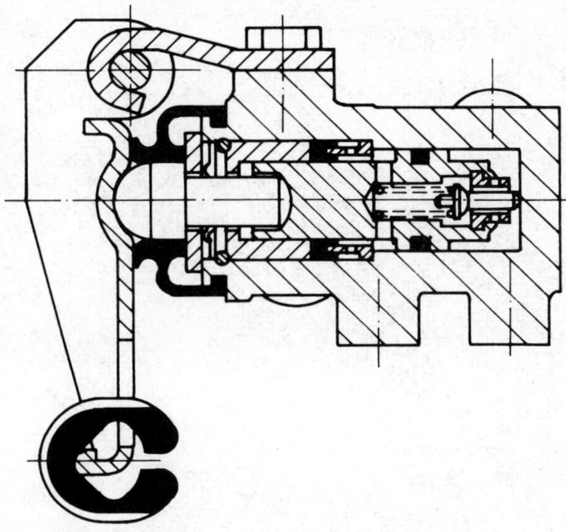

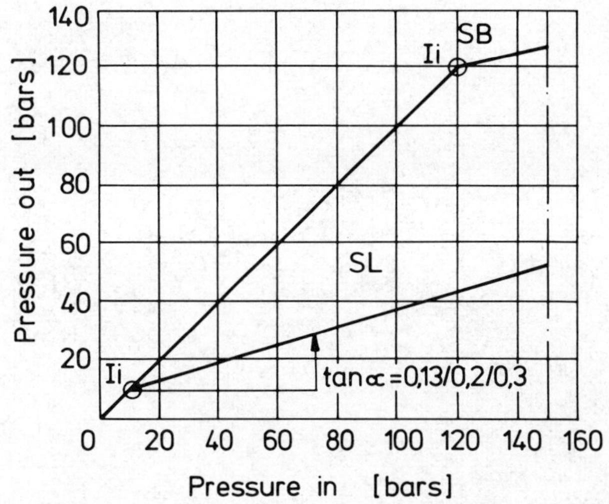

468 404 300 0

Fig. 20: Hydraulic Knee-Point Load Sensing Valve.

© IMechE 1976

Corrigenda

Page 7 Under Fig. 1: for 'Highway service tyres' read 'Patterned car tyres'
For 'Dual purpose tyres' read 'Highway service tyres', and for 'Patterned car tyres' read 'Dual purpose tyres'

Page 11 Col. 2, line 24, for 'ayg = μ and μ $R_F = B_F$' read '$a/g = \mu$ and μ $R_F = B_F$'

Page 13 Col. 1, line 6, for 'as the type I ECE/EEC 6 km – 7% grade' read 'as the type II ECE/EEC 6 km – 6% grade'

Page 14 Col. 1, line 27, for 'It was interesting to note from the test work that the average temperature recorded on the vehicle under equivalent fade test energy levels.' read 'It was interesting to note from the test work that the average temperature recorded on the dynamometer was only 7% higher than the average of front and rear brake temperatures recorded on the vehicle under equivalent fade test energy levels.'

Page 64 In caption for Fig. 6: for '(γ)' read '($\beta\gamma$)' and for '(a/a_o)' read '($\bar{a}/\bar{a}_o$)'
Col. 2, penultimate line for '($\beta\gamma$)' read '($\beta\gamma$ increased)'.

Page 65 Col. 1, para. 3.3, last line for 'frequence' read 'frequency'

Page 67 Col. 1, reference 7, for 'Kraftfahrzeugen Automobil' read 'Kraftfahrzeugen, Automobil'.
Col. 1, reference 9, for 'A Computer Simulation of Three-Dimensional Motions for Use in Braking Systems and of the Driving Task' read 'A Computer Simulation of Three-Dimensional Motions for Use in Studies of Braking Systems and of the Driving Task.'

Page 77 Fig. 2: legend for horizontal axis, for '(mile/mm)' read '(mile/$^1/_{1000}$ inch)'.

Page 90 Fig. 5: Column of figures on extreme right should be headed 'TD'.
Column of figures between the graphs should be headed 'Mean SRV at 50 km/h'

Page 109 Fig. 3: Curve running from zero to point 'A' should have a label 'Dynamic Brake-Force Distribution'

Page 149 The Figure at top of page and that at bottom should be transposed, the captions remaining unaltered.

Page 180 Para. 23, line 12 for 'doubling' read 'tripling'

Page 183 Equation (29) for '$(\frac{u + d^2u}{d\phi^2})$' read '$(\frac{u + d^2u}{d\phi^2})^2$'

Page 202 Transpose captions for Fig. 6: and Fig. 7: for Fig. 8: the abscissa should read $\frac{t_o - t}{t_o}$.

Page 213 Col. 1, para 2, line 18, for 'by a button.' read 'at random by the computer.'
Col. 1, para 4., line 4, for 'patient' read 'patients'.

Page 214 Col. 1, under DISCUSSION line 16 for 'In the case of the patients with the right inguinal hernias, R. foot, the mean time from completion of signal to pedal movement is significantly prolonged on the' read 'In the case of the patients with the right inguinal hernias, the mean time for the signal to completion of the pedal movement on the right is significantly prolonged on the'
Col. 2, line 14, for 'in both feet. On day 4 for the right feet but not' read 'in both feet, on day 4 for the right foot but not'

Page 239 Paper C47/76. The page numbers for 239 and 240 should be transposed, the paper starting on the latter.

Page 241 The expression (11) should read

$$P(E) = P(E_{KS}) \cdot P(E_{TS}) \cdot P(E_{B1}) \cdot P(E_{B2}) + P(E_{KP}) \cdot P(E_{TP}) \cdot P(E_{B2}) \cdot K_{fP} - P(E_{KS}) \cdot P(E_{TS}) \cdot P(E_{B1}) \cdot P(E_{B2}) \cdot P(E_{KP}) \cdot P(E_{TP}) \cdot K_{fP} \quad (11)$$

or

$$R_S = R_{B2}[1 - (1 - R_{KS} \cdot R_{TS} \cdot R_{B1})(1 - R_{KP} \cdot R_{TP} \cdot K_{fP})]$$

© IMechE 1977

Page 248 Para. 12., heading should read 'Adjust/Bleed Work as a proportion of Total Brake Work by frequency'
Para. 18, line 4, for 'present' read 'presents'.

Page 249 Para. 34, line 6, for 'have had to have extensive rebuild.' read 'have had to have an extensive rebuild.'

Page 251 Para. 59, line 6 and 9, for 'acheive' read 'achieve'

Page 255 Paper C50/76. The figures were omitted from this paper. They are as follows: –

International Legislation Activity on Trucks

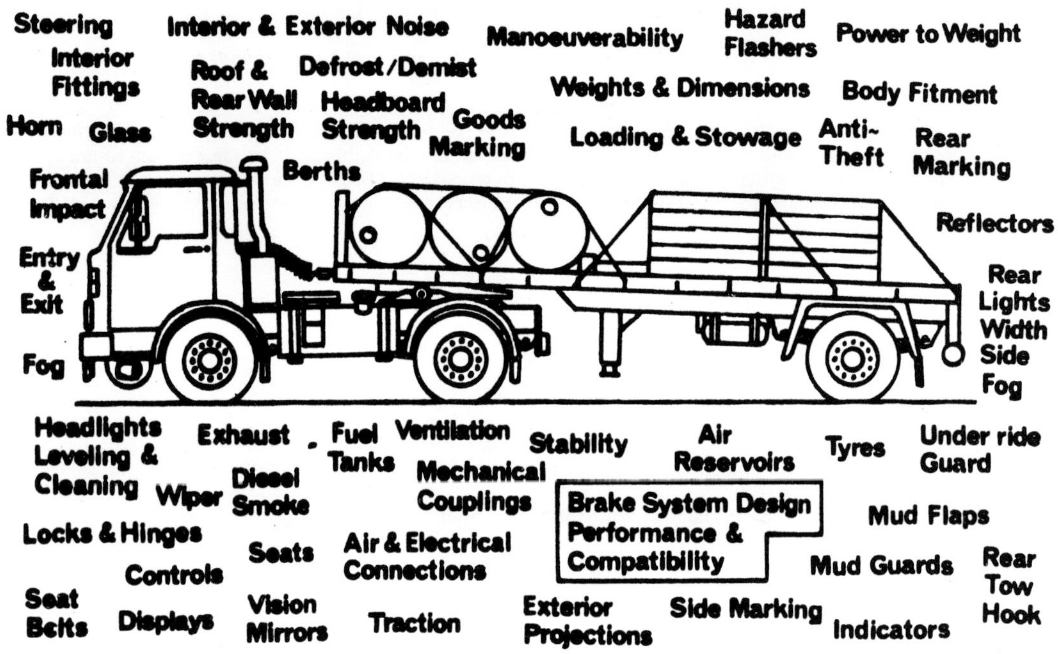

fig.1

ECE GENEVA

UN Gen. Assy. → Economic/Social Council → Inland TPT Cte → Road Tpt sub Cte → Group of Experts on Vehicle Construction (WP 29)

↓

Groupes Des Rapporteurs (GRs)

* PN	Tyres	* PA	(Pollution)
* F	Brakes	* B	(Noise)
* C	Seat Belts	* SG	(Safety)
* R	Truck Cabs	* SA	(Bus Safety)
* TRC	Containers	* LP	(Licence Plates)

fig.2

© IMechE 1977

EEC STRUCTURE

COMMISSION	COUNCIL OF MINISTERS	COURT OF JUSTICE
(Civil Servants)	(Politicians)	(Judges)
* Proposes Policies	* Decides Policy	* Decides Legalities
* Administers Decisions		* Ensures Observance
* Decides Amendments		

ECONOMIC/SOCIAL CTE	CTE PERMANENT REPS	ASSEMBLY
(Appointees)	(Diplomats)	(National MPs)
* Reviews Proposals	* Recommends Decisions	* Reviews Proposals
* Proposes Amendments		* Proposes Amendments
(ECOSOC)	(COREPER)	(EUROPEAN PARL)

fig.3

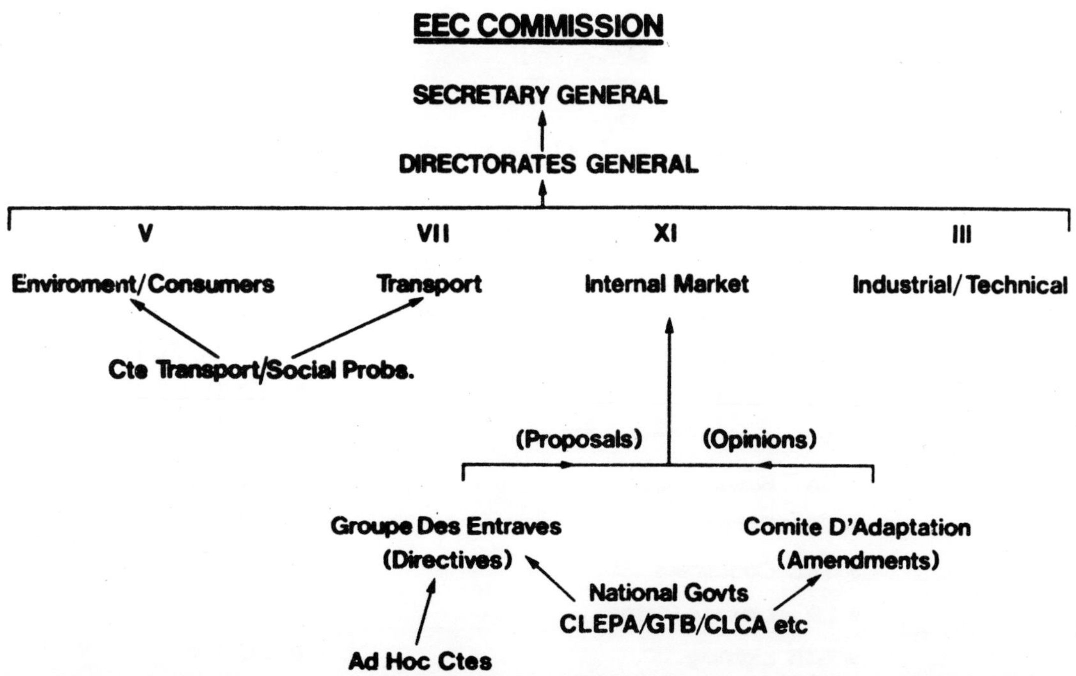

fig.4

© IMechE 1977

EEC DIRECTIVE DEVELOPMENT

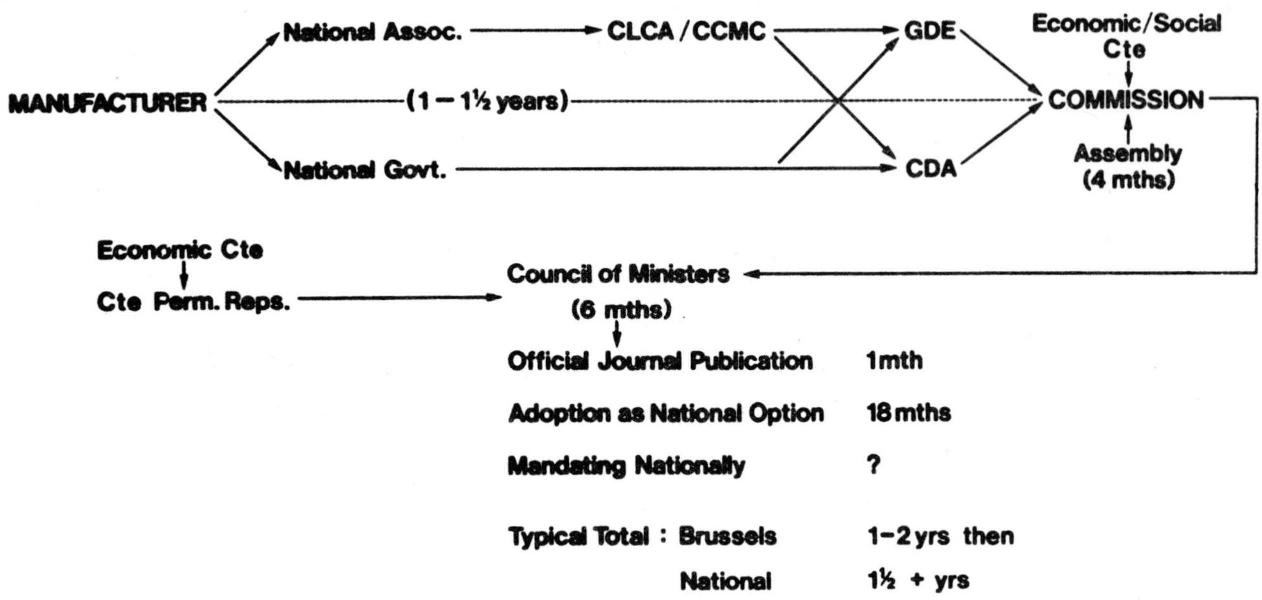

fig.5

FACING GROUPS

Government	Industry	Independant
ECE	BPICA	ISO
WP 29	TC	TCs
GRs	GEs (Groupes Des Experts)	SCs

* B	Noise	TC 43	
* C	Belts	TC 94 / SC 12	
* F	Brakes	● SC 2	
* PA	Pollution	TC 70 / SC 7	
* PN	Tyres	TC 31 / SC 9	
* R	Cabs	● SC 13 +	
* SA	Buses	● SC 18 +	
* SG	Safety	● SC 10 +	
* TRC	Containers	● SC 6	
* LP	Licence Plates	● SC 20	
* GTB	Lighting	● SC 8 +	● TC 22

fig.6

© IMechE 1977

'STANDARD' SEMI-TRAILER

fig.7

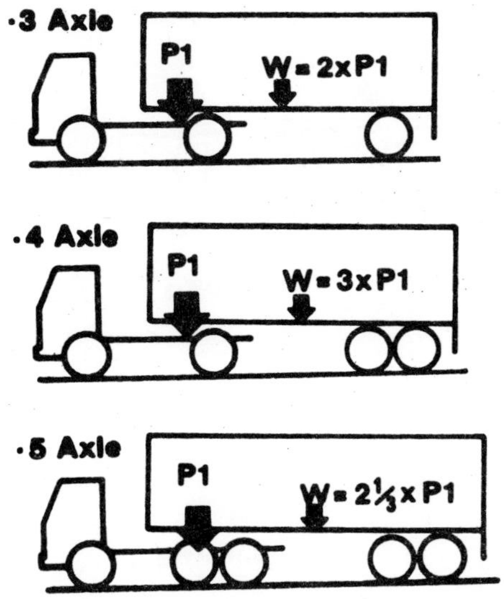

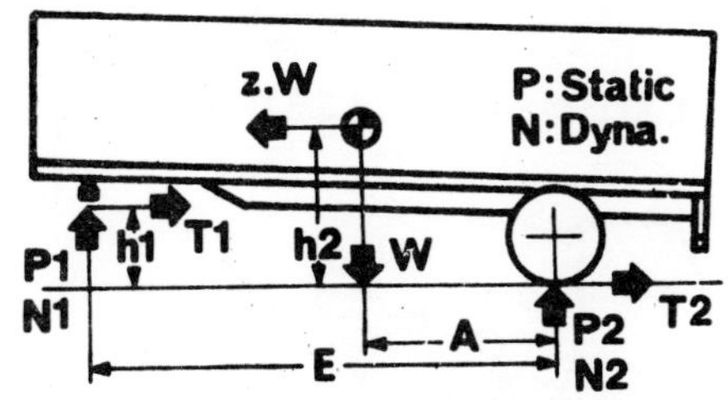

$$N1 = P1 + \frac{z.W.h2}{E} - \frac{T1.h1}{E}$$

Typical: $\frac{h1}{E} = 0{,}2$; $\frac{h2}{E} = 0{,}3$ & Obj.: $T1 = z.N1$

·3 Axle	$N1 = P1\,[1 + 0{,}37\,z]$	
·4 Axle	$N1 = P1\,[1 + 0{,}64\,z]$	$+0{,}45\,z$
·5 Axle	$N1 = P1\,[1 + 0{,}46\,z]$	

LADEN COMBINATION

fig.8

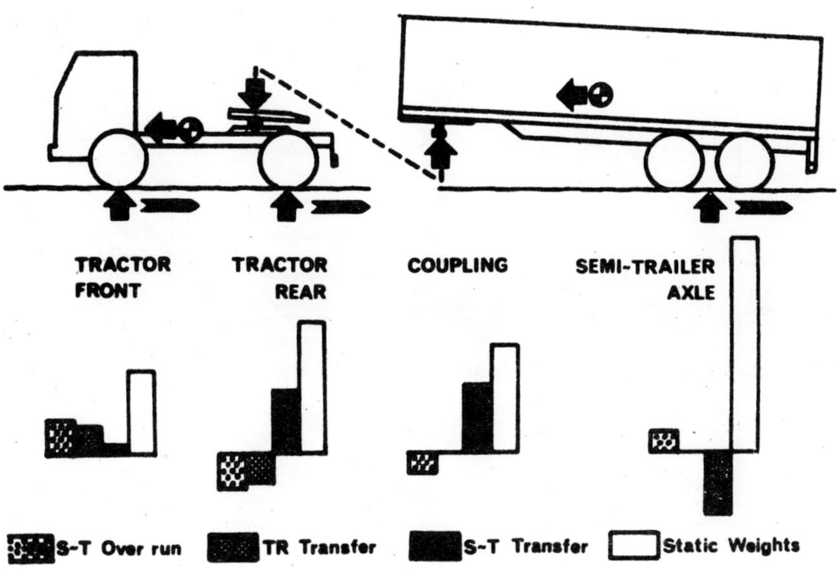

UNLADEN COMBINATION

fig.9

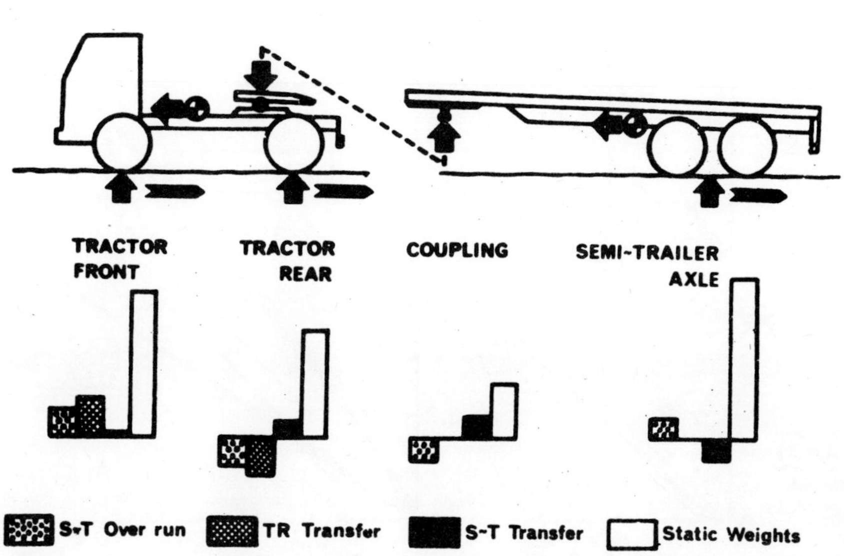

© IMechE 1977

ARTIC EXAMPLE

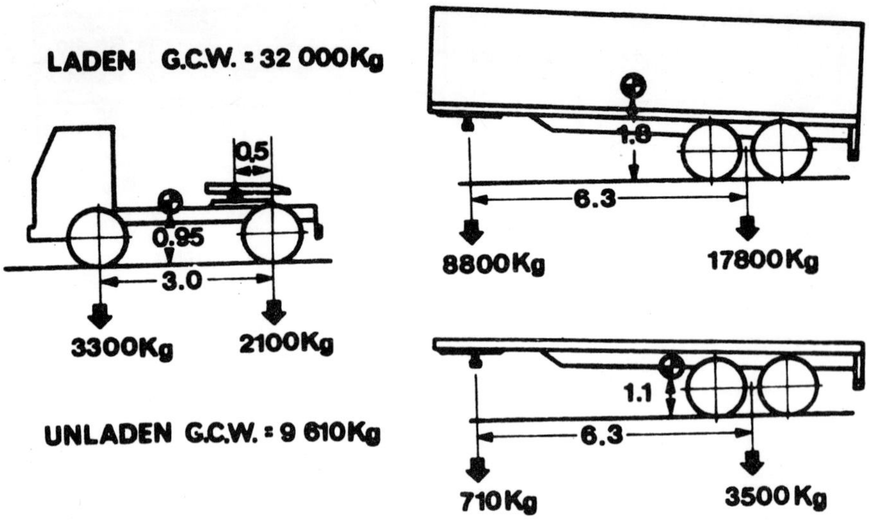

fig.10

LADEN ARTIC TO 'ANNEX 10'

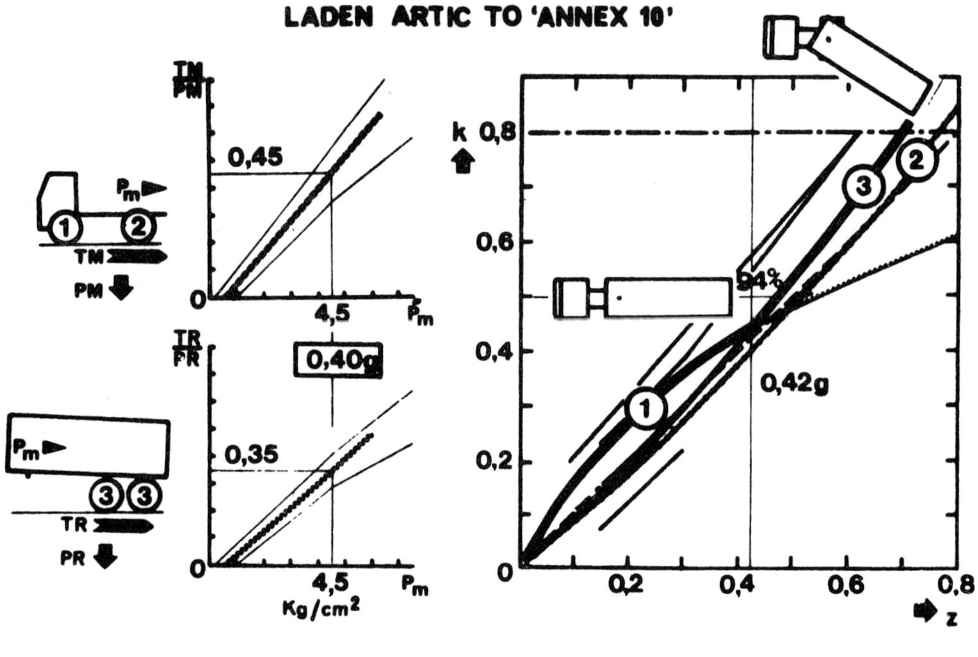

fig.11

UNLADEN ARTIC TO 'ANNEX 10'

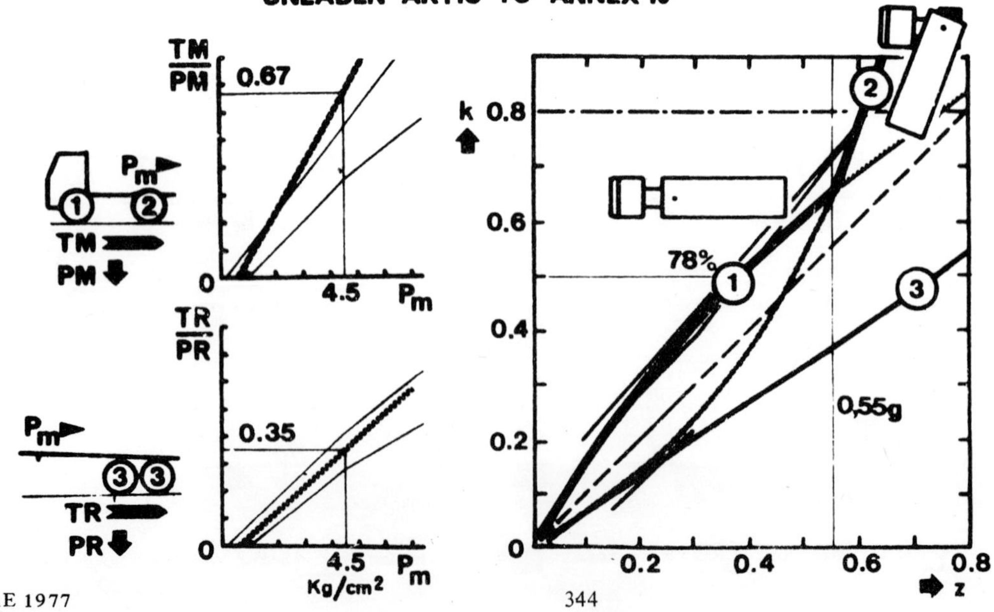

fig.12

© IMechE 1977

'ANNEX 10'+ FURTHER REDUCED REAR AXLE BRAKING

fig.13

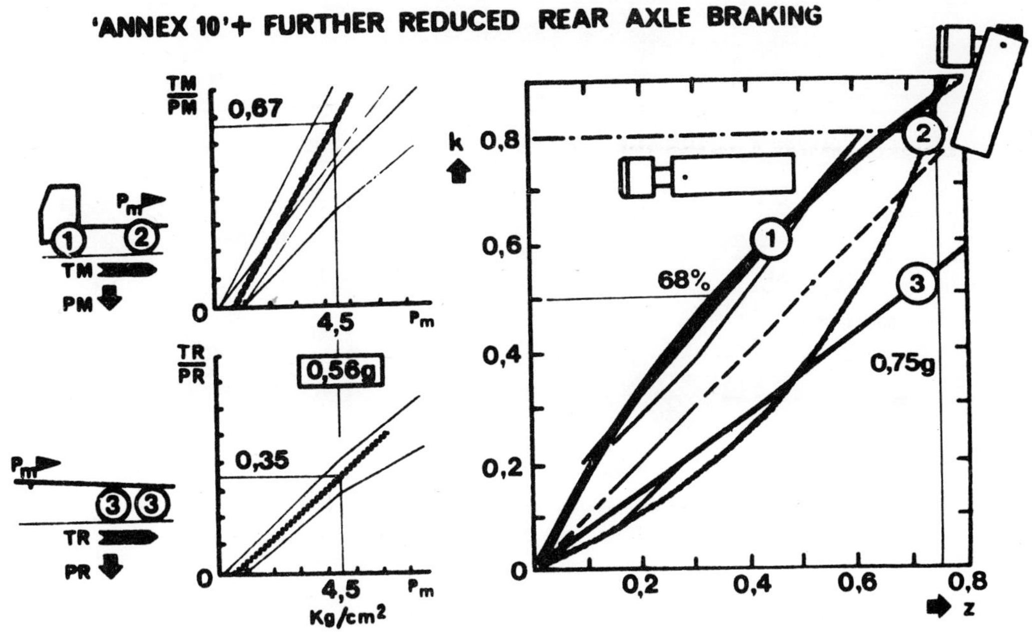

UNLADEN ARTIC TO 74/132/EEC

fig.14

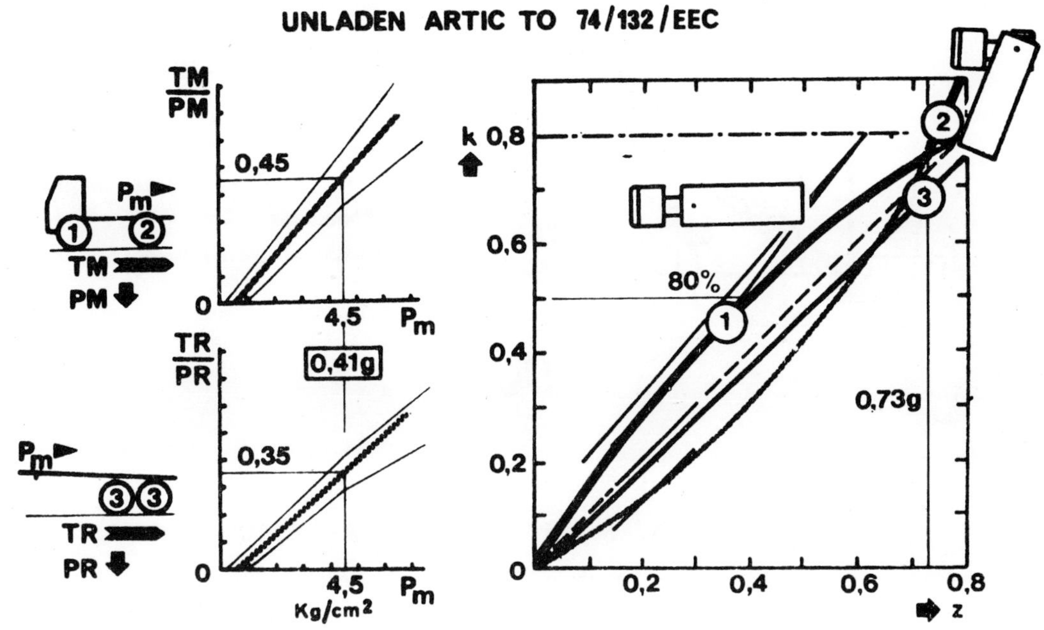

UNLADEN ARTIC TO 75/524/EEC

fig.15

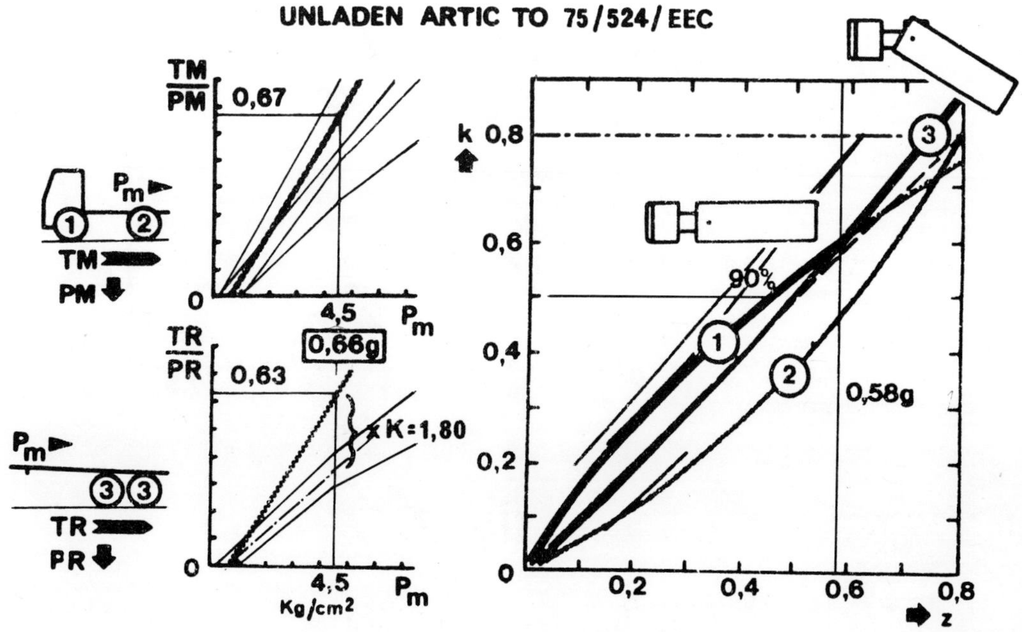

© IMechE 1977

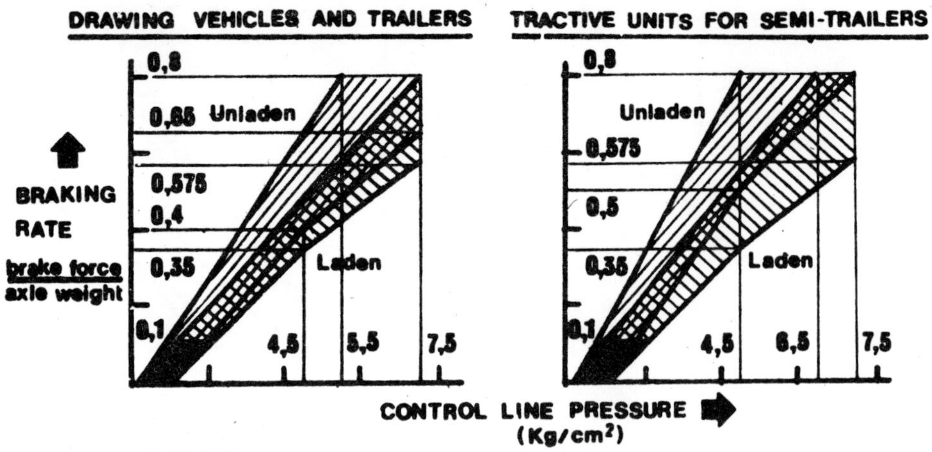

SEMI-TRAILERS

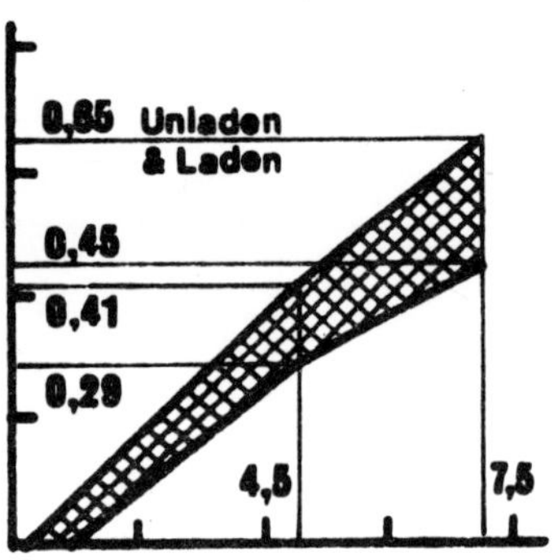

$$K = \left[1{,}7 - \frac{0{,}7\,PR}{PR_{max}}\right]\left[1{,}35 - \frac{0{,}96}{ER}\left(1{,}0 + (hr - 1{,}2)\frac{P}{PR}\right)\right] - \left[1{,}0 - \frac{PR}{PR_{max}}\right]\left[\frac{hr - 1{,}0}{2{,}5}\right]$$

where :

PR - Static semi-trailer axle weight

P - Static semi trailer total weight

EP - Wheelbase

hr - Center of gravity height

fig.16

75/524/EEC BRAKING ADHESION (not M1)

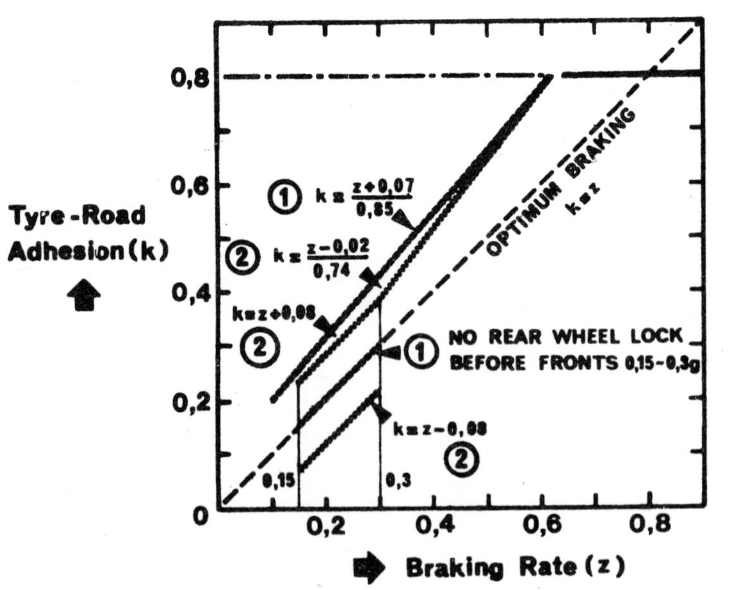

fig.17

© IMechE 1977

Page 299 Paper C79/76. Please note owing to a mistake on computing the conversion factor for lbf/in^2 to N/mm^2, an error of 10^6 has been made in the figures quoted for liner drag pressures. All values for 'specific liner drag' or 'drag pressure' should therefore be divided by 10^6. This applies to diagrams in Fig. 1b and 1c as well as to values quoted in paragraphs 4, 5 and 6 of section 3.2.

Page 312 Col. 1, line 9, for 'shows the fading behaviour' read 'shows the heat fading behaviour'.
In equation (5) for 'y' read 'μ'.
Equation (8) should read:–

$$\vartheta n = \frac{q}{c.y.v.} \cdot \frac{1 - e^{-n\frac{\alpha.F_K.t_s}{c.y.v}}}{1 - e^{\frac{-\alpha.F.t_s}{c.y.v}}} \qquad (8)$$

Page 313 Col. 1, second line following equation (13) for 'Hocke' read 'Hooke'
Col. 2, line 17, for 'Steel cast-iron' read 'cast steel'
Lines 23 and 28 for 'steel cast-iron' read 'cast steel'
Paragraph headed DISC BRAKE DESIGN, line 1, for 'With regard to thermal resistance' read 'With regard to the thermal behaviour'

Page 314 Col. 1, line 19, for 'steel cast-iron' read 'cast steel'

Page 331 Fig. 2: should appear as below.

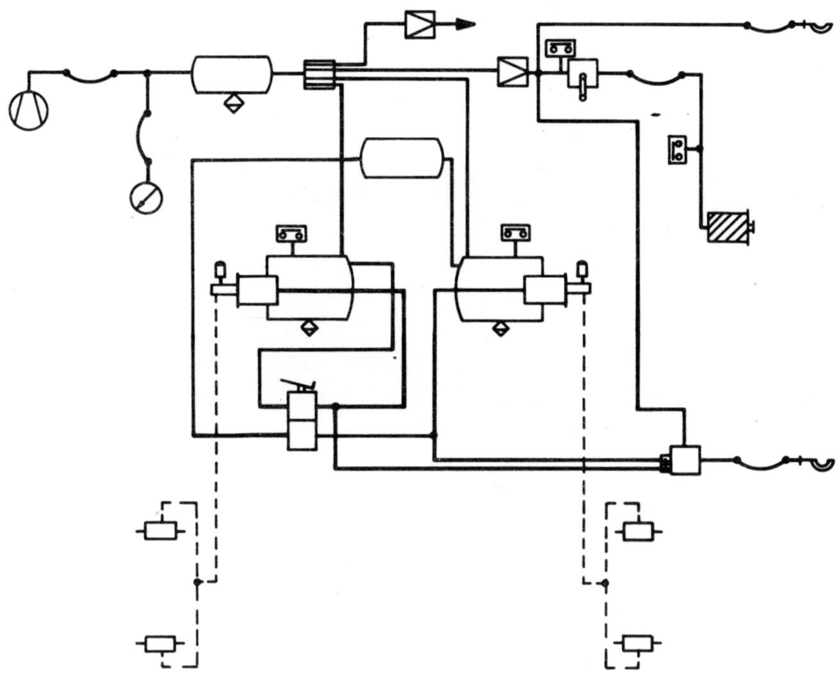

Page 324 Conclusion 8, instead of finishing at 'repairs.' should continue '67% of brake costs were labour and 23% replacement shoes.'

© IMechE 1977

© IMechE 1977

Discussion

Papers C22/76, C26/76 and C49/76

T. P. Newcombe Loughborough

I should like to pose a question to D. Peasley and ask him a little more about disc brakes for heavy vehicles. Although disc brakes obviously have some disadvantages on heavy-weights, I would like to ask whether work has been done on multi-disc brakes for commercial vehicles, and also ask him at what sort of weight level would one consider a disc brake not to be acceptable for use on vehicles.

D. Peasley Birmingham

Yes, we have done work on multiple disc brakes. Unfortunately, at the moment we are of the opinion that mechanical sophistication is not available to make multiple disc brakes a really satisfactory alternative. Girling, as a company has been associated with disc brakes on commercial vehicles for some years now, and we have a number of installations in production – some not very satisfactory – but we are of the opinion generally that we have not yet identified what are the correct operating features of disc brakes on trucks. Having said that, we think that you may very well have a satisfactory installation in running vehicles up to 12 tons, or possible slightly more than that, gross vehicle weight, but certainly at values above that we believe that there are associated problems in dissipating energy at that sort of level.

T. P. Newcombe Loughborough

Could I go further and ask whether that was using conventional discs or ventilated discs, and could you say whether there was any difference?

D. Peasley Birmingham

No, obviously there is an advantage in using ventilated discs, but at the moment the amount of improvement is still a subject for discussion and the disc brake equipment itself is fairly standard as opposed to piston calipers.

P. Oppenheimer Birmingham

I should like to comment on T. P. Newcombe's paper, if I may. Although it is fairly basic stuff, you do not seem to have given much regard to legislation in the actual paper, although in your speech today you have given a little more explanation. I should like to suggest that legislation and the regulations are probably of overriding importance in the design of any braking system now. The braking system can be as good as you like, but if it does not meet the regulations you will not get very far with it, unfortunately. In fact you can design a wonderful braking system that does not meet the regulations because many of the regulations have not been anticipated. For example, things like full-power hydraulics or anti-lock installations have a good chance of not passing the regulations at the moment. On the other hand, as mentioned by B. Ingram and D. Paisley in their paper, if you design a braking system that meets the regulations, it is not necessarily a good braking system. The latest European Directive 524 is a very good example of how you are allowed to have all four wheels locking in all conditions of road and all deceleration conditions, and the philosophy is all wrong, because people will design to the regulations rather than design a good braking system. I wonder whether you might give a little more attention to the regulations in the basic principles of design of a braking system. Thank you.

E. G. A. Elgeskog Göteborg, Sweden

I should like to put a question to all three speakers who have spoken so far. You refer to the road to tyre adhesion as a factor. We all know that this factor is very much dependent on many things, grip, slip, angles, speeds, loads etc., and they change very much according to conditions. Would this fact affect any of your arguments, do you think? I should like to hear your comments on that.

Chairman

The question is the varying value of μ under different conditions; does this affect your papers or arguments?

D. Peasley Birmingham

It does not affect my attitude because I think that in the written paper we made it perfectly clear that we were interested in performance of braking across the range. This is one of the important things which probably reflects on P. Oppenheimer's question about braking systems and legislation. What we are trying to achieve is a system which will utilize adhesion in the best possible way. This is what we were concerned about when writing our paper and it does not specifically alter my attitude, or our attitude.

B. R. Shilton Lincoln

When designing a braking system for a vehicle, one tries to take into account all possible ranges of tyre to road friction that will be presented. It is inevitably a compromise, between premature locking of wheels on one axle and one type of friction conditions, and premature locking of wheels on other axles. I think that views do differ somewhat from company to company, or from individual

© IMechE 1977

to individual on this particular aspect, but I think that all these things are considered in our papers. I think that how one would deviate from the ideal nominal value, may be influenced by personal preference or personal philosophy. I think that something more will be said about this later in the Conference.

T. P. Newcombe Loughborough

I accept all that P. Oppenheimer said. In my paper I just drew attention to what could be done in choosing a ratio, but obviously based on legislation, one now chooses one's optimum to ensure how to get a system to operate within certain bounds. To some extent, this now plays a large part in choice of the optimum parameter. I do not want to say too much about that at this stage, or I should not have anything to say at the end of the Conference. I accept that it is quite a difficult business, your choice is so restricted that you have no room to manoevre on parameters. Indeed, you could almost say that you can fix them quite easily if you are to use a fixed ratio system. In fact I think that this is the philosophy outlined in the second paper.

Concerning road-tyre adhesion generally, one gets to the stage where one has such a low value that one might just as well not be travelling on the road at all. There is nothing one can do in these circumstances but use common sense. Unfortunately, the motorist does not do this with the adhesion available. There is nothing one can do except to use an anti-lock system to cope with those situations, one has to do this; but should have reservations about anti-lock systems, as has been mentioned before. I can stick my neck out and say that I am an academic so no one will take any notice of what I say. Anti-lock systems have appeared on four wheel vehicles for a number of years, and they may well see the light of day for fitting on vehicles that have exreme changes in rear loading conditions, such as motor cycles.

R. D. Lister Slough

I think it is surprising from the simple basic statement on basic principles, how complicated they have already become. I should not like to think that P. Oppenheimer has put a damper on everybody, by thinking you must design to regulations. Of course you must do this, but it is the goal of designers and the purpose of meetings such as these to show the people who make the regulations what the different elements are. Legislation is man-made, and therefore may be man-altered if the need is demonstrated. On the other hand there are factors that you cannot change. You have the limiting factor of the road-to-tyre adhesion and the limiting factor of the input to the brakes, which you have to start from. One is the input and the other is the output. You are limited at one end and at the other.

Referring again to T. P. Newcombe's paper, it seems to me, in retrospect, that greater emphasis has been, and still is on the things that he mentioned in his introduction, the need to obtain consistent repeatable braking over long periods of time and also to achieve the maximum of braking in regard to control. This is quite right because all this makes for greater safety. Taking this consistency to a greater extent, should not the idea of uniformity extend to the whole traffic environment so that the differentials in braking performance, i.e., differentials between different vehicles, are reduced? This is being attempted, I feel, by trying to upgrade the commercial braking performance. This has been going on for a few years now, but at the same time improvements in car braking tend to maintain differentials so that we still end up with the car having a much better braking performance than commercial vehicles generally. Is there anything in the idea that this differential should be reduced by arbitrarily restricting the upper end of the car performance?

T. P. Newcombe Loughborough

I want to tackle the latter point first as I think it the most important. One can see how, unfortunately, we have got into this situation, and one really ought to blame the vehicle manufacturer for this. He originally designed a brake of a certain kind and got a satisfactory performance. Then what does he do? A short time after, he changes the performance and the last thing that he thinks of doing is to change the brakes because he thinks that they are settled. So what we get is a spread of parameters for different vehicles, over different tonnage rates and usage activities, producing temperature variations to quite a considerable degree, probably 200°C. So you ought to ask the manufacturers, once they have a vehicle with a good brake for the size of the vehicle, that when they intend changing things, at the same time to scale the brakes in accordance with the new performance. One would then find that they would have consistency over the whole range of vehicles.

D. Peasley Birmingham

I think that to start with you have a basic problem with commercial vehicles to fix the loads. I think this is one of the biggest problems in improving performance. I would hate to support the theory that we should reduce the performance of cars simple to make them more comparable with trucks, which is obviously one answer. It causes me much concern, when I go down a motorway, and I see an enormous vehicle behind me and I know that I am capable of stopping at a considerably higher rate of deceleration than he is. Before we can get a radical improvement in commercial vehicles, we shall have to develop a method of restraining loads.

B. R. Shilton Lincoln

May I add the comment that if all small vehicles on the road had virtually identical straight line braking performance, which at the moment could be achieved only be degrading that of passenger vehicles this might well produce a number of Newton calculations, the commercial vehicle being the rarer vehicle. However it could reduce very considerably the number of collisions between passenger cars and those going in the same direction.

T. P. Newcombe Loughborough

I think that we should pose this question to the car manufacturers.

R. W. Ladbrooke Wimborne

I should like to pose my two questions to the authors collectively. I should like to know about the tyre-road ratio. Someone mentioned 20 brake horse-power per ton. I must have missed out somewhere, but I think that it must be about half that. If you went down the M1 this morning I

© IMechE 1977

do not think you would find many commercial vehicles that would match that.

The other thing is the aircraft disc brake. A lot of work seems to have been done, at least by the American aircraft manufacturers, with some quite revolutionary disc brake aid materials, particularly carbon material. I wonder if the authors could say something on that.

D. Peasley Birmingham

We actually run a vehicle with over 20 brake horsepower per ton. We have a Volvo F89 rigid vehicle which is intended to be run with a trailer, but is actually run without, and I think that there are quite a number of vehicles with tyre to road ratios of over 15 horsepower per ton. The point about this is, that it is a radical change from the levels of some years age of around 5 to 8 horsepower per ton.

I am sorry but I do not know anything about the use of carbon fibres.

T. P. Newcombe Loughborough

Carbon fibres make an ideal brake, because they can withstand high temperatures and are very light. If R. W. Ladbrooke has a gold mine at the bottom of his garden to pay for these things, I am sure that he would find that on performance, and everything else, they would be acceptable. Meanwhile, for people with more modest means, I think that we shall still see the asbestos based material coming into use for many years for commercial vehicles. You can get a good performance from carbon fibre disc brakes, in fact it has been tried by Porsche in racing cars with quite good results.

F. D. Hales Loughborough

I should like to take up the point made by a previous speaker, concerning the application of brakes by a driver on the motorway with a large truck behind him. It is quite interesting that about 95 per cent of car drivers are not aware of this problem, and somehow thay ought to be made aware that the heavy vehicle cannot stop as quickly in this situation. You can deal with the situation, but what about those who cannot.

D. Peasley Birmingham

What I believe is the most critical thing, is that the truck driver be made aware of the limitations of his vehicle. When I am in this situation, what frequently strikes me is that the chap driving at 70 miles/h cannot possibly have tried to stop in an emergency.

T. Williams Crowthorne

I should like to come back to the disc brake problem, because you can get severe heating problems of components. Could you tell me whether this disc brake really meets the problems, and if you have done any measurements?

D. Peasley Birmingham

I am not in a position to answer that very ably. T. P. Newcombe, is very much more well versed in disc brake installations that I. I do not know whether he can deal with that?

T. P. Newcombe Loughborough

We have been conscious of the risks of overheating. From the work that we are doing on disc brakes, I do not believe that the disc brakes will make any difference to the temperature all of the time.

M. A. I. Jacobson Basingstoke

If I may first come back to one of the previous contributors, there is in fact an article to which we, in our publication Drive, tried to give a fair amount of publicity, about the danger on your tail. If the lorry driver is unable to judge his speed, this is perhaps as relevant as anything else. He always blames the poor motorist for being unprofessional, why should he be? He has other pursuits to follow.

I should like to come back to another problem which has not been mentioned yet. That is servo action. There seems to be a tendency to think that overseas manufacturers should offer servo to the braking system, so that you only just touch the brake then stop dead. Then the fellow behind cannot stop that quickly. That is one problem. What do the authors consider to be the limiting factors of the servo operation on cars? The factor that I would like to hear about from them is, have they any experience, or hope, to offer on the effect of flexible pipes giving that sensitivity, and of sponginess of the braking system under extreme pressure application. Thirdly, what are their views regarding having servos on a disc or drum brake system, such as for instance the latest cheaper range of motor cars coming from mass production factories? Is it a right step to fit a servo or not?

D. Peasley Birmingham

I will take the one about servos first. You must be in a very fortunate situation where you can put bigger servos on a vehicle, than normally you find in a PSV. That is a problem in itself and obviously we have commercial pressures on us, not unnecessarily to use equipment which is larger than actually required. The operating parameters for brake systems generally are fairly well defined. I know that there are some anomalies here, in that Citroën use different values; they have a rather different system anyhow. This, people take into account when driving a Citroën and getting used to them. But the basic levels on which braking systems operate are fairly well defined. We normally have problems in simply trying to get sufficiently large actuating equipment in, to bring the pedal effort down to the levels required by the EEC.

On the disc brake situation, obviously there are advantages. You were suggesting that there was an advantage in having disc brakes all round?

M. A. I. Jacobson Basingstoke

No, I was saying there was a problem of fitting a servo on a brake design and then dealing with the abruptness of stopping.

D. Peasley Birmingham

I think that if you have a system which actuates generally with pedal pressure you run the risk of having rather jerky brake application. I would be inclined to think, that if you have a servo it generally tends to dull the responses and

© IMechE 1977

therefore the feedback would not be so clear and well defined, and therefore the risks of having problems with the servo would be worse.

M. A. I. Jacobson Basingstoke

I think there is an obsession, as one would expect, with factory cars to ensure that the vehicle is fully laden, but I think that a factory car is fully laden about once a year.

Chairman

There was a further question on the sponginess of brakes. This has not been aswered.

B. R. Shilton Lincoln

Is M. Jacobson suggesting that spongy hoses are a good thing or a bad thing?

M. A. I. Jacobson Basingstoke

A bad thing.

B. R. Shilton Lincoln

In the paper we say 12½ per cent. I think it is an academic question, but one always assumes that the system is 100 per cent right and argue from that. It becomes a little hypothetical when it come to ordinary service. People do not believe this, but it is a fact that if you have hoses a little thicker, it is irrelevant.

M. A. I. Jacobson Basingstoke

May I ask a supplementary question? You find in practice that what you have is a gradual deterioration, becoming fairly rapid in time, and before you are much older in some systems you floor the pedal without reaching full locking anyway.

B. R. Shilton Lincoln

On some of the modern systems you have a length of flexible hose.

M. A. I. Jacobson Basingstoke

In making accessible pedal efforts you have reserves, and there is no substitute for adequate reserves, but people do not like to pay for adequate reserves.

R. S. Sharp Leeds

Do the three authors consider that brake design is a special design which needs a special activity?

B. R. Shilton Lincoln

I think that the answer must be no, they cannot be considered divorced. Obviously it is a question of the kinematic effect of a given suspension linkage on a given vehicle, to the given anti-lift and anti-drive characteristics depending on the distribution of the particular road vehicle. In addition, I think that with modern suspensions those particular members are complaint in this direction while some members resist. It may be possible to have design in compliance with the control system. I think that they must be considere together.

R. S. Sharp Leeds

Could I ask to what extent they are agreed?

B. R. Shilton Lincoln

Not as much as we would like, but possibly the vehicle manufacturer himself would be in a better position to answer that question.

T. P. Newcombe Loughborough

As regards the first question by M. Jacobson about servo action and its abruptness, we find from the results of tests, that people soon get used to the characteristics of servo and non-servo and it does not need many applications to adapt oneself to the system. We have made measurements of vehicles driven without and find little difference in braking behaviour. That is, of course, with drivers who are reasonably experienced.

Papers C23/76, C24/76, C27/76 and C28/76

D. McLean Loughborough

I wish to ask one question of F. D. Hales; your paper has concerned itself with the model characteristics of a linearised set and the assumption allowed you to linearize those equations so that the acceleration was constant. It is well known that if you monitor dynamic equations the stability is dependent to a marked degree on the initial provisions. Have your studies explicitly considered the effect on the trai stability of different provisions for braking other than small changes in speed?

My comment to R. S. Sharp is in connection with his remark when he was speaking of the cat's eye mode. He mentioned that this allowed one to tolerate an unstable mode because the time is perhaps less than four seconds. I have to remind him that such a performance applies only to aircraft pilots, who are pilots by virtue of three things, (1) they are selected for training because they have shown initial aptitude for that kind of task, (2) their physical condition is always beyond the level enjoyed by the majority of us, and (3) they are trained and tested continuously to ensure that their mental performance does not deteriorate. These facts may not apply to motorcyclists.

F. D. Hales Loughborough

I have not looked at any other technique than the one I outlined. The defence of it is that in reality changes of speed are small and the acceleration terms, although they introduce model realities, in effect do not affect the stability in this sense at all until you have back axles, which is a secondary effect of the acceleration. So, although these things are non-linear I think this is a reasonable assumption. I must confess that it would have been very interesting to use a much more sophisticated concept and perhaps some day you will either instruct me as to how, or actually do it for me.

R. S. Sharp Leeds

May I say thank you to D. McLean for his comment and say that often the traditional policy in motorcycle engineerir

© IMechE 1977

is to regard the motorcyclist as something similar to a pilot, although they are not selected by anyone; they select themselves. A lot of the things that they are expected to do by virtue of the design of the motorcycle is more than the general public should be expected to do, in my opinion. For instance they have the control of both brakes with the hand, which also works the throttle. When you are braking and changing gear, and at the same time you have gloves to keep your hands warm, you are expected to do the whole job properly. With the growing use of motorcycles by people who simply want to get around cheaply, a change in philosophy is desirable.

E. G. A. Elgeskog Göteborg, Sweden

I should like to comment on the first question about unusual conditions. As we have found in many states, you can easily get into a condition where anything you do will just make it worse. You cannot get out of the unstable situation, the forces are not there.

M. A. I. Jacobson Basingstoke

When F. D. Hales assumed his train of trailers, did he consider the possibility of doing something, which in the dim and distant past I did, when I worked on the signals of the San Francisco Bay project. That is to brake progressively from the rear. The only way that we could control a large number of vehicles; admittedly they were track vehicles, but they were automatic; was to have a light delay in action between each progressive unit. There was a much larger number of units than in the train you considered, but you can maintain stability, in a curve particularly, only if you consider the whole thing to be a piece of string with a conker at the end of it. If you keep the string taut, it will follow the path that you want it to follow. If you somehow brake in the middle or the front, the likelyhood is uncontrol and oscillation in it will vary considerably. We have recently carried out some tests with tractor and trailor units, a Land Rover, Range Rover and so on, and on some of our relay trailers we have found that you must sequence your braking in such a manner that you do not set up that slight oscillation, because you call on the skill of the operator to anticipate it and alter the braking mode to an extent that he may be able to respond to it. I wonder if you could respond to my suggestion of braking from the rear. I know that it could possibly lead to some theoretical loss of maximum braking on a dry road, but unfortunately that is not a situation we have to cope with very often.

F. D. Hales Loughborough

Thank you. I am not quite sure on the point that you mentioned about the instability that you experienced with guided vehicles was in longitudinal oscillations. If the motion that you experienced was longitudinal, I would accept that this could be a serious complication. Perhaps you could clarify that point?

M. A. I. Jacobson Basingstoke

We are concerned with guided units, but the sideways force on the rails was of considerable magnitude.

F. D. Hales Loughborough

The point you make about braking from the rear could have some advantages, but I think the thing that indicates that this is not a critical parameter, is that the evidence suggests that there is not too much change of stability. The evidence from the quasi steady approach is that there is not too much evidence of the acceleration as such producing the stabilizing effect. The stabilizing effect is largely associated with the actual locking. I think that it has been substantiated that in practice these vehicles change their behaviour in that sense, and it does not matter too much where you apply the braking. In one particular vehicle configuration you lose a great deal; that is in an articulated vehicle which of course is carrying part of its load on the tractor. If you brake the trailer first, clearly you are braking only half the mass. The other handicap, the big problem in commercial practice, is to overcome having application too far from the operating pedals. Of course, if you insist on putting brakes on the back first, the system does not start to slow down until it reaches the axle point. If you can show by experience that this has been done, on theoretical grounds there is no disadvantage in braking from the front axles first.

M. A. I. Jacobson Basingstoke

The actual delay period is not seconds, but probably of the order of less than one tenth of a second.

F. D. Hales Loughborough

On three axles, one tenth of a second represents about 9 metres travel, so I think we are talking about things that matter.

P. Oppenheimer Birmingham

Without wishing to expose my inexperience of eigenvalues too much, I understand from F. D. Hales that if you lock axle two, it becomes unstable. Some people say that, provided that you lock the front axle very soon after, it could remain stable. Does your mathematical analysis show that this is true and do you, or anyone in the audience, have any evidence to show that if you lock the rear first and the front after, it remains stable?

F. D. Hales Loughborough

The actual paper contains a table which shows what happens when the two axles lock together. This shows quite clearly that simultaneously locking the front and rear axles has a pronounced stabilizing effect. The point about locking the rear axles first and the front axle immediately afterwards is that the front axle can be locked before there is any significant deviation and of course this did not have to occur immediately when travelling in a straight line. On the point about whether this has been done experimentally, I remember a series of experiments we did with articulated vehicles when we managed to show that if you lock both the tractor and the trailer there is a tendency in a jack-knife situation for it to just slide along.

P. S. Fancher Ann Arbor, USA

I have a question directed to F. D. Hales and R. S. Sharp to do with the influence of longitudinal slip in deceleration. On the influence of longitudinal slip, E. G. A. Elgeskog pointed out that he had an optimum strike period of 15 per cent slip on the rear wheels of a vehicle. I can think of no work of a theoretical nature in progress, on the rules for slip on the rear wheels and slip on the front. I wonder if it is

© IMechE 1977

possible to include the influence of longitudinal slip and lateral force. I realise that this would mean complicated data, but it seems that things like that should be included, particularly in the analysis of deceleration.

R. S. Sharp Leeds

I can agree with that. It is very feasible to include it. In my case it would mean simple changes to the coefficients in the tyre behaviour. What has stopped us doing that is lack of information, particularly about motorcycle tyres. I think it reasonable to suppose that the characteristics are significantly different from those of car tyres.

F. D. Hales Loughborough

I think that the values of 20 per cent and 10 per cent are such that you are degrading the efficiency of the front wheels in that sense, rather than that of the back wheels. What you have to do is to require a most sophisticated braking system. I was looking at this family of vehicles and the braking upper and lower limits, and the tracking and longitudinal curves showed about 10 per cent and 20 per cent, in the absence of some mechanism in the performance tests for controlling slip because of the instability, or near instability, associated with braking ratios. I was looking at a family of vehicles, but you could look at a specific vehicle.

B. R. Shilton Lincoln

I would like to ask R. S. Sharp a question. I think we shall hear a lot in the next few days about automatic braking of front and rear axles. Probably the motorcyclist is unique here in having discretion to apply torque as he thinks fit. Does R. S. Sharp accept that a motor car in rear braking is generally unstable, and a motorcycle, if something goes wrong, capsizes? Can he advise how the motorcyclist ought to apply his brakes to slow down in the most stable manner possible?

If I may go on, M. B. Packer's analysis of releasing brakes, in and out of gear, and the resulting μ cohesion was very interesting, but I wonder if his analysis took account of the compliance in the transmission to the wheels themselves and what must be the major inertia component in the system? If this was not included in the analysis, what effect would he think compliance would have on response?

R. S. Sharp Leeds

I am afraid that I am a very unskilled motorcyclist and I am not the right person to answer. I should say that the motorcyclist should apply the brakes carefully, particularly carefully when you first start riding because, although it is a joke, you obviously learn an awful lot with a bit of practice. The big danger with motorcycles is that people kill themselves before they have learnt the value of experience.

M. B. Packer Leamington Spa

The effect of braking in gear in general is not tremendous on compliance. In gear, everything happens so slowly, that compliance is a secondary consideration in many ways. The biggest problem in compliance arises when you have excessive brake release with a low tension system, i.e. gears, and the wheel recovers extremely fast. What happens in those circumstances is that the inertia of the wheel compliance unfortunately extends to wheel lock. This is critical and in anti-locking one has to pay particular attention to this in dealing with the problem. In my paper I have tried to keep this simple and to study the basic cases, but unusally in an anti-lock situation you try to avoid complete brake release which is a disturbing force on the wheel operations.

B. R. Shilton Lincoln

Perhaps I worded my question rather badly. What I meant to say, was that the compliance in the driving channel between the rear wheels which were in a state of inertia in gear, would lead to the conclusion that everything expands very slowly. I would imagine that because of these factors and the compliance of the transmission system it would not behave in quite that manner.

M. B. Packer Leamington Spa

I accept your question. The theoretical model did not include this, but from practical experience – we have done similar exercises by locking up the wheels on commercial vehicles – one gets an oscillation superimposed on the wheel velocity, but it is of relatively low amplitude and can be filtered out. It becomes very significant if one senses on the crankshaft and you are in real trouble with this condition. In general people who use epoxy sensing have severe difficulty with a very sluggish system because you have to filter this component out.

R. S. Sharp Leeds

May I make one more comment to B. R. Shilton? If you lock the rear wheel on a motorcycle, you have the yawing stability that other vehicles have not. If you lock the front wheel, you have the rear wheel to consider because the balance is destroyed, and if you lock the front wheel and then the back wheel, you fall on your face!

D. A. Hoffman Bristol

You included in your summary, as well as in your paper, the question of stabilizing. I think that we need to know more about that than to take it at its face value. Can you provide anything?

F. D. Hales Loughborough

It is stabilizing in the sense that if you take a car as a simple example, and drive round in a circle, and apply brakes to the front axle, you will leave the circle at a tangent and proceed to travel in a straight line until you stop. It is clearly uncontrolled in that you cannot make it do anything else. One has to distinguish the idea of stability and controllability, doing what you want to do.

Papers C25/76, C29/76 and C30/76

N. F. Barter Nuneaton

I should like, first of all to ask a question of A. R. Williams concerning the classification of road surfaces: As everyone knows, for some time there has been a requirement by the Department for a skid number. I am a great admirer of people who ask what is the skid number of a proving ground? Investigation of this shows that a skid number is a figment

© IMechE 1977

of the imagination, in that, depending on the measuring device used, and the conditions on the day of measurement etc., so the answer varies. This being the case, it seems to me that there is very little to fear from the authorities in carrying out the test on any surface in that category, because in no court of law under circumstances where they required proof could this be justified. Therefore does this dry ASTM skid number have any value? Secondly, is it possible in the light of your talk about the influence of road surfaces on tyres, that no one could do much better than that in terms of specified standard surface? In anticipation that you will say yes, would it be feasible to do that?

A. R. Williams Birmingham

Thank you for the question. Whether or not I am the right person to answer it or not we shall see, but I think T. Williams here has certainly some part to play in the answer. I think it rather depends on what they are for. We have the question of climate and the different variations. There is the question of what you want the standard surface for, whether you are interested in testing tyres, surfaces, or trucks or anything else. If we are talking of weight, as T. Williams has mentioned on a number of occasions, there is not just one number. Speed is a most important parameter and one cannot describe surface characteristics at one speed and assume that it is the same for all speeds; or on water conditions, which will vary with the depth of water. I do not think that in our paper we have proposed any standard method for measuring skid conditions and I do not think that this is part of our description or task. I think that perhaps our colleague from TRRL, who has proposed devices like SCRIM, could probably answer the question better than I could.

T. Williams Crowthorne

I understand your problem, but, as I said at the end of my talk, unfortunately I am not working on this problem any longer. You cannot give any one number and expect people to carry out their tests. I quite agree with you; I think that we have to look at the details available, as I suggested, and find something that will measure the texture of the road. Some will take the speed and some the texture separately. We have to combine these with the results obtained from a car or a trailer device and then produce a matrix which you can use. I think that it is far more complicated than is suggested when using one number.

Chairman

I think that it would be useful to have your comments on this, Mr Ervin.

R. D. Ervin Ann Arbor, USA

We have certainly concerned ourselves with this problem for some time. I guess as we see it the problem will remain unsolved and the technical community will continue to thrash it over until someone comes up with some data to show what is actually taking place. In all other scientific disciplines in which some kind of process has contributed a foundation people can identify, the community finds itself in a position to describe the process and then says 'We know what matters and what does not matter'. Clearly friction on the surface is of a first order in determining the answer. This can vary as a consequence of the condition that is provided for the test and one can get to grips with the classification of the conditions for a correct mode. So as we see it, the technical community has applied itself, not at all rigorously, to the endeavour to normalize braking performance with the tyre traction performance, and so on until we have standard surfaces and new skid tests and we get a classical solution to the problem.

P. Oppenheimer Birmingham

Despite what N. F. Barter has said, I would like to pin down the authors a little more about skid number. Skid number is specified in America now and may be in Europe, especially as regards wet surfaces. If you have a certain surface, say a wet Bridport surface, which is very well known to you, it may be that it measures, say, 30 on a certain day. Do you believe it is necessary to make this measurement every hour, every day or every week, to satisfy yourself that it is still skid number 30? That is the first question.

The second question is, what roughly is the difference between skid knock up, which is the sliding μ value of a certain rim tyre, and the μ value of a representative truck tyre, measured at the same speed and same load and fitted into the ADTM level? What would be the representative μ value of a truck tyre and also of a representative car tyre? Can you please answer any of those questions?

T. Williams Crowthorne

I think that you have picked on the surface which is more easily produced to get a standard number, as shown in the lecture. In other words, it is not very speed dependent; it is a very common surface and once one has measured the surface it is unlikely to change. That particular surface, having reached the standard values, is unlikely to alter very much, perhaps it will remain in that condition for a number of years. But other surfaces are going to change over the years. They are going to change so I do not think you can cope by laying a surface down when you know that a year later it will alter. You cannot prevent it weathering and you cannot prevent the effect of the traffic.

P. Oppenheimer Birmingham

Is it enough to measure every month or every six months?

T. Williams Crowthorne

If you hope to keep a skid number within a boundary, how is this to be done? It is very difficult to do. You cannot take a surface like a Bridport, which will remain constant over the years. Most other surfaces will change in this time, depending on traffic and weather conditions, in winter or in summer and depending whether you are running a lorry or a car. All those things will change the characteristics of that surface.

Chairman

There is an argument for specifying the surface, and such a surface has a variation of temperature, weathering and polishing which is minimum. But, having done that, you still have to take your control measurements at the time you are doing the test.

© IMechE 1977

P. Oppenheimer Birmingham

But that measurement will take longer than the actual test!

Chairman

It depends on how far we pass this ASTM requirement.

A. R. Williams Birmingham

I think one point you might perhaps think about is the load test with trailer. That is the locked wheel configuration, which is fairly expensive on test tyres which I do not think are actually given away. I am not quite sure that the load configuration is the right one, particularly in the absence of more information or pending legislation or relevant tyre skid resistance. It may be that a SCRIM machine is better than a road test or a peak friction as opposed to locked wheels.

P. Oppenheimer Birmingham

Could you answer the second part of my question?

R. D. Ervin Ann Arbor, USA

We have already done a little work to endeavour to discriminate the relationship between the ASTM reading and truck tyre performance. The only thing that is observed is that there appears to be no clearly defined relationship between the ASTM measure and the resulting truck tyre performance. We measured one surface and it was 75 and another was 81. We did not necessarily observe that our truck tyre surfaces would all be 81 or 75, which is a shattering observation in itself. I think that so long as that can be the case, it is not clear how we shall talk to one another about how one differentiates between 81 and 75.

C. G. Giles Staines

I have been prompted to come into this mainly by some of the interesting things that P. S. Fancher said this morning and I think that I would choose one or two remarks on slip relationships. T. Williams very kindly acknowledged the fact that I used to have something to do with this.

I think that this afternoon we have been blinded by science and I congratulate those who have produced such detailed papers, but if some people are a little confused I can understand this, having been in that area quite a number of times myself. Everything that I say will be about wet surfaces. Roads are wet for 25 per cent of the time in this country, so you get some idea of the relative importance of this. I want to draw attention to the μ relationship. If you have a slippery surface the braking force coefficient is maintained very much up to about 80 per cent or 90 per cent slip, then it drops suddenly. I think that this is a very important point for people trying to make anti-locking devices, that you can get a very high coefficient from the point of view of braking even with 100 per cent slip. There is nothing imaginative about 50 per cent slip, but on peak value the important thing is how much the tyre runs up. There is nothing magical about this.

When you come to combining braking economy, I think we can talk about sideways force. People talk as if the whole thing was simple, but again I do not think this is true. When we talk of the locked wheel case I am pretty certain that more work is required and I hope that this will be done. One of the things that has not come out clearly in the papers is how to deal with a car suffering from sideways slip. This SCRIM machine which is talked about, always refers to a smooth tyre. The prime reason for that, is that this is the only way of producing a solid tyre for different road surfaces. So that when we talk about sideways force coefficient, we are talking about smooth surfaces.

That brings me to the second point that I want to make, and goes back to paper C29/76 and some work done by Locke about 10 years ago. He measured braking force coefficients by locking the front wheels at high speeds. We have always been trying to ask how you can characterize road surfaces by a single number. Do you do it by peak value or by lock value? The difficulty is that all these numbers are different and it is hard to find one that correlates to accidents. What I want to talk about is using the same pattern tyres on a very good surface at 60 miles/h. I think that this is quite sensible. The heat value, which should go up, was 0·6. That was a very good surface; not many surfaces would have a value as high as that. Then he locks the wheels and gets about 0·1, and if you test on ice you get about 0·2. So on a road like that with very high adhesion, you lock the wheels and they are safe.

I believe that this is the kind of condition on motorways because people lock the wheels. Someone asked earlier on, how do you stop the vehicle when driving along the motorway at 70 miles/h. If the road conditions are satisfactory it is only when they try to brake that they get into difficulty. You can judge whether the road is dry or not by your steering. A chap gets the feeling that is is safe, and only when he does something, and finds that he has to lock his wheels is he in trouble. This at high speeds is something that we need to think about quite a lot and I hope that those trying to make antilock devices will remember this, because to me it seems to be at the root of a great many of our difficulties in wet conditions.

Chairman

C. G. Giles said that our roads are wet for 25 per cent of the time, but of course the average is probably very much higher.

E. G. A. Elgeskog Göteborg, Sweden

I am happy to note that data are emerging on truck tyres. They have been scarce in the past. In that connection I ask R. D. Ervin, as he has measured the sideways force on truck tyres, if he has also considered the function of longitudinal slip?

R. D. Ervin Ann Arbor, USA

No, the machine cannot provide an answer, because the machines do not incorporate longitudinal force. We are very much interested in this, and are looking into it.

B. D. A. Phillips Coventry

I would like to ask A. R. Williams and T. Williams a question because it seems to me that their two papers were in conflict at one stage. I may have got it wrong, but I think that A. R. Williams implied that if we wanted to get large ground playing force, tractive or lateral, one would do well to have a large contact area, which implied the lower the contact area the better. T. Williams, at the end of his talk, showing the model identification devices, seemed to imply that the indentations in the tractive force were far better for locking work.

The second question has to do with the tie-up that A. R. Williams gave between a tyre's stiffness and its peak potential tractive force, referred to on page 72. To my way of thinking, the cornering stiffness of the tyre has to do with its elastic stiffness, that is spring mainly, yet it is a peak factor and more to do with the friction of the road, and the two may not be related.

A short quick question is to do with routing the peak braking force point further along the second slip graph. Perhaps this could be done if you are clever enough to style your tyres to have a high lateral stiffness, to give you a higher pulling stiffness and a low torsional stiffness, if you have a low longitudinal stiffness.

T. Williams Crowthorne

I will try to answer the first question. The curves I showed were of the greatest area you could put on the tyre and as you saw in the slide, if you take the higher speeds, it is apt to be the question of macrotexture and microtexture.

A. R. Williams Birmingham

You mentioned the point about tyre-road ratio. In the test we carried out there was little difference in the ratio within a reasonable range, say from 60 to 80 or 90. The balance of contact area and drainage was the second question. We believe that you need to get the maximum rubber contact with the road, bearing in mind that under wet conditions you need to balance that requirement with the requirement of adequate drainage, either with the macro design or the wet surface. What we were trying to do as far as the macrotexture is concerned was to find where the balance exists between the tyre and the contact area, yet giving adequate drainage for up to normal traffic speeds i.e. 70 miles/h. The suggested geometry is one which in theory and from practical results indicates something at least to investigate further, but you need the microtexture which is the level of roughness to a minimum scale of 5 to 10 μm. If you have that, I think that on the greater range of roads there is is a rupture of the film that is quite adequate.

On the point about peak braking force, I can say yes I agree with you. On cornering stiffness and braking stiffness, in the test we were looking at the construction variables and under those tests we were looking at the controllability where stiffness characteristics did affect controllability and the power and cornering braking.

Papers C31/76, C32/76 and C33/76

J. W. Davis Coventry

Today, we are not talking so much about the problems of making anti-skid systems, which was our prime concern earlier on, but how to make anti-locking systems, as they are now called, which will be reliable in performance under widely differing operating conditions. I think perhaps it is opportune that today we have put together two papers, one which is concerned with the computer program approach to assessments of anti-locking systems on vehicles, and one which is concerned with the practical testing of anti-locking systems, according to the type of test which we are likely to have to meet arising from legislation.

My question is aimed at both authors. Is it possible to lay down a standard computer program for evaluating anti-locking systems, so that throughput the EEC, if not throughout the world, a system when fitted to a vehicle can be approved for something which is exactly the same, at least in principle, in any part of the area that we are considering? Also I would like to have W. J. P. Adams' view on whether in his opinion this would meet some of the problems, especially in regard to energy consumption, on which over the years we have been arguing and discussing in various bodies throughout Europe and this country as to the nature this test should take.

To perform a practical test, one needs inordinately long surfaces to be able to keep an anti-locking system going for 16 seconds, and this is expensive. Is it cheaper to get work into the computer to assess this sort of aspect? Also for the proving of work on low coefficient surfaces, is it possible to devise a program which says that in principle an anti-locking system is capable of working on surfaces of something less than 0·3 μ without actually having to take to that surface?

P. S. Fancher Ann Arbor, USA

On the computer prediction capability, my feeling is that a certain amount of vehicle testing is absolutely necessary at the moment. Computer prediction has the dangers of all predictions. There are so many factors that come into play to indicate a zone, that it is difficult to be sure that you do not overlook something in your understanding of what constitutes program testing for vehicles. You always have to do some testing to make sure that you are on safe ground in your predictions. I look to the future in the hope that we shall be able to make better predictions, but I would be very hesitant to lean too much on computer predictions for anti-lock, although it can be used in combination with vehicle testing.

W. J. P. Adams Birmingham

I am quite convinced that in the years to come computers will be used to a much larger extent to evaluate performance predictions for vehicles fitted with anti-skid systems, but, there are so many factors involved in the overall performance of the vehicle, suspension stiffness, tyres and things like that, that while I think a computer could be used for doing some of the simpler tests, the proof of the pudding is in the eating. I prefer to get on with the test with a driver with a big hard boot and a brake pedal.

H. Hodkinson Leamington Spa

In paper C31/76, why have you considered only one loading? Secondly, why have you considered braking distribution and criticized the possibility which may arise under 61 system to a certain extent, whereas in fact the cases supply an agument which could probably be satisfied by a practical system and meet the law?

E. C. Glasner Stuttgart, GFR

I did mention that there was not very much difference between the load on distribution. As I said, it is assumed that there are no extreme loading differences.

© IMechE 1977

H. Hodkinson Leamington Spa

Could I come back on that and say that is a most unusual case in my experience of 25 years with vehicles.

E. C. Glasner Stuttgart, GFR

The loading distribution was 46 per cent between the unladen vehicle and 50 per cent on the loaded vehicle, so there was very little difference.

H. Hodkinson Leamington Spa

Again, I would repeat, this is a most unusual case and, as I said earlier, possibly we should design vehicles in that manner rather than control distribution between the axles.

E. C. Glasner Stuttgart, GFR

I wish to point out that that is because all these vehicles are all on these ratios.

H. Hodkinson Leamington Spa

I accept that. May I ask if the calculations are adhered to in general, or are they to a certain extent directed at a set of circumstances which one vehicle is expected to achieve?

E. C. Glasner Stuttgart, GFR

As far as I am aware, my company's vehicles are still within those percentages.

H. Hodkinson Leamington Spa

I am sorry to go on in this way, but what I am saying is that in my experience there are more vehicles which have substantial changes in rear axle laden braking than the peculiar narrow band represented in E. C. Glasner's case. I read the paper as being a general criticism of kinked braking distribution, rather than a representation of how to design vehicles.

G. Burridge Bristol

I want to go back to the question of practical engineering. First of all, the practical aspect of anti-skid. We have to build vehicles which will run on roads. Quite honestly, I learned very quickly and very simply that if you apply the front brake, or the rear brake too hard on a motorcycle, you fall into the road. I did not need to work that out on a computer.

Changing the aspect a little, we have seen in America many of the practical aspects of braking systems and anti-skid imposed by legislation across the board on motor vehicles and today many things have been said about our experience here. But the experience itself was something of a disaster. We do not want to get ourselves into that sort of situation in Europe. In the past I do not think that we have let our legislation get us into that position with new developments, and I do not think we want to get into that situation in the future. I believe that the practical aspect of using anti-skid on vehicles is to learn each step very carefully and very slowly on the system before we start to use it too widely. It might be nice to have the ultimate sophisticated system to reduce the stopping distance, but if that system has a tendency to be unreliable, then I am afraid it will have a negative effect. So I ask W. P. Adams, does he agree that the best way to proceed with anti-skid for the time being in the U.K. and in Europe, is to proceed with anti-skid on one axle first, and to learn about anti-skid on single axles before we spread it over every brake on every vehicle?

W. J. P. Adams Birmingham

I agree that what you say is absolutely right. One has to look at the figures I quoted in terms of usage value. Remember, 180 times in 20 000 miles when this vehicle could not be steered and 230 times when a jack-knife situation could have occurred if anti-skid systems had not been fitted on the vehicle. If an anti-skid system is to be fitted on one axle, it should be fitted on the drive axle if the vehicle is an articulated vehicle. I agree that in certain cases the very sensible way of proceeding slowly and deliberately leads to an improvement in road safety, and by adopting that attitude I am absolutely convinced the improvement will be considerable.

At the same time, there are a lot of accidents caused by vehicles that cannot be steered. A lot of these motorway accidents in fog would never happen if the vehicles could be steered. Purely and simply the vehicles are travelling too fast in reduced visibility and on surfaces they cannot possibly stop on once the obstruction has been sighted. A lot of these accidents would be prevented if anti-skid was fitted to the front axle, as we have proved. I take your point and agree with you absolutely. By all means let us go ahead with anti-skid on the drive axle of the tractor, but do not let us put the question of steerability out of our minds, because to a large extent it is just as sure to put an anti-skid system on the front axle as on the rear. It might cost you a little bit more, but there is an improvement in safety. I am not suggesting that you leave the question there. It is a debatable point. I could debate all evening on whether you put the second set on the front or the rear of the vehicle.

R. A. Turner Slough

Was the driver aware that there was an anti-skid set fitted to the vehicle? Obviously, if the vehicle was fitted with handbrakes and he was not aware that there was this new anti-skid device on it, that could artificially put up the figures.

W. J. P. Adams Birmingham

Yes, the driver did know that an anti-skid system was fitted. We took that precaution before we released the vehicle for service, to make absolutely sure that the driver knew what the counters were for before he went out. Those are the figures that were recorded. This is not a normal haulage vehicle, but is a vehicle used by a manufacturing company in the automobile industry. Therefore it is travelling light for 50 per cent of its journey. It takes manufactured goods on the outward journey, and comes back with empty pallets, at the most. So probably the usage value is exaggerated for those reasons, but I do not believe that the usage was exaggerated because the driver was showing off to his friends.

© IMechE 1977

I. Slack Leyland

I would like to ask a question regarding partial failure of the system. In America it is, I think, standard practice to fail each axle. I should like to know the authors' views on this. If it happens, do the axles remain in the anti-lock control condition?

P. S. Fancher Ann Arbor, USA

I am not sure of the question. In America the drive axle ability is unfortunately such a consideration.

I. Slack Leyland

What I was getting at, is that you put the data into the drive axle and the front axle is incapable of acting. If you have a lock which is continued, you cannot use it.

P. S. Fancher Ann Arbor, USA

I agree. The only hope is that the driver can steer and do something about that situation.

W. J. P. Adams Birmingham

Undoubtedly there are cases where with one axle failing, problems arise with instability for the remaining axle in circuit, but are you not arguing that one has to get a failure? But the failures experienced in the United States when one gets rid of the motion are few and far between. Our experience has been that the system is quite reliable when one talks in terms of production components. In my opinion they started anti-skid off on the wrong step altogether. They were putting in devices just because they were afraid that the system would fail. The right thing to do is to get a system which does not fail.

F. E. Webber Reading

Fifty per cent of the vehicles fitted with anti-skid equipment in our organisation, are giving trouble, We have now changed the system so that the light goes out when the system is working. The drivers then know that they do not have a light when the equipment is working. The first warning a driver gets is when he goes into a corner. With 32 tons, he would like a bit of warning.

W. J. P. Adams Birmingham

I am indeed sorry to hear of your experience. I am sure that we are all fortunate in that you are here today to tell the tale. The system I am experienced with, does not work quite like that. This possibly is a basis for a standard system to check out a failure warning, which could be applied throughout Europe. The system of which I have experience, automatically indicates to the driver when he switches on that the system is functioning correctly, by means of a light which comes on on the dashboard. This signals that the system is working. Having done that satisfactorily, the light on the dashboard goes out. While the vehicle is moving, whether the driver brakes or not, the system is continually checking itself out and, irrespective of whether he puts his foot on the brake pedal or not, the light will show or not show. It returns to perfectly standard braking reaction. The driver then has a perfectly standard vehicle without anti-skid.

W. Manning Basildon

I should like to ask P. S. Fancher, will all US manufacturers have anti-skid fitted?

P. S. Fancher Ann Arbor, USA

It is my experience that with relaxed stopping distances a number of manufacturers are going to remove the front axle brakes. Customers do not like brakes on the front axle of a vehicle.

Papers C34/76, C35/76 and C76/76

A speaker

I ask D. Hatch and B. J. Chapman if the benefits of their discovery are equally available on friction materials, or is the response limited to linings of particular composition and manufacture?

B. J. Chapman Chapel-en-le-Frith

Different friction materials do respond differently. If you plot friction or wear against a percentage of titanium, what you find is that at very low levels you get high friction wear; as the proportion increases the friction and wear drop. With some materials there is more trouble than with others, and some drift across. Some materials drift down about 0·01 per cent and reach a low point of about 0·03 and others drift down to 0·07. The shape of the friction and wear varies between titanium and other friction materials. It also varies to some extent with the service and duty under which friction materials operate.

M. S. F. Croissant London

I feel that B. J. Chapman and D. Hatch have produced the final piece to the puzzle of braking problems that we have experienced in the braking of fire appliances. They have pointed out the desire to achieve consistency in braking performance and the need to reduce complaints of pulling. This is particularly true for fire appliances.

We first experienced these problems in 1969 following reports from drivers of erratic braking. Investigation showed that this was due to pulling in stubbed or emergency stops, the disturbing feature being that a pull to one side was likely to be followed by a pull to the other. Since that time we have been testing brakes in order to identify the problem and to solve them by introducing changes in the braking system and steering design. The problem remained. To rectify a brake defect of this type it was sufficient to skim the drums and to profile new linings to match, to achieve an acceptable balance. It was only 9 000 miles before the defect reappeared.

Can I now assume that titanium in the cast iron is at the root of the problem? Secondly, may I ask the brake manufacturers or lining manufacturers, and drum founders whether the collaboration called for is forthcoming?

B. J. Chapman Chapel-en-le-Frith

First I must make it clear that all blame for braking faults are not due to titanium. There are cases where drums and discs are the cause of the fault, but there are cases where drums or discs are out of balance and this is the cause of the problem. If a vehicle fails the cause is not bound to be with the discs or the drums, but if it is due to titanium or niobium in the rotors, then no matter what you do it will

© IMechE 1977

eventually come back again. From the story that you told about getting the drums reground and other parts reground to fit precisely, which achieved a cure only to have the trouble return, makes it appear to be a classic case. We should not forget that there are plenty of other reasons for this kind of trouble.

On the second question, we have visited a number of motor manufacturers and we hope that as a result of the papers published at this conference, eventually we shall eliminate this problem. The trouble was that in the old days there was titanium in everything, and now they are going more and more to steel scrap and using electric foundries etc. Titanium has crept in as a persistent impurity and as multi-national companies have expanded their markets so the problem has escalated.

N. R. Gold Leyland

I would like to ask whether it would be necessary to move away from the titanium actuaters we normally get in brake systems, to more distant actuators to maintain the working stroke for the units.

O. Ehricke Hannover, GFR

I understand the question, but I am afraid that I cannot give you an answer. This is a question referring to the hardware which we have not crept into yet that far.

G. Burridge Bristol

I can easily say that I have not worked on hybrid systems, so I cannot answer it either. I think the paper showed that the actuation equipment would remain pretty well unchanged in the present proposals for high pressure air. The high pressure sytem is for storage only and then reduces to normal levels for operating the brakes, so it seems that the actuating system can still be used.

W. V. C. Batstone London

The systems G. Burridge described in his paper are not strangers to us on the operating side, but he mentioned that air test points were something new which had been incorporated in the systems. In 1968 we included in our vehicle specification the critical points for air pressure valves and it is gratifying to find that manufacturers are now catching up with the operators requirements.

On test procedures Mr Burridge showed a vehicle tied up with string. I cannot imagine such a vehicle going to a test station and being allowed to go through with a piece of string on it, but what it does highlight is the difficulty, apart from reliability and durability of valves, of testing with a load sensing valve. It would go some way to get new efficiencies. Obviously anti-lock systems are a much better bet. Perhaps Mr Burridge can give us some indication of his views on comparative efficiencies of load sensing devices and anti-lock systems.

G. Burridge Bristol

With regard to the test points, I was trying to illustrate, though perhaps I did not do so clearly enough, that test points are available and are widely used. The two examples I showed were of Swedish chassis. In Sweden test points have been widely used and encouraged, and are required by legislation. Moves are now on foot to ensure by legislation that test points are fitted in this country. Quite a number of manufacturers have already fitted test points and it is gratifying to know that quite a number of the bigger manufacturers fit test points.

The picture of a vehicle tied up with string was perhaps a little facetious but this does happen and F. Whalley's paper, to be presented in Session X, shows that 70 per cent of vehicles present for testing had valves that were disconnected. Presumably they are fixed in some way. Perhaps the string is a little more sophisticated than the type I use!

The point about air pressure valves is that they are relatively straightforward, simple devices. They are mechanical and pneumatic and they function if they are maintained properly. Anti-skid, as we discussed yesterday, is a more complex subject. It involves more equipment and components and much greater potential for malfunction. As illustrated, the American experience has shown that it is not a mechanical pneumatic problem. We have hi-fi interference and so on to contend with.

One would like to be able to say that anti-skid will solve lots of braking and maintenance problems, but I believe that we have a lot to learn yet, legislators, steel manufacturers, operators, everybody, on this subject. We are making progress in the right direction but we have to progress rather slowly and learn where the weaknesses are and cure them before we launch out and fit anti-skid to everything. Even when it is more widely used, it will need checking and maintenance. Anti-skid will require failure indication legislation. It still requires one to check the failure indication system at least daily and to have self-checking, as was described yesterday.

P. B. Murphy London

I am from one of the maligned ministries. I should like to give information about the position regarding testing. At last week's meeting in Geneva, it was agreed by the governments that more test points should be fitted. This would obviously include one on either side of the air pressure system.

G. Burridge Bristol

Good.

Papers C36/76 and C37/76

Mr Potter Birmingham

I would like to address a question to J. A. Odier and his team. I was extremely interested in your vehicle brake dynamometer and I can see that it must have a very wide application and be very useful to a development engineer. One thing I could not see clearly from your film, which I think is extremely important if you are to carry out brake testing, is how to achieve simulated brake cooling. I wonder if you could explain how you carry out simulated brake cooling?

J. A. Odier Paris

I agree with you that on this film there is no brake cooling, because it was made two or three years ago when there was no cooling, because it was no problem. But for two or three years now we have had cooling at the front of the car which delivers air by a servo at the same speed as the rollers, that is to say, at the same speed as on the road, and covers

© IMechE 1977

the front of the car. The experiment showed that it would give a good result at 5 or 10 per cent. If you compare the front axle and at the rear on the road, the latter is not well cooled, but there is no great difference either in a straight line at different speeds, or if we configurate a car going down a mountain, for example, in 20 minutes with a range of 40 or 50 brake applications with a temperature which increases by 300°C or 400°C at 50 miles/h.

R. S. Sharp Leeds

I would like to express doubts about J. A. Odier's rig. It reproduces accurately certain inertia facts, particularly to do with lateral movements of the vehicle and with regard to deceleration and acceleration of the vehicle, but I should like to know if the theoretical treatment of the inertia has been published and, if so, he could give the reference.

J. A. Odier Paris

I am not sure whether I understand the question. Were you asking 'is the inertia on the road the same as on the bench?' It is, because the inertia on the bench is adapted for each car, and for a given speed it is the same kinematic inertia on the road.

R. S. Sharp Leeds

I appreciate that longitudinal inertia is represented accurately, but when one is considering lateral motion there are certain terms which appear in the lateral acceleration, for instance, when the car has a forward motion, which, it seems to me, are not represented on the rig where the forward motion of the vehicle is not present.

J. A. Odier Paris

We can release a pedal for the engine brake and we can measure the holding resistance of the tyre. We have some losses in the machine because there is some friction in our machinery, but as far as we know them we compensate with an electric motor. The car is permitted lateral motion and other degrees of freedom which should be the same as in a car on the road. Does that answer your questions?

R. S. Sharp Leeds

Yes. Could you refer me to the theory? Has it been published?

Chairman

The question is, is there a theoretical set of equations of the treatment of this subject which R. S. Sharp can read?

J. A. Odier Paris

Yes. You have the bibliography at the end of the paper. If it does not satisfy you, you can write to me and I will try and answer.

F. D. Hales Loughborough

May be I can shed a little more light on this particular problem. Most people here will be aware that tomorrow there is to be a paper on these inertia effects and questions about the computer to which R. S. Sharp has referred occur equally on that. We should go into this in more detail to try to identify areas of deficiencies in the system. I find myself agreeing with R. S. Sharp that there are inaccuracies in this technique. You cannot represent the lateral motion exactly. You can approximate to it very closely, and from an engineering point of view I think one can get to the stage where the differences are negligible.

P. Oppenheimer Birmingham

I was interested to read that J. A. Odier has tested against the American regulation SS105 and your stopping distances on the bench correlate very well with those on the road test results, although they are not actually shown in your paper. How do you know that your rollers represent a skid mark of 75 or any one number? Also, I want to know if you could produce other skid numbers such as skid number 30, for example, by various codings.

J. A. Odier Paris

We have no skid number on the bench, but on the road you have no number until you make a test. On the bench you have a great advantage over the road, in that you can get reproducible answers. On the bench you can measure on each wheel, the ratio between the vertical force and the longitudinal force, so we have no use for the skid number. On each application of the brakes we can measure the efficiency of the braking. This can be seen in the paper.

The second question is about the coefficient of friction of 0·3. On the bench it is difficult, but we can change immediately for the difference between the dry and the wet, but it is not very reproducible. As we can measure the vertical force and the longitudinal force, it is not necessary to change the adherence to see whether the device is good or not.

Chairman

I think that the question is not whether you can measure the ASTM, but rather whether you are using the ASTM, not can you measure it, but what do you use to measure it?

P. Oppenheimer Birmingham

I think that J. A. Odier has explained how he measures it, but if the skid number is 30 and the coefficient for adhesion is 0·9, I suggest that makes it a lot easier to meet the stopping distances requirement.

J. A. Odier Paris

On the bench we can change the adherence without changing the position, but we cannot do that on the road. We can meet 0·3 with water. We have to put a film of water on, but that makes the machine very dirty. We can do so if necessary for the laboratory, not for car or aeroplane test braking. We have the methods, and there is good reproducibility. That is not necessary for the present task. The problem is difficult, but we hope to solve it in the future by continued work on the question.

© IMechE 1977

H. Hodkinson Leamington Spa

May I ask a question of W. R. Newton? In noting the development of disc brakes, one of the earlier objections to four-wheel disc brakes was the complexity and intricacy of systems at that time. Reviewing his assessment of present progress towards providing calipers which are more readily adaptable to front brake operation, how does he see future trends in relation to four-wheel brakes, bearing in mind that in the last 15 years we have seen a large number of companies adopt four-wheel disc brakes and then move away from them?

W. R. Newton Birmingham

I think you are quite right. There were several cars which went to disc brakes seven or eight years ago and subsequently they have come off them, and gone back to drum brakes. The difficulty is that disc brakes are doing only a third and drum brakes two-thirds, and we can get good performance with the linings we have. I think that the cost is the real factor in why we do not see a greater trend towards rear disc brakes. We as a company do not find that we can make a rear disc hand brake for anything near the price we can make a drum hand brake. Unless there is a breakthrough, I cannot see the trend altering at all.

Papers C71/76 and C79/76

Chairman

I already have some questions for this session from B. J. Chapman.

H. L. Schmidt Frankfurt, GFR

I have some corrections for B. J. Chapman. Owing to a translation error, the paper refers to 'cast iron'; this should be 'cast steel'. Secondly, 'organic linings' were 'cast steel discs' for some reason we do not know.

C. H. Curtis London

My questions are directed to I. Slack in the main. Referring to the short sentence 'Public service vehicles are also included', I think one or two comments may be of interest. I think that he has been very frank in his paper, but I should like to assure all my friends here that not all our buses suffer from all the troubles he has mentioned. Possibly he is quite right in saying that there has been a great deal of rationalisation in the various foundation braking systems. Perhaps I may quote from a famous statement: 'You have been too busy rationalizing and not had enough time to do some designing'.

Changes have taken place in vehicle design and in foundation braking systems particularly and this is a solution to some of our problems. We are trying to use the same type of braking. Vehicles have got heavier and we all seem to have forgotten the lightweight vehicle produced around the 1959 era, of which we have over 2 700 still running. The B15 is a step in the right direction, but I still think there is more to be done in design.

Dealing with the special requirements, it may be of interest to I. Slack to note that on different types of service we have found different conditons existing, particularly with brake linings. Where we co-operate with Greenline services we use an identical type of vehicle to the Routemaster, yet we are unable to run that with the same lining combination simply because of crazing at low temperature, etc., but I do not think that we have gone far enough with brake linings on foundation brakes and so on to meet the bill.

The question of dispersing the heat from the brakes is a problem. We changed from 16½" brake gear on the old RCs to 15½" brake gear and we found greater pull between the drum and the wheel, so most of our tyre problems have produced a change in our lining requirements. Automatic brake adjusters are essential to run a service because vehicles do not come back then over a four week period, the return time then being governed by topping-up of the batteries. Also on moulded brake linings we found a problem inasmuch as we get heat crazing and so on. A multi-combination seems to give a very good set-up. The cold friction, being given by the woven material, it fades very quickly with the heat, while the moulding makes the heat rise. By putting the woven material on the trailing shoe and the moulding on the leading shoe, we get a fairly small drop in braking efficiency; we do not get that cold efficiency which then fades right away.

I would question the order in which you put the troubles. I agree that safety is the first priority; it must be. I do not agree that environment is the second, I would put it third. I put the commercial aspect as the second because we are in business to make a profit. I think that we are being led astray by environmental conditions sometimes. I can see it coming now with the tendency to make vehicles smaller instead of taking steps to get people onto the buses.

I think that the solution to the whole problem is for the industry, that is the operators, to get together to decide the bus they want to run, and which they intend to run, and to get it endorsed, and to revert to some sensible vehicle so that our braking problems can stand still or recede over the present period. This would give us a breathing space to do some development to answer the question, will this brake succeed on buses or not?

I. Slack Leyland

Thank you C. H. Curtis. I think that you have said many of the things I would like to have said had I had a little more time. You have also given us the benefit of your intense operational experience.

Taking the first point about application of liners to specific functions, I think you made the point that on your type of operation, where you may wish to switch vehicles from one type of service to another without doing a reline as part of that change, you are looking for a more all embracing situation and you cannot pick your liners for that operation.

Taking the point about drum diameters, and going down from 16½" to 15½" and getting an improvement, I agree with you. My view of PSVs is that they should stay on 15½". On the heavy truck it is worth thinking about 16" drums as a possible way of reducing brake pressure, and therefore reducing the actual drum temperature. This in itself would probably give you an effective reduction in TAB temperatures as well, although I do not think that is part of your problem. Possibly the best thing to do is to get a compromise between the two.

Combination liners present another interesting question. There is quite a lot of debate on the philosophy as to which type of liner should fit each shoe. It seems that this is a solution which is not understood in an analytical way. Nevertheless, it is well proven and I think empirical results applying to a particular vehicle are those you have to stick

© IMechE 1977

to, but it is very difficult to generalize over the range.

The question of commercial pressures versus the environment. Yes, I would agree very much with your remarks. The order in which I took them was not primarily to indicate priority, although it does imply that to some extent.

On vehicle weights, it would solve all our problems if we could get them down a little. Another thing I did not have time to say, is that one way of reducing the level of loading on the brakes is by no other means than hanging an extra axle on, but that would not suit your operation. Reducing weights, or getting more braking area from somewhere would get us over many of our problems. Thank you for your comments.

Chairman

I have a request for a contribution to the discussion to be made by A. Spielfiedel, who would like to say a few words.

A. Spielfiedel Gennevilliers, France

I am acting in Europe on behalf of AC Delco France. Thank you Mr Chairman for allowing me to intervene in this discussion. To widen the information available to this audience, and because others have described systems for control of trucks with internal wheel sensors, I should like to say that there is a system of external wheel sensors also and I would like to say a few words about it.

There is a pick-up which is essentially similar to those that you have already seen, but it has a bigger output and can tolerate bigger areas. There is a drum which has slots and teeth around it and the wheel speed sensor is to be fixed on a little bracket somewhere on the axle itself. From this position it will mesh with the teeth on that drum. This is a very simple system to rotate the sensor. The drum takes up a little slotting on the drum edge to get this configuration. We are using that type of sensor, which is an axle by axle system including electronic control based on a well established technique which has been used extensively in the United States.

If we prefer this approach of an external sensor it is because we like to have two sensors from the axle. Other systems have a different arrangement, but we like this one because it allows the posssibility of having a full analysis of the wheel behaviour. As producers of the system, we believe that it is reliable and that experience in the United States will eventually prove that this is right. We emphasize that it is a very easy installation and makes it possible to do checking without the necessity of removing the machine. Of course we like this approach and believe that it has a future. Thank you.

A. Gurley London

May I make a plea to I. Slack concerning the need to provide plenty of air around brakes? When we retreated to 22½" rims, we knew that it would be only a few years before the chassis designers followed us. I was glad to note that I. Slack has a preference for 15½" drums and I want you to prevail on your masters not to go to 16" or even worse.

C. H. Curtis has dealt with particular aspects of PSVs and their problems and requirements. I want to reinforce and complement what he has said by mentioning one or two details. We have had sticking brakes since 1953, when particularly Leyland began to build two air brakes on a sliding scale. Although we have suggested rollers in front of the shoe and these came out in about September of last year, it took a long time before they were fitted. AEC, in particular, have made them standard practice for a number of years.

Could I. Slack make some comments on the apparent incapacity of designers ro control the distribution of braking? We have the most embarrassing situation that we are changing brake linings as often at one end as we are at the other. We even have situations where the passengers will not walk close to the hot front of the vehicle. It is a remarkable state of affairs.

I agree with C. H. Curtis in his contribution, but I detected in something that I Slack said, a little ingredient that I was not too happy about. I rather gathered that he was retreating from an air-brake system to a hand actuated one. I am sure that there is still life in the old dog of the air-brake system. I think that air-braking systems have some life ahead of them. We have an enormous amount of equipment and know-how invested in the air-brake systems and we should continue to make the most of the improvements that manufacturers have made at a great deal of expense.

Another aspect which is causing me concern is the lack of a serious offer from the manufacturers of brake fluid, to give assistance in reconditioning that fluid. Frankly, the cost alarms me with the large volumes that our vehicles carry.

Finally, a question to H. L. Schmidt. Has he considered operating both hand and foot brakes by using the mechanism of the handbrake?

I Slack Leyland

I do not think that you need have any fear about a move to larger drums on PSVs. To take your point about sticking brakes, yes, it is a long time. We have done it now and I think that we are moving quickly along the right lines with the roller failure brake, and there are many other things in the pipeline to deal with aspects of sticking brakes. So it is progressing.

Brake distribution; yes this is the old problem of differential wear, particularly between the front and rear axles, and the embarrassment of changing linings at one end while not bothering to do it at the other. It is not primarily a question of the capacity of designers to get round this problem. I think that it is a combination of a number of things. Differential wear or bad control of braking distribution arises from a number of different areas. The first I would put forward is that rating ideally is one way to keep wear in the right kind of relationship, but particularly in the case of the PSV, rating is not enough in simple terms of brake pressure because you have vast differences in cooling in different parts of a PSV due to the body, and even from one side to the other where you have an exhaust pipe passing in front of the brake, for instance. Nevertheless, I think that a better arrangement of brake rating in the future will produce some improvement there and the cooling aspect will be considered in relation to the kind of brake pressures which are aimed at for that sort of installation.

One of the most important aspects, one that I think is almost as important as that of friction materials, is the air stream. A lot of the difficulty in maintaining the correct brake distribution comes down to low pressure, and many of the brake applications on the PSV are only 10 to 15 psi pressure valves. Conventional air control valves are still not

© IMechE 1977

particularly brilliant at maintaining low differentials under these conditions. That is one aspect in which I think power hydraulics can give you some advantage because the pressures in the valves are high in relation to the friction forces involved. So potentially, I think power hydraulics can give a better balance even for pressures associated with the PSVs.

Beside the brake valve, the other thing about low pressure brake applications is what we call the drag point process, the minimum pressure at which you bring the linings into contact with the drum. I think we are becoming increasingly sensitive to equalizing these pressures on vehicles by the kind of force we apply externally to the brakes and by the type of foundation braking we match to the matrices.

All these points do significantly affect the differential wear on the liners between one axle and another. We are dealing with all these points and really it is time that the problem was solved.

Chairman

I think time is the problem. Have you any more points to make?

I. Slack Leyland

The other point was on the question of stability, which I think is quite important, but I am not qualified to answer that in detail.

H. L. Schmidt Frankfurt, GFR

It seems that the development of disc brakes for heavy commercial vehicles will go into two stages as has been done in the United States; first we shall have the hydraulically operated disc brakes, because as we know there are fewer problems associated with them than with the mechanically operated brakes.

W. V. C. Batstone London

To some extent I think that the last two speakers have been unusual in that they have suffered all the problems we have heard about. I should like to cast a pebble into the pool to see where the ripples go. During this Conference we have had a lot of very good technical papers on subjects ranging from actual situations, to hypothetical situations in computer programs indicating to me that there is extensive research going on into brake failures. But on what? Finally on brakes. I am concerned that little emphasis is placed on standards which I consider to be an absolute minimum. I would be seeking a much greater improvement in total standards to obtain greater reliability, durability and better performance so that I do not have to worry too much about legislation. This is where I find I am in a dilemma.

At this time in my organization we are actively engaged in specifying parameters for design and improvement of power units, transmission and suspension; but what of brakes? Do I still use the foundation design for stopping the vehicle introduced 40 years ago? I believe a case can be made for disc brakes. Certainly I would like the incorporate them in my vehicles by 1980. I think more research should be engaged in by the manufacturers to this end. I ask H. L. Schmidt if he considers that disc brakes will be available for me to employ in my braking system by that time, and will the cost be acceptable?

H. L. Schmidt Frankfurt, GFR

As I said in my paper, it seems that disc brakes will be definitely more expensive than drum brakes. This is accepted on the Continent. To what extent prices will increase, at present I cannot tell you.

I. Slack Leyland

I was stating a general attitude towards disc brakes rather than policies regarding disc brakes. They will be viewed with some scepticism mainly due to the difficulties of handling, but as soon as those difficulties are resolved and we are able to get disc brakes into the available space – or to make more space available to accommodate them – I think that they will be treated with more enthusiasm. Disc brakes are being treated as something which will be looked at very seriously in future.

The other point on which I should like to comment is your remarks regarding legislation and the suggestion that the manufacturers would build to the legislation as a maximum and a minimum requirement. I think I speak for all the manufacturers when I say that this is not the case. In designing the vehicle, particularly as regards the braking system, the manufacturer designs to a particular usage, a particular operation. Some of the comments made here, have indicated, perhaps, that he has not always managed that, but nevertheless that is his target. Maybe it is only a minimum standard, the vehicle manufacturer's standard.

W. V. C. Batstone London

May I come back on that? Manufacturers do not allow sufficient tolerance between legislation requirements and initial requirements.

I. Slack Leyland

All I can say is that I hope that we shall do better in the future.

W. V. C. Batstone London

Very good. I am glad to hear it.

W. Manning Basildon

May I ask H. L. Schmidt to refer to p217 on Operational Views and Requirements for Maintenance? If my interpretation is correct, the drum is delivering far less energy. To compete with the drums would not a caliper have to exert something like 80 kg?

H. L. Schmidt Frankfurt, GFR

So far as I understand your question, you are concerned about fading given on that diagram?

W. Manning Basildon

I am concerned about the comparisons that are being made. If the performance required of the vehicle, is actual braking performance, the caliper is not capable of pulling that vehicle up with the same performance. If you design a caliper brake to give the torque requirement of the vehicle manufacturer, I would consider the caliper itself a critical safety factor.

© IMechE 1977

H. L. Schmidt Frankfurt, GFR

What is given here is the energy input for both types of brake, also the fading behaviour.

W. Manning Basildon

If the drum brake had started at 550 tolerance, would it have faded so much?

H. L. Schmidt Frankfurt, GFR

Actually it was taken from two different tests and it should show an increased energy input and the fading is less.

W. Manning Basildon

It is not a fair comparison to say that it would not affect the fading.

H. L. Schmidt Frankfurt, GFR

There is a different torque here.

Papers C38/76 and C39/76

F. D. Hales Loughborough

I think I must start by saying that both authors are to be congratulated for their excellent papers. I personally feel that brake squeal is one of the most difficult aspects of braking, certainly the most difficult I have dealt with. Regarding M. R. North's paper on the disc brake question, I feel that it is a very elegant piece of analysis and I find it quite convincing. I would like to ask him, from this theory is there hope of being able to eliminate brake squeal, and if so, will a solution be found in the disc, the pad or the choker? This disc brake squeal is possibly a fascinating phenomenon, but one wants to get rid of it. This we accept and I think we have gone a fair way to understanding it and I would like to go on.

Coming to the drum brake squeal, the feature of many problems that occur in service, is that they do not happen all the time. I wonder if N. Milner would indicate how his theory covers the sort of situation where you have a braking operation with squeal on the side. I wonder whether he would like to indicate how his theory would be able to cope with using a more realistic representation of the brake drum in which the simple symmetry of the shell is lost?

M. R. North Nuneaton

I will start by answering as far as I can the questions specifically on disc brake squeal. The answer is that at this stage I think we have a model which broadly represents most of the problems of disc brake squeal. This is as far as we have got, and of course we would like to carry on and extend this model, particularly to the conventional caliper and twin piston caliper, where perhaps the major part of the problem lies.

As for avoiding the problem, this of course is more difficult. It is a practical engineering approach and perhaps in many respects it is more difficult than the aircraft approach to flutter whereby they make sure that the operating conditions of the aircraft are such that flutter is avoided. On a disc brake this is more difficult. You have a large number of disc modes. All the disc modes ally when they are coupled up to the pallets in a close area in the squeal frequency. How one avoids this by design, particularly within costing strengths, I do not know, but certainly there is one system at the moment which is built with the caliper and discs to make sure that they behave linearly. If they did this you could almost certainly design them so that they did not move into the region where squeal can occur. As I tried to indicate very briefly, the pad is a horribly non-linear component, particularly when it has been heated up to a hot condition and then cools off. This means that you have a very large range of pad surfaces and all sorts of potential squeal regions. So I am not convinced that the answer will lie in simple design because of the costing restraints which must inevitably occur.

In dealing with the last part of your question, whether the problem will eventually be found to be solved, I do not know. I suspect that it lies very close to the disc and pad relationship. Perhaps the caliper is a little further away from where the problem originates. However, we all know that calipers do have some effect. At this stage all one can say is that there are steps towards a solution, but where the solution is I do not know. That, probably, is the answer you expected.

N. Milner Chapel-en-le-Frith.

I fully agree with F. D. Hales that squeal can occur in one application and you can do another application under apparently identical conditions and you do not get squeal. I put this down to variations in the contact area and I have tried to show in the paper that if you get contact over the central portion of the lining the squeal propensity is much less than if you had heel and toe contact, provided that there is a certain amount of pressure to bring about fixed central line contact. I tried to account for the variations of speed propensity with contact area.

As regards the second part of the question, I think the theory could be applied to another system, provided it was a two-dimensional problem. In the model I used only linear and tangential components and if there were any axial movements I could not cope with them.

R. P. Jarvis Leamington Spa

N. Milner has most successfully used his drum brake theory to bring the mathematics of drum brake squeal up to the level of our knowledge of disc brake squeal. As one grows older and cynical, one ceases to be surprised that the requirements to avoid squeal are conflicting, for the shoe more mass but less stiffness and for the drum less mass but more stiffness. Lighter drums may conflict with thermal requirements. Everything that N. Milner says is in accordance with general experience. Perhaps an increase in the drum material internal damping offers the best hope of eliminating the main squeal, which his colleagues D. Hatch and B. J. Chapman, who spoke earlier today, may attend to.

Turning to M. R. North's paper, I think this is a very useful comparative study of the various theories. I would not have thought the measurement of the two modes of disc vibration was all that difficult, but it would be as well to be aware of the classic paper by Zenneck on the preferred modes of imperfect discs. I found in my own work that, if the frequency of the friction material support happens exactly to match the disc frequencies, peculiar effects are possible with the appearance of the modal pattern rotating with the disc and an amplitude variation with heat frequency.

I would like M. R. North to clarify what he means by harmonics, because the frequencies of higher modes of the

© IMechE 1977

elastic systems involved in brakes are not integral multiples of the fundamental as they are of strings and pipes. Random, discontinuous contact will excite the higher modes of the disc, but not give rise to 'harmonics' as such.

I agree with his assertion that the problem of a selection of a mathematical model is now not one of incorrect mathematics, but I would like to extend the point. The value of mathematical models does not lie in our ability to analyse them and come up with intriguing formulae called stability criteria. It is much more important that they should also give the designer an insight into what is happening so that he can take the creative step to eliminate the noise. The danger of a complex mathematical model is that it can become a strait-jacket to your thinking because once your computer is programmed, it is easier to vary the numerical values of your parameters rather than introduce a ninth or tenth degree of freedom. Many palliatives do take the form of an additional component inserted somewhere, adding usually both flexibility and hysteresis. I use the eight degrees of freedom model as a basis for analysing how known palliatives work, and why they sometimes prove ineffective would be a most useful way of continuing the work.

N. Milner Chapel-en-le-Frith

It is possible to increase the weight of the shoe without altering the stiffness.

As regards the second point, with my model, I agree that it is probably only approximate, I did some calculations where I increased the natural damping by about 30-fold and this did not bring about stability, although I agree that if increased further than that you are bound to get stability.

M. R. North Nuneaton

The first point was a question of harmonics. Yes, I think that I did use the term rather loosely. I was thinking of my own measurements of squeal where one certainly got a fundamental, and additional frequencies as well. I used the term harmonics just now and I think that I should have said additional frequencies.

I also agree that these sort of computer models could be used as an aid to understanding. It does not appear in the paper, but one could use them as an aid rather than an end in themselves. One can, of course, perhaps take mathematical modelling a bit further on the frequencies. There are techniques available which one should consider for this sort of work. I have in mind the use of finite elements where one can involve an engineering model, certainly a caliper. Those techniques, although they are time-consuming, might be very useful indeed when it comes to detailed design with a view to eliminating squeal.

M. A. I. Jacobson Basingstoke

I would like to ask the authors a very practical question. We have noticed in a number of repeated tests that two factors seem to come into operation, One is the change of humidity. Also, in the practical world, we find that the people who use brakes more frequently and drag their brakes seem to suffer this problem more than those who brake violently. This showed up in a cross-section of cars between eight weeks and five years old.

M. R. North Nuneaton

Perhaps I can make one comment. It is that squeal propensity tends to be fairly directly related to the friction levels of the lining material, certainly in disc brakes and changes in operating ability could change the friction levels. For example, early morning sharpness is a fairly well known effect, which is reckoned to be due to increasing friction which occurs on an overnight stop, perhaps due to early morning dew. This kind of brake usage may very well affect the brake friction. Maybe the lining manufacturers might like to comment on this.

N. Milner Chapel-en-le-Frith

I fully endorse what M. R. North said. I think humidity can affect friction levels and I think the figures in my paper show that they can rise very rapidly. As regards the drag braking problem, possibly it could be that after the brakes have cooled down you get this heel and toe contact which is part of the configuration of squeal.

R. S. Sharp Leeds

May I ask M. R. North to say something about the representation of inertia of the discs in his model?

M. R. North Nuneaton

That is really a nasty question. I have just been working on this. The method I used was to take the mass and stiffness that in combination would give me the free disc frequency, also the inertia and the rotational stiffness which would give me the same frequency in the disc mode, and also make sure that the mechanical impedence of this was the same as the mechanical impedance of a disc. Loosely, that was how I selected the pressure values of mass, inertia and the various frequencies needed. Another thing I used was the timing required to give the same DK rate of a disc.

Papers C40/76 and C41/76

I. D. Neilson Crowthorne

We have been wondering recently whether there is any difference between those drivers in normal driving situations and those in emergencies leading to accidents; do they react in the same way? A study has been made trying to find how drivers react to test situations. This has been carried out at Cranfield in conjunction with ourselves at TRRL. One of the measures has been the drivers' condition, their heart rate, which in ordinary driving is normally on a steady basis, but which in a stress situation becomes much faster. In any other test of subjects in driving situations it seems to us that there are at least three possibilities of situation for normal driving and for test subjects. They may be worried or stressed with a higher than normal heart rate; they may be calm and not flustered; or thirdly they may not be bothered in that they do not really react to the emergency at all.

I think that accidents occur suddenly and drivers who are usually calm and unflustered before the event have not had time to get worried or flustered until it is over. In experiments drivers may be in any of the states that I have described according to how they have been set up before they got going. In relation to the paper by D. McLean, I wonder in what category he would think the drivers were when performing the braking manoevres he represented and whether they were skilled or unskilled drivers? In other

© IMechE 1977

words, does he think they were worried about the test, or were they calm and not worried, or really not bothered about it at all and not really trying to act properly in a test situation?

R. T. Spurr Chapel-en-le-Frith

The drivers we used were mainly skilled test car drivers used to test driving. We also make measurements of people in the laboratory as a whole and other people, and we found that the results were qualitatively similar. The skilled test drivers gave consistent results. The unskilled and some of the less capable drivers gave the same pattern of results, but they were more scattered. In regard to the question of what happens in accidents, we have not done experimental work on that because we have only a limited supply of test car drivers.

H. Smales Cleckheaton

I should like to address my comments to D. McLean and R. T. Spurr. I notice that it is possible for a driver, a skilled driver, to close his eyes for up to three seconds. I hope this will not be published in the Press; that you can get yourself in an accident and close your eyes for two seconds.

I also notice that it is possible for a driver, to drive a vehicle reasonably well although the mechanical design departs from the ideal. Neverthless, there is a need to exploit the car's capabilities to the full and minimize the limitations. I wonder if the authors could indicate how they intend to use this work to achieve that objective.

Secondly I was interested to note the correlation obtained by using skilled drivers who were asked to judge deceleration. There appeared to be a very close correlation, but when they were asked to judge speed there did not seem to be such close correlation, although it is difficult to work out the actual correlation because of the difficulty of finding either the deceleration or the speeds used. Possibly at a speed of 30 miles/h there would be correlation errors of 16 or 17 per cent, whereas in deceleration much less error would be seen. Possibly you can comment on that.

R. T. Spurr Chapel-en-le-Frith

To answer the last two questions first, the deceleration range we used was between about 0·08 g, which for most people is the deceleration, and 0·8 g. Similarly, the speed recovered was about 10 miles/h, there were quite big errors, but at 10 miles/h and above it was surprising how accurately the driver could judge his speed. How he did it we do not quite know, but presumably he gathered together all sorts of clues, vibration, wind noise, how he pressed on the pedal and things like that, and the drivers were remarkably accurate.

M. A. I. Jacobson Basingstoke

I should like to ask R. G. Mortimer a particular question. When one compares American motor cars with European ones, there is one factor to remember above all. That is brake discipline and the fact that, by and large people manage to drive by observing the other driver from the front rather than constantly focussing attention on the rear via the mirror, so they observe other traffic lanes. There is also a tendency for people to apply maximum braking which is very much reduced because of the lay-out of the roads system. I wonder if R. G. Mortimer could give us his experience of driving here, where people drive more aggressively with high acceleration and deceleration in relation to other vehicles, and say if his findings might be moderated.

Secondly, I am not so surprised that people do not know the difference between the front and rear wheel locking up. I think this is a problem on which experience is very limited. I should like to challenge the boundary of 10 000 miles. I think that most people here have had their heart in their mouth more often than once every 10 000 miles if they are honest with themselves, but we do not always have accidents on these occasions. The remarkable thing is that people feel that they have reached the limit in braking at about 50 per cent of the absolute they can achieve, particularly when braking when going into a bend. I wonder whether the other authors carried out any tests in a straight line manoevre, or an operational straight line manoevre, and also at a bend. Our investigations, carried out some little time ago, showed that the judgement of speed and deceleration potential which a vehicle can stand is greatly affected by whether they are straight line tests on, say, an airfield or under road conditions, with a little camber or going into a bend. Even at MIRA we find that most of us have quite different levels of perception.

Thirdly, we should not forget that with the best system which we can invent and introduce into a motor car, it is still a matter of judgement by the individual which causes most of the accidents and near accidents. When one considers accident statistics, I think one should include the near accident, the narrow miss, which gives you a chance to survive and learn the lesson for next time. At 10 miles/h or 15 miles/h you will survive, but at 50 miles/h you may not.

I would like to ask one further question, whether the systems which have been evolved allow for the adaptability of the driver to handle the vehicle with virtually no braking system effective at all. We get used to it because of the driving situation, but can they go from a vehicle with a good braking system to a bad one, with an automatic change in response?

R. G. Mortimer Champaign, USA

First of all I would like to tackle your comment about the frequency with which emergency braking takes place. The data in the study we did in the States was obtained by instruments in a motor vehicle which we put into the motor pool at the university, and measured the deceleration which the drivers used in about 5 000 miles of driving and the number of brake applications. In fact we recorded the pressures for each of these decelerations and we were able to pick up the maximum deceleration they used for any kind of speed above 10 miles/h. About once in every 10 000 times the driver actually decelerated at more than about 0·5 g. I think this agrees with results from other similar tests. That might seem rare to you as an individual driver, but I once sat down and wondered how many times in America per day do we have 0·5 g stops and I think the number was something like 3 million per day. So it gets to be rather a different story when we start to look at accidents where there is severe braking. I still maintain that a driver does not get much practice in severe braking deceleration.

I think your second question referred to whether or not setting a standard on something like deceleration force gained is a reasonable thing to do, because, after all, drivers

© IMechE 1977

are valuable human beings and there is a limit to which we should go in designing systems that they have to control. I think that it is important to try to design the vehicle to make it as compatible as possible with the driver's skills, especially if it costs no more than it is costing today. I think that the kind of recommendations that we are making would probably not cost anything much more. In fact, as I mentioned, many cars do fall within those boundaries.

I am not sure whether you had another point to which I am supposed to respond.

M. A. I. Jacobson Basingstoke

The only other point was whether you noticed a difference in driving behaviour and driving discipline over here compared with that in the US because I think there is a need for more discipline in driving a large vehicle, in this country at any rate.

R. G. Mortimer Champaign, USA

There is obviously a different behaviour here compared with the US, but to a large extent I think this is due to the relative systems, and to the different densities in trafic. You also have different speed limits. Currently we have a 55 miles/h speed limit all over the country, whereas you are still driving at 70 miles/h on the motorways. Since I have been coming here for a number of years, I think that the lane discipline is much better than it was, based on my own observations, but I think that you have just as much need for good probing systems as we have in the States. I think that you are wrong in saying that your cars accelerate better. I am sure that they do not. They may stop better.

R. T. Spurr Chapel-en-le-Frith

Measurements were made on cars on the road and on the test track. The test track was the one shown on film this morning. A lot of measurements were made on the road when the drivers were aware of what was going on and they related very much indeed to measurements on the test track. In regard to deceleration in performance, we did not make any measurements, but we had some arrangement whereby we switched the servo in or out and found that the driver very quickly compensated for the change in conditions.

R. D. Lister Slough

I wonder if R. G. Mortimer is aware of the Sex Discrimination Act in this country. I have observed some discrepancy in his comment on women drivers. I think that possibly the main difference between men and women as far as braking is concerned is stature and body weight. You said that body weight did not give very good correlation to the brake pedal force. I agree with that and think it is right. On the other hand, you said females give only about 55 per cent of the male force. That does not seem quite to tie up. I wonder whether the ergonomic factor of the pedal and the seat, and the seat and the seat back, is perhaps a bigger factor. Perhaps it was not suitable for the female and for that reason apparently gave a lower figure. Many years ago we did some tests on both male and female drivers and found only a slight difference. We put that down to the tolerance action of the leg being just right for the female. This looks to me contradictory in the presentation of 55 per cent.

If I may also refer to the paper by D. McLean, R. T. Spurr and T. P. Newcomb, I am a little unhappy about this because the human operator is a very adaptable animal and can adjust to circumstances over a very wide range without any effort. The authors have illustrated four drivers with different braking responses, but if you had twenty drivers you would get twenty different responses. Perhaps given other conditions you could get a whole range of them. I wonder what is the best way of choosing these particular numbers. There seems to be a need for some explanation of why you have these equalization characteristics in equation 8 which are quite different from those in equation 3, which refers to quality, and throws doubt on the work if you justify your own numbers. Could we have a little explanation of that, please? I would be pleased to know how the simulation was used and what it is used for.

R. G. Mortimer Champaign, USA

I did not mean to devalue females, in any way. We had about 600 drivers, about half male and half female, whom we sat upon a specially designed wooden buck. This has a seat and a pedal which could be adjusted to any position. We adjusted the position of the pedal exactly for each individual based on their limb dimensions so that each was able to exert comparable maximum pedal force. The thing was positioned so that each person would be in the same relative position and have the same opportunity for exerting the force and we found that on average females exerted 55 per cent, and the relative force in the aircraft test was that females had about 55 per cent of the force exerted by men. So obviously sex is a very important factor and that is why we were looking at 50 per cent females and setting a pedal force which we thought reasonable in a braking standard.

C. B. N. Symonds Newbury

In regard to adaptability, we are in complete agreement that a driver can adapt to very bad circumstances, but we think that because he can adapt there is no reason to stretch him. There are still optimum conditions under which he can operate because, if he is driving not stretched, he can devote more attention to driving and other people on the road. If he has to cope with something which is rather off the optimum, a lot of his attention is already occupied.

On the question of adaptability, I think the generality of results is very interesting because we examined one driver over a range of speeds and found that all the curves for deceleration time traces fell on one line. When we examined the test drivers we found all their plots fell on this one line, showing remarkable generality in behaviour. That dimensional plot and that generality is the reason for much of this work.

D. McLean Loughborough

In regard to adaptability for operators, I think it must be clear that although a human being adapts to a situation, he does not in himself change his performance very greatly. The observed reactions change, but the actual general characteristics do not alter substance. The mathematical description given in this paper is the most general, but all those parameters of time constants, although changing, change only in very small ways. The principal change is in the gain of the operator. If the time is short, and the distance short, the gain will fall off rapidly, but adrenalin is pumped

© IMechE 1977

into the system so you get a faster response than you would otherwise obtain.

The fact is that there was a difference between the parameters in equation 3 and equation 5, I thought was explained in the paper. I tried to indicate in the talk that one must not expect results which are obtained for highly trained pilots in highly different, continuous situations, to obtain in a driving situation when doing different braking tests. Therefore you could not reasonably expect the conclusions which come up at the end of the day to be the same as in operations in a different situation. That is why there was the difference.

F. J. Jaksch Göteborg, Sweden

My first question is to R. G. Mortimer. In your paper you discussed only one of the brake control characteristics. pedal force versus heat collection, but there are more characteristics which are important. We have, for instance, pedal movement versus heat collection. We have found that this characteristic is of importance. Very high gains are acceptable, but there must be a lower limit for the deceleration if we have an improvement in the pedal movement in this direction. What is your feeling about that?

My second question is to D. McLean and R. T. Spurr. In your paper you have measurements of the driver's behaviour and you talk about stopping distances at different speeds. One can describe the pedal force deceleration and the speed. How do normal drivers react when they change the speed? The possibility is to hold the pedal force constant and moderate the distance or to have the distance constant and moderate the pedal force. Or you can have a mix of both pedal force and distance.

R. G. Mortimer Champaign, USA

In our study we did not evaluate the pedal displacement gain as such, but however, we did look at two levels of pedal displacement, one a maximum displacement of about 0·6 g, or it was a zero displacement. We ran tests with each of those configurations for each of the pedal force gain levels. I did not refer to the findings respecting the effect of pedal displacement. There were some, but they were probably minor. We did tend to have more occasions with brakes locking on the different surfaces, but they were not that more frequent than the preferred levels of displacement pedal. I could have described them to you, but I did not have time to do that. The drivers had to enter a lane and brake a vehicle to a stop when a signal was given, and they were not quite sure when the signal simulating stopping distance, lock-ups etc. would occur. The difference in the effect of the pedal displacement was really quite small compared with the effect of pedal force.

I am interested in what you found. I think what you are saying is that if you have a lot of displacement you do not need much force. It was partly objective, but this was an objective study with severe braking situations. This was not a subjective valuation of drivers and clearly their performance was compatible and good on all three surfaces when the brake system was operating within the pedal force gain which I outlined.

D. McLean Loughborough

The pedal force was taken as a linear function of the equation and that notion was in turn a piecewise linear function in the braking force, but we were not able to incorporate that in the paper. Recent results seem to indicate that we get higher levels of deceleration and much shorter times in severe braking, but these results have not yet been assessed quantitatively.

A. J. Wilson Birmingham

R. G. Mortimer appears to have assessed the limiting loads that a person can exert statically over all the dynamics in a dynamic situation. I think that if a person were put in one of these vehicles and directed to stop he would support himself with his foot on the pedal and it would be a more comfortable situation than the static pressing of a pedal. I think a person with one leg is likely to be able to apply foot force of 1·4 times his own weight on the brake.

R. G. Mortimer Champaign, USA

There have been studies done dynamically as well. Generally these studies were done in such a way that the drivers were told to drive round a track and enter a certain lane which occasionally would be sealed off, so they had to brake fairly hard. Then on one last circuit, the servo was failed and they still had to brake. In my interpretation of the results, I believe that there is general agreement between the Barker slip mock-up, such as we use, and a vehicle in which you can adjust the seat to give a comfortable position.

M. A. I. Jacobson Basingstoke

Maybe I can contribute some practical test results we have carried out. We have carried out tests at MIRA and other places and find, much to our discomfort, that our female drivers, cannot perform so well. They are most annoyed about it, my wife among them. Basically the problem is that we expect people to exert more pressure than they should legitimately be asked to, on some modern braking systems. I think we could all do with limiting the brake force to no more than 90 lb for dry weather road conditions.

T. P. Newcomb Loughborough

I should like to add one or two further comments to points raised on the paper. First, there was the question about adaptability of drivers. One must agree with R. T. Spurr, as we have found that drivers can brake quite well with the left or the right foot, and feel quite comfortable. The point made by R. D. Lister was that the human being is so adaptable that he can cope with almost any situation, but is he comfortable when doing this? The purpose of our exercise, is to check all the external parameters to see what we can do in regard to making them more comfortable at the wheel.

In answer to what M. Jacobson said about going into a bend, I should say that in our case our drivers had read the Highway Code and knew that they should complete the braking operation before they went into the bend.

R. T. Spurr Chapel-en-le-Frith

There are some comments I could make about the pedal element, because we carried out experiments on pedal pressure with a couple of hundred people. We found that all

© IMechE 1977

cars in this country are very much the same and the results of these measurements fit in very well with the results in the paper, not only qualitatively but also quantitatively. In general the pedal effort was two or three times the body weight for men, but there was very little correlation between pedal effort and body weight for women. We also found that age was not of great importance. Another interesting measurement is that in an accident some years ago, it was found that the driver braked very hard and actually bent the pedal. It was found that it had required a force of 450 lb to do this.

P. D. Livingstone Basildon

I wonder if R. G. Mortimer could say how this exercise applies to the truck situation as against that of the car, as there is some difference between the driver's relationship with the commercial vehicle and the car.

R. G. Mortimer Champaign, USA

Yes, I intended to preface my talk by a reference to commercial vehicles. Obviously we need to have similar data about truck operators. So far as I know there has not been a study done yet. It is very important that data should be available on all classes of drivers, servicemen or even schoolboys. But even if we had a sample of males and females of all ages, we could not use these data for truck drivers, because the cab and seating position in a truck is entirely different. It would be very nice to have a study of that sort.

D. McLean Loughborough

I can only reiterate what R. G. Mortimer has said. We too hope to start simulating on this very shortly.

H. Hodkinson Leamington Spa

I should like to ask R. G. Mortimer, whether in considering his recommended values for pedal effort against retardation, he has taken into account, in his calculations, the load condition of the vehicle concerned. Additionally, has any study been made by any of the authors on brake systems in which the pedal effort is of a non-linear characteristic of retardation, in so far as controllability of the brake system is concerned, particularly in the low retardation levels and also in regard to the overall speed factor in driving?

R. G. Mortimer Champaign, USA

Frankly, I am not clear what you mean in your first comment about gross vehicle weight being a consideration.

H. Hodkinson Leamington Spa

Obviously, with a limited number of passengers in the car, the gross vehicle weight and the retardation weight are usually different, from what you said. In legislation a vehicle in this condition has to be taken into account, but from the point of view of the great percentage of mileage and the incidents arising out of general driving, vehicles are not in that state. I am talking about passenger cars, where the pedal weight is related to gross vehicle weight condition, which is usually substantially different from when the driver is by himself.

R. G. Mortimer Champaign, USA

That probably is a reasonable way to run the testing of a vehicle. In principle, the pedal force relationship is clearly going to change, but from the ergonomic human factor standpoint the envelope does not change. The driver is concerned with how many are in the car at the time and it does not matter what the vehicle it is.

Your second point was on the question on non-linear relationships. As far as I know, they have not been accepted, as actual vehicles were used presumably with drum brakes fitted as opposed to discs. In our case the relationship was non-linear; we used servos, but I could not tell you whether or not the linear gains would be the same.

Papers C42/76 and C43/76

G. Jones Leyland

The subject of my contribution is loss of control as described in I. D. Neilson's paper. Our experience with cars which, in the lightly laden condition, have a high proportion of the weight on the front wheels has been, that to avoid a possible spin condition at high deceleration values and to ensure a reasonable adhesion on wet road surfaces a 'kinked' brake force distribution has advantages. We do, however, agree that such brake force proportioning devices are not ideal methods of preventing the locking of the rear axle, an approach to the ideal in terms of adhesion utilization provides conditions such that with normal variations there may be a higher proportion of unstable vehicles than may otherwise be the case. The compromise between stable, but possible dangerous front wheel lock, and unstable rear wheel lock conditions has been a difficult problem for manufacturers.

It is of interest to study the yaw motion of an unstable vehicle in relation to the equations of motion. Fig. D1 shows the yaw velocity/time history of a car spinning on a dry road surface.

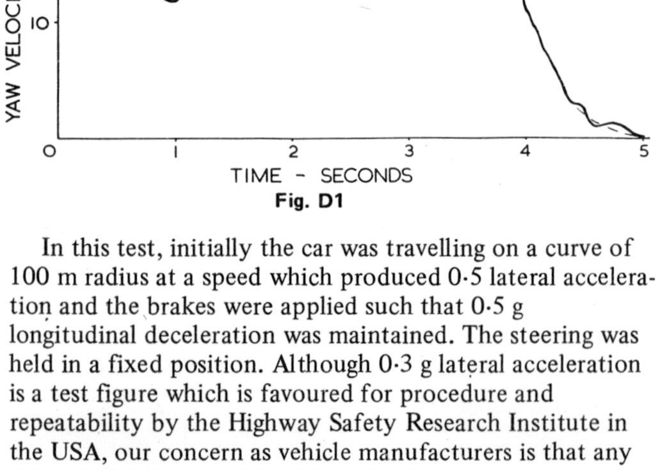

Fig. D1

In this test, initially the car was travelling on a curve of 100 m radius at a speed which produced 0·5 lateral acceleration and the brakes were applied such that 0·5 g longitudinal deceleration was maintained. The steering was held in a fixed position. Although 0·3 g lateral acceleration is a test figure which is favoured for procedure and repeatability by the Highway Safety Research Institute in the USA, our concern as vehicle manufacturers is that any future legislation on this subject should be appropriate to real safety requirements and be satisfactory in terms of repeatability. The present example is to illustrate the behaviour of a small car in the more extreme conditions.

Fig. D2 shows yaw angle against time as determined from an intergration of the velocity/time curve. It has been suggested that about 20 degrees is the maximum angle from which recovery is possible and this was reached in this case one second from the onset of brake application. This may in fact be beyond the limit for many drivers and so one can appreciate the marginal time available for human perception and response.

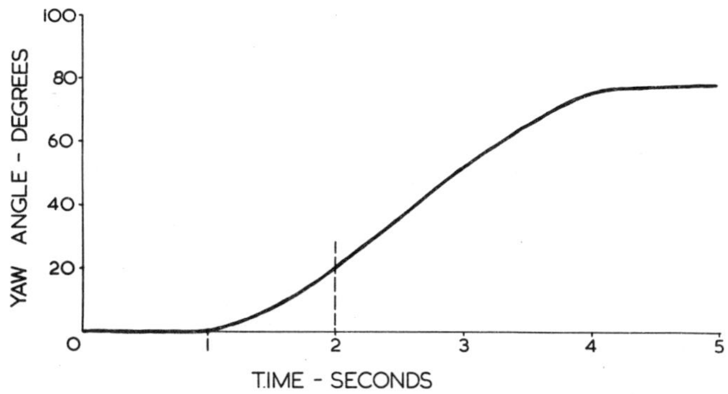

Fig. D2

From a step by step analysis of the yaw response in relation to the equations of motion for an idealized model proposed by Koiter and Pacejka (1), the reduction in rear side force necessary to produce the practical results was only a small proportion (around 10 per cent) of that associated with a total slide condition. Evidently this was due to the fact that only the rear wheel on the inside of the curve was locking to set off the unstable motion.

Fig. D3 shows recorded lateral acceleration at the centre of gravity in comparison with results determined from the model and calculated from the actual yaw response. Fig. D4 shows the yaw response from a similar car fitted with transverse apportioning of braking in addition to the normal fore and aft system.

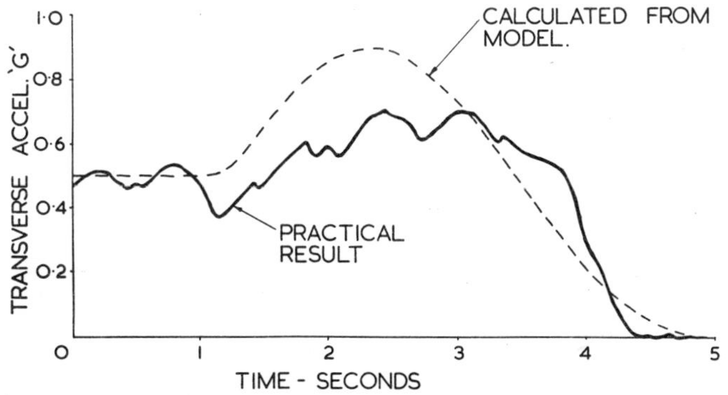

Fig. D3

It also shows the best result obtained on the dry road surface, braking from 0·5 g initial lateral acceleration and maintaining 0·6 g longitudinal deceleration without deviation from the desired path. Transverse apportioning was provided at both front and rear wheels in this case, the primary souce of benefit from the point of view of avoiding spin is of course the apportioning at the rear wheels. There are possibilities that developments on the above mentioned lines could produce a useful overall benefit in cases where this is necessary.

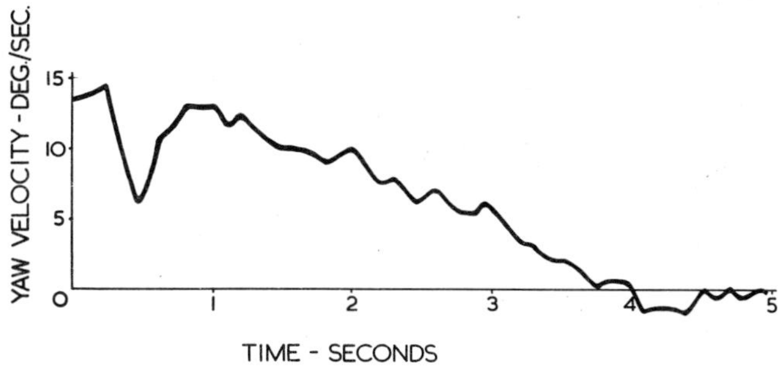

Fig. D4

REFERENCE

(1) KOITER and PACEJKA. 'Handling of vehicles under emergency conditions'. *Proc. Instn. Mech. Engrs.*, 1968–69, 183, (34).

B. R. Shilton Lincoln

I should like to ask I. D. Neilson a question and to make a comment on his paper. He deduces from the fact that approximately equal numbers of accidents were preceded by drift as by spin, that braking was perhaps not far off optimum. It may be suggested, I think, that a drift accident may well be caused by rear axle locking. So the deduction is not necessarily valid.

I should like to ask I. D. Neilson whether he has any information on the experience of people involved in accidents, either spin accidents or drift accidents, and the spin element in other accidents. Also on this question I should like his opinion on whether four-wheel anti-lock, which would result presumably in considerable reduction of spin and drift conditions, would reduce the number of spin accidents, although presumably it would do nothing towards reducing the drift.

I. D. Neilson Crowthorne

I think I certainly agree with your comment about drifting possibly being due to all wheels locking, but of course spinning being due to just the rear wheels locking. I am not absolutely certain which group of accidents was the worst in terms of injury. As far as I recollect there was not much in it, but I would have to check that up.

With regard to front wheel as well as rear wheel anti-lock, I gained the strong impression that about half the benefit from anti-lock would occur with rear wheel anti-lock only. The other half would occur from anti-lock of all wheels, but I should say beyond that, that this was for the sample of accidents I was able to look at. I was certainly left with the feeling after the study that possibly there were a number of other cases where braking had actually occurred, and we had not discovered it, and many of them could possibly have benefited from front wheel anti-lock because these were cars which had run quite a long distance when being braked before they met the obstruction. It was not easy to say they were braked with loss of control, but probably they were in the sense that if that there had been control the driver might have avoided the accident. So I think my feeling is that rear wheel anti-lock covers about half the cases, although it is probably less than half in reality.

© IMechE 1977

F. E. Webber Reading

May I comment on C. Wastell's paper? Your second study included non-drivers and you found much the same, little difference in the capacity of the left and the right leg. The availability of subjects was few. I think this referred to those who had never driven automatic cars. Some drivers may never have used that vehicle at all.

C. Wastell London

You are absolutely correct, of course. This is the obvious thing. We must compare clutch cars with automatic cars, but in this country it would still be quite difficult, I think, to find a group of people who had only ever driven automatic cars. It is this 'only ever' business which makes it difficult. You know that they have not had a licence, but unfortunately they have driven a brother's car and one does not include those people.

B. D. A. Phillips Coventry

May I ask I. D. Neilson a question? Does he think that if drivers when they are learning to drive had a compulsory session learning how to get out of emergency braking conditions, the number of accidents caused by lack of control would be greatly improved?

I. D. Neilson Crowthorne

One or two things are fairly obvious. One is that loss of control on braking occurs to people less experienced rather than fully experienced in driving, to a greater extent. Loss of control, like most accidents, does not seem particularly to occur when people are learning, but it is in the stage after they have learnt that accidents tend to occur. It might be a bit of help, I think, to have training in skidding, but, as I mentioned earlier in the Conference, I am not certain that skidding on very slippery surfaces would be greatly beneficial, because a lot of these accidents are on dry roads and nearly all the rest are on ordinary wet roads which are not particularly slippery. So the actual loss of control that is occurring, whether it is drifting or spinning, is mainly just on ordinary wet or dry roads and the training would not be on those conditions. I think training on really slippery surfaces is perhaps almost misleading.

C. Wastell London

I should like to ask I. D. Neilson a question. This relates to the fact that car accidents are an act of volition; to crash one's car one must do something. The number of accidents that are caused by driving error seem to be very much in excess of those caused by mechanical failure. Do you think, therefore, there would be profit in defining a group as you have already done with those over 25 years. One could be more exact about this defining a group in the population who are more prone to accidents, and secondly defining the psychological motives which result in an act of volition and lead to the crashing of a car?

I. D. Neilson Crowthorne

I do not know that I am in the best position to answer this question, but there are other members of the team, who are probably more familiar with a wide selection of data rather than just data on braking loss of control, who might answer. Perhaps C. A. Hobbs would like to make a comment?

C. A. Hobbs Crowthorne

My comment is that this is quite an involved field you are asking us to go into and it is not just related to brakes. If you are talking about driving errors, it is very involved and, as I will mention in discussing my paper later, there are many and varied factors concerned.

R. D. Lister Slough

I think one point which has not been brought out by I. D. Neilson, and which is worth bringing out, is the time limit involved. Anyone who has been working on this problem realises that the time element that is involved is crucial. In the case of front wheel locking or drifting, it is a relatively gradual process and the time is fairly long. In the case of rear wheel locking and rear wheel spinning, the time contracts dramatically. Once you have exceeded the steering angle of the front wheel beyond a certain degree you cannot return to the control situation, even if you were able to change it quickly enough. So the time element and the rate at which yaw is built up is really the problem.

I. D. Neilson Crowthorne

There is just one small point to add to that. When a driver locks his front wheels and has an accident it is quite often at a bend and he has passed the point of no return earlier. You can divide drivers approaching a bend rather fast, into two groups and this is related to experience. Those who are less experienced find that they have approached the bend too fast, do not do the right thing, end up by locking the front wheels and come off. There is nothing they can do about it; it had the same effect as rear wheel locking, but it is for a slightly different reason.

More experienced drivers, I think, experience this to an even greater extent because they have more ability and they are nearer to the actual speed at which they can or cannot get round the bend. Generally speaking, they get into trouble only when they are right up to the limit. In both cases there is, I think, a feeling of being incapable of doing much about it, or not doing the right thing.

W. Manning Basildon

Has there been any experience at all of measuring drivers' reactions with a car skidding? There was a suggestion that drivers could think faster.

I. D. Neilson Crowthorne

I do not think I have any useful evidence at present. As far as I am aware, drivers can do as well as anti-lock brakes, if they are experienced. Some work was done in Sweden which showed in relation to anti-locking that there were three groups of drivers. Drivers were put in a situation of driving on a test course and having to avoid an obstacle which suddenly appeared in front of them. There were drivers who did not react at all, which is quite common on the road. In other words, they just went straight into the barrier, which was of course flimsy and did not do them any harm. They were in the minority, if I remember they were under 20 per cent. At the other end of the scale there were drivers who could do as well as anti-lock whether or not they had anti-lock, it did not do anything for them. Those of the larger middle group, which was over half the total, benefited from anti-lock in the sense that they could get round the obstacle on anti-lock and they could not without it. It was not a very large study, but it does indicate the situation.

Papers C44/76 and C78/76

C. H. Curtis London

I cannot help but make just one comment on R. Denniss's paper: Thank goodness we do not operate commercial vehicles!

To return to the public service vehicle, much of what P. H. Wyke Smith has said I believe to be true in general, but I think we have quite a lot of answers to some of his problems at least. In 1959, some of you may have heard me say this, we introduced a somewhat revolutionary vehicle, the Routemaster, to the London fleet. This was based on some years' development work before it appeared as a production unit. We operated something like twenty five vehicles with various systems on, separately and together and I think we cured the majority of our braking problems, but the big change we made was the adoption of hydraulic power. By that I mean continuous flow of hydraulics. We were much taken by the development of this in the air context during the war and we took steps to try to fit it on to buses. It appeared that there was quicker build-up time, there was no bleeding, no corrosion, it was light in weight and did not require periodic maintenance and so on. Overall we saved about 1 cwt in weight of the vehicle. This meant that we could have either a lighter bus or we could carry one more passenger.

This went on, as you know, until that disastrous day in 1968 when everyone decided that we had reached the end, but I should like to say that the Routemaster as it stands today can easily meet the EEC Regulations without any real cost. So far as London is concerned, speeds have remained much the same, bar weight ratios, but what I find disturbing is the ridiculous increase in weight of public service vehicles. It amounts to something like 25 per cent, yet we are carrying no more passengers for that. I have quoted the 72-seat Bedford Routemaster which weighs less than 8 tons, but we have a Daimler Carrier with 72 passengers which comes out at 9½ tons.

It is no wonder that operators are complaining. I think this increase in weight has been one of the biggest problems we have to face. I look forward to the day when we shall return to vehicles with unladen weight of 8 tons at the most.

Regarding braking force on the air-brake system, yes we had this. We had to start braking the standard double-deck bus after the single-deck craze receded, and our problem was that they started off quite well but in time they deteriorated until we found that we were braking all on one axle. Let us face it, our brake pressure is not the maximum; it is about 25 per cent g in breathing time. I lay a little of the blame for this problem on the air-brake manufacturers who I think have not done their homework properly. We spent 10 years on split braking, unfortunately, and air-braking split just came in like that. I am glad to say that most of the air-brake people have now done their homework properly, but it involved us in making 200 changes in our Daimler two ton buses to get over the trouble.

I should just like to add my criteria for braking expectations. Foundation brakes must have automatic brake adjusters. There must be acceptable wear where wear checks are necessary. On the operating system brake pedal travel must be responsive and progressive and ideally should give a non-skid performance at all times. It should not be necessary to strip out braking systems under 160 000 miles. The dual system must maintain line pressure within 5 per cent of each other.

One last thing. P. H. Wyke Smith is operating some of our Routemasters. Just to show our good faith, would you agree to the purchase back at any time?

P. H. Wyke Smith London

I am open to offers on the Routemasters. There is a point on hydraulics, to answer three of C. H. Curtis's questions. The effectiveness of the hydraulic brakes is a matter of great interest to the operator. If you have a small batch of hydraulics in a big fleet, of course you have to be very careful where you place them because the delay time is negligible and to measure the brakes in the fleet is very considerable. On hydraulics and air the bulk of the industry is on air pressure because of the other usages of air in the vehicle, gear shift, door operation, wipers and many other facets.

I made the point in the paper that total consumption of air on the PSV is another factor that has been given a great deal of consideration by manufacturers and this is something we have to change. On the question of weight I believe C. H. Curtis has one particular vehicle appearing now in line and there are a number of EEC Directives and a number of DoE design considerations which are being built into these vehicles whether we like it or not.

R. Denniss Burton-on-Trent

I should like to make one comment. The solution is there, there are better brakes available, or will be available at a cost. I do not know whether it has been mentioned or not, but the Dunlop Aviation Department are looking very seriously at providing cheaper version of the Concorde multi-disc carbon fibre brake. There we have a brake which is compact and light which could well be maintenance free. So there is a 500 000 miles brake which is maintenence free. I said at the beginning in my paper that you have to keep your feet on the ground and look at the cost. The aircraft industry is prepared to pay £300 for one pound weight reduction. I ask C. H. Curtis what he is prepared to pay for a pound weight reduction? According to my calculations we could possibly afford to pay £2 for one pound weight reduction. The solution is there; it is just a matter of someone having development money and the customer being prepared to put his money where his mouth is.

G. Burridge Bristol

There are one or two comments I shall try to make very quickly, although this subject could occupy a whole day. It is very interesting to know that I shall have a job for the next 20 years. R. Denniss is setting us some very high targets for the future with the ultimate lightweight brake system which lasts for ever. I can assure him that we are determined to go this way, but although we also have some high targets, that is a long way off.

Air leakage has been mentioned. This is becoming a more important subject as time goes by. Air brake systems have

© IMechE 1977

always leaked, but they tolerate leaks unlike oil operated systems which will not tolerate leaks, and require much higher standards in assembly and maintenance.

The subject of variable load valve reliability was mentioned. The point should perhaps be made that the variable load valves should themselves be reliable. The problem we have, and I made that point in my paper, is that they require to be separated properly and need to be maintained when the vehicle is in service. This is brought about by the change in the dimension of the axle and the chassis, brought about by variations in the suspension system and in the suspension itself where we have failure, I think, rather than in the valve.

In his paper R. Denniss made a plea that we should increase our mileage based replacement service periods. I think I take this plea to mean increased reliability for our equipment. He illustrates very clearly in his paper, in paragraphs 5.6 and 5.7, that certain vehicles and certain types of service can do 100 000 miles without relining the brakes while, in paragraph 5.7 other types of vehicle have a brake life of only 100 000 miles. This illustrates very clearly that you cannot just relate the economy of brake to mileage, or maintenance period to mileage; the duty has to be taken into consideration.

I was interested in Mr Wyke Smith's introduction when he was preaching something like the same sermon that I preached yesterday, that we should slow down some of the proliferation if we can. Unfortunately, we do not have much choice about proliferation of valves and so on, but we can see that they are simple, reliable and maintainable. Is he aware that there are available air dryers and large compressers and individual valves in which the brake ratio can be set or varied?

R. Denniss Burton-on-Trent

The system will tolerate air leaks, but drivers do record these sort of defects. We are very pleased that they do because an air leak will not get any better on its own, or it is unlikely to. As I said in my paper, this is a time-consuming, frustrating pastime and in most cases not very serious.

P. H. Wyke Smith London

Westinghouse would be prepared to accept just two more valves, one on each axle in buses, just to prove the load sense. The 35 valves I mentioned on one single deck chassis would be left in a relatively unchanged state for, say, two years.

W. Manning Basildon

The incidence of failure in air brakes systems shown in R. Denniss's paper match the incidence levels of failures reported to F.M.Co via warranty and policy returns and reliability studies of fleet operators. However, in addition to air leaks and real failures, we have high incidence of failure of the brake low pressure warning buzzer switches. Air leaks are caused by incorrect fitting of the air pipe couplings and it is interesting to note that line diagrams of typical brake systems illustrated in Conference papers have around 300 threaded couplings to mate. About 25 per cent of these couplings require the use of 'blind end' female threads and these blind threads are nests for swarf and dirt particles which can become dislodged in service and contaminate the system.

My question to R. Denniss is, has he knowledge of the type of seal failure he is recording? Are they genuine wear problems or are they failures caused by swarf and dirt particles?

R. Denniss Burton-on-Trent

I am sorry, I just do not know. One has tried to analyse the reason for the seal failures. Not a lot of work has been done on that because it is not a major cost area. It is a frustrating, time-consuming area. From the foreign matter we have seen, it appears to be carbon matter rather than metal. That is all I can say.

R. D. Ervin Ann Arbor, USA

May I put another question to P. H. Wyke Smith? The double-deck bus is a curiosity for an American. My interest is in the safety experience you have with double-deck buses. There are two things here which interest me. One is the acceleration of brake wear risk. Is that to do with road bias and stability?

Do you find the safety experience on double-deck buses as high as in the braking losses suffered on other vehicles on average?

P. H. Wyke Smith London

Taking the first point first, we do not counteract what you call rear spin outs on double-deck buses. They are relatively slow moving and operate in high density areas where the opportunity for the driver is not so marked as it would be in other areas to get into the situation described. In the situation referred to, one can put a cantilever behind the rear axle, where there is concentration of heat from the body design of these vehicles, which is not conducive to good ventilation. It is where the difficulty arises and heat calculations of ratios which require one of these double-deckers to be relined four times on the rear compared with one relining on the front, produces the situation I referred to.

R. N. Kemp Crowthorne

I gather that R. Denniss is satisfied with the braking performance of the present day commercial vehicle. Bearing in mind that this is still far below that of a car, I wonder what constraints he has in mind in making that statement.

R. Denniss Burton-on-Trent

This statement is made on my own company operating experience. I think it fair to say that up to 1968–70 the articulated vehicle jack-knifing was quite prevalent. In our company we have an accident review committee in which we review all maximum growth vehicle accidents and any other accident of a major nature. It is fair to say that over the last four years we have not seen a real case of jack-knifing.

That is one parameter. The other is that there is a shortfall in the amount of accident statistics as they apply to commercial vehicles. I recall two years ago seeing a publication which stated that 60 per cent of commercial vehicle accidents involved other people, cars running into the back of commercial vehicles, usually when they were turning right. That basically is the reason for my statement. We do not appear to have accidents as a result of brake failure and

© IMechE 1977

nationally, from what statistics there are, the major cause of a commercial vehicle having an accident is when someone runs into the back of it.

Papers C45/76 and C46/76

I. Burgess Bracknell

First I should like briefly to comment on the paper yesterday on brake fluid reclamation. I do not think we know enough about this service.

My comment on C. A. Hobbs's paper is concerned with the statement that the 5 per cent water content is exactly the same as in brake fluid. Apparently this figure did not come from my company or any other company. I have no idea where it comes from.

As far as vapour loss is concerned, which is a parameter of conditions of safety, the paper considers that 2½ per cent to 3 per cent is already observed. In consideration of this point I ask if a more detailed breakdown of the water content figures in Table 6, p 227, is available.

My second point is that, as has already been mentioned, brake loss is a more relative measure of braking ability concerning safety. We would prefer more figures on this if they are available and I ask if TRRL would consider it further.

Thirdly, the paper says there were no significant differences found between samples taken from the master cylinders and wheel cylinders. This is contrary to my experience, where we have found a small difference. It has been to the effect that when we sample vehicles we sample from front wheel cylinder and also when desirable we sample from the rear.

Fourthly, concerning the occurrence of brake lock, I am not surprised. I am familiar with the area, having worked there, and conditions are those where brake lock is not likely to occur. In our experience it is more likely to occur in heavy town traffic, i.e. conditions of repeated very long brake applications. I ask if TRRL are currently taking research into brake lock? Concerning corrosion, it is our experience that it is mainly from the outside and not from the inside. We have examples where heavy seals break down rapidly.

C. A. Hobbs Crowthorne

On the first question you wanted a breakdown of water content against vehicle age. I tried to make it plain at the beginning when I was speaking that we have not looked at vapour lock as a major problem at the time when we were doing our analysis; the point being that we had no cases of suspected vapour lock and the analysis was done before the recent publicity about vapour lock. As you rightly say, the figure of 5 per cent was given; it was given to us by your organization. I think there is room for further work on this and that the TRRL will do some. R. N. Kemp is here and may be able to give more information about a future programme of work.

On your comment about vapour lock determination and water content determination, I entirely agree. At the time when the work was done, between 1972 and 1974, and the onset it did not appear that this was a feasible possibility. The question of the difference in water content between the master cylinders and the wheel cylinders is, I think, clear from our data. So far as we were concerned there was a big difference and you have the data that we have on that. I accept that the area we were considering is not hilly. It is flat and if you are going down an alpine pass you will have a rather different situation from that when driving through Berkshire. Perhaps some work can be done on this, but it is not part of the investigation we were doing.

I take your point about corrosion being from the outside rather than the inside. Most corrosion is from the outside. Perhaps corrosion in the internal system of components is not a great problem. I said that on the wheels it was relatively small. Whether that is caused externally or internally, I do not know. Perhaps you know more about that.

We did not find oil contamination a great problem, but there are obviously cases where people contaminate the fluid and on the few occasions when we have found oil contamination it appeared that it occurred through bad maintenance. Perhaps people are at least able to put brake fluid into their systems in our area if they are not able to do anything else with their brakes.

J. H. D. Harvey Barry

I would like to make a comment to reinforce what I. Burgess said. This 5 per cent of natural brake force has the effect of dropping the vapour lock from 240°C to approximately 130°C and it has the increasing effect on the internal braking, and corrosion of brake components. A vapour lock of 130°C compares with 175°C which I. Burgess has himself recommended in a comparatively recent paper, which corresponds to a moisture content of 2 per cent.

R. Ward Leeds

I have one question for C. A. Hobbs and a few general comments. I should like to know what criteria were used to assess his investigation. I also ask how they distinguished between surface corrosion and fitting, in other words, how these people did the work.

My general comment is, it is true, that relatively few accidents are caused by braking defects and obviously very few by corrosion. However, it has been estimated that some 40 fatal or serious accidents have been due to brake corrosion. Even the cost of measuring the 40 accidents was something in the region of £200 000, which is pretty small compared with the annual cost of inspecting brakes of vehicles on the road, which is estimated at £15 million a year. If inspection of brakes is considered necessary and useful surely it should be done as thoroughly as possible. The fact that C. A. Hobbs, in his paper, quoted 26 per cent to 27 per cent of vehicles with corrosion and the AA had a sample of 40 000 vehicles a few years ago and found that 13 per cent had corroded unions, and F. Whalley in his paper speaks of 13·7 per cent, and that accidents still occur, seems to indicate the difficulties of performing reliable inspection.

In conclusion, we should remember the old saying, 'Prevention is better than cure'. It would seem to make sense to spend £3 million to prevent to prevent a problem rather than £15 million to cure it.

C. A. Hobbs Crowthorne

Our assessment of brake surface corrosion unfortunately, gave only an indication that there was the presence of some corrosion. I do not have a breakdown of whether it was surface corrosion or fitting, the point being that the presence of some corrosion indicates a potential risk at some time in the future. Obviously the risk at the time is going to depend on more serious corrosion of the pipes. Unfortunately, I do not have those data, but that is the position at this stage. It might be worth commenting that in all our

© IMechE 1977

2 000 accidents we did not have one in which brake pipe failure contributed. We did pressure test the pipes and had no failures, so it cannot be a major problem; otherwise we would have had something happen.

R. Ward Leeds

May I add that we found that in 1969 thirty four accidents were due to brake corrosion and in 1970 fifty three were due to brake corrosion. I know this is a small number, but it does cost money.

F. Whalley Manchester

Category 61 includes master cylinders as well as pipes. We have no breakdown of individual component failures, but subjectively it is apparent that severe corrosion of brake pipes does exist in many cases. It is not unknown for pipes to burst in course of testing, presumably because pressure is heavier than in normal service.

J. P. A. Adams Birmingham

I have a question for F. Whalley, but first a comment. It has always been a source of amazement to me, this is a personal comment and it might not be agreed to by some of my colleagues, to see the leniency shown in investigation of some road operators. You put your machine in at the end of the operation, by which time the brakes have become cool, and therefore the test you do cannot possibly detect an operator who fitted spurious brakes which might be very poor in terms of performance, speed and sensitivity.

That is my comment. My question is, what attitudes would be taken, and by attitudes I mean would it be part of the brake test system, and what procedures would you adopt to test a vehicle if it came to be tested for an anti-skid system?

F. Whalley Manchester

A vehicle coming for test with an anti-skid system fitted would have to meet the same legal requirements as one without an anti-skid system. In other words, it would have to meet the minimum prescribed efficiencies. The fact that the system is designed to prevent the wheels locking up is, I think, irrelevant, provided enough load is applied to the vehicle to achieve 50 per cent, 55 per cent and 62 per cent respectively. This is no different in fact from a vehicle coming for test on a wet day perhaps with mud on the tyres. We have to apply sufficient load to ensure that we get the requisite efficiencies. I do not see that the provision of anti-skid affects it.

J. P. A. Adams Birmingham

I meant whether you would check that the anti-skid system was working.

F. Whalley Manchester

No, we would not check specifically that it was working. There is nothing in the Regulations saying that an anti-skid device must be fitted. It must be borne in mind that the conditions are built around the fundamental requirements.

F. D. Hales Loughborough

I take up the point made by C. A. Hobbs that a third of the cars with engines of under a litre had foot brakes which were out of adjustment, which appears to be due to the fact that small brakes are fitted to small vehicles. I do not find that to be a completely convincing explanation. Is he implying that small brakes are more difficult to adjust or that the ratio of reproducity to vehicle mass is deteriorating or is it for some other reason?

C. A. Hobbs Crowthorne

I can report only what we found, that there was a high incidence of brakes out of adjustment. I can only guess at the reason for this. I suspect that it is that the brakes required adjustment more frequently and that this sort of maintenance was not carried out with the regularity it should be on cars. Apart from that, your guess is as good as mine. It may be that people cannot afford to get their cars properly maintained. It is much of a guess.

M. A. I. Jacobson Basingstoke

If I can take up the point made by F. D. Hales, there is an answer. We have looked at 4 000 vehicles, including quite a few small ones, Minis and so on. They all use drum brakes and that is part of the reason.

I should like to put some comments and a question to C. A. Hobbs. As I said, we carry out roughly 4 000 inspections of cars among our members and we find that braking systems are not changed when they are in fact supposed to be checked by the garages selling them to customers. We still find a high incidence of brake system malfunction, often to the point of their being in need of immediate rectification, but they do meet the current DoE specification. Some go through without any brake linings at all and they are successful in the tests. In a sample we took last week there were 21 without any brake linings whatever and they did better than the other 55. They were submitted by the garage because they could not get the correct linings. It was quite a reputable garage.

One of the problems is the question of maintenance. Very few people, including those using garages, take any notice of the effects of fluid; they do not understand them. They do not do anything about it until it is too late. Rubber caps, we find, have a very finite life. Brake imbalance is a malfunction of heat rather than of mileage. The handbrake mechanism will fray and corode and some seize up quite early. There are two popular small cars which suffer from that, one a multi-national and the other from Leyland. We see far too many brake pipes with missing clips and corrosion due to mud. Surely that is a problem which should be ironed out.

I am surprised that one thing has not been referred to. That is flexible hoses which deteriorate with age. They expand more in the heat. Then we get a sinking feeling when we apply the footbrake. Perhaps you could comment on that. It may be of interest to see how rapid that deterioration is. If you put on 1 per cent up to four or five years old and then go up to 3 per cent, which is not uncommon for a motor car, 5 per cent can be detected as in need of attention or renewal.

The same is true of a gradual increase of deterioration in the mechanism. In the handbrake it starts off at 7 per cent in the first year and goes up to 22 per cent when it is six to eight years old. In general we confirm your evidence, but

© IMechE 1977

we would like to know what the TRRL and the DoE intend to do about it.

C. A. Hobbs Crowthorne

You have given me a number of questions again. Your comment about brake linings having a quality not complying with the Regulations, yes this does happen. It may be that this is why we have a high count for a number of brake defects. We were quite rigorous looking for defects, but it may be that a number of cars we found defective would comply with any brake regulation you would care to formulate. Your comment about rubber boots is covered by mechanical defects which we looked at. Yes, we did find self-adjusters not working. Missing clips were in the count for overall details. They do not occur very often. In regard to the question of brake hoses we had a practical problem because the brake examinations were carried out by one man in a garage, which is not an ideal situation. It was very nice of I. D. Neilson to show a picture of a typical accident in which it was raining and it was at night. You can almost guarantee that when you do this job the conditions will be all against you. Do you know a method of checking brake hose expansion with one man in a garage?

M. A. I. Jacobson Basingstoke

The answer is no. I wish I knew of one.

C. A. Hobbs Crowthorne

We had a system which was very difficult and it was difficult to do it in the laboratory.

F. Whalley Manchester

The only thing I want to comment on is the suggestion that vehicles can have braking efficiency without linings. This is absolutely true. If you look at Fig. 6, p 237 in my paper you will find that handbrake problems occur more often than footbrake problems. I should think, bearing in mind the human element, that if there were a brake lining problem it would occur more on the footbrake any way.

I also comment that it is always possible in the test procedure to ensure that the lining is in order and intact.

C. A. Hobbs Crowthorne

May I add something I omitted? Of the vehicles in accidents that we were able to test, all were able to pass the footbrake deceleration test, but we did have vehicles which were not able to pass the handbrake test.

Papers C47/76 and C48/76

R. T. Spurr Chapel-en-le-Frith

Some of the papers this morning have shown that things can and do go wrong in braking performing systems. J. Todovorić has gone one stage further and shown how things can be radically changed and improved on a systematic and scientific basis. I should like to ask J. Todovorić, does he think that the results to be obtained require a computer, or could it be done by a simpler process?

J. Todorović

I think it depends mainly on the number of items of information. Normally when we try to investigate a problem we collect a lot of information and it is necessary to have a computer to assess and calculate within the time allowed. Practically for a serious investigation it is advisable to have a computer.

J. D. W. Parsons London

I could add something to that which might help you. We in the Army have a computer and use it to process information to try to get reliable information. You have to be very sure of your definitions before you can start and to clothe things rather accurately. In handling it there are definite techniques which are quite easily learned, with a manual table. J. Todorović showed that it is fairly easy to gain an idea of the mean life from quite a little data. If you want a real answer you need a computer to enable you to do this.

J. D. H. Harvey Barry

I have a comment on J. D. W. Parsons's paper, on paragraph 46, which is headed Silicone Based Brake Fluids. I think it is important to realise that there are two types of silicone based brake fluids, the water tolerant type and those which are water intolerant. The water intolerant, or water resistant, are in fact water repellent in the system which contains them. Recent test results published by the US Army of a defence operation confirm this observation. Incidentally, it is entirely possible that water resistant fluids will be so developed that we will be able to envisage a system which is not only trouble free, but also maintenance free for the life of the vehicle.

J. D. W. Parsons London

Thank you for the information. Having read that particular paragraph again, it appears that I have combined two things rather than separating them as I should have done. The point I was trying to make was that it is possible to use these fluids but there is a price to pay. You do not get them for nothing. If you are working in heavy duty constant demand situations it may not pay you to do so except in specific areas.

I. Burgess Bracknell

There is a third type of fluid used as a sealer. It is based on mineral oil and is used in Chieftan tanks and similar vehicles. The extent of its use could be greater. It is non-hygroscopic, does not corrode and is available in reasonably small packages.

J. D. W. Parsons London

You also require a system designed to take them. We were examining data which exists already and looking for improvements of a practical nature. I could pose a comment from my co-author who reminded me that if I met awkward questions like this we could decline using the expertise of the Ministry of Defence and say that braking systems ideally should meet the requirements. This might cost ten or twenty times as much as a production item.

B. J. Chapman Chapel-en-le-Frith

In J. Todovorić's paper, in Fig. 4, p 245 there is only one leading shoe and one lining.

© IMechE 1977

J. Todovorić Belgrade, Yugoslavia

That is number 1 brake.

B. J. Chapman Chapel-en-le-Frith

I thought you would have two shoes.

J. Todovorić Belgrade, Yugoslavia

One leading and one trailing. The table in Fig. 4 gives only one brake as an example.

Papers C50/76, C51/76, C52/76 and C72/76

P. R. Oliver Bristol

I have some short questions to put to the authors. O. Ehricke mentioned the friction valve. The point I make is that we have never had this valve on British vehicles in the past. It enables vehicles to operate with a valve in the system which is thought to be a good thing when operating in Europe, whereas obviously the undeveloped countries have to get along without them.

My second point is again associated with O. Ehricke's paper and concerns parking control. I do not think we can expect drivers when they park vehicles on a slope always to ensure that the vehicle can be maintained with the brakes. Therefore it could be considered that the security of the vehicle could be endangered.

Finally, this concerns W. I. Baxter; he covered it very briefly in his opening. When the three-line was introduced in Great Britain a few years ago, we were led to believe that this sort of legislation would be general throughout the world. That is, where we have secondary braking on the vehicle. Could he say when he thinks we may go over to this?

O. Ehricke Hannover, GFR

With regard to your first question, I think the rule has to be understood so that in case a failure occurs to one of the service brake circles, the separation and continuation of the feeding of the circles is to enable the driver to continue in emergency conditions just to the next workshop, or parking area or whatever it is. It would be the function of the rule, I am sure, that you can continue to drive in that condition every morning for weeks.

Referring to the second question, the test position in the hand control and handbrake. The background of this is a different philosophy in regard to vehicle train braking on roads in Germany, France and Italy, and I think, in Belgium as well. In Germany it was the old national law; hence the custom that a vehicle train was held on a slope with all brakes, that means the tractor and also the trucks. In France it was the custom, and I think also the law, that a vehicle train must be held on a slope by the whole of the vehicle brakes only. The French were used to that and the Germans were used to that, so the question was, what to do? The respective governments agreed that the requirement of the rule calling for braking of the train by mechanical means only shall be considered complied with, also if the brakes of the tractor are in the condition that the driver has the possibility of checking whether the motor vehicle parking brakes alone are able to hold the train.

W. I. Baxter London

Dealing with the question of three lines or two, there is no doubt in my mind that three lines are better than two. The EEC Directive permits two or three lines and so does the ECE Regulation 13. Whether continental European countries will in fact go for three lines is a matter for them to decide. It is for each country to do whatever it wants to do in this. Three lines or two lines are acceptable under EEC rules. In this country, if you have a two-line EEC drive tractor and trailer and it has only two lines, that is acceptable. I heard a speaker say yesterday that he was working out a programme on how to deal with the mixed marriage problem. I think there is a problem of three lines or two lines and it is up to the operator to decide which he prefers, but my feeling is that three lines are better than two lines.

P. R. Oliver Bristol

The point is that we have done away with the requirement for a second brake and the safety of the trailer vehicle is not what it was under EEC regulations.

W. I. Baxter London

You prefer three lines. When you become a member of a family, the Common Market, you have to get compatibility among yourselves and perhaps accepting only two lines was the only solution.

O. Ehricke Hannover, GFR

Would you permit me to answer W. I. Baxter? I have not the rule in front of me, but I think the rule was that it must be a multiple line, preferably two lines.

W. I. Baxter London

So it says at least two lines?

O. Ehricke Hannover, GFR

Well it does not absolutely prohibit more lines.

W. I. Baxter

'At least two lines' does not mean 'prefer two lines'. The actual wording is 'the pneumatic link must be of the type with two or more pipes'.

G. Jones Leyland

I should like to echo P. Oppenheimer's concern about when legislation is talked about, accepted and enforced. What we want to know is when Directives are accepted and when they are to be in force because there is a distinct difference and a lot of confusion arises, not only on the subject of brakes, but on all the regulations. We do not know whether the legislators realise the confusion this can cause throughout the industry. Perhaps W. I. Baxter can use his influence on this?

W. Baxter London

Of course this question of dates is always cropping up. In the United Kingdom we always promulgate what we intend to do and the dates of application and not only inform manufacturers in Britain but also importers. What other countries do I do not know. I think this question should be directed to your agents in the countries to which you want

© IMechE 1977

to export. The agents in this country are very well aware that the manufacturer needs to know the dates of application. This question has been raised in Brussels already, but it is difficult to get agreement to codify.

G. Jones Leyland

Even our own agents get notice of regulations being in force in nine months time. It is ridiculous.

P. R. J. Harding Cleckheaton

I direct my questions to Mr Baxter. I wonder if he would agree to elaborate a little on the point of brake lining classification, and particularly what he considers to be the purpose. We have seen this operating in West Germany a few years ago, with a certain treatment. I wonder what his view on that type of lining is.

W. I. Baxter London

The reason why we are interested in the brake lining classification system is that under the EEC, approval tests have to be carried out on the first vehicle of the team and on the test report one has to state what make and type of brake linings are fitted. The vehicle will then be sold and put into operation and the question then arises about replacement marketing. One could argue that if owners could only fit linings which have been approved that is a barrier to trade and not a good thing to have. We do not want to be too precise in saying that a Ford Escort must have a certain lining. That is not good for other manufacturers and there are very small brake lining manufacturers we would not like to see go out of business. So we are anxious to avoid a trade barrier. To overcome that one needs to classify brake linings and one needs to have some method of grading brake linings, possibly according to friction level. Some operators also go into the rates of wear. I would not go so far as that, but this is the reason why we want a brake lining classification.

As to the German method, frankly that did not get off the ground. It was done by one of the universities in Germany, but we thought we would go along with a full-scale method of measuring brake lining coefficients and work is being done on this at the moment. Ideally, the classification should cover every aspect of brake lining technology, but this is just not practical. If we start with friction levels we shall get somewhere and we are very anxious to avoid a trade barrier in legislation.

B. J. Chapman Chapel-en-le-Frith

May I remind W. I. Baxter that it is pretty pointless to classify the friction if you do not classify the material?

W. I. Baxter London

Thank you very much for the reminder. Yes, we do appreciate that this is a problem, but we have to start somewhere.

D. A. Hoffman Bristol

W. I. Baxter, you indicated that one of the things to be discussed in Geneva in July was regulations. Can you and the other panel members give us the possible dates and what form the initial legislation will take?

W. I. Baxter London

I do not think I would like to go into the detail of what the specification will prescribe. In the international discussions the representatives range from people looking for promotion, people not particularly worried about anti-lock, and people who want something practical, useful and sensible and not to hinder developments in this very important field of technology. We cannot say what requirements will be listed in the regulation when it eventually emerges.

On the question of timing, anti-lock will be given top priority. One would imagine that a document could be finalized at the July meeting which would be submitted to the parent group, W29, in October and, if two governments sponsor that, it could be available in the early part of next year. It is an argument between the perfectionists in the group and the people who want some sort of progress, so the time scales depend on two camps.

F. D. Hales Loughborough

I thought, listening to P. Oppenheimer and some of the remarks he made on other papers, that he felt the lines of communication between the engineers in the various European countries and the rule makers in Brussels were somewhat tenuous and not functioning in the way they might. I wonder if he has a plan or a way in which the situation could be improved so that the difficulties could be overcome and they could solve these problems more quickly?

P. Oppenheimer Birmingham

One of the biggest problems we see from the industry side is that there are very few industry people allowed at these government meetings, perhaps for perfectly good reasons! We tend very often to have opposing views, or different views on what the legislators are trying to do. In Brussels in particular there tends to be only one representative from the vehicle manufacturers and one representative from the equipment manufacturers. There are names for these organisations and they are named in the paper. Both of these happen to be French, which is an additional problem to the industry because the French, I do not want to offend anyone, tend to look after their own industry in their own country, as U.K. people might look after the U.K. industry. It is a problem on which perhaps W. I. Baxter will have something to say. The meetings in Brussels tend to be fairly small, held at short notice and the agenda is never circulated in time and it becomes very difficult for people to go to these meetings. Very often they discuss braking, headlights and other things all in one day and you may go to Brussels to find that braking is not discussed on that day.

Perhaps I am over-stating our case, but that is the reason why we do not do very well in Brussels. We do better in Geneva because things are better organized and planned and there is more representation, usually five representatives from the industry, and this gives a chance for someone from England, from France and Germany.

W. I. Baxter London

I do not want to get into controversy on the politics of trade organisations the P. Oppenheimer has been talking about, but in fairness it should be said that they invited the manufacturers to be represented. The secretary who attends for that group of the EEC is in fact an Englishman.

© IMechE 1977

M. A. I. Jacobson Basingstoke

When one listens to the complexity of the new regulations introduced in Geneva, and in Brussels, one wonders who is to pay for all that. One is left with the impression that some regulations are rushed into existence by various countries who have no intention of going beyond the stage of adopting to that of implementing them. I think we should have learned by what the Americans have had to pay for by way of legislation. Would it not be better for the industry to catch up with the hardware and finish it before placing it on the file before the poor unsuspecting English-speaking ratepayer?

W. I. Baxter London

This is a question I have raised in all seriousness, the question of who is to pay for it? You and I pay for it in our subscriptions to the United Nations and the Common Market. I think we should look at it by thinking what an awful mess it would be if we did not have them. The way to look on Geneva and Brussels is to think what an awful mess there would be if they did not exist. They may be slow and ponderous, but if they did not exist you would have many more regulations and many diverse types sprouting everywhere, whereas now this acts as a safety valve and people float their ideas, discuss, ridicule or accept them and there is a very important forum for the introduction of regulations.

P. Oppenheimer Birmingham

May I add that basically the American standards are designed to upgrade the performance of the vehicles because they think they are not safe. That is not the case in the EEC. They are trying to harmonize the regulations, which is quite different. This could mean that instead of having a great chart we shall eventually have one set of regulations for Europe.

B. R. Shilton Lincoln

I would like to make two comments and ask one question. I think the first question and answer was, where one should be allowed to brake a car. P. Oppenheimer has spoken of unanimity of interpretation. We must avoid design being bound to regulations. I think a particular case could exist where, for example, the cheapest lay-out for a draw-bar truck could be made to meet the compatibility requirements of the EEC. I should like to ask O. Ehricke if he suggests that the front axle sensing bar on a draw-bar trailer with two axles which are independent, should be a dynamic axle. I wonder if he can explain why the front axle should be non-responsive?

O. Ehricke Hannover, GFR

That is because of the load transfer. Because of the load transfer, the valve on the front axle gets the signal that the vehicle is getting empty. That is why it is recommended to make l.a.v. static to avoid a decrease in pressure to the cylinders, because of the way the transfer on the front axle works. It is explained in the paper.

B. R. Shilton Lincoln

If it is explained in the paper, thank you very much.

W. Manning Basildon

Firstly a question to O. Ehricke, trailers last probably twice as long as tractors and with the new Regulation 524 we no longer instrument to a load variable, but give what the driver likes, that is for the trailer brakes to come on slightly ahead of the tractor brakes. Is there any answer for the operators to be able to follow 75/524, for trailers or tractors?

My second question is to P. Oppenheimer. With the proposed changes 71/2/20 to include anti-lock braking systems, is it the intention that the Directive will be changed so that other braking systems will also have the facility of meeting the regulations applied to anti-lock brakes?

My third question is this: P. Oppenheimer included in the EEC Directives which have been accepted, number 70/156. As explained, that includes axles. Has W. I. Baxter any information about hard-line axles?

O. Ehricke Hannover, GFR

I am not sure whether I understood your first question. Does it refer to compatibility? Is it a question of translating pressure in a motor vehicle, transforming the pressure in the motor vehicle cylinders to a higher level to get the vehicle into the band?

W. Manning Basildon

The initial service line.

O. Ehricke Hannover, GFR

Yes, you mean there should be a pressure level difference between the pressure coming out of the brake pedal and the trailer brake coupling head? Perhaps I could explain by referring to the diagram, service line E where the trailer reaction variable gives a lead for the first 1/1/5e.

W. I. Baxter London

On W. Manning's point about weights harmonization, I would not like to raise his hopes very much. There has to be another meeting in May when the question of weights will be on the agenda, but I do not think there will be very much progress in that area for some time. The only change which has taken place since two years ago is the change by the Federal Republic of Germany who said they would accept 11 tonnes on an axle and now it is 10 tonnes.

P. Oppenheimer Birmingham

The industry's intention is that it should be an annexe to Regulation 13 and a Directive in such a way that it is an alternative to 2/5/24. Instead of meeting this, an anti-lock system would have to meet other prescriptions. That is the industry's intention and I hope that W. I. Baxter sees it in the same way.

W. I. Baxter London

The intention of the people in Geneva and in Brussels is to have a separate annexe to cover anti-locking brakes and, of course, amendments may have to be made to the main document 75/524 and EEC Regulation 13 to tie up with the annexe.

O. Ehricke Hannover, GFR

In answer to B. R. Shilton, it is just the other way round. If you have an empty vehicle and a load transfer to the front axle, the l.a.v. gets the signal that the vehicle is getting laden. That would mean giving too high a pressure to the cylinders and complete braking of the trailer.

© IMechE 1977

Written Contributions

P. R. J. Harding Cleckheaton

Concerning choice of brakes, T. P. Newcomb argues that it is easier to maintain the braking ratio with a disc brake all round system than with a disc/drum combination and goes on to say that there is a trend towards fitment of disc brakes on all wheels. With regard to disc brake trends, there are a number of examples where designers have returned from full disc systems to ones which have rear drum brakes. There are a variety of economic and technical reasons for this, the main being cost, rationalization and handbrake.

Regarding the question of brake ratio, it is often supposed that fitment of the same type of brake and friction material to all axles virtually ensures that brake balance will be maintained because all brakes will have similar performance/temperature characteristics. However, as T. P. Newcomb points out, the cooling rates of front and rear brakes are different, particularly so under the wide range of heating and cooling cycles imposed by normal road usage. This does mean, even with the linear brake factor of the disc brake and the advanced properties of modern pad formulations, that there is still a difficult problem which is not made any worse by matching a rear drum brake of other than duo-servo design to a front disc brake. In fact, with modern high braking ratios and with modulation devices, it can be fairly argued that the drum brake is well able to cope with and very well suited to rear axle braking requirements. Measurement of temperatures at which brakes run in service always reveals the relatively cool values at which rear drums operate, and within which the drum brake performance is very stable.

A feature described as significant in the paper by B. Ingram and D. Peasley is the improvement in a friction material's fade behaviour with successive testing.

It is a characteristic which often leads to discrepancies in test results and I endorse the authors' reference to the care which needs to be exercised in pre-judging results without knowledge of the total work history.

All too often, when evaluating lining or pad materials insufficient attention is paid to control of work done during bedding-in and to use of vehicles prior to test and this can lead to apparent test result variance. A related effect is the apparent difference in fade performance of a particular friction material under different test procedures or on differing brake installations. This is illustrated in Fig. D5 which shows the results of two sets of the same lining material subjected to different procedures at similar energy input levels on the same vehicle.

Repeated tests confirmed the characteristic of sharp in-stop fade around stops 5, 6, or 7 followed by a rising friction and marked over recovery in the 30 second time cycle sequence. The 45 second cycle gave progressive fade with in-stop fade throughout, some delayed fade at the

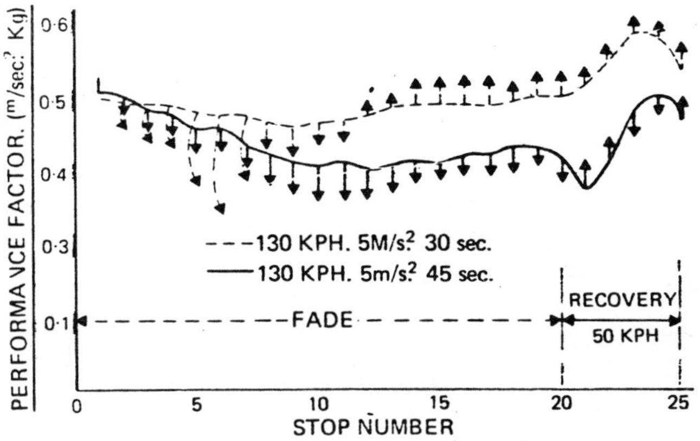

Fig. D5

commencement of recovery and some frictional 'knock-down' at the end.

Another point of interest relevant to the author's findings on rubbed area of drum is that the car concerned was equipped with amply proportioned drum brakes. The maximum temperatures recorded were 460°C and 510°C respectively and yet the lining used was a medium duty type which gives much greater, often unacceptable performance drop on more conventional drum brakes under lower energy/temperature conditions.

One important combination of brake lining characteristics mentioned by W. R. Newton and A. C. W. Wright in connection with 'one-shot' automatic adjustors is that of compressibility and heat swell. It is generally supposed that linings should not increase their thickness as temperature increases, or decrease their thickness as pressure increases since these effects may lead to dragging brakes or excessive pedal travel. Significant amounts of residual heat swell cannot be tolerated in a lining but some compressibility and thermal expansion are essential to optimize the operational efficiency of automatically adjusted brakes.

The reasons for this are:—

(a) to ensure pressure distribution around the arc of contact between lining and drum when higher temperatures cause the drum radius to expand away from shoe radius, and

(b) to minimize this radius differential by maintaining drum to lining clearance.

If the lining is incompressible, deflections of both shoe and drum under high temperatures and pressures will offset the lack of lining compression and cause over adjustment of 'one-shot' mechanisms. The same may apply to incremental adjustors, but only under very high brake application frequencies such as those associated with Alpine Pass descents, rallying or heavy duty PSV city operation.

Lining thickness expansion and contraction, with increase

© IMechE 1977

and decrease in temperature, will tend to maintain clearances and thus avoid over adjustment. Table D1 illustrates these effects, predicted for an 8″ drum brake having typical 'one-shot' adjustment. The final column shows the expected number of adjustment units when the drum temperature is 250°C and high actuation pressure is applied. Experimental work has confirmed the trend shown.

Table D1
Automatic Adjustment relative to lining expansion and compression under high temp./pressure operation

Lining	Therm. Expans. m/m/°C/10³	Compressive Strn. m/m/bar/10³	Units of Aut. Adjust
A	10·43	9·00	2
B	8·69	5·50	2
C	7·39	5·00	2
D	3·26	6·25	3
E	2·61	5·75	3
F	≃0	4.00	3

With reference to the paper given by B. J. Chapman and D. Hatch, the mechanism of many tiny particles giving rise to frictional stability and acceptable wear rates, whilst few (and presumably larger) such particles cause higher, less stable friction and accompanied high wear is in line with basic tribological concepts. One can easily appreciate that a pre-requisite of stable friction and wear is a smooth mating surface made regenerative by a fine abrasive action. Some electrical brushes are plated with silver containing a fine abrasive and in such a way the wear and reliability is improved by better maintenance of the commutator surface. The most surprising and exciting discovery of D. Hatch and his co-workers is that the tiny hard particles are so necessary in the mating iron itself when in the past it has been usual to rely solely on the friction element for desirable braking properties and to blame it completely for all undesirable braking properties.

It seems to be, so far as brake rotors are concerned, that due to re-cycling of scrap etc., the majority of irons do contain sufficient titanium or accumulations of such trace elements, as to ensure adequate refination of the carbo-nitrides. This assumes, and perhaps D. Hatch will confirm this or otherwise, that the higher titanium contents, say > 0·025%, are to be preferred since this will give good wear and stable disc friction, albeit at a lower level. Attempts to control the trace elements at a very low level in order to achieve high disc friction (or to incur high disc wear) would seem to be inadvisable since frictional stability appears to be very poor in this area. Even with irons which are suspect, there are some friction materials which are tolerant and which will give acceptable friction, stability and wear life. I know D. Hatch will insist that a wide range of materials behave with similar deficiency when used against a 'bad' iron; but there are good examples of high line/low line versions of the same car which use similar discs but different pad materials, and where pull or high wear incidence is highly correlatable to one pad material. This does not imply that the friction material which is the most tolerant can be readily substituted for the one which is not, because regulations and test standards try to ensure that the increased weight/performance potential of larger engined models is matched by their brake performance. When the same design and size of brake is fitted throughout a model range, this has to mean that at one end or the other, there is either an under-braked or an over-braked condition. The friction material is called upon to provide the solution to this problem since it is a component which is most readily changed. Thus pad materials of different friction and duty levels are required.

Specifically concerning the wear properties of discs, it does seem possible that the effect of the trace elements as innoculants in promoting a fine graphite structure may have a significance and I wonder if the experiments in which abrasive grit particles replaced the carbo-nitrides showed conclusively that disc wear, as distinct from pad wear, was similarly affected. If so, do the authors think that the mechanism for improved disc wear is some sort of dispersion hardening effect which gives a generally harder disc surface, or a more complex mechanism in which the heavier duty type of friction material is continuously honed to present a smoother contact with the disc over a larger true area of contact?

One must admire the dedication and scientific application of J. A. Odier and his colleagues in their development of a highly sophisticated laboratory system for testing car brakes. But in even attempting to argue a case for such a machine as an alterantive to the test track they do themselves a disservice. The cost of a track is certainly much higher but it can be used to carry out various types of test on many different vehicles simultaneously. The vehicles may be cycles, cars, tractors, buses and even large trucks with trailers.

A great deal of progress in road test techniques has been made in the past 25 years and the work of various authors bears some, but certainly not all of the evidence.

Early in their paper the authors are at pains to state that the results from their bench tests are identical to those obtained from the road. This itself underlines the importance of the road test. Regarding test reproducibility, the graphs of standard deviation against % population (Fig. 1 page 166) do not give a fair representation of general road test accuracy. The bench test distribution implies a platykurtic error curve with approximately half the results within 2 st. devs. and half spread up to 8 st. devs. The road test graph shows an even wider dispersion. With only 3 sets tested on both road and on bench, it would seem that any statistical significance merely indicates a deficiency in the fifth wheel equipment used by comparison with the roller dynamometer, or demonstrates the unsuitability of total stop distance as a legislative performance parameter.

In development of brakes and linings, the performance parameter is usually torque or deceleration as a function of input. With such parameters, road tests carried out by properly trained crews of highly skilled specialist personnel using appropriate instrumentation, are found to give repeatability as good as 1¾% coefficient of variation within set, within crew; 2½% c.o.v. between set, within crew and 3½% c.o.v. between set between crew. Analysis of variance and other statistical techniques have been used to investigate sources of error in brake testing. The greatest single source of variance in vehicle tests is the driver. Even he (or she) can be assisted, or even part replaced, by engineering devices in order to increase accuracy, but I am not advocating here that this is either necessary or desirable. I am stating that the repeatability of road testing can be satisfactory and comparable to that of bench testing.

Fig. D6 shows a typical distribution of performance results from 12 sets of disc brake pads tested on the same vehicle. The coefficient of variation is 3·6% and this may be compared with 2·6% obtained from 5 similar sets evaluated on a single brake type inertia machine. Table D2 gives further examples of road test repeatability on various vehicles/brakes.

With reasonably consistent friction materials and brake/

© IMechE 1977

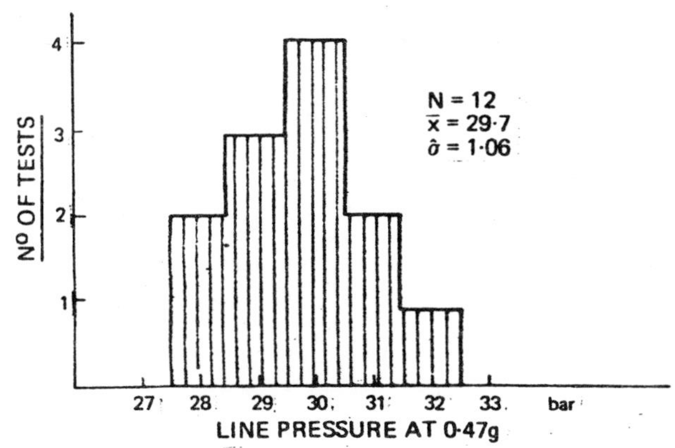

Fig. D6 Distribution of performance from road test results.

vehicle systems free from intrinsic variability, a c.o.v. of 3½% to 4% may be expected at 0·5 g deceleration level.

Taking the average value of $v^2/2pd$ quoted as 110, the average bench test c.o.v. (Fig. 1, page 166) is 4·4% and that of the comparable road tests is 6·6%. Perhaps J. A. Odier could give some idea of the c.o.v. to be expected from his dynamometer over a large number of similar performance tests.

Table D2
Road test repeatability

Brake type	No. of vehicles	Mat. sets	No. of tests	Crew	Coefficient of variation
Drum	1 (C.V.) }	1	4	Same	3·56
Drum	1 (C.V.) }	17	17	,,	4·35
Drum	1 (P.C.) }	1	6	Same	3·00
Drum	1 ,, }	1	6	Various	3·32
Disc	1 ,,	10	10	Same	2·79
Disc	1 ,,	9	9	Same	3·75
Disc	3 ,,	27	27	*	6·78
Disc	1 ,, }	4	4	Same }	2·42
Disc	1 ,, }	6	6	Various }	

* Most tests involved some crew
(Stat: significant difference between cars.)

With reference to paper C38/76 by M. R. North and C39/76 by N. Millner, the research work and mathematics of both authors supports and explains accepted brake design practice and development experience particularly in regard to the effects of stiffness of the friction material carriers; the brake shoe and the pad backplate. It has always struck me as odd that the effect of the stiffness of the friction element carrier is usually opposite in the respective cases of drum brake squeal and disc brake squeal, whilst the effect of conformability of the friction material itself is usually beneficial in both cases. Perhaps the authors could give further comment on this.

In the case of drum brake noise, the main problem concerns commercial and public transport vehicles and complaints arise mainly from an external source, i.e. from people unconnected with the vehicle operation but who live near to bus stops or road junctions. In the case of disc brake noise the complaint is invariably the owner or driver of the passenger car concerned, since the problem is almost exclusively in the passenger car field. Economically, this separates the two problems and is perhaps one reason why different types and sub-types of brake noise and characterising frequency ranges are much better defined in the case of disc brakes, where we have been concerned, for many years, with squeal, squeak, squelch (or scroop) and wire brushing as identifiably separate but related phenomena, all which may be accentuated in nuisance value by sometimes being present when the brakes are not being applied. This latter effect is perhaps the only real difference between disc brake and drum brake noise since all the different types of noise mentioned can be identified with drum brakes as well as discs.

It would seem to be important to recognize the different noises and the conditions under which they occur rather than to regard them all as brake squeal or as harmonics or modes of the same basic vibration, although I agree that it tidies up the theory and mathematics if one does so. Some of these aspects of brake noise seem best explained by and correlatable with the μ-velocity characteristics of the friction pair whilst this rather old fashioned theory is not reconcilable with the more sophisticated modern analyses. I wonder if this apparent non-reconcilability is really due to the tendency for the advanced researchers to treat all brake noise as squeal. Certainly it would seem from observation and experiment that such noises as wire brushing and squelch are much more to do with the nature of the friction pair interface and not much to do with the mass-elastic/damping characteristics of the main hardware. What is required from future research in this area is a better understanding of the factors which influence generation and damping of the different types of noise because it seems certain that the disc or drum, brake or caliper, friction material and the vehicle suspension all have a unique contribution as well as a collective one in the overall brake noise problem.

Regarding braking defects related to engine size and vehicle mileage, it would seem worthwhile to examine reasons for significant differences in more detail.

C. A. Hobbs suggests that the most likely explanation for smaller engined cars having more brake defects is that the brakes are smaller. It is also true that the smaller engined cars (below 1 litre) are nearly all operated in the privately owned sector whereas larger vehicles are spread into fleet and company owned sectors where maintenance is perhaps more regularized. Since company vehicles tend to cover a greater mileage per year than privately owned vehicles, their usage can be more conducive to better condition of the exposed brake components. I have certainly observed this to be so, and two similar vehicles which have both covered 12 000 miles, one in 12 months and the other in 3 years will usually show widely different faults or developing faults. It is also a fact that below 1 litre most vehicles are equipped with drum brakes all round, whereas the larger capacities have disc/drum systems or discs all round. This alone could explain why more of the 1 litre cars had front brakes out of adjustment, than was found on cars over 1 litre.

There could be some revealing statistical interactions or correlations if a more detailed analysis were made. For example, a large number of observations of disc brake pad wear in service from a survey carried out on one particular model available in both 1250 cc and 1500 cc versions, showed no significant difference in pad life between vehicles operated in the privately owned sector and those run by company representatives and fleets. These two facts were, in themselves, surprising in view of the widely held belief that the greater performance potential provided by the larger engine will lead to reduced pad life and that company operated or business usage is more arduous than privately owned/pleasure usage and will, therefore, lead to a similar

© IMechE 1977

result. However, when the incidence of fleet and private sector cars was examined in relation to the incidence of 1250 cc and 1500 cc cars in the sample, a very high significance was discovered between fleet/company representative operated 1500 cc vehicles and all other combinations.

In the case of brake defects relative to engine size and mileage, it may be that the true correlation is with other, perhaps compounded factors rather than simply brake size.

Referring to paper C47/76 by J. Todorović, work on predicted reliability of linings and pads, in terms of their wear life, has been carried out by the writer over many years. The basis of this work has been wear life distributions obtained from service and the estimation of their various percentiles and dispersion parameters. A particular interest in the 1st percentile arises because most car manufacturers operate warranty schemes under which claims exceeding 1% give rise to economically unacceptable situations.

Figure D7 illustrates the results of a pad wear life survey involving 160 cars of a particular make and model. They are plotted as Weibull probabilities and the slight non-linearity of plots <5% indicates the speculative nature of extrapolation to the distribution tail. The high slope parameter (β) shows the 'peaky' normal type of distribution and is comparable in this respect with that of the author's car disc brake results. (Fig. 5, p. 246).

In order to simulate the lower percentile of usage for car brakes, and thereby predict reliability of pads and linings in meeting warranty requirements, a 'sample' road route has been developed. It is designed to be typical of roads and terrain in the U.K. and also sufficiently low in traffic density to allow cars to be run to a controlling schedule of time and brake application. The route is, of necessity, long enough to obtain adequate measured wear per run, as well as to ensure representative proportions of the different types of road and terrain. The combination of running schedule and route composition has been determined experimentally to give as near as possible the 1st percentile wear life. Because of good repeatability, it is very cost effective and one or two runs are usually sufficient to give a reliable estimate of the pad life to be expected from service. For linings, which have a lower rate of thickness loss than pads, at least 3 runs are desirable. The technique is well known to vehicle and brake manufacturers in the U.K. as the Mintex RC7s method and has been used successfully since 1964.

From Figure D7, it can be seen that the 1 percent and 50 percent intercepts are 13 000 and 28 000 miles respectively. Four years prior to carrying out this survey, a 1st percentile life of 16 000 miles (and a 50th percentile of 30 000) had been predicted from RC7s test runs made on a prototype simulator car. Service experience by the vehicle manufacturer during the intervening time had confirmed that there was less than 1% complaints of pad life not achieving 12 000 miles, and also that average life was 25 to 35 000 miles.

Also plotted in Fig. D7 is the result of a lining trial on 49 public transport buses. This is not dissimilar from J. Todorović's bus brake distribution which shows a slope parameter $\beta = 2{\cdot}05$; indicative of the Rayleigh type of distribution.

In my experience this form, which is the classical 'wear out' type of failure pattern, was more usual when woven linings were in common use, and is perhaps more typical of the drum brake than the disc brake. The latter is usually found to give normal type distributions with positive skew and long tails with $\beta > 3{\cdot}5$ when submitted to a Weilbull analysis.

H. Hodkinson Leamington Spa

The paper C31/76 by E. C. Glasner draws the conclusion that 'kinked' brake distributions do not provide improved safety in braking and implies that the EEC directive, 75/524/EEC, which may require their adoption, can have a detrimental effect on braking safety.

In electing to consider one vehicle loading condition only, the author advises that the vehicle concerned does not have a significantly different weight distribution between laden and unladen condition. Such a vehicle, particularly one with a low C of G ht./wheelbase relationship, is unlikely to benefit from a 'kinked' distribution and in fact the EEC directive conditions could be satisfied easily with a 'fixed' distribution.

However, there is a large percentage of the vehicle population, particularly those directed at providing a compact economic package, which show a marked difference in ideal braking distributions between 'driver only' and 'fully-laden' cases. These conditions, together with a high C of G/wheelbase relationship, are typical of smaller vehicles, which can benefit from 'kinked' distributions in providing an effective and simple compromise over a wide range of limiting adhesion conditions.

Whilst the author has demonstrated the adverse effect of engine braking, it will be appreciated that this effect is beneficial in the case of front-wheel-drive which is a popular feature of the small economy vehicle.

Similarly, disc/drum brakes are widely used in this category of vehicle and the effect in distribution of the differing threshold pressures of such brakes can be offset by 'kinked' brake distribution. Graphs showing comparative distributions on typical front-wheel and rear-wheel drive vehicles are enclosed.

From the foregoing, it may be assumed that many vehicles may benefit from 'kinked' distributions and the greater utilization of tyre adhesion in the lower retardation levels will offer improved brake performance.

R. N. Kemp Crowthorne

Referring to paper C23/76 by F. D. Hales I should like to mention that in tests at TRRL with an unladen tractor/semi trailer/dolly/semi trailer train progressive braking on a 50 m curve on a low μ surface ($\mu \simeq {\cdot}25$) from a speed of 40 km/h resulted in lateral instability at axles 2, 4 and 5. The tractor jack-knifed about 10° anti clockwise, the dolly swung through 90° until the drawbar was parallel to the rear of the 1st semi trailer and the rear semi trailer also swung out of the curve; the combination coming to rest spreadeagled over 12 m of carriageway width. Without braking the train negotiated the curve without difficulty.

Does F. D. Hales think that this and other departures from his analytical results may arise from non-linear accelerations already mentioned by D. McLean or to other theoretical strategy not being observed in practice? Identical tyre characteristics at each axle are possible but in general use improbable. Braking matched to axle loading is not always possible with load apportioning values limited to a maximum ratio of about 305:1. In view of this would he not consider anti-lock brake systems on each vehicle of the train a wise precaution?

© IMechE 1977

WEIBULL DISTRIBUTION

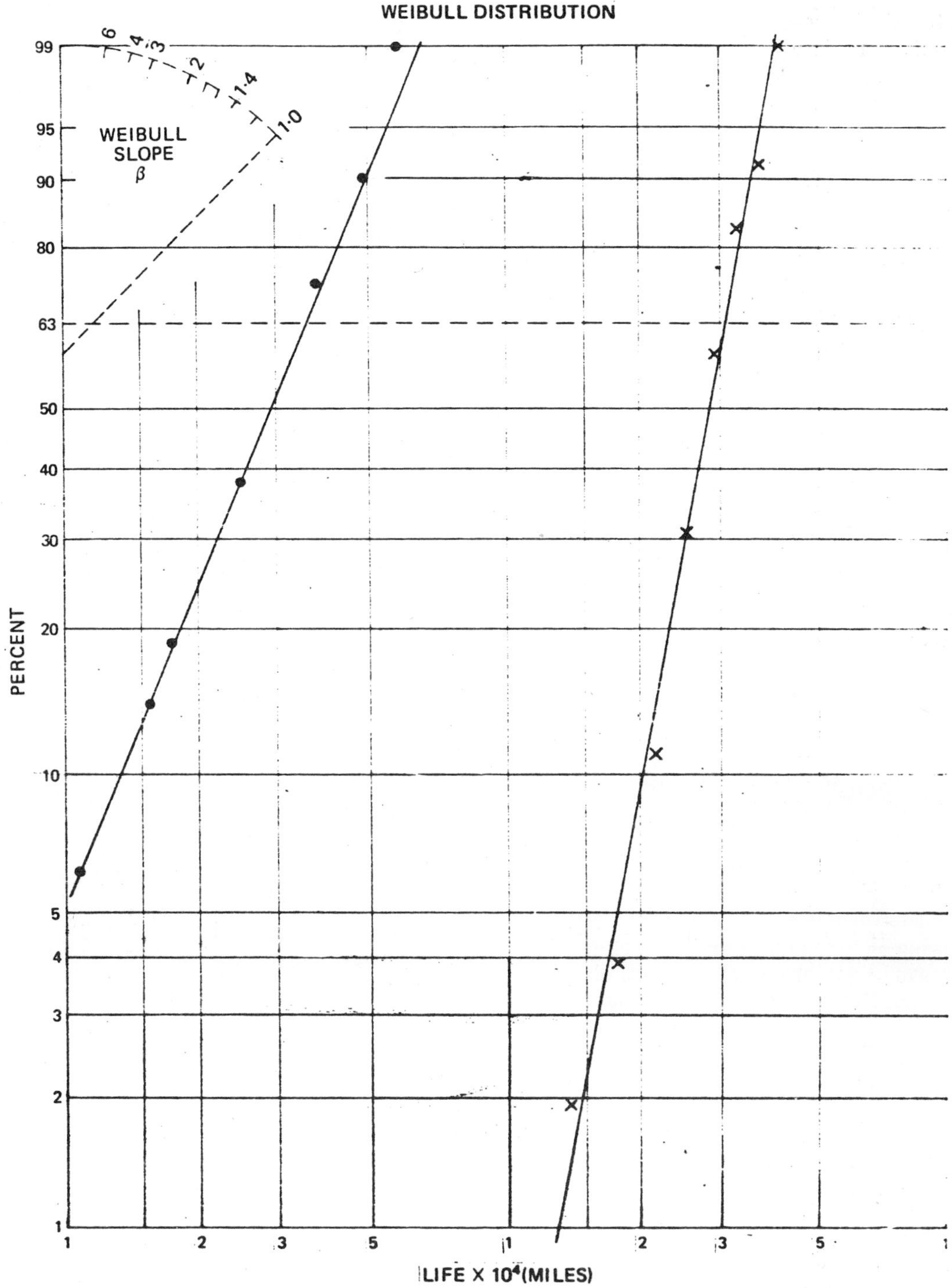

Fig. D7: Wear life of pads and linings.
X —— X Car disc pads $\beta = 5$
• —— • Bus linings (woven) $\beta = 2{\cdot}3$

T. P. Newcomb

The paper C37/76 by J. Odier contains some very interesting data on the braking performance of cars. Fig. 2, which shows average changes in the braking ratio of a car fitted with disc brakes on all four wheels, is particularly useful. These figures can be compared with similar earlier results obtained by Lister (1) mostly on drum-braked cars. The present author's results show much less change in braking ratio and indicate the improved stability which is attained by using an all-disc-braked vehicle.

One of the advantages of the authors' machine is that it readily enables transient results to be obtained in safety while braking from high speeds. I wondered if the authors could show some such results on all disc and front disc/rear drum-braked cars to illustrate the possible spread in variation in brake effectiveness that can occur in practice, while making individual stops from high speeds.

Fig. 6 shows a particularly treacherous behaviour of a brake correcting device and I wonder if the authors could say whether the behaviour observed was due to poor initial design and setting or malfunctioning of the components and how often have such results been obtained.

(1) LISTER, R. D. Symposium on Control of Vehicles During Braking and Cornering (Inst. Mech. Engrs.) 1963, p. 44.

© IMechE 1977

G. M. Ranger Wrexham

This I intend as a contribution to the discussions following Session 11 on Legislation. It was evident that there are some discrepancies between interpretations of EEC Regulation 71/320, particularly in relation to articulated vehicles and the way in which the brakes should be applied when they are parked.

In his paper C72/76, O. Ehricke outlined a two line braking system, which featured parking of the tractor unit conventionally on spring brakes, while the trailer service brakes were left on, and the trailer was technically parked 'on air'. The hand control/parking valve of Fig. 9 has a check position which allows the vehicle driver to momentarily release the trailer service brakes in order to demonstrate that the tractor spring brakes are capable of holding the whole combination. However in paragraph 3.2.3 of his text, O. Ehricke correctly states that on gradients with unfavourable friction properties and, it should be added, under conditions of unfavourable load distribution, it is not possible for the spring brakes of the tractor to hold the combination. Hence there will always be the occasion, in practice, where the driver will use the check and merely demonstrate the inability of the tractor spring brakes alone to hold the combination. He will then rely on air applied trailer service brakes to park the vehicle, a most unsatisfactory condition.

Of course, this point was raised during the discussion by B. R. Shilton and it was evident that certainly as far as the UK is concerned, any form of parking 'on air' is regarded as taboo.

Unfortunately, EEC braking systems currently proposed by relevant bodies in the UK do not offer any solution to the parking combinations problem and in fact systems so far published completely ignore it, since trailers are left unbraked and hanging on to the spring braked tractors, even though there will be not infrequent occasions when the tractor will be incapable of holding the combination.

It is my opinion that faced with such a dilemma, the lead should be shown by the EEC Directive itself, in that it should lay down recommended codes of practice in braking system design, leading eventually to the adoption of a standardized European braking system, which, one assumes, is what the Directive is working towards anyway. The Directive would offer a real contribution to vehicle safety if it recommended a system which included mechanical application of the brakes on all axles of a tractor/trailer combination when parked. Fig. D8 shows a two-line braking system which would achieve this, in principle. For clarity the full complement of quick release valves, relay valves, anti compounding valves etc., have been omitted, and it is not suggested that this system in its present form complies with all the requirement of EEC Directive 71/320. However, it does incorporate the following features:–

(1) Split service braking on the tractor unit in the usual fashion.

(2) Trailer service braking triggered by either the front tractor axle or rear tractor axle or both via the double check valve (1), which should be of the type to allow simultaneous feed from both inlets if they are at equal pressures. The relay valve (2) may also be triggered by the hand control valve (3) to give independent trailer braking, if required.

(3) The trailer emergency line is fed from the tractor spring braking system which may in turn be fed by either or both of the trailer/parking reservoirs (5) and (6) via the double check valve (4), which should be of the same type as DCV (1).

(4) On the trailer, the conventional relay emergency valve has been replaced by a simple relay valve, the trailer breakaway feature being met by spring brakes which are held off by emergency line pressure. Hence in the event of breakaway, the loss of emergency line pressure will lead to automatic spring brake application. Similarly, when parking the combination, the operation of the simple on/off parking valve (7) on the tractor, will result in total exhausting of the tractor and trailer spring brake hold off pressures and the brakes will be mechanically applied on all axles.

(5) The valve (8) on the trailer, which may be termed a shunting valve, usually allows the free passage of emergency line pressure to provide spring brake hold off. To allow yard shunting of the trailer without connecting the trailer air lines, the button on the shunting valve (8) is depressed and air from the trailer reservoir admitted to the spring brakes. The shunting valve will incorporate a reset feature, whereby on reconnection of the emergency line, the line pressure is sufficent to knock back the plunger on the shunting valve and restore the spring brake hold off pressure feed from the emergency line. The double check valve (9) is to prevent compounding of service and spring braking.

Compatibility

(a) *Trailer*
A Trailer equipped with the spring braking system of Fig. D8 would be entirely compatible with both 2 and 3 line tractor units, according to EEC Regulations, since secondary trailer braking is not a requirement. Obviously the automatic trailer brake application would still be obtained on breakaway, since emergency line severance gives spring braking. Unfortunately the all axle parking feature for the combination would be lost, since the majority of 2 and 3 line tractor units would not exhaust the emergency line when parking brakes were applied. In this respect the spring braked trailer would behave exactly as a conventional 2 line trailer; and simply hang on to the braked tractor. A bonus is that a slow leak from anywhere in the emergency line system would cause gradual trailer spring brake application.

(b) *Tractor*
Service braking with the tractor of Fig. D8 presents no compatibility problems when connected to conventional 2 or 3 line trailers.

Secondary braking is also no problem since in the event of either tractor axle losing brakes, full trailer service braking is preserved via the DCV (1) together with the service brakes of the remaining tractor axle.

Parking of the combination using the parking valve (7) results in exhausting of the emergency line, leading to automatic trailer service brake application with the conventional 2 or 3 line trailer equipped with a relay emergency valve. Hence the tractor will be parked on spring brakes and the trailer on air. As stated in the beginning, this is not really desirable, but apparently is permissible under EEC 71/320.

© IMechE 1977

When considering 3 line C + U trailers, it is assumed that the secondary line is left open, and in this mode it behaves as a conventional two line trailer.

Hence Fig. D8 shows the rudiments of a two line braking system which, while preserving the basically sound service and secondary braking requirements of EEC Directive 71/320, gives a definite improvement in performance and safety over current systems which comply with the too vaguely specified parking requirements.

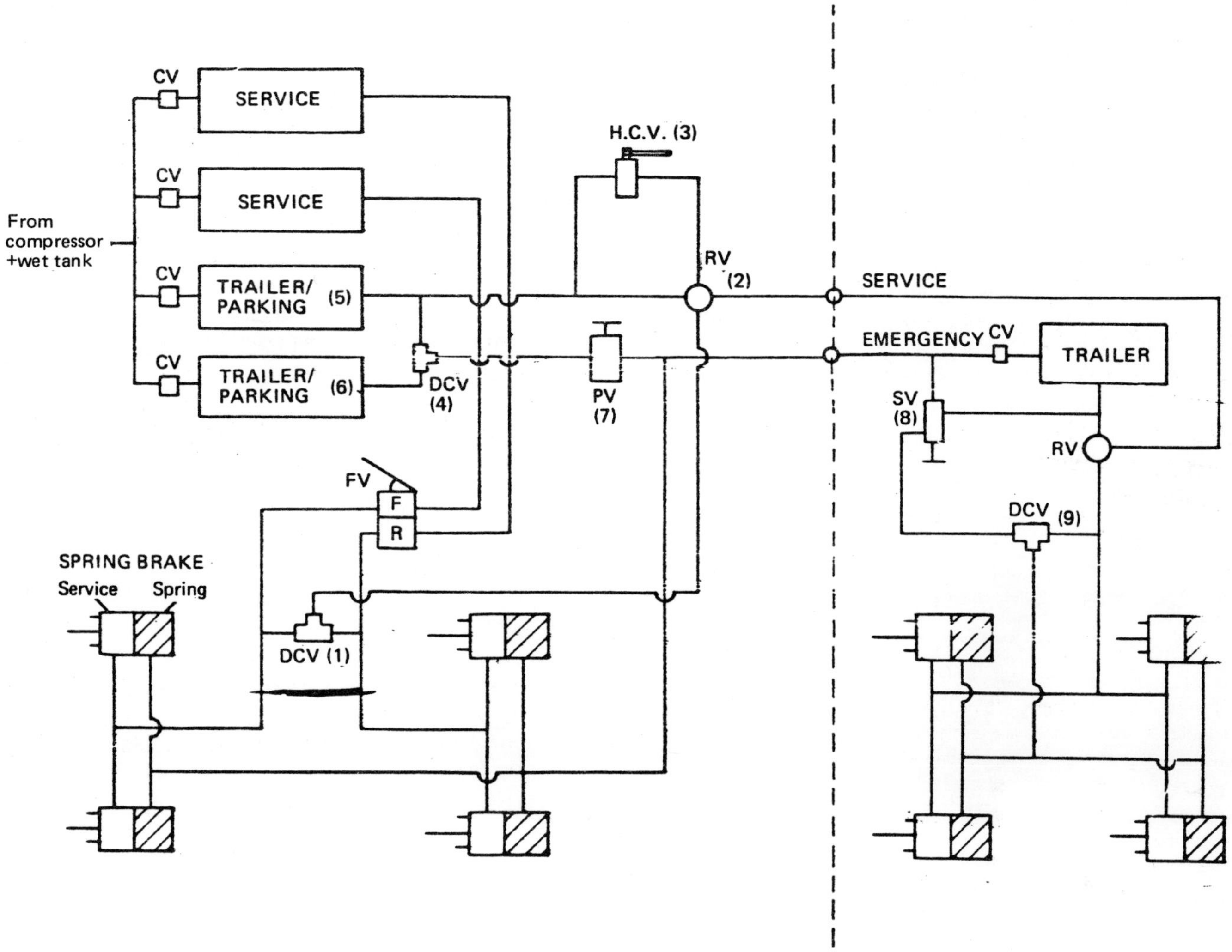

Fig. D8. Schematic of 2-line air braking system with all axle parking feature.

© IMechE 1977

Author's Reply

J. A. Odier Paris, France

The two questions which have been asked by P. R. J. Harding and by T. P. Newcomb are important and interesting. As far as the first question is concerned, the study which is being proposed has not been carried out systematically up to now. As far as the curve of Fig. 6 is concerned, it should be pointed out that it is not a typical curve of a general nature; it is only an example, chosen from other examples, just because the curve is complex, and therefore long and difficult, when established by road testing whereas on a fixed-stand test bench, it is much more convenient to determine using tests which are easy to carry out and very short in duration.

© IMechE 1977

List of Delegates

Adams, J. P. A.	Girling Ltd., Birmingham
Anderson, J.	British Leyland Bus & Truck Division Ltd., Leyland
Avent, R. J.	Bendix Westinghouse Ltd., Bristol
Bacher, M. B.	SAF du Ferodo, Saint Ouen, France
Bainbridge, W. N.	Automotive Products Ltd., Leamington Spa
Baines, R. G.	Clayton Dewandre Co. Ltd., Lincoln
Baker, A. K.	Ferodo Ltd., Chapel-en-le-Frith
Baker, D. E.	Department of the Environment, Nottingham
Ballinger, F. J.	Michelin Tyre Co. Ltd., Stoke-on-Trent
Barker, C. A.	Wichita Co. Ltd., Bedford
Barson, C. W.	Dunlop Ltd., Birmingham
Barter, N. F.	MIRA, Nuneaton
Barton, J. R.	Massey Ferguson Manufacturing Co., Coventry
Bath, M.	Clayton Dewandre Co. Ltd., Lincoln
Batstone, W. V. C.	National Freight Corporation, London
Baxter, W. L.	Department of the Environment, London
Bayliff, R.	B.F. Goodrich Brake Systems, Heveadorp, The Netherlands
Beasley, S. A.	Dunlop Ltd., Coventry
Bergdahl, B. G.	National Swedish Road and Traffic Research Institute, Linköping
Berin, I.	Saab-Scania AB, Södertälje, Sweden
Blase, H.	Wabco Westinghouse GmbH., Hannover, GFR
Booth, G. R.	Brake Linings Ltd., Buxton
Brearley, M.	Girling Ltd., Birmingham
Brimaud, G. B.	SAF du Ferodo, Saint Ouen, France
Broad, H. J.	Goodyear International Corporation, Brussels, Belgium
Brocklehurst, P.	Rolls-Royce Motors Ltd., Crewe
Brodd, S.	AB Volvo, Göteborg, Sweden
Brown, S. J.	Southdown Motor Services, Brighton
Bryant, C. G.	Trent Motor Traction Co. Ltd., Derby
Buckby, K. N.	York Trailer Co. Ltd., Northallerton
Burckhardt, M. H.	Daimler-Benz AG, Stuttgart, GFR
Burgess, I.	Burmah-Castrol Co., Bracknell
Burridge, G.	Bendix Westinghouse Ltd., Bristol
Burrows, C. W.	Rolls-Royce Motors Ltd., Crewe
Burrows, N.	Eastern British Road Services Ltd., Derby
Burton, R. T.	Girling Ltd., Birmingham
Carr, R. N.	Mintex Ltd., Cleckheaton
Caton, P. B.	Mintex Ltd., Cleckheaton
Cattaneo, C. M.	Fiat S.p.A., Torino, Italy
Cattley, R.	Crane Fruehauf Trailers, Toftwood
Cena, B.	Fiat – DCR, Torino, Italy
Chapman, B. J.	Ferodo Ltd., Chapel-en-le-Frith

Clark, D. J.	Clayton-Dewandre Co. Ltd., Lincoln
Cramp, G.	Dunlop Ltd., Birmingham
Croissant, M. S. F.	London Fire Brigade
Curtis, C. H.	London Transport Executive
Czinczel, A.	Robert Bosch GmbH, Heidelberg, GFR
Davis, J. W.	Dunlop Ltd., Coventry
Denman, T. H.	Weeks Trailers Ltd., Hessle
Dennis, R.	Bass Production Ltd., Burton on Trent
Dijks, A.	University of Technology, Delft, The Netherlands
Dimond, B. R.	Bendix Westinghouse Ltd., Bristol
Dore, R. D.	Wabco Westinghouse (UK) Ltd., Milton Keynes
Easton, J. M.	Bendix Westinghouse Ltd., Bristol
Ehricke, O.	Wabco Westinghouse GmbH, Hannover, GFR
Elgeskog, E. G. A.	AB Volvo, Göteborg, Sweden
Emmerson, J.	Don International Ltd., Manchester
Ervin, R. D.	Highway Safety Research Institute, University of Michigan, Ann Arbor, USA
Estmar, G.	Saab-Scania AB, Södertälje, Sweden
Fagerlund, G.	Saab-Scania AB, Södertälje, Sweden
Fairbank, J. M.	Stephen James (Air Brake), Leeds
Fancher, P. S.	Highway Safety Research Institute, University of Michigan, Ann Arbor, USA
Frith, H. A.	Brake Linings Ltd., Buxton
Fuke, W. A.	Ford Motor Co. Ltd., Basildon
van Genugten, J. T. M.	DAF Truck BV, Eindhoven, The Netherlands
Gibson, R. W.	British Friction Materials Council, London
Giles, C. G.	Staines (formerly TRRL)
Glasner von Ostenwall, E. C.	Daimler-Benz AG, Stuttgart, GFR
Gold, N. R.	British Leyland UK Ltd., Leyland
Grainger, H.	HQ REME Data Centre, London
Green, M.	Girling Ltd., Birmingham
Gurley, A.	National Bus Company, London
Hackman, P. W. H.	Ford Motor Co. Ltd., Basildon
Hales, F. D.	University of Technology, Loughborough
Hallett, R. C.	Bendix Westinghouse Ltd., Bristol
Hancock, M. G.	Redland Technology, Horsham
Harding, P. R. J.	Mintex Ltd., Cleckheaton
Harper, H. M.	ERF Ltd., Sandbach
Harries, D. A.	Girling Ltd., Birmingham
Harrison, A. W.	Girling Ltd., Birmingham
Harvey, J. H. D.	Dow Corning Ltd., Barry
Hatch, D.	Ferodo Ltd., Chapel-en-le-Frith
Hawker, M. J.	Clayton Dewandre Co. Ltd., Lincoln
Heger, K.	Adam Opel AG, Rüsselsheim, GFR
Hemingway, D. J.	Girling Ltd., Birmingham

© IMechE 1977

Hiett, L. M.	Girling Ltd., Birmingham
Hobbs, C. A.	Transport and Road Research Laboratory, Crowthorne
Hodgson, T. E.	Mintex Ltd., Cleckheaton
Hodkinson, H.	Lockheed Brakes Division, Automotive Products Group, Leamington Spa
Hoffman, D. A.	Bendix Westinghouse Ltd., Bristol
Holbrook, W.	Dow Corning Corporation, Midland, Michigan, USA
Hughes, T. J. A.	United Kingdom Land Forces, Wilton
Huusom, P. H.	Roulunds Fabriker Ltd., Hjallese, Denmark
Ingram, B.	Girling Ltd., Birmingham
Jacobson, M. A. I.	The Automobile Association, Basingstoke
Jaksch, F. J.	AB Volvo, Göteborg, Sweden
Jefferies, J. K.	Rockwell International Automotive Operations – Europe, Alcester
Jepson, J. L.	National Travel (NCB) Ltd., London
Joensson, L.	AB Volvo, Göteborg, Sweden
Johnson, B.	Eaton Axles Ltd., Darlington
Jones, D. V.	Department of the Environment, London
Kapoor, J. K.	High Commission of India, London
Karlsson, J. A.	SAB Thulinverken, Landskrona, Sweden
Kateb, F. J.	Girling Ltd., Birmingham
Kemp, R. N.	Transport and Road Research Laboratory, Crowthorne
Ladbrooke, R. W.	Wimborne
Laland, C.	Chrysler UK Ltd., Coventry
Lewis, J.	Eaton Ltd., Manchester
Lister, R. D.	Slough (formerly TRRL)
Livingstone, P. D.	Ford Motor Co. Ltd., Basildon
von Loewis of Menar, A.	Robert Bosch GmbH, Stuttgart, GFR
Ludvigsson, R. L.	Hesselman Bil – Aero AB, Stockholm, Sweden
Lundberg, J. G.	AB Volvo, Göteborg, Sweden
McHattie, M. A.	Ministry of Transport, Ottawa, Ontario, Canada
McKay, D.	Conoco Ltd., London
McLean, D.	University of Technology, Loughborough
Mair, J.	Clayton Dewandre Co. Ltd., Lincoln
Manning, W.	Ford Motor Co. Ltd., Basildon
Mathias, N.	Girling Ltd., Birmingham
Meadows, R. W.	Fodens Ltd., Sandbach
Millner, N.	Ferodo Ltd., Chapel-en-le-Frith
Mizon, R.	Mintex Ltd., Cleckheaton
Molinier, P. M.	SAF du Ferodo, Saint Ouen, France
Morris, A. L.	MVEE, Chertsey
Morson, P. A.	Dunlop Ltd., Coventry
Mortimer, R. G.	University of Illinois, Champaign, USA
Murphy, P. B.	Department of the Environment, London
Neilson, I. D.	Transport and Road Research Laboratory, Crowthorne
Nelander, N. A.	SAB Thulinverken, Landskrona, Sweden
Newcomb, T. P.	University of Technology, Loughborough
Newton, W. R.	Girling Ltd., Birmingham
Nichols, S. A.	Vauxhall Motors Ltd., Luton
Nielsen, O. K.	A/S Roulunds Fabriker, Hjallese, Denmark
North, M. R.	Motor Industry Research Association, Nuneaton
Oakley, R.	Consumer Association, Gosfield
Odier, J. A.	SAF du Ferodo, Paris, France
Old, J. D.	Road Transport Industry Training Board, High Ercall
Oliver, P. R.	Bendix Westinghouse, Ltd., Bristol
Oppenheimer, P.	Girling Ltd., Birmingham
Ouwerkerk, H.	University of Technology, Delft, The Netherlands
Packer, M. B.	Automotive Products Ltd., Leamington Spa
Page, W. M.	Clayton Dewandre Co. Ltd., Lincoln
Parker, R. C.	Watford
Parkinson, D. R.	British Leyland Bus & Truck Division Ltd., Leyland
Parry, G. W.	BP Chemicals International Ltd., Penarth
Parsons, B. J.	Bendix Westinghouse Ltd., Bristol
Parsons, D.	Automotive Products Ltd., Leamington Spa
Parsons, J. D. W.	REME Data Centre, London
Pearce, D. A.	Michelin Tyre Co. Ltd., Stoke-on-Trent
Peasley, D.	Girling Ltd., Birmingham
Phillips, B. D. A.	Lanchester Polytechnic, Coventry
Plumley, D. C.	Dennis Motors Ltd., Guildford
Pratt, B. H.	Don International Ltd., Manchester
Prior, C.	Trist, Draper Ltd., Bristol
Ranger, G. M.	Rubery Owen Rockwell Ltd., Wrexham
Rath, H.	Girling GmbH, Koblenz, GFR
Rauscher	Knorr-Bremse GmbH, München, GFR
Ray, F.	Department of the Environment, London
Richardson, A. J.	SAB Brake Regulator, Darlington
Riddy, F. T.	Girling Ltd., Birmingham
Rouse, J. A.	Girling Ltd., Birmingham
Rutter, B. E.	Stephen James (Air Brake), Leeds
Sadler, W.	Koblenz, GFR
Scharff, T. E.	Telma Retarder Ltd., Slough
Schlösser, L. H. M.	Institute for Road Safety Research SWOV, Voorburg, The Netherlands
Schmidt, H. L.	Alfred Teves GmbH, Frankfurt, GFR
Sebley, C. H.	Dunlop, Ltd., Coventry
Sharp, R. S.	University of Leeds
Shilton, B. R.	Clayton Dewandre Co. Ltd., Lincoln
Simonsen, E.	Roulunds Fabriker Ltd., Hjallese, Denmark
Sivill, A. D.	Lucas (Industries) Ltd., Solihull
Slack, I.	British Leyland UK Ltd., Leyland
Smales, H.	Mintex Ltd., Cleckheaton
Smales, P. M.	Chrysler UK Ltd., Coventry
Smith, W. W.	Dunlop Ltd., Coventry
Southorn, R. V.	MVEE, Chertsey
Spence, D. R.	Girling Ltd., Birmingham
Spielfiedel, A.	AC Delco France, Gennevilliers, France
Spurr, R.T.	Ferodo Ltd., Chapel-en-le-Frith
Stott, G. W.	CBS (Automotive & Industrial) Ltd., Newbury
Sumira, A	APG Gandy, Wallasey
Symons, C. B. N.	CBS (Automotive & Industrial) Ltd., Newbury
Tanguy, C. T.	DBA, Drancy, France
Tarlton, F. H.	Rolls-Royce Motors Ltd., Crewe
Tee, N. E.	Kirkstall Forge Engineering Ltd., Leeds
Teunissen, H.	DAF Truck B.V., Eindhoven, The Netherlands
Thirion de Briel, J.	SAF du Ferodo, Saint Ouen, France
Thomas, M. P.	Mintex Ltd., Cleckheaton

© IMechE 1977

Thoms, E.	Saab-Scania AB, Södertälje, Sweden
Thrower, A. J.	B.F. Goodrich Engineered Systems, Troy, Ohio, USA
Tiggeler, J. A. M.	B.F. Goodrich Aero Products and Brake Systems, Heveadorp, The Netherlands
Todorović, J.	Belgrade University, Yugoslavia
Tompkin, J. B.	Girling Ltd., Birmingham
Törnqvist, Y.	Volvo Europa NV, Ghent, Belgium
Toyne, C. C.	Department of the Environment, London
Treyer, A. A.	Bendix Poids Lourds, Drancy, France
Turner, R. A.	Calor Gas Ltd., Slough
Turton, J.	Mintex Ltd., Cleckheaton
Unsworth, D.	Rubery Owen Rockwell Ltd., Wrexham
Vergoz, J. M.	ITT Corporation, New York, USA
Ward, R.	Yorkshire Imperial Metals Ltd., Leeds
Wardle, W. B. D.	H.K. Porter Co. (GB) Ltd., Stirling
Wastell, C.	Westminster Hospital and Medical School, London
Watanabe, Y.	Honda Motor Co. Ltd., Tokyo, Japan
Weaving, B.	Trist, Draper Ltd., Bristol
Webber, F. E.	Metal Box Ltd., Reading
Whalley, F.	Department of the Environment, Manchester
Wienecke, F. J.	Alfred Teves GmbH, Frankfurt, GFR
Wildig, A. M.	University of Technology, Loughborough
Wilkes, M. J.	Kirkstall Forge Engineering Ltd., Leeds
Williams, A. R.	Dunlop Ltd., Birmingham
Williams, E. P.	Department of the Environment, Nuneaton
Williams, T.	Transport and Road Research Laboratory, Crowthorne
Willmer, P. F.	Automotive Products Group, Leamington Spa
Willmott, G. F. A.	MVEE, Chertsey
Wilson, A. J.	Girling Ltd., Birmingham
Wood, J. F.	Girling Ltd., Birmingham
Wouters, C.	DAF Truck BV, Eindhoven, The Netherlands
Wyatt, J.	British Leyland UK Ltd., Coventry
Wyke Smith, P. H.	National Bus Company, London
Young, H. J. T.	Society of Motor Manufacturers & Traders, London
Zeiträger, G.	Adam Opel AG, Rüsselsheim, GFR

© IMechE 1977